Modern Classical Optical System Design

Fundamentals, techniques, tips, and tricks

Online at: https://doi.org/10.1088/978-0-7503-6059-3

IOP Series in Emerging Technologies in Optics and Photonics

Series Editor

R Barry Johnson, a Senior Research Professor at Alabama A&M University, has been involved for over 50 years in lens design, optical systems design, electro-optical systems engineering, and photonics. He has been a faculty member at three academic institutions engaged in optics education and research, has been employed by a number of companies, and has provided consulting services.

Dr Johnson is an IOP Fellow, an SPIE Fellow and Life Member, an OSA Fellow, and was the 1987 President of SPIE. He serves on the editorial board of *Infrared Physics & Technology* and *Advances in Optical Technologies*. Dr Johnson has been awarded many patents, has published numerous papers and several books and book chapters, and was awarded the 2012 OSA/SPIE Joseph W Goodman Book Writing Award for Lens Design Fundamentals (second edition). He is a perennial co-chair of the annual SPIE Current Developments in Lens Design and Optical Engineering Conference.

Foreword

Until the 1960s the field of optics was primarily concentrated in the classical areas of photography, cameras, binoculars, telescopes, spectrometers, colorimeters, radiometers, etc. In the late 1960s optics began to blossom with the advent of new types of infrared detector, liquid crystal display (LCDs), light emitting diode (LEDs), charge coupled device (CCDs), laser, holography, and fiber optics along with new optical materials, advances in optical and mechanical fabrication, new optical design programs, and many more technologies. With the development of the LED, LCD, CCD, and other electro-optical devices, the term 'photonics' came into vogue in the 1980s to describe the science of using light in the development of new technologies and the operation of a myriad of applications. Today optics and photonics are truly pervasive throughout society and new technologies are continuing to emerge. The objective of this series is to provide students, researchers, and those who enjoy self-education with a wide-ranging collection of books, each of which focuses on a topic relevant to the technologies and applications of optics and photonics. These books will provide knowledge to prepare the reader to be better able to participate in these exciting areas now and in the future. The title of this series is *Emerging Technologies in Optics and Photonics*, in which 'emerging' is taken to mean 'coming into existence', 'coming into maturity', and 'coming into prominence'. IOP Publishing and I hope that you will find this series of significant value to you and your career.

A full list of titles published in this series can be found here: https://iopscience.iop.org/bookListInfo/emerging-technologies-in-optics-and-photonics.

Modern Classical Optical System Design

Fundamentals, techniques, tips, and tricks

Ronian Siew

inopticalsolutions.com, Vancouver, British Columbia, Canada

IOP Publishing, Bristol, UK

ISBN   978-0-7503-6059-3 (ebook)
ISBN   978-0-7503-6057-9 (print)
ISBN   978-0-7503-6060-9 (myPrint)
ISBN   978-0-7503-6058-6 (mobi)

DOI   10.1088/978-0-7503-6059-3

Supplementary material is available for this book from https://doi.org/10.1088/978-0-7503-6059-3.

Version: 20240201

IOP ebooks

British Library Cataloguing-in-Publication Data: A catalogue record for this book is available from the British Library.

Published by IOP Publishing, wholly owned by The Institute of Physics, London

IOP Publishing, No.2 The Distillery, Glassfields, Avon Street, Bristol, BS2 0GR, UK

US Office: IOP Publishing, Inc., 190 North Independence Mall West, Suite 601, Philadelphia, PA 19106, USA

To the optical adventurer

Contents

Foreword   **xiii**

Preface   **xiv**

Acknowledgements   **xvii**

Author biography   **xviii**

Abbreviations   **xix**

Symbols   **xxi**

1   Imaging   **1-1**

1.1   An introduction to the real world   1-1

    1.1.1   Real lens systems take in all the light until rays hit a 'STOP'   1-3

    1.1.2   Shoot for the 'minimum viable product' (make it 'perfect' in steps)   1-4

    1.1.3   But even 'perfect' cannot be perfect: limitations due to physical laws   1-4

    1.1.4   Realistic product development operates between these two limits   1-10

    1.1.5   What to do about ambiguous requirements   1-10

    1.1.6   Your list of 'things I still need to understand' will only grow (which is fine)   1-11

    1.1.7   Everything you really want to do will likely take place only after office hours   1-12

    1.1.8   Why books are still necessary for your knowledge   1-12

    1.1.9   You become good at something by doing it over and over for a very long time   1-13

    1.1.10   How to bug people for help   1-14

    1.1.11   Truth = the best estimate $\pm$ uncertainty   1-14

    1.1.12   Are you ready for this?   1-15

1.2   Optical system design using Ansys Zemax OpticStudio®   1-16

    1.2.1   Set the aperture   1-18

    1.2.2   Set the fields   1-21

    1.2.3   Set the wavelengths (I will tell you how many you need)   1-22

    1.2.4   Create a paraxial thin-lens equivalent system   1-24

    1.2.5   Use 'solves'   1-25

    1.2.6   Check the MTF and defocus sensitivity and create defocus invariance   1-29

1.2.7   Creating a real lens model of the thin-lens model   1-36

1.2.8   Lens MTF, spatial frequency, field curvature, distortion, and relative illumination   1-39

1.2.9   Why it is not always about MTF in real life (it depends on your application)   1-44

1.2.10 Amazing OpticStudio features you may not know about (which we will use)   1-45

1.2.11 Tilted and decentered components and assemblies   1-51

1.2.12 Optimizing a lens   1-61

1.2.13 Tolerancing analysis for a lens   1-74

1.2.14 Create and use a 'black box file'   1-81

1.2.15 Nonsequential modeling and analysis   1-83

1.2.16 Dealing with 'ray trace noise' in nonsequential modeling   1-86

1.2.17 Deciding between nonsequential and sequential approaches   1-87

1.2.18 OpticStudio's hybrid nonsequential mode (this is a powerful tool)   1-88

1.2.19 A wrap-up; get set to use OpticStudio for the rest of this book   1-95

1.3   Practical concepts for optical system layout and analysis   1-96

1.3.1   First-order: all you really need is $1/f = 1/s + 1/s'$ 1-96

1.3.2   Example: a microscope tube lens using commercial off-the-shelf lenses   1-98

1.3.3   Example: a xenon arc lamp with an elliptical reflector   1-104

1.3.4   Notes on designing with commercial off-the-shelf components   1-108

1.3.5   If you master the concept of conjugate planes, you can go very far   1-109

1.3.6   Example: collimation at an intermediate plane and its application   1-111

1.3.7   Example: conjugate planes in modern microscope condensers   1-113

1.3.8   Example: a simple modern digital microscope using commercial off-the-shelf lenses   1-116

1.3.9   Example: locating and modeling dust artifacts in imaging systems   1-124

1.3.10 Conjugate planes in a classical projector   1-127

1.3.11 Object and image conjugates at the same location   1-129

1.3.12 If you master the concept of pupils, you will understand what detectors 'see'   1-132

1.3.13 Example: relay lenses (you will often need them)   1-132

1.3.14 Example: pupil and scene visibility in a terrestrial telescope   1-134

1.3.15 More on pupils: Max Berek's 'forgotten' formula   1-135

1.3.16 The optical center of a lens (you have probably never heard of this) — 1-139

1.3.17 Locating and optimizing the optical center of a lens system — 1-141

1.3.18 The application of the optical center and pupils to depth sensing — 1-142

1.3.19 Approximate analogies: eyepieces, tube lenses, and scan lenses — 1-146

1.3.20 Approximate analogies: condensers as eyepieces in reverse — 1-149

1.4 Practical lens design and aberration management — 1-151

1.4.1 In rapid product development, just 'manage' the aberrations (we show you how) — 1-151

1.4.2 Heuristic lens design theory — 1-152

1.4.3 Why are mobile phone lenses not used as high-aperture laser scan lenses? — 1-177

1.4.4 Do not be afraid of 'aplanatism' (it is just a term for an optimized lens, except...) — 1-185

1.4.5 The optical sine theorem is not the same as the Abbe sine condition — 1-208

1.4.6 Analogous imaging systems: aspheric aplanatic singlets and Ritchey–Chrétien mirrors — 1-209

1.4.7 Heuristic color correction theory — 1-220

1.4.8 Conrady's D-d method for achromatizing — 1-236

1.4.9 Do commercial off-the-shelf achromats satisfy D-d? — 1-239

1.4.10 Example: achromatizing a monochromatic four-element lens — 1-242

1.4.11 Example: an apochromatic microscope tube lens design — 1-251

1.4.12 Example: secondary color in a high-aperture double-Gauss lens — 1-254

1.5 Preparing drawings for optical fabrication — 1-259

1.5.1 What an optical design drawing for production looks like — 1-259

1.5.2 The relation between ISO 10110 specifications and tolerance operands — 1-260

1.5.3 Modeling the centering process of cemented elements — 1-264

1.5.4 Alternatives to design drawings: communicating with suppliers — 1-269

References and further reading — 1-270

2 Illumination — 2-1

2.1 The illumination problem — 2-1

2.2 Essential radiometry for illumination problems — 2-3

2.2.1 What type of source is being modeled in sequential ray tracing? — 2-3

2.2.2 What is different about sources in nonsequential ray tracing? — 2-4

2.2.3 Flux, radiance, and étendue in illumination design   2-4

2.2.4 From radiance to radiant intensity: modeling sources   2-8

2.2.5 The concept of source spread functions and the irradiance of images   2-9

2.2.6 If a source radiates and nobody is there to see it, does it shine?   2-12

2.2.7 Why is the full width at half maximum often the width of a distribution?   2-15

2.2.8 Is the image of a Lambertian source a Lambertian source?   2-15

2.2.9 Is chromatic aberration important in illumination?   2-16

2.2.10 The radiometry of LEDs and the use of source files in nonsequential ray tracing   2-16

2.2.11 There is no free étendue   2-21

2.3 The concept of ray density in illumination design   2-23

2.4 The concepts of global and local uniformity   2-24

2.5 The concepts of étendue division and superposition   2-25

2.6 'First-order' illumination design   2-27

2.6.1 Illumination using paraxial thin lens models   2-27

2.6.2 How to correct the radiance problem (because paraxial thin lenses are fake lenses)   2-30

2.6.3 Relative illumination is called *critical illumination* in illumination design   2-31

2.7 How to design for uniform relative illumination   2-33

2.8 Relative illumination in direction cosine space   2-40

2.9 The phase space viewpoint of relative illumination   2-44

2.10 Aplanatism and the relative illumination in the pupil   2-49

2.11 Regions of uniformity in collimated light: the searchlight optical layout   2-53

2.12 The specification of flashlights and searchlights based on the ANSI FL1 Standard   2-57

2.13 Searchlights, critical illumination, and Köhler illumination: a comparison at equal flux and track length   2-58

2.14 How to lay out light pipes for uniform illumination   2-69

2.15 How to lay out fly's eye arrays for uniform illumination   2-71

2.16 Fly's eye arrays that have negative-focal-length lenslets   2-77

2.17 Uniform oblique illumination   2-78

2.18 Point spread function illumination   2-80

2.18.1 The coherent case: Gaussian to top-hat laser beam shaping   2-80

2.18.2 The incoherent case: LED Lambertian to top-hat beam shaping   2-87

2.19 A summary of the approaches used in illumination   2-91
2.20 Tips on optimization and tolerancing in nonsequential ray tracing   2-93
       References   2-106

3   Optical system product development   **3-1**

3.1   Lights at the ends of tunnels (not light pipes, but a personal story)   3-2
       3.1.1 Gratification and enlightenment   3-2
       3.1.2 Challenges in academic life   3-2
       3.1.3 Transition to the real world and product development   3-3
       3.1.4 The light at the end   3-4
       3.1.5 The systems perspective on optical system product development   3-5
3.2   An example of a complex optical system: virus detection using   3-6
       real-time quantitative PCR instruments
       3.2.1 What is the minimum viable product for such a device?   3-6
       3.2.2 Why you cannot be 'just' a lens designer when designing   3-10
              real-time qPCR instruments
3.3   Statistical principles for optical system product development   3-12
       3.3.1 The 'expectation value' is not the value you should expect   3-12
              to get
       3.3.2 The difference between the standard deviation of a function   3-14
              and the standard deviation of random values
       3.3.3 Statistical principles related to error analysis, error bars,   3-14
              sensitivity analysis, and optical tolerancing analysis
       3.3.4 In optical tolerancing analysis, a merit function is a *function* of   3-16
              random variables
       3.3.5 The std dev of a merit function has a std dev   3-21
       3.3.6 The different ways in which engineers and statisticians solve   3-21
              problems
3.4   The concept of the signal-to-noise ratio   3-24
       3.4.1 What exactly is the signal, and what do you mean by 'noise'?   3-24
       3.4.2 How does the signal-to-noise ratio scale with the size of a   3-24
              region of interest?
       3.4.3 How does the signal-to-noise ratio scale with integration time?   3-28
       3.4.4 Does camera 'gain' increase the signal-to-noise ratio?   3-29
       3.4.5 What is 'charge conversion efficiency'?   3-29
3.5   The concept of the limit of detection   3-30
       3.5.1 The noise in the background after subtracting the background   3-30
       3.5.2 The limit of detection is like an 'apparent nuisance signal'   3-30

3.5.3 The relationship between the limit of detection and the signal-to-noise ratio 3-32

3.6 Remarks concerning the tolerancing of complex optical systems in product development 3-33

3.7 Monte Carlo tolerancing as a means to justify an alignment philosophy 3-34

3.8 Some nuances of optical systems in product development 3-44

3.8.1 When a lens images an intermediate transmissive or reflective surface 3-44

3.8.2 The eternal challenge of stray light analysis and control 3-45

3.8.3 Ghosts and 'narcissists' (I mean narcissus) effects 3-47

3.8.4 The spectral 'blueshifts' of thin-film filters and beam splitters 3-50

3.8.5 Optical density and the transmittance of stacked filters 3-55

3.8.6 Optical fibers versus free-space components for illumination . 3-57

3.8.7 Fundamental limitations to illumination in microscopy 3-58

3.8.8 Drift: an enemy of statistics and the reason for calibration . 3-58

3.8.9 Should manufacturing processes be easy or hard? 3-59

3.8.10 Optical design for robustness 3-60

3.8.11 Innovation tip—ask the question: 'How bad is it?' 3-61

3.9 Simple conceptual case studies 3-61

3.9.1 A compact optical system for virus detection 3-61

3.9.2 The Texas Instruments DLP® chip (DMD) projection optical system 3-68

3.10 Wrap up, README, and I wish you all the best! 3-82

References and further reading 3-83

Appendix A: Further notes on imaging **A-1**

Appendix B: Further notes on illumination **B-1**

Appendix C: Further notes on optical system product development . **C-1**

Appendix D: Notes on some advanced topics **D-1**

Foreword

There are many books on lens design, optical fabrication, optical testing, optical instrument design, optical system modeling, etc. Unlike these books, Ronian Siew's book, *Modern Classical Optical System Design: Fundamentals, techniques, tips, and tricks*, provides a different perspective on developing optical systems. Having skill and knowledge in the myriad of disciplines involved in developing optical systems is invaluable to the client or company for which the engineer/scientist works. Mr Siew earned his undergraduate and graduate degrees in optics from the University of Rochester's Institute of Optics, and he has a total of over two decades of experience with a decade of it as an independent optical consultant. During this period, he experienced challenges in developing various projects and learned how to productively develop optical systems while mitigating common problems often plaguing projects. His book presents fundamentals that a designer should understand, a useful coverage of how to use Ansys Zemax OpticStudio for lens design and analysis, lens design techniques, preparing drawings for optical fabrication, numerous examples of practical concepts for optical systems, and a comprehensive discussion of illumination and how to avoid mistakes often made. I feel that many readers will find chapter 3, which discusses optical system product development, to be an extremely valuable portion of the book. The book ends with four appendices that provide additional information on imaging, illumination, optical system product development, and several advanced topics. Once the three chapters of this book have been digested, I suggest gaining an understanding of the appendices will further provide the reader with improved knowledge valuable in achieving project success. An optical system engineer is far more than a domain expert in lens design, or optical fabrication, or detectors, or testing, etc. This person has broad knowledge and experience in numerous disciplines needed to successfully manage simple to complex projects. Mr Siew's book is an important contribution to the optics community and the broader device and instrumentation field. Optics is pervasive throughout the plethora of products available today and tomorrow, and learning the content of this book will improve the reader's ability to achieve project success.

Dr R Barry Johnson, FInstP, FOptica, FSPIE, HonSPIE
Senior Research Professor, Alabama A&M University
Institute of Physics Publishing Series Editor,
Emerging Technologies in Optics and Photonics
Past-President SPIE - The International Society for Optics and Photonics
Huntsville, Alabama

Preface

Experienced optical designers and engineers often develop personal 'bags of tricks' in the course of their careers. These tricks may, for example, involve special techniques to design lens systems, or perhaps they may be simple methods to align components. A designer's bag of tricks saves time in typical modern fast-paced product development for commercial projects. If you are an early-career designer, perhaps the hardest part of product development is knowing how to get started in applying the fundamentals to a real-world design within a tight schedule, and a recurring problem is knowing how to use a modern optical design program for that task. You may turn to quick sources, such as lessons on YouTube or tips from various notes posted online. If, like me, you have come across David Shafer's illustrative optical design tips on SlideShare, you will have been drawn to his way of presenting useful material in short, simple layouts of lens and mirror systems. This can motivate you and get you started on a design. You may have also come across Mark Nicholson's channel on YouTube®, 'Design Optics Fast!', which is also an excellent source of highly practical tips that are short, simple, accessible, and accurate. If you have enjoyed those videos, as I do, you may have been inspired to search for further material. As you pore over textbooks, you may feel overwhelmed by the necessary mathematical rigor. But in the near term, you need to deliver. You are in a company, you have a deadline, and you may feel that you have too little time. So now what to do?

Modern Classical Optical System Design (MCOSD) is a book that can hopefully save you some needed time. MCOSD shares this author's bag of tricks to help get you started in optical system design in a modern, fast-paced product development context. Overviews of basic design techniques, tips, and tricks are written with simple explanations and illustrations, but some in-depth discussions with mathematical rigor are also provided for further study (you can procrastinate, but you ought to eventually know this material well, because deep knowledge is the ultimate power of a designer). The use of the term *modern classical* in the title of this book is meant to indicate that well-established principles are applied to the design of optical systems by way of using specific tools and features of a modern optical design program (in this book, I have used the Ansys Zemax OpticStudio® program for Microsoft® Windows®). Also, some well-known concepts and formulas are reinterpreted (i.e. proven and derived differently or expressed in a different form) and specific examples are provided to show how they can be applied to the design of optical systems. Optical prescriptions are provided for many of the design examples, and their lens files are made available as supplementary material at IOP Publishing's website for this book.

In this book, the term *optical system* generally refers to the integration of optical components into a full instrument/product. These components can be commercial off-the-shelf parts (lenses, mirrors, sensors, light sources, etc) or they may be custom-designed assemblies. Where indicated in this book, an isolated assembly that is used either for imaging or illumination may also be called *the optical system*. Topics include imaging principles, elements of lens design, illumination, modelling, analysis, tolerancing of optical systems, and detection. Due to this book's rather quick-paced delivery of practical concepts (and the assumption that the reader is already working in industry), MCOSD does not dwell on derivations of some basic formulas (such as Snell's law) and on reintroducing certain basic definitions (such as what stops and pupils are). You should also know a little bit about basic wave (physical) optics, at least enough to know what is meant by interference, phase, and the Airy disk. Thus, MCOSD does not provide an introduction to optics, nor does it focus solely on lens design. Rather, MCOSD is a book for the *optical system designer*—individuals who integrate, model, optimize, analyze, and develop entire instruments (and may perform occasional lens designs).

In some sense, MCOSD may loosely be considered a second volume to *Perspectives on Modern Optics and Imaging* (a book I independently published in 2017), featuring material not covered in its predecessor. It follows the spirit of *Perspectives* in the sense that it nitpicks specific applied and fundamental concepts that I have found useful in industry, and as a practical resource book, it will help remind you of some basics within relevant practical contexts, give further insights, and show how to apply them to practical design. References and bibliographic information are provided where necessary. Some fine examples include books in this IOP Publishing Series. In this book, unless stated otherwise, lens dimensions are in mm, polarized light is not involved, sources are mostly incoherent and Lambertian, and all optical systems are for visible light. Further, systems are assumed to be immersed in air with refractive index = 1, and they operate on the surface of the Earth. When certain terms or phrases require emphasis, they may either be italicized (if it is required that the reader gain some familiarity with that term/phrase, such as *minimum viable product*, or *conic constant*) or shown in boldface (if a term/phrase possesses high technical significance, such as the **Lagrange invariant**). Rays are traced from left to right (but this convention is not always adopted in this book). Where needed, it may be indicated that specific topics that do not fit directly within the context of the discussion (but are nonetheless interesting and have pedagogical value) are provided as notes in the appendices. Scientific notation is either written in the form of $\times 10$ to the power or 'E' followed by the power (e.g. 1×10^{-2} or, equivalently, 1E−02).

I have often said that I am just a guy who has not stopped searching for answers. Consider this book a sample of knowledge gained from my learning journey. I took this path—this very long road that sometimes did not seem to have light at the end—when I was eleven. It was not easy. It took much study, asking dumb questions, many trials, many errors, doing lots of exercises (in optics, not aerobics), and doubting myself. Of course, it also took many years of real-world experience. The more topics I began to understand, the more questions I had. Then, I found out that much of what I knew was only what I thought I knew. But you do not only discover this in real-world product development settings (in fact, in industrial practice, it is possible to get something to work even if you do not understand it). Rather, I often found out my weaknesses through writing—which I did a lot. Writing taught me. Writing checked my mistakes. Writing exposed my deficiencies. Consequently, writing progressed me, and it continues to nudge me forward. So, I wrote this book, and I hope that it will help you too in your own search for answers and success in your development of optical systems.

Ronian Siew
Vancouver, BC, Canada

Acknowledgements

I am grateful to professor R Barry Johnson for his support and encouragement throughout the course of writing this book, including the proposal period, for which he gave me valuable advice. Barry has been both mentor and friend, and he has provided significant motivation for me to keep on pursuing the rich field of optical design. I thank Mrs Marianne Johnson for her patience while Barry—being the series editor for the IOP Series in Emerging Technologies in Optics and Photonics— had to spend time reading through my manuscript, which I do hope did not include too many mistakes, whether in prose or in the fundamentals. I am grateful to professor Kedar Khare of the Indian Institute of Technology Delhi for educating me in the area of noise statistics and Wiener filters as they are applied towards deconvolution in imaging, and for helping me to polish this very topic in appendix D.3. I thank IOP Publishing's senior commissioning editor, Ashley Gasque, and editorial assistants Erika Radzvilaite, Isabelle Defillion, and Rory Weaver, as well as IOP Publishing's editorial board, for their patience, support, and encouragement, without which I would not have been able to convince myself that I could actually complete this book. I thank IOP Publishing's legal and rights advisor, Cameron Wood, for his assistance in advising me on the rights processes. I am also grateful to the IOP Publishing eBooks Production Team for their careful attention to detail and their hard work in making all of the necessary edits to this book's manuscript. I would like to thank the reviewers of this book's initial proposal, whose feedback helped me make this book better. I also thank the nice folks at Ansys Zemax (Sanjay Gangadhara, Matt Ramos, Nicholas Herringer, and Mike Steffensmeier) in helping to get my free trial of OpticStudio up to speed and for permissions to re-use some of the graphics in the program. Appreciation also goes to Andy Skoogman and the board of directors at PLATO for letting me use symbols from the ANSI/PLATO FL1-2019 standard. I thank my friend, Ching Hu, for letting me see and refer to his unfinished book on robotics and autonomous system architectures. I thank my parents and brothers for their support throughout my life, having to listen to my annoying theories and letting my room get messed with optics experiments. Finally, my deepest thanks go to my wife, Claire, for her patience and support in letting me sit and write this significant volume of work, whilst also reminding me to keep exercising and take breaks.

Author biography

Ronian Siew

Ronian Siew obtained his degrees in optics and physics at the University of Rochester. He is an optics consultant with more than 25 years of professional experience in designing optical systems, with a recent focus on bioimaging and sensing. He has authored scientific peer-refereed papers and four books. He holds two patents, one related to an optical system in a digital PCR instrument that has been applied to COVID-19 surveillance and detection. Ronian also serves as an associate editor in the area of optical design for SPIE's Spotlight book series.

Abbreviations

2D	Two-dimensional
3D	Three-dimensional
ANOVA	Analysis of variance
ANSI	American National Standards Institute
AOI	Angle of incidence
AR	Anti-reflection (coating)
BFL	Back focal length
BRDF	Bidirectional reflectance distribution function
BSL	Beam shaping lens
CCD	Charged coupled device
CCE	Charge conversion efficiency
CG	Computer generated
Cg	Conversion gain (used synonymously with CCE)
CMOS	Complementary metal–oxide–semiconductor
COB	Chip on board (e.g. a COB light-emitting diode module)
COTS	Commercial-off-the-shelf
CLT	Central limit theorem
CPC	Compound parabolic concentrator
DLP®	Digital Light Processing (by Texas Instruments)
DMD	Digital micromirror device
DN	Digital number (in the context of conversion of photoelectrons into digital values)
DNA	Deoxyribonucleic acid
DOE	Design of experiment
EFL	Effective focal length
$F/\#$	F-number (unless stated otherwise, it is the infinite conjugate f-number, given by dividing the EFL by the system's entrance pupil diameter)
FFT	Fast Fourier transform
FOV	Field of view
FWHM	Full width at half maximum
GBI	Geometric bitmap image (in the context of the analysis feature/tool provided by OS)
GD&T	Geometric dimensioning and tolerancing
GIA	Geometric image analysis
H	When used in context of examining the graphics for an optical layout in OS, this is the horizontal coordinate in the layout.
HNSC	Hybrid non-sequential
IES	Illumination engineering society
Index	Refractive index
ISO	International Organization for Standardization (e.g. ISO 10110 for optical specifications)
LCD	Liquid crystal display
LCoS	Liquid crystal on silicon
LED	Light-emitting diode
LES	Light-emitting surface
LLG	Liquid light guide
LOD	Limit of detection
lp/mm	Line-pairs per millimeter

Max (or max)	Maximum
MCOSD	Modern Classical Optical System Design (this book)
MIL	Military (i.e. United States MIL standard for optical components)
Min (or min)	Minimum
MTF	Modulation transfer function
MVP	Minimum viable product
NA	Numerical aperture
NEMA	National Electrical Manufacturers Association
NEP	Noise equivalent power
NSC	Non-sequential (ray tracing)
OC	Optical center
OPD	Optical path difference
OPL	Optical path length
OPTICA	Formerly, the Optical Society of America
OS	OpticStudio® (from Ansys Zemax OpticStudio)
OSC	Offense against the sine condition
OST	Optical sine theorem
OTF	Optical transfer function
PCP	Product commercialization process
PCR	Polymerase chain reaction
POC	Point of care
POP	Position of peak irradiance
PSF	Point spread function
Pupil	Either the entrance pupil, or the exit pupil, used when there is no ambiguity in context
qPCR	Quantitative polymerase chain reaction
RMS	Root mean square (square root of the average of the sum of the squares)
RNA	Ribonucleic acid
ROI	Region of interest
RPD	Relative partial dispersion
RSS	Root sum square (square root of the sum of the squares)
rad	Radians (i.e. the unit for an angle, such as π rad)
SAC	Spherical aberration compensator
SC	Sequential (as in sequential ray tracing)
SLE	Spectral luminous efficacy (i.e. the photopic sensitivity of the human eye)
SNR	Signal-to-noise ratio
SPIE	International Society for Optics and Photonics (formerly, Society of Photographic Instrumentation Engineers)
SSF	Source spread function
Std dev (or std dev)	Standard deviation (the plural is std devs)
Stop	Aperture stop
TEMoo	Transverse electromagnetic zero-zero (transverse mode of a beam)
TI	Texas Instruments
TIR	Total internal reflection
TIRun-out	Total indicator runout
USAF	United States Air Force
V	When used in context of examining the graphics for an optical layout in OS, this is the vertical coordinate in the layout.
VOC	Voice of customer
vs	Versus

Symbols

ϕ	Radiometric optical flux (e.g. watts, milliwatts, etc.)
ϕ_ν	Photometric luminous optical flux (e.g. lumens)
L	Radiance (flux per unit projected surface area, per unit solid angle)
E	Irradiance (flux per unit surface area)
F or $\overline{F}$	Photon flux or mean photon flux, respectively (# photons per unit time)
I	Radiant intensity (flux per unit solid angle)
$\mathcal{E}$	Total étendue
τ	Transmittance (e.g. $\tau = 0.1$ means 10% transmission)
$\tau(\lambda)$	Spectral transmittance
$V(\lambda)$	Spectral luminous efficacy
T	Time period (e.g. $1/T$ is frequency)
Φ	Lens or surface power (e.g. the reciprocal of a focal length, such as $\Phi = 1/f$)
n	Refractive index (or simply, index; where appropriately specified, this symbol may also be used to indicate 'the number of' something)
θ	Any angle, defined and used within a specific context in a discussion
λ	Wavelength
H	Lagrange invariant
$\mathscr{W}$	Wiener filter transfer function
$\mathscr{F}$	Fourier transform operation
y	The y-coordinate of a paraxial marginal ray at any surface
$\overline{y}$	The y-coordinate of a paraxial principal ray at any surface
u	The angle from the axis subtended by either a paraxial or real marginal ray at a surface in object space, depending on the context in which it is used in a discussion. A superscript prime symbol is used to denote the quantity in image space.
$\overline{u}$	The angle from the axis subtended by either a paraxial or real principal (or chief) ray in object space, depending on the context in which it is used in a discussion. A superscript prime symbol is used to denote the quantity in image space.
U	The marginal ray at the full entrance pupil zone
U'	The marginal ray at the full exit pupil zone
h	The height of the object for a real ray (i.e. not a paraxial quantity)
h'	The height of the image for a real ray (i.e. not a paraxial quantity)
$>$	In the context of discussing the location of a program feature (or a file in a folder in a directory) in the user interface drop-down menus of Zemax®, this symbol indicates the directory path towards that feature (e.g., c drive > Zemax > Documents > Samples).
# or #	'Number' (e.g. part #, or $F/\#$, or $F/\#$)
$F/\#$:	F-number (unless stated otherwise, it is the infinite conjugate f-number, given by dividing the EFL by the system's entrance pupil diameter)
$F/\#'$:	Working f-number (the f-number in finite conjugate imaging, defined in section 1.3.15)

Parallel Rays

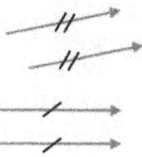

In figures, parallel rays that are significant are either indicated with short black double lines or with short black single lines that intersect the rays, as shown here.

K	Conic constant (note that Zemax OpticStudio's help file uses the small letter k)
k	Wave number, $k = 2\pi/\lambda$
μ	Mass density (mass per unit volume)
$\gg$	Much greater than
$\equiv$	Defined as (e.g. $A \equiv B$ means A is defined as B)

Chapter 1

Imaging

This chapter is devoted to a variety of imaging principles and practices that I have found useful in my career. The material is organized into five sections, which begin by introducing the reader (assumed to be an early-career optical engineer/designer) to the *real world*. Here, I tell what it is like to transition from academia into industrial practice in a product development setting. The second section provides an overview of the most essential and basic features of the Ansys Zemax OpticStudio program, which sets the reader up to use this program for all of the design examples provided in this book. As a natural progression towards optical modeling, analysis, and design, the third and fourth sections respectively provide practical concepts for modeling optical systems in terms of first-order layouts and then include the effects of aberrations and how to *manage* them effectively in the design of optical systems. The final section provides an overview of how to communicate with manufacturers in order to get parts fabricated for your optical designs.

1.1 An introduction to the real world

This book is about getting things done. So, let us get right to it. Figure 1.1 shows a typical modern product development *phase gate* flowchart. In this example, a one-year timeline has been given to a project to finish developing an *alpha prototype*, which is a common term used for a company's first working prototype of a product under development (an improved version in a next phase could be called the *beta prototype*). Today, a one-year timeline is not an uncommon timeframe for the development of a complex product, such as an analytical instrument for virus detection based on the *polymerase chain reaction* (PCR) method [1]. During a pandemic, the timeline for such an instrument could be halved (or reduced even further).

A company often uses phase gate flowcharts to plan, manage, and execute projects in its product commercialization process (PCP) [2]. The idea is to reduce risk from phase to phase. Logically, risk is reduced if, at each phase, problems are identified, questions are answered, ideas are proven, and doubts are eliminated. A phase gate approach is also a

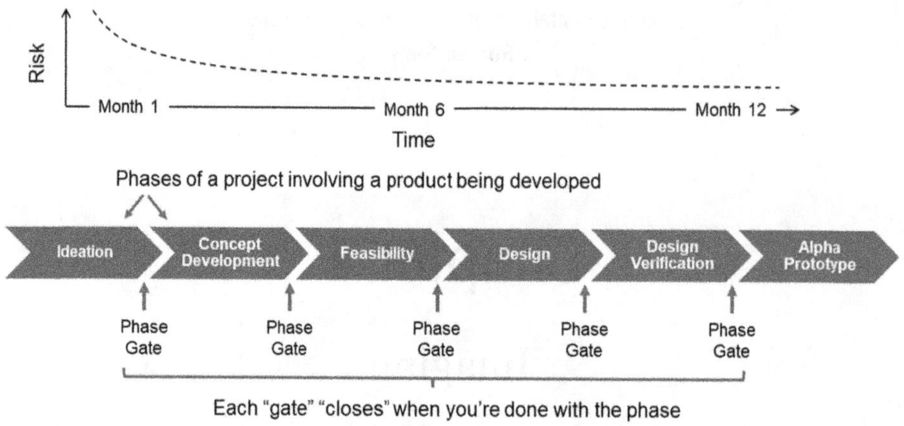

Figure 1.1. A modern product development phase gate flowchart.

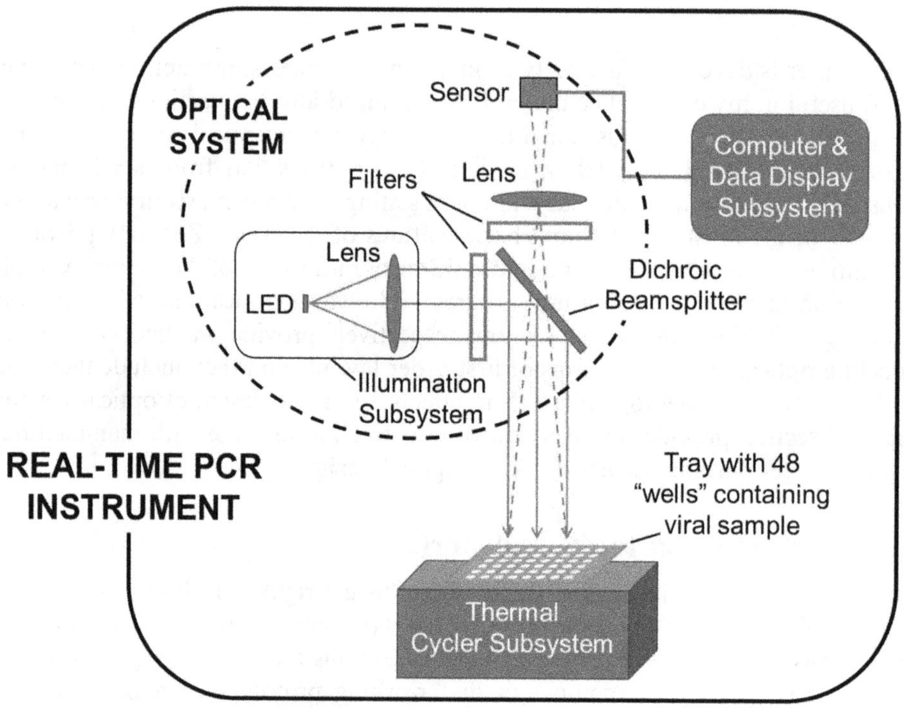

Figure 1.2. A system—in this case, a *real-time PCR* instrument [1]—broken down into constituent parts called subsystems. The optical system, for example, is a subsystem of the instrument, and the illumination components combine to form a subsystem of the optical system.

means to break down the complexity of a project into parts. Similarly, a complex instrument (a 'system') may be broken down into parts (subsystems), as illustrated in figure 1.2. However, note that a system can often involve more than the instrument [3]. Among the subsystems of an instrument, we may regard the combined optical parts as 'the optical system' of the instrument. However, there is much flexibility in this definition. For instance, if an instrument is primarily optical (such as a microscope), then the instrument is itself the optical system (or optical instrument).

In order to accomplish the feat of developing a product within a short timeline, we will often be concerned with the question of how detailed an optical system model has to be in an optical design program. The more detail is put into a model, the more time is needed to work on it. Early-career designers often want to model everything. Experienced designers know that this is not always necessary. In fact, it is often unrealistic to create an exact model of everything within a short project timeline. So, in the next section, let us talk about what it means to model a lens system.

1.1.1 Real lens systems take in all the light until rays hit a 'STOP'

Figure 1.3 shows a comparison between two conditions being modeled for the same lens system. Do you see the difference? In real life, rays from the object fill the lens system (figure 1.3(a)). But in an optical design program, it is implicitly assumed that some of those rays are perfectly blocked by spacers, so that only rays that make it through the aperture stop are displayed (figure 1.3(b)). Is this realistic? Generally, this is not 100% realistic. This is because those points at which rays are thought to be blocked by spacers may still scatter the rays, resulting in stray light. But that is a piece of reality that can be included later, if needed. Meanwhile, the assumption that we have perfectly light-absorbing spacers and walls in lens housings enables efficient lens design optimization. When the optimization is completed, you just need to refine your model if or when further details (such as imperfectly light-absorbing spacers and walls) are needed. **The point is that, when we model an optical system, we create a temporary approximate representation of reality for that optical system, which is meant to be refined in steps where needed**. Remember Taylor series? We often begin studying a complicated function by representing it by the lower-order terms in a Taylor series expansion for that function. When greater accuracy is needed, we include the higher-order terms. This is no different from modeling optical systems: begin with the simplest model sufficient for the task at hand (e.g. figure 1.3(b)), and when you are done with that, involve more aspects of reality (e.g. figure 1.3(a)).

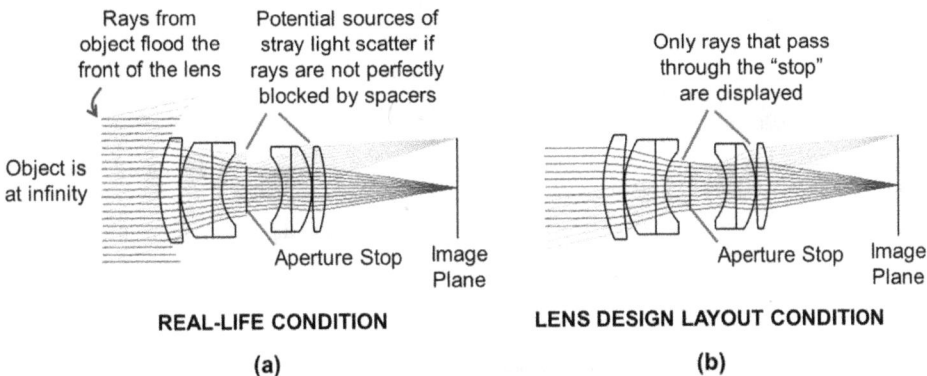

Figure 1.3. A comparison between two models of an imaging lens: (a) a realistic model involving rays flooding the front of the lens system. (b) A simplified model displaying only those rays that make it through the aperture stop.

1.1.2 Shoot for the 'minimum viable product' (make it 'perfect' in steps)

If we remember that a model of an optical system is an approximate representation of its real-life version, then it must be the case that the optical system model is not perfect. So, an alpha prototype is not perfect either. In fact, this is often the case and we would expect this result. Otherwise, we would never reach the finish line. After this, we would normally expect to build a beta prototype, which is an improved version of the alpha. In this way, the alpha prototype is often called the *minimum viable product* (MVP).

In a PCP, the MVP is a version of the product that meets a minimal set of requirements (the first edition of this book, for example, is an MVP). Sometimes, even the beta prototype can be considered an MVP. Your company's program managers would decide this, based perhaps on customer feedback. The point is that deadlines are reached by way of focusing on meeting the minimal requirements rather than targeting perfection in the first attempt. Similarly, your optical system model and its first built version can be an MVP. This is how you meet a deadline. The next version of your model can be more 'perfect' than its predecessor.

1.1.3 But even 'perfect' cannot be perfect: limitations due to physical laws

In the course of improving the alpha and beta versions of a product, we naturally cannot break the laws of physics. To give an obvious example, suppose your alpha prototype comprises a luminaire (i.e. a light source assembled with lenses or mirrors to project rays onto a screen for illumination) with 25% flux efficiency. This means that it can make use of 25% of the flux emitted from a light source to illuminate a screen. Perhaps a beta version for this luminaire may attain a flux efficiency of 50%, which is clearly an improvement, but it cannot ever exceed 100%, as this would violate flux conservation. This is clearly a trivial example, but other optical conservation laws are not so simple, such as the conservation of **étendue**.

You will learn more about étendue in chapter 2. But let me give you a heads-up. Figure 1.4 depicts a small source whose rays are captured by a circular aperture, yielding a cone of captured rays with a half-angle u through the aperture.

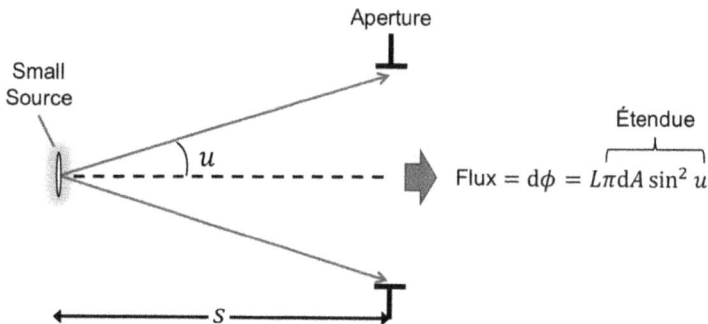

Figure 1.4. A small source of surface area dA emitting rays captured by a circular aperture. In this figure, the source and aperture are in a lossless medium.

The flux in the rays, as we know, should be constant, due to the law of conservation of energy. For a small source area dA, this flux may be expressed as

$$d\phi = L\pi dA \sin^2 u, \tag{1.1}$$

where $d\phi$ is the flux and L is the **radiance** of the source, which is known to be independent of the angle u if the source is *Lambertian* (think of a diffuse scattering sheet of paper, or the surface of the Sun) [4]. In addition, it is known that L remains constant as rays pass through apertures and lenses [3, 4]. In other words, radiance is a conserved quantity and is invariant through lossless media. Even if there is loss by absorption, we can include a transmittance factor τ and write equation (1.1) as

$$d\phi = \tau L\pi dA \sin^2 u. \tag{1.2}$$

In equation (1.2), other than τ, nothing changes the magnitude of the transmitted flux. So, the quantity $\pi dA \sin^2 u$ is a conserved quantity, and it is known as the étendue for an optical system with a circular stop.

How does knowledge of the étendue help us in designing an optical system? Let us take as an example the design of a refractive telescope at two magnifications, whose layouts at those magnifications and prescription (at the lower magnification) are shown in figures 1.5 and 1.6, respectively. Figure 1.6 is displayed as it would appear in the *Lens Data Editor* of Ansys Zemax OpticStudio® (OS) professional, version 22.3. In the prescription (which is for the layout shown in figure 1.5(a)), the lens material entered is manufactured by Schott. To produce the system shown in figure 1.5(b), simply set the radius of surface 3 to 246.342 mm and the thicknesses of surfaces 5 and 12 to 373 and 26.32 mm, respectively. The wavelengths are at the F, d, and C atomic lines, weights = 1.

Telescope objectives and eyepieces can usually be designed separately, followed by combining them. If needed, a bit of re-optimization may be performed. Neither the objective nor the eyepieces need to be designed first, but the ray properties in each must be considered. For instance, there is the étendue. As will be made clear later, once the objective's diameter and the field of view (FOV, given by the half-angle subtended by the distant scene of interest) has been defined, all ray properties following the objective are determined and fixed. In particular, any increase in the effective focal length (EFL) of the objective reduces the numerical aperture (NA) of rays behind the objective. That is, the half-angle for the cone of converging axial (on-axis) rays behind the objective gets smaller. Accordingly, the size of the

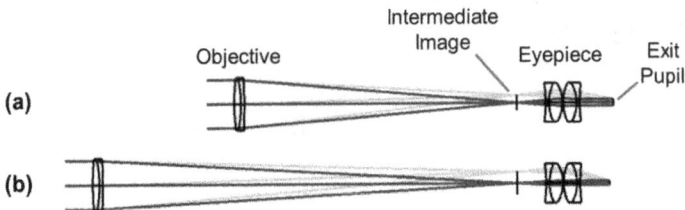

Figure 1.5. The impact of étendue conservation in a refractive telescope: (a) the telescope at 7× magnification. (b) The telescope at 10.7× magnification.

Lens Data

Update: All Windows

Surface 0 Properties — Configuration 1/2

	Surface Type		Comment	Radius	Thickness	Material	Coating	Clear Semi-Dia
0	OBJECT	Standard		Infinity	Infinity			Infinity
1		Standard	"Dummy"	Infinity	25.00000			21.65465
2	STOP	Standard	Stop	Infinity	0.00000			21.00000 U
3	(aper)	Standard	Objective	156.38900	6.00000	N-BAK1		23.00000 U
4	(aper)	Standard		-107.30700	3.00000	N-SF5		23.00000 P
5	(aper)	Standard		-523.78200	245.30000			23.00000 P
6		Standard	Int Image	Infinity	26.00000			6.56190
7	(aper)	Standard	Eyepiece, 1	-116.64800	3.00000	N-SF10		17.00000 U
8	(aper)	Standard		39.34000	12.00000	N-BAF10		17.00000 U
9	(aper)	Standard		-31.09700	0.50000			17.00000 U
10	(aper)	Standard	Eyepiece, 2	38.15600	12.00000	N-BAF10		17.00000 U
11	(aper)	Standard		-33.16700	3.00000	N-SF10		17.00000 U
12	(aper)	Standard		144.51100	29.94000			17.00000 U
13	IMAGE	Standard	Exit Pupil	Infinity	-			2.94416

Figure 1.6. Lens prescription for the telescope in figure 1.5(a) (units in mm).

intermediate image in front of the eyepiece (see figure 1.5(a)) gets larger, so that in consideration of this effect, an eyepiece must either be selected (or designed) to match the objective in the sense of having sufficient size to collect the oblique (off-axis) rays. Moreover, the half-angle FOV defines the *direction* of the oblique rays, so that the eyepiece must account for this angle. Otherwise, once the oblique rays have been refracted through the eyepiece, the exit pupil may not be properly located for the observer who will eventually use the telescope.

Note that the objective's diameter (and therefore, the stop diameter) has been kept the same in both magnifications (due perhaps to a requirement to maintain the telescope's weight and the diameter of its barrel, which is a reasonable design constraint). This restricts the étendue of the telescope system at both magnifications. Consequently, as shall be shown in due course, the size of the exit pupil has shrunk in the case of the telescope with higher magnification. This is why it is often more uncomfortable to look through eyepieces when the magnification for a 'scope' is increased for a fixed size of the objective. There is no other way around this discomfort at high magnification unless the diameter of the stop (and therefore, also that of the objective) is increased in proportion to the increase in magnification. This results in the need to increase the étendue of the telescope system, which is permissible as long as there is available étendue from the source for the telescope's aperture stop to collect (hold that thought—we will return to that point near the end of this section).

Let us now gain some insight into how, exactly, the telescope's étendue impacts the size of the exit pupil in the current example above. Recall from figure 1.3 that the

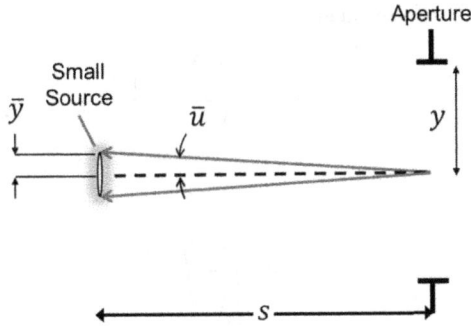

Figure 1.7. Definition of the angle $\bar{u} \approx \bar{y}/s$, based on the same source and aperture geometry as that shown in figure 1.4. For simplicity, $\bar{u}$ is *unconventionally* considered positive here.

stops control the passage of rays through an optical system. In the case of the telescope example, the stop (whose semi-diameter is 21 mm, according to the prescription in figure 1.6) is at the front vertex of the objective. Once rays from the object have been limited by the stop, the étendue for the telescope has been defined, and we shall now see how to express it and how it is related to the size of the exit pupil. To that end, have a look back at figure 1.4. There, you may note that, for large distances s, the sine of the angle u may be approximately given by the ratio of the aperture's semi-diameter to s. If we let y be the semi-diameter of the aperture, then equation (1.2) may be written as

$$d\phi \approx \tau L \pi dA (y/s)^2. \tag{1.3}$$

If the source is a small disk with semi-diameter $\bar{y}$, then $dA \approx \pi \bar{y}^2$. Substituting this into equation (1.3), we have

$$d\phi \approx \tau L \pi^2 \bar{y}^2 (y/s)^2 = \tau L \pi^2 [(y\bar{y})/s]^2. \tag{1.4}$$

In equation (1.4), the quantity $\pi^2 [(y\bar{y})/s]^2$ is approximately the étendue, and the ratio $\bar{y}/s$ can be identified as the absolute value of the angle $\bar{u}$ shown in figure 1.7. Since the étendue is conserved, the absolute value of the quantity $(y\bar{y})/s$ is constant. Therefore, for simplicity, regarding all quantities as absolute values, we have the following expression:

$$(y\bar{y})/s = y\left(\frac{\bar{y}}{s}\right) = y\bar{u} = \bar{y}\left(\frac{y}{s}\right) = \bar{y}u, \tag{1.5}$$

where u is the angle shown in figure 1.4.

In equation (1.5), the equality given by $y\bar{u} = \bar{y}u$ is one way to express the so-called **Lagrange invariant** of first-order (paraxial optics) ray tracing [5, 6]. Think of it as the one-dimensional paraxial version of étendue. We can therefore apply this invariant to explain how the étendue of the telescope impacts the size of its exit pupil. To do this, we may regard y as the semi-diameter of the telescope's aperture stop and $\bar{u}$ as the half-angle (in radians) subtended by the object at infinity. Since it

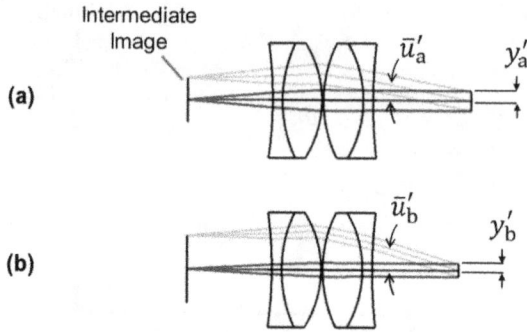

Figure 1.8. Definitions of the angles $\bar{u}_a'$ and $\bar{u}_b'$ and heights y_a' and y_b' at the exit pupils of the telescope systems of figure 1.5: (a) the eyepiece from figure 1.5(a). (b) The eyepiece from figure 1.5(b).

was mentioned earlier that this half-angle is 1.5°, then $\bar{u}$= 0.026 18 rad. Based on the prescription in figure 1.6, y= 21 mm. So, the Lagrange invariant of the telescope is $y\bar{u}$= (21)(0.026 18) ≈ 0.55 (we may ignore its units in the present discussion as long as we maintain the length and angle units as mm and rad, respectively). At the exit pupils of both telescope systems (figure 1.8), there are corresponding angles $\bar{u}_a'$ and $\bar{u}_b'$ and corresponding heights (semi-diameters) y_a' and y_b' for the systems in figures 1.5(a) and (b), respectively. If you model the telescopes using the prescription from figure 1.6, you would obtain $\bar{u}_a'$= 0.189 76 rad and $\bar{u}_b'$= 0.292 68 rad. Since the invariant quantity is 0.55, solving for the exit pupil heights, we obtain y_a'= 0.55/0.189 76 ≈ 2.9 mm, and y_b'= 0.55/0.292 68 ≈ 1.9 mm (we are purposely not being strict with significant figures).

Allow me to be long-winded in this section, because I want to show that there are many ways to arrive at the same result, which is a good way to check your answer. For instance, we could have applied first-order ray tracing to show that, since the exit pupil is the image of the aperture stop (formed by the eyepiece), it makes sense that its size in figure 1.5(b) should be smaller than in figure 1.5(a). Clearly, at equal stop diameters, when the objective has a longer EFL, it has to be shifted farther away from the eyepiece, so its image should be smaller. Further, we could have noted that the ratio given by $y_a'/y_b' ≈ 1.53$ is actually equal to dividing the higher magnification (10.7×) by the lower magnification (7×), yielding 10.7/7 ≈ 1.53. Finally, since $y\bar{u} = \bar{y}u$, and since in the current example we have only computed $y\bar{u}$ at the *pupil conjugates* (i.e. at the stop (which is also the entrance pupil) and at the exit pupil), you might perhaps want to know what $\bar{y}u$ is at the *object and image conjugates*. In this case, we would let $\bar{y}u$ be the Lagrange invariant at the object and $\bar{y}'u'$ be the invariant quantity at the intermediate image (figure 1.9). However, since the object is at infinity, then what is $\bar{y}u$ at the object plane? Well, infinity is not necessarily located at the edge of the universe. Rather, it is just a very distant location—sufficiently distant that its intermediate image formed by the objective is negligibly shifted from where it would have been if the object was truly at the edge of the universe. If you could follow the present exercise, you would find that by inserting a thickness value of 10^6 mm at surface 0 in the prescription shown in

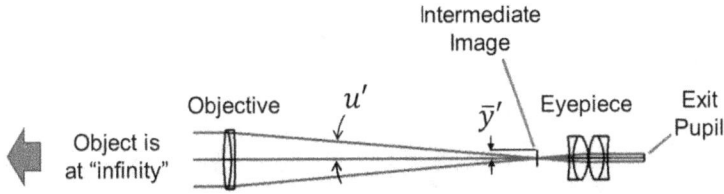

Figure 1.9. The definitions of the angle u' and height $\bar{y}'$ at the intermediate image plane of the telescope system in figure 1.5(a). Corresponding quantities may be defined for the system in figure 1.5(b).

figure 1.6, there would indeed be negligible difference in the location of the intermediate image. Given that $\bar{u} = 0.026\ 18$ rad, an object that is at 10^6 mm away would have a height $\bar{y} \approx 0.026\ 18 \times 10^6$ mm. Since the stop's height is 21 mm, then $u \approx 21/10^6 = 21 \times 10^{-6}$ rad. Thus, the Lagrange invariant at object–image conjugates is $\bar{y}u = (0.026\ 18 \times 10^6\ \text{mm})(21 \times 10^{-6}\ \text{rad}) \approx 0.55$. Hey, look! This is the same as the invariant computed at the pupil conjugates! Thus, $y\bar{u} = \bar{y}u$, as expected.

In a paragraph just prior to equation (1.3), I mentioned that it would be possible to increase the size of the exit pupil at high magnification by increasing the size of the telescope's aperture stop, provided that there is *available étendue* for the telescope's stop to receive from the source. What does this mean? It means that the half-angle u in figure 1.4 is defined by the stop's height y only if the source emits rays over an angle larger than y/s (e.g. if the source is truly Lambertian and emits over a hemisphere). As an extreme example, someone could have *imposed an aperture* on the rays from the source by placing the telescope inside a building (such as an observatory) and forgetting to slide open a window wide enough to accommodate a larger angle for u. In another example, perhaps the source might be a Lambertian emitter with some physical structure that limits its rays such that they span a finite angle. In either of these cases, even if you were to increase the size of the telescope's stop (and objective lens), you would not see an increased exit pupil size *if there is no light to produce it*. Another way to understand this is by virtue of the Lagrange invariant $y\bar{u} = \bar{y}u$. Here, we can see that y is fixed if one limits the quantity $\bar{y}u/\bar{u}$.

One last point. You may be familiar with the Lagrange invariant H being expressed as a surface-by-surface *difference* given by

$$H = \bar{y}_i u_i - y_i \bar{u}_i, \tag{1.6}$$

where the subscript i denotes the ith lens surface in the ray trace [5, 6]. I want to show that this formula is consistent with the expression shown in equation (1.5). In equation (1.6), there is ordinarily a sign convention that $\bar{u}_i$ is negative. Hence, the right-hand side of this equation is actually an *addition* (did you know this?—it is not obvious). In other words, the difference is really given by $H = \bar{y}_i u_i - \left[y_i(-|\bar{u}_i|) \right] = \bar{y}_i u_i + |y_i \bar{u}_i|$. If we consider that $i = 0$ at the object plane, then the Lagrange invariant at the object is $\bar{y}_0 u_0$. At some ith plane that is slightly beyond the object, we would have $\bar{y}_i < \bar{y}_0$ (a ray from the tip of the object points towards the center of the entrance pupil, so the height of this ray would be less than $\bar{y}_0$). According to the sign convention that $\bar{u}_i < 0$, the difference $\bar{y}_i u_i - y_i \bar{u}_i$ ensures that the reduction of $|\bar{y}_0 u_0|$ to $|\bar{y}_i u_i|$ at the next

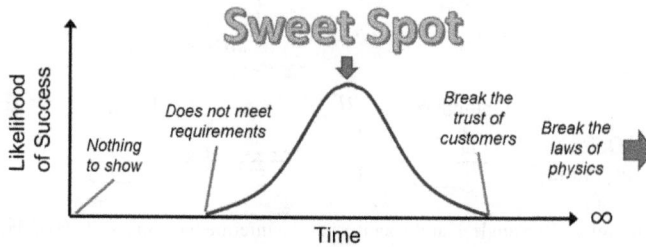

Figure 1.10. The limiting conditions of product development and the 'sweet spot.'

nearby plane would be compensated for by the addition of $\left| y_i \bar{u}_i \right|$ at that next plane, thus, satisfying the invariance. Incidentally, if the ith plane is the image plane, then $H = \bar{y}_0 u_0 = \left| y_i \bar{u}_i \right|$, since $\bar{y}_i = 0$ at that plane, which is precisely the expression in equation (1.5).

1.1.4 Realistic product development operates between these two limits

On one hand, if a product does not meet any requirements but is shipped to the customer earlier than expected, then it would be unlikely to be well received by the customer. On the other hand, a perfect product—one that far exceeds customer expectations—may take an excessive amount of time to complete, which may result in a loss of trust in your product on the part of the customer. Figure 1.10 depicts such limiting conditions in product development. In between these limits is a 'sweet spot' of realistic operation. This sweet spot can be the MVP, or it can be something better, but it cannot break the laws of physics. Shoot for the sweet spot. That is how to make realistic products.

1.1.5 What to do about ambiguous requirements

Some customers may not know how to specify what they want, but they may have a fair idea of what they want to see. Sometimes, they may not even know what they want to see, and they may be open to some suggestions. If you enjoy innovating, then chances are, you may actually like these sorts of customers (I do!), because such customers give you freedom to explore the creation of a variety of solutions. On the other hand, if you prefer clear specifications and requirements, you may need to look elsewhere for more 'savvy' customers (because those who are not able to tell you exactly what they want are unlikely to tell you exactly what they want, no matter how much you press them—because they simply do not know exactly what they want, and they do not know how to know what they want).

When customers may not know what they want, all you need to do is to offer options. I'm sure you could do this, even if you may prefer having clear specifications and requirements. However, if the customer does provide clear specifications and requirements, you may find it hard to meet them. If you want to negotiate, then you have to be prepared to 'lose' the bid for the project if someone else claims to be able to meet the requirements. Then, you might as well prefer that your customer is one of those who do not know exactly what they want. The point is that when there are ambiguous requirements, just offer suggestions and options. Your customer will love you for it. Further, if the requirements require clarification, then clarify them. If your customer may not understand technical jargon, then you have to ask and express your questions in a clear comprehensible fashion, with illustrations. If you cannot do this, then it may become quite difficult to have and retain customers.

One last thing. In some organizations, specifications and requirements can mean different things. Requirements may be thought of as 'wish lists' expressed in the *voice of the customer* (VOC), while specifications may be thought of as 'numbers' related to specific technical quantities (such as effective focal length, modulation transfer function, image distortion, etc.) However, it can be said that there is not necessarily any general agreement on the difference between specifications and requirements. Ultimately, it is just a definition that has to be agreed upon among stakeholders of a product development project, including the customer.

1.1.6 Your list of 'things I still need to understand' will only grow (which is fine)

If you are reading this book, then chances are you may have a list of things you still need to understand. This is good, because who knows where they may lead to in addition to learning new things about designing optical systems. This book, for example, is a consequence of my own list of things I needed to understand. And even when I have completed writing this book, I will still have a long list of more things I need to understand.

So, do not worry if your list of things you still need to understand keeps on growing. This does not mean that you are not making progress, and it does not mean that you are a lesser person than individuals who may appear to know everything, which is impossible in any case (maybe they do not even know that they do not know everything). College professors, for example, always have a list of things they still need to understand. Otherwise, their careers would end, as they would no longer be making discoveries from their list of unanswered questions, and they would not be publishing any new research findings.

But what does a list of things you still need to understand have to do with engineering and product development? The answer is that if your list does not grow, then neither does your knowledge. You do not need to be a competitive person to realize that this means you may lose out on employment opportunities. The point is that if you continue to have a growing list of things you still need to understand, then provided that you do not stop searching for answers, you will keep learning and be able to offer new solutions to new problems, thereby sustaining your employability. (However, I should point out that this is not the main reason that I have a growing list of things I still need to understand. For me, I just enjoy finding things out and using new knowledge to come up with new ideas.)

1.1.7 Everything you really want to do will likely take place only after office hours

In the real world, you will often be hired to do the things a company needs you to do. Sometimes, you may have a new idea, and you may think about asking for time to work on it. If you are lucky, you get to have that time. Sometimes, a company may not be able to give that time to you. As long as you still want to work for the company that you are in, then very often, the only way for you to realize that new idea of yours is to do it after office hours and perhaps even on weekends. In my career, I have often done this. It was not easy, but I figured that in those situations, it would just be a matter of making temporary sacrifices of my personal time. From time to time, I would take breaks, then charge forward once again. This was how I approached the problem of the work–life balance—by making it my personal decision on when to go above and beyond the call of duty and when to stop to recharge.

Ex-colleagues asked me why would I bother doing things that may not be appreciated, since they were not part of my job, and that I would not be getting paid for it. I did not have an answer. Somehow, I just wanted to find out if my ideas would work. And in some of the cases in which I did research and investigations at a company after office hours, I did get my ideas to work, followed by reporting on them. This was enough to satisfy my curiosity. Years later, when I had already left one company, a senior manager (who was still at the company) who remembered what I had done after office hours made a recommendation to another company that I would be a good optics consultant. And the rest, as they say, is history.

There is never any guarantee that we will be so lucky as to get something back in return. You have to do it because you want to do it, not because you hope that someone will recognize your work. Similarly, in the context of learning modern classical optical system design, you may not be given free time within company hours to read this book and try the examples. Often, you would have to make time for it after office hours. Do it whenever you can. It does get harder when you get older, not necessarily because of age but because of added responsibilities and commitments (new bills, a mortgage, a family, extended families, etc.)

1.1.8 Why books are still necessary for your knowledge

Many years ago, when I was employed as a full-time lens designer, I would undergo an activity I called 'insanely rapid research and development.' This was an activity that often began with customer specifications for a lens that I had never designed. But I had to design it nonetheless. Often, I would have just a week or two to come up with something that the business development manager could use to quote the price for that lens. So, I had to read books and book chapters on how to design the lens, including a review of the relevant fundamental aberrations of the typical design form of the lens. I spent nights and weekends doing research, doing exercises, and doing many optimization trials to come up with a rough design. This was the only way to meet the deadline for pricing the lens.

Today, I continue to read and study books and research papers because my customers and my employer's customers constantly challenge me with problems that

I may have never solved, with questions that I may have never answered, and with design requests that I may not have experience in. I cannot tell a customer that just because I solved a problem or designed a system in a certain way in a previous company (and that it worked), then I would be able to do the same for their company. Not only could that possibly involve intellectual property issues, but it could also make me appear uncreative. Moreover, there would be no way for me to prove that what I did had worked, as everything I did for a prior company would be under a nondisclosure agreement. I need to justify my design approaches and solutions not just by words and expressions of prior experience but with clear rationale, models, calculations, and predictions and by citing published work. In fact, this is no different from the ability of professors and researchers to produce new findings based on fundamental knowledge. Experience alone cannot help to grow and sustain my knowledge. How many companies must I work at in order to acquire sufficient experience? Is it impressive if I say that I had worked in such-and-such company with a big branded name? No, I cannot rely on this. I need to keep reading and try to produce new material based on what I read.

1.1.9 You become good at something by doing it over and over for a very long time

Before retiring, David Shafer spent over fifty years designing lenses [7]. This is why he is good at what he does. Similarly, if you have watched Netflix's *The Last Dance* documentary on the Chicago Bulls basketball team of the 1997–98 NBA season, you will have found out how hard Michael Jordan worked at perfecting his basketball skills over many years. It is a famous story that he was not selected to be part of his high school's varsity basketball team. But over time, his hard work ethic won him two Olympic gold medals and six NBA championships.

There is no short cut to being good at something. Even though this book shares bags of tricks to hopefully help save you some needed time in your work as an optical system designer, these tips and tricks can only help save you time in getting started. You still need to put in hard work and many long years of designing optical systems before you become an expert. I started 'designing' optical systems when I was eleven (and because I was not the smartest kid in class, I am glad that I started doing optical design early!) Sure, as an eleven-year-old, my designs were simple and crude instruments made out of card boxes. Sometimes, they did not work. But over time, I became better at it by trial and error, by self-studying from textbooks, by asking teachers and authors 'dumb' questions, by going to college (and asking professors more dumb questions), and by working as an optical system designer for 25 years. Today, I am writing this book. It took quite some time for me to get here, yet I am still learning new things. And in the real world of deadlines, it does not get any easier. So, if you want to be good at optical systems design, then you have to endure many long hours of designing optical systems. Solving brainteaser puzzles and winning chess games—while stimulating for the brain—are not going to help you become a good optical system designer. The only way to be good at designing optical systems is to design optical systems.

Today, modern optical design programs make it possible to for you to do design exercises by creating 'mock' designs of optical systems. You never have to actually build them but just build them virtually in the program. As you will see throughout this book, many of the simulation features of an optical design program will help you understand and check your design. Moreover, if your company has an optics lab with off-the-shelf components, you can design your mock optical system using models of those components in an optical design program, then just try them out in the optics lab (we will talk about how to do this in section 1.3). Nobody will pay you to do these mock design exercises. So, is it worth it? Absolutely. Sometimes, instead of investing in the stock market, it can be more worthwhile to invest in yourself—to invest in your own time to develop new and special skills (optical design is a much-needed special skill).

1.1.10 How to bug people for help

You will never be alone in your quest to learn optical system design and to further your knowledge in this field. Reading this book and other books is a good start. Asking people for some help can be a bonus. Who would be so willing to give you free advice? Actually, many would. But getting their attention and advice works best if you begin by asking them questions within some context. A good context is to begin by asking them questions about something from their work, such as a book or book chapter that they wrote, or a paper that they published. They would appreciate that. Consequently, they would be pleased to help you.

1.1.11 Truth = the best estimate ± uncertainty

Measured quantities are associated with uncertainties, and any quantity that has been computed from data is destined to vary each time new data is collected with which to compute that quantity. This is due in part to measurement error, and, where applicable, quantum mechanics. Perhaps quantum fluctuations can be considered negligible at macroscopic scales, but in optics, you can see them often when you capture images at low light levels (such as fluorescent and luminescent bio-samples) using a digital camera. You can see it in the pixels, where a histogram of pixel intensities within a region of interest (ROI) has a spread of values. This is Poisson noise—the behavior of light quanta striking the pixels of the image sensor randomly but at an average rate, and the conversion of fractions of lumps of the energy into an accumulation of electric charge. When you have to report a number associated with that detected radiation, you find that each time you take a measurement, the number has changed a little.

So, what do you do? You must take more than a single measurement. Each measurement is an *estimate* of the *true* value, and the average is that true value's *best estimate*. So, you compute the average. You also need to compute a quantity that estimates the spread (i.e. the error, uncertainty, or variability) of your computed average. This is known as the standard deviation of the average. Note that this is not the same as the standard deviation of the measured values that were used to compute the average. The standard deviation of the measured values represents the error of a single measurement, while the error in the average is the standard deviation divided by the square root of the number of measurements. This is a consequence of the **central limit theorem** (references to this and other statistical concepts are made in chapter 3). So, by this theorem, you would report your computed quantity as $\bar{x} \pm \sigma/\sqrt{n}$, where $\bar{x}$ is the average, σ is the standard deviation, and n is the number of measurements. In science and engineering, *truth* is, at best, a best estimate (the average) plus or minus its uncertainty (the standard deviation times $1/\sqrt{n}$, which represents a sort of *degree of reliability*). And that is not all. Even the standard deviation has a standard deviation, but we will get into that in chapter 3.

In some aspects of the *real world*, where the media and world politics are involved, it is possible for people to form different perspectives on what is meant by *truth*. Sometimes, truth can appear as *fact from an authority*, such as perhaps, an expert. But a good expert knows that there is no absolute truth in anything, other than a best guess and a measure of its reliability. An effective approach to science and engineering is to regard all fact as temporary truth. This is the reason that academic peer-refereed research papers are published and continue to be reviewed, critiqued, and perhaps even criticized. When new knowledge is found, textbooks are rewritten. Knowledge progresses by this endless cycle of discovery, review, and rewritten facts. This cannot be any different in product development, which is the reason that even some engineers (like this author) perform research and publish books. If you write what you have learned, you learn even more (somehow, it just works that way).

As you progress in your learning journey in optical system design, you will face situations in which a measured or computed quantity does not remain the same each time you make the measurement or computation—which is something you need to get used to. One example is in the analysis of your designed optical system under manufacturing conditions, where all the quantities that specify the optical components must be associated with **tolerances**. Another example is what is called nonsequential (NSC) ray tracing, which we will discuss in a number of sections in this book. NSC ray tracing is used by optical design programs to simulate rays for illumination, and the rays are spat out randomly, governed by a probabilistic distribution. In real life, measured and computed quantities are best guesses that can vary. Put simply, truth equals a best estimate plus or minus uncertainty.

1.1.12 Are you ready for this?

If you have a list of things that you still do not understand, if you are willing to pick up and read not only this book but many other books by authors with different methods and perspectives, and if you are willing to put in the effort and time to design many optical systems (even virtually, using an optical design program), then I think you are ready to embark on your journey to become an effective optical system designer for product development. The format of this book is such that sections 1.2 and 1.3 provide the foundation for subsequent sections and chapters. While, in the context of this book, the word 'design' in *optical system design* mostly involves the selection and integration of commercial off-the-shelf (COTS) components, it is often beneficial for optical system designers to know how to perform lens design optimization. For this reason, sections 1.4 and 1.5 provide some basic foundations and tips for practical computer-aided lens design and the creation of optical drawings for manufacture. We will not dwell on details that can be found in texts that specialize in lens design. After all, this book is about getting things done in a fast-paced product development context, so we ought to get to the point of the subject matter efficiently. Thus, main points are highlighted and topics are often illustrative. Those who have read *Perspectives on Modern Optics and Imaging* [8] have seen a familiar format (except I do not have any cartoons to show here).

As mentioned in the Preface, in MCOSD, we will be using the Ansys Zemax OpticStudio® (OS) optical design program. The next section introduces you to this program, but it is not intended to be a thorough course on the program. Rather, it provides a brief overview (and sort of a refresher) on the program. It is assumed that you will refer to many of the online resources available from Ansys® and Zemax® (such as videos on YouTube, knowledge base articles, courses, etc.) As such, in some cases, it is purposely taken for granted that you are familiar with specific terms and features of the program. Feeling good? Good! Then let us forge ahead.

1.2 Optical system design using Ansys Zemax OpticStudio®

Figure 1.11 shows what the user interface in OS (professional edition, version 22.3) typically looks like. By default, the program opens up in sequential (SC) ray tracing mode. The *Lens Data Editor* (the leftmost tab inside the light blue circle) is likely to be displayed as shown, with three surfaces (OBJECT, STOP, IMAGE). The concept is that you enter data (such as lens radii of curvature) into *data editors*. You need to play around a bit with how the tabs and individual windows are displayed in this user interface to get used to how to make the windows appear as shown in figure 1.11. You can even left-click, hold, and drag each tab/window around.

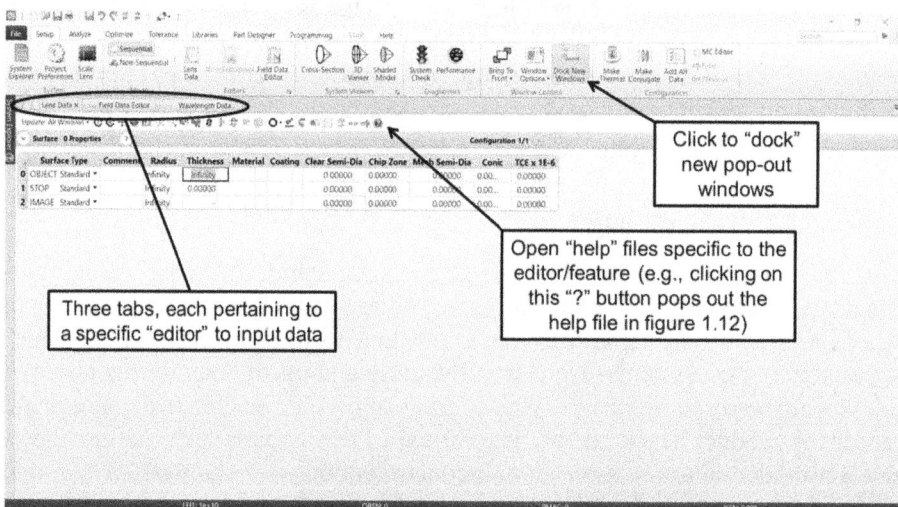

Figure 1.11. The user interface of OS (professional edition, version 22.3) for a *blank* new file.

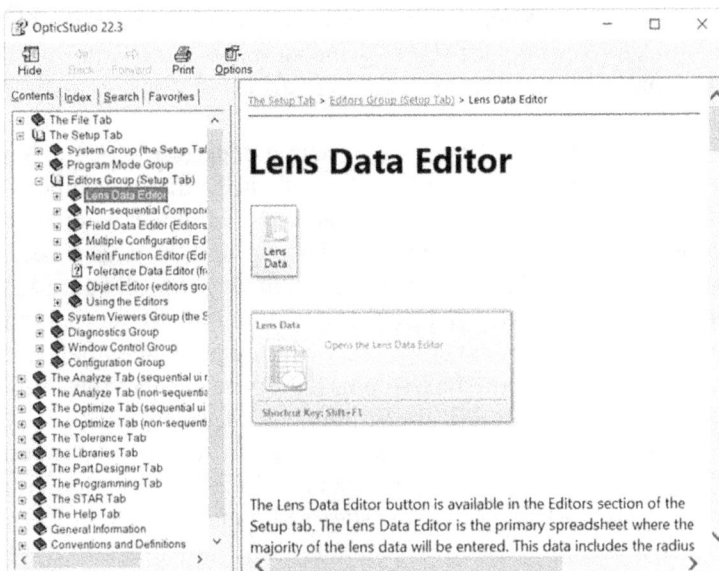

Figure 1.12. This help file appears upon clicking the '?' button for the Lens Data Editor. Reproduced from [9], with the permission of ANSYS, Inc.

Personally, I like to dock all the windows related to data editors and display layouts and analysis graphics (such as spot diagrams, MTF, distortion curves, etc.) as panels below and to the right of the screen.

At the top of any of the program's windows, there is a row of buttons with icons or symbols; clicking on any of these triggers a program feature. Scroll to the right of these icons and you will see a button with a '?' symbol. Clicking on the '?' symbol pops out a help file for that specific editor or feature. All of the program features, editors, and windows in OS come with this '?' button to click on, which pops out specific help files that describe elements of that program feature, editor, or window. For instance, in figure 1.11, clicking on the '?' button of the Lens Data Editor pops out a separate window as shown in figure 1.12, which is a help file specific to the Lens Data Editor.

Scrolling down the menu to the left of this help file (figure 1.12) also points to various other help files. Further, clicking on the Help menu provides a variety of options for guidance. This book assumes that you have either taken an introductory course on using OS or browsed a variety of online resources. There are also many helpful sample OS lens and optical system files you can open to get a rough feel for how optical systems are put together in this program. To see this, simply left-click on the File menu, select 'Open,' and you should see, by default, a list of folders; one of them is called 'Samples,' in which you should be able to navigate and see many lens examples. If you are not pointed to that path when you click on File > Open, then simply locate the file path for Documents > Zemax > Samples. By the way, in the Samples folder, you may note that there is a folder called 'Design Applications,' in which there should be a folder for various lens examples from one of my earlier books [8]. If you keep opening sample files and simply play around with the system, trying different program features and clicking various buttons, you will get a good feeling for the user interface of OS.

How can you begin building an optical system model from scratch? In particular, let us suppose we were starting from the blank file in figure 1.11. Where do you go from here? In any optical design program, **you first need to tell the program how you want rays to enter the optical system that you intend to model**. This is because in real life, light can fill an optical system in many different ways (recall figure 1.3 in section 1.1.1). For example, if the object to be imaged is at a finite distance from the optical system, and if it emits (or scatters) light over a wide angle such that all rays flood the optical system, then it is the **aperture stop** in the optical system that limits the rays that pass through it (e.g. figure 1.3(b)). In this case, there has to be a way to *set the aperture* of the system in the optical design program to mimic this condition. In other cases, perhaps an existing aperture has already defined the rays entering your optical model; in this case, the *system aperture* setting in the optical design program should create the defined rays, and you should not let any aperture in your optical system model cut off (vignette) those rays. In the following section, we shall discuss how this is done in OS.

1.2.1 Set the aperture

To begin, let us reduce clutter and make more room on your computer screen. Right-click anywhere across the top row of the program's feature icons and select 'Minimize Ribbon.' Next, right-click on the row for surface 1 and select 'Insert Surface.' This produces a new row for surface 1, resulting in four surfaces. Enter 10 for the thickness of surface 1, and enter 250 for the thickness of surface 2. You will not see anything yet (wait for it). Next, navigate towards the top left of the menus and select 'Setup,' then select 'System Explorer.' This opens up a left panel with a drop-down menu of parameters (Aperture, Fields, Wavelengths, Environment, etc.) Select 'Aperture,' then 'Entrance Pupil Diameter,' then enter 25 for the 'Aperture Value.' Next, navigate to the Setup menu, left-click on the 'Cross-Section' icon, and out pops a layout of your optical system, which is just a hole 25 mm in diameter. Dock this window somewhere (say, at the bottom of your screen). You should now see the view shown in figure 1.13, which is essentially a model of a 25 mm diameter hole, letting rays pass through it over a distance of 250 mm (but displaying 10 mm of space in front of the hole), where rays begin their journey from infinity. Lens designers often like to see where rays came from before they entered a lens system. It it is a good idea for optical system designers to do the same.

By default, the units are in mm. In OS, the mm unit is often called *lens units*, perhaps because most lenses and optical components have sizes on the scale of millimeters, which seems like a common natural size relative to humans. You can switch units (scroll to the bottom of the System Explorer menu and you will see it). But we will stick to the default. Another default is the wavelength of the rays, which is set to monochromatic at 0.55 microns (we will return to wavelength settings later).

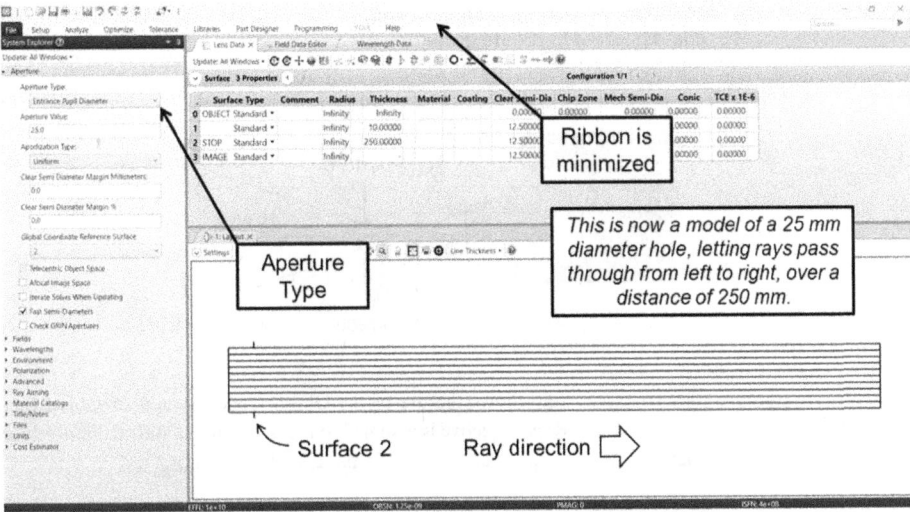

Figure 1.13. The layout of rays propagating through a hole that has a 25 mm diameter in OS.

At this point, it is also a good idea to take note of the *x*, *y*, and *z* coordinate system of the ray-trace layout. By hovering your cursor over any space on the layout (without clicking), you can have the horizontal (H) and vertical (V) coordinates displayed. By default, these are the *z*-axis and *y*-axis coordinate values, respectively, on the layout, and the *global coordinate* (i.e. the point at which *x*, *y*, and *z* equal zero) is usually located at the stop. To check this, put your cursor on any cell in the STOP row (i.e. surface 2 in figure 1.13). Then, at the top left corner of the Lens Data Editor, left-click on the ∨ symbol (towards the left of *Surface Properties*) and you will notice a drop-down menu that lists options for Type, Draw, Aperture, etc. Look at the Type menu, navigate right, and you will see a tick mark on 'Make Surface Global Coordinate.' If the tick mark is not on that surface, then it is on another surface, and you can easily see which surface it is on by clicking on the < and > buttons (which navigate you to the next surfaces up and down, respectively). Alternatively, you can return to the ray trace layout and hover your mouse cursor over the space in the ray trace layout until the *H* and *V* coordinates equal zero at a surface's vertex.

Now, I mentioned that light can enter your optical system in many different ways. In figure 1.13, the hole is the only aperture in the optical system. So, it is the aperture stop of the system, and it is also the entrance pupil. In OS SC mode, there is no such thing as not having an aperture stop. In fact, this is generally true in any optical design program that performs SC ray tracing. There is always a stop. If you have studied lens design, you already know this, but let me repeat the concept here for completeness: the stop is the smallest aperture in an optical system and it is the aperture that defines the NA (equivalently, the *f*-number) in the image space of the optical system. The stop also controls aberrations of the optical system by letting

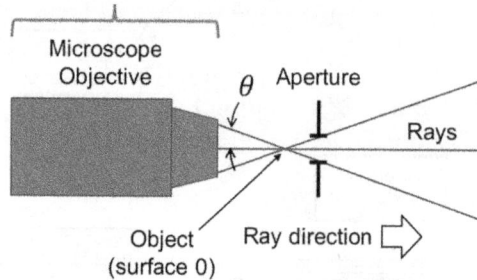

Subsystem not being modelled, and it is assumed to create a perfect focused spot at the object plane

Figure 1.14. An example of a condition that allows the use of the Object Cone Angle θ or Object Space NA = $\sin \theta$ as the Aperture Type. A microscope objective is assumed to focus rays into a diffraction-free point source, so this source is modeled as a point object at surface 0 in the Lens Data Editor.

axial and off-axis *field rays* (i.e. rays from field points, sometimes also called oblique rays) travel specific paths through lens and mirror systems.

So, in the current setup (figure 1.13), the model of an optical system with a single aperture is a model comprising the stop as the aperture, which has a specified entrance pupil diameter. Clearly, this lets you model a condition in which rays strike an optical system comprising only a single aperture (the stop), and all rays outside of the aperture are blocked, leaving only rays traveling through the aperture. If you scroll down the options for the aperture setting (under the Aperture Type drop-down menu), you will see other ways to let rays in, which include the following settings: Image Space *F/#*, Object Space NA (numerical aperture), Float by Stop Size, Paraxial Working *F/#*, and Object Cone Angle. The first and third of these are not reasonable choices for a single-aperture optical system (i.e. figure 1.13), because there is no lens or mirror to create an *F/#* (*f*-number, which is defined in OS as the ratio of the lens EFL to the *paraxial* entrance pupil diameter). However, the rest are perfectly reasonable options for various conditions. For instance, if you made your object distance finite (i.e. by inserting a nonzero value into the thickness of surface 0 in the Lens Data Editor), then you can model conditions in which there is a point source at the object emitting rays with a specified NA or cone angle entering an aperture. One example could be that you may have a diffraction-limited microscope objective lens on the left-hand side of your aperture, and the objective is assumed to focus axial rays to a perfect point source in front of the aperture (figure 1.14). Perhaps an approximation you are making is to regard the focused spot as a diffraction-free point source, so that you can avoid modeling the objective lens and instead create a point source for the object, which possesses a diverging cone of rays with half-angle θ, or NA = $\sin \theta$. Finally, perhaps you wish to determine an appropriate aperture diameter at some finite distance to the right of the point source, so you let θ define that size. Note that when you select either Object Space NA or Object Cone Angle for the aperture setting, your stop size then possesses the value that results from the specified NA or cone angle for the object.

1.2.2 Set the fields

If you play around with the layout given in figure 1.13, it will probably not be long before you realize that you can create a model of a pinhole camera (without diffraction effects). You simply need to adjust the size of the pinhole by changing the Aperture Value for the Entrance Pupil Diameter setting. Alternatively, select 'Float by Stop Size' for the Aperture Type, then adjust the stop size directly by entering any value for the Clear Semi-Diameter column on surface 2 in the Lens Data Editor. Let us create an aperture 10 mm in diameter (this aperture will eventually become the stop for a lens in section 1.2.7). So, using the file from figure 1.13, set a value of five for the Clear Semi-Diameter on surface 2. You may then wonder how to generate rays from across a field of view (FOV). To do this, navigate towards the Field Data Editor. If this editor has not shown up on your user interface, then go to the System Explorer drop-down menu and double-click on 'Fields' (see the left-hand side of figure 1.13), then right-click to insert more fields and enter 6° and 12° for the Y Angle, as shown in figure 1.15.

By default, the field settings are given in terms of angle, in degrees. Under the Field Properties menu, you may notice a parameter called 'Normalization,' which has been set to 'Radial.' That setting has particular significance with regards to how rays are traced through the entrance pupil of the optical system (but do not worry about this for now, as you will get to know *field normalization* in the relevant sections of this book). Your system should now appear as shown in figure 1.16. Generally, the term *fields* can mean either angles or heights (and *height* means a length above the optic axis). Note that there are *weights* for each field, but they are only applicable to certain types of **merit function operands** in OS (you will get more familiar with the merit function and operands later—so do not worry about them now).

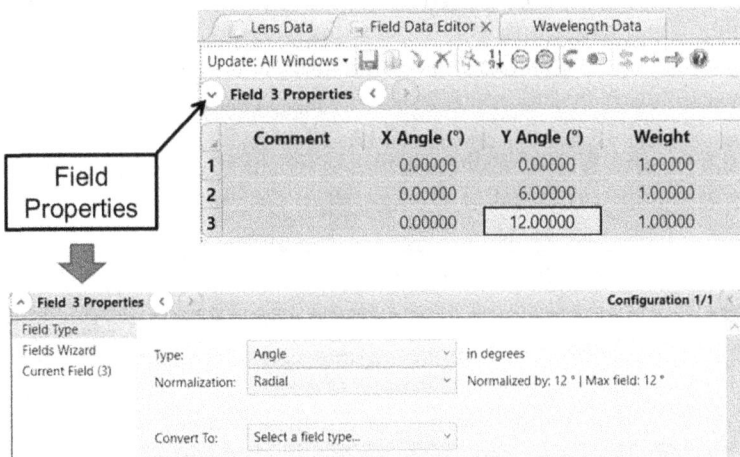

Figure 1.15. Field settings and properties for the system in figure 1.13. The intent is to set this system up to become that shown in figure 1.16.

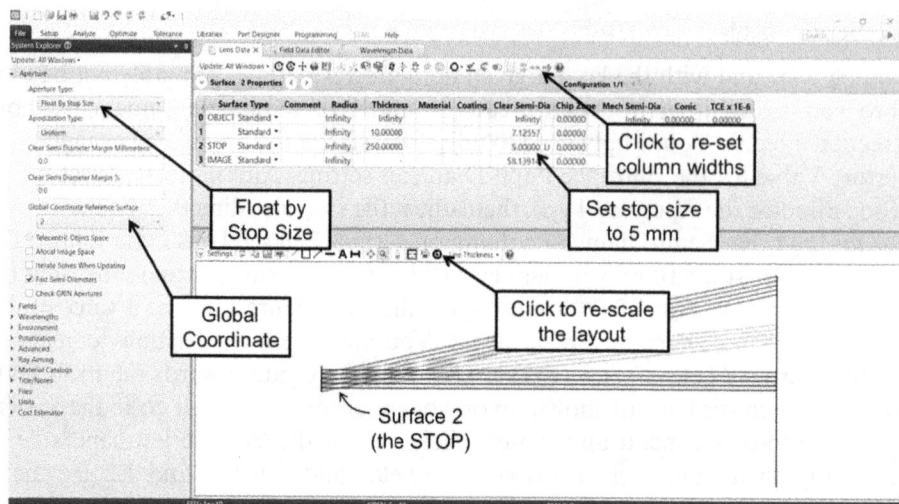

Figure 1.16. The user interface from figure 1.13 after the field settings from figure 1.15 have been applied.

In figure 1.16, note that the Global Coordinate can be set within the System Explorer menu, as indicated by the arrow pointing towards that setting below the Aperture Type (so, you have the option of setting the Global Coordinate either here or in the Surface Properties of the Lens Data Editor). It is also worthwhile to note that there are plenty of graphics options. For instance, clicking on the black button with a curved arrow re-scales the layout to fit everything nicely into the graphics window. You can also hide the display of specific surfaces. For instance, perhaps you may not like the appearance of the *dummy surface* on surface 1, since, in real life, there may be nothing there (but you still want to see rays before they enter the stop). Just navigate to the Surface Properties of surface 1 and select 'Draw,' then select 'Do Not Draw This Surface.' In the Lens Data Editor, you may sometimes see that the column widths are not optimally set, and instead of manually adjusting each column (by dragging it), you can click on the double horizontal arrow, which resets all of the widths such that all numerical values in cells are fully displayed. Finally, you may note that the rays from each field angle are color coded in the layout. This is my default setting. You can change this by clicking the settings button for the layout display window (located at the top left of the window).

1.2.3 Set the wavelengths (I will tell you how many you need)

I will provide the theoretical justification for setting *spectral weights* in an optical design program in section 1.2.12. In the current section, let me give you the gist of it. In the file for figure 1.16, if the Wavelength Data Editor is not already displayed in one of the tabs of your editors on the screen, then double-click on 'Wavelengths' under the System Explorer menu. I have included three wavelengths by ticking checkmarks on the left boxes and entering the values (in microns), as shown in figure 1.17. The choices of *weights* roughly span the visible spectrum, *assuming that*

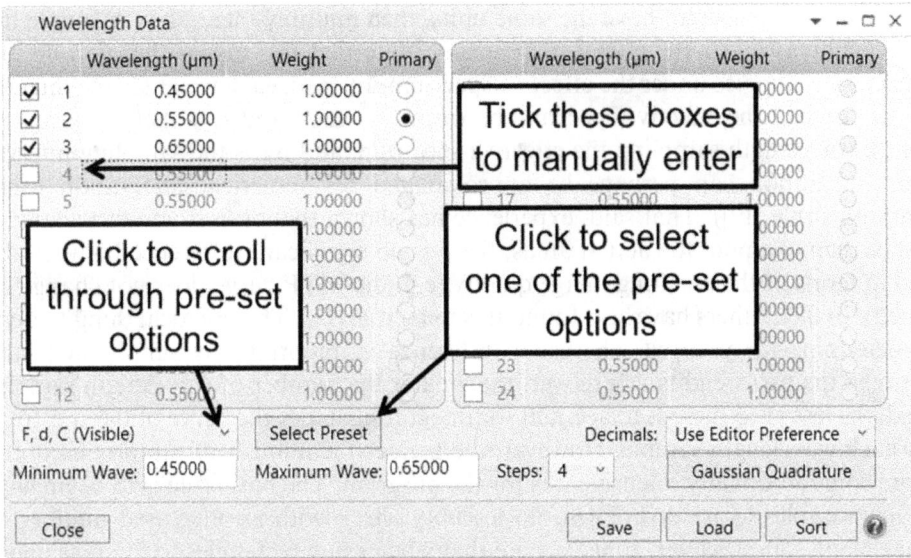

Figure 1.17. Three wavelength settings in the Wavelength Data Editor.

the object emits a flat spectrum across the visible spectrum, and assuming that there is a detector at the image plane with that same flat spectral response and with the same choice of units for the object's spectral function. If this detector had been the human eye, we would have needed to use a different set of wavelengths and weights for each, such as the so-called *Photopic* spectrum [4] (you can see this selection by scrolling down the preset options). In addition, if the object emitted some other spectrum, then the wavelength settings (and weights) would have had to sample the *product* of the object's spectrum and the detector's spectrum. We may as well call this spectrum the *effective spectrum* of the optical system, as it comprises the object (source) and the spectral response at the image plane (detector). Further, if there are any bandpass filters or optical coatings on the lenses in the ray path, their spectra must also be accounted for in the effective spectrum.

For an optical system comprising just a source and an aperture, the weights may be thought of as discretized sampled *areas* for a narrow wavelength band (centered on that wavelength) divided by the total area of the effective spectrum (so that the sum of the weights equals one). However, as I will show in section 1.2.12, in general cases, you would not divide each sampled area by the total effective spectrum's area. The usual approach is to use the sampled areas as the weights [10, 11]. Note that for a flat effective spectrum, there is no reason to use '1' as the weight for each (because the spectrum is uniform). For a non-flat spectrum, the weights are the sampled areas of the spectrum, so each weight represents the relative *importance* of each wavelength, regardless of the vertical scale used for the intensity of the spectrum. This means that if you had plotted the source and detector spectra using, say, Microsoft® Excel®, you would not need to normalize each spectrum's intensity, neither would you need to normalize the effective spectrum's intensity. You would only need to ensure that both spectra have the same units, then multiply them, then calculate the sampled areas under the effective spectrum. Note that this means that you do not simply fit rectangles under the effective spectrum. Instead, each weight is the integral of the curve of the effective spectrum between $\lambda_i - \Delta\lambda/2$ and $\lambda_i + \Delta\lambda/2$, where λ_i is the ith wavelength sampling the effective spectrum, and $\Delta\lambda$ is the sampling interval (this integration can perhaps be approximated by numerical integration using Simpson's rule [12]). That said, experience has shown that even if you fit rectangles rather than compute the actual areas, there is no significant difference between the two in terms of the resulting image quality (e.g. the MTF curve does not change by much). In fact, others have also found this to be true [11]. The practical thing to do is to determine for yourself whether a difference is important by simply switching between the two weights. Moreover, the greater the number of wavelength samples used, the less need there is to account for integrating under the curve of the spectrum for each wavelength sample. However, the greater the number of samples used, the more computationally intensive it is for the program. The basic rule is to begin with as few samples as needed. Then, do a sanity check with an increased number of wavelength samples later. Finally, note that when using rectangles to fit areas under the effective spectrum, the relative weights are simply given by the points on the vertical scale of the effective spectrum. This is because the wavelength intervals $\Delta\lambda$ for each rectangle are the same. At any rate, do not worry—you will learn more about spectral weights and see how all this is done in section 1.2.12.

1.2.4 Create a paraxial thin-lens equivalent system

Using the lens file from figure 1.16, first make some room on your screen by closing the System Explorer panel. Then, create the thin-lens system shown in figure 1.18 by entering the prescription that is in the Lens Data Editor of figure 1.18. To create the

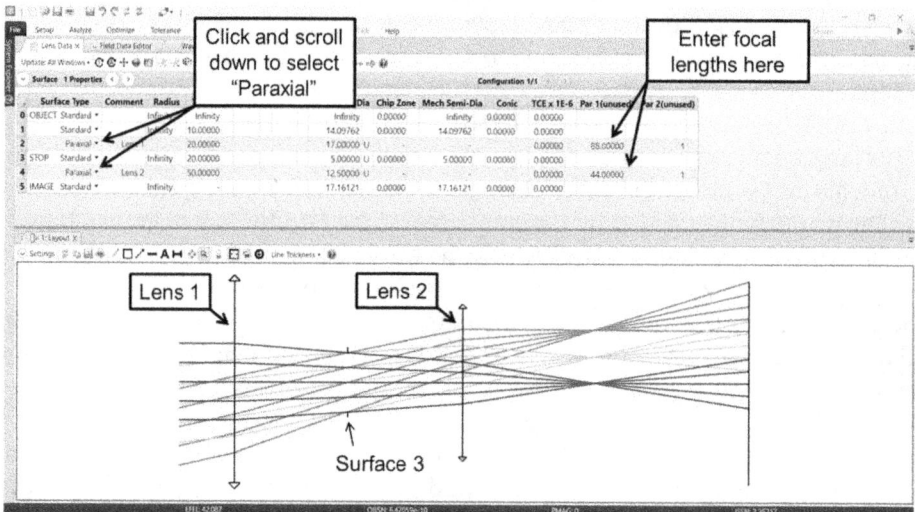

Figure 1.18. Paraxial thin-lens layout of a lens system with an EFL of 42.087 mm.

two thin lenses there, you need to left-click on the arrows of the surfaces as shown, then scroll down to select 'Paraxial.' Enter 88 and 44 for their focal lengths in the columns shown. Enter 17 and 12.5 for their Clear Semi-Diameters, respectively. Enter 20 for the thicknesses of surfaces 2 and 3. Enter 50 for the thickness of surface 4. The combined two thin lenses should now yield a system EFL of 42.087 mm. It does not matter what wavelength this EFL is currently valid for, because these thin lenses are not dispersive. At an image distance of 50 mm, as shown, we can clearly see rays focusing and then diverging (we will focus the rays in the next section, using a 'solve'). Notice that I have written 'Lens 1' and 'Lens 2' in the Comment column for surfaces 2 and 4. Try it. It is often helpful to label your components.

1.2.5 Use 'solves'

Now, we will make use of *solves*, which is a common function or feature in any optical design program. The concept of a *solve* function is to let the program literally determine (solve) a specific numeric value for a component's variable (such as the thickness of a surface or the surface's radius of curvature) that satisfies a specific constraint. Here, we are going to let OS solve for the best focus of the lens system. To do this, left-click on a button just to the right of the cell for the thickness of surface 4 (figure 1.19), then select 'Marginal Ray Height,' and select zero for the choice of Height and also for the Pupil Zone.

The system should now be focused and appear as shown in figure 1.20. The thickness of surface 4 is now the focal distance from that surface to the image plane, and its value should be 22.956 52 mm. You may have noticed that I have been using five rays per field to display the rays traced through the lens system in the layout graphics. You can display as many rays as you wish. Just right-click anywhere on the layout screen or left-click on Settings, and select the number of rays. Personally, when either the x-fan or y-fan of rays is selected for display, I tend to enter an odd number of rays, because this displays the central ray along the optic axis for the axial

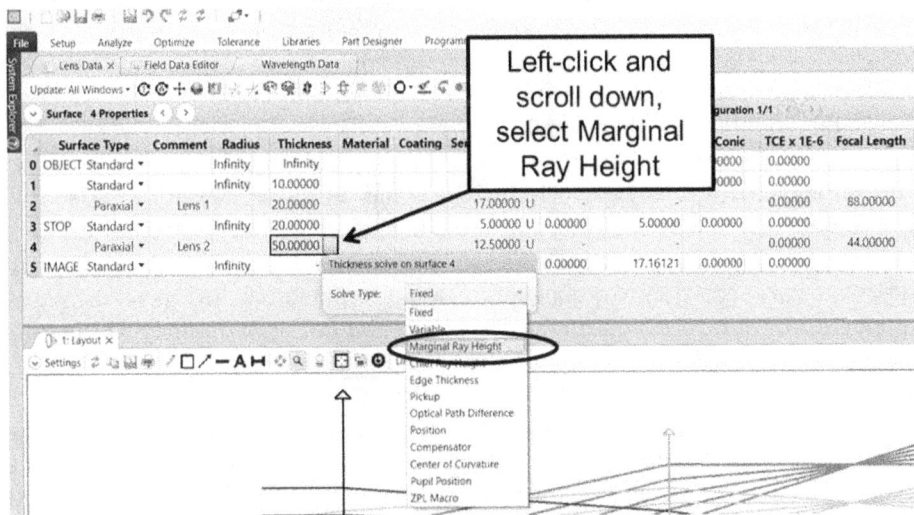

Figure 1.19. Setting the Marginal Ray Height on surface 4 for the system in figure 1.18.

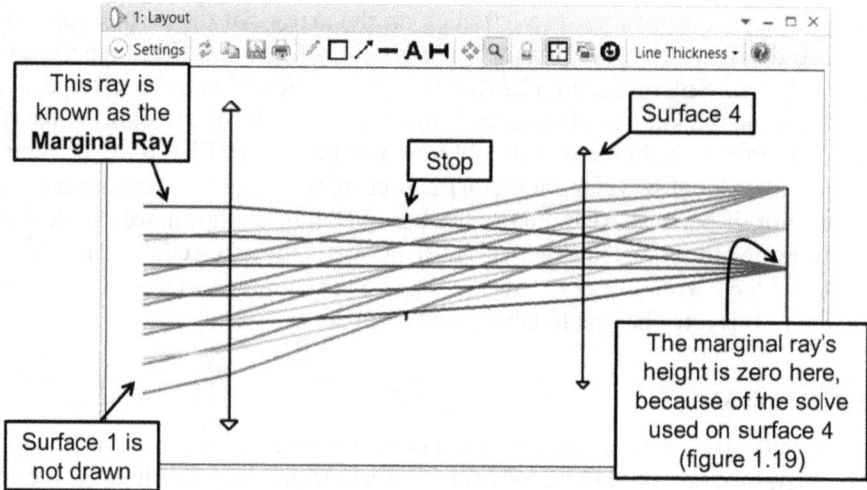

Figure 1.20. The focused lens system from figure 1.19 after applying the solve function for marginal ray height = 0 and pupil zone = 0.

bundle, leaving equal numbers of rays above and below the axis. The same is done for the oblique (off-axis) rays. This helps me visualize the passage of the central ray as the ray bundle progresses from surface to surface, especially when tilts and decenters of lens elements are imposed on the optical system (we will discuss element tilts and decenters in a later section). By now, you may have also guessed that you can play around with the settings for any window, whether it is for displaying graphics, analyses, or in any of the data editors. Also, you should realize that you can click on the '?' button to read more about the features present in any window.

Zero is chosen for the marginal ray height solve to constrain the program to determine the thickness of surface 4 that yields a height of zero for the **marginal ray** (i.e. the ray at the rim of the stop). This is equivalent to telling the program to focus the axial bundle of rays. The choice of zero for the pupil zone selects the paraxial region for focus (the pupil here refers to the entrance pupil). That said, since we have a system of paraxial thin-lens models, there would have been no difference between selecting zero or a nonzero value for the pupil zone. Now, you may have noted that when you scroll down, the choices for solve type include 'Fixed' and 'Variable.' Selecting a fixed solve type removes any solves and *fixes* the value in the cell of the Lens Data Editor to whatever you enter. Selecting a variable solve type places a 'V' symbol on that cell and enables it to vary according to design optimization commands called **operands**. We will now enter some basic operands in the *Merit Function Editor* to optimize this paraxial system of lenses to achieve an EFL of 50 mm.

The Merit Function Editor is the 'heart' of an optical design program (apart from the *Tolerance Data Editor* and *Monte Carlo* tolerancing feature, which we will discuss in section 1.2.13). This editor is the program's workhorse, and it enables us to design optical systems through a process of optimization. Open this editor by navigating to the Optimize menu, select Merit Function Editor (note that if you want to locate a

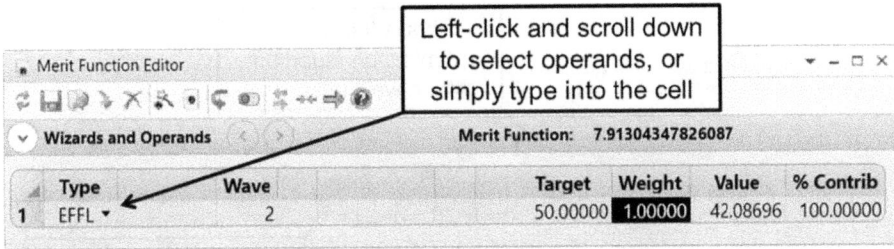

Figure 1.21. The Merit Function Editor (the 'heart' of an optical design program), showing the EFL operand used to optimize the lens system to achieve a target value of 50 mm.

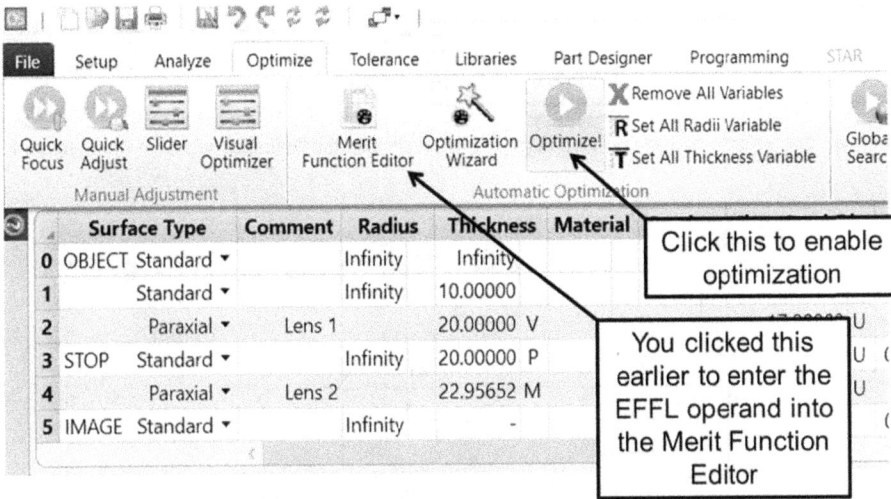

Figure 1.22. The Optimize! button under the Optimize menu.

specific program feature, you can also search for it by entering that feature name in the search box located at the top right of the screen). Dock the Merit Function Editor anywhere (I usually dock it at the top right of my screen), then enter the operand shown in figure 1.21. Enter 50 in the Target cell and weight = 1. This is done to set the target EFL of the system of thin lenses to 50 mm. The weight lets the program know that this operand should not be ignored. When more than a single operand is used, different weights for each operand indicate varying degrees of relative 'importance' for the operands (you will get used to entering them in due course throughout the book).

Next, set the solve for the thickness on surface 2 to variable 'V,' and set the solve for thickness on surface 3 to 'Pickup'; then, enter two for 'From Surface,' Scale Factor = 1, and Offset = 0. This Pickup Solve on surface 3 will assume the same value for the thickness on surface 2 (so, it 'picks up' the value from surface 2). If the Scale Factor were some other value, then the pickup would multiply the value from the selected surface by that factor. The Offset value adds a numeric value to the Pickup. Now, go to Optimize, then left-click the Optimize! button (figure 1.22). This should open up the *dialog box* shown in figure 1.23.

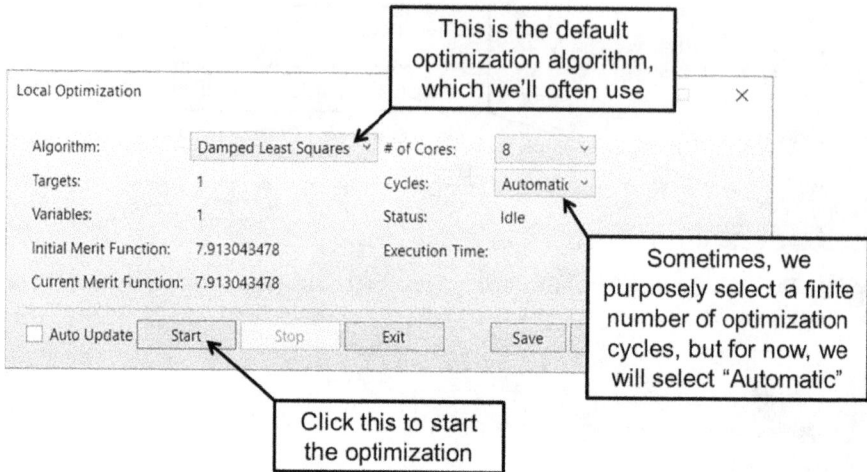

Figure 1.23. The optimization *dialog box* that pops up after clicking the Optimize! button shown in figure 1.22.

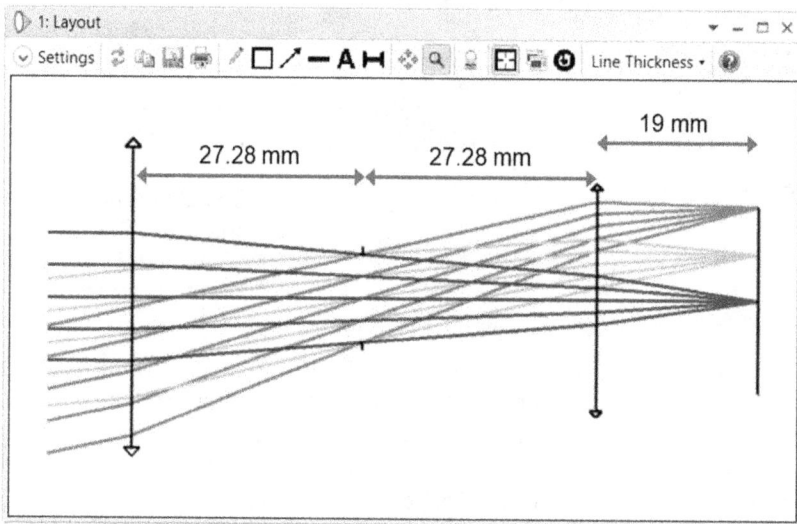

Figure 1.24. Layout for the optimized lens system, with EFL = 50 mm.

When the dialog box for optimization pops up (figure 1.23), left-click on 'Start,' which begins the optimization. At the end of optimization, click on Exit in the dialog box, and the result should be the layout shown in figure 1.24, whose system EFL is now 50 mm. Further, the thickness of surface 4 should be 19 mm, which is the **back focal length** (BFL) for this specific lens system.

You have just experienced the power of optimization in optical design. There are many different operands to choose from in order to target a variety of quantities, including the size of a point spread function (PSF), geometric spot size, and even *control operands* that limit the values of other operands (such as the central thickness of a lens, the air spaces between lenses, etc). You can view the list of choices for operands by clicking on the '?' symbol in the Merit Function Editor (or the Merit Function Editor Toolbar), which opens the associated help file and indicates clickable links to help topics such as Optimization Operands Summary, Optimization Operands by Category, Optimization Operands (Alphabetically).

1.2.6 Check the MTF and defocus sensitivity and create defocus invariance

Using the file from figure 1.24, navigate to the Analyze menu, then select FFT MTF (see figure 1.25). This outputs the Fast Fourier Transform MTF plot, as shown in figure 1.26 (note that the font displayed in the FFT MTF plot is Arial bold, because I have set this as my preferred font under the Project Preferences menu, which can be found under the Setup menu at the top left, next to the File menu).

Depending on your FFT MTF plot settings, the graphics may appear different from the output shown in figure 1.25. Clicking on the Settings for my FFT MTF plot yields the output shown in figure 1.27.

Note in figure 1.27 that I have put a tick mark on 'Show Diffraction Limit,' which means that the diffraction-limited MTF curve is also displayed in figure 1.26, but it cannot be seen because it has been covered by the other curves from all of the fields. But if we defocus the image plane a little, we should be able to see the diffraction-limited curve, as all of the other MTF curves will drop. To see this, insert a surface after surface 4, and input 0.026 mm. You should see all MTF curves drop slightly, thereby revealing a black solid curve, which is the diffraction-limited MTF curve. By

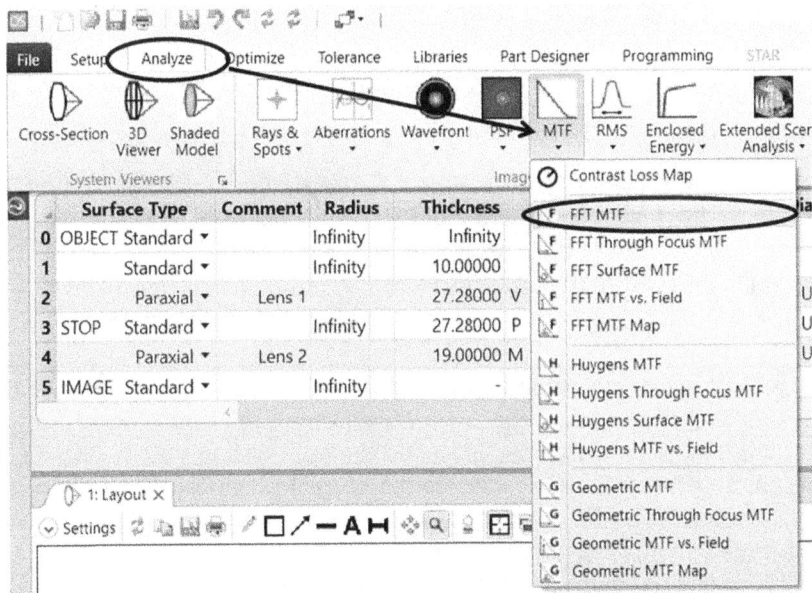

Figure 1.25. Selecting the FFT MTF plot from the Analyze menu.

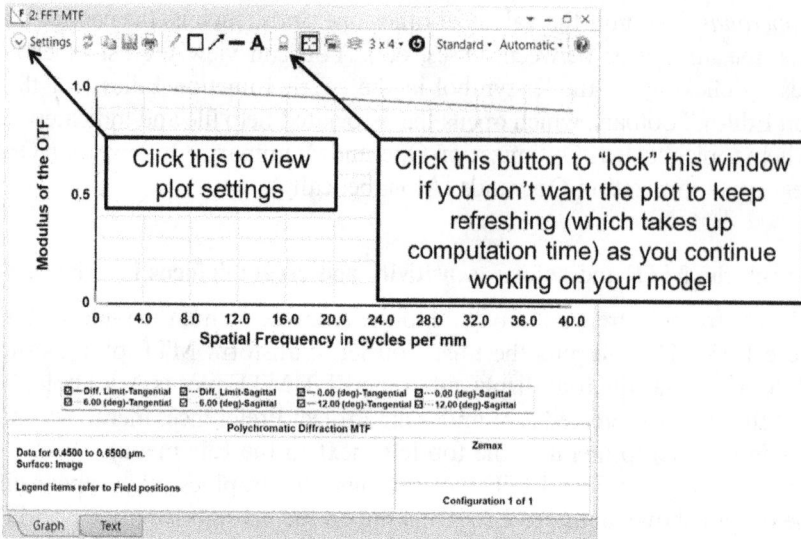

Figure 1.26. The FFT MTF plot output from the selection made in figure 1.25.

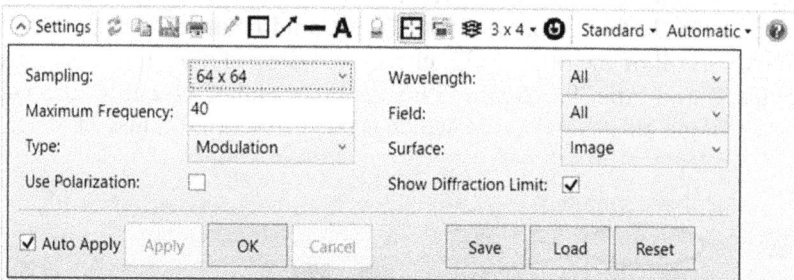

Figure 1.27. The settings menu for the FFT MTF plot in figure 1.26.

the way, why did I choose to defocus the image plane by the specific value of 0.026 mm? It has to do with knowing that the depth of focus Δz for a diffraction-limited imaging system may be expressed as

$$\Delta z \approx \pm 2\lambda(f/\#)^2, \tag{1.7}$$

where λ is the wavelength being considered (if the system is achromatic, such as a paraxial thin-lens system, then the *central* wavelength of the spectrum is all that is needed), and $f/\#$ is the *f*-number given by the lens system's EFL divided by the entrance pupil diameter [13]. We can use the Merit Function Editor to compute Δz given by equation (1.7). To do this, enter the operands shown in figure 1.28.

You can read about the operands entered in figure 1.28 and the structure in which they have been entered in the Merit Function Editor using the associated help file for the Merit Function Editor's operands. Note that in figure 1.28, I have reset the

Figure 1.28. Operands for computing Δz given by equation (1.6).

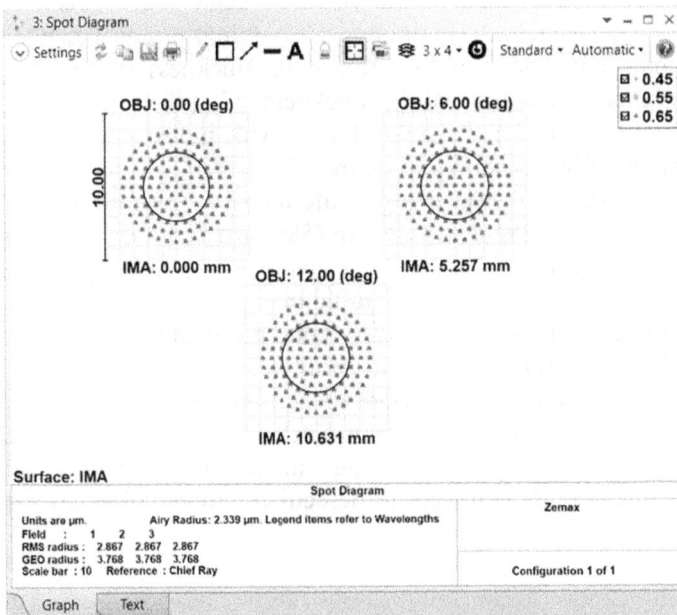

Figure 1.29. A standard spot diagram at a +0.026 mm defocused image plane for the paraxial thin-lens system shown in figure 1.24.

weight of the EFL operand (EFFL) to zero. In figure 1.28, Δz is estimated to be roughly 13 microns. This implies that if we defocus by 0.013 mm, then there should be a negligible drop in the MTF curves (try this and you will see). So we need to defocus by more than 13 microns to see a larger effect on the curves, which is why I simply doubled the value and suggested 0.026 mm for the defocus. If you defocused the image plane by up to 0.1 mm, the curves would drop significantly. You can see from figure 1.25 that there are many options for analysis features, such as Rays & Spots, Aberrations, Wavefront, and so on. Clicking on them will reveal their plots. For example, click on the Rays & Spots menu and select Standard Spot Diagram. This will output the window shown in figure 1.29 for the case of an image plane defocused by + 0.026 mm. Here, the scale is 10 microns. Note the size (diameter) of the Airy diffraction spot, which is indicated by the solid black circle overlaid with the spots. Note that the spots are shown for all wavelengths, and there is no indication of chromatic aberration (which we would expect).

Since we have a *perfect* lens system model here (which is naturally free from chromatic aberration), there seems to be no point in going any farther than we already have. However, we can do something interesting and perhaps even somewhat surprising. In particular, I will show that **the introduction of spherical and chromatic aberration can extend the depth of focus for this lens system**. Further, I will show that this is closely related to an observation in optical design that **whenever an imaging system has bad image quality, then the rate of change of *badness* from its bad state tends to be *less* than the rate of change of *badness* from its good state**. This is sometimes seen to be a compromise between achieving reduced sensitivity to manufacturing tolerances and obtaining peak nominal performance.

To begin, remove the marginal ray solve on the thickness of surface 4, and instead place the marginal ray solve on the thickness of surface 5 (height = 0, pupil zone = 0). Set a *Material Solve* on surface 4 with index = 1.8 and an Abbe V_d (dispersion) value of 30, as shown in figure 1.30.

This creates a slab of glass whose material is a fictitious model that has a refractive index of 1.8 at the d wavelength (588 nm) and a dispersion value that is typical of some flint glasses. The resulting layout is shown in figure 1.31, provided that you use the layout settings shown in figure 1.32.

In the layout shown in figure 1.31, the second paraxial thin lens is placed directly at the left surface of the slab of glass. The result is added chromatic aberration due to dispersion at the glass and also spherical aberration due to the nonlinear characteristic for the refraction of rays arising from Snell's law. If you examine the standard spot diagram at a scale of 10 microns, you will see that the spot at each field angle has grown significantly. If you were to defocus the image plane by ±0.1 mm, you

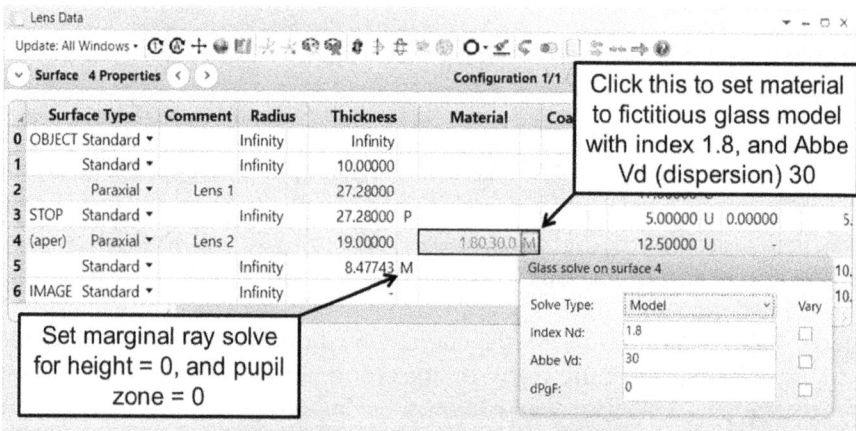

Figure 1.30. Creating a fictitious glass model for an index of 1.8 and an Abbe dispersion of 30.

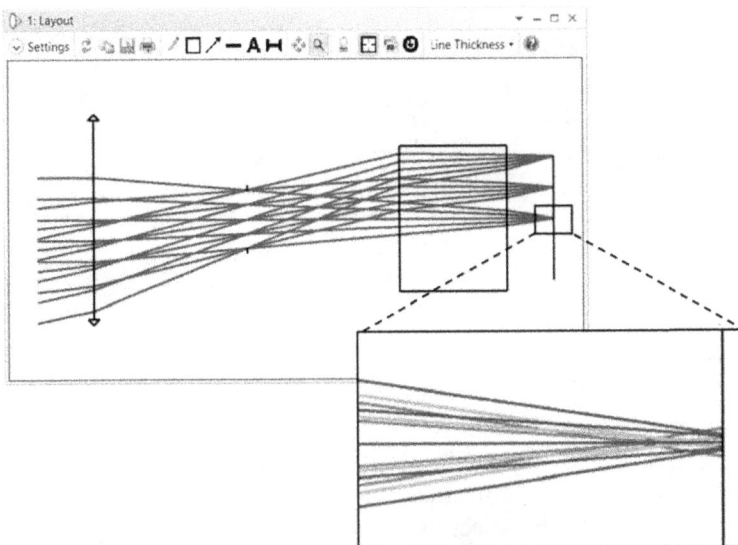

Figure 1.31. The layout resulting from the prescription in figure 1.30. The colored focusing rays indicate chromatic aberration and spherical aberration.

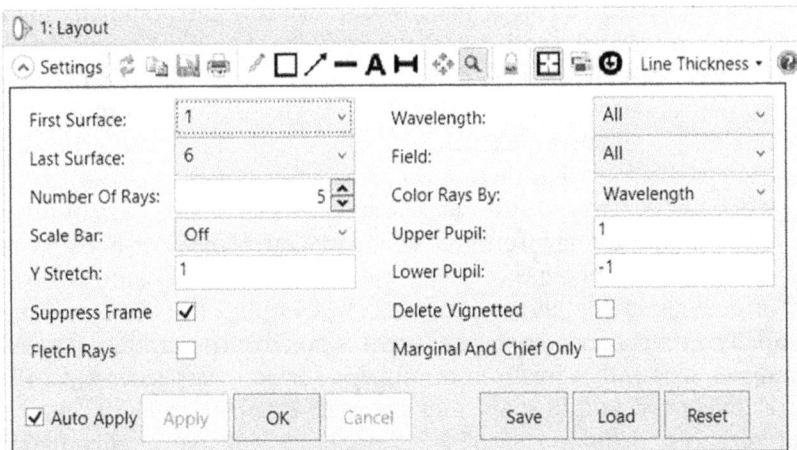

Figure 1.32. The settings used to view the FFT MTF plots in figure 1.33. Note that the wavelength selection (which has been set to 'All') is that of figure 1.17.

would obtain the FFT MTF plots shown in figure 1.33 (be sure to increase the sampling in the settings menu to 128 × 128).

In figure 1.33, note that the MTF curves are not very different across the full 0.2 mm focal range. This behavior is significantly different from the behavior you observed for the system in figure 1.25. In that system, had you defocused by ±0.1 mm, you would have seen a significant drop in MTF from the nominal state of no defocus. The behavior of the MTF curves shown in figure 1.33 is due to the

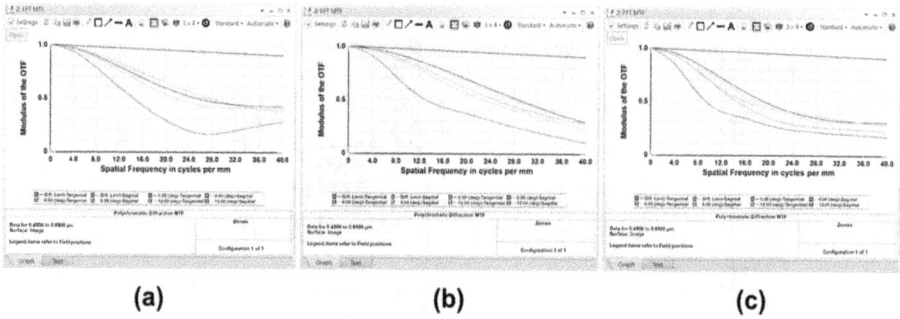

(a)   **(b)**   **(c)**

Figure 1.33. FFT MTF plots for the system in figure 1.31 with various amounts of image plane defocus. (a) −0.1 mm defocus. (b) No defocus. (c) +0.1 mm defocus.

addition of spherical and chromatic aberration from the thick glass slab, but the larger effect comes from the chromatic aberration (specifically, the *primary axial color*). You can check this by selecting a monochromatic wavelength in the settings menu for the MTF plots. You will then see that the effect of defocus on the monochromatic MTF curve is greater than its effect on the polychromatic MTF curve. However, the spherical aberration is non-negligible, so it can be said that the combined effect of spherical and axial color aberration has increased the depth of focus for this lens system, albeit at the cost of sacrificing the nominally high MTF performance when these aberrations are not present.

The characteristic of an extended depth of focus due to the addition of aberrations is not uncommon and is well documented [14–18]. The basic idea is rather simple, as it has a completely geometric interpretation. If you look closely at the enlarged box in figure 1.31, you will see that there is no plane beyond the defined image plane at which rays cross with zero spot size. So, the spread of a congregation of rays at any plane beyond (but not too far from) the defined image plane does not vary rapidly. This spread of rays is, of course, a geometric PSF, whose size changes slowly as a function of defocus in the current example. In contrast, the PSFs in figure 1.29 change rapidly with defocus. In some cases, it is possible to introduce a *phase mask* (which can be a lens with a freeform or aspheric surface) that generates a PSF that remains relatively constant in shape and size over a defocus range far greater than the range considered in figure 1.33. This is the concept behind extending the depth of field/focus by way of *wavefront coding* [19, 20]; the action of tailoring the PSF to achieve the invariance is often called *PSF engineering* [21]. However, in wavefront coding, a final clear image is recovered by way of *deconvolving* the blurred image from the PSF. Incidentally, if all of the MTF curves in figure 1.33 were somehow engineered to remain constant over the defocus range, then it could be implied that their PSFs had been made fairly invariant over that range. This would enable the method of deconvolution that is performed in wavefront coding. However, there are important aspects of the deconvolution technique to be considered for practical applications. In particular, for this method to work, one should have some knowledge of the *average* of the Fourier transform of the noise in the detected signal as

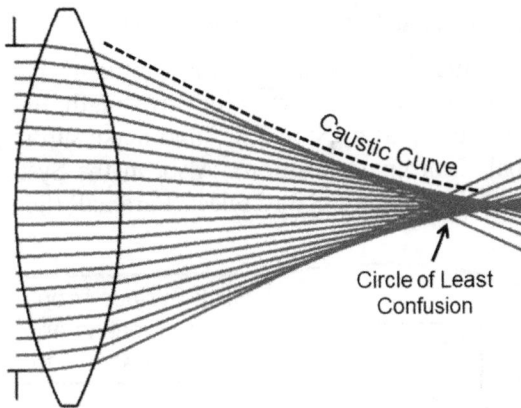

Figure 1.34. A caustic curve formed by spherical aberration in a focusing beam.

well as some knowledge of the Fourier transform of the desired signal to be detected (see a brief note on this topic in appendix D.3).

A topic that is closely related to defocus invariance is the *desensitization* of an optical system to tolerances. The observation that the MTF curves in figure 1.33 do not drop significantly over the defocus range of ±0.1 mm implies that the system has —in some sense—been somewhat desensitized to that amount of focal variation. But the resulting MTF curves are also not high. In fact, experienced optical designers have often encountered trade-offs between the nominal performance of an optical system and its sensitivity to tolerances [22–24]. However, the situation is not so simple. For example, the same designers have also mentioned cases in which it was possible to desensitize a lens and still achieve *satisfactorily* high performance, with the provision that the lens designs were *well optimized* [24], which means that aberrations were corrected without high degrees of cancellation. In the language of lens design, this cancellation is better described as *aberration balancing* (or, as Sasián puts it, *aberration propagation* [24]). Indeed, the approximate invariance of the PSF with defocus in figure 1.33 can be considered to be a balance between defocus and aberrations from spherical and axial color. In fact, this is how you determine the best focus when spherical aberration and defocus are present [5, 13]. In this case, the degree of aberration balancing is not high (otherwise, defocus and spherical aberration would cancel to near zero), which is one way to understand why the presence of spherical aberration and axial color in figure 1.33 results in a fair amount of desensitization to defocus. Another way to think about the slow variation of the MTF with defocus in this figure is to consider that whenever a focusing beam takes the form of a *caustic* [13, 25], the size of the PSF—which is defined by a roughly asymptotic behavior of the caustic curve—has a rate of change with axial distance that decreases near the *circle of least confusion* (figure 1.34). So, in general, it can be said that *the rate of change of badness from a bad state tends to be less than the rate of change of badness from a good state.*

1.2.7 Creating a real lens model of the thin-lens model

Create a new file and enter the prescription shown in figure 1.35. Enter 0°, 6°, and 12° for the Y angle fields. Enter 0.55 microns for wavelength. In the System Explorer menu, set the Aperture to 'Float by Stop Size,' and make surface 7 the stop (set the stop to be the Global Coordinate position). Also, in the System Explorer menu, scroll down and locate 'Ray Aiming,' and select 'Paraxial' (figure 1.36).

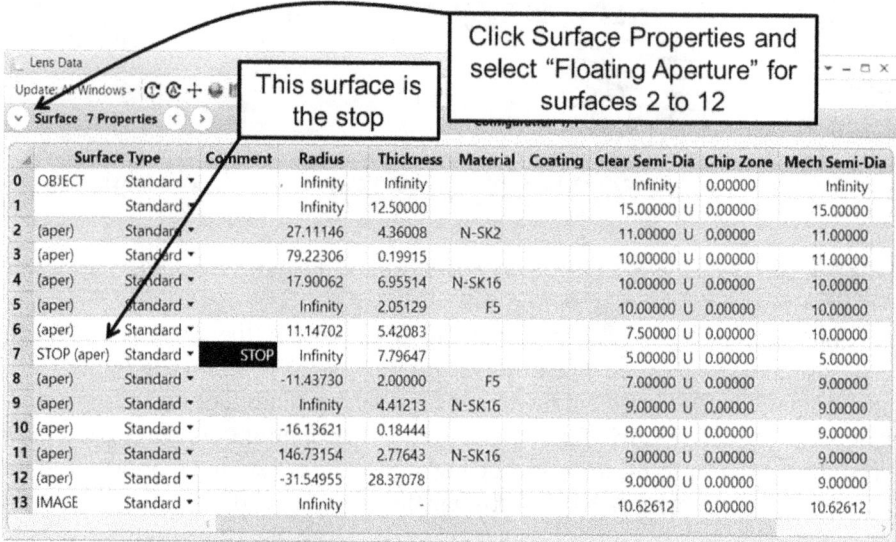

Figure 1.35. The prescription for a real lens version of the system in figure 1.24.

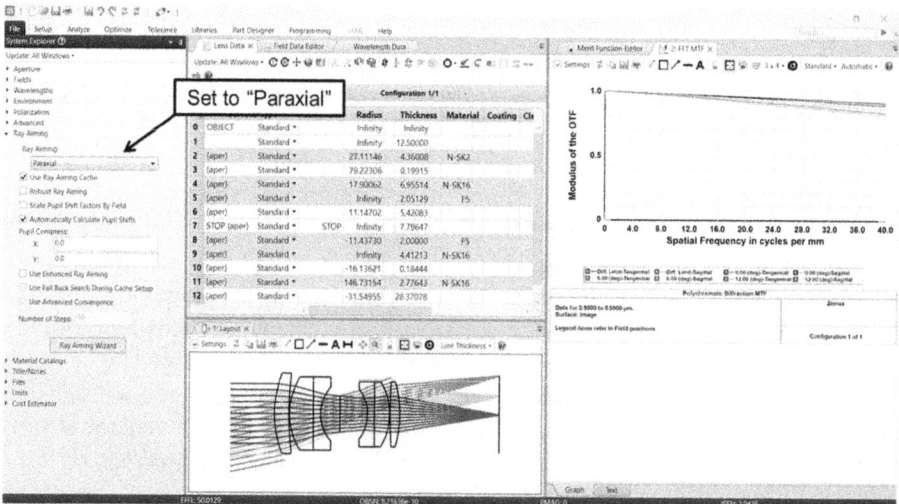

Figure 1.36. Setting Ray Aiming to 'Paraxial' in the System Explorer menu.

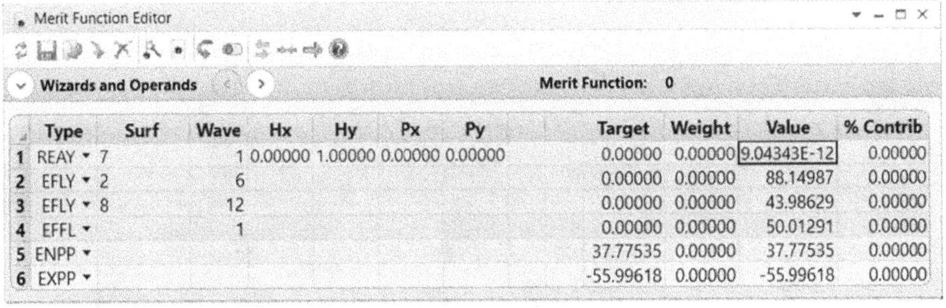

	Type	Surf	Wave	Hx	Hy	Px	Py	Target	Weight	Value	% Contrib
1	REAY	7	1	0.00000	1.00000	0.00000	0.00000	0.00000	0.00000	9.04343E-12	0.00000
2	EFLY	2	6					0.00000	0.00000	88.14987	0.00000
3	EFLY	8	12					0.00000	0.00000	43.98629	0.00000
4	EFFL		1					0.00000	0.00000	50.01291	0.00000
5	ENPP							37.77535	0.00000	37.77535	0.00000
6	EXPP							-55.99618	0.00000	-55.99618	0.00000

Figure 1.37. Setting operands in the Merit Function Editor for the system in figure 1.36.

Setting Ray Aiming to Paraxial instead of 'Off' is a crucial step in the optical design and analysis process, as it lets the program ensure that the **principal ray** (the central ray in the bundle of rays from the largest field point) strikes the center of the stop. You can check whether the principal ray does indeed strike the center of the stop by entering the REAY operand shown on the first row in figure 1.37 (I have also added other operands, which we will discuss soon).

The REAY operand tells the program to trace a ray from the largest field point (Hy = 1), at the first wavelength given in the Wavelength Data Editor (Wave = 1), towards surface 7, which spits out the resulting Y height for that ray at surface 7 (Value = 9.043 443E−12). This is saying that the principal ray's height at surface 7 is $\approx 9.043 \times 10^{-12}$ mm, which is essentially zero. If you select 'Off' for Ray Aiming, you will see that this value changes and is no longer correct (the principal ray should always cross the center of the stop, because it is required to *aim* at the center of the entrance pupil and exit from the center of the exit pupil, regardless of the presence or absence of aberrations).

Now, set surface 1 to be the stop. You should now see that the lens model is the same as that shown in figure 1.3(a), except that we have one extra field angle here, which is at 6° (you may remove it if you wish, but we will use it again very soon). But by setting surface 1 to be the stop—and letting this surface have a large semi-diameter—we can better visualize what happens in real life when rays from all over the field strike a lens system. You may note that rays fill the entire lens system, and they are meant to be blocked by spacers and the inner walls of the lens barrel that mounts the lenses. But this is assuming that the rays are 100% blocked. Each point where a ray does not strike the lens is, in theory, a potential source of stray light. Setting up a lens system in the manner shown here will remind you of this. In theory, when we set surface 1 to be the stop, there is no need to set the Ray Aiming (i.e. we may leave it as 'Off'). When a surface is not *buried* (i.e. when it does not lie between lenses in a lens system), it is perfectly valid to not set the Ray Aiming, because there are no lenses in front of the stop. In this case, the principal ray does not need to be aimed towards the image of the stop in object space (which is the entrance pupil) in order to strike the center of the stop. The pupils and stop are *conjugates* of one another (this is a term we will revisit in further detail in section 1.3). By this, we mean that they are images of each other. Evidently, a ray that strikes the center of the

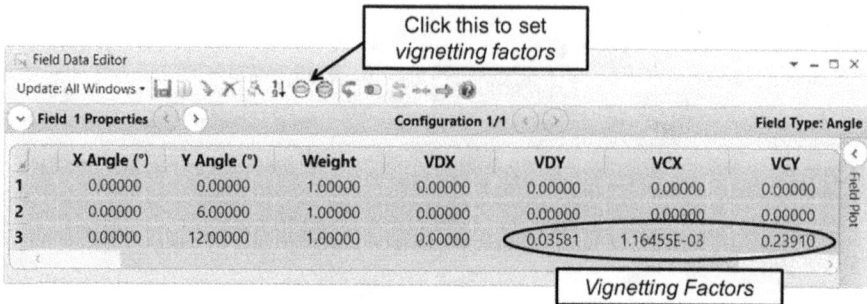

Figure 1.38. Setting vignetting factors automatically in the Field Data Editor.

entrance pupil should also strike the center of the stop and that of the exit pupil (when points are conjugates of one another, rays from a point should end up crossing its other conjugate points—this is a basic premise of imaging theory, and we will delve further into this in section 1.3). When a stop is buried, the lenses in front of it form an image (virtual or not) of it elsewhere, and this image (which is the entrance pupil) may be aberrated. Setting Ray Aiming to Paraxial (or Real) lets the program account for the *pupil aberration* and ensures the proper crossing of all rays (including the principal ray) through the stop.

Now reset surface 7 to be the stop, and ensure that Ray Aiming is set to Paraxial. Let us account for vignetting of the off-axis rays by clicking on the button shown in figure 1.38. This automatically computes appropriate *vignetting factors* for the fields so that only rays that pass through the lens system are displayed. The decentration factors (VDX, VDY) and compression factors (VCX, VCY) effectively shift and squeeze the off-axis rays to fit through the stop. If you now look back at the merit function, you will see that the value of REAY is equal to 0.179 06 mm. If you hover your cursor across the layout screen and let it point towards the height of the **chief ray** (the central ray in the farthest off-axis vignetted ray bundle) at the stop, it is indeed about 0.179 mm.

Now return to the Merit Function Editor and notice the other operands below REAY. Rows 2 and 3 compute the EFLs for the respective lens groups for the *double-Gauss* lens system being modeled, while row 3 computes the total lens system's EFL (we could have also set the operand to EFLY for surfaces 2–12 and arrived at the same result). Notice that the EFL for the first lens group is about 88 mm, and the EFL for the second lens group is about 44 mm. Remember these? I used them for the paraxial thin-lens model in the previous section (see figure 1.18). But it is important to note that the current real lens system is not an accurate representation of the paraxial thin-lens model from figure 1.18, and vice versa. I would have needed to include more paraxial thin-lens surfaces to mimic the condition for the real lens system here. You can see what I mean by examining the operands ENPP and EXPP on rows 5 and 6 in the Merit Function Editor. These are the entrance pupil and exit pupil positions, respectively, for the current real lens system, measured relative to the current Global Coordinate setting, which is at the stop. If you had entered these operands into the Merit Function Editor for the paraxial thin-lens model in figure 1.18, you would have seen that they compute different values. But we will not mind this minor issue here. Our next task is to check the performance of this real lens system in order to get acquainted with the common metrics to be examined for imaging systems.

1.2.8 Lens MTF, spatial frequency, field curvature, distortion, and relative illumination

Figure 1.39 shows the FFT MTF plot for the lens system in figure 1.36 at all field angles, displayed at a monochromatic wavelength of 0.55 microns, at a maximum spatial frequency of 40 *cycles* per mm, and with *sampling* set to 64×64. If you have studied lens design, you know that there are curves for *sagittal* and *tangential* rays, in view of astigmatism.

For almost all of the image quality analyses you are likely to perform with the MTF plot, you will generally use the FFT MTF computation among all of the choices provided in OS. This is because, when aberrations are minimal—which is what we would expect when we begin to examine MTF during optimization (otherwise, when aberrations and spot sizes are large, the MTF would have many zero crossings, which is senseless for imaging)—the FFT computation is fast and accurate. The maximum spatial frequency of 40 cycles/mm used in figure 1.39 is interesting and has some significance. Let us suppose that the lens system forms images onto a 35 mm format image sensor (whose size is 36 mm×24 mm). When

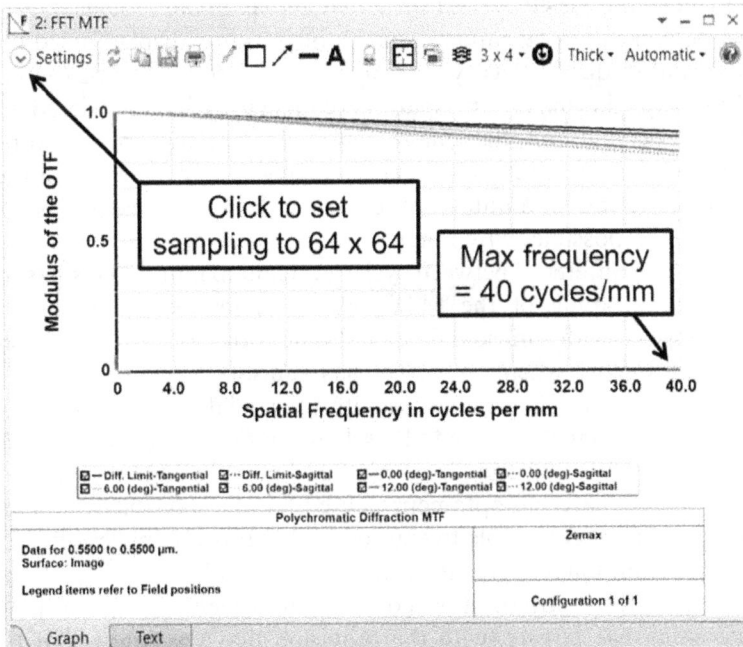

Figure 1.39. The MTF for the lens system in figure 1.36, at a wavelength of 0.55 μm.

the full image on this sensor is enlarged to roughly that of A4 paper, the enlargement factor is about eight times. Under the assumption that a human observer will view this enlarged image at a *near distance* of 250 mm, the human eye is often considered to have a resolution limit of about 5 *line pairs* (lp) per mm at the target. Applying the 8× multiplicative factor to this resolution yields 40 lp mm^{-1} at the size of the 35 mm format image sensor [26, 27]. So, a lens to be designed for such an image sensor format may be regarded as having to yield good MTF at a maximum spatial frequency of 40 cycles/mm. At any larger spatial frequencies, the 8× enlargement factor would yield resolution that is beyond the 5 lp mm^{-1} limit for the human eye.

You may have noticed that we used the terms *cycles per mm* and *line pairs per mm* interchangeably. We will return to this soon. Meanwhile, the above example is clearly meant for a 35 mm format image sensor, whose final image is considered to be viewed on printed A4-sized paper. Therefore, it is not necessary to adhere strictly to the example above. Different conditions and customer requirements can change the spatial frequencies one has to consider for designing an imaging lens. Also, other approximations and assumptions can be made. For instance, Walker's model of the human eye indicates that, at the fovea, the limiting resolution is roughly 110 cycles/mm [11]. If we then consider that this eye model has a BFL ≈ 17 mm [11], then when it is viewing a scene at 250 mm from the cornea, the resolution that is projected onto the scene is roughly 110 cycles/mm × (17/250) ≈ 7 cycles/mm. So, for a 35 mm format image sensor, applying an 8× enlargement factor to the 7 cycles/mm limiting resolution yields 56 cycles/mm, which is a higher spatial frequency than the 40 cycles/mm stated above. In fact, for such *large-format* image sensors, you often see camera lens manufacturers displaying MTF data at maximum spatial frequencies varying roughly from 20 cycles/mm to 80 cycles lp mm^{-1}. Ultimately, the maximum spatial frequency to be considered for a lens being designed depends on the image sensor format it is meant for and the enlargement factor to be applied. The enlargement factor depends on the display size you are considering for a human observer—unless your observer is not human (hey, that is indeed possible, why not?)

Now, what is the difference between cycles/mm and lp mm^{-1}? This has to do with the fundamental meaning of the MTF, which represents an imaging system's response to objects with sinusoidal spatial intensities [5, 28]. Technically, the MTF only makes sense when the object has sinusoidal transmittance (and it is back-illuminated) or if it is self-luminous with sinusoidally varying spatial intensity (we are using the term *intensity* here to loosely mean *flux per unit area of the emitted radiation*). But many test targets—such as the popular USAF bar target (and the famous alphabetical chart that optometrists use for eye tests)—have dark and bright *bars*. It turns out that it is possible to extract the MTF from images of bar targets if appropriate computational corrections are employed [28]. Further, even if these corrections are not employed, it is common to *roughly* gauge a lens's MTF performance using bar targets (with the provision that you know that this is not technically accurate). For example, recall from above that Walker's model of the human eye yields a limiting resolution of about 7 cycles/mm at an object whose

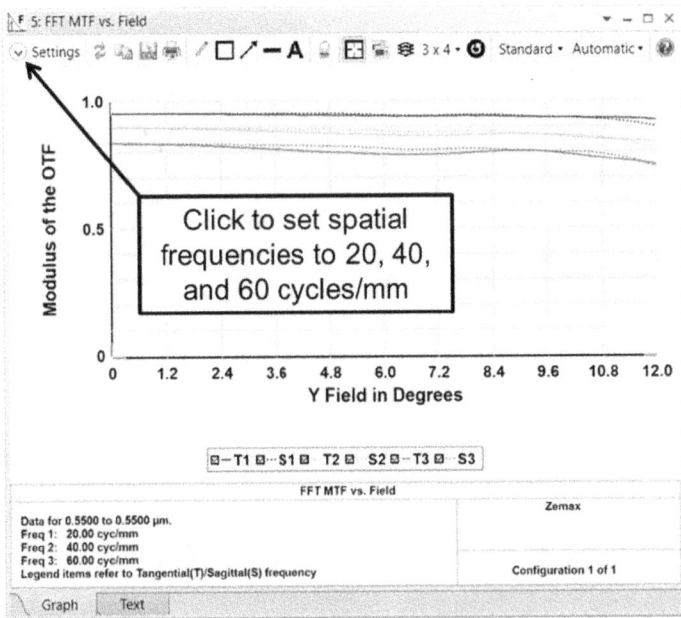

Figure 1.40. The MTF vs Field output for the lens system in figure 1.36 at a wavelength of 0.55 μm and at 20, 40, and 60 cycles/mm.

distance from the front of the eye is about 250 mm. The reciprocal of 7 cycles/mm is roughly 0.143 mm. Divide this by 250 mm, take the arctangent, and we have about 0.033°. Convert this to arc minutes (multiply by 60) and we obtain 2 arc minutes. This angle subtends the distance between a bright and dark bar for the letter 'E' if this letter were placed at 250 mm from the eye. If we took half of the angle, we would obtain 1 arc minute, which is the angular limit of resolution subtended by a single dark bar in the letter E [11] or the letter H [29]. The point here is that in this example, we were able to estimate the angular limit of resolution for the human eye (whose object is made of bars in alphabets) using the limit of resolution (in cycles/mm) from a model of the human eye.

The above discussion highlights that the MTF (which we will take to roughly mean the *contrast* in the image) is one of the most common and important performance characteristics to check. Other common metrics are as follows: MTF versus (vs) Field (figure 1.40), Field Curvature and Image Distortion (figure 1.41), and Relative Illumination (figure 1.42). If you work as a lens designer, you will be analyzing far many more plots than these. But here, we are not going to be full-time lens designers. Figures 1.39–1.42 display the most basic plots to check if you are an optical system designer and normally spend most of your time selecting off-the-shelf lenses (but may sometimes optimize a lens). The MTF vs Field plot in figure 1.40 shows how the MTF varies across the field. I have arbitrarily chosen to display the MTF at 20, 40, and 60 cycles/mm. Here, you generally want to see that the sagittal and tangential curves are as close to each other as possible across the full field (which

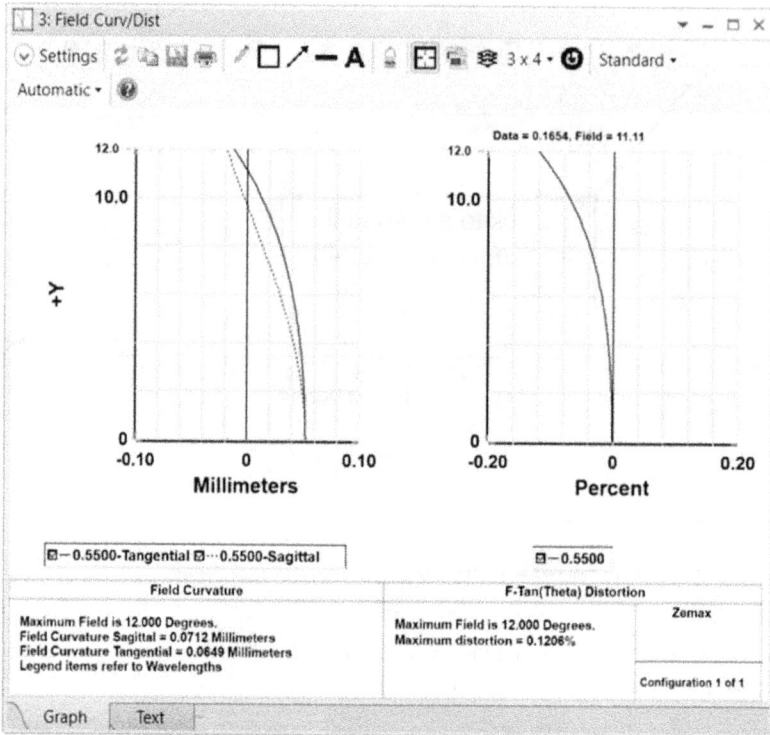

Figure 1.41. Field Curvature and Image Distortion outputs for the lens system in figure 1.36 at a wavelength of 0.55 μm.

would imply that there is minimal astigmatism across the field of view). In many cases, practical imaging lenses have an MTF that can drop to 0.5 (or less) at full field.

By default, the field curvature plot in OS is displayed in the same window as image distortion. The main thing you check in the field curvature plot is the variation of focal position as a function of field for the sagittal and tangential rays. As a reminder, the sagittal rays are the rays in the x–z plane, while the tangential rays are the rays in the y–z plane. To see them, go to Setup > 3D Viewer to display the layout. Then, go to its settings and select X Fan or Y Fan (or both) for the 'Ray Pattern.' The X Fan displays sagittal rays, and the Y Fan displays tangential rays. The image distortion plot (or simply 'distortion' plot, which should not be confused with *pupil distortion*) in OS is determined by computing

$$D\% = \left[\left(h' - h'_\mathrm{p} \right) / h'_\mathrm{p} \right] \times 100\%, \tag{1.8}$$

where $D\%$ is the percent distortion in the image, h' is the height of the image determined by a *real* ray (i.e. it is the height of the *chief* ray—but hold that thought), and h'_p is the height of the *paraxial* image. Note that h'_p is used synonymously with the variable $\bar{y}$ that was used in section 1.1.3 (just consider h'_p as $\bar{y}$ at the image).

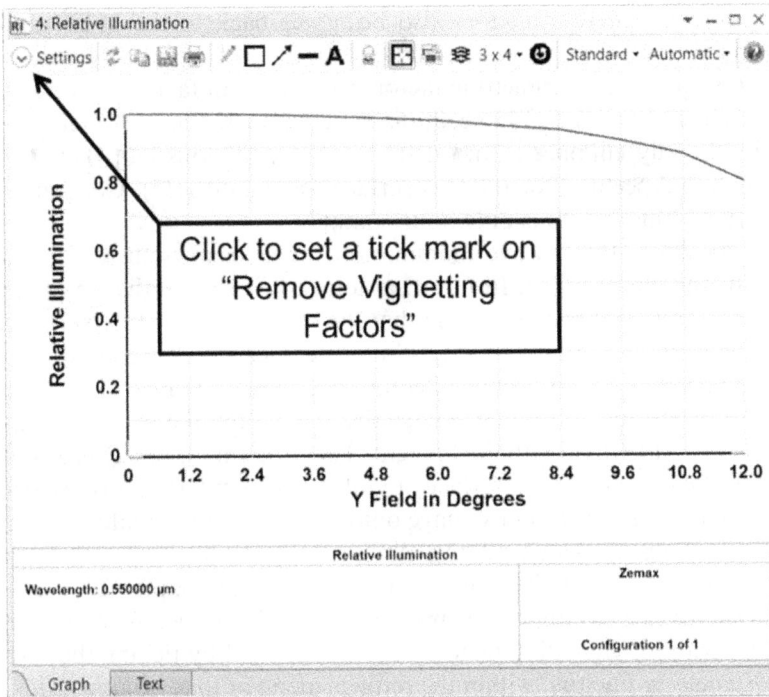

Figure 1.42. The Relative Illumination output for the lens system in figure 1.36 at a wavelength of 0.55 μm and at the default ray and field density settings of 5 and 10, respectively.

Now, since h' is the height of the chief ray at the image, then vignetting factors should be employed. But I have not done this (because, in this specific case, there would be a negligible difference, though generally you would need to use vignetting factors by clicking on the button shown in figure 1.38). You can check the impact of vignetting more clearly if you use the DISG operand in the Merit Function Editor to compute the distortion. Try it and click/un-click the automatic setting for vignetting factors. Incidentally, you can also manually compute equation (1.8) using the REAY and PARY operands to compute h' and h'_p, respectively. Further, you can compare that calculation with the result obtained using the CENY operand in place of the REAY operand. Which is the correct operand? Neither is wrong nor right. The REAY operand is be more conventional, as it provides the y-height of the chief ray when vignetting factors are set. But the **centroid** of the y-rays is located using the CENY operand. In reality, which operand yields the *actual height* of the image? Both. And yet, neither. If you really want to know the size of an image, then you must use an analysis feature of the program that simulates an image, and then you must determine for yourself what is considered the *edge* of the image. Here, you may ask yourself whether edge-detection is essential in your application. It would be, if you were designing a machine vision system to measure object sizes. Still, it is entirely up to you and your colleagues to justify how object and image size are defined for your intended application.

The subject of vignetting factors also brings us back to the MTF (and it also brings us to the topic of relative illumination). In the MTF plots of figures 1.39 and 1.40, I did not specify that vignetting factors were used. In fact, they were purposely not used. Generally, the MTF needs to be computed when vignetted rays are modeled by literally clipping them off from the calculations [30]. This is done by setting 'Floating Aperture' for the lens surface (see figure 1.35). Compared to this, the use of vignetting factors is only approximate.

Loosely speaking, the relative illumination for a lens system is the brightness profile across the image plane. In technical terms, it is the irradiance (flux per unit area) across the image plane, assuming that the source is a flat Lambertian emitter. Under unvignetted conditions, relative illumination can be designed to be uniform across the image plane, as it is controlled by several factors, including the apparent size of the entrance and exit pupils, image distortion, and the instantaneous rate of change of image distortion with field height. Therefore, if an object is a source with uniform flux per unit area at its surface, and if it has Lambertian ray emission, then using an imaging lens to project its image onto a screen enables uniform illumination on the screen when the lens's relative illumination is tuned to be flat across the screen. We will discuss relative illumination in detail in chapter 2, where we will make many references to important prior work on this topic. When OS computes relative illumination, vignetting factors may be removed by ticking the box labeled 'Remove Vignetting Factors' within the settings menu of that window. This lets you examine relative illumination properly even if vignetting factors are being used for the fields. Many practical imaging lenses with high fields of view have relative illumination that drops considerably near the edge of the field. Finally, the reader should be cautioned that the default values for the Ray Density and Field Density in the Settings menu of the Relative Illumination plot may not necessarily yield accurate results. So, we may need to adjust these values accordingly. Fortunately, there is a way to determine the settings that are sufficiently accurate, as discussed in section 1.2.10.

1.2.9 Why it is not always about MTF in real life (it depends on your application)

Even if a lens's MTF is reported in its supplier datasheet, it may not mean that you have to qualify a lens by its MTF when you use it in your product. There are a couple of reasons for this. First, why repeat the supplier's work? This would require replicating the supplier's test, which is difficult and time-consuming. Second, if your object to be imaged is a biological cell, then just get a supplier to etch structure onto a medium and make it into a model of the cell, and let that be the gold standard test target. The point is to qualify your imaging system in the manner in which you would actually use it.

For example, in one project, our team used a lens to image an array of *wells* in a manner similar to that shown in figure 1.2. In our application, optical crosstalk was the primary concern, in which adjacent well images could receive stray light from each other, partly due to the blurring of the well images (i.e. the spread of light due to image blur from the edge of one well image could creep into the image of the next well). Of course, a high MTF implies less image blur. But MTF is not that easy to measure accurately in volume manufacture. So, we decided to measure the lens's crosstalk (using a customized test fixture on a lab bench) and specify a maximum for it. We had also determined that our crosstalk values were directly relatable to the instrument's end performance (and the end users of our instrument tended to compute crosstalk in the same way that it was defined for our test fixture). It also made our lens supplier happy, as they too did not want to measure the MTF in mass production.

One is right to ask: 'But how would you know whether a lens's MTF is good enough for your intended application, such as in the case of satisfying the crosstalk requirement?' One way is to tabulate the MTF data against crosstalk values and, by way of correlation, determine the point at which the MTF for a lens failed the crosstalk test. Such data for a population of lenses may become available in the next (beta) phase of your project. Certainly, correlation is not necessarily causation. But correlation can be quite sensible if the quantities being compared (e.g. crosstalk vs MTF) are theoretically and mathematically relatable. In other words, one can in theory express their relationship as a formula (see appendix A.1 for a derivation and discussion of this relationship). Moreover, consider that correlation is what we often do in many cases of practical lens design, lens testing, and use by the end customer. For instance, eyepieces and magnifiers are designed by tracing rays backwards, i.e. from the eye's pupil (which serves as the aperture stop) to the field. So, the image quality that is ascertained at the eyepiece's field stop in the optical design program (and at the test facility) has to be correlated with the experience of the end user of the eyepiece and magnifier, who provides us with feedback. (It is the same when an optometrist specifies a spectacle lens after testing your eye!) Today, of course, we also have plenty of advice and guidance for this correlation, based on the experiences shared by many authors of optical design books [5, 10, 11, 31, 32].

1.2.10 Amazing OpticStudio features you may not know about (which we will use)

In section 1.2.8, I mentioned that if you wanted to know how to define the edge of an image, you would need a means to simulate an image. As an example, using the lens file from figure 1.35 (or, equivalently, from figure 1.39) go to the Analysis menu and select Image Simulation from the Extended Scene Analysis menu (figure 1.43). Next, under its Settings menu, enter the settings shown in figure 1.44. You should see the output displayed in figure 1.45. Note that the Field Height setting is the full field angle (i.e. $2\times12°$) if the fields in the Field Data Editor are expressed as angles.

The Image Simulation feature computes the *convolution* of the PSF with the image that is in the file type used for simulation (in this example, the file type is a bitmap image, whose filename is 'Demo picture–640×480,' found in the Documents > Zemax > IMAFiles folder). You may recall that the image formed by a lens is a convolution operation [34] (see also, e.g. appendix A.1), where the flux per unit area at each position across the image is proportional to the volume under the PSF. However, note that in the settings shown in figure 1.44, we have selected '3' for the PSF-X Points and PSF-Y Points. This roughly means that the convolution is only

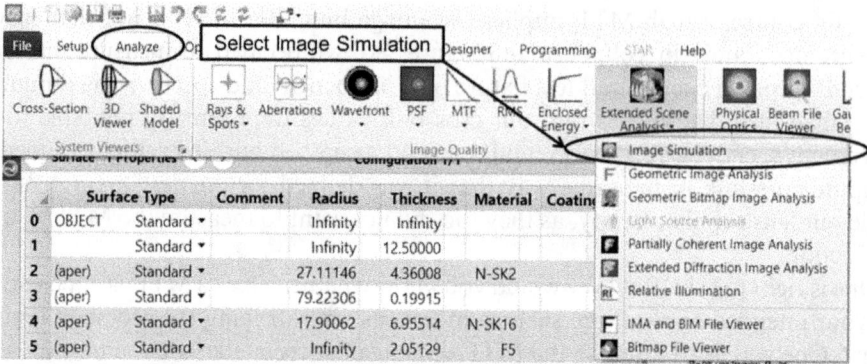

Figure 1.43. The image simulation analysis feature in OS.

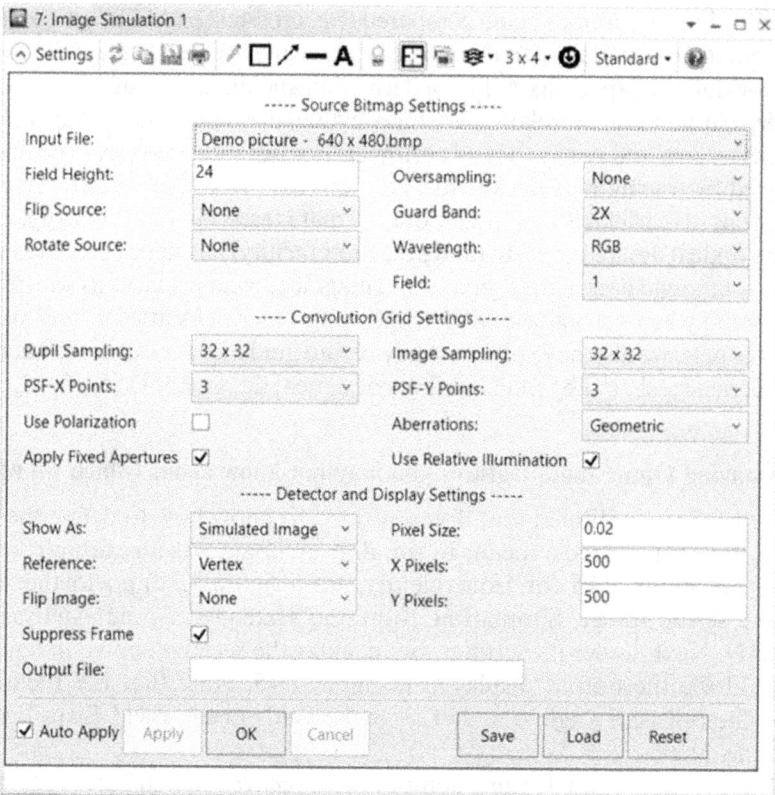

Figure 1.44. Settings to use in the Image Simulation to output the scene shown in figure 1.45.

performed over three annular areas (with interpolation in between), where the PSF is *space-invariant*. Still, this lens's spot diagram appears to show approximate space-invariance across the full field (the MTF vs Field plot also implies this). Alternatively, you could just check how different the image looks by selecting one

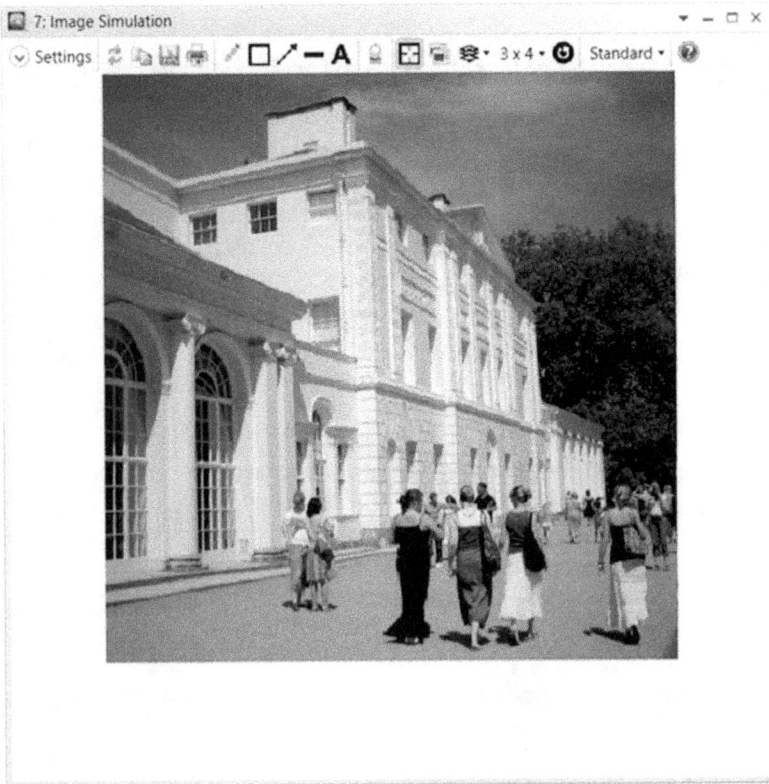

Figure 1.45. The output of the Image Simulation produced using the settings in figure 1.44. Reproduced from [33], with the permission of ANSYS, Inc.

or more PSF points for the settings. Note that we have placed a tick mark on Use Relative Illumination, which results in some edge darkening.

For the current lens, you may also see that since it is a near-diffraction-limited lens (at the wavelength of 0.55 microns), we may select 'Diffraction' under the 'Aberrations' setting in figure 1.44. This applies the Airy PSF for the convolution operation. In addition, you may check how much defocus would impact the image quality by shifting the image plane by ±0.026 mm (or less, or more) as was done in figure 1.29. As you can tell, there is plenty that can be done by playing around with this analysis feature of OS (incidentally, this analysis feature was used to simulate the images in an OPTICA presentation by the author in relation to a freeform machine vision lens system [35, 36]). You may consult the help file for the Image Simulation feature for further details on this analysis tool (as usual, by clicking on the '?' button of that window).

Another highly useful image analysis tool in OS is the Geometric Image Analysis (GIA) feature (see the option just below Image Simulation in figure 1.43). As we will see in chapter 2, **this analysis feature enables one to perform approximate illumination design and analysis in sequential ray-tracing mode**. Let us take an example by way of

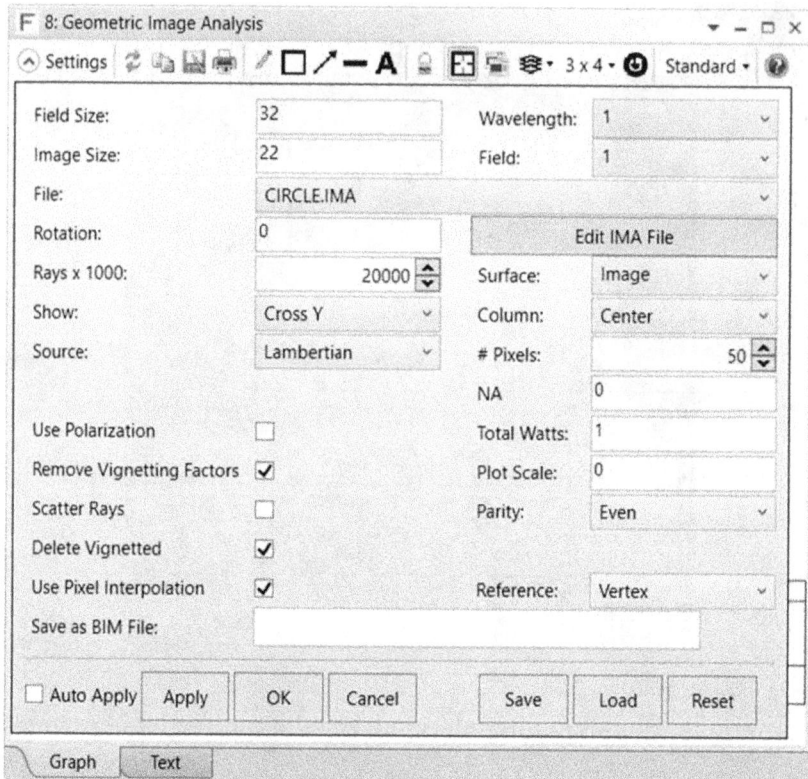

Figure 1.46. The settings used to obtain the GIA plot shown in figure 1.47.

comparing a GIA plot across the image plane and the relative illumination for the lens in figure 1.36. Once you have selected the GIA feature from the Extended Scene Analysis menu, enter the settings shown in figure 1.46, which should yield the output plot shown in figure 1.47. The object is a disk Lambertian source subtending a 32° full field angle from the vertex of the entrance pupil. Now return to the relative illumination plot of figure 1.42, open its Settings menu, and enter 24 for the Ray Density. The output should be as shown in figure 1.48. Compare this to the GIA plot shown in figure 1.47. They should be similar.

You may notice that the computation time was significantly longer for the GIA plot. In this plot, 20×10^6 random rays (weighted by a cosine factor that is consistent with the way in which a Lambertian source emits rays) were traced *from across the entire circular field*. The result is that one obtains a full irradiance distribution with some *ray trace noise* across the image, but we have only selected a two-dimensional (2D) column display. If you tilted and decentered one or more elements of the lens system (which we will do in the next section), then the resulting irradiance distribution could be analyzed as a surface plot, contour plot, false color plot, gray scale, etc. In a relative illumination plot, only $\pm x$ and $\pm y$-scan plots may be displayed under tilts and decentered conditions, and minimal ray tracing is

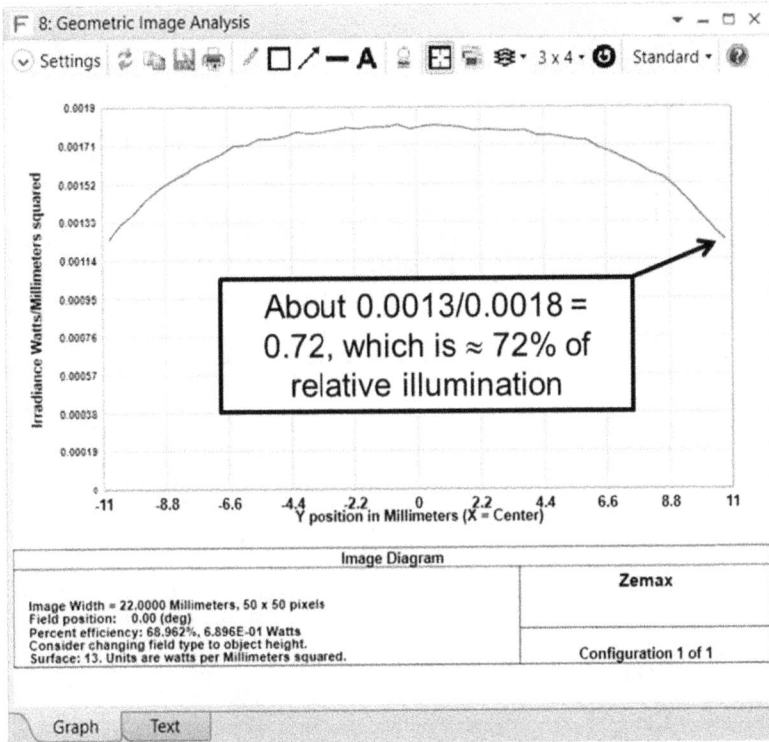

Figure 1.47. The GIA plot produced as a result of the settings used in figure 1.46.

performed, in which only a finite sample of field points (ten in the case of figure 1.48, which is the default) is considered. Further, for each field point, only a finite grid of rays sampling the exit pupil is traced.

There is one other point to be cautious about when using GIA to analyze the illumination across a plane. Although the rays are emitted in a random fashion, it is not done in the same way as in a pure NSC ray trace (whose rays from a source are emitted in a random, Monte Carlo fashion and weighted by a specified **radiant intensity** profile). In particular, the GIA plot in figure 1.47 has a slight error in that ray aiming was on, and it is unclear how *random* rays are *aimed* at the entrance pupil (in fact, the OS help file on GIA states that *pupil distortion* may occur if ray aiming is on—which is not a precise description of what is going on computationally and algorithmically). But if random rays in GIA are aimed in some fashion, then this could potentially *tilt* the off-axis ray bundle from the field towards the entrance pupil, because the chief ray would be aimed there. In a pure NSC ray trace, rays would simply be emitted in accordance with radiant intensity profiles defined by the source model (you will become further acquainted with—or at least reminded of —what is meant by radiant intensity in chapter 2). Thus, in GIA, it is often the case that it is more correct to remove ray aiming and set a first dummy surface (i.e. just a plane in front of the lens system, such as surface 1 in the prescription shown in

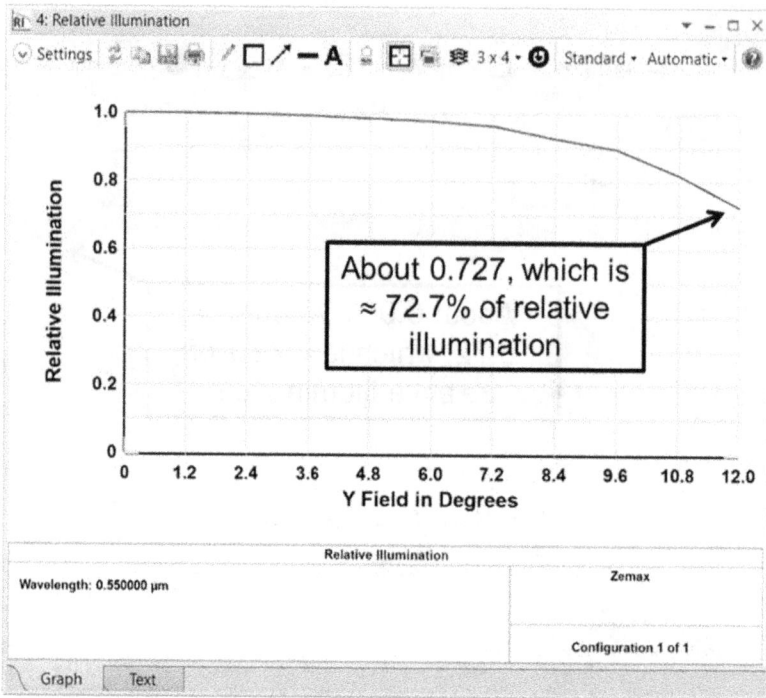

Figure 1.48. A Relative Illumination plot produced using Ray Density = 24.

figure 1.43) as the aperture stop. This makes the first surface into the entrance pupil for rays to fill (as in figure 1.3(a)).

If ray aiming is turned off, the computation is also significantly faster, as the program no longer has to spend computational time to aim rays (including the chief ray) towards the aberrated entrance pupil when the stop is buried in the lens system. However, in many cases, since the GIA is used (by the author of this book) as a first approximation to illumination analysis while designing an illumination optical system in SC mode, it is often not necessary to impose all of these settings. As always, a combination of tests using different settings is all that is needed to check results. For illumination problems and designs, the final test in the model is done in a full *pure* NSC ray tracing mode.

Many other useful OpticStudio (OS) features specific to the design examples in this book (i.e. those clickable icons across the Analyze menu, as shown in figure 1.43) will be discussed in relevant sections of the book. But before we end this section, one other useful feature to highlight is the Reports data feature output (figure 1.49). If you select Prescription Data from the Reports menu, then this outputs not just the surface data, but also the edge thicknesses of the lenses (which are helpful in checking manufacturability), element volume, refractive index, and more. Further, the Reports data include System Data and Cardinal Points, which are useful for quick reference to the first-order properties of the optical system (some of this information may also be obtained using operands in the Merit Function Editor).

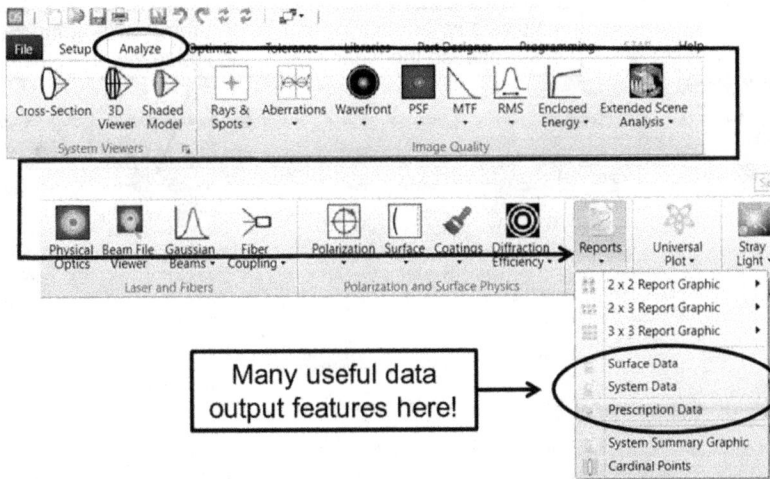

Figure 1.49. The Reports data feature under the Analyze menu.

1.2.11 Tilted and decentered components and assemblies

In optical design, the term *elements* generally refers to a single refracting (or reflecting) piece of an entity assembled into an optical system. For example, the lens system shown in figure 1.36 has six elements. It also has two *lens groups*, one to the left of the stop and the other to the right. Here, each lens group has a singlet and a doublet. A doublet is not generally regarded as an element (but it can be a *component*, since two elements are cemented into a single entity). The lens system as a whole may also be considered *a lens*. It would be called an *assembly* if it were mounted into a barrel (which is also often called a *housing*). However, this terminology is not strict. After all, the lens in figure 1.36 (which has six elements) is an assembly of *lens elements*. Over time, the usage of these terms has become somewhat self-explanatory within the context of discussion among optical designers and engineers.

Since elements that are mounted into assemblies may suffer from tilts and decentering due to tolerances, we should get acquainted with modeling the tilts and decentering of elements, components, and assemblies. Using the same lens file from figure 1.36, enter new rows above surface 8 and below surface 12 in the Lens Data Editor, select the Coordinate Break surface types at the new surfaces 8 and 14, and enter the data shown in figures 1.50 and 1.51. This data decenters the second lens group by 2 mm in the $+y$ direction and tilts it by $+7°$ about the x-axis. Figure 1.51 is a continuation of the Lens Data Editor scrolled to the right. Note that in figure 1.50, some new entries have been used in the prescription, such as values in the Chip Zone column. These allow annular clearances (for mounting spacers) such that the surface area they enclose is the clear aperture, and the values in the Mechanical Semi-Diameter column are half the outer diameters of the lens elements.

This sets an annular clearance of 1 mm surrounding the surface, so the clear aperture is inside this annulus.

#		Surface Type	Comment	Radius	Thickness	Material	Coating	Semi-Diameter	Chip Zone	Mech Semi-Dia
0	OBJECT	Standard ▾	Obj	Infinity	Infinity			Infinity	0.00000	Infinity
1		Standard ▾	Dummy 1	Infinity	12.50000			15.00000 U	0.00000	15.00000
2	(aper)	Standard ▾	1	27.11146	4.36008	N-SK2		11.00000 U	1.00000	12.00000
3	(aper)	Standard ▾		79.22306	0.19915			10.00000 U	1.00000	12.00000
4	(aper)	Standard ▾	2	17.90062	6.95514	N-SK16		10.00000 U	1.00000	11.00000
5	(aper)	Standard ▾	3	Infinity	2.05129	F5		10.00000 U	1.00000	11.00000
6	(aper)	Standard ▾		11.14702	5.42083			7.50000 U	1.00000	11.00000
7	STOP (aper)	Standard ▾	STOP	Infinity	7.79647			5.00000 U	0.00000	5.00000
8		Coordinate Break ▾	+TETY 8 12		0.00000			0.00000	-	
9	(aper)	Standard ▾	4	-11.43730	2.00000	F5		7.00000 U	0.70000	9.50000
10	(aper)	Standard ▾	5	Infinity	4.41213	N-SK16		9.00000 U	0.50000	9.50000
11	(aper)	Standard ▾		-16.13621	0.18444			9.00000 U	0.50000	9.50000
12	(aper)	Standard ▾	6	146.73154	2.77643	N-SK16		9.00000 U	0.50000	9.50000
13	(aper)	Standard ▾		-31.54955	-9.37301 T			9.00000 U	0.50000	9.50000
14		Coordinate Break ▾	-TETY 8 12		9.37301 P			0.00000	-	
15		Standard ▾	Dummy 2	Infinity	28.37078			9.41871	0.00000	9.41871
16	IMAGE	Standard ▾	Img	Infinity				12.50000 U	0.00000	12.50000

Pickup Solve, which lets the ray trace continue at surface 15

Position Solve, which, after rays have travelled through surfaces 9-13, lets the lens group un-do the tilts and decenters that were done at surface 8

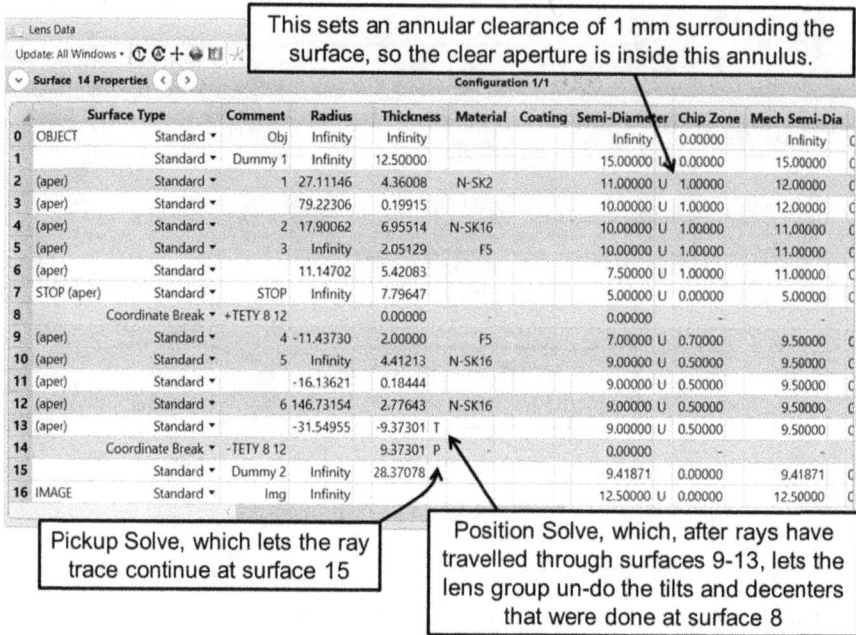

Figure 1.50. Prescription with Coordinate Break surface types for the lens in figure 1.36. This figure shows data up to the Mechanical Semi-Diameter column.

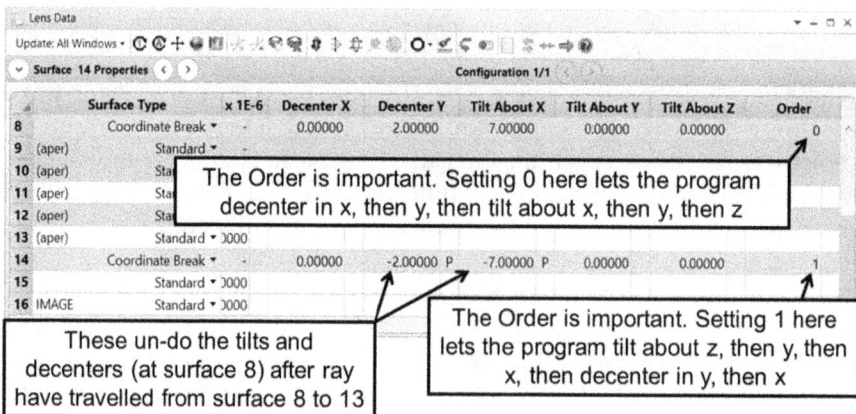

#		Surface Type	x 1E-6	Decenter X	Decenter Y	Tilt About X	Tilt About Y	Tilt About Z	Order
8		Coordinate Break ▾	.	0.00000	2.00000	7.00000	0.00000	0.00000	0
9	(aper)	Standard ▾	-						
10	(aper)	Sta							
11	(aper)	Sta							
12	(aper)	Sta							
13	(aper)	Standard ▾	)000						
14		Coordinate Break ▾	.	0.00000	-2.00000 P	-7.00000 P	0.00000	0.00000	1
15		Standard ▾	)000						
16	IMAGE	Standard ▾	)000						

The Order is important. Setting 0 here lets the program decenter in x, then y, then tilt about x, then y, then z

These un-do the tilts and decenters (at surface 8) after ray have travelled from surface 8 to 13

The Order is important. Setting 1 here lets the program tilt about z, then y, then x, then decenter in y, then x

Figure 1.51. Continuation of the prescription with Coordinate Break surface types from figure 1.50, showing the columns from Decenter X to Order. Note that the value 1 has been entered for the Order column on surface 14.

Thus, the diameter of the lens element with surfaces 2 and 3 is 24 mm, and its clear aperture diameter is 22 mm. Element numbers have been entered under the Comment column, and the 'Do Not Draw This Surface' option is selected for surface 15.

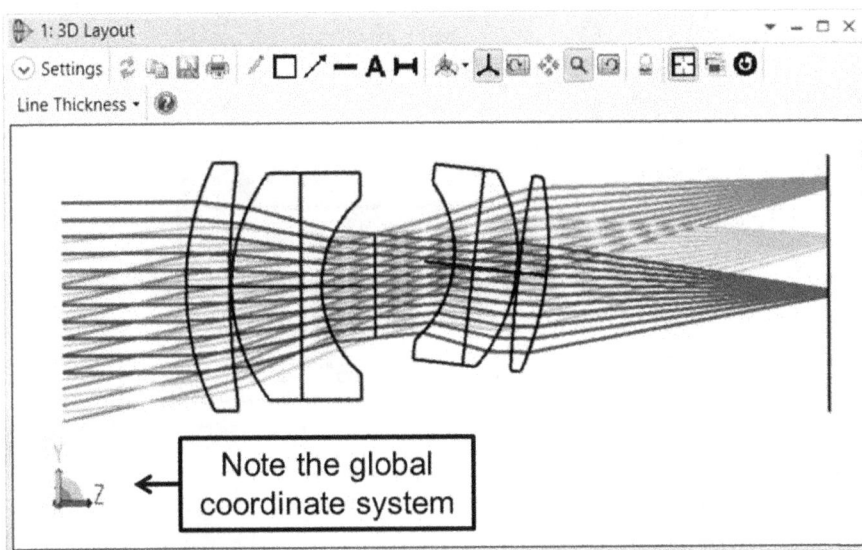

Figure 1.52. The resulting layout, shown in the 3D Viewer option from the Setup (or Analysis) menu, based on the prescriptions from figures 1.50 and 1.51.

Under the Decenter Y and Tilt About Y columns in figure 1.51, the values +2 mm and +7° have been entered for the Coordinate Break on surface 8. These set the amount of decenter in the y-dimension to 2 mm and specify a +7° tilt about the y-axis, respectively, for the second lens group. The resulting layout is shown in figure 1.52. This layout is from the 3D Viewer feature, which is found in the Setup (or Analysis) menu. It should be noted that OS's coordinate axes are such that the $-x$ axis points out of the page, so the $+x$ axis points into the page. The $+z$ axis points right, so $-z$ points to the left. The $+y$ axis points up, so $-y$ points down. These are the global coordinate axes, which are currently defined to be at the aperture stop (surface 7).

At the image plane, due to the asymmetry in the y-dimension, there is a change in irradiance distribution, as shown in figures 1.53 and 1.54. Note that the tilt and decenter has increased the relative irradiance in the top half of the image. This is also indicated by the relative illumination plot, as shown in figure 1.55. The image quality is expected to suffer, as implied by the aberrated rays along the y-axis in figure 1.52. Indeed, an image simulation (figure 1.56(a)) displays a degraded image, compared with the undegraded image (figure 1.56(b)) obtained using the nominal lens design from figure 1.35. (Note that the source bitmap file for this simulation is named 'StanleyLostLagoon20151102_144844' and is provided in the supplementary material for the current chapter. This file must be placed into the IMAFiles folder of Zemax. It is a picture taken by the author at the Lost Lagoon in Vancouver's beautiful Stanley Park.) In regard to the decentering and tilting of elements and groups in OS SC mode, this is all there is to it. The concept is that the Coordinate Break surfaces are always placed before and after the group of elements being

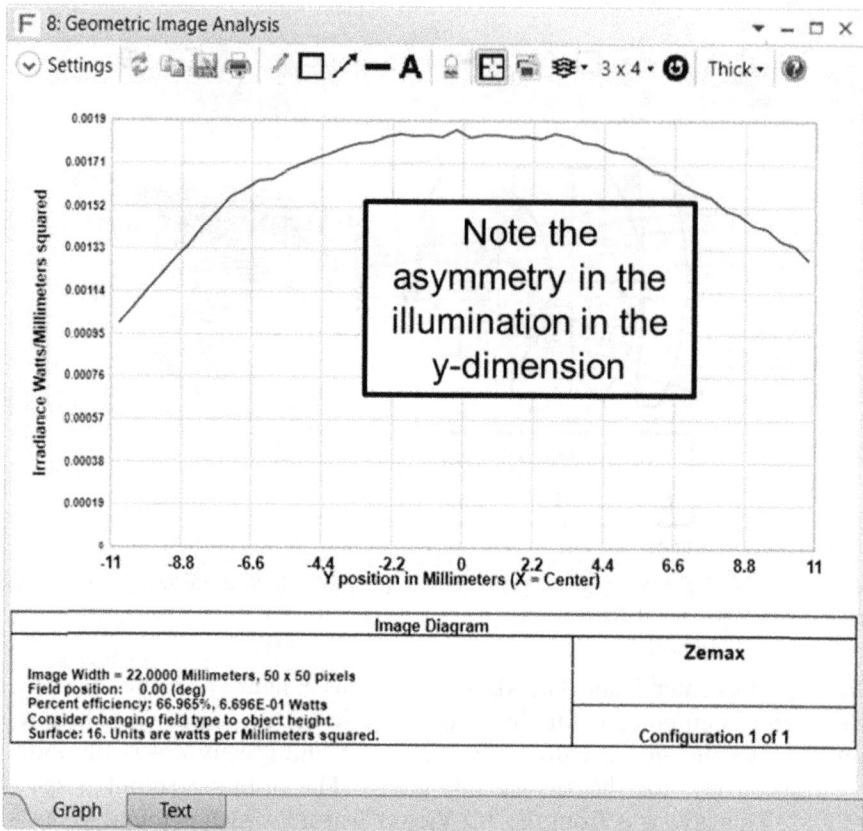

Figure 1.53. GIA plot in the *y*-dimension for the lens in figure 1.52, using the same settings from figure 1.46.

decentered and tilted. After the rays have traveled through the intended element/group, the second Coordinate Break surface must undo the decentering and tilts at the origin of where they occurred.

This is why a position solve has been placed for the thickness of surface 13. After the decentering and tilts are undone, rays must continue their journey from the last surface they had propagated through, which is surface 13. This is why a pickup solve has been used just after surface 13. Finally, a dummy surface is placed at surface 15, and the 'Do Not Draw This Surface' option has been selected for that surface's Draw property. Otherwise, a plane would appear at surface 15, which might not be visually aesthetic.

It is not only elements and groups that can tilt. The surfaces of lenses and mirrors can tilt too, as well as the image and object planes. One way to perform such surface tilts is to make their surface types into the Irregular type, as shown in figure 1.57, where the Irregular surface type has been selected for surface 1. Then, scroll right on the Lens Data Editor to the Decenter and Tilt columns, and enter −5° as shown in figure 1.58. The result is as shown in figure 1.59. When OS performs tolerancing, this

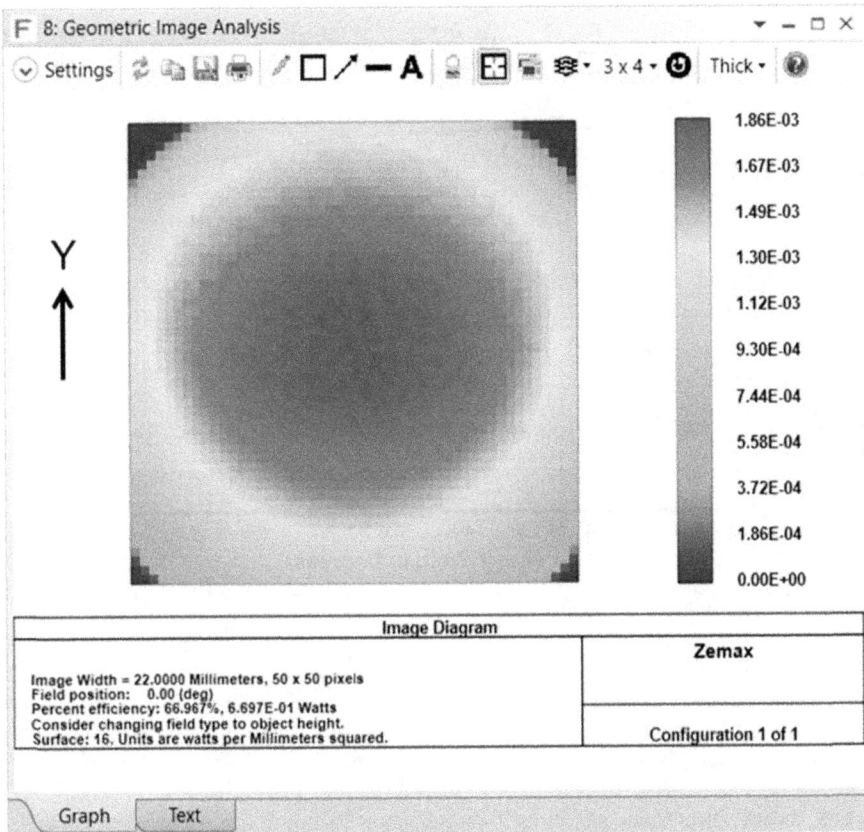

Figure 1.54. GIA false color plot for the lens in figure 1.52. The scale on the right-hand side should be regarded as relative, because the flux in the image is a fraction of the flux through the entrance pupil, which has been set to 1 watt (see the settings in figure 1.46).

is how it performs tilts and decenters on surfaces (and the coordinate breaks are used to tilt elements, components, and assemblies). You may notice that a positive angle rotates the entity clockwise if the *negative* axis points out of the page (or rather, when it points towards you). Accordingly, if you were *looking* towards the *positive y* axis from the top of the page so that this axis points towards you, a positive angular rotation would turn the entity *counter-clockwise*.

Let us now add a mirror tilted at + 45° about the *x*-axis. To do this, enter the surfaces and data shown in figures 1.60 and 1.61. Note that a mirror is always sandwiched between two Coordinate Break surfaces. Note in figure 1.61 that the pickup solve in the Tilt About *X* cell of surface 18 has been given the same angle as in the first tilt given to the mirror at surface 16. This lets the local axis tilt by an additional + 45°, which is as it should be, because we expect rays to reflect and continue orthogonally upwards after reflecting off a mirror at 45°. Click on the surface properties for the mirror, select Aperture, then Aperture Type, then select Rectangular Aperture, and enter 7 and 14 for the *X*-Half Width and *Y*-Half Width,

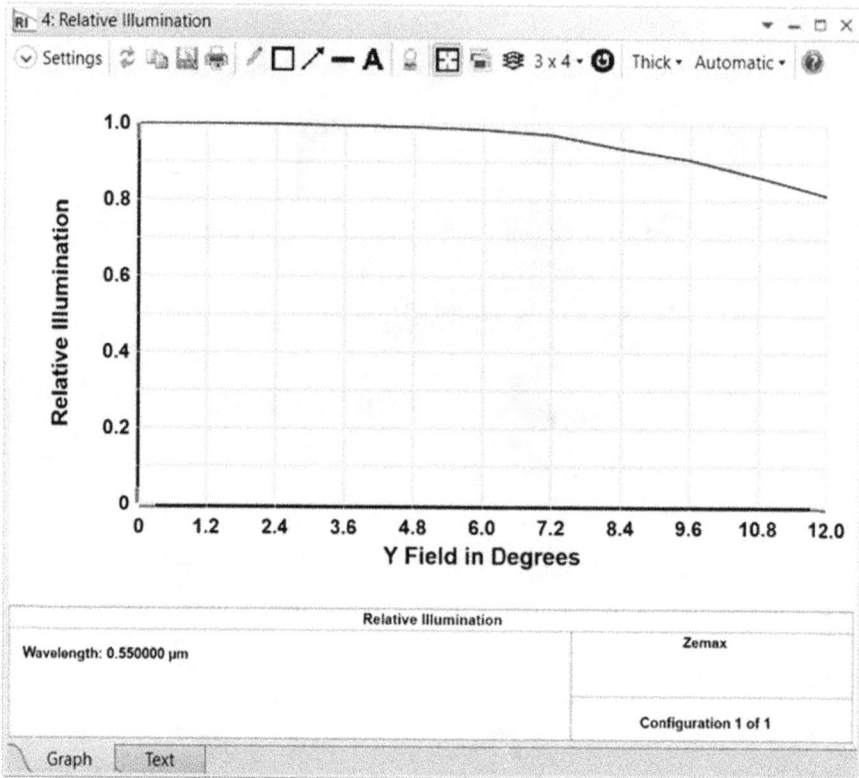

Figure 1.55. Relative illumination plot in the *y*-dimension for the lens in figure 1.52, using the same Ray Density settings as those of figure 1.48.

respectively. This makes the mirror into a rectangular element with those dimensions. Now, scroll up through the surface property menu and select Draw, then enter 2 for the Thickness. This makes the mirror appear to have a 2 mm thickness. You should now see the layouts shown in figures 1.62 and 1.63. The shaded model layout in figure 1.63 is obtained by selecting the Shaded Model icon under the Setup (or Analysis) menu.

You may have noticed that the mirror has vignetted the largest off-axis rays. We can make the mirror larger to account for those rays, but the vignetting is also a demonstration of one possible way to remove problematic oblique rays if they have high aberration. You could do the vignetting with an aperture. But if you happen to need a *fold mirror* to fold the ray path between the last element and the image plane, then the mirror could do the work of both vignetting and folding.

Perhaps now is a good time to introduce another amazing OS feature: the Universal Plot. You will find this feature on an icon just to the right of the Reports icon in the Analyze menu (see figure 1.49). Click it, then select 1-D, then 'New,' and enter the values shown in figure 1.64 under Settings. Click 'OK' when done, and you should see the plot shown in figure 1.65. This is a plot of the on-axis (axial) root

(a)   (b)

Figure 1.56. GIA image simulation displays using the same settings as those of figure 1.44 but for a different bitmap source file. (a) Simulation result obtained using the lens in figure 1.52. (b) Simulation result obtained using the lens in figure 1.35. *The source bitmap file is provided by courtesy of Ronian Siew.*

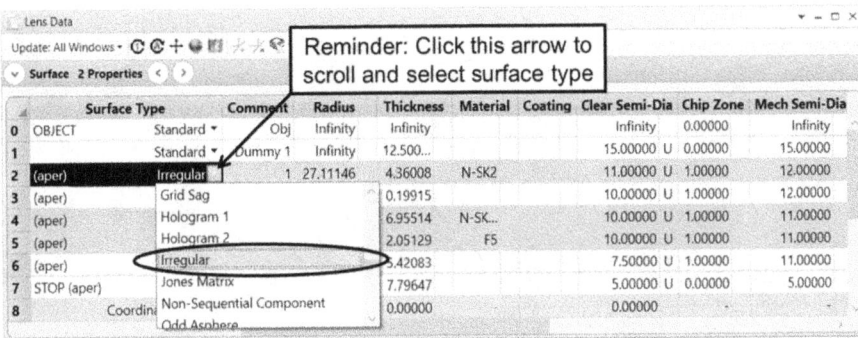

Figure 1.57. Selection of the Irregular surface type for surface 1 of the current lens system.

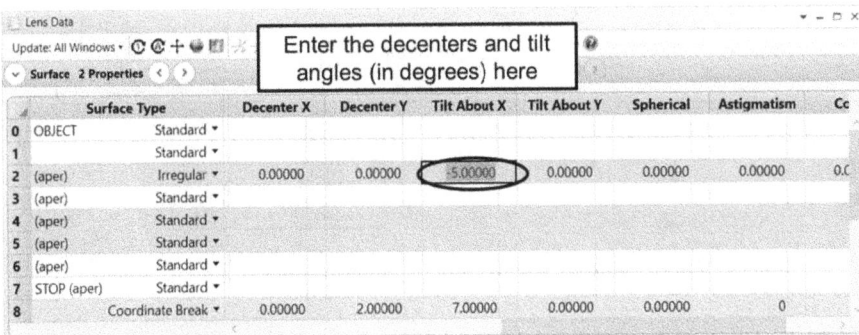

Figure 1.58. Entering a $-5°$ tilt about the x-axis for surface 1.

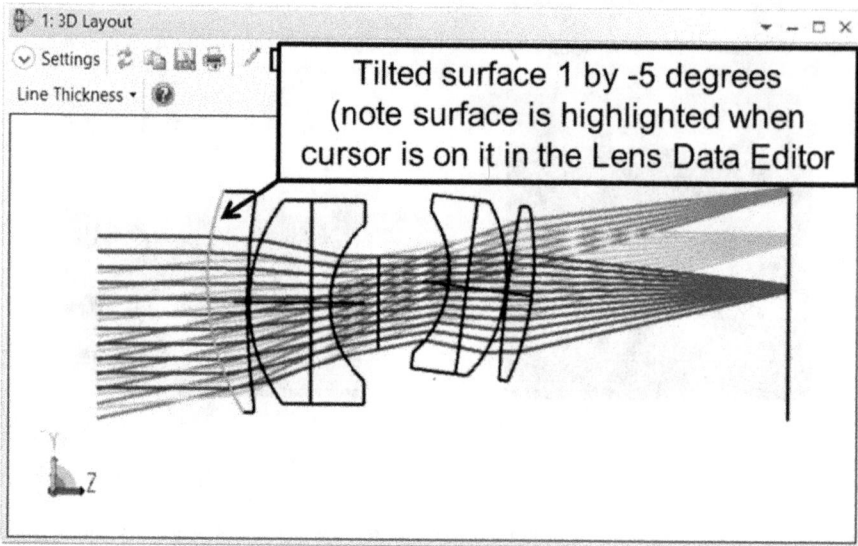

Figure 1.59. The resulting tilt on surface 1, based on data entered in figures 1.57 and 1.58.

Figure 1.60. A prescription that adds a mirror to the lens system from figure 1.59. This figure only shows the columns up to Mechanical Semi-Dia. The rest are shown in figure 1.61.

Figure 1.61. Data continued from figure 1.60, where a pickup solve has been entered for the Tilt About X cell of surface 18. This lets the local axis 'go up' after reflection.

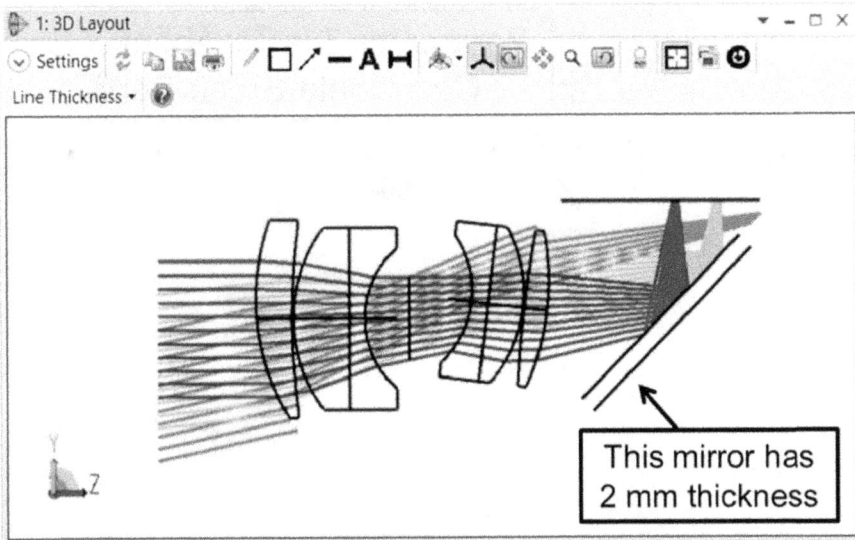

Figure 1.62. 3D layout for the lens system, obtained using the prescriptions from figures 1.60 and 1.61.

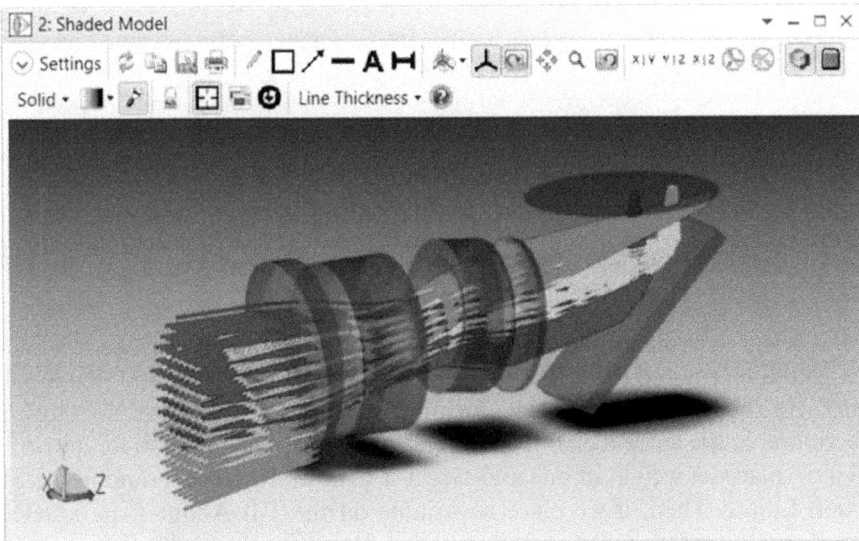

Figure 1.63. Shaded model layout of the lens system in figure 1.62.

mean square (RMS) spot size vs tilt of the second lens group (i.e. the tilt about the x-axis of surface 8 for the current lens system).

In figure 1.64, you entered RSCE for the Operand, which is the merit function operand for RMS spot size (referenced to the *centroid* of the spread of the axial bundle of rays at the image plane). We could have actually selected 'merit' under the

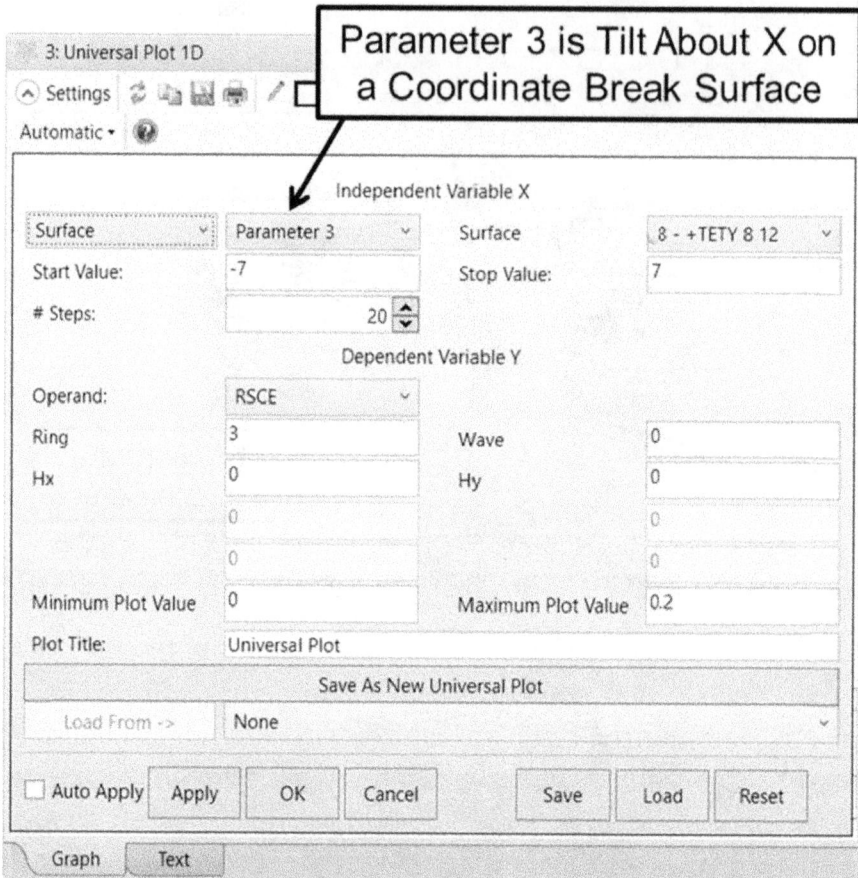

Figure 1.64. Settings to enter in order to output the plot shown in figure 1.65.

Operand setting (scroll to find it), entered RSCE directly into the Merit Function Editor, and then entered the line number in the editor for this operand (the input for the line will show up in the settings menu if you select 'merit' for the Operand). Furthermore, in the current example, instead of plotting the graph in figure 1.65, we could have placed a weight of one and target $= 0$ for an RSCE operand in the Merit Function Editor. Then, if we place a variable on the Tilt About X parameter on surface 8 and optimize the lens system, we can determine the x-tilt angle that would yield the minimum spot size shown in figure 1.65, which looks to be somewhere between 1.5° and 2°. But the current exercise provides a good reason to showcase the Universal Plot feature of OS. Isn't this neat? The resulting output in figure 1.65 lets you graphically see the impact of the tilt about the x-axis of the second lens group on the spot size of the axial field. Note that the minimum is not at 0°, because there remains the surface tilt of $-5°$ on surface 2 and the $+ 2$ mm Y decenter on surface 8. If you remove those perturbations, then the minimum is indeed at 0°. On the other hand, another way to see the result in figure 1.65 is that in the presence of other fixed

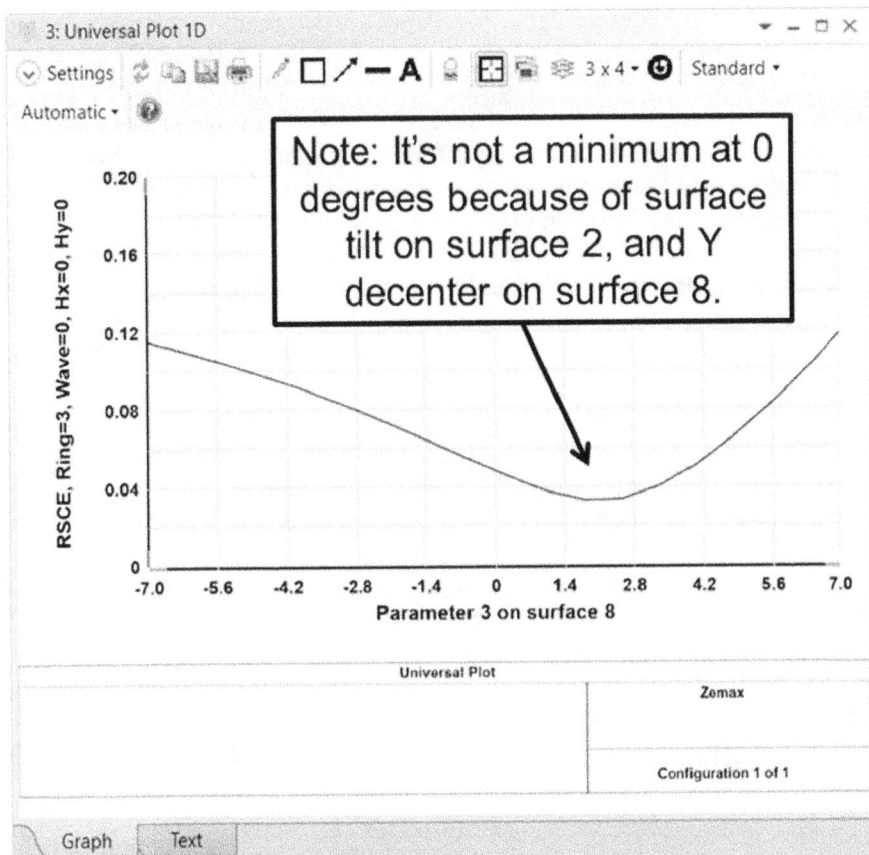

Figure 1.65. Axial RMS spot size vs tilt for the second lens group of the current lens system.

perturbations, if you have at least one *compensator* to adjust in the optical system, then you may be able to use it for *tuning* so as to improve the quality of your product. This is precisely what happens when you make adjustments to your optical systems that are in factory production. Those adjustments are called compensators. Focusing is a common compensator. Tilts are used as compensators if you suspect that there are other tilts and decentering of elements, components, and assemblies that result from manufacturing tolerances. It is not practical to provide adjustments for all tolerances, as this would be laborious and therefore costly. So, when those residual tolerances (perturbations) are fixed, you select a number of compensators to tune your optical system.

1.2.12 Optimizing a lens

Taking the same lens from the previous section, if you now include wavelengths at 0.45 and 0.65 microns, each weighted at 1 (set the primary wavelength to 0.55 microns), you will see that the MTF drops. This is because the lens has not been optimized for a broad spectrum. We will now do this. If you are new to lens design, then you may wonder whether this is how lenses are designed (i.e. by taking someone else's prescription and optimizing it). Yes, it is—mostly. Especially today. Perhaps not in the early days of lens design and development. But in modern times, lens design begins by taking a reference lens and re-optimizing it (or sometimes, even simply just re-scaling followed by slight re-optimization, as will be shown later in this section). Even in cases in which an experienced lens designer is designing from scratch, the designer is relying on their memory of what was done previously, whose origin is rarely the designer's. Today, there are even powerful programs that can start from scratch for you [37]. Certainly, experienced and knowledgeable designers also study the intrinsic and induced aberrations of a lens in some detail. But, by and large, there are still plenty of iterative optimizations, trial and error, and tweaks, based on a prior reference design.

Since our lens optimization exercise requires us to set the right wavelength weights for a broad spectrum, let us revisit the topic of how the weights for a spectrum are selected. Suppose that the object to be imaged either emits or scatters a spectrum $S(\lambda)$, and, as a consequence of the light passing through an optical system with transmittance $\tau(\lambda)$ and reaching an image sensor with response $R(\lambda)$, the detected effective spectrum is $S(\lambda)\tau(\lambda)R(\lambda)$, which is given in some arbitrary units (e.g. it could be volts per nm, or amps per nm, etc.) Now suppose it is somehow known—through some separate measurement—that the total flux collected by the optical system from the object is ϕ_{tot} in units of watts. Then, regardless of the intensity units used for the spectrum $S(\lambda)\tau(\lambda)R(\lambda)$, we may normalize it by its area and multiply the result by ϕ_{tot} to obtain the object's *spectral flux* (i.e. its flux per unit wavelength):

$$\phi(\lambda) = \phi_{tot}\frac{S(\lambda)\tau(\lambda)R(\lambda)}{\int_{-\infty}^{+\infty} S(\lambda)\tau(\lambda)R(\lambda)d\lambda}. \tag{1.9}$$

So, if an optical system has just a polychromatic source, an aperture (or filter), and a detector, then equation (1.9) has the proper flux units for the light being detected. That is, integrating both sides of the equation with respect to wavelength yields ϕ_{tot}. When modeling this condition, the *continuous* spectral weights for the program may be considered as the quantity $S(\lambda)\tau(\lambda)R(\lambda)/\int_{-\infty}^{+\infty} S(\lambda)\tau(\lambda)R(\lambda)d\lambda$. Evidently, for just the source (or object) alone and an entrance pupil, the source's spectral flux $\phi_S(\lambda)$ in the optical system may be expressed as

$$\phi_S(\lambda) = \phi_{Stot}\frac{S(\lambda)}{\int_{-\infty}^{+\infty} S(\lambda)d\lambda}, \tag{1.10}$$

where ϕ_{Stot} is the total flux of the light entering the optical system from the object, prior to any influence from the optical system's spectral transmittance and the detector's spectral response. Once again, integrating both sides of the equation with respect to wavelength yields that total flux. Note that generally, ϕ_{tot} is not necessarily equal to ϕ_{Stot} because of the product $\tau(\lambda)R(\lambda)$, which can remove flux at specific wavelength bands.

For arbitrary intensity units, an incoherent *quasi-monochromatic* PSF at the image plane may be given by the product $P(x, y, \lambda)S(\lambda)\tau(\lambda)R(\lambda)$, where x and y are spatial coordinates at the image plane and $P(x, y, \lambda)$ is the PSF's spatially varying function, while $S(\lambda)\tau(\lambda)R(\lambda)$ represents its spectral content. The wavelength that appears in $P(x, y, \lambda)$ signifies that the PSF's width may be wavelength dependent, which is the case if there is axial color aberration to any order (e.g. primary and secondary axial color, etc). If there is lateral color aberration, then even the spatial variables x and y are wavelength dependent. It is reasonable that the polychromatic PSF may be obtained by integrating $P(x, y, \lambda)S(\lambda)\tau(\lambda)R(\lambda)$ over all wavelengths. However, the computation of the polychromatic PSF is generally performed by including normalization [34, 38]. This means that the polychromatic PSF, which we denote by $p(x, y)$, may be expressed as

$$p(x, y) = \frac{\int_{-\infty}^{+\infty} P(x, y, \lambda)S(\lambda)\tau(\lambda)R(\lambda)\mathrm{d}\lambda}{\int_{-\infty}^{+\infty} \int_{-\infty}^{+\infty} \int_{-\infty}^{+\infty} P(x, y, \lambda)S(\lambda)\tau(\lambda)R(\lambda)\mathrm{d}\lambda \mathrm{d}x \mathrm{d}y}. \tag{1.11}$$

In some cases of practical significance, the polychromatic PSF given by equation (1.11) may be expressed in a manner analogous to equations (1.9) and (1.10) in that a total flux can be made to multiply the right side of the formula. For example, if we let $\phi_p(x, y)$ be the flux per unit area of the polychromatic PSF, then we can express it as

$$\phi_p(x, y) = \phi_{ptot}\frac{\int_{-\infty}^{+\infty} P(x, y, \lambda)S(\lambda)\tau(\lambda)R(\lambda)\mathrm{d}\lambda}{\int_{-\infty}^{+\infty} \int_{-\infty}^{+\infty} \int_{-\infty}^{+\infty} P(x, y, \lambda)S(\lambda)\tau(\lambda)R(\lambda)\mathrm{d}\lambda \mathrm{d}x \mathrm{d}y}, \tag{1.12}$$

where ϕ_{ptot} is the total flux in the polychromatic PSF. Its practical significance lies in the determination of the irradiance (flux per unit area) that is received by a photosensitive surface at a focused spot [34, 39]. In the current discussion, however, our purpose is to understand the origin of spectral weights in an optical design program and how they are related to an imaging system's performance. For this purpose, ϕ_{ptot} is not needed in the PSF, but the normalizing factor in the denominator of equation (1.11) is, as you will now see. Since the optical transfer function (OTF) is the normalized Fourier transform of the PSF [34], and since the MTF is the modulus of the OTF, then the polychromatic MTF (denoted by PMTF) is given by the Fourier transform of equation (1.11), expressed as

$$\mathrm{PMTF} = \left| \int_{-\infty}^{+\infty} \int_{-\infty}^{+\infty} p(x, y) \exp\left[-i2\pi(\nu_x x + \nu_y y)\right]\mathrm{d}x \mathrm{d}y \right| \tag{1.13}$$

$$= | \mathcal{F}\{p(x, y)\} |, \tag{1.14}$$

where $\mathcal{F}\{p(x, y)\}$ denotes the Fourier transform of $p(x, y)$, $i = \sqrt{-1}$, ν_x and ν_y are spatial frequencies in cycles per unit length, and $p(x, y)$ is given by equation (1.11). Because the Fourier transform operation is wavelength independent (i.e. it does not involve integrating over wavelength), the PMTF expressed in equation (1.13) may be written as an average of monochromatic MTFs, weighted by the effective spectrum. To see this, substituting equation (1.11) into (1.13) gives

$$
\begin{aligned}
\text{PMTF} &= \left| \ \mathcal{F}\left\{ \frac{\int_{-\infty}^{+\infty} P(x, y, \lambda)S(\lambda)\tau(\lambda)R(\lambda)\mathrm{d}\lambda}{\int_{-\infty}^{+\infty}\int_{-\infty}^{+\infty}\int_{-\infty}^{+\infty} P(x, y, \lambda)S(\lambda)\tau(\lambda)R(\lambda)\mathrm{d}\lambda\mathrm{d}x\mathrm{d}y} \right\} \right| \\[2mm]
&= \frac{\int_{-\infty}^{+\infty} | \ \mathcal{F}\{P(x, y, \lambda)\} \ | \ S(\lambda)\tau(\lambda)R(\lambda)\mathrm{d}\lambda}{\int_{-\infty}^{+\infty}\int_{-\infty}^{+\infty}\int_{-\infty}^{+\infty} P(x, y, \lambda)S(\lambda)\tau(\lambda)R(\lambda)\mathrm{d}\lambda\mathrm{d}x\mathrm{d}y} \\[2mm]
&= \frac{\int_{-\infty}^{+\infty} \text{MTF}(\nu_x, \nu_x, \lambda)K(\lambda)S(\lambda)\tau(\lambda)R(\lambda)\mathrm{d}\lambda}{\int_{-\infty}^{+\infty} K(\lambda)S(\lambda)\tau(\lambda)R(\lambda)\mathrm{d}\lambda},
\end{aligned}
\tag{1.15}
$$

where:

$$
K(\lambda) = \int_{-\infty}^{+\infty} \int_{-\infty}^{+\infty} P(x, y, \lambda)\mathrm{d}x\mathrm{d}y.
\tag{1.16}
$$

Just in case this last step was not clear, note that the OTF is conventionally normalized [34]:

$$
\text{OTF}(\lambda) = \frac{\mathcal{F}\{P(x, y, \lambda)\}}{\int_{-\infty}^{+\infty} \int_{-\infty}^{+\infty} P(x, y, \lambda)\mathrm{d}x\mathrm{d}y} = \frac{\mathcal{F}\{P(x, y, \lambda)\}}{K(\lambda)}.
\tag{1.17}
$$

By virtue of Rayleigh's theorem (sometimes called Parseval's theorem) [34], the denominator in equation (1.17) is actually the Fourier transform of $P(x, y, \lambda)$ at the spatial frequencies $\nu_x = 0$ and $\nu_y = 0$. That is:

$$
\int_{-\infty}^{+\infty} \int_{-\infty}^{+\infty} P(x, y, \lambda)\mathrm{d}x\mathrm{d}y = \int_{-\infty}^{+\infty} \int_{-\infty}^{+\infty} P(x, y, \lambda)\exp\{-i2\pi[(0)x + (0)y]\}\mathrm{d}x\mathrm{d}y.
\tag{1.18}
$$

So, redefining either side of equation (1.18) as the quantity $K(\lambda)$ in equation (1.15) makes the expression in equation (1.15) more compact. Equation (1.15) is analogous to expressions for polychromatic MTF that have been published by several authors [38, 40–42]. It may be interpreted as a *spectrally weighted average of monochromatic MTFs*, which is a consequence of the normalization step in equation (1.11). This leads us to an understanding of the origin and determination of spectral weights that are used for optical design, first introduced in section 1.2.3. I will now show you how to make practical approximations for these weights.

In an optical design program, the polychromatic PSF and the PMTF are computed by numerical integration. In the case of the PMTF (and ignoring, for convenience, the fact that numerical Fourier transforms are generally performed efficiently via the FFT algorithm), this means that equation (1.15) can be approximated by discretizing the integrals, which, for n samples of equal intervals, we may express as

$$\text{PMTF} \approx \frac{\sum_{i=1}^{n} \text{MTF}(\nu_{xi}, \nu_{yi}, \lambda_i) K(\lambda_i) S(\lambda_i) \tau(\lambda_i) R(\lambda_i) \Delta\lambda}{\sum_{i=1}^{n} K(\lambda_i) S(\lambda_i) \tau(\lambda_i) R(\lambda_i) \Delta\lambda}, \qquad (1.19)$$

where $S(\lambda_i)\tau(\lambda_i)R(\lambda_i)\Delta\lambda$ (for which $i = 1, 2, 3, \ldots, n$) are the spectral weights for an optical design program. In OS, these are entered into the Wavelength Data Editor. Note that they are area intervals, not the peaks of the spectral curve. This means that, in theory, the wavelength intervals $\Delta\lambda$ should be small and plentiful, or they can be not that small and not that plentiful if Simpson's rule for numerical integration is applied [12]. Yet, if we can accept certain approximations (as mentioned in section 1.2.3), then we can just fit rectangular areas under the peaks of the effective spectrum given by $S(\lambda)\tau(\lambda)R(\lambda)$. In the case of fitting rectangles under the curve of the effective spectrum, $\Delta\lambda$ cancels between the numerator and denominator in equation (1.19), so, that leaves just the peaks of $S(\lambda)\tau(\lambda)R(\lambda)$ sampled at every ith interval. Furthermore, if you should decide to use spectral weights that have been normalized by the total area under the curve of the effective spectrum, then that normalization factor would also cancel in equation (1.19). Therefore, it is often sufficient to select spectral weights by sampling the peaks of the effective spectrum, as depicted in figure 1.66. I acknowledge that the wavelength intervals $\Delta\lambda$ in figure 1.66 do not cover the entirety of the widths that span to the left and right of the sampled wavelengths λ_i. This was deliberate for two reasons: one, the intervals are more clearly represented this way; two, as a consequence of normalization, it really does not matter how wide they are if we are fitting rectangles under the peaks of the curve. In optical system design, making such approximations (which are eventually experimentally verified by a prototype) is practical and helps save time.

Before moving on to optimizing a lens, I want to show that although the scales chosen for the vertical axes of the spectra are inconsequential (due to normalization), the choice of *relative units* among the spectra $S(\lambda_i)$, $\tau(\lambda_i)$, and $R(\lambda_i)$ is important. For instance, note that a sensor's spectral response $R(\lambda_i)$ is not the same as its quantum efficiency. The quantum efficiency is defined as the number of photoelectrons out divided by the number of photons in, per unit wavelength. But by virtue of quantization, if the spectral flux is $\phi(\lambda)$, then the number of photons at any wavelength is $\phi(\lambda)/[hc/\lambda] = \phi(\lambda)\lambda/hc$, where h is Planck's constant and c is the speed of light in a vacuum. If $n_e(\lambda)$ is the number of electrons per unit time detected from the sensor, then $n_e(\lambda)t$ is the number of photoelectrons collected at integration time t. So, if $\eta(\lambda)$ is the quantum efficiency, then $\eta(\lambda) = n_e(\lambda)t/[\phi(\lambda)\lambda/hc] = n_e(\lambda)hct/[\phi(\lambda)\lambda]$. The spectral response (often also called **responsivity**) is the current out divided by the flux in, per unit wavelength, so it is given by $\eta(\lambda)\lambda e/(hct)$, where e is the electron charge [43, 44]. In other words, spectral responsivity $R(\lambda_i)$ (either discrete or continuous) is proportional to $\eta(\lambda)\lambda$, and if you do use responsivity rather than quantum efficiency from a sensor's datasheet for the purpose of determining the effective spectrum and spectral weights, then you must also select units that are proportional to flux rather than to the number of photons for the object (or source) and the transmittance of the optical system.

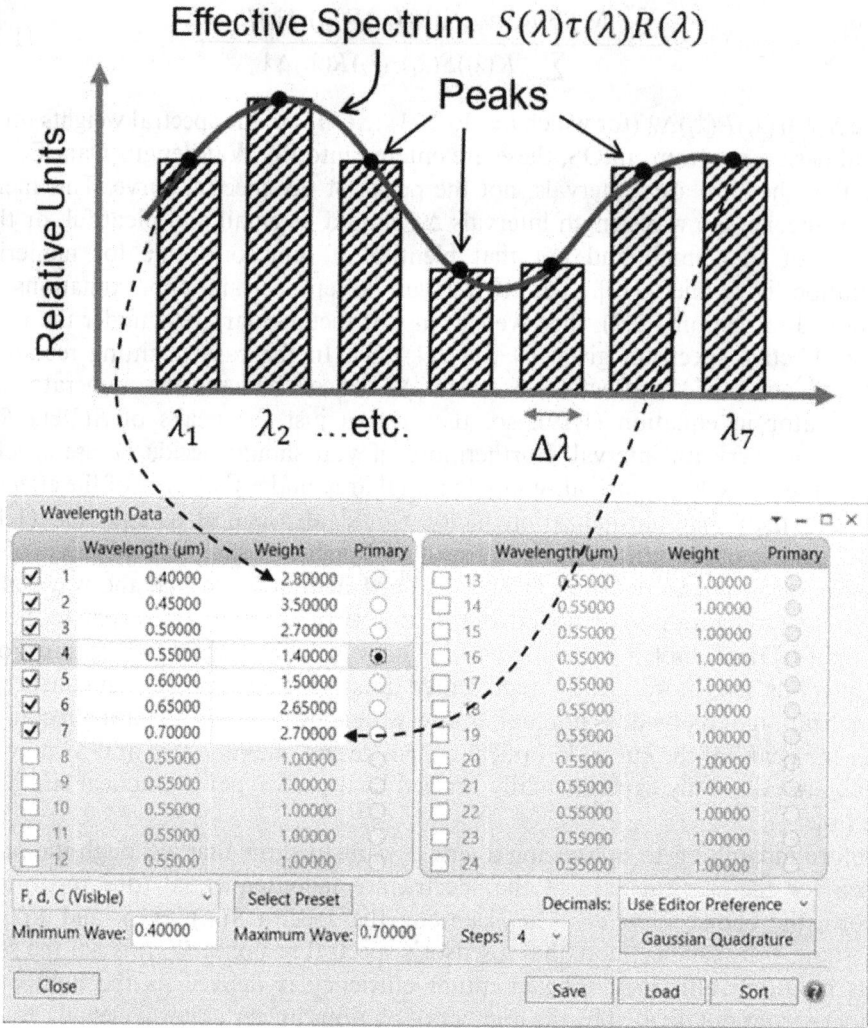

Figure 1.66. Spectral weights, sampled using peaks of the effective spectrum.

The above discussion is helpful for us in our discussion of lens optimization, and, as you will see, makes things very simple and convenient. Let us suppose that the object's spectrum $S(\lambda)$ and the optical system's spectral transmittance $\tau(\lambda)$ are flat across visible wavelengths and that they have been taken from datasheets that provide units proportional to the flux per unit wavelength. Let us suppose that the sensor's *quantum efficiency* is not flat. However, the visible light quantum efficiencies of most monochrome CMOS image sensors are blue-shifted. That is, they are more sensitive to blue than to red (to confirm this, you just need to search for this online and examine available datasheets, such as the Sony IMX183 monochrome image sensor). Since spectral responsivity is proportional to $\eta(\lambda)\lambda$, the multiplication of the

Figure 1.67. Lens data editor for the current lens system, with variables.

quantum efficiency by λ tends to shift the peak of the spectrum towards the middle, resulting in an approximately even spectrum across the visible range between about 450 and 650 nm. So, for simplicity, it seems quite practical to assume a flat effective spectrum for our lens optimization example. Sometimes, this means having weight = 1 at three wavelengths given by the F ($\approx$486 nm), d ($\approx$588 nm), and C ($\approx$656 nm) atomic wavelengths. Alternatively, we may use weight = 1 at 450, 550, and 650 nm. For convenience, we shall use the latter.

Using spectral weights = 1 at wavelengths of 0.45, 0.55, and 0.65 microns and taking the lens from the previous section, let us remove the fold mirror, tilts, and decenters. Also, remove coordinate breaks and eliminate the third field point so that you only have the axial field (0°) and the 6° field point. Set variables for the radii of curvature and thicknesses for surfaces 2–12. The Lens Data Editor should look like that shown in figure 1.67. Enter the operands with targets and weights shown in figure 1.68. Now, go to the Optimize menu and click Optimize! When the optimization dialog box pops out (figure 1.69), just click Start, and wait till the execution stops. The resulting lens data I obtained are shown in figures 1.70–1.73. This demonstrates an optimization run in lens design for a relatively simple lens system (it has a low field angle that is not generally useful for many modern camera lenses, but it serves its purpose here as an example).

Let me explain the operands used in figure 1.68. The first, the EFFL, constrains the system's effective focal length. Some designers use the last lens surface's radius of curvature to constrain this (e.g. in the current lens system, by placing an f-number solve on the radius of surface 12 and maintaining a specific entrance pupil diameter

Merit Function Editor ▾ – ☐ ✕

ˇ ⊟ ▷ ↘ ✗ ↖ • ᄃ ▭ ↕ ⇠ ⇢ ❶

▾ **Wizards and Operands** ◁ ❭   **Merit Function:** 0.00862448108155068

	Type	Wave			Target	Weight	Value	% Contrib
1	EFFL ▼	**2**			50.00000	1.00000	50.01291	36.15813
2	RSCE ▼ 3	0 0.00000 0.00000			0.00000	2.00000	6.30647E-03	17.24825
3	RSCE ▼ 3	0 0.00000 0.70000			0.00000	0.00000	7.74025E-03	0.00000
4	RSCE ▼ 3	0 0.00000 1.00000			0.00000	2.00000	8.97145E-03	34.90578
5	RWCE ▼ 3	0 0.00000 0.00000			0.00000	0.00000	0.45977	0.00000
6	RWCE ▼ 3	0 0.00000 0.70000			0.00000	0.00000	0.74432	0.00000
7	RWCE ▼ 3	0 0.00000 1.00000			0.00000	0.00000	0.94981	0.00000
8	BLNK ▼							
9	AXCL ▼ 1	3 0.70000			0.00000	0.20000	-0.05887	0.00000
10	BLNK ▼							
11	MNCG ▼ 2	12			2.00000	0.20000	2.00000	0.00000
12	MXCG ▼ 2	12			12.00000	0.20000	12.00000	0.00000
13	MNEG ▼ 2	12 1.00000	1		1.00000	0.20000	1.00000	0.00000
14	MNCA ▼ 2	11			0.20000	0.20000	0.18358	11.68784
15	MXCA ▼ 5	8			20.00000	0.20000	20.00000	0.00000
16	MNEA ▼ 5	8 1.00000	1		1.00000	0.20000	1.00000	0.00000

Figure 1.68. Merit function editor for the current lens system, with operands, targets, and weights.

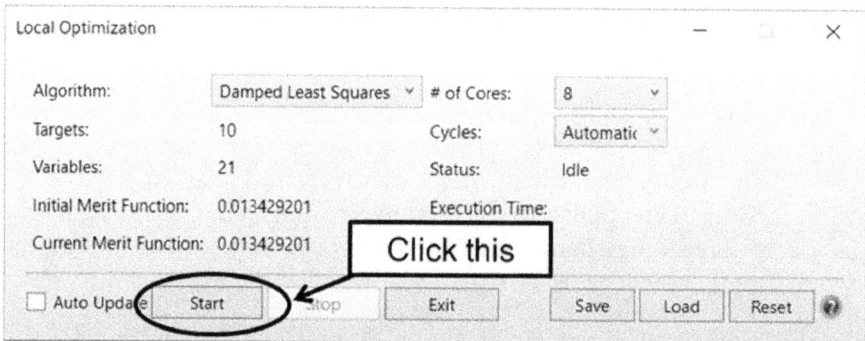

Local Optimization   —   ☐   ✕

Algorithm:	Damped Least Squares ▾	# of Cores:	8 ▾
Targets:	10	Cycles:	Automatic ▾
Variables:	21	Status:	Idle
Initial Merit Function:	0.013429201	Execution Time:	
Current Merit Function:	0.013429201	**Click this**	

☐ Auto Update   Start   Stop   Exit   Save   Load   Reset ❶

Figure 1.69. Click Start when this optimization dialog box pops up.

or stop semi-diameter). Operands 24 constrain the root mean square (RMS) spot size at the *three* fields. Here, the middle field point has been set at 70% of the field angle, but I have left it uncontrolled (no weight is placed on that operand). Generally, it should be constrained, but it need not be at the 70% field. The process of modern automatic lens design optimization is so quick that you can try different field positions (and numbers of field points) to see whether they improve anything. When a middle field is used at 70% of full field for a finite object distance, the field area coverage between the axial and full field is a circle of half the area of a circle covering the full field. Similarly, when using a 0.7 zone in the entrance pupil to analyze lens performance, this zone covers half the area of the full lens aperture and gives some indication of how much influence half of the flux entering the lens would have on the PSF, which is sensible. However, in other cases, it is beneficial to correct aberrations (such as spherical and coma aberrations) at the full aperture (i.e. pupil

Figure 1.70. The author's result in the lens data editor after performing a single optimization run of the lens system from figure 1.67.

	Surface Type		Comment	Radius	Thickness	Material	Coating	Clear Semi-Dia
0	OBJECT	Standard ▾		Infinity	Infinity			Infinity
1		Standard ▾	Dummy 1	Infinity	12.50000			15.00000 U
2	(aper)	Standard ▾	1	26.77480 V	4.41344 V	N-SK2		11.00000 U
3	(aper)	Standard ▾		77.36275 V	0.20089 V			10.00000 U
4	(aper)	Standard ▾	2	17.90949 V	6.47651 V	N-SK16		10.00000 U
5	(aper)	Standard ▾	3	-796.19923 V	2.27833 V	F5		10.00000 U
6	(aper)	Standard ▾		11.11568 V	5.49907 V			7.50000 U
7	STOP (aper)	Standard ▾	STOP	Infinity	7.97679 V			5.00000 U
8	(aper)	Standard ▾	4	-11.48863 V	2.02972 V	F5		7.00000 U
9	(aper)	Standard ▾	5	-63.35601 V	4.45257 V	N-SK16		9.00000 U
10	(aper)	Standard ▾		-16.12572 V	0.20167 V			9.00000 U
11	(aper)	Standard ▾	6	148.43735 V	2.59709 V	N-SK16		9.00000 U
12	(aper)	Standard ▾		-31.38900 V	28.98876 V			9.00000 U
13	IMAGE (aper)	Standard ▾	Img	Infinity	-			10.62600 U

Figure 1.71. The author's result in the merit function editor after a single optimization run of the lens system from figure 1.67.

Merit Function: 0.00234634620798509

	Type	Wave			Target	Weight	Value	% Contrib
1	EFFL ▾	2			50.00000	1.00000	50.00000	4.26865E-06
2	RSCE ▾ 3	0 0.00000 0.00000			0.00000	2.00000	2.81572E-03	45.00335
3	RSCE ▾ 3	0 0.00000 0.70000			0.00000	0.00000	2.92493E-03	0.00000
4	RSCE ▾ 3	0 0.00000 1.00000			0.00000	2.00000	3.10304E-03	54.65642
5	RWCE ▾ 3	0 0.00000 0.00000			0.00000	0.00000	0.16469	0.00000
6	RWCE ▾ 3	0 0.00000 0.70000			0.00000	0.00000	0.20417	0.00000
7	RWCE ▾ 3	0 0.00000 1.00000			0.00000	0.00000	0.23398	0.00000
8	BLNK ▾							
9	AXCL ▾ 1	3 0.70000			0.00000	0.20000	7.74197E-04	0.34023
10	BLNK ▾							
11	MNCG ▾ 2	12			2.00000	0.20000	2.00000	0.00000
12	MXCG ▾ 2	12			12.00000	0.20000	12.00000	0.00000
13	MNEG ▾ 2	12 1.00000	1		1.00000	0.20000	1.00000	0.00000
14	MNCA ▾ 2	11			0.20000	0.20000	0.20000	0.00000
15	MXCA ▾ 5	8			20.00000	0.20000	20.00000	0.00000
16	MNEA ▾ 5	8 1.00000	1		1.00000	0.20000	1.00000	0.00000

zone = 1), as discussed in section 1.4. Ultimately, some trial and error here can be helpful.

Operands 5–7 are *just-in-case* controls for the RMS wavefront error. By habit, I use them just in case I need to switch to using these rather than RMS spot size

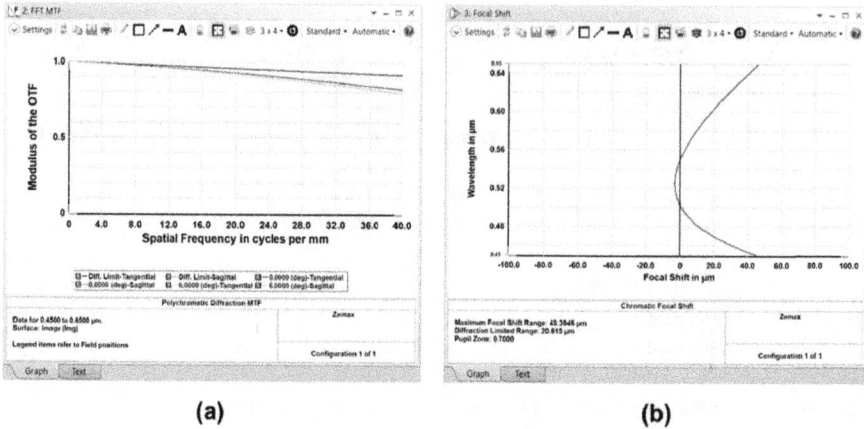

Figure 1.72. The author's lens's performances resulting from a single optimization run of the lens system shown in figure 1.67. (a) MTF. (b) Chromatic Focal Shift.

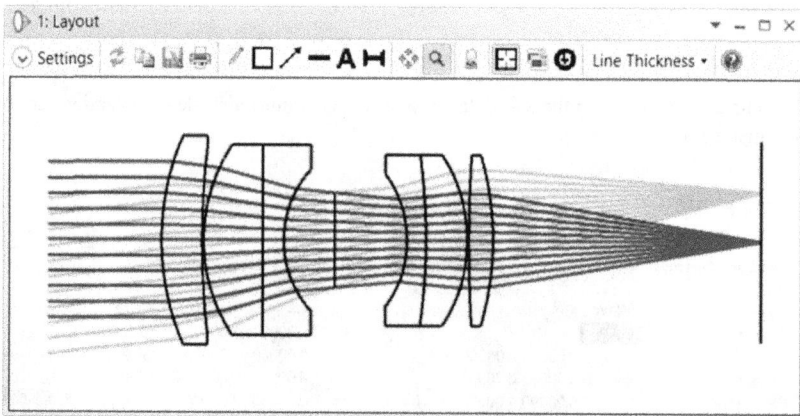

Figure 1.73. The author's lens layout resulting from a single optimization run of the lens system shown in figure 1.67. The prescription is provided in figure 1.70.

operands. In more complex lens systems with more elements, larger field angles, and higher apertures (lower f-numbers), switching to RMS wavefront error helps to improve the MTF, but requires some sacrifice in the other metrics, such as axial color and EFL. In the current example, the lens is quite simple, so there was no need to use RMS wavefront error. I have placed the operands there out of habit.

Operand 9 controls the primary axial color at the 0.7 pupil zone (as mentioned, this is a somewhat arbitrary choice at this stage). This operand does the achromatizing, but it does not *know* anything about achromatizing. The six-element lens system (whose origin was one of the double-Gauss lens samples provided by OS in their sequential lens samples directory) already has glass choices selected that *enable* the achromatizing (see sections 1.4.7–1.4.12 for a further discussion of color correction). The thing we wish to highlight here is the *chromatic focal shift* plot in figure 1.72(b),

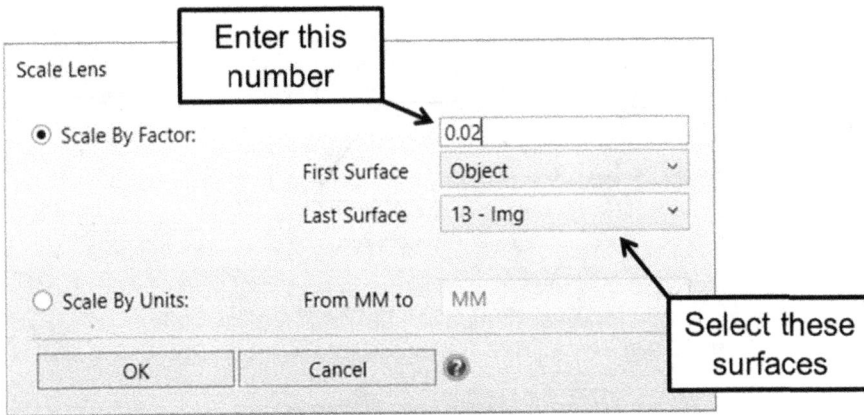

Figure 1.74. Dialog box for scaling a lens. Select 0.02 to scale the EFL from 50 mm to 1 mm.

in case this is the first time you have seen it. It gives a plot of the focal position in image space vs wavelength for the chosen pupil zone. The curve you see in the current figure is characteristic of a lens that has been corrected for primary axial color, which is the difference in focal positions (but not the EFLs) of the shortest and longest wavelengths used for the lens system, which is near zero in the current lens system. The nonzero residual is the difference in focal positions between the central wavelength and these two wavelengths. This residual is called the *secondary color* aberration.

Operands 11–16 control the mechanical structural properties of the lens elements. For instance, they prevent the central thicknesses and edge thicknesses of the elements from becoming zero (which is obviously not realistic), and they prevent air spaces from going negative. Reading through the help files on these operands will tell you very clearly what each of them does. The only useful thing for me to tell you here is that these operands generally do not require high values for their weights, relative to the operands for controlling aberrations and EFL.

Now, let us say that we want to have a lens at a smaller EFL (say, 1 mm) and at a larger EFL (say, 100 mm). It will become apparent why these two extreme values for target EFLs are chosen. To do this, first scale the current lens's EFL to 1 mm by selecting the Setup menu, then locating and left-clicking on Scale Lens. At the dialog box, enter the scale factor and surfaces displayed in figure 1.74. The scale factor of 0.02 is entered because we are scaling the lens's EFL by 1/50 of its current value.

The result of scaling down to a 1 mm EFL is shown in figure 1.75. Note that the MTF is now diffraction limited. The maximum spatial frequency for a diffraction-limited lens is given by $1/(\lambda F/\#)$. Since the f-number of this lens is roughly $F/3$,[1] its maximum spatial frequency is ≈ 600 cycles/mm. The reason that this scaled-down

---

[1] The f-number is given by the EFL divided by the entrance pupil diameter, which is twice the semi-diameter of the ray height at surface 1 (just hover your cursor to that ray height or use the REAY operand and determine the Y ray height at surface 1, then multiply by 2).

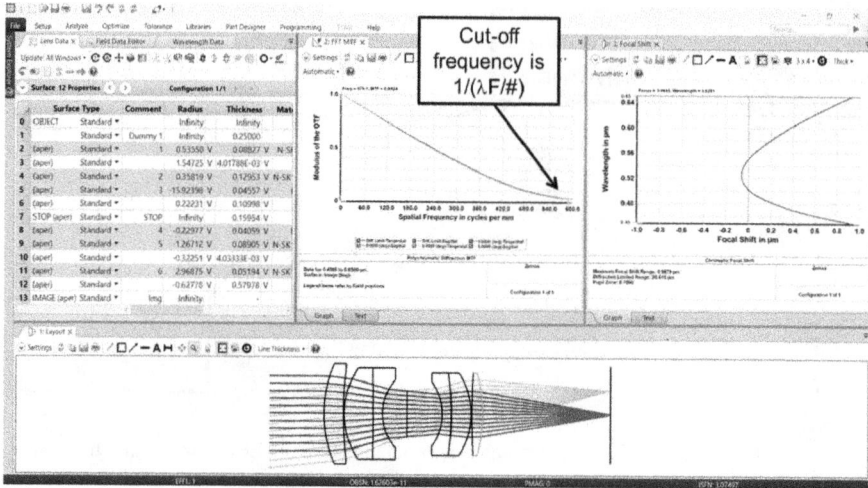

Figure 1.75. OS lens system for the lens scaled to 1 mm EFL. Note that the variables have been left on the surface radii and thicknesses, but no optimization was performed.

lens has improved MTF is because aberrations scale with size (unless the wavelength scales too, which cannot happen), with the exception of percentage distortion of the image. Therefore, in the context of aberrations, it is not difficult to design miniature lenses (of course, mobile phone lens design presents further challenges, due to packaging constraints). However, there are other nuances.

First, from a practical standpoint, the challenge in the miniaturization of lenses is in the manufacturing aspects. Second, there is the problem of the *information capacity* of a lens. By this, we basically mean the amount of detail that can be captured by a lens when it scales in size. One way to quantify this is to use the so-called **space–bandwidth product** [45, 46]. It sounds esoteric, but it is not. To compute it, just do as you normally would when you want to know how many lines can fit in a fixed length: you would divide that length by a pitch for the lines, right? But this is the same as multiplying the length by one over the pitch. The latter (one over the pitch) gives line pairs per unit length (or, if you used sinusoids, then it gives cycles per unit length), which is the spatial bandwidth (as opposed to temporal bandwidth, which is given in units of cycles per unit time). If the pitch is now defined by the smallest spot size you can get in a lens system, then that spot size is given by a combination of the PSF from the Airy disk's size and the contribution of perhaps RMS spot size from aberrations. Let us call this the *effective PSF spot size* for an imaging lens. So, the space–bandwidth product is simply a measure of how many effective PSFs can sample your image. If you square the ratio of length to effective PSF spot size, you get the *official* definition of the space–bandwidth product [45]. When we did the scaling for the lens just now, we maintained the lens's *f*-number. This fixed the diffraction-limited spot size given by the Airy disk (which also fixed the cutoff spatial frequency). Now, the Airy disk size is fixed, while the lens gets

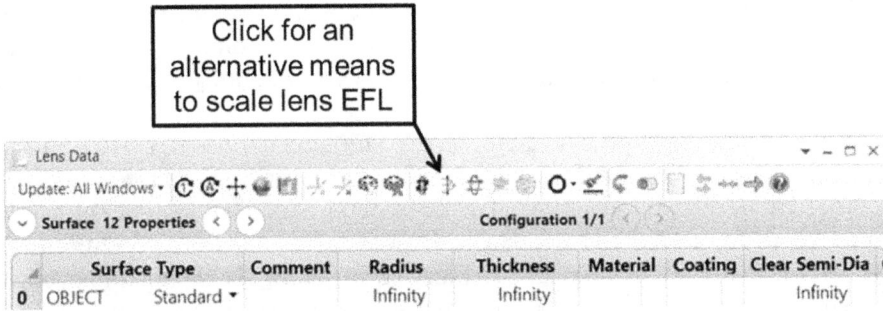

Figure 1.76. Another way to scale a lens's EFL, found in the Lens Data Editor.

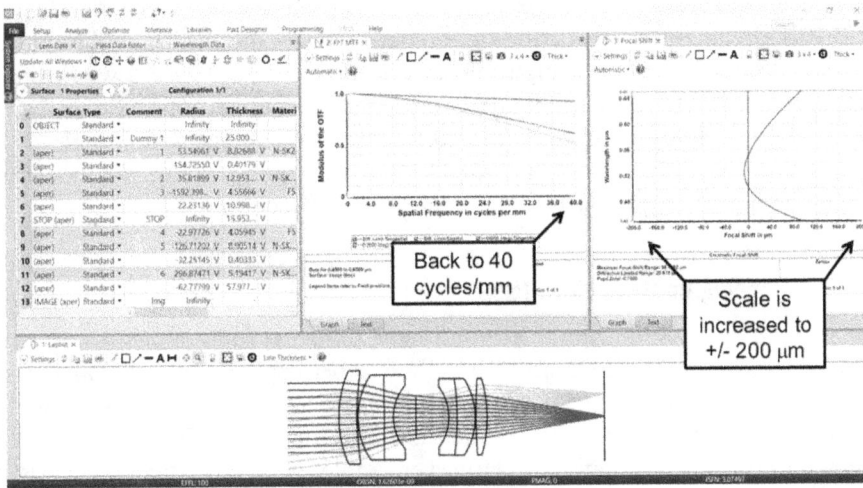

Figure 1.77. OS lens system for the lens scaled to an EFL of 100 mm. Note that the variables have been left on the surface radii and thicknesses, but no optimization was performed.

smaller and smaller. So, two Airy disks next to each other (whose origins are at a fixed separation of point sources from the object plane) get closer and closer at the image plane, due to the fixed angle subtended by the two object point sources at the lens's entrance pupil. Therefore, the *angular resolution* of a lens decreases for small lenses at a fixed *f*-number, even though the aberrations decrease. Astronomical telescopes are large, but so are their *f*-numbers, so their size does not help this issue (their larger size collects higher flux for distant point objects).

Now take the lens prescription from figure 1.70 again and scale it to an EFL of 100 mm. By the way, another method for scaling this lens's EFL is to click on the button shown in figure 1.76. Try this, and set the EFL to 100 mm. The result I obtained is as shown in figure 1.77. Note that the MTF has dropped and is less than the MTF for the 50 mm EFL lens at 40 cycles/mm. The aberrations have been magnified by scaling to a larger EFL. This is the scaling problem for lens design optimization when the lens gets larger. You can try re-optimizing this larger lens using the same merit function from figure 1.71, but you will not get far. The problem is in the secondary color in this lens.

Actually, the MTF at 40 cycles/mm shown in figure 1.77 is not bad and quite reasonable. Depending on the viewing conditions of the final image, you do not always need to be diffraction limited in order to obtain sufficient image contrast. For instance, recall from section 1.2.8 that if the image is formed on a 35 mm format image sensor, and if we assume an 8× enlargement factor for the final image to be projected onto a screen (such as a sheet of A4-size paper), then a human observer looking at that screen at a near distance of 250 mm should be able to see a sufficiently clear image. However, I did not mention the MTF that is needed for the image to be considered to have sufficient contrast. It turns out that, according to a curve called the *Aerial Image Modulation*, this MTF should be >20% [11]. The current lens already meets this criterion. Under these conditions, there should be no significant benefit in having a higher MTF, even though the image would indeed appear clearer at a higher MTF. Under a different set of conditions, you might need a higher MTF at 40 cycles/mm. And if so, for the current lens with a high EFL of 100 mm and an aperture of roughly $F/3$, due to the secondary color, you would need new glass choices and, in general, more surfaces (see section 1.4.12).

1.2.13 Tolerancing analysis for a lens

When you are done optimizing a lens and have fixed the radii of curvature, central thicknesses, and sizes of the elements, it is time to perform a tolerancing analysis for the lens, which is an exercise that involves not only determining the sensitivity of the lens system to tolerances but also estimating the yield (or likelihood of meeting a specification) of the lens system. However, quite generally, tolerancing is not necessarily an activity saved for the last step of optical design. Some designers go back and forth between lens optimization and tolerancing. I do this whenever I suspect that a specific element or number of elements have shapes that—based on experience —may have the potential to be sensitive to tolerances. Some examples of such elements include thin air-spaced doublets and highly curved meniscus elements. Other designers may include tolerance sensitivity directly in the merit function [22–24], which can be quite beneficial, as the optimized lens often does possess reduced sensitivity to tolerances (we touched briefly on this topic in section 1.2.6). OS has at least two merit function operands that can be used to achieve this reduction in tolerance sensitivity, which are the TOLR and HYLD operands (see appendix C.1). In this section, we will simply perform a tolerance analysis of the lens given by the prescription in figure 1.70.

First, note that when you are fixing the surface radii and center thicknesses of elements to the intended nominal values for production, you do not normally use all of the decimal places available in the Lens Data Editor. Most manufacturers require a specific finite number of decimal places. From experience, the surface radii may be set to three decimals, and the central thicknesses can be given with up to two

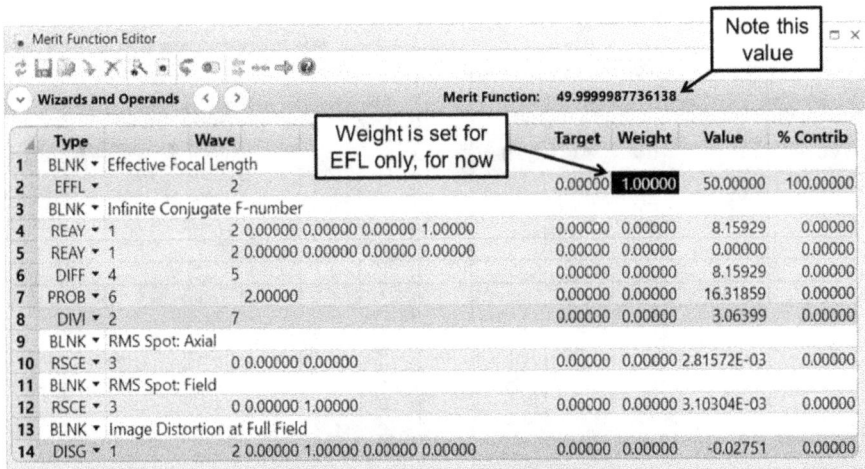

Merit Function Editor

Note this value

Wizards and Operands ⟨ ⟩   Merit Function: 49.9999987736138

Weight is set for EFL only, for now

	Type	Wave		Target	Weight	Value	% Contrib
1	BLNK ▾ Effective Focal Length						
2	EFFL ▾	2		0.00000	1.00000	50.00000	100.00000
3	BLNK ▾ Infinite Conjugate F-number						
4	REAY ▾ 1	2 0.00000 0.00000 0.00000 1.00000		0.00000	0.00000	8.15929	0.00000
5	REAY ▾ 1	2 0.00000 0.00000 0.00000 0.00000		0.00000	0.00000	0.00000	0.00000
6	DIFF ▾ 4	5		0.00000	0.00000	8.15929	0.00000
7	PROB ▾ 6	2.00000		0.00000	0.00000	16.31859	0.00000
8	DIVI ▾ 2	7		0.00000	0.00000	3.06399	0.00000
9	BLNK ▾ RMS Spot: Axial						
10	RSCE ▾ 3	0 0.00000 0.00000		0.00000	0.00000	2.81572E-03	0.00000
11	BLNK ▾ RMS Spot: Field						
12	RSCE ▾ 3	0 0.00000 1.00000		0.00000	0.00000	3.10304E-03	0.00000
13	BLNK ▾ Image Distortion at Full Field						
14	DISG ▾ 1	2 0.00000 1.00000 0.00000 0.00000		0.00000	0.00000	-0.02751	0.00000

Figure 1.78. Merit function operands used for tolerancing analysis of the current lens.

decimals. But you would need to discuss the manufacturer's preferences with them (normally, there would be a lot of back-and-forth communication prior to commencing production). Further, while you are fixing the values, you will notice that your nominal lens design performance changes slightly. The usual thing to do is to fix the central thicknesses of the elements first, as the lens performance is usually not as sensitive to element thickness as it is to surface radii and air spaces. So, you fix as many element central thicknesses as possible, followed by re-optimization if needed. After that, you fix the air spaces, and re-optimize as needed. Finally, fix the surface radii one by one and re-optimize one by one, until it is all done. In this section, we will not bother with the decimals, as the point is just to demonstrate how basic tolerancing is done.

Now, taking the lens with the prescription of figure 1.70, remove all variables. Then enter the operands, weights, and targets shown in figure 1.78. Note that the targets are all zero, and only a weight (which can be any value if only one operand is used) is given to the EFFL operand. At the top of the Merit Function Editor, note that there is a *merit function value* of 49.999...etc. This is the result of taking the square root of the sum of the squares (i.e. it is a RMS computation) of the *differences* between an operand's measured *Value* and its *Target* value in the Merit Function Editor, provided that there is a weight on that operand. In this case, since only the EFFL operand (which is for the EFL of the lens system) has a weight, the merit function value should be 50 mm. Numerical computations and round-offs make that value slightly different. If the Target cell in row 2 is given some other value (such as 50), then the merit function value should be zero (or a really small number). In the current example, since I intend to analyze the tolerance of the lens's EFL, I have only set a weight for that operand. Otherwise, OS would have measured the RMS merit function value for all operands that had weights set. If more than a single merit function operand is needed for evaluation, then a *tolerance script* may be written to report each value (an example is provided in section 3.7).

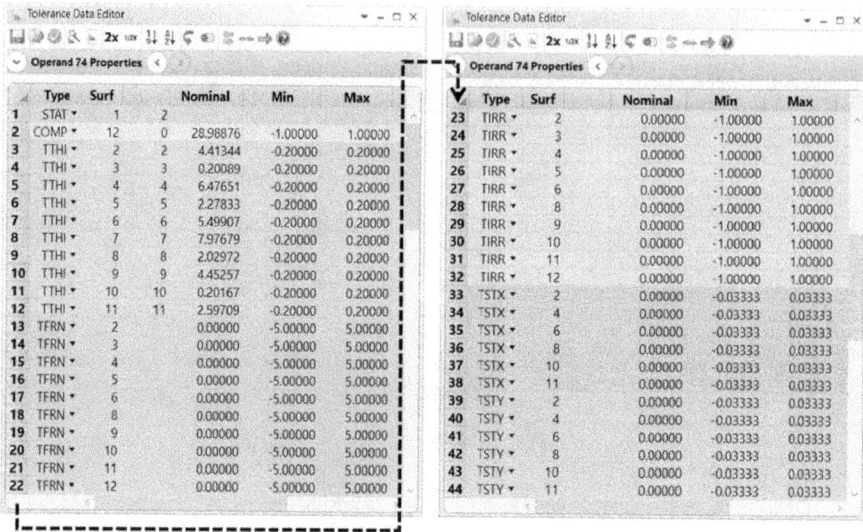

#	Type	Surf		Nominal	Min	Max
1	STAT	1	2			
2	COMP	12	0	28.98876	-1.00000	1.00000
3	TTHI	2	2	4.41344	-0.20000	0.20000
4	TTHI	3	3	0.20089	-0.20000	0.20000
5	TTHI	4	4	6.47651	-0.20000	0.20000
6	TTHI	5	5	2.27833	-0.20000	0.20000
7	TTHI	6	6	5.49907	-0.20000	0.20000
8	TTHI	7	7	7.97679	-0.20000	0.20000
9	TTHI	8	8	2.02972	-0.20000	0.20000
10	TTHI	9	9	4.45257	-0.20000	0.20000
11	TTHI	10	10	0.20167	-0.20000	0.20000
12	TTHI	11	11	2.59709	-0.20000	0.20000
13	TFRN	2		0.00000	-5.00000	5.00000
14	TFRN	3		0.00000	-5.00000	5.00000
15	TFRN	4		0.00000	-5.00000	5.00000
16	TFRN	5		0.00000	-5.00000	5.00000
17	TFRN	6		0.00000	-5.00000	5.00000
18	TFRN	8		0.00000	-5.00000	5.00000
19	TFRN	9		0.00000	-5.00000	5.00000
20	TFRN	10		0.00000	-5.00000	5.00000
21	TFRN	11		0.00000	-5.00000	5.00000
22	TFRN	12		0.00000	-5.00000	5.00000

#	Type	Surf	Nominal	Min	Max
23	TIRR	2	0.00000	-1.00000	1.00000
24	TIRR	3	0.00000	-1.00000	1.00000
25	TIRR	4	0.00000	-1.00000	1.00000
26	TIRR	5	0.00000	-1.00000	1.00000
27	TIRR	6	0.00000	-1.00000	1.00000
28	TIRR	8	0.00000	-1.00000	1.00000
29	TIRR	9	0.00000	-1.00000	1.00000
30	TIRR	10	0.00000	-1.00000	1.00000
31	TIRR	11	0.00000	-1.00000	1.00000
32	TIRR	12	0.00000	-1.00000	1.00000
33	TSTX	2	0.00000	-0.03333	0.03333
34	TSTX	4	0.00000	-0.03333	0.03333
35	TSTX	6	0.00000	-0.03333	0.03333
36	TSTX	8	0.00000	-0.03333	0.03333
37	TSTX	10	0.00000	-0.03333	0.03333
38	TSTX	11	0.00000	-0.03333	0.03333
39	TSTY	2	0.00000	-0.03333	0.03333
40	TSTY	4	0.00000	-0.03333	0.03333
41	TSTY	6	0.00000	-0.03333	0.03333
42	TSTY	8	0.00000	-0.03333	0.03333
43	TSTY	10	0.00000	-0.03333	0.03333
44	TSTY	11	0.00000	-0.03333	0.03333

Figure 1.79. The tolerance data operands used for tolerance analysis of the current lens.

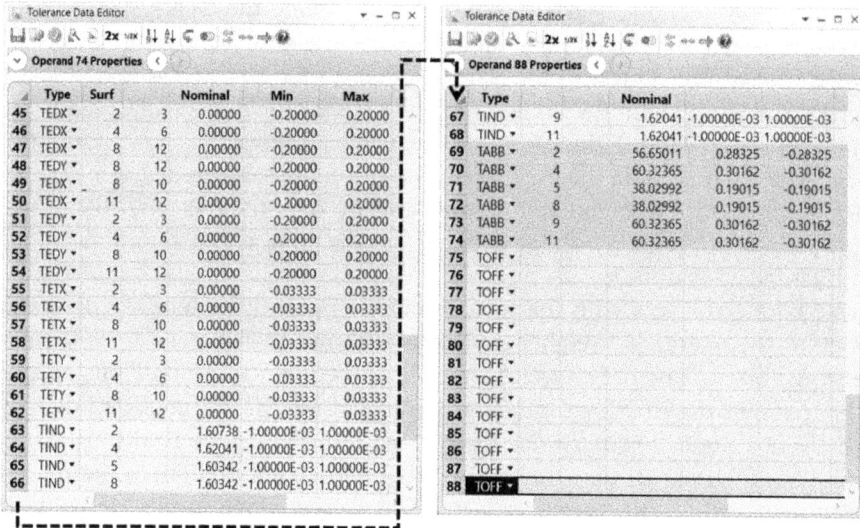

#	Type	Surf		Nominal	Min	Max
45	TEDX	2	3	0.00000	-0.20000	0.20000
46	TEDX	4	6	0.00000	-0.20000	0.20000
47	TEDX	8	12	0.00000	-0.20000	0.20000
48	TEDY	8	12	0.00000	-0.20000	0.20000
49	TEDX	8	10	0.00000	-0.20000	0.20000
50	TEDX	11	12	0.00000	-0.20000	0.20000
51	TEDY	2	3	0.00000	-0.20000	0.20000
52	TEDY	4	6	0.00000	-0.20000	0.20000
53	TEDY	8	10	0.00000	-0.20000	0.20000
54	TEDY	11	12	0.00000	-0.20000	0.20000
55	TETX	2	3	0.00000	-0.03333	0.03333
56	TETX	4	6	0.00000	-0.03333	0.03333
57	TETX	8	10	0.00000	-0.03333	0.03333
58	TETX	11	12	0.00000	-0.03333	0.03333
59	TETY	2	3	0.00000	-0.03333	0.03333
60	TETY	4	6	0.00000	-0.03333	0.03333
61	TETY	8	10	0.00000	-0.03333	0.03333
62	TETY	11	12	0.00000	-0.03333	0.03333
63	TIND	2		1.60738	-1.00000E-03	1.00000E-03
64	TIND	4		1.62041	-1.00000E-03	1.00000E-03
65	TIND	5		1.60342	-1.00000E-03	1.00000E-03
66	TIND	8		1.60342	-1.00000E-03	1.00000E-03

#	Type		Nominal		
67	TIND	9	1.62041	-1.00000E-03	1.00000E-03
68	TIND	11	1.62041	-1.00000E-03	1.00000E-03
69	TABB	2	56.65011	0.28325	-0.28325
70	TABB	4	60.32365	0.30162	-0.30162
71	TABB	5	38.02992	0.19015	-0.19015
72	TABB	8	38.02992	0.19015	-0.19015
73	TABB	9	60.32365	0.30162	-0.30162
74	TABB	11	60.32365	0.30162	-0.30162
75	TOFF				
76	TOFF				
77	TOFF				
78	TOFF				
79	TOFF				
80	TOFF				
81	TOFF				
82	TOFF				
83	TOFF				
84	TOFF				
85	TOFF				
86	TOFF				
87	TOFF				
88	TOFF				

Figure 1.80. Tolerance data operands continued from figure 1.79.

Now open up the *Tolerance Data Editor* by going to the Tolerance menu. Then enter the long list of *tolerance operands* and data shown in figures 1.79 and 1.80. These are the lens variables that are being assigned the tolerances (Min and Max values). Many of these operands can be understood by reading the OS help files associated with tolerancing (some are discussed in section 1.5.2). Here, we wish to highlight that the order of the rows for the TEDX, TEDY, TETX, and TETY

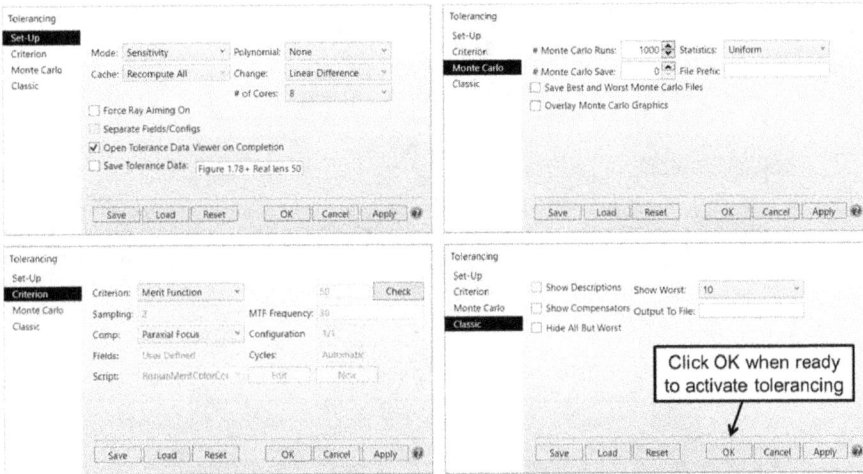

Figure 1.81. Selections used for the dialog box of the Tolerancing feature.

operands matters. In particular, in figure 1.80, note that **TEDX** and **TEDY** on rows 47 and 48 are entered *prior* to those on rows 49 and 50. This ensures that decenters and tilts for the second lens group (i.e. surfaces 8–12) are performed first, followed by the decenters and tilts of the individual elements in that group. Next, click the Tolerancing icon under the Tolerance menu, and enter the selections shown in figure 1.81 when the dialog box shows up for this feature. Click OK when you are done with all selections shown. This activates (starts) the tolerancing of the lens (by this, it is meant that the program makes perturbations according to the Min and Max values entered in figures 1.79 and 1.80, and the output is the result of the tolerancing action, which I will explain). Wait for the program to terminate the tolerancing 'run,' and then wait a while longer for the tolerancing status window to disappears when it is done tolerancing. You should see a new window pop up called *Tolerancing Results*, as shown in figure 1.82. This window pops out because we placed a tick mark on *Open Tolerance Data Viewer on Completion* in the Setup tab of the dialog box (see the top left of figure 1.80). You always want this window to pop up after completion of a tolerance run, as it provides all of the results of the run, which I will explain shortly.

Go to the Tolerance menu and click on Histogram. The result should be an output window (figure 1.83) displaying a histogram of the EFLs listed in the second column of figure 1.82. At a glance, you can see the statistical distribution of EFL values from 1000 random samples of lens systems modeled by the Monte Carlo tolerancing simulation that we selected in the tolerancing dialog box (see the top right of figure 1.80).

Now go back to the Tolerance Results output window. In figure 1.82, reading the columns from left to right, the first column lists the name of each Monte Carlo sample generated out of the population of 1000 lens samples. Each lens sample possesses elements with random amounts of perturbations of the tolerances listed in

There are THREE tabs. We are currently displaying results from the Monte Carlo tab

Tolerancing Results: 1

Settings

Monte Carlo | Sensitivity | Summary

	Merit Function	TTHI	TTHI	TTHI	TTHI
	Statistics	Statistics	Statistics	Statistics	Statistics
	Field: 0	Surf: 2	Surf: 3	Surf: 4	Sur
	Config: 0	Adjust: 2	Adjust: 3	Adjust: 4	Adjus
	Nominal: 49.999998773613	Nominal: 4.4134421243198	Nominal: 0.2008940464852	Nominal: 6.4765106949971	Nominal: 2.2783313771
	Comp: Paraxial Focus	Min: -0.2	Min: -0.2	Min: -0.2	Min:
		Max: 0.2	Max: 0.2	Max: 0.2	Max
		Comment:	Comment:	Comment:	Comme
MC_1	49.66215	4.39330	0.04707	6.56109	2.18
MC_2	50.15252	4.23879	0.03005	6.44216	2.47
MC_3	49.80474	4.30451	0.00441	6.62430	2.22
MC_4	50.39482	4.42230	0.18601	6.29823	2.40
MC_5	50.26016	4.37945	0.19377	6.30012	2.11
MC_6	49.89479	4.45816	0.36942	6.35791	2.16
MC_7	50.19053	4.52767	0.30623	6.47153	2.19
MC_8	50.06068	4.35212	0.24969	6.30212	2.44
MC_9	49.81683	4.35772	0.33970	6.61593	2.34
MC_10	49.43116	4.24128	0.02762	6.40787	2.12
MC_11	49.84990	4.27900	0.14242	6.56793	2.12

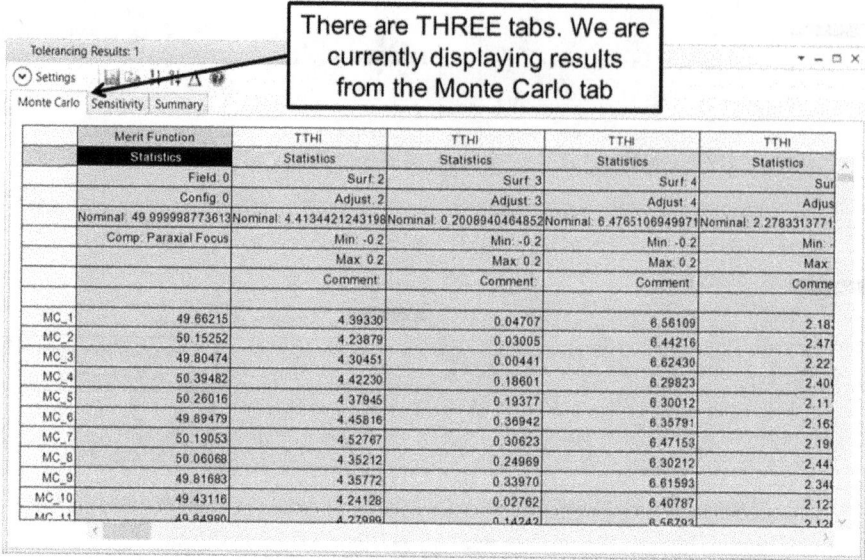

Figure 1.82. Tolerancing results window (partially displayed) from the first 'tolerance run.'

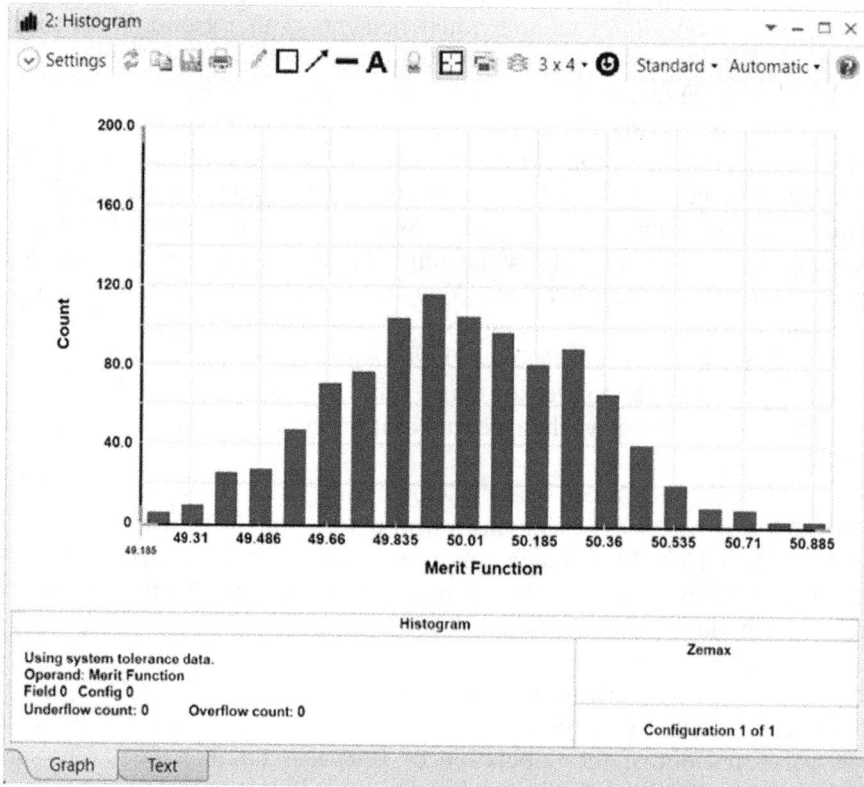

2: Histogram

Settings   3 x 4   Standard   Automatic

Histogram

Using system tolerance data.
Operand: Merit Function
Field 0   Config 0
Underflow count: 0   Overflow count: 0

Zemax

Configuration 1 of 1

Graph | Text

Figure 1.83. A histogram of EFLs from the first 'tolerance run.'

in the Tolerance Data Editor of figures 1.79 and 1.80 (we will talk a little about those operands soon). The second column lists the operand being tolerance analyzed (often, we will simply say that this operand or figure of merit or quantity is being *toleranced*). This is the EFL operand (row 2) in the Merit Function Editor, where a weight has been entered for this operand alone. The rest of the columns list the randomly perturbed amount of each operand listed in the Tolerance Data Editor for each Monte Carlo sample. The amount of perturbation is bounded by the Min and Max values entered into the Tolerance Data Editor, and the statistical distribution that governs this bound is the uniform distribution, which was selected in the tolerancing dialog box (see the top right of figure 1.80). However, this selection had already been pre-determined by our use of the STAT tolerancing operand on row 2 of the Tolerance Data Editor.

The listing of all perturbations in the columns to the right of the first two columns in figure 1.82 is highly beneficial. You can use these data to determine whether there is any correlation between any of the tolerances and the *figure of merit* being toleranced. For example, in the current tolerancing results output, we can analyze whether there is any correlation between the EFL and the thickness of surface 9 of the prescription shown in figure 1.70. You could do this by cutting and pasting these two columns into a separate spreadsheet. You would see that in this example, the EFL and the thickness on surface 9 are uncorrelated, which is rather interesting, because if you click on the Summary tab in the Tolerancing Results output (i.e. the third tab from the left in figure 1.82) and scroll down until you reach Worst Offenders, you will note that the thickness of surface 9 is the top *offender* (has the most influence) for the EFL. This list of top ten *worst offenders* is provided because we selected '10' for Show Worst in the tolerancing dialog box (see bottom right in figure 1.80). This shows that the worst-offending tolerance is not necessarily correlated with the changes in EFL.

There are some reasons why, in the current tolerance run, the EFL is not correlated with the worst offender. First, the worst offender is not the only offender. Every perturbed variable is an offender, so it is difficult to draw correlations when everything is being perturbed at once. Second, there may be several of the top ten worst offenders that have roughly equal amounts of influence on the EFL. Third, in OS, these offenders are determined by examining changes in the *parameter* under test (i.e. the EFL) with respect to **independent** changes in the tolerancing variable (e.g. the thickness of surface 9). This is known as a *sensitivity analysis*, but it only considers independent sensitivities. In general, when random perturbations are assigned to lens variables (e.g. surface radii, central thicknesses, etc.), the resulting variation in the parameter can be a function of higher-order mixed derivatives called **interactions** between the variables [47] (we will discuss this topic a little more in section 3.3).

While OS does not provide computation of these higher-order terms at this stage, they are accounted for in Monte Carlo simulations. Of course, the individual derivatives of higher-order terms cannot be extracted from the Monte Carlo simulations, so you cannot determine the relative contributions of each term to the parameter being tolerance analyzed. However, note that the *difference* between the variance of the parameter from the Monte Carlo run and the squared total change from a *root sum square* (RSS) computation can give an indication of the presence of higher-order terms. What do I mean? For instance, if you scroll down the Summary listing and go past the list of worst offenders, you may note that there is a summary called *Estimated Performance Changes based upon Root-Sum-Square method*. Under this, you will find the *Estimated Change*, which is the square root of the sum of the squares of the deviations for the toleranced parameter (i.e. the EFL in this case), based on perturbing each variable (i.e. each of the operands in the Tolerance Data Editor) one variable at a time. Statisticians call variables *factors* (so, an RSS analysis is like a so-called one-factor-at-a-time analysis, as often referred to by statisticians and proponents of the *Six Sigma* paradigm of product development). In OS, a deviation is defined as the difference between a nominal value and the value attained as a result of a perturbation. In the current example, in which the EFL is the parameter, OS calculates the change in the EFL from its nominal value that results from perturbing a variable. OS creates these perturbations as many times as there are tolerancing operands in the Tolerance Data Editor, but it also does this twice—once for the Min values of the operands and once for the Max values of the operands. OS takes the sum of the squares of the parameter changes that result from perturbing the variables to their Min values, and it does the same for the Max values; it then sums both squared deviations and divides by two; finally, it takes the square root. This is like taking the square root of the sum of variances in error analysis. This computation therefore does not account for interactions between variables. In contrast, the standard deviation that is computed from the list of EFL values in the Monte Carlo tab is a consequence of combinations of deviations (to all orders, but truncated in the sense that the tolerances are truncated) and interaction terms. So, taking the difference between the variance from the EFL values in the Monte Carlo run and the variance from the RSS gives some indication of the presence or absence of interaction terms. If the difference yields a small number, then there is unlikely to have been any interaction. But if the difference yields a non-negligible value, then there may be some interactions. That said, note that there are other nuances involved in statistical calculations, some of which are discussed in section 3.3.

Returning to figure 1.78, you may note that there is a list of other merit function operands that can be toleranced if weights are entered for them. In particular, the operands on rows 8, 10, 12, and 14 could individually be assigned weights (i.e. not all at once) to analyze how much they would vary as a result of all of the tolerances entered in the Tolerance Data Editor. We will not be performing that tolerance. Rather, I have placed them into the Merit Function Editor in order to show you how you would proceed if you wanted to analyze several parameters in the merit function (just note that they cannot all have weights) and to also show you how careful you need to be to compute certain of these parameters. For instance, row 8 computes the infinite conjugate *f*-number, given by dividing the EFL by the diameter of the entrance pupil. I could have used the EPDI operand to obtain the entrance pupil's diameter, but I did not, because that operand computes the *paraxial* entrance pupil diameter. In real life, we would need the *real* diameter, which is given by twice the ray height on surface 1. This is why operands 4–8 exist in the current Merit Function Editor. Note also that, in order to get the Y ray height at surface 1, I did not just use a single REAY operand. Rather, I took the *difference* between the ray height and the position of the axial ray. Take a good guess why. The reason is that when tolerancing is performed, surface and element decenters and tilts can shift and tilt the axial ray, which is equivalent to a perturbation of the optic axis. In turn, the Y height of the *rim ray* (i.e. REAY for Py = 1 in the Merit Function Editor) shifts and tilts as well. Therefore, to determine a diameter for the entrance pupil, I need to take the difference between the rim ray and the axial ray Y positions. In fact, in theory, I need to sample around the entrance pupil and take an average. But you get the idea.

One last point. Do you know why the ray height at surface 1 is nominally the rim ray for the entrance pupil? This has to do with geometrical optics principles. The entrance pupil is the *window* into the lens system as seen by rays entering the system, and it is the image of the aperture stop. So, the rim ray for the aperture stop must also be the rim ray for the entrance pupil. We have set Ray Aiming to 'on,' which means that rim rays entering the system properly touch the aperture stop's rim. So, since the ray Y height at surface 1 grazes the rim of the aperture stop, then its height at surface 1 must be the semi-diameter of the entrance pupil. In real life, this is measured in reverse. A point source is placed at the back focal point (the axial point on surface 13), and rays travel towards the front of the lens [48]. If you are an optical system designer who works in a company that develops instruments (i.e. not an optics supplier), you will normally not need to care much about how the f-number is measured (just as with the MTF, you would not want to spend too much time doing optics-centric subassembly tests that can be performed better by the optics supplier). But if you work as an optical engineer or a lens designer in an optics-centric company, then it is almost always necessary to know how optical components and assemblies are tested.

1.2.14 Create and use a 'black box file'

In OS, a *black box* (BB) file is an encrypted version of the lens system that possesses all the properties of the lens system without revealing its prescription. Some lens suppliers provide such files, as these files enable us to integrate some selection of COTS lenses into our own optical system models for analysis. So, let us do this for the lens from the previous section. Go to the File menu, then locate and click on Zemax Black Box file, as shown in figure 1.84. Then select surfaces 2 and 6 as the First and Last surfaces, respectively. This creates a BB file for the first lens group (because OS does not permit a range of surfaces with a buried stop to be selected). Ensure that you add a tick mark to *Create and load test file after black box export.* Click OK when done, and OS will prompt you to name your first BB file. I named mine 'figure 1.84 BB1,' which saved the BB file with the '.ZBB' prefix (you do not need to add that prefix when you name your file, as OS will do it for you). After that, OS saves your original file and loads up your new file with the BB file, and the whole OS file is renamed with a underscore BB '_BB.' You then literally see a box that replaces the first lens group.

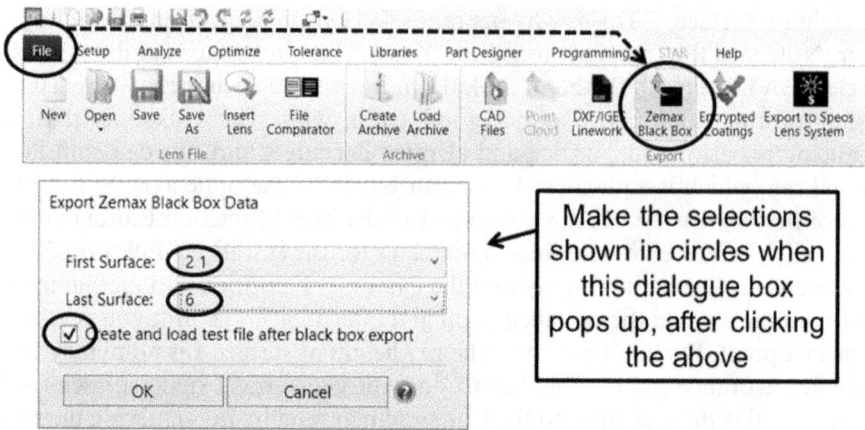

Figure 1.84. Creating a first **BB** file in front of the stop for the lens from the previous section. Go to the File menu, then click on Zemax Black Box.

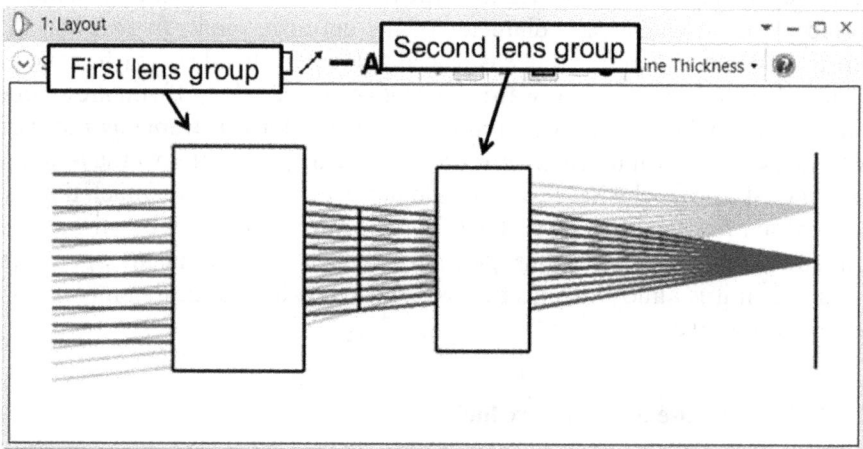

Figure 1.85. Two **BB** files with the stop in between for the lens from the previous section.

Do the same for the second lens group, and you will end up with the system shown in figure 1.85. Note that the names of the two **BB** files are displayed in the comment column of the Lens Data Editor. When you want to load other existing **BB** files that are in their proper Zemax directory, you create a blank surface, then enter the name of the **BB** file (with the .ZBB prefix included) in the comment column for that surface.

Now that you have created your **BB** file/s, you can share them with someone who is requesting a design from you to integrate and analyze in their own OS model. And you can do so without revealing the prescription of your special lens. Some may feel that a **BB** file reveals significant information, and may therefore feel hesitant to share their **BB** files with others. For instance, the current lens system reveals that there is a stop between two lens groups. Some may feel that this is a piece of significant information about the design. However, revealing where the stop is located is not very significant. Optical designers know that, according to aberration theory, some lenses need to have a stop in between lens groups if coma, distortion, and astigmatism are to be more easily controlled and corrected, such as in the design of high-performance microscope objectives [49]. It is also known that the stop location can control relative illumination (we will cite most of the significant references to this piece of knowledge in chapter 2 of this book). And it is also known that the stop location can influence the position of the entrance and exit pupils relative to the principal planes of an optical system. In particular, when the stop is positioned at the **optical center** of a lens system, then the entrance and exit pupils are located at the front and rear principal planes, respectively [11, 50]. Admittedly, perhaps this last piece of information is not as well known, but it is published and therefore accessible to avid readers of optical design. Moreover, it will soon become known in section 1.3 of this book that we can make specific use of the optical center of a lens for applications involving depth sensing. In some other applications, it is important for the customer to know where the stop is located in a lens being considered for integration. For instance, in microscope tube lenses, it is important to know the *pupil distance*, which is the distance from the exit pupil of a microscope objective (which is the stop for a tube lens) to the front entrance of the tube lens. Microscope objectives that are intended for phase contrast must have an accessible stop, because that is where a phase plate would be mounted. Microscope condensers for illumination must have an accessible field stop, as that is the object conjugate for the *substage condenser*. Laser scan lenses must have an accessible stop because that is where the scan mirror is located. And so on. Therefore, the stop location is not much of a *secret* recipe for a lens. Rather, the essential ingredients in a lens design are in the choice of materials and the precise values for the surface radii and central thicknesses, which are all hidden by a **BB** file.

1.2.15 Nonsequential modeling and analysis

When you create a new lens file in OS, it is in sequential ray tracing mode by default. The term *sequential ray tracing* refers to the way in which an optical design program computes the path of rays from surface to surface. In SC ray tracing, element surfaces must be located in an ordered consecutive (sequential) fashion. This is why you see that surface numbers are ordered sequentially from top to bottom in the Lens Data Editor. Everything we have done in this book until now has used the sequential ray tracing mode. This is fine for analyzing aberrations and image quality, but in illumination and stray light analysis, one often uses the NSC ray tracing features of a program, in which the optical components do not need to be placed in an ordered fashion in an editor, with the exception of some *nesting* rules that the program uses to decide which surfaces are considered significant when one or more surfaces are in contact in the NSC model.

To create an NSC model, you need to convert your new blank OS file into a NSC file, and to do that, simply go to the Setup menu and click on the Non-Sequential

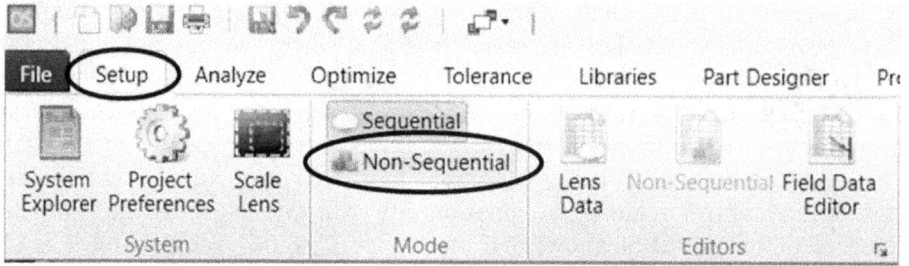

Figure 1.86. Locating the Non-Sequential button to convert a blank file into NSC mode.

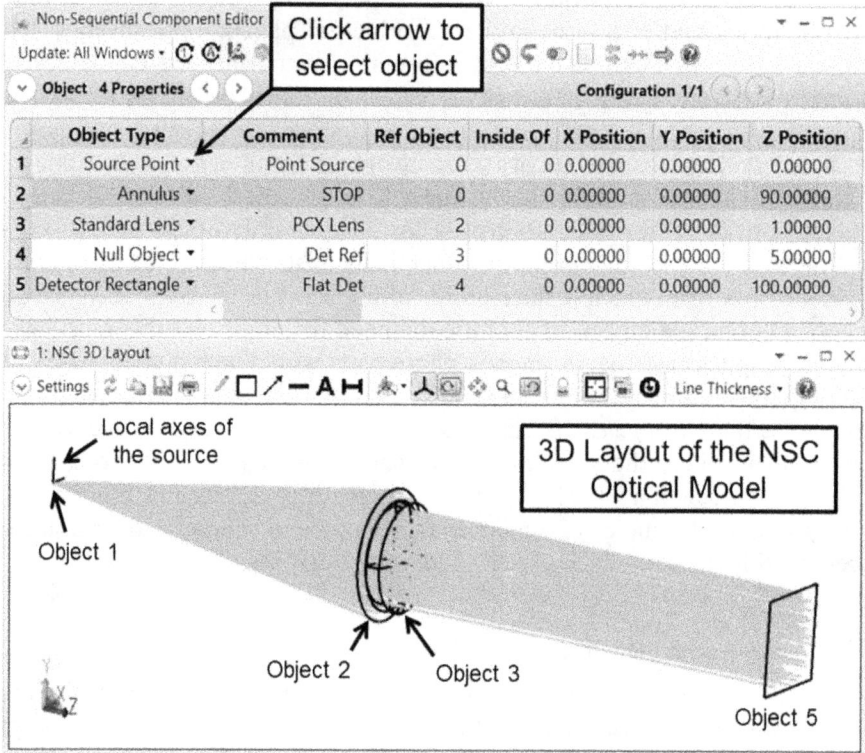

Figure 1.87. A simple NSC optical model.

button (figure 1.86). A dialog box will appear. If you are starting from a blank file, then you do not need to save any current data. At any rate, click on one of the choices in the dialog box and the file will be converted into a NSC *environment*, with a blank *Non-Sequential Component Editor*. The NSC Editor in NSC ray tracing mode is the counterpart of the Lens Data Editor in SC ray tracing mode. The NSC Editor lets you enter all of the *objects* that you want to model. In NSC mode, you enter objects rather than surfaces. As usual, you right-click and select the appropriate options to insert objects before or after any object, at any row of the NSC Editor. Try entering the five objects shown in figure 1.87. Left-click on the down-arrow of the Object Type column to select the objects. For each object, scroll rightwards in the NSC Editor to enter more object properties, such as material, radii of curvature, etc. I will tell you in a moment what properties to use, but first, note that in the current model, the source is placed at position $(x, y, z) = (0, 0, 0)$, which is the global coordinate.

In the current model, I used a point source (Object 1) with a half-cone angle of 10°. I used 200 for the # of Layout Rays (note the random nature of the ray output). The annulus (Object 2) has a 15 mm maximum half-width and an 11 mm minimum half-width. To make the annulus absorb rays, just type Absorb into the Material column for it (do this for any object whose material should be an ideal absorber). The lens is plano-convex and its flat side faces the source. Its material is Schott N-BK7, and its central thickness is 5 mm. The radius is −50 mm on the right side, and the clear semi-diameter is 12.5 mm. Object 4 is a *Null Object*, which means it is not used as an object. Instead, it allows me to enter values in specific cells for specific purposes. In this case, I enter 5 mm at the Z position and the value 3 for the Ref Object (i.e. the object in the NSC Editor that serves as the reference location for subsequent objects in that editor). This allows me to make the right vertex of Object 3 a reference location for the next object, which is the detector (however, note that any other subsequent objects can reference Object 4). The location of the detector (Object 5) is then set at 100 mm from the right vertex of the lens. You can see that the rays are roughly collimated, which was my intent. Note also that the word Absorb has been entered for the material of the detector. As with all OS features, you must read the help files to become further acquainted with NSC modeling. We will be using more NSC modeling of optical systems in chapter 2. However, you will see in chapter 2 that most illumination designs involve fundamental theory, and in fact, some illumination designs can be produced by using SC ray tracing and then checking the final performance using NSC modeling.

In OS, there is another way to build NSC models of existing lens systems that have been created in SC mode: first, return to the lens whose prescription is given in figure 1.70 and save it as a new file (to protect the original), then remove all the variables. Go to the File menu and click on Convert to NSC Group. You will see a dialog box—do not click OK yet. Note that if you do not select specific surfaces, the conversion is done for all surfaces. If you only want to convert surfaces 2–12, you can do so by exiting from this dialog box (click Cancel), then select surfaces 2–12 by keeping the Shift key down and left-clicking on all those surfaces. Then return to the Convert to NSC Group function and convert them. If you do the conversion this way, then the result is a *hybrid-nonsequential* model, in which surfaces 2–12 become an NSC group of components within your current SC optical model, which is a very powerful feature of OS, as you will see in section 1.2.18. Meanwhile, if you just convert the entire system into an NSC model, you will see that the conversion is done very quickly and you will obtain the model shown in figure 1.88.

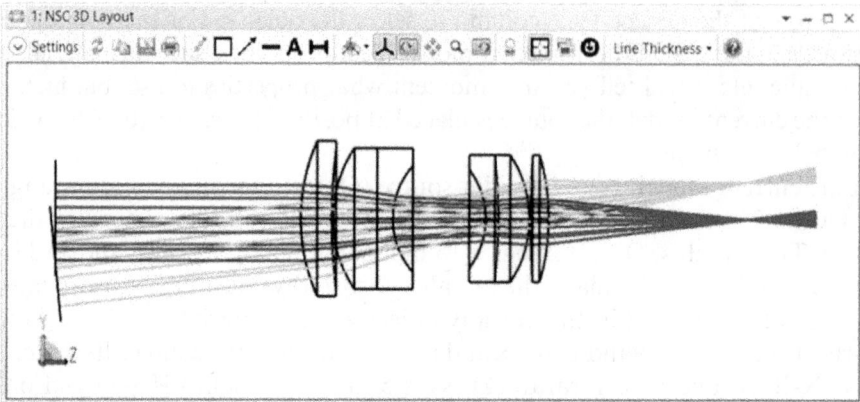

Figure 1.88. The lens from figure 1.70 following conversion into pure NSC mode.

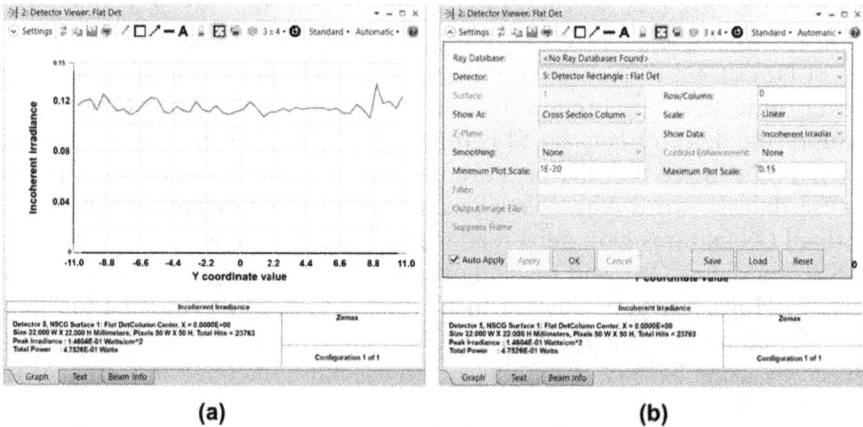

(a)   (b)

Figure 1.89. Detector output for the NSC model in figure 1.87. (a) The irradiance for the column of pixels. (b) The settings used to display the output irradiance.

1.2.16 Dealing with 'ray trace noise' in nonsequential modeling

Let us return to the lens shown in figure 1.87. Go to the Analyze menu and click on Detector Viewer. This outputs what you would see on Object 5. If you go to its settings, you can select the way in which you want the detected rays to be displayed (Grey Scale, False Color, etc.), but you will not see any irradiance profile until you trace those rays. To do this, first enter the number of random rays to be traced into the # Analysis Rays cell for Object 1. Enter 50 000 (just type 5E4), then go to Analyze > Ray Trace. A dialog box appears. Leave all boxes blank except for Ignore Errors. Click Trace, and the output should be as shown in figure 1.89(a) if its output settings are those shown in figure 1.89(b).

It is apparent in figure 1.89 that noise is present when rays are traced in NSC mode; this is commonly known as *Monte Carlo ray trace noise* or simply *ray trace noise* in NSC ray tracing, which resembles the Poisson nature of photons reaching a detector. The rays are randomized using a probability distribution governed by the source's **radiant intensity** (flux per unit solid angle). Why does the program not display beams like those we see in computer-generated (CG) animated films? The reason is because that would not only require a very powerful computer, but there are some special algorithms, interpolation, and artistic *rendering* involved [51–53]. For the purpose of optical system design, interpolation is indeed possible (e.g. the *Lighting Trace* feature in the Premium edition of OS does some interpolation), but it is also often possible to gain quantitative measures of performance using a finite but large number of rays [54]. The idea in NSC ray tracing is to either trace a large number of rays once, or to trace a small finite number of rays but do it many times and take the average. Both approaches are precisely the same. In the current simulation, since the source is a point source (whose radiant intensity is uniform in angular space), the randomization is distributed within this small cone, so tracing a large number of rays does not take a lot of time (try increasing the # Analysis Rays to 1E+06 and you will see that the result is very quick, with a fairly smooth curve). Furthermore, one can sometimes reduce the number of pixels at the detector, which is equivalent to doing a moving average. Finally, some smoothing algorithms are provided by the detector setting (see the Smoothing selection on the left side of figure 1.89(b)).

In fact, in the current ray trace, there is a default setting for the source's property setting that makes the output not precisely random and somewhat smoothed. It is called *Sobol Sampling*. If you go to Object 1's properties and look under Sources in the left pane, you will see that under the *Raytrace* column, Sobol Sampling is selected by default. OS's help file for this feature provides some good reasons for using this sampling method to trace *pseudo-randomized rays*. The idea is to lessen the amount of ray trace noise, and it does help (try turning it off and perform the ray trace again—you will see a noisier signal). I often turn off Sobol Sampling, but I sometimes use it if I do not want to trace too many rays, or if I do not want to perform the ray trace many times. Because NSC ray tracing involves Monte Carlo noise, quantitative analyses and assessments made in NSC modeling actually involve making statistical estimates of parameters based on the ray-trace results.

1.2.17 Deciding between nonsequential and sequential approaches

Generally, if you want to analyze the consequences of stray light and the irradiance distributions produced by illumination optics, then NSC ray tracing is the way to go. For image quality assessment and aberration correction in lens design, SC ray tracing is the approach. However, as will be shown in chapter 2, SC ray tracing may also be used for some illumination problems, because SC ray tracing can be significantly faster. You may wonder how it is that SC ray tracing may be applied for illumination design. The answer is that while the analysis of irradiance profiles generated by illumination optics is best done through NSC ray tracing, the *act* of designing the illumination optics does not necessarily have to be performed by NSC ray tracing. It takes a combination of deep knowledge of the fundamentals of optical radiometry and application of the tools of SC and NSC ray tracing to design illumination optics.

The same is true of stray light analysis. In this case, it is not just a matter of applying a scatter model to a surface and seeing where rays go upon striking that surface. Rather, it is also about knowing where the source is and where its rays strike. Recall from figure 1.3 that in SC ray tracing, identifying potential sources of stray light is part of the action of including more aspects of reality in your optical system model. Clearly, if a surface does not scatter, then there is no stray light that you need to analyze in NSC mode.

So, what do I personally do? I use both SC and NSC modes of ray tracing in my analyses. I often begin by modeling an optical system in SC mode, even if the system is primarily designed for illumination. In SC mode, provided that the entrance pupil is properly sampled (see sections 2.2.1 and 2.3), I can quickly see where rays go and make qualitative assessments of irradiance distributions. Moreover, note that the relative illumination plot is in fact an illumination analysis performed in SC mode (see sections 2.2.5 and 2.6–2.9). Further, recall from section 1.2.8 that the GIA plot in OS is a simple and quick means to assess an illumination profile. A final illumination analysis may then be performed in NSC mode. Of course, it is simpler to model the integration of optical and mechanical components in NSC mode. In practical optical system design, we try to be as creative as possible by exploring all possible routes to perform a thorough analysis of our optical models. Speaking of this, now is a good time to introduce a very powerful *mixed mode* of SC and NSC analysis: OS's hybrid nonsequential mode. We discuss this in the following section.

1.2.18 OpticStudio's hybrid nonsequential mode (this is a powerful tool)

Let us insert a Pechan prism into the lens given by the prescription in figure 1.70 and simulate image rotation by such a prism. First, save the lens from figure 1.70 into a new file and enter two new surfaces after surface 7, as shown in the prescription in figure 1.90. Note also that the radius of surface 4 and the thickness of surface 14 have changed from the prescription in figure 1.70. Finally, change the largest half-field angle to 3°. In other words, you have just two field settings for this lens: the axial field at 0° and the oblique field at 3°.

Note that the aperture setting is Float by Stop Size. In figure 1.90, surface 8 is now a so-called Non-Sequential Component surface type. This means that at that surface, OS is giving you the means to insert a NSC object. In doing so, we create a hybrid nonsequential (HNSC) optical model (in other words, a *mixed* model of SC and NSC objects). The next step is to set the Z positions for the *entrance port* and *exit port* for the NSC object being inserted into the SC model. The concept here is that the NSC object enters into the SC prescription by way of setting a surface as a *port* at which OS begins the NSC ray tracing and another port that ends the NSC ray trace. These two ports are known as the entrance and exit ports, respectively. In the current prescription, surface 8 serves as the entrance port and surface 9 is the exit port. Since we have defined the location of the entrance port (surface 8), we now need to allow space along the z-axis so that we can insert the NSC object (which is a

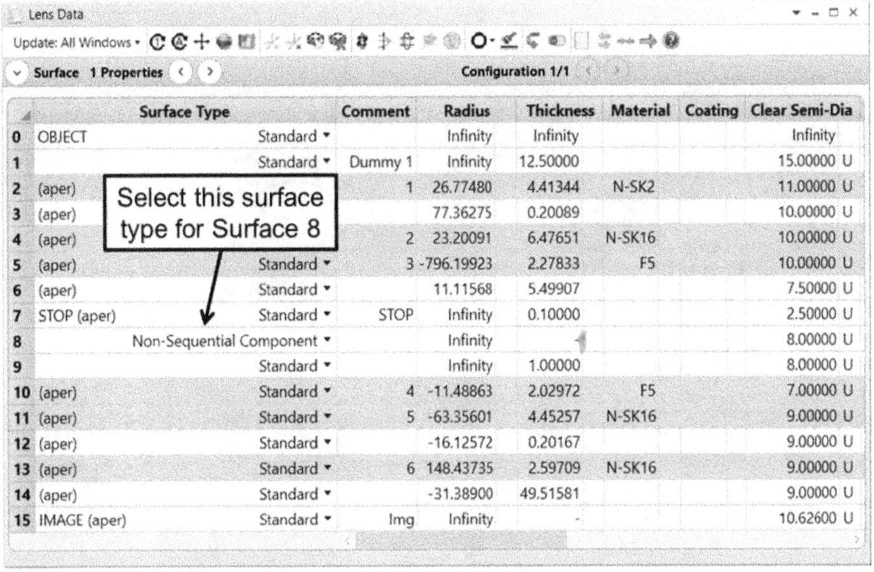

	Surface Type		Comment	Radius	Thickness	Material	Coating	Clear Semi-Dia
0	OBJECT	Standard		Infinity	Infinity			Infinity
1		Standard	Dummy 1	Infinity	12.50000			15.00000 U
2	(aper)		1	26.77480	4.41344	N-SK2		11.00000 U
3	(aper)			77.36275	0.20089			10.00000 U
4	(aper)		2	23.20091	6.47651	N-SK16		10.00000 U
5	(aper)	Standard	3	-796.19923	2.27833	F5		10.00000 U
6	(aper)	Standard		11.11568	5.49907			7.50000 U
7	STOP (aper)	Standard	STOP	Infinity	0.10000			2.50000 U
8		Non-Sequential Component		Infinity				8.00000 U
9		Standard		Infinity	1.00000			8.00000 U
10	(aper)	Standard	4	-11.48863	2.02972	F5		7.00000 U
11	(aper)	Standard	5	-63.35601	4.45257	N-SK16		9.00000 U
12	(aper)	Standard		-16.12572	0.20167			9.00000 U
13	(aper)	Standard	6	148.43735	2.59709	N-SK16		9.00000 U
14	(aper)	Standard		-31.38900	49.51581			9.00000 U
15	IMAGE (aper)	Standard	Img	Infinity	-			10.62600 U

Figure 1.90. The starting prescription used to create a hybrid nonsequential model of a Pechan prism inserted into the imaging lens of figure 1.70.

Pechan prism). To do this, go to surface 8 and scroll to the right of the Lens Data Editor until you see parameter entry cells called *Draw Ports*, followed by short-form terms for *Exit Locations* in *X*, *Y*, and *Z*. Enter 21 (mm) for the *Z* location. You can now see two surfaces separated by 21 mm, each of which has a semi-diameter of 8 mm. These are your entrance and exit ports for NSC ray tracing. Enter 0 for the Draw Ports, and the port surfaces disappear (which is convenient for visualizing the optical system without annoying port surfaces that do not really belong there when you present your work to customers). Now go to the Setup menu and navigate *past* the Non-Sequential button that you used in section 1.2.15 until you see the Non-Sequential button within the *Editors* 'frame.' Click that button and enter the values shown in figure 1.91.

After entering everything shown in figure 1.91, the resulting 3D layout should be as shown in figure 1.92 (use whatever ray display layout settings you wish).

You can visualize the reflections of rays in the prism by displaying only a single ray and selecting the *Fletch Rays* option under the Settings menu of the 3D layout. The result is shown in figure 1.93.

Now, let us have some fun with image rotation. Activate the Image Simulation feature (as described in sections 1.2.10 and 1.2.11 and used to create figures 1.45 and 1.56), and enter the settings shown in figure 1.94. The result should be as shown in figure 1.95(a).

Next, return to the NSC Editor and enter 22.5° for the Tilt About *Z* parameter for the prism. The result should be as shown in figure 1.95(b). Note that the image has rotated by 45°, even though the prism was rotated by 22.5°. This is because the internal reflection of rays in the prism rotates the image by twice the rotation of the prism.

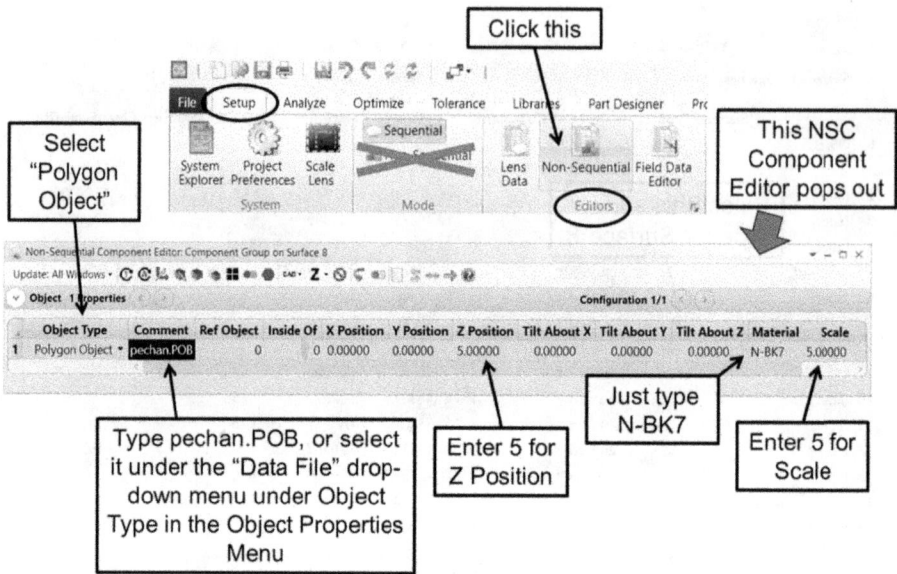

Figure 1.91. Clicking on the Non-Sequential button within the Editors frame of the Setup menu displays the NSC Editor for HNSC mode.

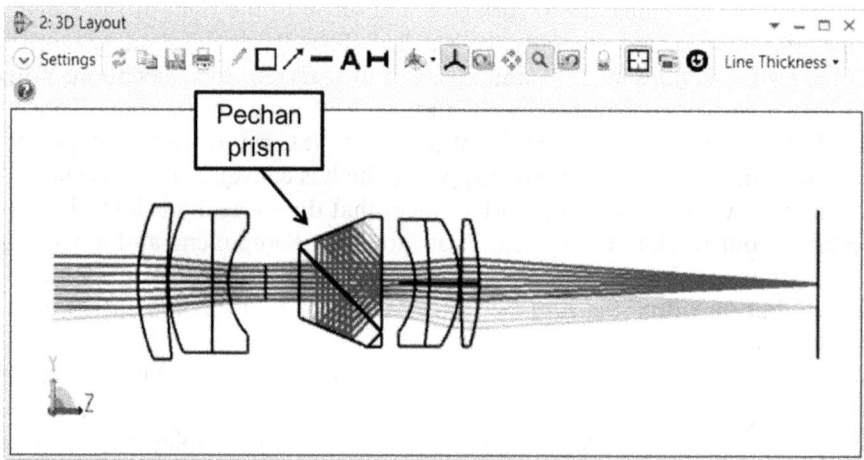

Figure 1.92. A Pechan prism between lens groups in HNSC mode.

The Pechan prism is one of many prisms used for image rotation [55]. You can see that this action is quite easily simulated in OS. You can use any of the Extended Scene Analysis features in OS to visualize the rotational effects of the current prism model. Note that the pechan.POB object file comes with the list of *polygon object* file samples in OS. You may try selecting other types of prism samples to visualize their effects on rays and images.

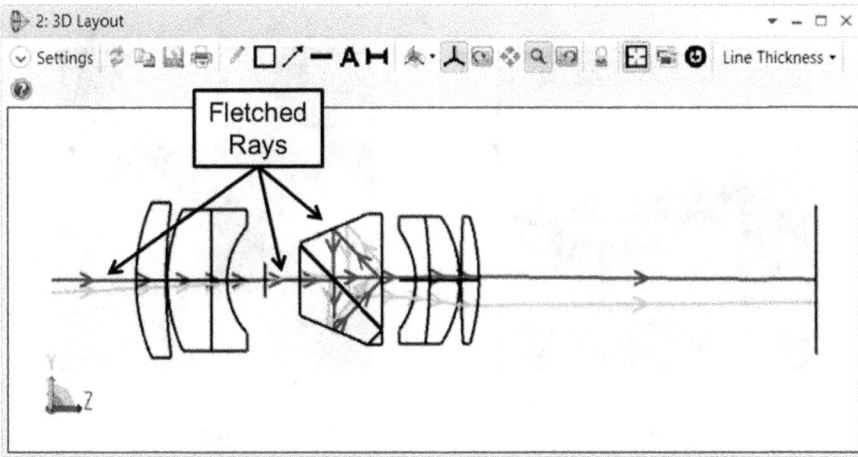

Figure 1.93. Arrowed 'fletched' rays showing their path in the prism.

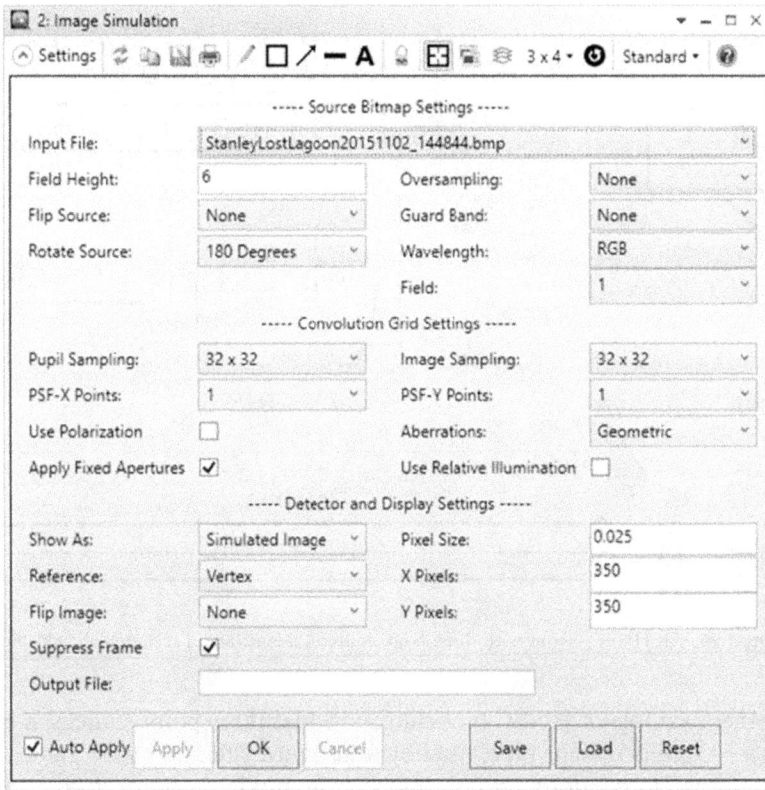

Figure 1.94. The settings for Image Simulation used to produce the display shown in figure 1.95(a).

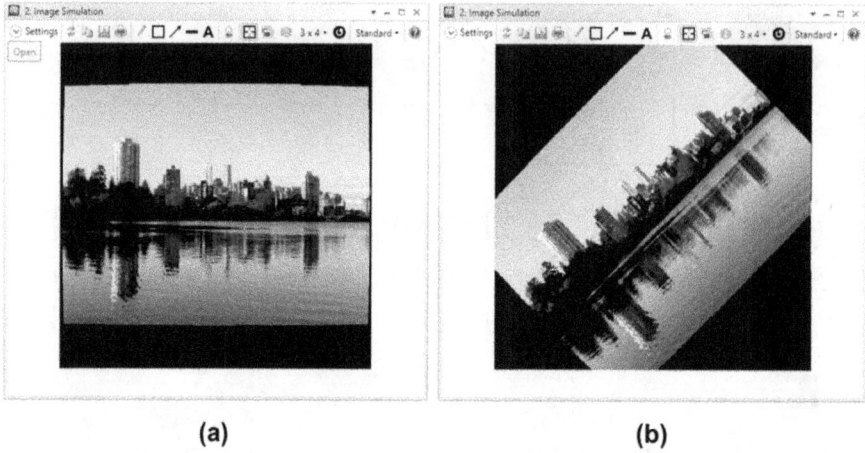

(a)   (b)

Figure 1.95. Image rotation by the Pechan prism. (a) Unrotated. (b) Rotated by 45° due to the 22.5° setting for Tilt About Z.

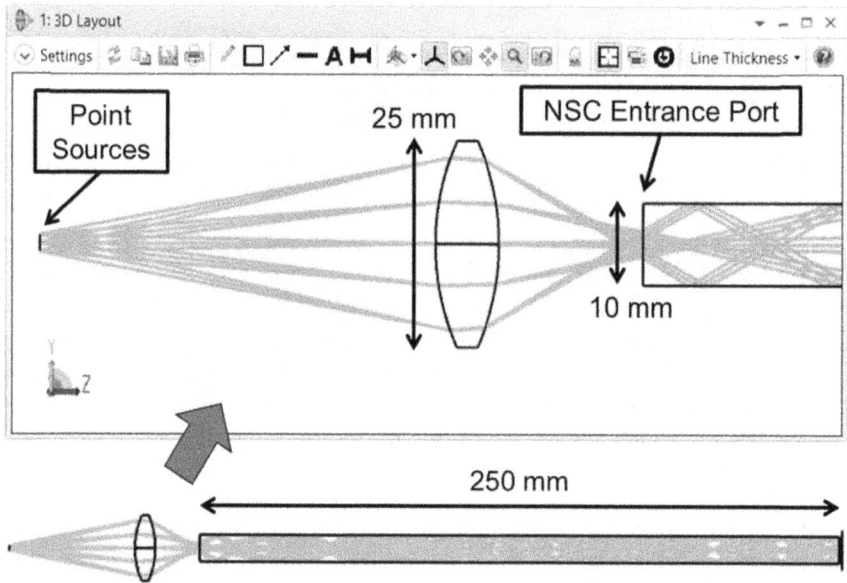

Figure 1.96. An HNSC layout with three point sources, a lens, and a rectangular light pipe.

As another example of HNSC modeling, consider three point sources, a biconvex lens, and a rectangular light pipe, as shown in figure 1.96.

The OS Lens Data Editor prescription for the model in figure 1.96 is shown in figure 1.97.

The system aperture is set to Float by Stop Size. In the Lens Data Editor, the NSC object's exit location in Z is 251 mm, because this light pipe has a length of 250 mm and you cannot let the NSC object surface's exit port (surface 5) be the

Figure 1.97. OS lens data prescription for the model in figure 1.96.

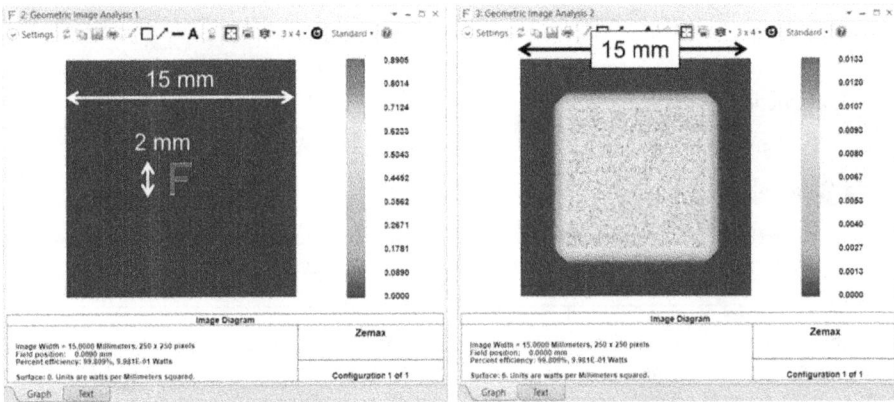

(a)   (b)

Figure 1.98. GIA for a letter F object as a Lambertian source. (a) Object distribution at surface 0 of the Lens Data Editor. (b) Irradiance at surface 6.

location of an NSC object. The three point sources are separated by 1 mm in the vertical, totaling a full height of 2 mm. This is modeled using Object Height rather than Angle in the Field Data Editor. As will be shown, this 2 mm dimension shall serve as the size of a *letter F* Lambertian source that the biconvex lens focuses into the light pipe.

The NSC prescription consists of a single object, which is a Rectangular Pipe whose material type is Mirror (i.e. just as you typed in 'N-BK7' for the Material type in the NSC Editor in figure 1.91, you should type 'MIRROR' for the Rectangular Pipe model here). The rectangular pipe's entrance and exit dimensions are both 10 × 10 mm (so their X and Y half-widths are 5 mm), and its Z Length is 250 mm. If a letter F object (at surface 0) is used as a Lambertian source in GIA (figure 1.98(a)), then the resulting irradiance distribution near the exit of the light pipe (surface 6 in the Lens Data Editor) is as shown in figure 1.98(b).

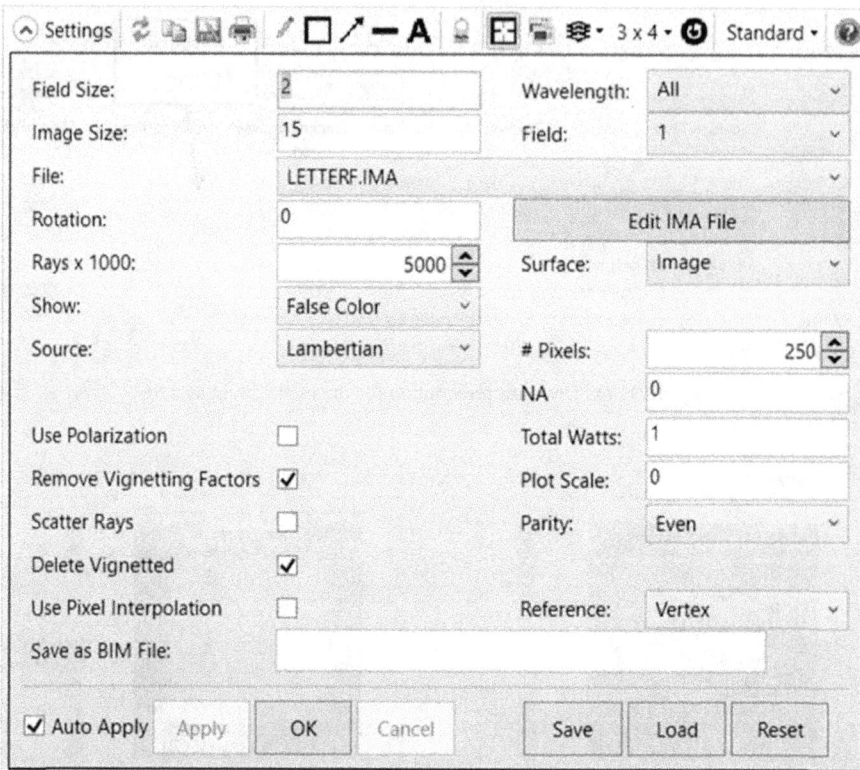

Figure 1.99. Settings for the GIA analysis displayed in figure 1.98(b). To display the distribution in figure 1.98(a), select surface 0 instead of Image.

The GIA settings used to display figure 1.98(b) are shown in figure 1.99. You may be interested to know what the distribution looks like at surface 4, the entrance of the light pipe, which is, of course, also the focal plane of the biconvex lens. This is shown in figure 1.100.

The distribution shown in figure 1.100 is quite recognizable as the inverted and aberrated image of the object, which has been demagnified by roughly 0.4 times. At a plane near the exit of the light pipe, this distribution has been virtually homogenized and spread over the full aperture of the pipe, despite the fact that we have placed the physical exit face of the light pipe 1.1 mm from the plane of the GIA's detector surface. While this demonstrates the homogenizing power of a light pipe, real light pipe systems do not end here. More optics is required beyond the exit face of a light pipe in order to *relay* the homogenized plane onto some desired plane of illumination. This is discussed in section 2.14.

There is one thing I have not yet explained about the GIA settings. In figure 1.99, note that the Flux setting (which is called Total Watts at the middle right side of the settings dialog box) is 1 Watt. *This refers to the flux that you estimate will enter the entrance pupil*, which is the aperture stop (surface 1) in the current model. You have to enter this number. OS will not compute it for you. In the current example, I have

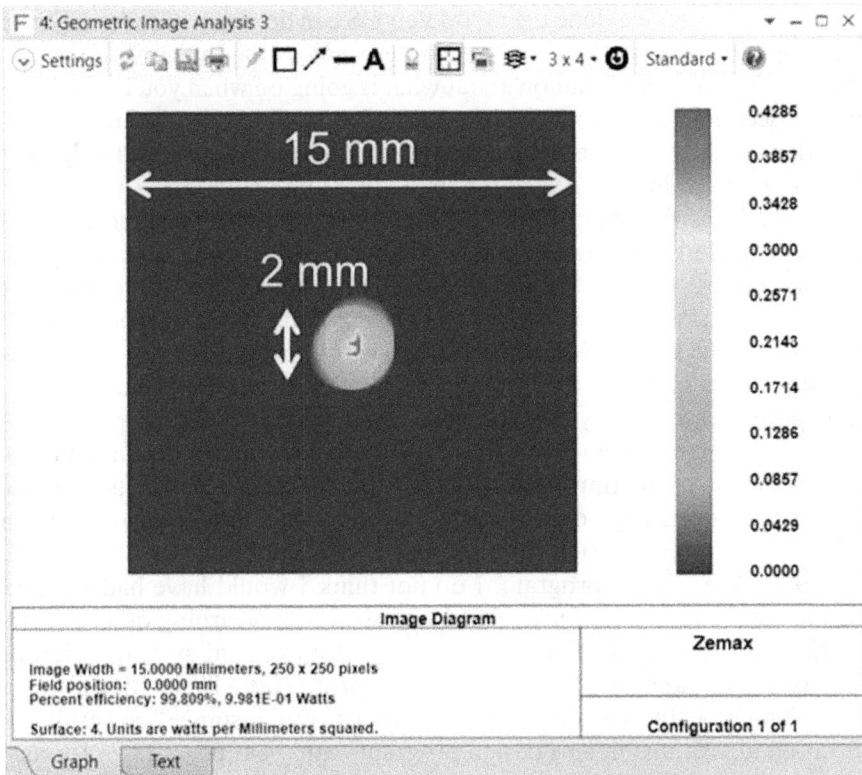

Figure 1.100. Irradiance distribution at surface 4, the entrance of the light pipe.

entered an arbitrary value, but in section 2.13, a specific estimated value is applied in the GIA feature to perform an illumination analysis (and I will show you how to estimate this flux).

1.2.19 A wrap-up; get set to use OpticStudio for the rest of this book

By now, you should have rudimentary knowledge and experience in using OS. In this program, by default, the SC ray tracing mode is the *ray tracing environment* that you are in when you open up a new blank OS file. You then begin a model by setting the system aperture (see section 1.2.1). This system aperture could be the semi-diameter of the stop, or it could be the NA in object space, and so on. The idea is to tell the program how rays should enter the optical system. Begin there, and then set up the fields, the wavelengths, and everything else, such as populating the Lens Data Editor with surfaces. Once done with the above, you may perhaps realize that it takes many playful exercises with an optical design program to attain sufficient familiarity with its features, nuances, and even the fundamentals of optical system design. If you have checked out appendix A.1, then you have seen how deep knowledge of the MTF and PSFs leads to an understanding of the relationship between optical crosstalk and MTF. This has nothing to do with using OS, yet you may have realized (having done the math) that OS can do many of the computations for you when you use specific program features, such as the Image Simulation tool. This helps you gain better intuition about what is going on when you use those tools. Further, in section 1.2.12, you saw how an understanding of polychromatic MTF computations leads to a deeper understanding of how and why spectral weights in an optical design program are selected for simulation and analysis.

You need not restrict your journey in optical system design to the sole use of OS. You can apply the knowledge gained in the current section to other programs. I just happen to use OS for my work. Someday, I look forward to using other programs, which are equally excellent for design. It is just that when I am busy delivering optical system designs, apart from using a program, I also spend a lot of time doing mathematical derivations and analyses and studying books and research papers. I normally do not have the time to try other programs. By sticking with a single program that has most of the features I need for my work, I have been able to spend time developing new ideas and new products. If a specific analysis feature is missing from the program I am using, then instead of searching for a different program, I often try to come up with a different way to analyze the problem using the current program. I do not think I would have had the time to pursue the creative aspects of my job if I had spent time learning other programs. I do believe that many of the creative aspects of an optical system design are a function of the designer, while the program is just a tool. This is not to say that programs cannot help provide creative solutions. For instance, powerful global optimization algorithms in lens design programs can yield new design forms. But global optimization is a program feature that is most helpful to lens designers. In contrast, my work involves optical system design, in which balancing aberrations is just one aspect of the job. The bulk of my work involves creating new products and modeling the sum of all components and subassemblies in a complete instrument. This may not be the case for you. So, while I am currently stuck with using OS, you do not have to be. Perhaps you may follow the ideas laid out in this book and apply them to any program of your choice.

1.3 Practical concepts for optical system layout and analysis

1.3.1 First-order: all you really need is $1/f = 1/s + 1/s'$

In many optical system design situations, you begin with a thin-lens paraxial layout, such as that shown in figure 1.20. In the *old days*, we used to sketch this on paper. Today, we either sketch a layout electronically (for example, using some electronic drawing application on a computer or tablet) or we lay out a first-order design by way of a paraxial thin-lens model in an optical design program, such as OS. In OS, you would use the *Paraxial* surface type that was described in sections 1.2.4 and 1.2.5. The governing equation for such a surface type is the well-known first-order thin-lens equation, expressed as

$$1/f = 1/s + 1/s',$$
(1.20)

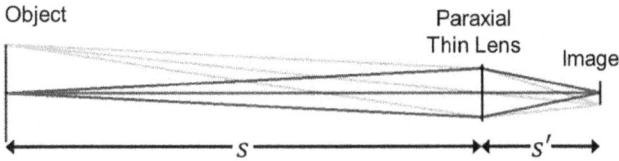

Figure 1.101. A first-order (paraxial) thin-lens layout in OS, with dimensions relating to equation (1.20).

where f, s, and s' are the EFL, object distance, and image distance, respectively. You know this formula well, but for clarity, let us put it in its usual graphical context (figure 1.101).

Note that in figure 1.101, the layout was created using the Paraxial surface type for a thin lens in OS, but the graphics for the window and its frame are not displayed. In section 1.2, the window frames (including the icons for clickable buttons) for all graphics were displayed for the purpose of showing how they actually appear in OS. In the current and subsequent sections and chapters, I may sometimes not display the full graphics, unless under specific circumstances. You can just assume that, unless stated otherwise, all optical models and layouts are done using OS. You may wonder what sign convention holds for the paraxial thin-lens model in OS. You can find out by making f negative or by bringing s closer to the lens until a virtual image of the object is created (hint: you need to set a solve for the marginal ray height to zero at the surface for thickness s', as discussed in section 1.2.5). You will eventually discover that OS's paraxial thin-lens model satisfies the usual convention that virtual images are located at $s' < 0$ and that a negative-powered lens has $f < 0$.

Some rules that are worth knowing at all times—based on familiarity with applying equation (1.20) to paraxial thin-lens models—are as follows: (a) the image is virtual, magnified, and upright when $0 < s < f$. (b) The image is real, magnified, and inverted when $f < s < 2f$. (c) The image is real, equal in size to the object, and inverted when $s = 2f$. (d) The image is real, demagnified, and inverted when $f < s < \infty$. (e) And of course, the image is virtual, upright, at infinity, and subtends a tangent half-angle given by h/f (where h is the object's height above the optic axis) when $s = f$. If you have not already mastered these rules, then master them now! Use a combination of exercises to do this, including entering numbers into equation (1.20), playing with the paraxial thin-lens model in OS, and by hand sketching those handy basic ray geometries for thin-lens layouts that you have often drawn in undergraduate physics and optics classes, such as that shown in figure 1.102.

Based on the geometry of the rays in figure 1.102, you should be quite familiar with the basic derivation of equation (1.20), but let us review it as a simple exercise in algebraic manipulation, which is often useful in analyzing optical formulas. From figure 1.102, by similar triangles, $h/s = h'/s'$, and $h/f = h'/(s' - f)$. So, eliminating h and h' by taking h/h' and combining the equations, we obtain $s/s' = f/(s' - f)$. Cross-multiplying gives $ss' - sf = fs'$. Dividing both sides by ss' yields $1 - (f/s') = f/s$. Finally, after dividing both sides by f, we have $1/f - 1/s' = 1/s$, which is algebraically equivalent to equation (1.20).

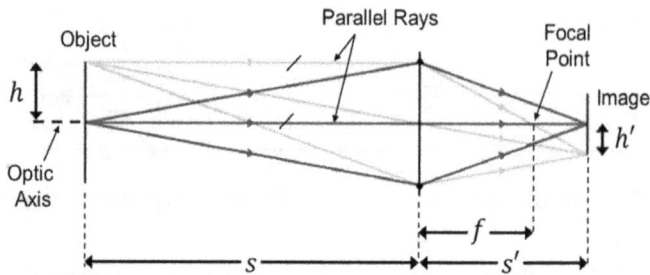

Figure 1.102. A basic ray geometry for a thin-lens graphical layout.

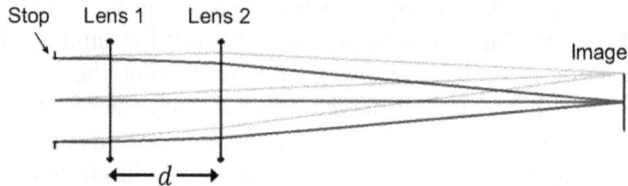

Figure 1.103. An air-spaced double paraxial thin-lens system.

1.3.2 Example: a microscope tube lens using commercial off-the-shelf lenses

Equation (1.20) clearly does not apply to thick lens systems (which would involve the use of so-called *y-nu ray-tracing formulas* to determine the image position) but if you are starting a rough optical model, then you will not have already included glasses in your layout. Besides, once you involve glass and element thicknesses, you should not be doing any more hand calculations! As long as you have combinations of paraxial thin lenses in your model, then equation (1.20) is all you need. For example, a common lens system comprises a combination of two thin lenses separated by an air space d, as shown in figure 1.103.

By way of the principle that the image formed by a first lens becomes the object for the second lens (and recalling what principal planes are), equation (1.20) can be applied to the system in figure 1.103 to obtain:

$$1/f = 1/f_1 + 1/f_2 - d/(f_1 f_2),\qquad(1.21)$$

where f_1 is the EFL of Lens 1, f_2 is the EFL of Lens 2, and f is the EFL of the double thin-lens system. There are many practical uses for this formula. By making $f_2 < 0$, we obtain a basic telephoto arrangement and can make quick estimates for all of the variables in the formula to *pre-design* a simple telephoto lens. By placing a stop in between the two lenses, we can realize a basic Petzval-type lens. As yet another practical example, consider a situation in which we reduce the stop diameter, place it in front of the two-lens system, and shift it leftwards some distance from Lens 1, as shown in figure 1.104. The resulting lens system resembles a modern microscope *tube lens*, which is used to focus rays exiting the back of an *infinity objective* (i.e. a microscope objective lens whose rays from the specimen

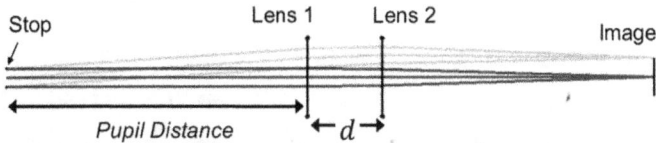

Figure 1.104. An air-spaced pair of paraxial thin lenses used as a microscope tube lens.

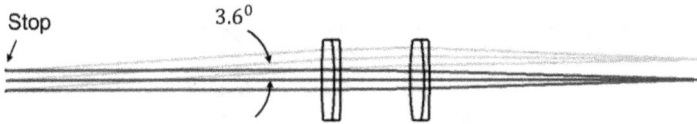

Figure 1.105. An air-spaced pair of achromatic doublets used as a microscope tube lens.

plane are collimated) onto the image sensor of a digital camera [56, 57]. The long *pupil distance* between the stop and Lens 1 is an intentional space allowance for the insertion of auxiliary optics, such as beam splitters and filters, between the back of the objective (which is where the stop is) and the tube lens.

There are some benefits to assembling a microscope tube lens comprising two thin lenses. For instance, the separation d between them enables us to vary the lens's EFL, which can either serve as a magnification compensator if microscope objectives from different manufacturers are used or vary the magnification for a fixed objective. In addition, if COTS lenses are used for the lens pair, then the separation between them provides some degree of correction for astigmatism and field curvature. Let us take a real lens example. A common numerical aperture (NA) for a microscope objective with a 20 mm EFL is 0.3. When we trace rays from infinity towards the specimen, this results in an objective lens *entrance pupil* diameter (see equation (1.43) in section 1.3.15) of $2 \times$ NA $\times$ EFL = 12 mm, where NA = 0.13 and EFL = 20 mm. In figure 1.104, the stop diameter has indeed been purposely set to 12 mm. A common magnification in modern microscopes is 10 $\times$, and a common *image circle* or *field number* is 25 mm, which is the diameter of a circle enclosing the full spatial field of view of the projected image onto a camera's image sensor. At 10x magnification, this means that the full spatial field of view at the focus of the objective is about 2.5 mm in diameter, and the half-angle subtending this field is arctan(1.25mm/20mm) $\approx$ 3.6°. At this half-field angle, oblique (off-axis) rays from the field pass through both lenses of the tube lens without vignetting if the lenses have diameters of roughly 50 mm. Indeed, the two thin lenses shown in figure 1.104 have been set to 50 mm diameters. To achieve 10 $\times$ magnification, an objective with a 20 mm EFL must be paired with a tube lens with a 200 mm EFL. Applying equation (1.21), if we select a first thin lens with a 500 mm EFL and a second lens with a 300 mm EFL, then the separation required to achieve an EFL of $f = 200$ mm is $d = 50$ mm. According to the current Edmund Optics lens catalog, part numbers 45-271 and 45-181 fit the choices for Lens 1 and Lens 2, respectively. Using these, we arrive at the lens system shown in figure 1.105, whose prescription is given in figure 1.106.

	Surface Type	Comment	Radius	Thickness	Material	Coating	Clear Semi-Dia
0	OBJECT Standard ▾		Infinity	Infinity			Infinity
1	STOP Standard ▾		Infinity	200.00000			6.00000 U
2	(aper) Standard ▾	45-271	305.74000	8.00000	N-BK7		25.00000 U
3	(aper) Standard ▾	45-271	-223.20000	4.00000	N-SF5		25.00000 U
4	(aper) Standard ▾		-663.82000	43.86832 V			25.00000 U
5	(aper) Standard ▾	45-181	173.11000	9.00000	N-BAK4		25.00000 U
6	(aper) Standard ▾	45-181	-164.03000	3.50000	N-SF10		25.00000 U
7	(aper) Standard ▾		-709.83000	172.83527 M			25.00000 U
8	Standard ▾		Infinity	-0.01037 V			12.53884
9	IMAGE Standard ▾		Infinity	-			12.53865

Figure 1.106. The prescription for the lens system shown in figure 1.105.

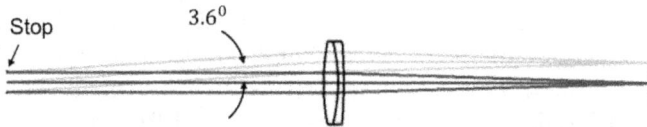

Figure 1.107. A single achromatic doublet used as a microscope tube lens.

In figure 1.106, note that the separation between the two doublets is not 50 mm, because the elements are not ideal thin lenses. Two variables are shown in the prescription. The thickness of surface 4 was used to optimize the system's EFL (the target was 200 mm at the monochromatic wavelength of 550 nm), and the thickness of surface 8 was used to optimize the RMS spot size of the axial field alone in order to achieve best axial focus. The marginal ray solve at surface 7 is an optical designer's *old trick*, which is to maintain the location of the paraxial focal plane whilst varying the final focal plane position (which is surface 9) relative to the paraxial focus. Before displaying its resulting performance (in terms of MTF, field curvature, and image distortion), let us consider the alternative, which is to use a single achromatic doublet, as shown in figure 1.107, whose prescription is given in figure 1.108. This doublet uses part number 45–179 from Edmund Optics. Now, only a single variable is used, which is the thickness of surface 6 used to optimize the axial RMS spot size.

Figure 1.109 compares the monochromatic MTF curves at 550 nm for the lenses in figures 1.105 and 1.107. Note the significant improvement for the air-spaced double-lens system compared to the single lens (a comparison of their polychromatic MTFs at 450–650 nm would also indicate better performance for the double-lens system). The field curvature, astigmatism, and image distortion curves (at 550 nm)

	Surface Type	Comment	Radius	Thickness	Material	Coating	Clear Semi-Dia
0	OBJECT Standard ▾		Infinity	Infinity			Infinity
1	STOP Standard ▾		Infinity	200.00000			6.00000 U
2	(aper) Standard ▾	45-179	130.48000	9.00000	N-BAK4		25.00000 U
3	(aper) Standard ▾	45-179	-99.36000	3.50000	N-SF10		25.00000 U
4	(aper) Standard ▾		-320.20000	193.82519 M			25.00000 U
5	Standard ▾		Infinity	-0.01635 V			12.53344
6	IMAGE Standard ▾		Infinity	-			12.53296

Figure 1.108. The prescription for the lens shown in figure 1.107.

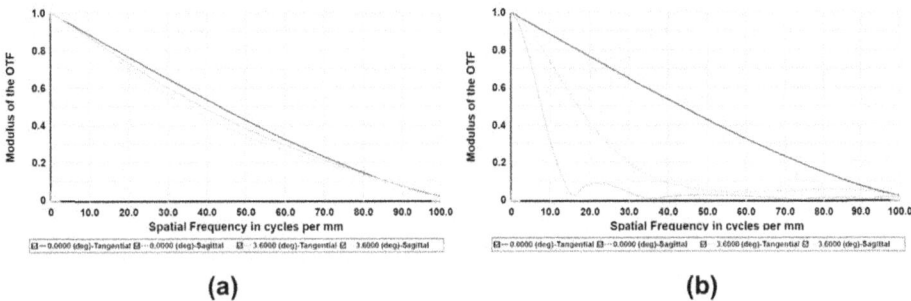

(a)   (b)

Figure 1.109. MTF curves at 550 nm for tube lens models. (a) The lens from figure 1.105. (b) The lens from figure 1.107. The blue solid curve is the axial MTF, which is diffraction limited. The solid and dotted green curves denote the tangential and sagittal planes, respectively.

for the two cases are shown in figure 1.110, which also indicate better performance for the double-lens system. This improvement is not due to a reduction in spherical aberration. At low aperture ($F/16.7$) and field, both lens systems have negligible spherical aberration, coma, and image distortion. The only monochromatic aberrations left are field curvature and astigmatism, both of which are functions of the surface power and the spaces between elements. It is the separation d for the two-lens system as well as the use of different surface curvatures for each doublet that provide the right degree of correction for astigmatism and field curvature.

We will discuss aberrations further in section 1.4, but let me share something interesting here. From the point of view of the oblique ray bundle (i.e. the rays from the 3.6° field angle), it can be said that the oblique rays *see* a sort of *effective lens system* that is different from the one *seen* by the axial ray bundle. Figure 1.111 illustrates this idea; it depicts that the oblique rays effectively travel a slightly larger gap distance d' between the doublets. Furthermore, the oblique rays travel through a different portion of the two doublets, and in that portion, there is somewhat a different focal length than the one experienced by the axial ray bundle. As such, it

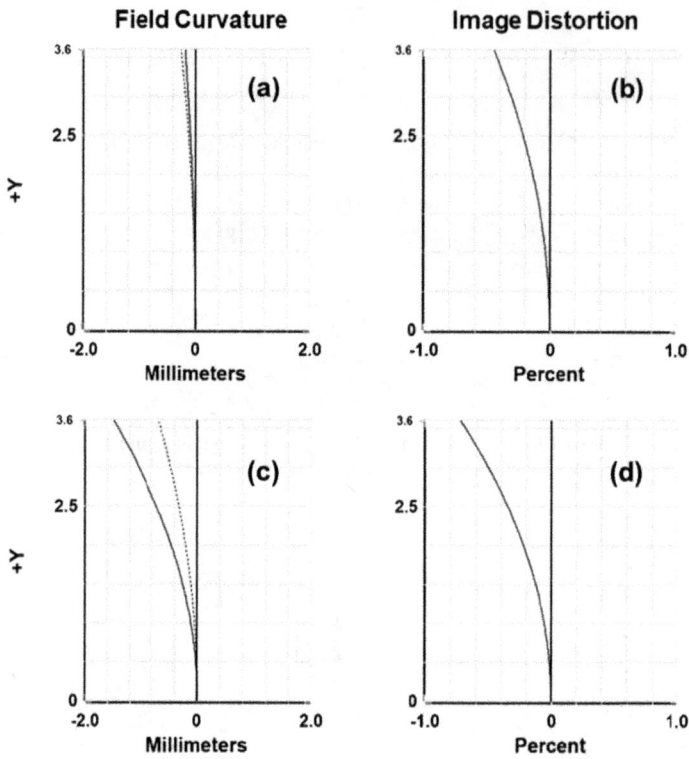

Figure 1.110. Field curvature/astigmatism and image distortion curves. (a) and (b): The lens from figure 1.105. (c) and (d): The lens from figure 1.107. The dotted curve in the field curvature plots denotes the sagittal rays.

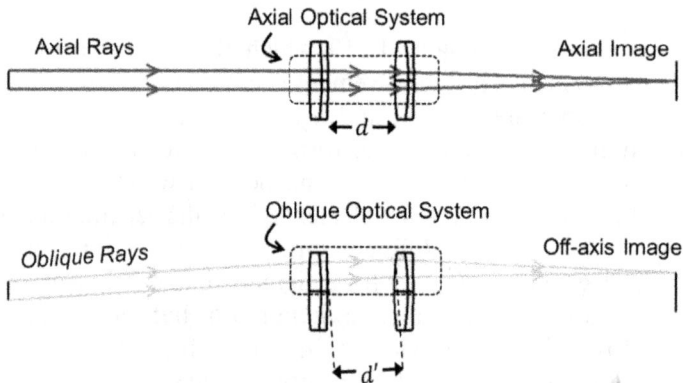

Figure 1.111. An illustration of effective optical systems for axial and oblique rays for the air-spaced tube lens design in figure 1.105.

can be roughly said that according to equation (1.21), there are different values of f, f_1, and f_2, and of course a different separation d' for the oblique rays relative to the axial rays. If we allow the application of equation (1.21) to be stretched to the extent of associating the oblique rays with an effective oblique relative EFL, then the larger distance d' for the oblique rays can be said to aid in the reduction of the refractive power imparted to those rays. This results in an extension of the EFL for those rays, thereby *equalizing* the focal lengths between oblique rays and axial rays, which is not the case for rays passing through a single doublet. This is one way to understand why the doublet pair in figure 1.105 has better imaging performance than the single doublet in figure 1.107.

The concept of a focal length for oblique rays is nothing new—it is, in fact, the central idea for understanding astigmatism and for deriving what is sometimes called *Thomas Young's astigmatic formulas* [58] but otherwise called *Coddington's equations for astigmatism* [5, 25]. These formulas apply to a local area of an optical element that focuses an oblique *ray pencil* (a narrow bundle of rays). The design of spectacle lenses (including progressive powered spectacles) is based on applying these *Young-Coddington* equations [59]. In one study by this author, these equations were applied to the design of a freeform lens that provided variable focus across its surface without moving parts [35, 36, 60]. Oddly, a subsequent study by others introduced the concept of *field focal length* [61], even though this had been implied by the equations of Young and Coddington mentioned above. Moreover, it is known that when image distortion is present, then it can be said that the EFL varies with field [5, 62].

When you are in a fast-paced project involving microscopy, perhaps you may at times need tube lenses with some specific focal lengths that are not available from any manufacturer, and you may not have sufficient time to have them custom made. In such cases, through a judicious selection of COTS lenses, the two-lens design approach that we have discussed here may be right for you. You may wonder why a monochromatic MTF plot was displayed in figure 1.109, despite the fact that we used achromatic doublets. The reason is that the design remains unsuitable for broadband white light. However, the two-lens doublet tube lens design performance can be quite good within narrow bands whose spectral width is roughly 20 nm. Such narrow bands are common in fluorescence microscopy [56, 57], in which multilayer narrowband filters are mounted between the objective and the tube lens.

As a final remark, note that instead of applying equation (1.21), we could have played with a double thin-lens paraxial model in OS to determine the EFLs of the two doublets in figure 1.105. But that type of trial and error would waste time if you were not already aware of equation (1.21). In many cases, knowing a formula helps to accelerate the optical modeling done in a program. In fact, this was how I arrived at the paraxial double thin-lens model in OS shown in figure 1.104. I first used equation (1.21), then modeled two thin lenses in OS. In the following section, we will return to equation (1.20) and apply it to the conceptual model of an elliptical reflector for a xenon arc lamp.

1.3.3 Example: a xenon arc lamp with an elliptical reflector

One example of a product line of xenon arc lamps is supplied by Excelitas Technologies; these lamps are sold under the brand name of Cermax® Xenon [63]. At the cited website, you may find datasheets for lamps with elliptical reflectors, and in those datasheets, information is provided for the dimensions of the semi-major and semi-minor axes of ellipses for lamps with elliptical reflectors. So, given that we know the semi-major and semi-minor axis lengths (figure 1.112), when modeling the ellipse in OS sequential mode, what is the *base radius of curvature* of the elliptical surface that focuses the rays from the arc source located at Focus 1 onto Focus 2? Furthermore, what shall we use as the reflector's *conic constant*?

To begin our model, we apply our knowledge of the geometry of ellipses and basic reflective optics. If the arc source (which, for simplicity, we shall consider to be a point source) is located at Focus 1, then its image is located at Focus 2. So, to the first order, the object distance to the curved reflective surface on the left is s, and the image distance, s', is given by $s' = s + 2c$. Therefore, applying equation (1.20) gives:

$$1/f = 1/s + 1/(s + 2c). \qquad (1.22)$$

From the theory of ellipses and optics, we know that the hypotenuse that forms the triangle given by the dimensions a, b, and c has the dimension a, because for an ellipse, when light travels the distance from Focus 1 to Focus 2 by first reflecting off the left reflective surface, that distance is equal to the path taken from Focus 1 to Focus 2 by reflection off the top (i.e. by traveling the distance $2a$). So, $2(a - c) + 2c = 2a$. Therefore, $c^2 = a^2 - b^2$, and this is the way to determine c, given a and b. Clearly, $s = a - c$, so we have all the means to plug values into equation (1.22) in order to determine the EFL of the left-hand reflective surface of the mirror. Knowing this, the base radius of curvature of that reflective surface is just $2f$, according to our knowledge of reflective optics (figure 1.113).

In figure 1.113, θ_i is the angle of incidence (AOI) and θ_r is the angle of reflection. Since $\theta_i = \theta_r$, then to the first order (i.e. in the limit $\sin \theta_i \approx \tan \theta_i \approx \theta_i$ and $h \to 0$), we have $\theta_i \approx h/R$ and $2\theta_i \approx h/f$. So, combining these, we find that $f \approx R/2$. Why am I referring to a radius of curvature for a spherical mirror when our current design

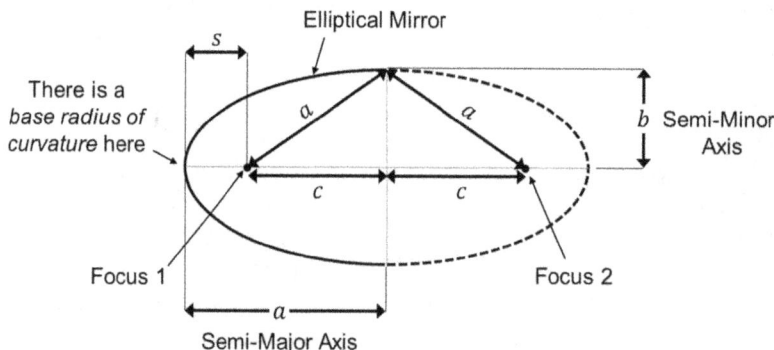

Figure 1.112. An illustration of an elliptical reflector geometry for an arc lamp. The arc is not a point source, but we shall assume that it is for simplicity.

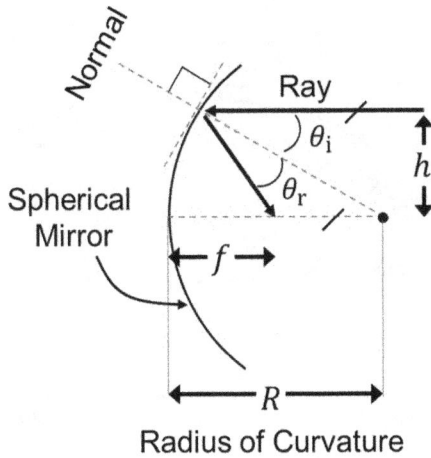

Figure 1.113. The geometry for a ray from infinity that is focused by a spherical mirror. We are purposely ignoring the convention that light has to travel from left to right in a layout.

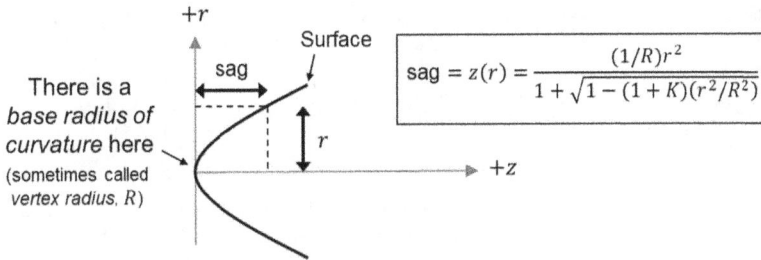

There is a *base radius of curvature* here (sometimes called vertex radius, R)

$$\text{sag} = z(r) = \frac{(1/R)r^2}{1 + \sqrt{1 - (1 + K)(r^2/R^2)}}$$

Figure 1.114. A geometry describing the sag of a surface commonly used in optical design.

problem concerns an ellipse? The reason is as follows: many classes of surface shapes in optical design may be

expressed by a *sag* formula, given by

$$z = \frac{(1/R)r^2}{1 + \sqrt{1 - (1 + K)(r^2/R^2)}}, \tag{1.23}$$

where R is the *base radius of curvature*, r is the surface's radial coordinate (figure 1.114), and K is the so-called *conic constant* that governs the *eccentricity* of the shape [6]. If you remember the concept of conic sections in analytic geometry, this is where the conic constant comes from. In other words, K determines whether the shape is a circle (when $K = 0$), a parabola (when $K = -1$), a hyperbola (when $K < -1$), an *oblate ellipse* (when $K > 0$), or a *prolate ellipse* (when $-1 < K < 0$). So, the radius of curvature in figure 1.113 is R in equation (1.23) for the case of $K = 0$. For an ellipse, this radius must still exist, by virtue of its presence in equation (1.23). Since the ellipse that is depicted in figure 1.112 is that of a prolate ellipse (or rather, in the case of 3D geometry,

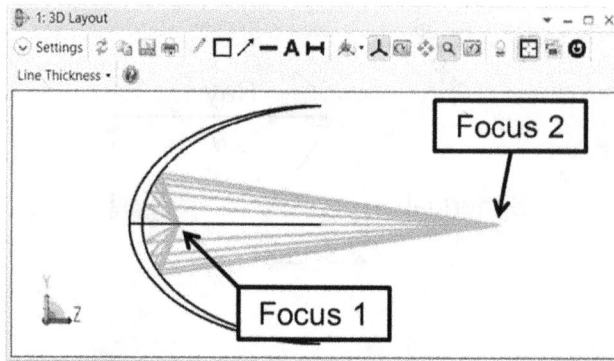

Figure 1.115. The layout in OS for the optimized model of the elliptical reflector given by part number PE300C-10F of the Cermax® Xenon arc lamp from Excelitas Technologies [63].

	Surface Type		Comment	Radius	Thickness		Material	Coating	Clear Semi-Dia		Conic	
0	OBJECT	Standard		Infinity	-5.86300				0.00000		0.00000	
1	STOP	Standard		10.61000	0.00000				15.00000	U	0.00000	
2	(aper)	Standard	Ellipse	10.61000	55.74053	M	MIRROR		18.00000	U	-0.65554	V
3	IMAGE	Standard		Infinity	-				4.57054E-10		0.00000	

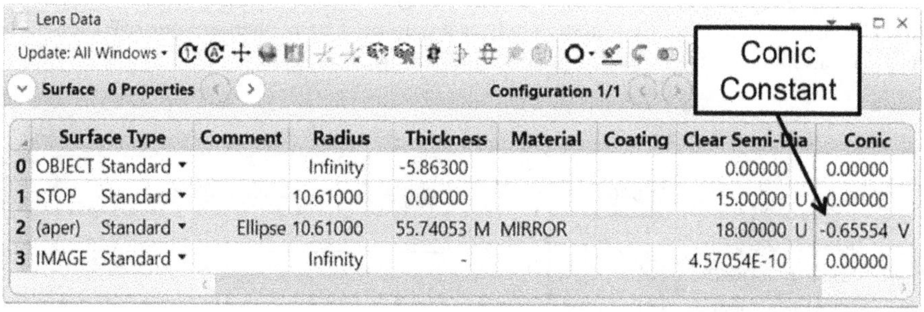

Figure 1.116. The prescription in OS for the layout in figure 1.115. The stop has been given a radius equal to the ellipse's radius. This is done for display purposes in the layout.

a *prolate ellipsoid* [6]), its conic constant should have a value somewhere between negative one and zero, while its base radius is $2f$, where f is given by equation (1.22).

Let us now take a numerical example and model it in OS. For one current model of the Cermax® Xenon arc lamp from Excelitas Technologies (model number PE300C-10F), it is specified that the semi-major axis length is $a = 1.2136$ inches, and the semi-minor axis length is $b = 0.712$ inches [63]. In mm units, these are $a \approx 30.825$ mm and $b \approx 18.085$ mm. Applying the formulas above, this means that $c = \sqrt{a^2 - b^2} = \sqrt{30.825^2 - 18.085^2} \approx 24.962$ mm. The object distance to the mirror is $s = a - c = 30.825 - 24.962 = 5.863$ mm. Applying these values for s and c in equation (1.22), we obtain $f \approx 5.305$ mm. Since $R = 2f$, the base radius is $R = 2(5.305) = 10.610$ mm. Therefore, if we let 10.61 mm be the radius of curvature for a mirror surface in OS, and if we let the object distance (the thickness of surface 0) be 5.863 mm, then we can obtain the image distance by setting a marginal ray height solve to zero for the thickness after the mirror surface. Finally, we may set an arbitrary value for the conic constant somewhere between 0 and -1, and let this be a variable. If we then set the target axial RMS spot size to 0 in the Merit Function Editor, we may set the conic constant for the ellipse to be a variable, thereby obtaining its value through optimization. The result is the layout shown in figure 1.115, whose prescription in OS is given in figure 1.116.

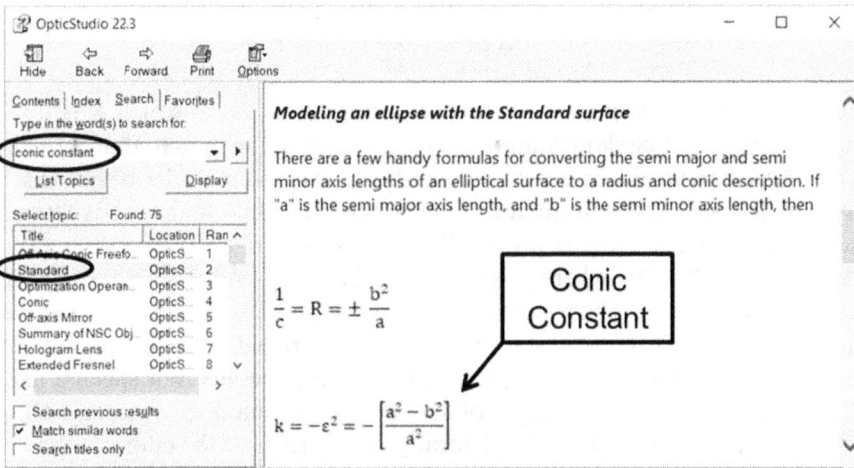

Figure 1.117. OS's help file on modeling elliptical surfaces. Reproduced from [33], with the permission of ANSYS, Inc.

Note that we expected the image distance to be $s + 2c \approx 55.787$ mm, but OS gave the result as $55.740\,53$ mm using the marginal ray solve. This is perhaps due to some rounding errors in our prior calculations. The conic has been determined to be $-0.655\,54$ (it is unitless). Let us compare this with the theoretical value for the current ellipse. According to the OS help file for elliptical surface shapes (see figure 1.117), the conic constant should be $K = -(a^2 - b^2)/a^2$. Applying the values in the current exercise, we obtain $K = -(30.825^2 - 18.085^2)/30.825^2 \approx -0.655\,784$, which is slightly off from the value obtained through optimization in OS, as displayed in figure 1.116. No worries. It is just a matter of checking rounding errors and confirming that the resulting performance does not significantly change when the chosen values for all variables are varied within some agreed tolerances. Also, you could try entering the theoretical value computed above for the conic constant into OS. Consequently, you would just need to re-optimize the base radius, targeting an axial RMS spot size of zero. You would observe negligible differences in the axial MTF and spot size. Finally, as with any application involving illumination, a last check and finishing touches have to be done through NSC ray tracing, which we shall leave for the reader to explore.

There is one thing left to be said before closing this section. You may already know that the sag formula provided by equation (1.23) describes an *aspheric* surface when $K \neq 0$. More generally, such surfaces (which are rotationally symmetric about the optic axis) are called *quadrics of revolution* [6] (this is similar to rotating a conic section function about an axis in calculus). Furthermore, you may also know that departures from these conic sections are also called *aspheres*, which are conventionally expressed as higher-order terms added to the right-hand side of equation (1.23). However, equation (1.23) may itself be expanded into a Taylor series [6], which we can express as:

$$z = \frac{1}{2}\frac{r^2}{R^2} + \frac{1}{8}(1 + K)\frac{r^4}{R^3} + \frac{1}{16}(1 + K)^2\frac{r^6}{R^5} + \ldots \qquad (1.24)$$

An examination of equation (1.24) shows that the condition at which $K = 0$ (which we know represents a circle) results in a first-order term for the sag that is quadratic with r. Under conditions where $K \neq 0$ (with the exception of the trivial case of a parabola when $K = -1$), the quadratic term persists, while higher-order terms are included. Therefore, to the first order, all quadrics of revolution are parabolas. And for all parabolas, $R = 2f$, where f is the parabola's EFL and R has several different names: it is the parabola's base radius, vertex radius, or *directrix*. **This means that, to the first order, any asphere that does not possess additional terms in** equation (1.23) **has a base radius given by R in** equation (1.24)**, which results in a specific paraxial EFL.** This is why, in our example of modeling the xenon arc source's elliptical reflector, I was first concerned with determining the EFL of the ellipse, followed by determining its base radius, which has to be entered into the column for the surface radius of the ellipse when modeling it in OS. It is worth keeping equation (1.24) in mind, because the common aspheric formula given by including further terms in equation (1.23) *disrupts* the base radius of a lens, which is the reason that experienced designers and lens manufacturers caution against the inclusion of additional quadratic and fourth-order terms (due to the factor $1 + K$ that multiplies r^4/R^3 in equation (1.24)) in optical designs that use aspheres.

1.3.4 Notes on designing with commercial off-the-shelf components

Clearly, the benefit of using COTS components is that it saves time, since the components are (mostly) readily available from the supplier. Therefore, in many practical cases of product development with compressed schedules, the use of COTS components is a good approach to designing and constructing a proof-of-concept optical system, which possibly leads to a first (alpha) prototype of the product. Based on data gathered from the COTS-based optical system, we may custom design the optics for the next (beta) prototype, and so on. Of course, this would not apply to those circumstances in which the product is a precision optical instrument whose specifications call for a high MTF.

In the previous two sections, you saw examples in which some COTS components were selected for use in optical design. In section 1.3.2, COTS achromatic doublets were selected for a microscope tube lens design after modeling them as paraxial thin lenses in OS. In section 1.3.3, we approached the reverse—we first selected a COTS xenon arc lamp, then applied some first-order theory to compute the EFL of the elliptical mirror, then modeled it in OS sequential ray tracing mode by performing a final optimization for one variable (the conic constant). In the case of the tube lens example, the use of COTS lenses was primarily made possible because we had a system operating at a low field angle and a low aperture. Further, the tube lens system provided a single variable for optimization, which was the separation between the doublets. Unfortunately, that separation may only be used to optimize one of two parameters: (a) the EFL of the tube lens system, or (b) the astigmatism and field curvature.

This is the drawback of using COTS components—they have significantly fewer variables for optimization. In fact, the only variables for optimization are the air gaps and finite choices of different elements with different EFLs from a supplier of COTS lenses (also called *stock* lenses). However, with patience and care, it is indeed possible to apply these two variables to obtain a completely COTS multielement lens design with reasonable performance, such as a four-element triplet operating at the F, d, and C wavelengths [64]. **The process is iterative, involving the sequential optimization and replacement of elements. This is the key to designing optical systems using COTS components.** In other, simpler conditions, we can follow certain practical rules [65]. For instance, when using a singlet at low field and high aperture with an object at infinity, it can be beneficial to select a plano-convex element and orient the convex side towards the source (e.g. see section 1.4.4). Such rules are derived from a combination of sound knowledge of aberration theory and experience, and they are well suited to optical developers who do not have access to optical design programs. With the use of modern optical design programs, checking the performance of a COTS component through modeling it in the program can be rather straightforward—if facing one side of a COTS lens towards the source does not work, then flip it and see! Still, it is worthwhile to know some aberration theory, which is a prerequisite for optical design (we will discuss some tips on aberrations and optical design in section 1.4). Further examples of the use of COTS lenses are provided throughout specific sections in this book.

1.3.5 If you master the concept of conjugate planes, you can go very far

Looking back at figures 1.112 and 1.115, the points labeled Focus 1 and Focus 2 are known as **conjugate points** or simply **conjugates** of an optical imaging system. Put simply, **object and image points are conjugates of an optical imaging system**. Accordingly, planes containing those points are **conjugate planes**. So, object and image planes are conjugate planes. There is good reason for you to master this concept because many optical systems utilize more than a single conjugate plane *simultaneously*. In other words, in all optical systems, there are multiple conjugates *doing work* for the optical system. For instance, a telescope has to have four conjugate planes: (a) the scene to be magnified; (b) the image of the scene; (c) the entrance pupil (i.e. the objective lens, or a plane near it, which is the aperture stop); (d) the exit pupil, which is where you would place your eye, as described in section 1.1.3. Now, hold that thought—did I say four conjugate planes? Actually, there are more.

First, let us state the obvious: All optical imaging systems possess an infinite number of conjugate planes, because, according to equation (1.20), for any object distance s, there is an image located at s'. And since any plane can serve as the object for each and every element in a lens or mirror system, there are even more object–image conjugates than those associated with the total optical system. All of this stems from the basic premise of imaging, which is that **an image of a point is**

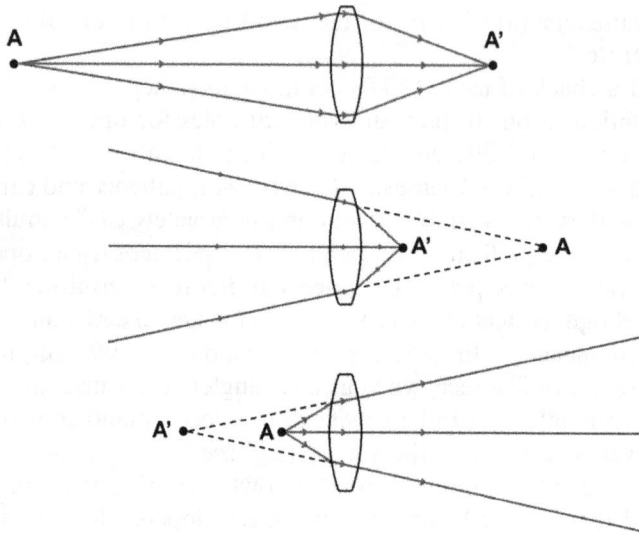

Figure 1.118. An object (A) and image (A′), which are conjugate points of a lens.

formed when rays from that point intersect after passage through an optical system, as illustrated in figure 1.118. For any object point A, if there is an image A′, then all rays from A intersect at A′, and when this happens, it means that all rays from A take the same time to reach A′. This is a consequence of **Fermat's principle** [66]. For axial conjugate points, certain special surfaces satisfy Fermat's principle exactly, and they are the quadrics of revolution for which the conic constant is nonzero. For spherical optics, where the conic constant is zero, conjugate points and planes are defined within the first-order approximation.

The point to take from the above is that **whenever you see a pair of rays intersect at any intermediate plane within the path of all rays in an optical imaging system, then that intersection may either be a conjugate image or a conjugate object**, which may either be real or virtual (and we will discuss applications of this important concept over the next few sections). As a consequence of this, there are the so-called six (actually seven) **cardinal points** of an optical imaging system [5, 65], which are the first and second principal points, the first and second nodal points, and the first and second focal points (figure 1.119).

Of the six cardinal points mentioned, the first four are conjugate points. That is, the first and second principal points are conjugates of one another at unit transverse magnification, while the first and second nodal points are conjugates of one another at unit angular magnification (which is why a **nodal ray** at an angle u passing through the first nodal point emerges from the second at the same angle). When the medium is air on both sides of the optical system, then the principal points coincide with the nodal points. Extending lines orthogonally from those points gives the principal planes, as depicted in figure 1.119.

Take a close look at how the blue and red rays enter the optical system from the left and right sides, respectively. Reversing the red ray from the first focal point

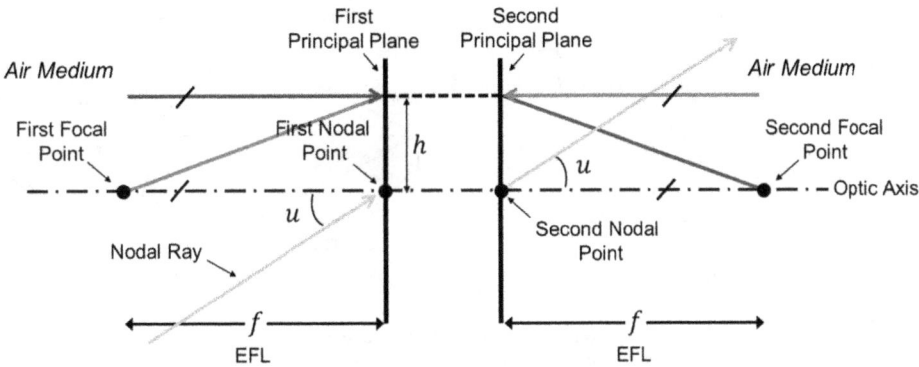

Figure 1.119. The cardinal points and planes of an optical imaging system.

would make it intersect with the blue ray at a point on the first principal plane at height h above the optic axis. Then, on the right side, they would appear to emerge at height h from a point on the second principal plane. In accordance with the concept described above (that describes how conjugate points are identified by way of locating pairs of rays that begin at an intersection and end at another intersection), it becomes clear that the principal planes are conjugate planes, and the fact that the intersecting blue and red rays enter and emerge the principal planes at the same height h means that the principal planes are at unit transverse magnification. Similar logic applies to intersecting rays for the nodal ray at an angle u to the optic axis. That is, the nodal ray and a ray collinear to the optic axis intersect at the first nodal point, and they emerge together from the second nodal point. Thus, the two nodal points are conjugates, and the fact that the entering and exiting oblique nodal rays are at an equal angle u with the optic axis means that they are at unit angular magnification. This makes four conjugate points out of the six cardinal points. There is actually a *seventh cardinal point*, which is itself a conjugate of the two nodal points. Somewhere along its path, the nodal ray intersects the optic axis at a third location, and that intersection is called the **optical center** [5, 50]. Johnson [50] has pointed out that since it is a conjugate of the nodal points, the optical center should accordingly be regarded as a **seventh cardinal point**. In sections 1.3.16–1.3.18, we discuss this seventh cardinal point and its practical applications in more detail.

1.3.6 Example: collimation at an intermediate plane and its application

Figure 1.120(a) shows collimated rays from an axial point source, while figure 1.120(b) shows collimated rays from an off-axis field point of an extended source.

For extended sources, the angle θ defines a half-angle divergence for the collimated oblique rays. Consider figure 1.121, which depicts a second thin lens is mounted at distance f_2 from the first thin lens. If f_2 is the EFL of the second thin lens, specific sets of rays exiting the second thin lens can be considered to be collimated. This is what is meant by collimation at an intermediate plane.

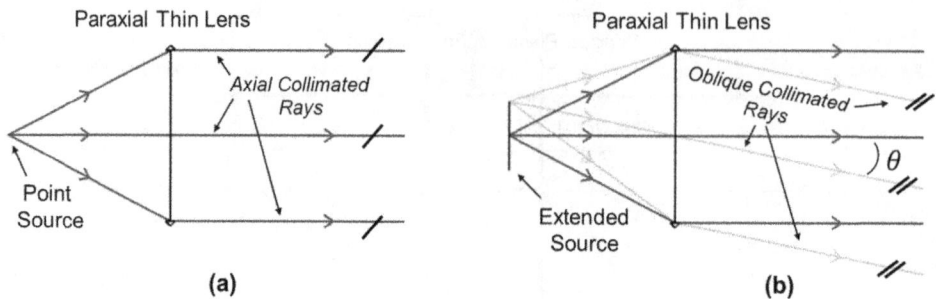

Figure 1.120. Collimation. (a) For an axial point source. (b) For an extended source. The distance between the source and the thin lens is the EFL of the thin lens.

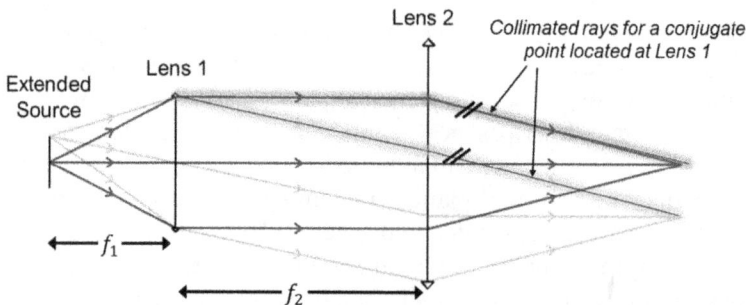

Figure 1.121. The collimation of rays from an intermediate plane (Lens 1).

In figure 1.121, Lens 1 is an intermediate plane from which Lens 2 collimates rays. Applying the concept of conjugate points and planes from the previous section, it can be seen by examination that the highlighted (glowing) pair of rays that originate from the top of Lens 1 are indeed parallel to each other. This is at least part of what meant by collimation, as we commonly know it.

But there is another aspect of collimation that deserves attention. Due to the concept of conjugate planes, collimation should really be regarded as a matter of perspective, and it is also a *relative* condition. It is a matter of perspective in the sense that if I did not highlight those glowing parallel rays for you in figure 1.121, then perhaps you might not have noticed their parallelism. Instead, you might have just seen that there are two focusing cones of rays: the green cone from the off-axis field point of the extended source and the blue cone from the axial field point. But now that I have highlighted the glowing parallel rays, you can clearly see that Lens 2 is actually collimating rays originating from a point at the top of Lens 1. Yet, that is not all there is to it. Collimation is also a relative condition in the sense that a first conjugate plane must be located at the front focal plane of a lens, and a second conjugate plane must be located at the back focal plane of the next lens. For instance, in figure 1.121, if there were a third lens to the right of Lens 2, then the glowing parallel rays would only be considered collimated if the image plane of the

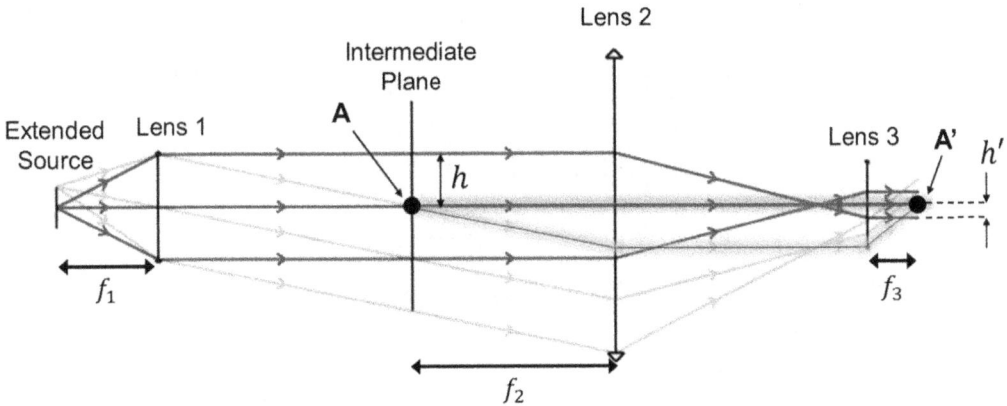

Figure 1.122. The collimation of rays from an intermediate plane to the right of Lens 1.

third lens were placed at the back focal plane of that lens. Furthermore, if Lens 2 were shifted some distance farther to the right of Lens 1, then Lens 2 would be collimating rays from a plane that is to the right of Lens 1, even if there is nothing at that plane other than the rays originating from the extended source placed at distance f_1 from Lens 1. This condition is illustrated in figure 1.122, where a pair of rays highlighted in glowing blue and red now form a conjugate pair at A and A′. Further, the height labeled h is measured along a conjugate plane at A and is imaged at A′ with height h'. For convenience, the intermediate plane in this example has been placed at a point where the oblique ray from the top of Lens 1 intersects with the axial ray along the optic axis. However, the conjugate points A and A′ can exist even if the intermediate plane is located anywhere to the right of Lens 1, by virtue of maintaining the distance f_2 between the intermediate plane and Lens 2 and f_3 to the right of Lens 3.

Now, consider figure 1.123, where the intermediate plane is back at Lens 1. Under this condition, for all cases a, b, and c, no matter where Lens 3 is placed to the right of Lens 2, a perfectly sharp image of Lens 1 is found at the image plane of Lens 3.

On the other hand, if the image plane of Lens 3 were shifted farther to the right, then that plane would see a focused image of some other plane that is to the right side of Lens 1 by virtue of applying equation (1.20) (i.e. the image formed by Lens 2 would become the object for Lens 3). Among the three conditions shown, Lens 3 would take on the smallest size if it were positioned at the back focal plane of Lens 2 (figure 1.123(a)). However, figure 1.123(b) depicts a more common situation in a practical application, namely the layout of a typical microscope condenser, in which a biological specimen would be placed at the image plane of Lens 3.

1.3.7 Example: conjugate planes in modern microscope condensers

Referring back to figure 1.123, all three conditions represent suitable layouts for a modern transmission *brightfield* microscope condenser system, in which the extended object is a light source (typically an LED), while Lens 1 is often called

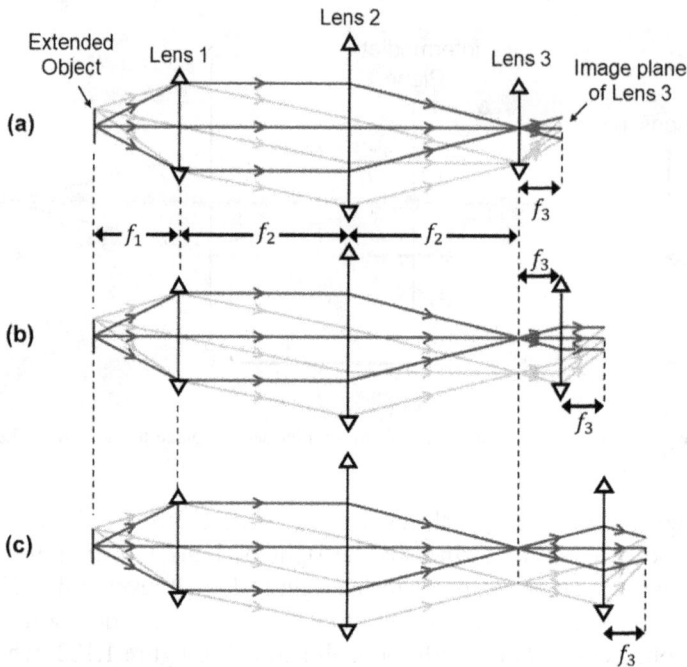

Figure 1.123. A third lens focuses rays from the infinite conjugate image of Lens 1 formed by Lens 2. (a) Lens 3 placed at the back focus of Lens 2. (b) Lens 3 shifted right by one focal distance. (c) Lens 3 placed at an arbitrary distance from Lens 2.

the *flux collector*, Lens 2 is often called the *field lens*, and Lens 3 is often called the *substage condenser*. In fact, these terms are more commonly used by microscope manufacturers (such as Nikon and Olympus) than in academic texts. And among those manufacturers, the layout shown in figure 1.123(b) is the most common in transmission brightfield condenser systems because the ray cone generated by placing Lens 3 at one focal length from the focal plane of Lens 2 can be made to match an *acceptance cone of rays* of a microscope objective, whose NA should be either greater than or equal to the NA of the ray cone produced by the substage condenser. This is illustrated in figure 1.124, where a simple modern microscope condenser system layout is assembled using COTS lenses (including the objective) and whose prescription is provided in figure 1.125. The design wavelengths are 450, 550, and 650 nm (all weights = 1), but the layout is displayed at the 550 nm wavelength. The NA of the condenser (denoted by NA_c) is 0.15, while the NA of the objective is designed to be 0.25. As shown, rays from the substage condenser are nicely transmitted through the objective and can be said to be well *matched* with the NA of the objective. In transmission brightfield microscopy, NA_c is normally less than or equal to the objective's NA. The distance between the substage condenser and the specimen plane is called the substage condenser's *working distance*.

We now discuss the conjugate plane relationships in the system above as they relate to figure 1.123(b). In figure 1.124, the plano side of the flux collector (surface 2

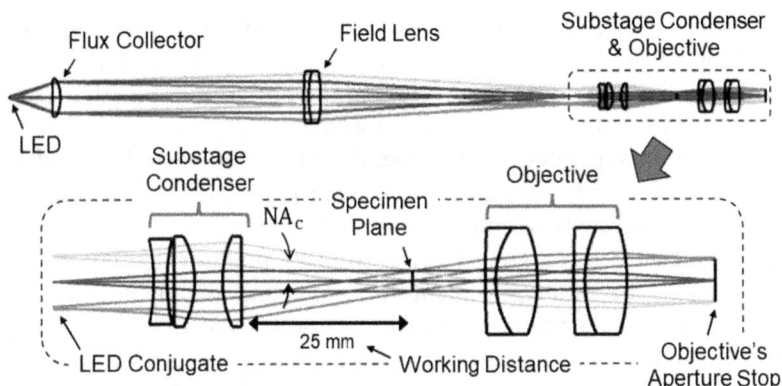

Figure 1.124. A simple modern transmission brightfield condenser system layout with an objective lens. All lenses are COTS components.

	Surface Type		Comment	Radius	Thickness	Material	Clear Semi-Dia	Chip Zone
0	OBJECT	Standard ▾		Infinity	21.21195		0.70700	0.00000
1	STOP	Standard ▾		Infinity	0.00000		7.00000 U	0.00000
2	(aper)	Standard ▾	49-901	Infinity	4.00000	N-SF11	9.00000 U	0.00000
3	(aper)	Standard ▾		-19.62000	118.89000		9.00000 U	0.00000
4	(aper)	Standard ▾	47-642	162.43000	2.40000	N-SF5	12.50000 U	0.00000
5	(aper)	Standard ▾	47-642	54.44000	6.00000	N-BK7	12.60000 U	0.00000
6	(aper)	Standard ▾		-76.28000	122.53896 M		12.50000 U	0.00000
7		Standard ▾	Int Plane	Infinity	14.25250		3.60851	0.00000
8	(aper)	Standard ▾	48-695	-25.14000	2.00000	N-BK7	6.00000 U	0.00000
9	(aper)	Standard ▾		25.14000	1.00000		6.00000 U	0.00000
10	(aper)	Standard ▾	62-564	Infinity	3.00000	N-BK7	6.35000 U	0.00000
11	(aper)	Standard ▾		-13.13000	4.00000		6.35000 U	0.00000
12	(aper)	Standard ▾	38-300	15.51000	2.70000	N-BK7	6.35000 U	0.00000
13	(aper)	Standard ▾		Infinity	25.00000		6.35000 U	0.00000
14		Standard ▾	Specimen	Infinity	10.63762		1.45656	0.00000
15	(aper)	Standard ▾	47-700	128.44000	1.50000	N-SF10	7.50000 U	0.00000
16	(aper)	Standard ▾	47-700	13.17000	6.00000	N-BAF10	7.50000 U	0.00000
17	(aper)	Standard ▾		-20.72000	5.36000		7.50000 U	0.00000
18	(aper)	Standard ▾	47-701	205.03000	1.60000	N-SF10	7.50000 U	0.00000
19	(aper)	Standard ▾	47-701	17.64000	5.98000	N-BAF10	7.50000 U	0.00000
20	(aper)	Standard ▾		-26.97000	12.97177 M		7.50000 U	0.00000
21	IMAGE	Standard ▾		Infinity	-		3.01178	0.00000

Figure 1.125. The prescription in OS for the layout in figure 1.124. Numerical values in the Comment column refer to Edmund Optics part numbers for the COTS lenses.

in the prescription provided in figure 1.125) is placed at roughly the *back focal plane* of the field lens, as the field lens is an achromatic doublet with the flint glass on the left side. This collimates the stop (surface 1 in the prescription) so that the substage condenser can project the stop onto the specimen plane. Thus, the stop and the specimen plane are conjugates. Meanwhile, the LED (surface 0) is conjugate with a plane to the left of the substage condenser. That plane is the entrance pupil of the condenser; thus, if an iris is mounted there, then the iris can control the NA of the condenser (i.e. it can make $NA_c \leqslant 0.15$). At this point, you may perhaps have noticed that this type of arrangement for projecting the light source onto the entrance pupil of the substage condenser is generally known as **Köhler illumination**, which, in the current example, has been applied to illuminate a microscope specimen plane to avoid seeing *hotspots* and *cold spots* from the light source [56] (we will discuss this illumination arrangement a little more in section 1.3.10 and in greater detail in chapter 2). Finally, note that the LED has another conjugate, which is at a plane to the right of the objective. That plane is the objective's aperture stop, which is also its entrance pupil (note that the objective would be designed the other way around, from stop to specimen plane). In summary, by examining where pairs of rays intersect, you can determine where their conjugate points are located throughout the optical system. In a first-order layout, whenever the source is the starting conjugate in OS (i.e. when it is located at surface 0), then using marginal ray solves (for which the marginal ray height and the pupil height are set to zero) at appropriate surfaces will indicate all of that source's respective conjugates, as is the case for surfaces 6 and 20 in the prescription provided in figure 1.125. Accordingly, setting the *chief ray solve* (for which the chief ray height is set to zero) on appropriate surfaces indicates the location of the pupils. For instance, if I had used a chief ray solve on surface 13, then the thickness at that surface would have been close to 25 mm (it would not be exactly 25 mm because of aberrations produced by the COTS lenses).

1.3.8 Example: a simple modern digital microscope using commercial off-the-shelf lenses

Referring back to figures 1.124 and 1.125, note that the marginal ray solves on surfaces 6 and 20 are actually not at a pupil height of zero. Rather, they are at a pupil height of 1. This is a little trick I use when I select COTS lenses quickly and know that the aberrations—especially spherical aberration—have not been fully corrected. By selecting the marginal ray height solve to be at zero at a pupil height of 1 and then displaying the ray trace in a layout using only three rays per field point (as shown in figure 1.124), the ray crossings at focal positions appear exact and therefore rather intuitively pleasing for illustrative purposes. When more ray numbers are selected for display, the spherical aberration becomes apparent, as shown in figure 1.126 using five rays per field point.

For the purpose of illuminating the bio-sample at the specimen plane in a transmission brightfield microscope system and achieving sufficient image resolution, the aberrations from the illuminator are not particularly significant.

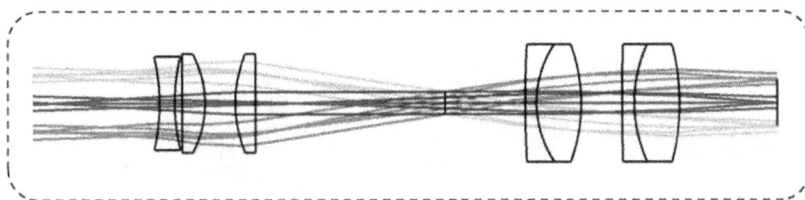

Figure 1.126. The layout of the substage condenser and objective portion of figure 1.124 using five rays per field point, which makes the residual spherical aberration apparent.

For instance, Frolov [67] expresses the view that theorists have offered contrasting opinions about the influence of the aberrations from a condenser on image resolution: some say that the aberrations have virtually no effect, while others say that a condenser's aberrations should at least be comparable with that from the pairing objective. Further, Born and Wolf [68] express the opinion that the condenser's aberrations have no impact on the resolving power of a microscope. In this author's opinion, all of the thoughts and studies above are well-justified in the following sense: in microscopy, a significant number of techniques are available to the user for viewing specimens at various degrees of image contrast, and many of these techniques involve different ways of illuminating the sample. These include such things as covering one side of the cone of rays so as to obtain just one set of rays at an oblique angle (often called *oblique illumination*) or masking the central rays such that an annulus or 'ring' of rays strikes the specimen (often called *amplitude contrast imaging*) [56, 57]. A microscopist tries all available methods to view the specimen, and the best images are selected for further study. So, there is no reason to say which technique is correct, or whether or not the aberrations in the illumination should be eliminated. The point is to get the most information out of the specimen that can possibly be offered by the illumination technique. In this way, it is therefore understandable that the layouts shown in figures 1.123(a) and (c) also provide acceptable illumination for transmission brightfield microscopy, with the exception that in some cases, rays at angles greater than that given by the objective's NA may vignette and cause some stray light. Yet, it is particularly significant to note that there should always be some rays at an angle, rather than having only incident rays that are all parallel to the optic axis of the objective. There is a good reason for this, as follows: according to the Abbe and Fourier concept of imaging, an object is like a diffraction grating (see, for example, [34, 68]). Different amounts of structure in the object diffract rays at different angles. Smaller structures diffract rays into larger angles, while larger structures diffract rays into smaller angles. So, if only rays that are parallel to the optic axis of an objective are provided as illumination (figure 1.127(a)), then high-angled scattered (diffracted) rays from fine structure in the specimen are not captured by the objective. But if the illumination is angled (figure 1.127(b)), then this enables the objective to capture those high-angled scattered rays. This is a graphical means to understand why a microscope condenser should provide some angled rays, which is enabled by providing a numerical aperture in its illumination. Thus, in cases in which there are aberrations in the

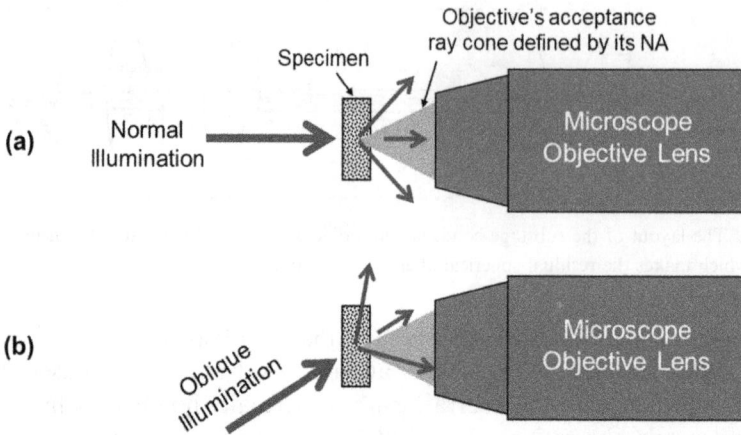

Figure 1.127. Illumination in transmission brightfield microscopy. (a) Rays arriving with normal incidence result in high-angled rays scattered (diffracted) by fine structure in the specimen, which are not captured by the objective. (b) Rays arriving with oblique incidence enable the objective to capture those high-angled scattered rays.

illumination beam, this just means that rays at various angles are present in the specimen. In view of the fact that angled rays provide *viewability* of high-angled scattered rays from fine structures in the specimen, then perhaps some degree of aberration in the illumination may be acceptable.

As explained above, I was not too concerned with the aberrations in the illumination, so I began my selection of COTS lenses by starting with a rough paraxial thin-lens layout of the flux collector and field lens. A choice had to be made for the size of the light source, which I decided should be a small LED. A common emitter size is 1×1 mm, so this was chosen, and the field heights used for the layout were at the axial position and a height of 0.707 mm, which is the half-diagonal length of a 1×1 mm square. Even if aberrations are not very important, I wanted to minimize them as much as COTS elements would allow, so I used a small aperture stop diameter of 14 mm at the left side of the flux collector. This size also allowed me to minimize the size (diameter) of the condenser system. I initially used a 25 mm back focal length for the flux collector, as this is a common dimension in lens systems (an inch is about the right dimension for many reasonably sized optics). A choice then had to be made for the EFL of the field lens. At this point, I proceeded to lay out a paraxial thin-lens model of the substage condenser. From practical experience, I knew that I eventually wanted to have the condenser designed to integrate with a 10× magnification objective (i.e. an infinity objective with a 20 mm EFL paired with a 200 mm EFL tube lens) at 0.2–0.3 NA, which are manageable NAs for COTS lenses. I also knew from experience that, in practice, it is often desirable for the NA in the illumination to be less than that of the objective (in one experience, a customer said that they preferred a lower condenser NA rather than the higher NA I had provided for them in my design, despite the fact that my original intent was to maximize resolution using high NA for the condenser).

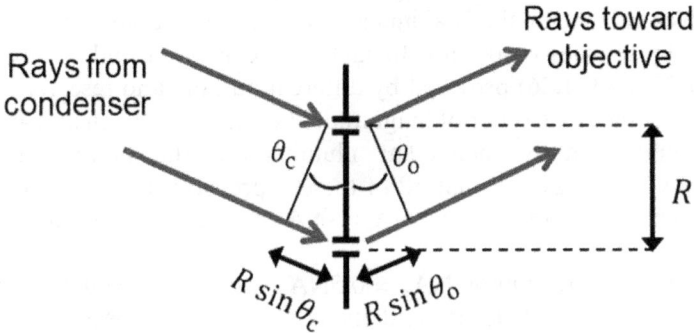

Figure 1.128. The ray geometry used to derive the relationship expressed in equation (1.25).

This point deserves some attention. A rule of thumb in microscopy for maximizing two-point resolution at the specimen plane is given by

$$R \sim \lambda/(NA_c + NA_o),\qquad(1.25)$$

where R is the smallest resolvable separation between the two points, λ is the central wavelength of the light for illumination, NA_c is the condenser's NA, and NA_o is the objective's NA [56]. The symbol $\sim$ is taken to mean that the formula is approximate and *in the order of magnitude* given by the right-hand side of the equation. This formula is based on Abbe's theory of imaging and can be easily derived by regarding the specimen as a diffraction grating such that a pair of points are like thin slits separated by a grating pitch R, as illustrated in figure 1.128.

According to the ray geometry shown in figure 1.128, it is apparent that $NA_c = \sin \theta_c$ and $NA_o = \sin \theta_o$. By interference, maximum contrast is achieved if $RNA_c + RNA_o = m\lambda$, where $m = 1, 2, 3...$etc. What value should be selected for m? It should be a value that minimizes R at the largest combination of numerical apertures. This means $m = 1$. Based on this and solving the above for R, we obtain equation (1.25). However, an in-depth analysis provided by Born and Wolf [68] for the case of coherent illumination from a condenser actually yields $R \sim 0.82\lambda/NA_o$. In contrast, equation (1.25) implies that when $NA_c = NA_o$, then $R \sim 0.5\lambda/NA_o$. Moreover, equation (1.25) provides no *guidance* nor any *hint* on the bounds for the maximum value of NA_c (so, it is a rather heuristically derived relation). Also, this formula is not the only relation for estimating two-point resolution. Hopkins [69] determined that under the condition of partially coherent illumination produced by an incoherent source, the illumination from a condenser can yield a limiting resolution given by

$$R \sim 0.58\lambda/NA_o.\qquad(1.26)$$

A discussion of the derivation of equation (1.26) (based on the work of Hopkins) is also given by Born and Wolf, which highlights that the relation given by equation (1.26) is met when the condenser's NA is roughly 1.5 times the NA of the objective [68]. This is interesting and is likely true if one sets conditions in an experiment that perfectly matches those assumed by Hopkins. But in practice, there would be far too many variables influencing the final image resolution. So, equations (1.25) and (1.26) are, at best, loose rules of thumb. In fact, you can often find different forms of equations (1.25) and (1.26) provided by different authors and researchers, and they all yield roughly the same order of magnitude. For instance, Born and Wolf [68] also show that under purely incoherent illumination, the limiting resolution is $R \sim 0.61\lambda/NA_o$. Yet despite what all of these expressions imply, my customer had expressed a preference for having $NA_c < NA_o$ (under the conditions set by this customer).

So, I eventually decided upon $NA_c = 0.5NA_o$, where $NA_o = 0.25$, partly because I was going to use COTS lenses and therefore did not expect to achieve nicely focused rays from the condenser. I then proceeded to lay out three paraxial thin lenses for the substage condenser, two of which were positive elements (to split powers) and one of which was negative (to reduce field curvature). Perhaps you would also be surprised to know that when I laid out my substage condenser *pre-design* using paraxial thin lenses, I did it *backwards* (from the specimen plane towards the field lens and flux collector). You will learn why I did this in section 1.3.20. Next, some amount of trial and error was needed to adjust the EFL of the field lens in order to match the height of the chief ray with the semi-diameter at the back (i.e. the surface that would face the field lens) of the substage condenser. When I was ready, I selected three COTS lenses whose EFLs were as close as possible to the paraxial thin lenses and optimized the air spaces to achieve the results you see at surfaces 8 to 13 in figure 1.125. I then integrated that selection with the paraxial thin-lens model of the flux collector and field lens, then selected the COTS lenses you now see for surfaces 2 to 6 in figure 1.125.

Finally, there is the objective (and eventually, the tube lens, which we will integrate using the lens from figure 1.105). As described by Zhang and Gross [70], a simple design (but perhaps not meant for very high resolution) is given by the form of the Lister or Petzval type, in which the objective comprises two doublets separated by an air space and the stop is placed in front. So, I began with two paraxial thin lenses separated by an air space, just to obtain an EFL of 20 mm. I then selected the two COTS lenses given by the part numbers at surfaces 15–20 in figure 1.125. The layout I initially used for inspecting lens performance is as shown in figure 1.129(a); it traces rays from infinity towards the specimen, yielding the MTF shown in figure 1.129(b). Note that the original stop position was 19.2 mm from the left vertex of the first doublet, but it was shortened in the condenser system layout shown in figures 1.124 and 1.125 (see the thickness of surface 20) due to aberrations in the illumination beam.

Based on the MTF shown in figure 1.129, the objective can resolve two points separated by 25 microns (reciprocal of 40 cycles/mm) at the specimen plane (because there is sufficient modulation of > 30% at 40 cycles/mm). Let us pause for a moment to consider this MTF. Upon further analysis, you would find that a non-negligible amount of third-order spherical aberration exists. Perhaps you may have heard of *spherical aberration compensator plates* [71], which are aspheres with no base radius; they have a fourth-order aspheric coefficient in order to remove a certain degree of spherical aberration from an imaging system. The concept is not new, as such

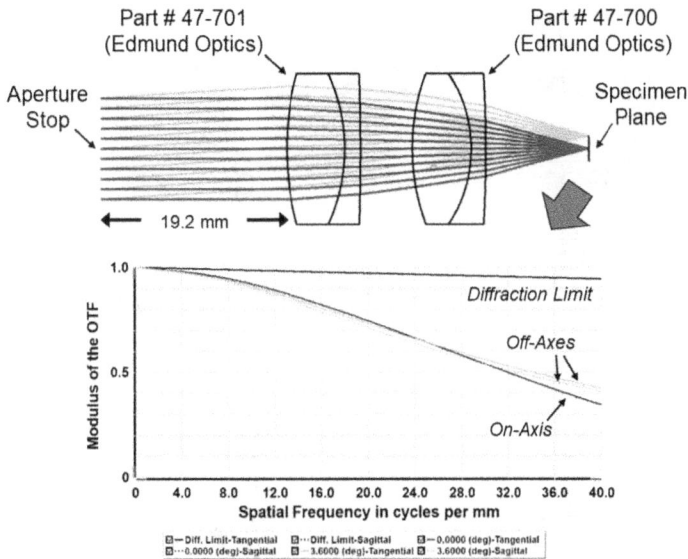

Figure 1.129. Layout for only the 10× objective and its MTF, based in part on surfaces 15 to 20 in figure 1.125. The two oblique field points are at 3.6° (half-angle). The MTF is based on wavelengths of 450, 550, and 650 nm, with weight = 1 for each.

compensator plates were already employed in the Schmidt camera as early as 1932 [5]. In theory, we can apply a customized part of such a compensator near the stop of the objective, as was done by Schmidt. Doing so yields a good improvement in the MTF, as shown in the layout and MTF plot in figure 1.130. The thickness of the plate is 3 mm and its fourth-order coefficient is $-9.749\,86 \times 10^{-06}$ at the left surface. Unless there are other coefficients present to correct further aberrations, the compensator plate needs to be near the stop (otherwise, aberrations other than spherical would be imparted to the oblique/off-axis rays, in accordance with *stop-shift* effects [6]). As an alternative, a meniscus lens can take the place of the aspheric, as is done in a *Bouwers-Maksutov* system [62]. At any rate, since we are using COTS lenses in this example, we shall forgo the use of such compensators.

When the COTS objective from figure 1.129 is integrated with the COTS tube lens from figure 1.105, the result is as shown in figure 1.131.

Note that the distance between the objective and the first element of the tube lens is 219.2 mm (i.e. the sum of the stop distance in figure 1.129 and the 200 mm pupil distance of the COTS tube lens). Note also that the MTF at the camera sensor plane at 4 cycles/mm (i.e. 10× reduced in spatial frequency due to the 10× magnification of the objective-tube lens system) matches that of the MTF at the specimen plane of the objective, which is quite good considering that all elements are COTS components. Finally, note that the conjugate planes displayed in figures 1.129–1.131 so far refer to the object–image conjugates of the microscope system. If we were to integrate the system from figure 1.131 with the condenser system from figure 1.124, how would we

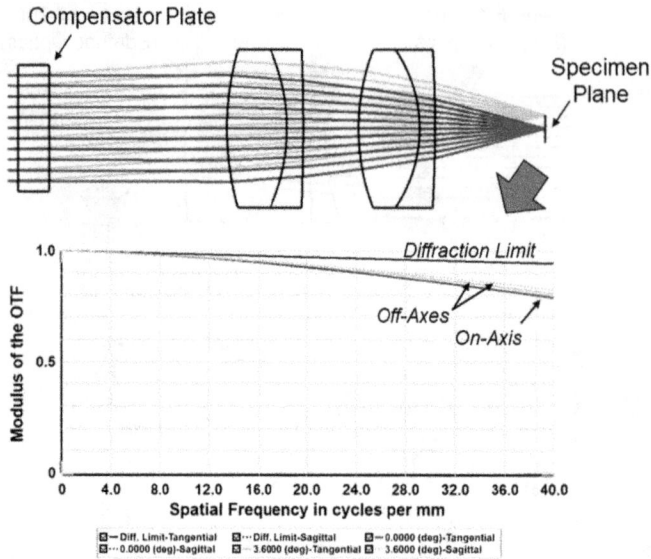

Figure 1.130. The addition of a fourth-order-only aspheric plate to reduce spherical aberration in the objective of figure 1.129, thereby improving the MTF.

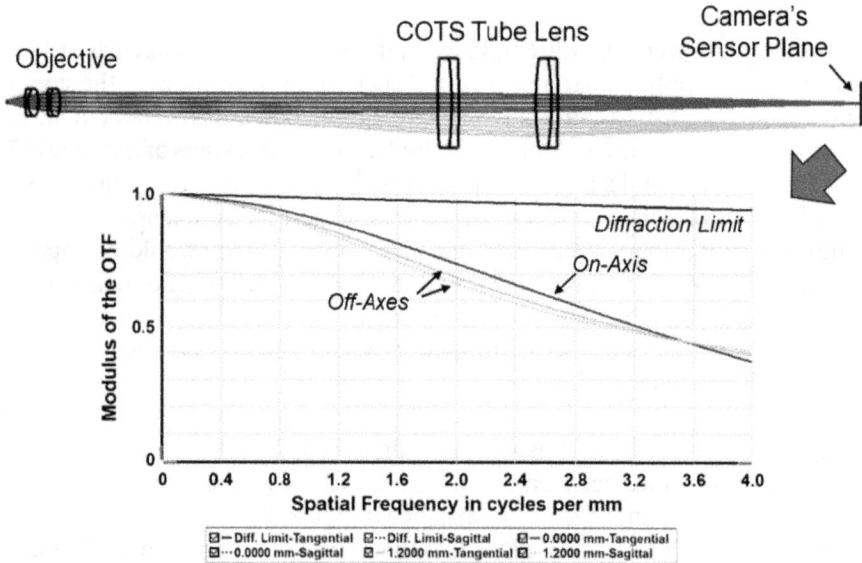

Figure 1.131. The objective from figure 1.129 integrated with the tube lens from figure 1.105, and the resulting MTF at the same wavelengths as those used for the MTF in figure 1.129.

display the object–image conjugate planes in the full layout, since, in this case, the object conjugate plane for the system in figure 1.124 is the light source, not the specimen? The answer is to optimize the surface thickness (the distance from the second doublet of the tube lens to the final image plane) such that the chief ray

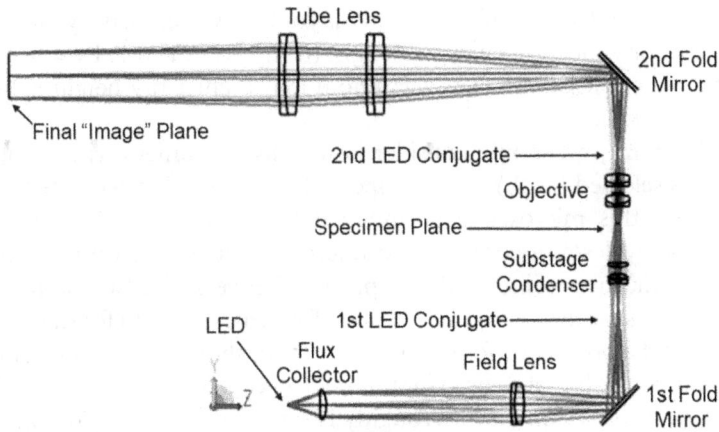

Figure 1.132. The objective-tube lens system from figure 1.131 integrated with the system from figure 1.124, with two mirrors to *fold* the path, yielding a shortened system length.

height is zero at the final image plane. The integrated layout—using two *folding (or simply, 'fold') mirrors* to reduce system length—is shown in figure 1.132.

In figure 1.132, as the layout has been set up in OS so that the LED is used as surface 0 (and the aperture stop is placed at the left vertex of the flux collector), rays begin their journey from the LED surface as point sources across a starting conjugate object plane. At surface 0, there are three field points: one at the central (axial) location and two at a height of 0.707 mm above and below the optic axis. Each field point is displayed in different colors, so wherever you see rays (three per field) of the same color converge at a plane, then that plane is a conjugate plane of the LED. If you look at the final image plane, the word 'image' is placed in quotes because, in actual use of the microscope system, while it is indeed the conjugate image of the specimen plane, it is not the image plane of the current optical layout in the sense that the object is the LED, not the specimen. Thus, from the perspective of illumination in this system layout, the final image plane is a plane of the *true field* (i.e. it should be regarded as a true conjugate of the specimen plane, which is also what we are considering as the *field*, where a field of view resides). As such, the aperture stop in this specific layout may also be considered to be the *true field stop* or perhaps the *effective field stop* in actual practice, even though the field settings used in the layout correspond to the LED, which is the source. Are you confused yet? I hope not. This is the reason why I mentioned in the title of section 1.3.5 that *if you master the concept of conjugate planes, you can go very far*. That is, if you master this concept, then you will never be confused about what goes on in a semi-complex optical system such as this all-COTS modern digital microscope. Let us not get confused by converging rays at other conjugate planes. Recall that I mentioned that rays of the same color in this layout converge at planes that are conjugate to the LED, which are indicated by the first and second LED conjugate planes labeled in figure 1.132. Note that *different colored rays* can also be seen to converge at specific planes, such as the final image plane. At such planes, they are conjugate with the *true field stop* and the specimen plane. In OS, the distance between the last tube lens element and the final image plane in this layout was obtained by optimizing the thickness there for which the chief ray height (using REAY = 1, Py = 1, and Px = 0 in the Merit Function Editor) was zero (note that a chief ray height solve was not used, due to aberrations).

As a final remark, we should consider how a digital camera with an appropriate image sensor is selected for this microscope system. To do this, we note that the two-point resolution this microscope is intended for is roughly 25 microns (i.e. the reciprocal of 40 cycles/mm converted to microns). Therefore, we need at least one pixel at one of the two points and one pixel in between the two points (this is an application of the so-called Nyquist sampling frequency [10, 26] for digital cameras). This means that the pixel pitch for the chosen camera should be roughly half of the 25-micron two-point resolution desired for this microscope system, which is 12.5 microns. This implies the use of a camera with roughly 12-micron pixel sizes. Anything less than this size would not be helpful for this current microscope design, as it would just help you to sample a PSF of this system, which you should not be interested in (unless you are). On the other hand, many modern CMOS sensors today have pixel sizes less than 10 microns and a reasonable price, so you might as well not worry too much about having smaller pixels (in case you do become interested in the PSF at some point). The field of view at the final image plane is a circle of roughly 25 mm, so, assuming that we are fitting a square into this circle, the sides of this square are roughly $25/\sqrt{2} \approx 17.68$ mm. Since 17.68 mm divided by 0.012 mm (the size of a pixel in mm) is roughly 1473 pixels, the required camera resolution in terms of the number of pixels is roughly 1473 × 1473 or 2.2 megapixels. Different cameras have different aspect ratios (i.e. the ratio of the long side to the short side of the image sensor dimensions), so you just need to pick a sensor that fits inside this image circle and has roughly 2.2 megapixels.

1.3.9 Example: locating and modeling dust artifacts in imaging systems

In many situations, dust and other artifacts may appear in the image of an optical imaging system. Dust, for instance, cannot wholly be avoided if the imaging system is not assembled in a perfectly clean room (which would be costly) and if the imaging system is not fully enclosed in a tight housing. In some cases, dust artifacts may reside in the illumination subsystem of the imaging system, such as the condenser assembly in figure 1.124. In fact, due to the location of the field stop at the left surface of the flux collector (which is a conjugate plane of the specimen area), any dust settling on that surface will appear at the specimen plane. If this dust is inaccessible and cannot be cleaned out, then a temporary fix could be to simply defocus the illumination such that the dust is located at a different conjugate. Fortunately, in the condenser system of figure 1.124, this can be accommodated by shifting the field lens towards the left, as shown in figure 1.133, where the field lens has been shifted by half its original length, yielding a image of the field stop defocused by about 2 mm relative to the specimen plane (whilst maintaining the focal distance between the specimen plane and the objective). This is one benefit of the current condenser system design, in which the LED is collimated by the flux collector such that its first conjugate plane always remains at the left focal plane of

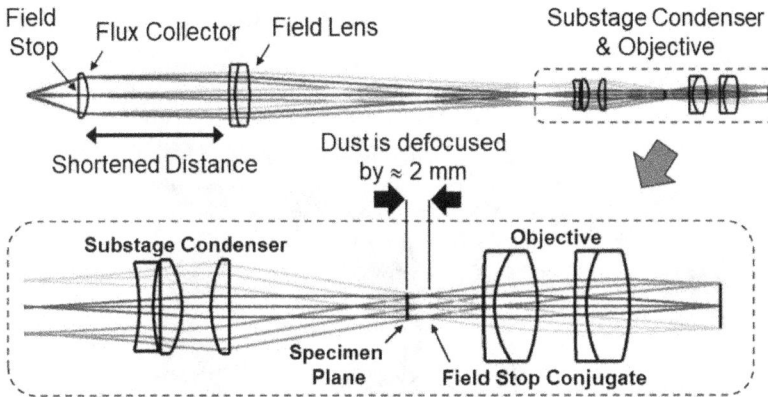

Figure 1.133. Shifting the field lens to the left by half of its original length has the result of defocusing the field stop by ≈ 2 mm to the right of the specimen plane.

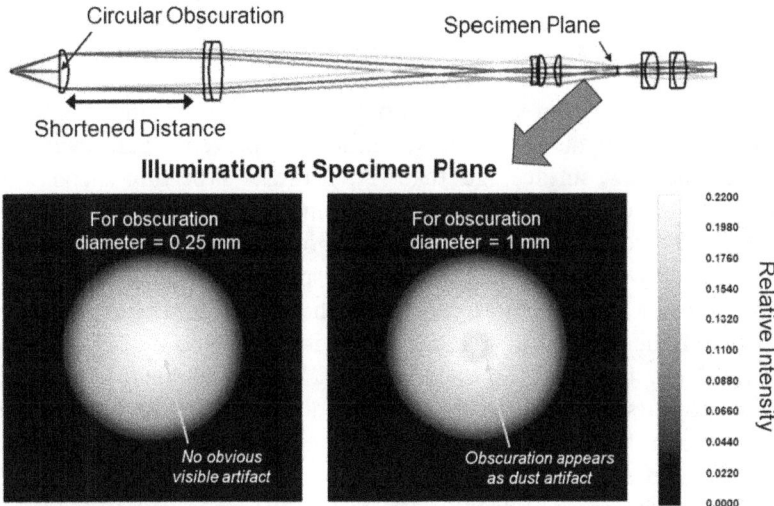

Figure 1.134. GIA simulation in OS for the system in figure 1.133, in which a central circular obscuration is inserted at the left vertex of the field stop to model a dust particle.

the substage condenser, thereby allowing the geometry of rays to maintain their convergence and *acceptance* by the objective.

An inspection of the objective alone will reveal that its depth of focus is approximately ±0.2 mm (i.e. defocusing the specimen plane by up to this amount results in a drop in the objective's MTF to near-zero modulation), so the dust would not be observable if it were defocused by 2 mm to the right or left of the specimen plane, depending of course on the size of the dust. What maximum size is permissible? There are no definitions for this, other than constraints imposed by the needs of your application. Still, we can model this size effect of a dust-like particle using the GIA feature in OS (see section 1.2.10) and inserting a *circular obscuration* at the left vertex of the field stop, as illustrated in figure 1.134. To do

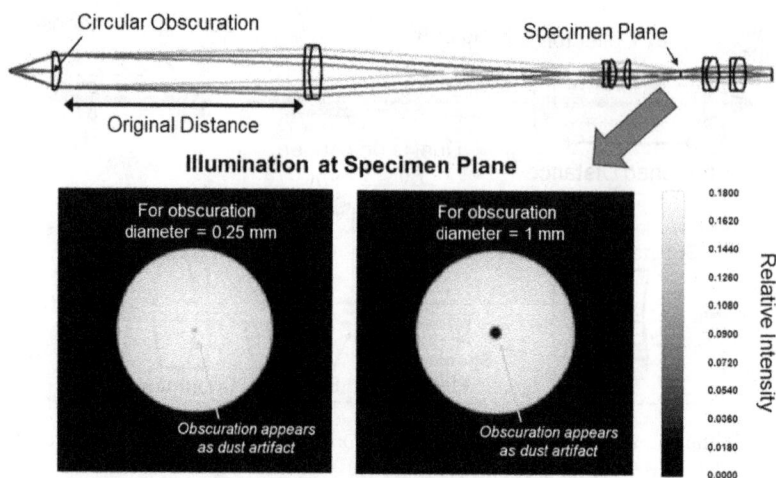

Figure 1.135. GIA simulation in OS for the original system in figure 1.124, in which a central circular obscuration is inserted at the left vertex of the field stop to model a dust particle.

this, insert a surface between surfaces 1 and 2 in the prescription in figure 1.125 (of course, first ensure that the thickness of surface 3 is halved), then open the surface properties of the new surface, go to > Aperture > Aperture Type > Circular Obscuration, enter 0.125 (mm units) to obtain the effect shown on the left of figure 1.134, and enter 0.5 to obtain the effect shown on the right. As this is a purely geometrical ray-trace simulation, the effects displayed do not account for diffraction. As a means of comparison, figure 1.135 shows the simulation result when the distance between the flux collector and field lens is restored to the original distance of 118.89 mm. Note that for the case of a 0.25 mm diameter obscuration, the illumination appears dust-free in the defocused system (figure 1.134), whereas in the original system (figure 1.135), the dust artifact remains. However, it is acknowledged that the uniformity is worse in the defocused system, which, in the current example, may be considered a trade-off effect.

In reality, any surface in the optical ray path may be subject to the settling of dust particles of various sizes, resulting in some level of artifacts that may be observable at the camera plane, regardless of whether or not that element's surface is conjugate to that plane. To assess the impact of such conditions, circular obscurations of various sizes may be inserted next to any surface of interest along the ray path. If other shapes for obscuration are not provided in OS's SC mode, then one may convert the entire optical model into a NSC model, then insert obscuring objects of various shapes at those surfaces to see how they may appear at the camera plane (where you would create a model of a detector). Moreover, in NSC mode, the obscuring objects may be made into sources. In this way, one can analyze the impact of bright light-scattering (or fluorescent) artifacts rather than just light-absorbing artifacts. In fact, some dirt or foreign particles are fluorescent when excited with light at some wavelength. As such, the use of a *fluorescence detection system* [72]

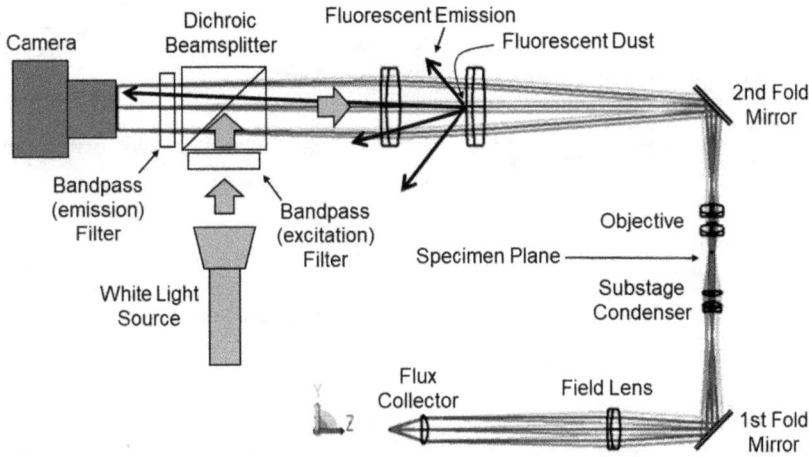

Figure 1.136. The insertion of a white light source, dichroic beam splitter, and bandpass filters to serve as an external fluorescence-excitation system for inspecting dust.

coupled into the optical system in an appropriate way would enable such particles to be detected. In the lab, one may inspect for the presence of such particles by inserting a white light source and filters in the manner shown in figure 1.136. Of course, this example may be specific to the current microscope system being modeled, but the principle can be applied to other general optical systems as long as a suitable space can be identified for the insertion of an external fluorescence-excitation system for the inspection of potentially fluorescent dust and other foreign particles.

1.3.10 Conjugate planes in a classical projector

Here, a classical projector is taken to mean an optical system that projects *slides* onto a screen [62]. If you are of the current generation of early-career designers but did not receive formal training in optics, then you may not know what slides are. They are *transparencies*—but you may not know what these are either. Okay. In the old days, long before the iPhone®, the iPad®, digital cameras with CCDs and CMOS image sensors, and digital office projectors that connect with a personal computer, there was the camera and a thing called *film* [73, 74]. If you are a fan of photography, you already know this, but otherwise, the thing called film may not be immediately obvious to you. Cameras for the masses used to have film in place of digital image sensors (but Hollywood and avid photographers still use film). Just think of the film as some plastic-like transparent material with images captured on it from the camera lens. Such film can be made into a thing called a photographic *slide*, which is itself transparent and is inserted into a *slide projector* (which I am calling a classical projector here) that comprises a light source, a condenser lens, and the projection lens responsible for forming the image of the slide on a screen, as illustrated in the first-order layout of such a system in figure 1.137, whose prescription in OS is provided in figure 1.138.

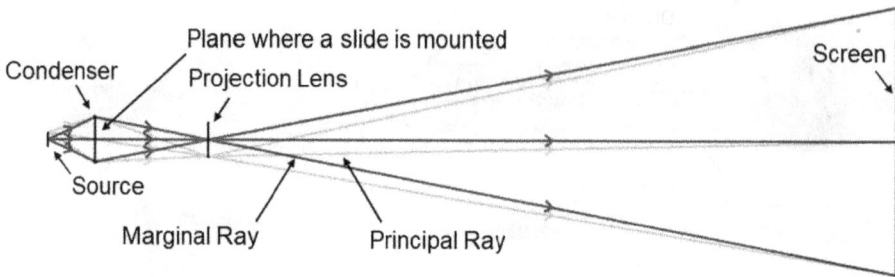

Figure 1.137. The first-order layout of a classical projector.

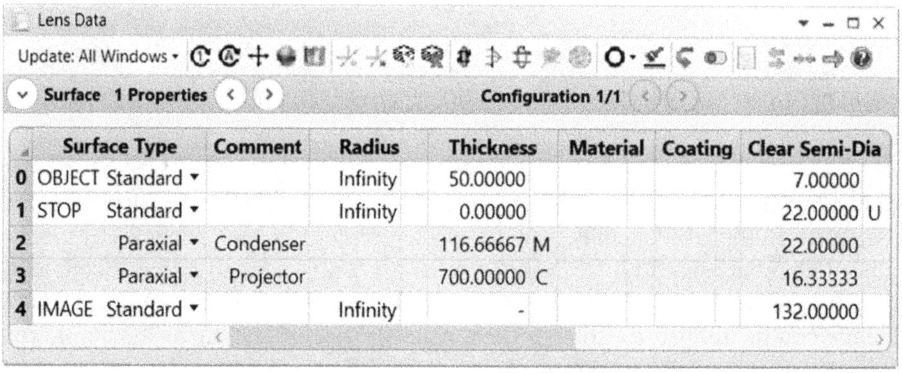

	Surface Type		Comment	Radius	Thickness	Material	Coating	Clear Semi-Dia
0	OBJECT	Standard ▾		Infinity	50.00000			7.00000
1	STOP	Standard ▾		Infinity	0.00000			22.00000 U
2		Paraxial ▾	Condenser		116.66667 M			22.00000
3		Paraxial ▾	Projector		700.00000 C			16.33333
4	IMAGE	Standard ▾		Infinity	-			132.00000

Figure 1.138. The prescription in OS for the layout shown in figure 1.137.

In this model, the field setting is at an arbitrary height of 7 mm. The paraxial thin-lens models are at EFLs of 35 mm and 100 mm for surfaces 2 and 3, respectively. As shown in the prescription, a marginal ray solve (ray height = 0, pupil zone = 0) is used on surface 2, and a chief ray solve (ray height = 0) is applied to surface 3. By now, you may have already become somewhat acquainted with my use of these two types of solves in OS. You saw the marginal ray solve being used in previous sections, but I only mentioned the chief ray solve without using it in section 1.3.7. Here, in figure 1.138, you can see the chief ray solve at work for the thickness of surface 3. I am hoping that as I reiterate the use of these solves in multiple examples involving the concept of conjugate planes in optical system layouts, you will become more and more familiar with this fundamental concept and its application. Let me try to drive you further towards a major trick in optical system layout, as follows. In the case of the classical projector, the condenser is the name given to the projector's flux collector, and it forms the image of the source on the entrance pupil of the projection lens. Hence, it follows the principle of illumination utilized in figures 1.123 and 1.124, which is often called **Köhler illumination** [75]. The key point to this arrangement of elements for illumination is that the plane of the slide is not conjugate to the light source, so you do not see the source focused together with the image of the slide containing the scenery and images captured by the camera. This is a good thing to do because the light source may have some structure (such as the filament of a tungsten bulb) that you do not want to see on the screen. This is another example of the application of the concept of conjugate planes. In a layout of an illumination system based on the Köhler principle, there is the opportunity for me to highlight that there are often two main pairs of conjugate planes of interest in any optical imaging system: a conjugate pair for the system's object (which is the source in the current example) and its image and a conjugate pair for the pupils of the system. In the Köhler illumination method, the pupil conjugates are the slide and the screen. Since the principal ray (which is an unvignetted chief ray) always crosses the optic axis at the pupils (i.e. the stop, the entrance pupil, and the exit pupil) of an optical imaging system, the chief ray solve provides the means to determine the location of the screen, which is the exit pupil of the system. Hence, this solve is used on the thickness at surface 3. In such a system, the source (object) is conjugate with the *entrance pupil of the projection lens* (note that this pupil is not regarded as part of the pupils *of the system*), hence, the marginal ray solve is used to determine the distance between the object (the source) and its image (the entrance pupil of the projection lens). Master these concepts, and you will be able to design complex optical systems.

1.3.11 Object and image conjugates at the same location

As another useful example of applying the concept of conjugate planes, consider the layouts of the so-called *Bravais* optical systems shown in figures 1.139(a) and (b).

Bravais systems are characterized by having the object and the image at the same location [62]. For instance, in figure 1.139(a), since the central glass thickness is equal to the radius on the right-hand side, axial rays from the object are coincident with themselves when they exit the lens (projecting the rays backwards yields the image conjugate, which is the location of the object). In figure 1.139(b), axial rays converging from the left-hand side are aimed towards an axial point at the far right,

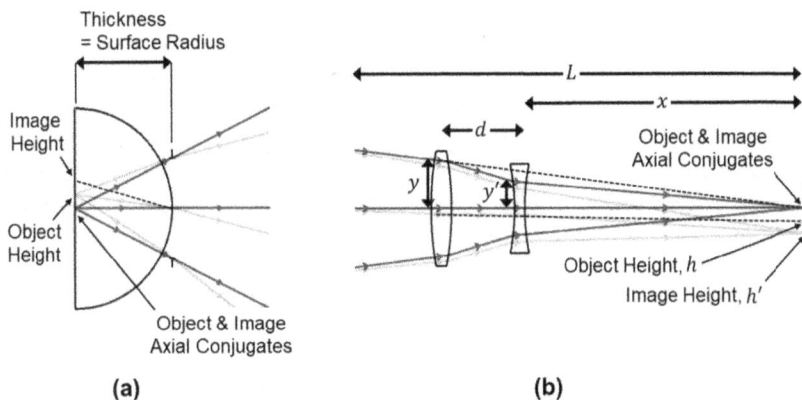

Figure 1.139. Two examples of Bravais optical systems. (a) A Bravais hemispheric lens. (b) A Bravais telephoto lens system.

which is the object conjugate (given by extending the top ray, as illustrated by the dashed black line). But after refraction through the positive and negative lenses, these rays converge again at the same axial location, which is the image.

Bravais systems are also seen to magnify the image. For the hemispherical lens in figure 1.139(a), the magnification is given by the lens's refractive index *on the way out* (i.e. if the object is in the glass). This relation is derived by noting that if h is the height of the object in the lens, h' is the height of the image, n is the index of the lens, and R is the radius of the hemisphere, then applying Snell's law in the paraxial approximation gives $n(h/R) = h'/R$. Solving for the index yields $n = h'/h$, which is the magnification.

For the system in figure 1.139(b), assuming thin lenses, the governing first-order equations can be presented as

$$\frac{1}{f_1} = \frac{x(m-1)}{md(d+x)},$$ (1.27)

and

$$\frac{1}{f_2} = \frac{(1-m)(d+x)}{xd},$$ (1.28)

where f_1 and f_2 are the EFLs of the positive and negative thin lenses, respectively, d is the separation between them, x is the distance from the negative lens to the image, and m is the magnification of the image (note that $f_2 < 0$) [62]. However, there is actually an additional nonobvious constraint given by

$$m = \frac{y}{y'}\left(\frac{x}{d+x}\right),$$ (1.29)

where y and y' are given by the heights of the marginal ray at the surfaces shown in figure 1.139(b). Further, L has to be maintained, otherwise we would not have a Bravais system. Equation (1.29) is derived by noting that, to the first order, the transverse magnification h'/h of the image is equal to dividing the angle $y/(d+x)$ by the angle y'/x (think of a telephoto lens). In these formulas, there are too many variables preventing a closed-form solution if it is necessary to make specific choices for all variables. A practical way to apply equations (1.27)–(1.29) is to select numerical values for three of the variables, optimize the rest, and iterate as necessary. For instance, if you should decide on specific values for f_1, f_2, and m, you must let x, d, y, and y' be variables for design optimization in a program, whilst maintaining L. A suitable rule is begin with a value for f_1 that is larger than $|f_2|$. To see this, we combine equation (1.27) with (1.28) and solve for m to obtain

$$m = -\frac{f_1}{f_2}\left(\frac{x}{d+x}\right)^2.$$ (1.30)

Clearly, in order to obtain an enlarged image, m has to be greater than 1. Since $x/(d+x)$ is necessarily less than one, we must have $f_1 > |f_2|$. Note also that equations (1.29) and (1.30) indicate the constraint that if one has selected values for f_1, f_2, and m, then x, d, y, and y' must be free. For a pair of thick positive and negative lenses, the distances x and d can become the air spaces between the lenses, and the total length L must be constrained to include the lens thicknesses (a simple design example is provided in appendix A.2).

A Bravais system of the type just described is useful if magnification is needed in the image without changing the location of the original image. **However, this added magnification is actually traded off against reduced contrast and resolution, due to the reduction in the NA of the image**. Some call this *empty magnification* [56]. Still, note that the reduction in the resolution applies to the diffraction-limited case in the image. In some cases, the image's MTF may not be diffraction limited, and in such cases there is neither loss nor gain in the image resolution at the higher magnification, other than the possibility of some reduction in image quality due to added or residual aberrations from the Bravais system that may remain uncorrected.

The hemispherical Bravais lens type shown in figure 1.139(a) can also be quite useful. For one thing, it is free of spherical aberration and third-order coma (so, the image is **aplanatic**), and it is also free of axial color [5]. Note that an LED whose emitter is encapsulated in a hemispherical dome lens is an approximate example of such a Bravais system, though in this case, the application is only incidental. The main purposes of the dome lens for LEDs are to encapsulate the die and wire bonds so as to prevent breaking of the wires, to aid in light extraction efficiency, and to shape the output distribution of rays [76]. Still, it is helpful to note that such encapsulated LEDs are Bravais-like systems, because when you want to model the output of rays from the LED but do not have access to the so-called *ray files* obtained from goniophotometers, you can simply model an emitter (without the dome) whose size is given by n times the actual emitter's size (where n is the assumed refractive index of the dome) and apply the radiant intensity data provided in the LED manufacturer's datasheet to that modeled emitter (we will talk more about this in chapter 2).

At this point, you may realize that I like citing some old references, dating, for instance, back to Rudolf Kingslake's classic text called *Optical System Design*, published in 1983 [62]. The reason for this is that many old texts discuss some ingenious methods and concepts for designing and analyzing optical systems. Many of these have either been forgotten, or they are being used in optical shops that have little visibility to modern designers who are too busy to pay attention to them. In addition, today's powerful optical design programs make it very easy to model optical systems without having to apply old techniques that were developed to aid in design (e.g. with certain exceptions, the so-called Delano diagrams—sometimes also called *y, y-bar* diagrams [77]—have become virtually nonexistent in a modern designer's bag of tricks). Some old principles are worth noting, if for no other reason than to gain further insight into the workings of optical systems. For instance, there is the quantity called the *hiatus* of a lens system, which is the distance between the principal planes [62]. In the cited reference, Kingslake describes a formula for the hiatus of a singlet. If a singlet is sufficiently thin, then for a 'typical' lens index of ~1.5, the hiatus is roughly a third of the lens thickness. This is something worth remembering because it means that for any thin lens (i.e. no matter whether it is biconvex, plano-convex, biconcave, plano-concave, positive meniscus, or negative meniscus), having determined the location of one principal plane, one knows that the next principal plane is near the first. Other interesting techniques involve the determination of the EFL of a lens on a lab bench and mechanisms for maintaining image focus while varying the magnification of the system [62].

1.3.12 If you master the concept of pupils, you will understand what detectors 'see'

The key point to note from the previous few sections is that whenever an SC ray trace model in a program begins with the object as the starting surface to be imaged, then the use of marginal ray solves in the program determines the conjugate positions of the object and image, while the use of chief ray solves determines the conjugate positions of the pupils (including, of course, the stop position). But in cases in which the starting surface is a light source for illumination, then provided that the Köhler illumination technique (which is by far the most common technique) is employed to illuminate the object to be imaged, then the marginal ray solve determines the conjugate positions of the pupils, while the chief ray solve determines the conjugate positions of the object to be imaged. This reversal of the roles of the marginal and chief ray solves is key to modeling the complete chain of rays for complex optical systems that integrate the illumination source with the imaging system. The digital microscope system modeled in section 1.3.8 was an example of this.

Once you have mastered the above first-order ray tracing technique of using marginal and chief ray solves, many other types of optical systems become fairly straightforward to model in a first-order layout. It is particularly significant to have some idea of what a detector plane 'sees' in the ray path of an optical system. For example, the camera mounted on the left-hand side of the digital microscope model in figure 1.136 sees just the image of the specimen at the plane of the image sensor, as it is a camera without its own lens (imaging is provided by the tube lens). But some detectors, such as your eye, a digital camcorder, a digital photographic camera, or a mobile phone camera, have lenses. Often, such 'eyes' must view a scene provided by an optical instrument, such as a telescope or an *afocal attachment* (for example, the use of a mobile phone's camera to photograph the image produced by a micro-scope). What does the camera/detector see? How are lens systems combined in such a way that a suitable image of the scene of interest is captured? These questions are addressed in the following sections.

1.3.13 Example: relay lenses (you will often need them)

Figure 1.140 shows a system of two lenses with equal EFLs; the first lens (labeled the 'imager') forms an intermediate image, and the second lens (labeled the 'relay') projects the intermediate image onto the final image plane. This is what a relay lens does, but, as can be seen in the system of figure 1.140(a), some off-axis rays have been vignetted. For this system, reducing the stop of the imager would not help, but if the stop were shifted left by one focal length, the principal ray would be aimed

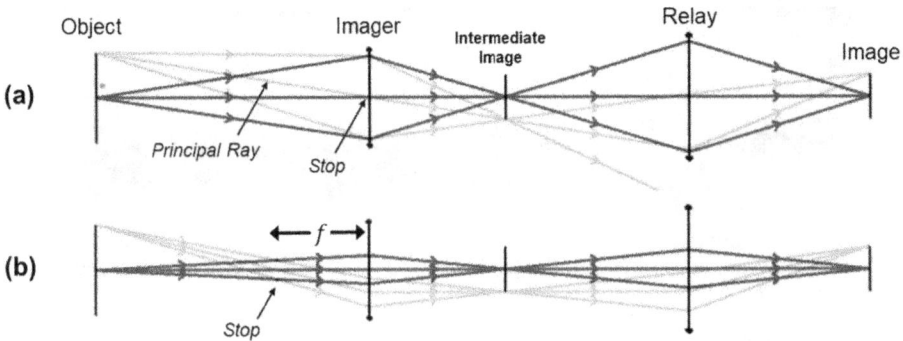

Figure 1.140. A lens system comprising two lenses, where the first forms an intermediate image, and the second relays this image onto the final image plane at the far right. (a) Off-axis rays are vignetted (clipped off) at the relay. (b) The imager is made telecentric in its image space such that all rays pass through the relay and onto the final image.

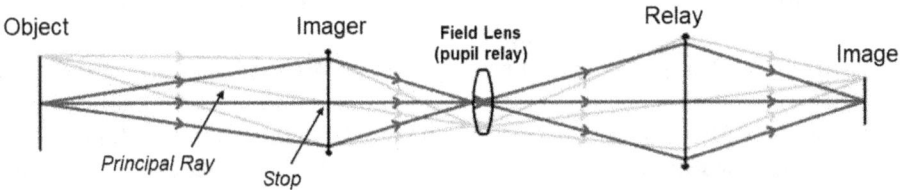

Figure 1.141. Inserting a field lens at the intermediate image of the imager in figure 1.140(a). The system has been slightly lengthened by the field lens.

towards its center, yielding an imager that is *image space telecentric*, which helps to let through all the rays, as illustrated in figure 1.140(b). However, in many cases, you may not have the luxury of customizing the imager to make it telecentric. Further, perhaps the imager's aperture must be maintained large in order to obtain a sufficiently bright image. The solution in this case is to insert a *field lens* at the intermediate image (figure 1.141) such that the principal ray is bent towards the center of the relay's entrance pupil (i.e. the center of the relay's entrance pupil is a conjugate point to the center of the imager's exit pupil). Note that the field lens is itself a relay lens in that it relays the exit pupil of the first lens into the entrance pupil of the second. Hence, it is a *pupil relay*.

What does the final image look like in all three cases above? Another way to ask this is: what does the imaging relay (not the field lens/pupil relay) see, and why should we be interested in this? The answer to the latter question is that in many practical cases of using *visual optical systems* (such as in using telescopes and binoculars with our eye), the placement of the imaging relay in the above systems is analogous to the condition of placing our eye at the eyepiece of a telescope or any other visual optical system. I want to show that when it comes to the *viewability* of the intended scene of interest in the final image, it is required that we place our eye at the exit pupil of the system in front of us (so it follows that in the case of using a camera with a lens rather than our eye, we

Figure 1.142. GBI analysis of the scene from figure 1.56(b) at the final image plane for (a) the system in figure 1.140(a), (b) the system in figure 1.140(b), and (c) the system in figure 1.141.

must place the camera's lens at the exit pupil also). To answer the former question of what the final image would look like, we may use OS's Geometric Bitmap Image (GBI) analysis feature/tool. In contrast to the Image Simulation tool discussed in section 1.2.10, the GBI tool geometrically traces rays from pixels that sample the scene to be simulated rather than performing a convolution of PSFs. In some cases (especially when analyzing the effects from vignetting), the GBI is more realistic than the Image Simulation tool. Also, the difference between the GBI and GIA tools is that the latter cannot display simulations of bitmap image files. So, using the GBI tool, figures 1.142(a)–(c) display a model scene at the final image plane for the respective systems in figures 1.140(a) and (b), and 1.141. You may find it fascinating that figure 1.142(c) displays the scene within an enclosed circle, despite the fact that we used a field lens to relay the off-axis rays into the second lens. This is due to the aspect ratio of the bitmap file used, whose horizontal field is beyond the current field of view of the layout in figure 1.141. Thus, the final image of this extended scene has included the finite diameter of the field lens, which is clipping off the field rays (this is often what you see if you hold up a magnifying glass with your arm extended such that a real but inverted image of a scene is formed as an intermediate image between the magnifying glass and your eye). The problem in figure 1.142(c) can be fixed by increasing the field lens's diameter, as shown in figure 1.143.

Often, the addition of a field lens introduces field curvature, especially if the field angle for the principal ray is large (note that in the case of some unavoidable vignetting, the *chief ray* is the ray to be relayed by a field lens—see section 1.2.7 for a review of the difference between the principal ray and the chief ray). The reason for this is that a field lens is primarily a positive-powered element or lens system, and field curvature is a function of the sum of the surface powers. The solution to this is to include a negative-powered element near the final image plane to serve as a *field flattener* [5, 64, 78]. An example is provided in section 1.4.2.

1.3.14 Example: pupil and scene visibility in a terrestrial telescope

Seeing through visual optical systems can be analyzed using a similar approach to that used for imaging through pupils in the previous section. Figure 1.144 shows a

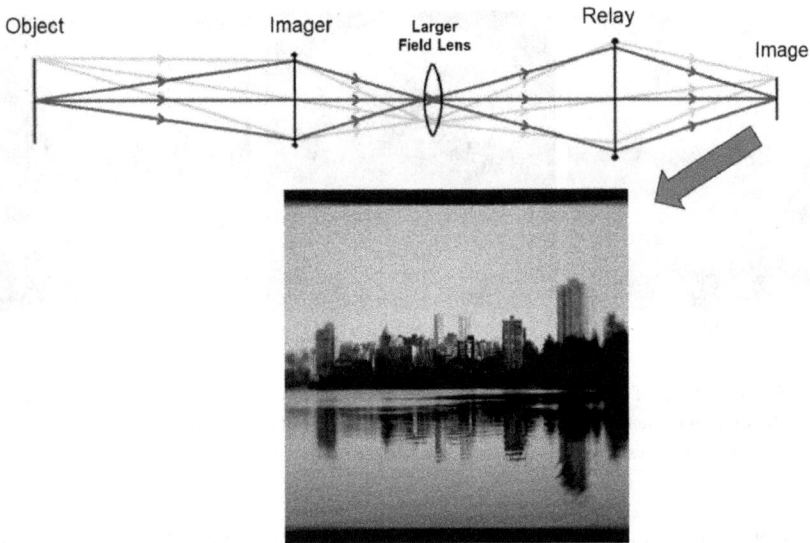

Figure 1.143. A larger field lens for the system in figure 1.141 yields a full field of view.

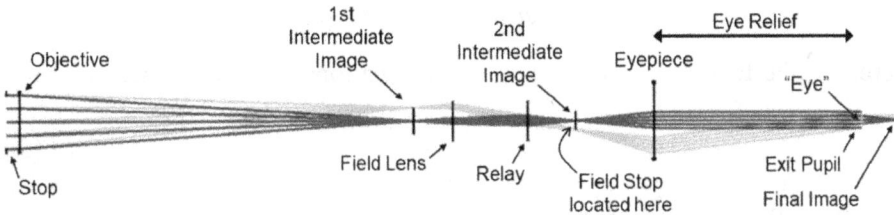

Figure 1.144. A first-order layout for a terrestrial telescope.

first-order layout for a terrestrial telescope of the type often used in a riflescope [79]. This model yields a magnification of roughly 3.5× and an erect image. There is also a field stop, modeled as a floating aperture in OS. Figure 1.145(a) shows a GBI-simulated image of a scene (based on a photograph taken using my mobile phone camera along a trail at Vancouver's beautiful Stanley Park) as viewed by just the human eye, which is modeled as a paraxial thin lens with a 4 mm diameter pupil (the iris). This same eye model is used in figure 1.144. Figure 1.145(b) is a simulation of the scene at the final image plane of figure 1.144. In this figure, the eye is at the exit pupil of the telescope, which is where it should be. In figure 1.145(c), the eye has been shifted 25 mm away from the exit pupil, resulting in vignetting.

1.3.15 More on pupils: Max Berek's 'forgotten' formula

In an insightful article concerning the irradiance (flux per unit area) of images, Vladan Blahnik of Carl Zeiss AG [80] shared a formula that he said was attributed

Figure 1.145. Simulations of a scene from a photograph. (a) The naked-eye view. (b) The view through the telescope of figure 1.144 when the eye is placed at the exit pupil. (c) The view when the eye is shifted 25 mm to the right of the exit pupil, resulting in vignetting. Courtesy of Ronian Siew.

to Max Berek, who had derived it in a book published in 1930 [81]. This formula relates the EFL and **pupil magnification** of a lens to the object and image distances *measured from the entrance and exit pupils, respectively*, which may be written as

$$-\frac{1}{\beta_p q} + \frac{\beta_p}{q'} = \frac{1}{f},$$
(1.31)

where f is the EFL and β_p is the pupil magnification, which is defined as:

$$\beta_p = \frac{D_{ex}}{D_{en}}.$$
(1.32)

Here, D_{ex} is the diameter of the exit pupil, and D_{en} is the diameter of the entrance pupil. The dimension q is the distance between the object and the entrance pupil, and q' is the distance between the image and exit pupil. Note that these should not be confused with the distances in equation (1.20), which are measured from the first and second principal planes, respectively.

Blahnik remarks that equation (1.31) 'has fallen into oblivion in many modern optics texts' [80], which seems to be the case, as not much can be found on it other than an application by Blahnik for estimating the impact of defocus on spot size and other optical properties in a smartphone [82]. This application of the formula is quite helpful, as many non-first-order-related performance characteristics of imaging systems are quantified in terms of the pupils of optical systems rather than the principal planes. This includes image irradiance, relative illumination, and aberrations (which, in the context of wavefront error, are optical path differences at the exit pupil). Moreover, as I will show in section 1.3.18, knowledge of β_p is helpful for applications involving depth sensing (distance determination).

Given the significance of equation (1.31), it makes sense to know how to derive it from first principles. To do this, we refer to figure 1.146, where certain approximations are implied concerning the **aplanatism** of two image positions, one at point F

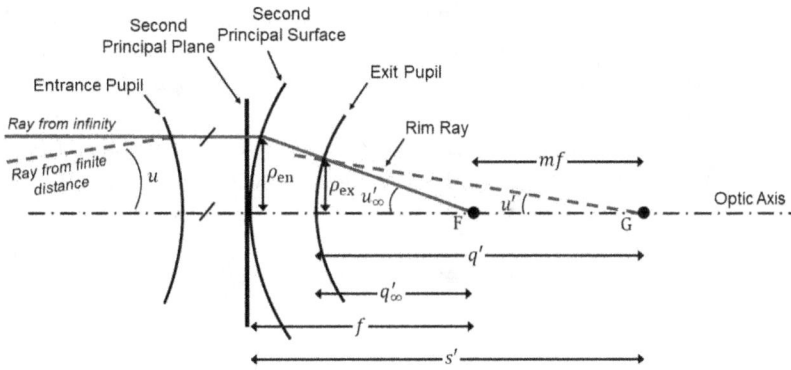

Figure 1.146. The construction used for deriving equation (1.31).

(the second focal point, which is the image of a point object at infinity) and the other at G (the image of a point at a finite object distance). An aplanatic image is free from coma and spherical aberration. However, since no two objects at different locations along the optic axis can be imaged perfectly [66], it follows that there cannot be two aplanatic image locations (aplanatism is discussed in further detail in section 1.4). But if a lens system can function sufficiently well over several object distances (which is the case for most well-designed lenses), then we can make the approximation that there are roughly aplanatic images at various locations in image space. Using this assumption, the magnification m of the image at G may be expressed as

$$m = \frac{\sin u}{\sin u'} = \frac{\text{NA}}{\text{NA}'} = \frac{s'}{s}, \qquad (1.33)$$

where $\text{NA} = \sin u$ is the numerical aperture in object space, $\text{NA}' = \sin u'$ is the numerical aperture in image space, and s is the distance between the object and the first principal plane, which is not shown in the illustration. If q (not displayed in the figure) is the distance between the object and the entrance pupil, then due to the aplanatism in the image, the marginal ray is a **rim ray** from the exit pupil that converges towards G, and we can regard the image magnification as being given by $m = (\rho_{en}/q)/(\rho_{ex}/q') = (D_{en}/q)/(D_{ex}/q')$, where ρ_{en} and ρ_{ex} are the semi-diameters of the entrance and exit pupils, respectively. Applying this result to equation (1.32) gives:

$$\beta_p m = \frac{q'}{q}. \qquad (1.34)$$

It shall become clear that equation (1.34) is useful. Next, we note that the distance between the exit pupil and the image at G is

$$q' = q'_{\infty} + mf, \qquad (1.35)$$

1-137

where mf is the displacement of the image from F to G when an object at infinity is shifted to a finite location indicated by the angle u for the dashed blue ray in figure 1.146. To justify that this displacement is the correct amount, we combine equations (1.33) and (1.20) by eliminating s in both equations to yield $1/f = (1 + m)/s'$. Solving for s' we have $s' = f(1 + m) = f + mf$. Thus, the displacement from F (which is at a distance f from the second principal plane) to G is mf. Next, due to the aplanatic condition for the ray from infinity, we note that

$$\sin u'_\infty = \frac{\rho_{en}}{f} = \frac{\rho_{ex}}{q'_\infty}. \tag{1.36}$$

Combining equation (1.32) with (1.36) gives

$$\beta_p = \frac{q'_\infty}{f}. \tag{1.37}$$

Substituting equation (1.37) into (1.35) gives

$$q' = \beta_p f + mf. \tag{1.38}$$

Dividing both sides of equation (1.38) by $q'f$ gives

$$\frac{1}{f} = \frac{\beta_p}{q'} + \frac{m}{q'}. \tag{1.39}$$

Finally, substituting equation (1.34) into (1.39) for the image magnification yields

$$\frac{1}{f} = \frac{\beta_p}{q'} + \frac{1}{q\beta_p}. \tag{1.40}$$

Equation (1.40) would be algebraically equivalent to equation (1.31) if the sign convention that $m < 0$ were applied, but we are leaving it out of the formula because, in the present exercise, we are mostly concerned with the magnitudes of the geometric dimensions (care would then need to be taken when applying the formula in a strict manner).

The derivation just performed is also useful for arriving at a formula that relates the f-number of a lens system at finite conjugate imaging (often called the *working f-number*) to the infinite conjugate f-number that is given by dividing the EFL by the entrance pupil diameter. Such a formula has been provided in some texts (for example, [44, 83]) but the derivation has not been given. The formula is expressed as

$$F/\#' = F/\#\left(1 + \frac{m}{\beta_p}\right), \tag{1.41}$$

where $F/\#'$ is the working f-number, defined as

$$F/\#' = \frac{1}{2\sin u'} \tag{1.42}$$

and $F/\#$ is the infinite conjugate f-number, defined as

$$F/\# = \frac{1}{2 \sin u'_\infty}. \qquad (1.43)$$

To obtain equation (1.41), begin by noting that $\sin u' = \rho_{ex}/q'$, where q' is given by equation (1.35). Substituting this into equation (1.42) gives

$$F/\#' = \frac{q'_\infty + mf}{2\rho_{ex}} = \frac{\beta_p f + mf}{D_{ex}} = \frac{f(\beta_p + m)}{D_{ex}}. \qquad (1.44)$$

Finally, since $\sin u'_\infty = D_{ex}/(2q'_\infty)$ and $q'_\infty = \beta_p f$, substituting these into the denominator of equation (1.44) and applying equation (1.43) yields

$$F/\#' = \frac{f(\beta_p + m)}{2q'_\infty \sin u_\infty'} = F/\#\frac{(\beta_p + m)}{\beta_p} = F/\#\left(1 + \frac{m}{\beta_p}\right). \qquad (1.45)$$

In chapter 2, it will be shown that the irradiance in the image is proportional to $1/(2F/\#')^2$ for any object distance (note that $m = 0$ for an object at infinity). Equation (1.45) provides some further insight in that it states that the irradiance is a function of three quantities: m, β_p, and—now hear this—don't forget about $F/\#$. This means that even at a fixed image magnification and unit pupil magnification (i.e. $\beta_p = 1$), the irradiance in the image can be varied by adjusting the sizes of both the entrance and exit pupils equally, as this amounts to varying the infinite conjugate f-number, $F/\#$. This makes sense, as increasing the size of the entrance pupil would let more flux through the lens. Further, the size of the exit pupil is increased proportionately. This situation is easily visualized for the case of an ideal thin lens for which, clearly, $\beta_p = 1$.

1.3.16 The optical center of a lens (you have probably never heard of this)

In section 1.3.5, I mentioned the optical center (OC) of a lens and said that Johnson [5, 50] has pointed out that it should be labeled as the seventh cardinal point. A close look at figure 1.147 reveals this cardinal point, which is a conjugate of the two nodal points.

In figure 1.147, the left-hand surface has a radius of 25 mm and the right-hand surface has a radius of −50 mm. The central thickness is $t = 25$ mm. The lens's refractive index is 1.5 at 550 nm (it is a fictitious model glass). If t_1 is the distance from the left vertex to the OC (where $t_1 > 0$) and t_2 is the distance from the right vertex to the OC (where $t_2 < 0$), then the OC is given by [5]:

$$t_1 = \frac{t}{1 - (R_2/R_1)}, \qquad (1.46)$$

where R_1 is the radius of the left surface, and R_2 is the radius of the right surface (note that these radii follow the usual convention such that, in figure 1.147, $R_1 > 0$ and $R_2 < 0$). To derive equation (1.46), we look to figure 1.148, which shows that the

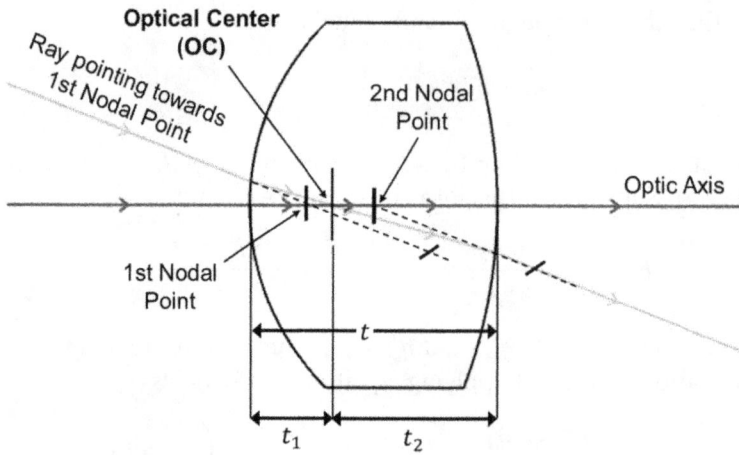

Figure 1.147. The optical center (OC) of a single-element lens.

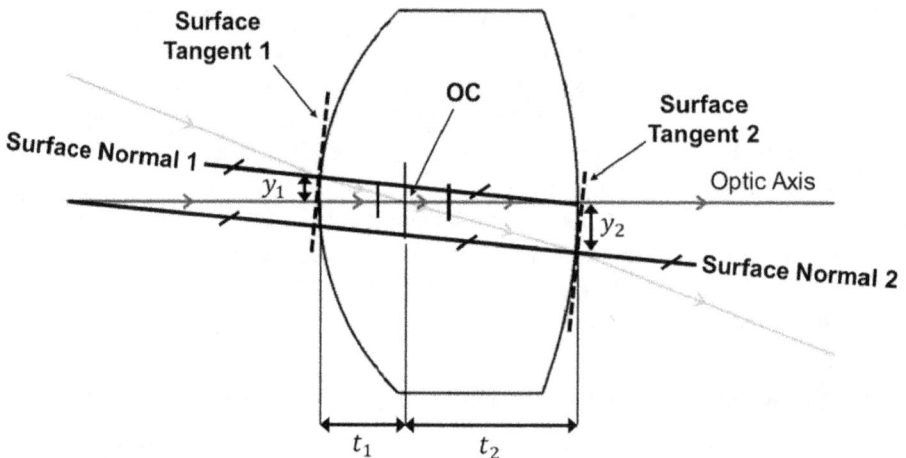

Figure 1.148. The construction to derive the location of the OC, based on the same lens as that shown in figure 1.147.

dashed lines labeled Surface Tangent 1 and Surface Tangent 2 are parallel to each other. This must be so, because the oblique ray (which points towards the first nodal point prior to entering the lens) emerges from the lens at equal angles relative to the optic axis. This is analogous to a ray entering a slab of glass whose left and right surfaces are parallel, which, in the case of the lens in figure 1.148, is represented by the surface tangents on both sides of the lens.

Based on the geometry in figure 1.148, if y_1 is the height from the optic axis to the intersection of the oblique ray with Surface Tangent 1 on the left surface, and y_2 is the height from the optic axis to the intersection of the oblique ray with Surface Tangent 2 on the right surface, then note that

$$\frac{y_1}{R_1} = \frac{y_2}{R_2},\tag{1.47}$$

and

$$\frac{y_1}{t_1} = \frac{y_2}{t_2}.\tag{1.48}$$

Equation (1.47) is realized when one observes that the line connecting Surface Normal 1 from Surface Tangent 1 to the optic axis is the radius of the left surface, and the line connecting Surface Normal 2 from Surface Tangent 2 to the optic axis is the radius of the right surface. Thus, by combining equations (1.47) and (1.48) through the elimination of y_1 and y_2 (and noting that $t = t_1 - t_2$), one obtains equation (1.46). The reader may enter the prescription for the lens in figure 1.147 to verify that equation (1.46) does indeed compute the OC for this lens, which is $t_1 = 8.33$ mm. The lens's index does not enter into equation (1.46), which tells us that the OC is wavelength independent! However, the angular magnification of the OC (defined by the ratio of the angle of the oblique ray at the first nodal point to its angle at the OC) is a function of the refractive index, so the magnification of the OC is generally wavelength dependent [5]. The significance of knowing the OC of a lens shall be made evident in the next two sections.

1.3.17 Locating and optimizing the optical center of a lens system

Equation (1.46) enables us to locate the OC of a singlet, but it would surely be more complex to derive a formula for the OC of a multielement system of lenses. To locate the OC for a lens system using an optical design program, the idea is to note that **when the aperture stop is located at the OC, then the first and second nodal points are located at the entrance and exit pupils, respectively** [5]. Take, for example, the lens system in figure 1.73, whose prescription is provided in figure 1.70. At its current stop position, the first nodal position is located at 43.8841 mm from surface 1, and the second nodal position is located at −49.9768 mm from the last surface (this means it is located to the left of the image surface). In OS, this is determined using two methods. The first is to go to the Analyze menu, then click on Cardinal Points under the Reports menu (see figure 1.49). When using this feature of OS, make sure that the data covers the region from surface 1 to the image plane by selecting these surfaces in the settings menu for the cardinal points. The other method is to use the CARD operand in the Merit Function Editor. Next, at the Analyze > Reports menu, we may use the System Data information to note that the location of the entrance pupil is 37.4594 mm from surface 1, and the exit pupil is −57.3486 mm from the image surface. Note that we can also use the ENPP and EXPP operands in the Merit Function Editor to obtain the entrance and exit pupil locations, respectively. By definition, OS measures these relative to surface 1 and to the image plane, respectively.

Based on the information above, we can roughly tell in which direction the OC is located, relative to the current stop position, by changing the stop location. For instance, if we purposely make surface 6 into the stop, then we find that the exit pupil location is −82.38 mm from the image surface, which is farther to the left from its original position and is therefore more distant from the second nodal point. This means that the OC is located somewhere to the right of the original stop position, because shifting the stop to the left resulted in a larger difference between the position of the second nodal point and that of the exit pupil. The OC is then eventually determined by iterating the change of position of the stop until the position of the second nodal point is equal to the position of the exit pupil (and thus, the position of the first nodal point is equal to the position of the entrance pupil). For the lens system in figure 1.73, the OC lies somewhere between the current stop position and surface 8.

In OS, in cases in which you have the freedom to design a lens whose OC is located at the stop, it would be a straightforward matter of including operands for, say, EXPP and CARD (for which Data = 5) and targeting their difference at zero (or their ratio at one). For example, let us again take the lens from figure 1.73. Suppose we only wish to locate the stop at the OC, and we do not care about optimizing all other parameters for the lens. We found above that the OC should be somewhere to the right of the current stop position. In fact, if we were to switch the stop position to surface 8, we would find that the exit pupil location is closer to the image than the second nodal point, which implies that the OC lies somewhere between the current stop position and surface 8. So, we can make the thickness of surface 6 a variable, set a pickup solve for the thickness of surface 7 such that the total thickness of surfaces 6 and 7 remains constant, and enter the EXPP and CARD operands into the Merit Function Editor to set their target ratio to one. Further, we can reduce the stop size just to be sure that no rays vignette. To maintain approximate image focus, we can add a marginal ray solve for the thickness of surface 12 to set the image to lie at the paraxial focal plane (by setting pupil zone = 0, ray height = 0). Then, optimize. The resulting layout, prescription, and merit function are shown in figure 1.149.

1.3.18 The application of the optical center and pupils to depth sensing

Imagine an optical system that has a spot projected towards a distant diffuse planar surface and an imaging lens to capture the image of the projected spot, as depicted in figure 1.150. Such a basic system may be used for distance measurement [84, 85]. The concept is that if the height h is fixed and known, then by simple geometry, the distance s to the target surface (i.e. the diffuse planar surface) may be determined if one can measure the image height h' and image distance s', provided that there is negligible image distortion (otherwise, if there is non-negligible image distortion, then the distortion can be measured through a calibration process and accounted for in the calculations). Further, this concept does not require the spot's image to be in sharp focus, because it is only necessary to know the center of the spot image, which is located at h' in the image. This helps to save the cost of an autofocusing mechanism for the imaging lens. Thus, the imaging lens may remain fixed in position (so it is a *fixed-focus lens*). Under these conditions, $h/s = \tan\theta = h'/s' = \tan\theta'$. Therefore, upon determining h', then for a fixed known dimension h and fixed known distance s', s can be determined.

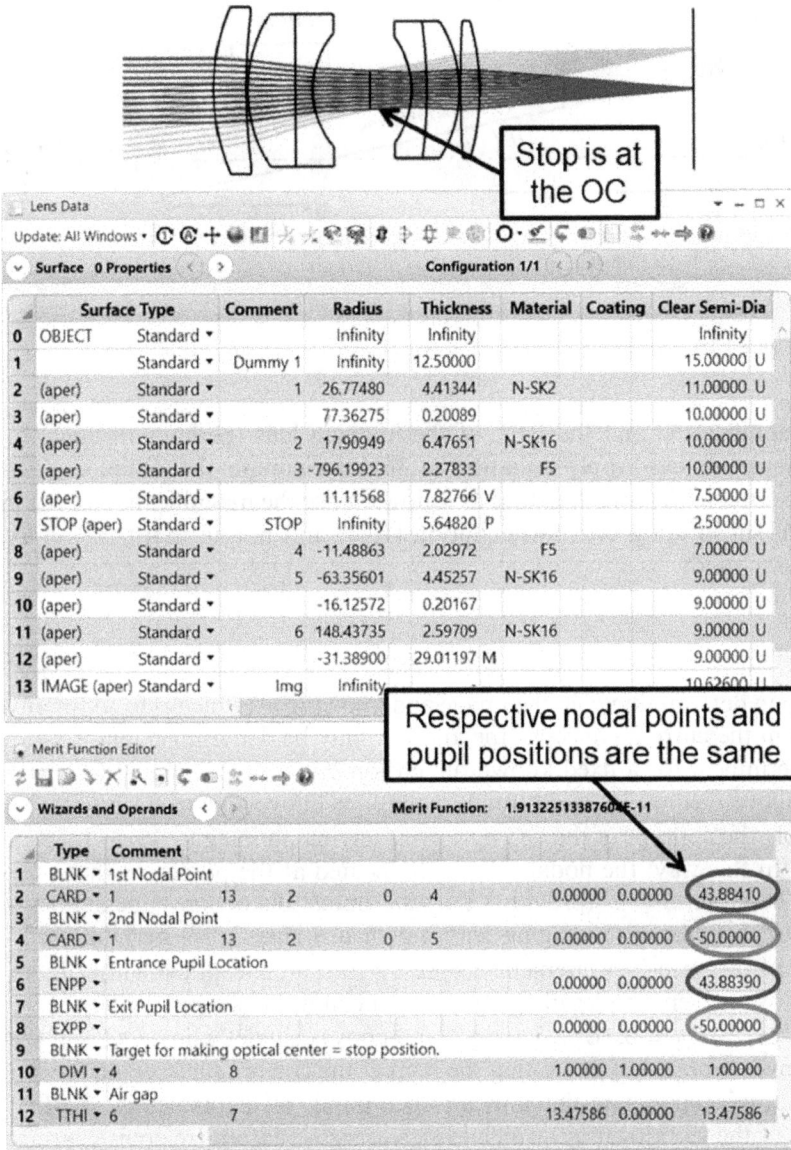

Figure 1.149. The result of optimizing the location of the stop to be at the OC for the lens in figure 1.73. The layout (top), prescription (center), and merit function (bottom) are shown.

Upon closer inspection, the system in figure 1.150 is rather simplistic in that it would only apply if a thin lens were used for the imaging lens, as, in this case, the distances s and s' would be measured relative to a single plane at the lens. For multielement lenses, the distance s would have to be replaced by q (the distance from the target surface to the entrance pupil), and the distance s' would have to be replaced by q' (the distance from the exit pupil to the image plane). If we know the

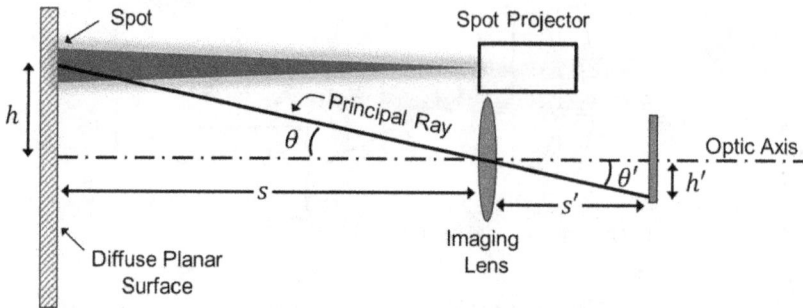

Figure 1.150. An example of an optical system layout for distance (depth) measurement.

pupil magnification and the EFL of the imaging lens (perhaps through an initial calibration), we can apply equation (1.34) to estimate the distance q from the entrance pupil to the target surface as follows: since the magnification of the image is $m = h'/h$, substituting this into equation (1.34) and noting that $\tan \theta' = h'/q'$ and $\tan \theta = h/q$ yields

$$\beta_{\mathrm{p}} = \frac{\tan \theta}{\tan \theta'}. \tag{1.49}$$

Thus, upon determining h' in the image and knowing q' (which is fixed for any object distance in the current concept), the angle θ' may be determined, and from this, we may determine q for a fixed known dimension h.

The above method for determining q requires knowledge of β_{p}. Alternatively, we may apply the concept of the OC by way of having an imaging lens whose stop is at the OC. In this way, the nodal points are located at the pupils, so that $s' = q'$ and $s = q$. Therefore, $\theta = \theta'$ (provided that the image distortion is negligible) so that $\beta_{\mathrm{p}} = 1$. In this case, the imaging lens shown in figure 1.149 may be used for the purpose of distance measurement according to the concept shown in figure 1.150, but there is an even simpler imaging lens that can do this. Applying equation (1.46), if $R_2 = -\infty$, then $t_1 = 0$, so we have a plano-convex singlet whose OC is at the vertex of its convex surface. By mounting the stop at the convex surface and having it face the target, we can ensure that the entrance pupil (or, equivalently, the first nodal point) is at the stop location, which is an accessible datum (reference surface) from which s may be measured. Figure 1.151 illustrates such a system, using part # 48–244 from Edmund Optics for the lens.

In figure 1.151, surface 4 in the prescription is not the plane at which we have a focused image. As such, is it not strictly the 'image plane.' Rather, we have maintained this surface at a fixed distance from the plano surface of the lens, and we regard this distance as s'. We further assume that at the target surface's height h, there is a spot projected by an external source, as depicted by the 'spot projector' in figure 1.150. Thus, at the image plane of the plano-convex lens, there is a defocused image of that spot, whose center determines the image height h'. We have ignored image distortion, and we have used the operand REAY on row 13 of the Merit

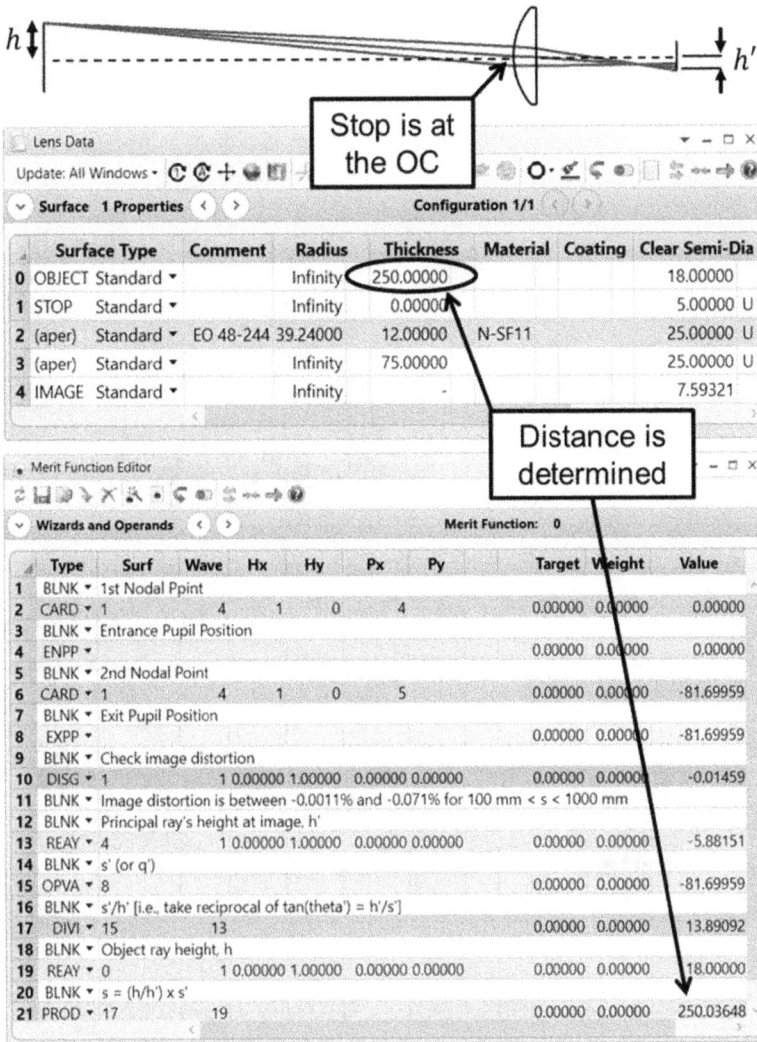

Figure 1.151. Example of a singlet used for depth sensing, based on the method shown in figure 1.150.

Function Editor to compute the height h' of the defocused spot image. In reality, there would be an image sensor at the defocused spot, so, some means would be needed for an algorithm to interpolate between pixels to estimate h'. The resulting estimated distance on row 21 of the Merit Function Editor is off by 0.036 48 mm, due perhaps to ray aberrations, as the locations of the nodal points and the OC are, after all, first-order (paraxial) parameters.

If a pinhole were used in place of the imaging lens, then no aberrations would be present. Although a pinhole may result in low spot brightness at the camera, the use of a laser [85] for spot projection might help to provide a sufficiently bright spot at the target surface. A laser's beam can result in speckle when it scatters off a diffuse

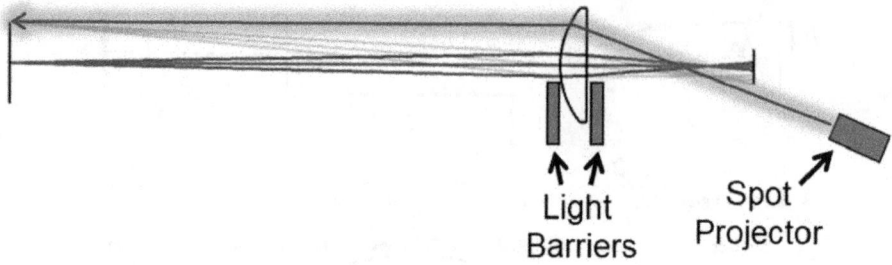

Figure 1.152. A concept for inserting a spot projector *inside* the lens of figure 1.151.

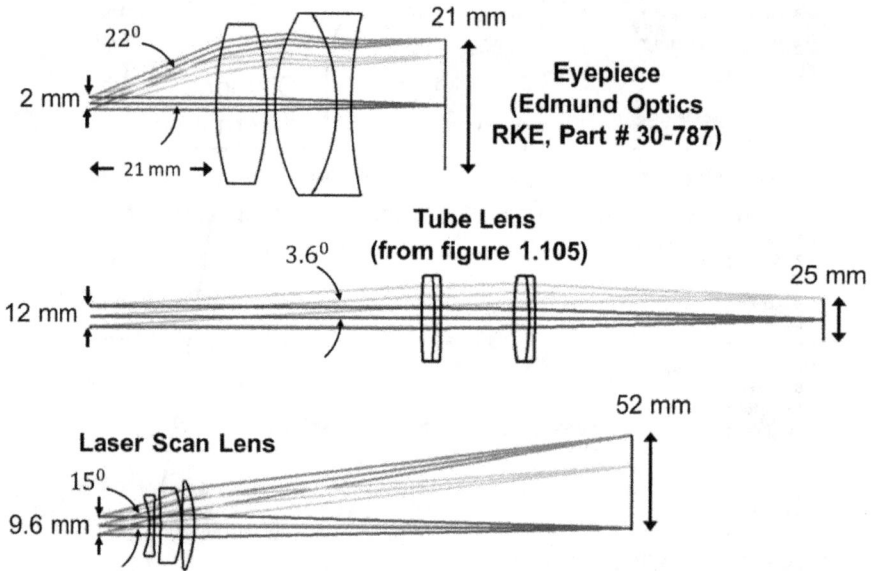

Figure 1.153. Comparison of layouts for an eyepiece, tube lens, and laser scan lens. The layout dimensions are not scaled relative to one another.

surface (though it is not clear whether this would be a problem), so we could use a de-speckling component (though this would add cost). These are the types of ideas to think through in product development—comparing the pros and the cons of various approaches. Incidentally, if we fixed the size of the image sensor and added 'light barriers' in front of and behind the lens, then the stop diameter for the lens in figure 1.151 could be opened up, which might allow for the use of a different optical concept for spot projection, such as that shown in figure 1.152. This concept might be feasible, provided that its pros and cons were evaluated and a prototype fabricated and tested.

1.3.19 Approximate analogies: eyepieces, tube lenses, and scan lenses

Figure 1.153 shows a comparison of layouts for an eyepiece, tube lens, and scan lens. With the exception of scale, first-order properties (field angle, including the

Lagrange invariant), and aberration content (or perhaps, aberration *intent*), they all share similar characteristics in that rays begin from infinity, they then hit the stop (which is also the entrance pupil), and finally, they end at the *image* plane (which can be a field plane, depending on the application). The eyepiece is the RKE® type (part # 30-787), whose Zemax file (with full prescription) is available for download from the Edmund Optics website. It has an EFL of 28 mm. The tube lens is from figure 1.105 of this book. The laser scan lens is based on a design by Laikin (starting with OS's Zebase file # P_013), but I have re-optimized it to have an EFL of 200 mm for a wavelength of 633 nm and a smaller field angle. The point is that all three lens systems look similar but have very different aberration contents and different first-order parameters. Yet, because they are so similar, it is possible to stretch their limits by interchanging them for a variety of applications.

Let us take an example. Suppose we want to quickly build a miniature digital microscope with 10× magnification, and we do not care about standardizing the field of view nor the size of the image at the camera (which is ordinarily at a *field number* of about 25 mm). We would not need to spend extra time and money to customize one. As a first prototype, we could use the RKE® eyepiece from Edmund Optics shown in figure 1.153 as a tube lens. For the objective, we begin by scaling down the COTS lens from figure 1.129 (whose prescription is provided on surfaces 14–20 in figure 1.125) to an EFL of 2.8 mm (because the RKE® eyepiece in figure 1.153 has a 28 mm EFL, and we want a 10× magnification microscope). After scaling down the objective from figure 1.129, we find that the individual doublets have an EFL of nearly 4 mm (but slightly larger) at a wavelength of 550 nm (we are using 450, 550, and 650 nm for the wavelength settings, with a weight of one for each). Hence, we may select Edmund Optics part # 84-125 and use two of them for the objective. Each doublet of part # 84-125 has an EFL of 4 mm, but if we space them apart at the right air thickness (recall equation (1.21)), we can make the total EFL equal 2.8 mm, so that the system magnification is 10×. Upon doing this, it is found that the air gap for the two doublets is 1.056 48 mm. We also find that collimated light is obtained from the objective when the distance between the specimen plane and the front vertex of the objective is 0.117 75 mm. Further, when we scaled down the prior objective, the distance from the second doublet to the stop was about 0.75 mm, so we shall use this length. Upon integrating the resulting doublet combination with the RKE® eyepiece of figure 1.153, we obtain the layout and MTF shown in figure 1.154. Note that since the distance from the right vertex of the objective to the stop is 0.75 mm and the distance from the stop to the left vertex of the eyepiece is 21 mm, therefore the distance between the objective and the eyepiece is 21.75 mm. The MTF is plotted using broadband wavelength settings of 450, 550, and 650 nm and weights of one for each. The field height at the specimen plane is 0.24 mm, but better image quality is achieved between the axial point and about 0.7 × the field height, which is 0.168 mm. The MTF implies that two points spaced 5 microns apart at the specimen plane are resolvable at a modulation of greater than 20% at the full field of view. This is because the maximum spatial frequency shown in the figure is 20 cycles/mm, so 10 × demagnification at the specimen plane yields 200 cycles/mm. Take the reciprocal of that, which gives a pitch of roughly 5 microns. Finally, if a suitable choice of image

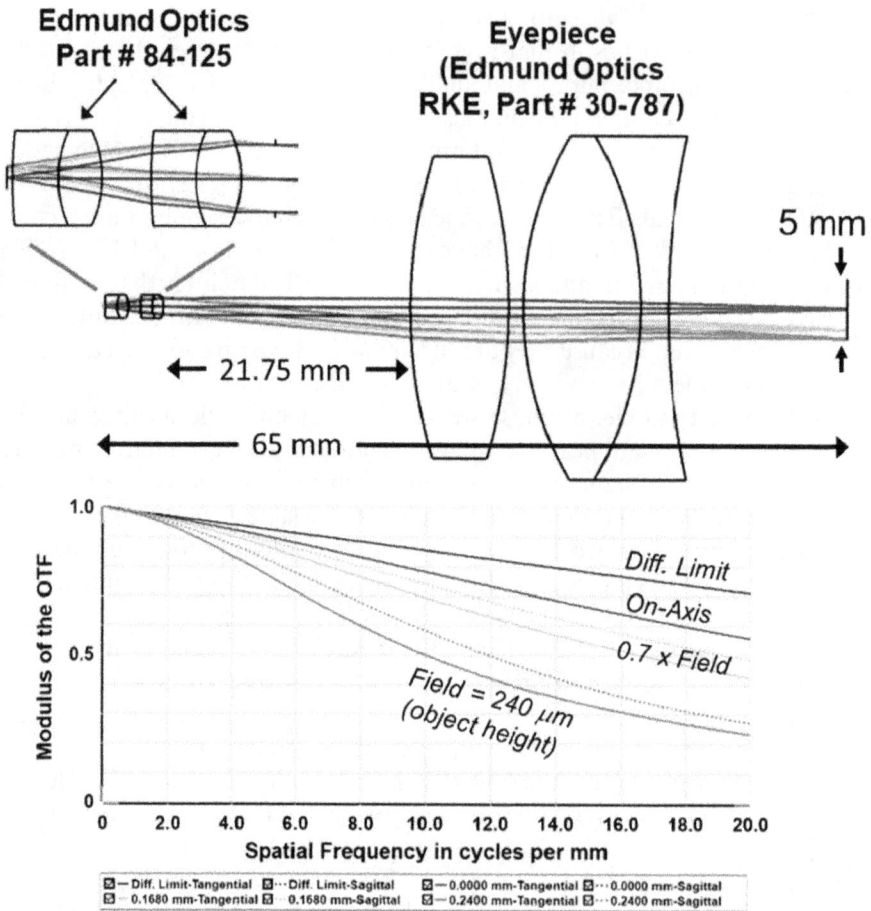

Figure 1.154. Layout and MTF of a prototype miniature digital microscope using COTS lenses. It combines a scaled-down version of the objective from figure 1.129 (but using a different choice for the doublets) with the eyepiece shown in figure 1.153.

sensor at roughly 5 mm × 5 mm were found, then integrating it with the system in figure 1.154 would yield a miniature digital microscope. The pixel size would have to be roughly 2.5 microns (half of the 5 micron pitch between two object points) or less. A popular sensor that fits this requirement is the 5 megapixel CMOS image sensor from ON Semiconductor, part # MT9P031 (https://www.onsemi.com/products/sensors/image-sensors/mt9p031), which has a 2.2 micron pixel size.

Of course, the miniature digital microscope described above has a limited field of view, given by a diameter of about 0.48 mm at the specimen plane. So, this microscope needs to scan over the ROI at the specimen plane. Furthermore, there is an obvious mismatch in sizes between the objective and the eyepiece-based tube lens. But you get the idea here, which is to illustrate how to recognize analogous lens systems and push their limits in applying them to practical applications. This is especially useful when there is a short project timeline and you need to develop a prototype system. Incidentally, considering that scan lenses are also roughly analogous to eyepieces and tube lenses, note that the supplier Thorlabs provides a tube lens (part # TTL200MP) that can be used as a laser scan lens.

1.3.20 Approximate analogies: condensers as eyepieces in reverse

Let us take the Edmund Optics RKE® eyepiece of figure 1.153 from a different point of view. Applying the concept of conjugate planes shown in figure 1.121, we can identify a different conjugate pair of points beginning with the point at C in figure 1.155, which can be considered a point at the specimen plane of a substage condenser (see figure 1.124). This point C could form a conjugate pair with a point somewhere beyond the eyepiece's field stop if a second lens system were to focus a pair of rays from C. For instance, a conjugate for C could be a point on the plane at the flux collector of the condenser system in figure 1.124. Therefore, reversing the eyepiece would make it into a substage condenser, and we could then replace the previous COTS substage condenser used in figure 1.124 with a reversed RKE® eyepiece, as shown in figure 1.156.

In figure 1.155, in order for the eyepiece to possess similar characteristics to those of the COTS-based substage condenser used in figure 1.124, the entrance pupil (i.e. the stop) diameter was increased slightly—from a diameter of 2 mm (the original pupil size of the RKE® eyepiece) to 2.5 mm (the field of view at the specimen plane of the condenser system in figure 1.124). When the eyepiece is reversed for use as a substage condenser, the sine of the half-field angle for the eyepiece becomes the NA of the substage condenser. If the NA is limited to roughly 0.15 (as was done for the

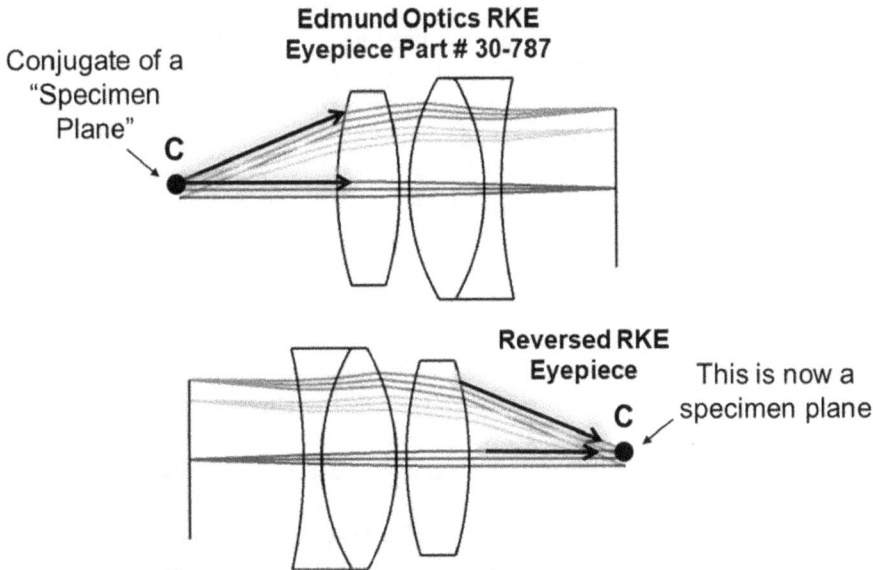

Figure 1.155. Employing the Edmund Optics RKE® eyepiece as a substage condenser by using it in reverse. The conjugate point C becomes a point at the specimen plane.

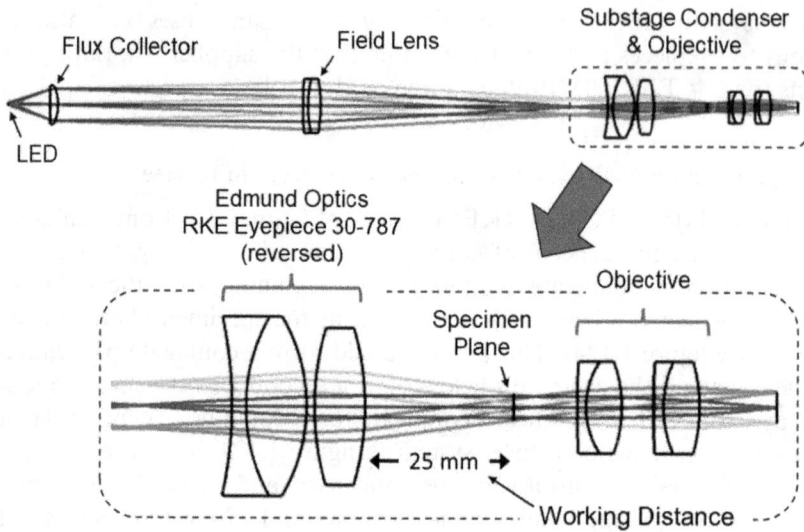

Figure 1.156. Inserting the RKE® eyepiece into the condenser system from figure 1.124.

substage condenser in figure 1.124), then its half-angle is about 8.6°, which is less than the half-field angle of the eyepiece. Therefore, widening this eyepiece's stop diameter to 2.5 mm does not cause much of an issue, as the field angle has also been reduced (which is like using the eyepiece at a lower $F/\#$ and a lower field angle). The eyepiece's eye relief (which was originally at around 21 mm) has purposely been pushed out to 25 mm in order to match the working distance for the substage condenser in figure 1.124. The result is that when the eyepiece is reversed, it is a good substitute substage condenser for the one used in figure 1.124, as it integrates well with the flux collector and field lens, as shown in figure 1.156.

In figure 1.155, when the eyepiece is used as an eyepiece, it functions as a lens with a small aperture (entrance pupil) but a large field angle. On the other hand, when we recognize that the point C may be regarded as the focal position for two rays (one being the upper ray of the oblique ray pencil, and the other being the upper ray of the axial ray pencil), then reversing the eyepiece makes it into a lens with a high aperture but a low field angle. This is a consequence of the Lagrange invariant, as implied by figures 1.4 and 1.7 and equation (1.5). That is, figure 1.4 is analogous to the condition of an eyepiece if the 'source' in this figure is considered the pupil and the angle u is the half-field angle. In contrast, figure 1.7 is analogous to the condition of a reversed eyepiece if the 'source' is now the focal area for rays traveling from the aperture (at the right side of figure 1.7) towards it. **Thus, as a consequence of the Lagrange invariant, we can make a basic rule that a lens operating at a low aperture and a high field angle has a counterpart that can operate with reasonable quality at a high aperture and a low field angle. This counterpart can be found by reversing the lens and its pupil/image conjugate planes such that the field becomes a pupil and a pupil becomes the field.** As shown by the practical example in this section involving the reversal of an eyepiece to become a substage condenser, this rule can be rather useful.

1.4 Practical lens design and aberration management

1.4.1 In rapid product development, just 'manage' the aberrations (we show you how)

If you are not a lens designer and your optical engineering job primarily involves selecting COTS components for system integration, then you might as well just buy a COTS lens. If no COTS lens meets your requirements, then you just have to contract out a custom lens design task to an optical consultant or a small optical design firm. However, a custom lens does not always need to be very complex, and knowing just enough about designing lenses can get you to a solution. Moreover, even if your job as an optical system engineer/designer does not usually involve custom lens design work, it is rather beneficial to know how to design basic lenses on your own, using an optical design program. In modern optical design, a successful design is often achieved by way of starting from an existing optimized lens and **patiently** tweaking it through iterative optimization in an optical design program until the new specifications are met (see, for example, section 1.2.12). Throughout this process, you will be guided by a combination of some basic knowledge about aberrations and their role in lens design. This is what is meant by **aberration management**. My approach is to divide design and aberration management into four parts:

1. **Check the first-order architecture:** identify the first-order lens architecture that best suits the present product requirements and specifications.
2. **Identify the lens form:** take suggested lens designs available in the literature based on the required field and aperture. Scale the lens and tweak it using iterative optimization.
3. **Manage monochromatic aberrations:** if the starting design is well chosen, then one normally reaches an optimized solution fairly quickly (see, for example, section 1.2.12). However, if needed, first check residual aberrations (are they monochromatic or chromatic?) Correct the astigmatism by adding more meniscus elements, stop-shifting, and optimizing element shapes, air spaces, and element thicknesses (add or remove elements where necessary). Minimize the Petzval sum by adding negative-powered surfaces. Minimize spherical aberration and coma by splitting elements where necessary and optimize iteratively with all available variables. During optimization, interchange between minimizing RMS spot size and RMS wavefront error (I often do not need to use MTF operands to achieve high MTF, as I find that minimizing RMS wavefront error does the job well). Note that additional merit function operands are always included to control other parameters, such as system length, element edge thicknesses, and so on.
4. **Manage chromatic aberrations:** the first instance of achromatization is often assumed to be addressed by the initial choice of available lenses in the literature. Subsequently, appropriate merit function operands are used to maintain the lens's achromatic state. Subsequent improvements can be made by combining an iterative optimization process involving trial and error with the automated selection of new glass choices (see sections 1.4.7–1.4.12).

In the process above, the first part—which is to consider first-order architectures—attempts to satisfy basic constraints for imaging the scene of interest. For instance, if there is a need to image a wide field of view, then a wide-angle lens is needed. The first-order layout of such a lens must have a front negative-powered lens and a back positive-powered lens; we call such an architecture a reverse-telephoto or *retrofocus* lens. Further, such an architecture results in a BFL that is longer than the EFL, which is helpful in cases in which space is required between the positive lens group and the image plane. If, on the other hand, one requires scene magnification at the image plane of a camera, then an extended focal length lens is needed (and usually, one would like to fit this lens within a reasonably compact space). A telephoto lens architecture would be suitable for this. Such architectures can then be explored by way of laying out a system of paraxial thin-lens models in the optical design program, followed by replacing them with actual lenses (if designing from scratch) or replacing the complete model with an appropriately scaled lens from the literature, which is the second part in the list above for aberration management. In this part, you search for lens design forms and any advice given in the literature (books or other sources of information on lens design). A useful chart has been provided by Warren J. Smith [31], in which design forms are sorted by half-field angle and system aperture (*f*-number). The third and fourth parts of the list involve knowledge of aberrations and their role in practical lens design. This shall be the topic of the next few sections.

1.4.2 Heuristic lens design theory

Imagine a glass sphere with a radius of 50 mm and an index of 1.5 at 550 nm (figure 1.157).

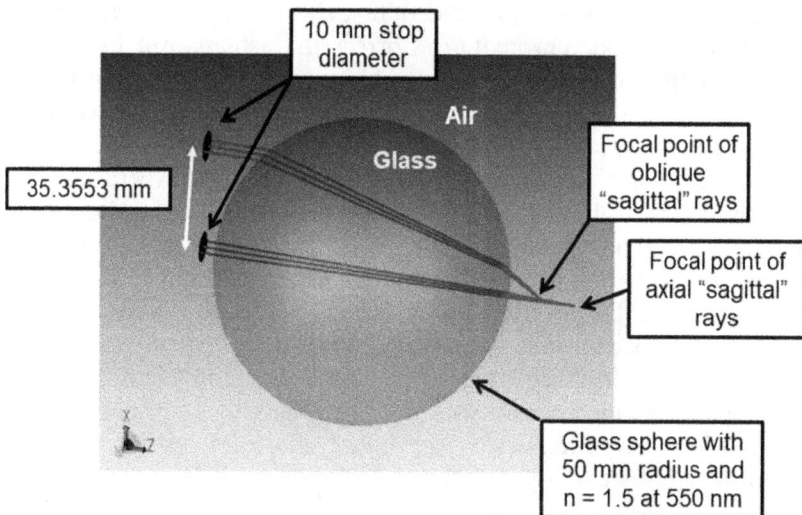

Figure 1.157. A glass sphere with two entrance pupils letting 'sagittal' rays through. The +*y* dimension points out of the page.

The glass sphere in figure 1.157 is immersed in air. Two entrance pupils are shown, one at the axis, the other situated 35.3553 mm above. This separation is somewhat arbitrary (it so happens that it is the height that makes the angle of incidence of the central ray in the upper pupil 45° to the normal at the sphere). Although the axes indicate that the $+x$ direction is vertical, the rays being traced are here considered 'sagittal' because they are horizontal (i.e. they span the Y–Z plane). In this figure, I ask that you temporarily dispense with the usual definition that the sagittal is in the x–z plane. It so happens that when I created this model in OS, I applied a quick trick to focus the rays, namely a marginal ray solve (pupil zone = 1). In OS, the marginal ray solve places the **tangential ray of the Y–Z plane** (for the defined pupil zone) at the optic axis of the system. Since, in this example, I wanted the sagittal rays to come to a focus for the upper pupil, I decided to shift that pupil in the $+x$ direction whilst tracing three tangential rays. This effectively makes those tangential rays into 'sagittal' rays under the conditions shown in figure 1.157.

Note that the oblique rays (i.e. the 'sagittal' rays from the upper pupil) are focused at a location on the axis between the right side of the sphere and the focal position of the axial rays. So, the oblique rays experience a higher optical power while passing through the sphere than the axial rays, even though they are going through the same sphere. Note that the upper pupil is not at an angle. That is, the upper pupil is in the same x–y plane as the axial pupil, and their central rays are parallel prior to entering the sphere. Why do the oblique rays focus closer to the sphere than the axial rays? The answer is illustrated in figure 1.158.

In figure 1.158, two horizontal discs are drawn inside the glass sphere of figure 1.157. The disc at the axial position has a radius equal to that of the sphere, while the upper disc (which is at the height of the upper pupil) has a radius equal to 35.3553 mm, which is smaller than the sphere's radius. Thus, the pencil of rays from the upper pupil experiences refraction through a smaller radius of curvature (but it is not exactly the radius of the upper disc—rather, due to the curvature of the sphere, the 'effective' radius that the rays from the upper pupil 'feel' can be considered to lie somewhere between the radii of the upper disc and the axial disc). If we let the

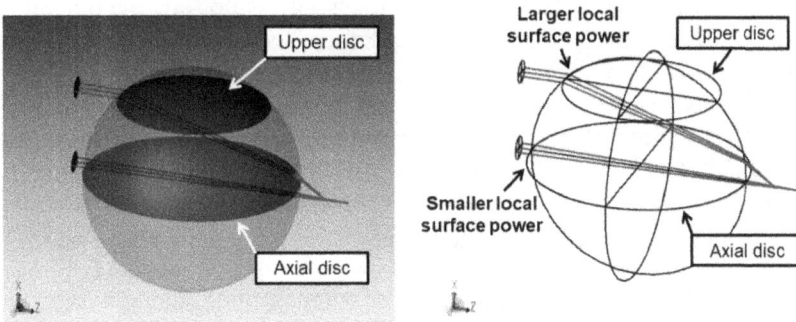

Figure 1.158. Two horizontal discs with different radii of curvature in the sphere. The left-hand figure shows a 3D solid model. The right-hand figure shows a 'skeleton' of the model.

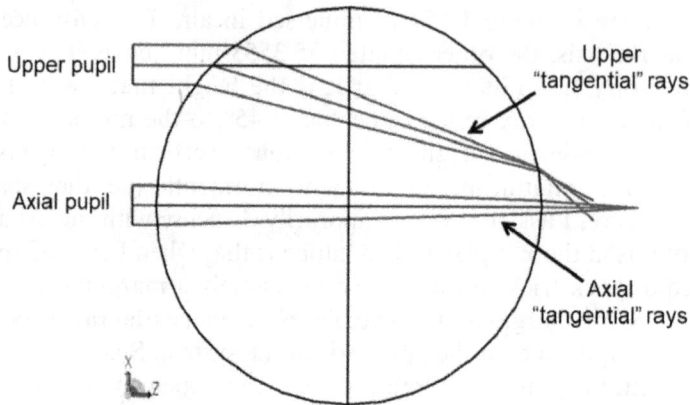

Figure 1.159. 'Tangential' rays pass through the sphere of figure 1.157. The $+y$ direction points out of the page.

'tangential' rays through the upper pupil, they are focused at a different location from the 'sagittal' rays (figure 1.159). In fact, they are even farther from the axial focal position (closer to the sphere). In essence, this is what causes **astigmatism** in a lens. The equations describing the focal positions for the oblique tangential and sagittal ray pencils are those derived by Young and Coddington [5, 25, 58] that were mentioned in section 1.3.2.

Let us gain further insight into the *local power distribution* that exists across the left-hand side of the sphere's surface. Have a look at figure 1.160, where, instead of letting rays through both surfaces of the sphere, we let them pass through the left-hand surface and focus inside the glass.

In figure 1.160, rays enter from infinity from the left-hand side at two field angles: zero (blue rays from the axial field) and a half-field of 2.5° (green rays). Thus, the green rays focus at a height above the axis. The left-hand surface of the glass is the left hemisphere of the sphere of figure 1.157, which has a radius of 50 mm. The rays are then allowed to focus (using a marginal ray solve for zero ray height at the full pupil zone, where the pupil semi-diameter is 5 mm). In OS, due to symmetry about the optic axis, when the medium that rays focus into is not air, the optical element appears as if it is a capsule. At the full pupil zone, the axial sagittal rays focus at a position about 149.666 mm from the left-hand vertex of the hemisphere. If the pupil zone were zero (paraxial), the focal distance would have been 150 mm, which is indeed the EFL for a surface of radius 50 mm and index = 1.5 for a wavelength at that index (here, I have created a model glass with index 1.5 at 550 nm, and the rays are traced at 550 nm).

There are two ray pencils from the upper pupil: the axial oblique pencil (due to rays entering the upper pupil parallel to the sphere's axis) and the field oblique pencil (due to rays entering the upper pupil at a 2.5° half-field about the y-axis). When a marginal ray solve (at full pupil zone) is used in OS, the axial distance from the dotted vertical line towards the focal point at the right is about 130.7637 mm. This is

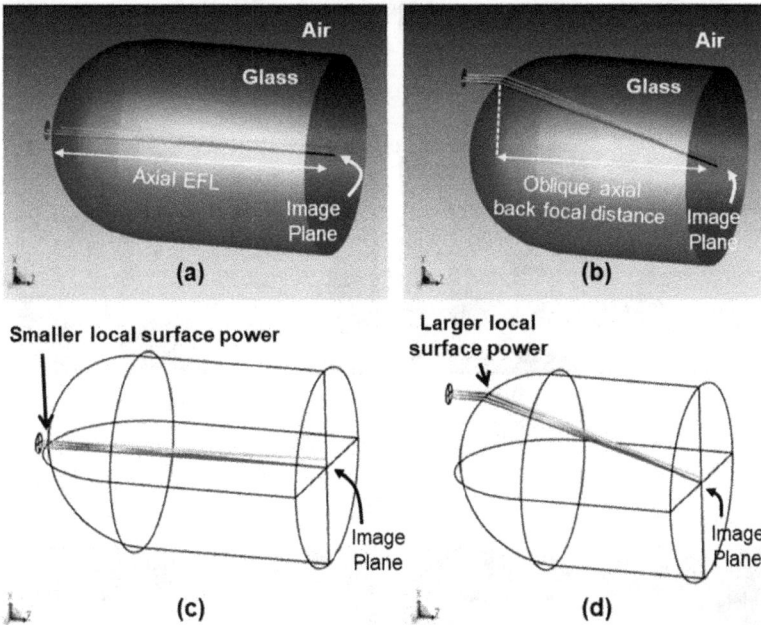

Figure 1.160. Focusing the 'sagittal' rays into the glass of the sphere of figure 1.157. (a) and (b): Solid models of the axial and upper pupils, respectively. (c) and (d): 3D 'outline' models for the axial and upper pupils, respectively. Blue rays are from infinity and axial, while green rays are from infinity at 2.5° half-field.

not the EFL of those rays. Rather, it is simply an *axial oblique back focal distance* associated with the sagittal rays at full pupil. The EFL associated with the ray pencils from the upper pupil may be determined by way of comparing the size of the *local image* formed by those rays and the image formed by the axial pupil. Due to the asymmetry of the ray pencils from the upper pupil, there is no single definition for the size of the local image formed by those rays. For instance, at the image plane, the height of the green rays from the upper pupil is roughly 2.6 mm in the $+x$ dimension. The ray height for the 2.5° half-field for the axial pupil is about 4.4 mm. Since the axial EFL is 150 mm (for paraxial rays on-axis), the EFL for the oblique ray pencil in the $+x$ dimension is roughly 2.6/4.4 times 150 mm, or 88.6 mm. If we use OS's POWF operand to compute this EFL, we obtain a value of roughly 87.5 mm, which is close to the above estimate. (Indeed, the POWF operand in OS computes the EFL associated with a field point!) If a 2.5° half-field is given about the x-axis, then the height of the focal point in the $+y$ dimension is about 3.5 mm. So, taking 3.5 mm divided by 4.4 mm times 150 mm, we obtain an EFL of about 119.3 mm in the $+y$ direction for the oblique field ray in the upper pupil. Using the POWF operand to compute this EFL, we would get 121.4 mm, which is not too far from the above estimate.

OS offers a way to visualize the effect of associating a smaller EFL with the oblique rays: make use of the Image Simulation feature in OS (see section 1.2.10) for the two pupils. Doing this gives the result shown in figure 1.161, where it can be seen

Figure 1.161. Simulated images: (a) for the axial pupil; (b) for the upper pupil. Here, all rays (sagittal and tangential) are allowed through the pupils.

Figure 1.162. Simulated images for the upper pupil. (a) For comparison, this is the image from figure 1.161(b). (b) The image when the pupil is an ellipse that mostly lets through the sagittal rays.

that the axial pupil forms a larger image (figure 1.161(a)) than the upper pupil (figure 1.161(b)) at the image plane. However, as mentioned above, there is a distortion in the image formed by the upper pupil related to the asymmetry of its ray condition. Moreover, the upper pupil has formed a blurred image, due to the astigmatism in the image. If we make the upper pupil into an elliptical stop with a semi-major axis length of 5 mm in the y-dimension, and a semi-minor axis length of 0.1 mm in the x-dimension (thereby letting only most of the 'sagittal' rays through), then we obtain the result shown in figure 1.162(b). Note that the image is clearer. For comparison, figure 1.161(b) is replicated in figure 1.162(a). In one study [35, 36], I corrected the astigmatism by way of using a pair of freeform surfaces. Aspheric shapes can also perform this correction, but it is not necessary to use aspheres and freeform surfaces. In fact, the whole point of classical lens design is to use a combination of spherical elements. Clearly, as indicated by the figures just described as well as by

Young and Coddington's equations, there is a **localized transverse optical power distribution** in the sagittal and tangential dimensions across the surface of any spherical element. **This implies that a ray pencil traversing a well-designed lens system can be made to refract through a combination of local surface powers that combine in such a way as to focus tangential and sagittal rays equally towards the image plane—in fact, this is essentially how optimized lens designs work**. Hold that thought. We will return to it after we discus in further detail how the spherical surfaces of lenses combine with other spherical surfaces to correct the astigmatism and the field curvature in the image.

Since, from figures 1.157–1.160, we have seen that there is a **localized transverse refractive power distribution across a spherical lens's surface**, we may wish to know whether two or more surfaces can combine in power in such a way as to bring oblique sagittal and tangential rays to a focus at the same plane as that of the axial rays. In this way, all focal points from across the field would lie on the same plane. Such a focal plane is called a flat field in the image. We shall approach this problem *piecewise*, by first flattening the tangential rays across the field, and we shall ignore distortion. Also, we shall avoid most (not all) of the spherical aberration and coma by reducing the stop size. Let us begin with a biconvex singlet, which we will *bend* (i.e. its surface curvatures shall be made to vary whilst maintaining the lens's EFL). Further, I will begin this exercise by creating a *multi-configuration* system that has two vertical pupil locations, similar to the conditions shown in figures 1.157–1.160. You will soon see why this is done. To this end, have a look at the layout shown in figure 1.163, whose prescription and multi-configuration settings are shown in

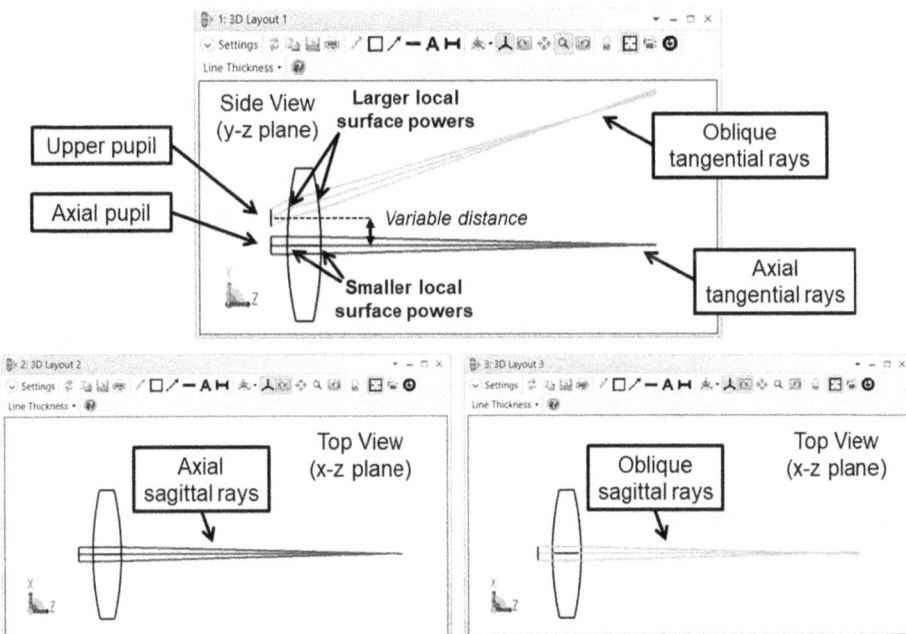

Figure 1.163. A layout for a biconvex lens with N-BK7 glass. The prescription and multi-configuration settings are provided in figures 1.164 and 1.165, respectively.

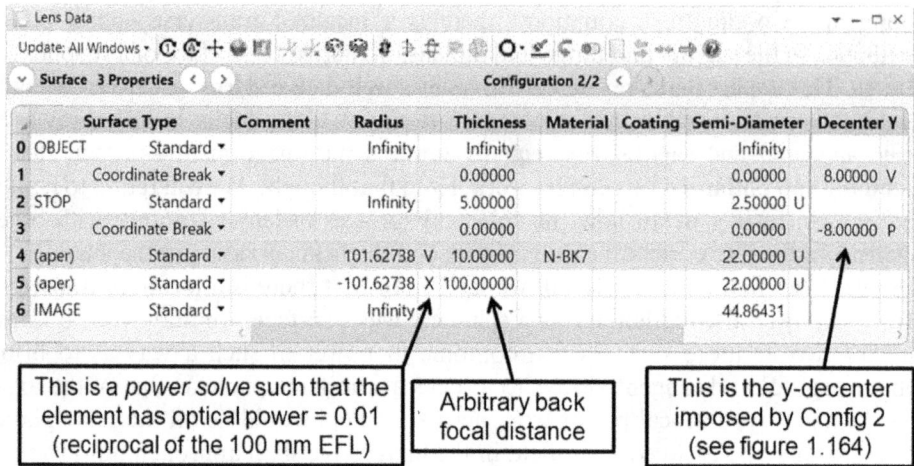

Figure 1.164. The prescription for the layout in figure 1.163 but using the configuration imposed by Config 2 in figure 1.165, which is the Multi-Configuration Editor in OS.

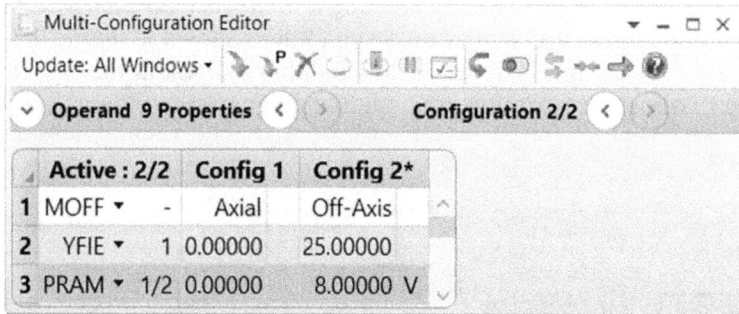

Figure 1.165. The Multi-Configuration Editor that imposes settings on the prescription in figure 1.164. The asterisk symbol on Config 2 denotes that the current prescription reflects the settings in configuration 2, which are controlled by the operands YFIE and PRAM.

figures 1.164 and 1.165, respectively. The wavelength is monochromatic at 588 nm (the d wavelength), and the system aperture is Float by Stop Size. Also, ray aiming is used (aimed at the paraxial entrance pupil). Note that in figure 1.163, I am now using the 'normal' labels for the tangential and sagittal rays.

In the top layout of figure 1.163, the upper pupil is at a y-height of 8 mm from the axial position. This displacement is a starting point and shall be made into a variable during optimization. This y-decenter of +8 mm is achieved using the coordinate breaks shown in figure 1.164, but the values for the decenter are controlled by the second configuration (Config 2) of the Multi-Configuration Editor (figure 1.165). In OS, the Multi-Configuration Editor lets you set up multiple versions of your lens system controlled by operands entered into rows of the Multi-Configuration Editor. In the current system, I have used two operands, YFIE and PRAM. The YFIE sets

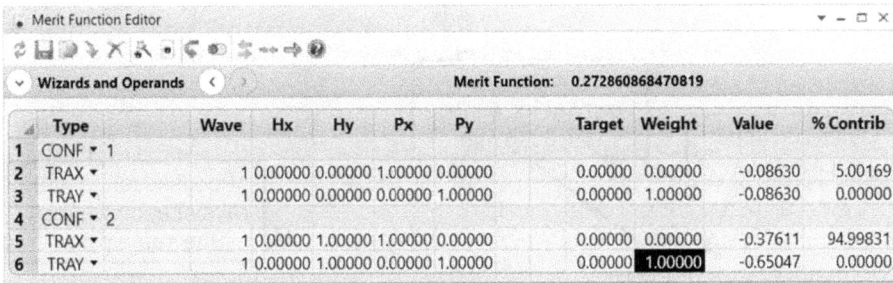

	Type	Wave	Hx	Hy	Px	Py	Target	Weight	Value	% Contrib
1	CONF ▾ 1									
2	TRAX ▾		1	0.00000	0.00000	1.00000	0.00000	0.00000	0.00000	-0.08630	5.00169
3	TRAY ▾		1	0.00000	0.00000	0.00000	1.00000	0.00000	1.00000	-0.08630	0.00000
4	CONF ▾ 2									
5	TRAX ▾		1	0.00000	1.00000	1.00000	0.00000	0.00000	0.00000	-0.37611	94.99831
6	TRAY ▾		1	0.00000	1.00000	0.00000	1.00000	0.00000	1.00000	-0.65047	0.00000

Figure 1.166. The Merit Function Editor for the system in figures 1.163–1.165.

the field angle in the Field Data Editor. You can see that the field angle is 0° for the first configuration (Config 1), and 25° (half-field angle) in the second configuration (Config 2). This sets up the system so that it has the axial pupil shown in the bottom left layout of figure 1.163 (Config 1) and the upper pupil shown in the bottom right layout of figure 1.163 (Config 2). Thus, the top layout in figure 1.163 shows both Config 1 and Config 2 together. The PRAM operand controls 'Parameter 2' (the y-decenter) on surface 1 (the coordinate break). As described in section 1.2.11, coordinate breaks are used in pairs when a surface is decentered relative to other elements. Thus, the coordinate break on surface 3 ensures that the lens is centered. Note that the global coordinate is placed on surface 4.

When using multi-configuration setups in OS, double-clicking on any of the Config tabs in the Multi-Configuration Editor toggles between the two configurations. For example, upon double-clicking on the Config 1 tab, the asterisk marks that tab, and the prescription reflects the operand settings in the Config 1 column of the Multi-Configuration Editor (so, the stop is re-centered, and the field angle is 0°, such that only axial rays can enter the pupil). You just need to spend time playing with the editor to get acquainted with it.

The lens shown in figure 1.163 is our starting point in optimizing the focus for the oblique tangential rays and the axial tangential rays to lie on the same plane. The Merit Function Editor for this system is shown in figure 1.166, where operands for both the tangential and sagittal rays in both configurations have been inserted to prepare for a series of optimizations. The TRAX and TRAY operands compute the transverse ray errors in the x and y dimensions, respectively, at the image plane. These errors are the differences between the x (if TRAX) or y (if TRAY) ray positions relative to the chief ray (which is also the principal ray in the current case, as there is no vignetting). Ordinarily, when minimizing transverse ray errors (even at a single field point), many more of these operands (TRAX, TRAY, TRAC, etc.) are used in the merit function, and different weights are given to rays at different pupil zones such that a root mean square (RMS) computation of the differences between the target values and the current values is performed in the Merit Function Editor. You can see a *best* selection of weights by using OS's default merit function tool and selecting either x weights only (to minimize the RMS transverse ray error for sagittal rays only) or y weights only (to minimize the RMS transverse ray error for

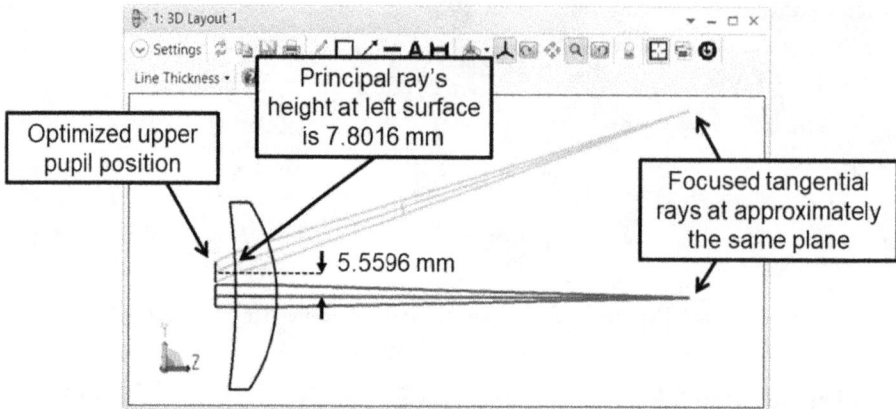

Figure 1.167. A layout showing focused tangential rays for the axial and upper pupil, which are the results of optimization using the operands and weights in figure 1.166.

tangential rays only), or both x and y weights (to minimize the RMS spot size of the focused rays, which is equivalent to the RSCE operand). A common approach to determining the appropriate rays to trace (and their weights) is the so-called *Gaussian quadrature* method introduced by Forbes [86]. In our current exercise, we shall make an approximation and simplification by using only one transverse ray error operand for each field point.

We will begin by only optimizing the focus for the tangential rays at the full pupil. Hence, only the TRAY operands have been given weights in the merit function of figure 1.166. Note that the displayed numbers under the Value column of the Merit Function Editor are the ray errors. You can estimate them graphically by pointing the cursor towards the rays at surface 6 on the layouts in OS and taking the differences in the x or y positions for those rays, relative to the chief ray. We want to minimize these errors, which is why the targets are zero for the operands. To do this, optimize the system (see section 1.2.12), and the resulting lens design layout is shown in figure 1.167. The left and right surface radii are −158.578 64 mm and −39.8149 mm, respectively. The upper pupil has shifted to a height of 5.5596 mm. The tangential rays are now focused at approximately the same plane for the axial and oblique rays. So, the current example in figure 1.167 has shown that there is a local vertical position near the lens for oblique rays to enter the lens such that their tangential rays are focused at the same plane as the axial tangential rays. At that vertical location (5.5596 mm for this lens), the oblique tangential rays are refracted through a pair of surfaces with local surface powers that combine to focus those rays towards a plane shared by the axial rays. Now, in reality, a lens can only have a single entrance pupil (here, it is also the aperture stop). But if we can only have a single entrance pupil to achieve this, then is there another way of letting the oblique rays pass through that vertical location of the lens? The answer is to shift the aperture stop towards the left at an appropriate position, as shown in figure 1.168.

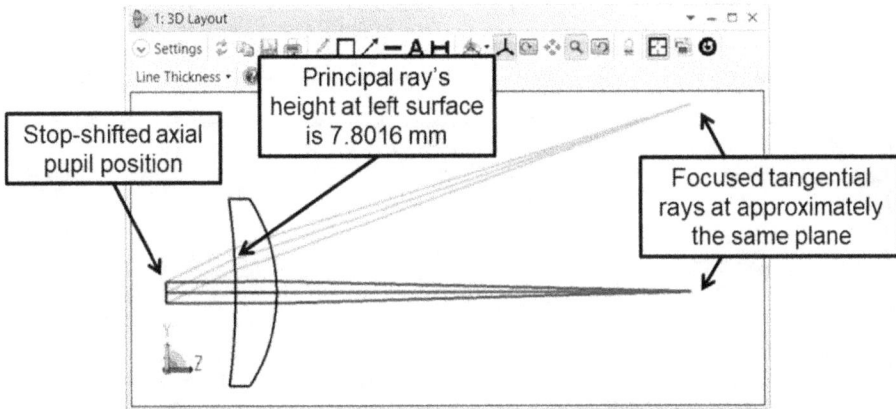

Figure 1.168. A leftward-shifted position of the stop yields a geometry for the oblique rays to pass through the same vertical location of the lens as that of figure 1.167.

Figures 1.167 and 1.168 provide a graphical means to understand the basis for so-called *stop-shifting* to control astigmatism in lens design. Let us now use the field curvature plot (such as the type of plot in figure 1.41) to examine the flatness of the tangential focus across the image plane for the lens in figure 1.168. However, to do this, we need to re-optimize this lens because the field curvature plot displays the focal position for **parabasal rays** (i.e. rays at small angles to each other) of the tangential (and sagittal) fields at the image plane. One pair of axial parabasal rays, for instance, consists of a ray that passes along the optic axis and a paraxial marginal ray, while another pair of oblique parabasal rays from the field consists of the principal ray and another field ray at a small angle approximation to the principal ray. To optimize parabasal tangential rays for both fields for the lens in figure 1.168, use a small value for the Py pupil zone (e.g. let Py = 0.1 in the Merit Function Editor of figure 1.166) for the TRAY operands. We must also place the image plane at the paraxial marginal ray focal position (using a marginal ray solve with height = 0 and pupil zone = 0 for the last thickness). After making these changes and re-optimizing the left surface radius (whilst maintaining the element power at 0.01 using the element power solve on the right surface) as well as the stop's axial position, we obtain the lens prescription shown in figure 1.169, whose layout and field curvature plot are shown in figures 1.170 and 1.171, respectively.

Figure 1.171 clearly shows a flattened tangential image field. This is not necessarily a bad state for a lens with residual third-order astigmatism and negligible spherical aberration and coma, which approximates the state of the current lens at low aperture. In fact, in classical lens design, if we have a lens in such a state of aberration, we would find it rather beneficial to have a flattened tangential image field at the image plane [87]. Let us go further and see what it would take to flatten only the sagittal image field. We can achieve this by removing weights from the TRAY operands, and, instead, using weights only for the TRAX operands. Further, we continue to use low values for the pupil zone (such as letting Px = 0.1 in the Merit

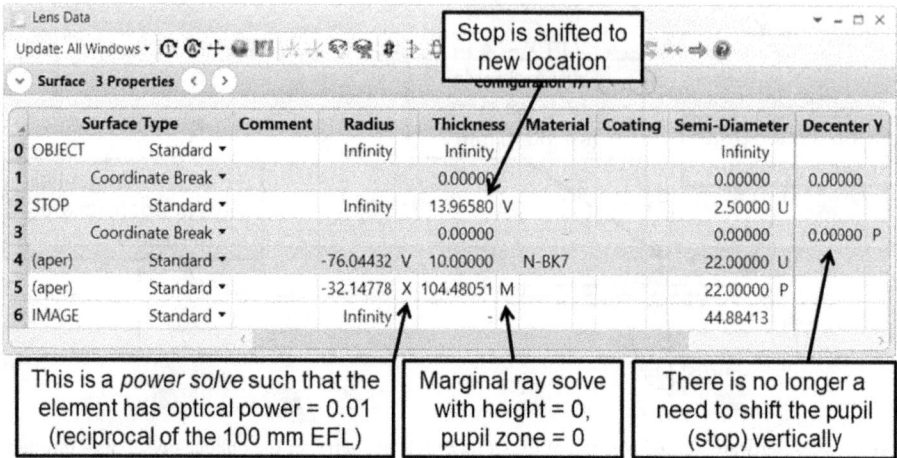

This is a *power solve* such that the element has optical power = 0.01 (reciprocal of the 100 mm EFL)

Marginal ray solve with height = 0, pupil zone = 0

There is no longer a need to shift the pupil (stop) vertically

Figure 1.169. A prescription for a singlet optimized for a flat paraxial tangential ray field.

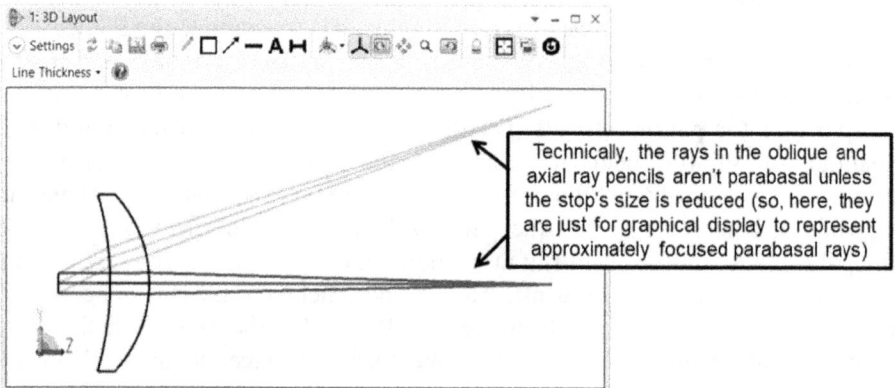

Technically, the rays in the oblique and axial ray pencils aren't parabasal unless the stop's size is reduced (so, here, they are just for graphical display to represent approximately focused parabasal rays)

Figure 1.170. The layout for the lens given by the prescription in figure 1.169.

Function Editor of figure 1.166) for the TRAX operands. We let the thickness of surface 2 and the radius of surface 4 be variables, we maintain the lens's EFL through the use of an element power solve for the radius of surface 5, and we continue to use the paraxial marginal ray solve for the thickness of surface 5. By doing so, we obtain the layout shown in figure 1.172, whose prescription and field curvature plot are shown in figures 1.173 and 1.174, respectively. For reference, the current state of the Merit Function Editor is shown in figure 1.175.

Note that the lens's edges in figure 1.172 are unrealistic, as they cross and have negative thickness due to the absence of an operand to constrain the edge thickness (if we had included such a constraint, it would have been the MNEG operand). For the purpose of the current exercise, we shall ignore this minor issue. The point is to note that we have successfully flattened the sagittal field, as shown in figure 1.174.

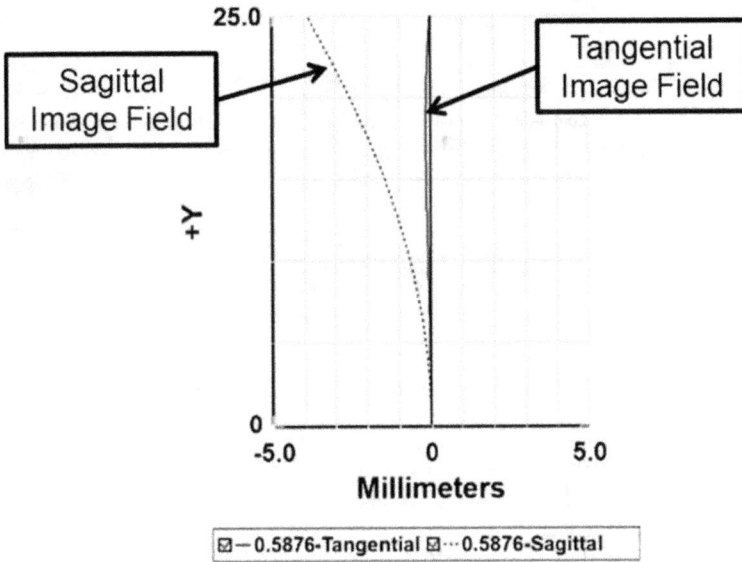

Figure 1.171. The OS Field Curvature plot for the lens in figure 1.170. The $+Y$ axis units are degrees.

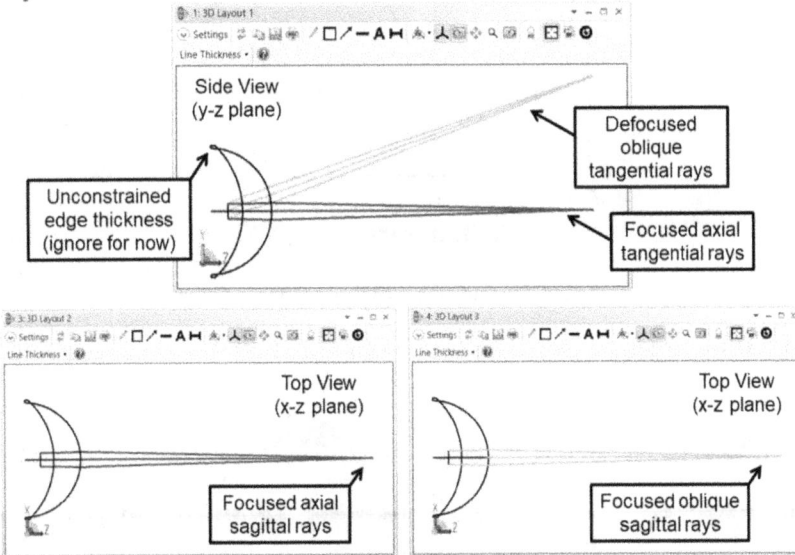

Figure 1.172. The layout resulting from optimizing the focus of parabasal sagittal rays.

Moreover, we note that the optimized radii for the lens surfaces are different from those required to flatten the tangential field, and the distance between the stop and the left-hand vertex of the lens is also changed. This implies that **for the current lens thickness,** there can be no solution that yields both a flat tangential field and a flat

Figure 1.173. The prescription for the layout shown in figure 1.172.

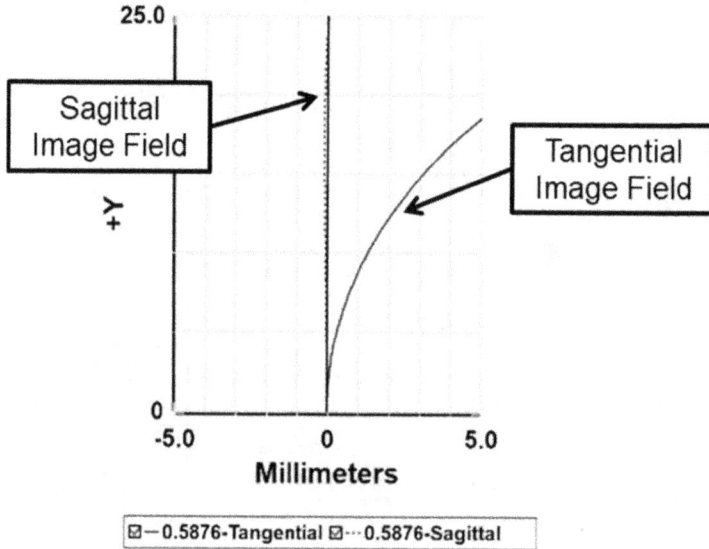

Figure 1.174. The OS Field Curvature plot for the lens of figure 1.172. The $+Y$ axis units are degrees.

Figure 1.175. The OS Merit Function Editor for the state of the lens shown in figure 1.172.

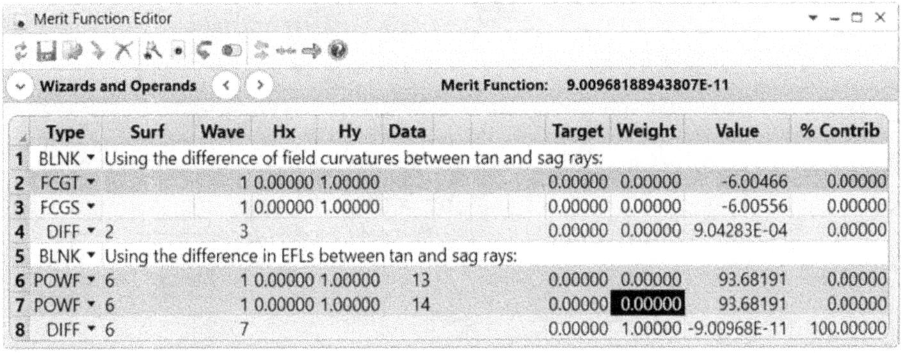

	Type	Surf	Wave	Hx	Hy	Data	Target	Weight	Value	% Contrib
1	BLNK	Using the difference of field curvatures between tan and sag rays:								
2	FCGT		1	0.00000	1.00000		0.00000	0.00000	-6.00466	0.00000
3	FCGS		1	0.00000	1.00000		0.00000	0.00000	-6.00556	0.00000
4	DIFF	2	3				0.00000	0.00000	9.04283E-04	0.00000
5	BLNK	Using the difference in EFLs between tan and sag rays:								
6	POWF	6	1	0.00000	1.00000	13	0.00000	0.00000	93.68191	0.00000
7	POWF	6	1	0.00000	1.00000	14	0.00000	0.00000	93.68191	0.00000
8	DIFF	6	7				0.00000	1.00000	-9.00968E-11	100.00000

Figure 1.176. The OS Merit Function Editor for the optimized state of the lens in figure 1.177.

sagittal field simultaneously—but hold this thought, as we will soon return to this point and determine a solution for this singlet that can indeed flatten both the sagittal and tangential fields, which provides a first example with which to introduce the concept of the *Petzval theorem*.

The next question is: 'Can a singlet reach an optimized state that yields zero astigmatism but without any regard to field flattening?' Yes, it can. To achieve this, we need to use a different set of operands from those of figure 1.175. This is because achieving zero astigmatism is not the same as targeting both the tangential and sagittal rays to achieve zero transverse ray errors at a specified field and at a defined back focal plane (i.e. surface 6 in the prescription of figure 1.173). Rather, the target *difference* between the focal positions of the tangential and sagittal rays must be set to zero. I will use two sets of operands for you to have a look at (but I will only use one of the two sets for optimization). As a first set, we can use the FCGT and FCGS operands for the farthest field, then set their target difference to zero. As a second set, we can use the POWF operand. Upon using these and performing a single optimization, the merit function of figure 1.175 is modified to that shown in figure 1.176. The resulting lens layout, prescription, and field curvature plot are shown in figures 1.177–1.179, respectively.

In figure 1.176, notice that a weight was applied only to the operand on row 8, which targets the difference in the EFLs (computed using the POWF operand) of the oblique tangential and the sagittal ray pencil. The OS help file for the POWF operand tells you that setting '13' and '14' for the 'Data' values of the POWF operand results in computing the tangential and sagittal ray EFLs of the field points defined by Hx and Hy. The reader may try to use the FCGT and FCGS operands instead, which are shown on rows 2 and 3. In this case, the weight on row 8 should be removed and placed on row 4 instead. You will find that a very similar (perhaps even identical) optimized lens is the result.

The term 'field curvature' can be ambiguous if one does not specify what one means by it. There are actually *four* 'image field curves' in the type of field curvature plot displayed in figure 1.179, but only two curves are normally displayed: the sagittal and the tangential image field (often also called the sagittal and tangential

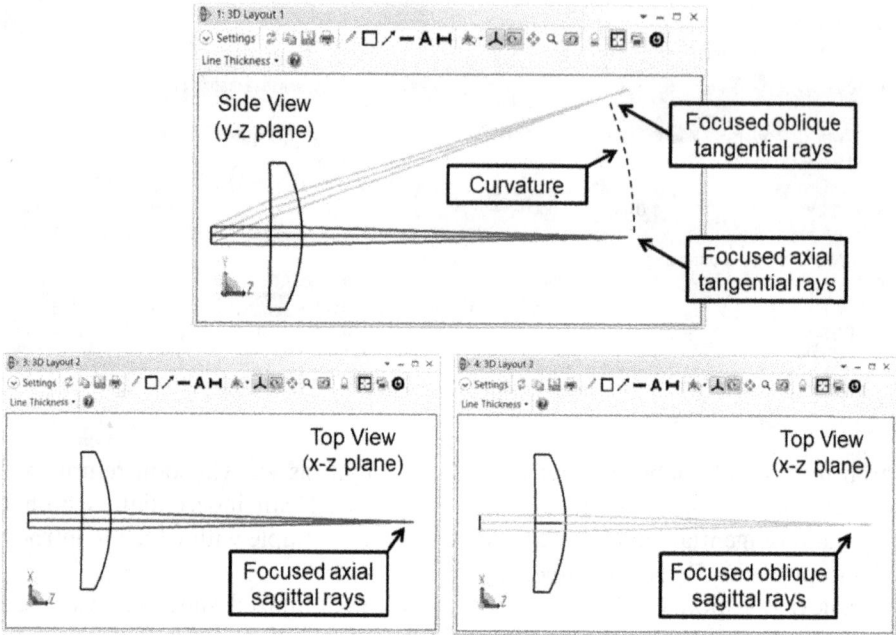

Figure 1.177. Layouts for a singlet optimized for zero astigmatism using the merit function in figure 1.176.

Lens Data

Update: All Windows

Surface 3 Properties — Configuration 1/1

	Surface Type		Comment	Radius	Thickness	Material	Coating	Semi-Diameter	Decenter Y
0	OBJECT	Standard		Infinity	Infinity			Infinity	
1		Coordinate Break			0.00000			0.00000	0.00000
2	STOP	Standard		Infinity	18.23816 V			2.50000 U	
3		Coordinate Break			0.00000			0.00000	0.00000 P
4	(aper)	Standard		-607.51594 V	10.00000	N-BK7		22.00000 U	
5	(aper)	Standard		-47.89548 X	100.56084 M			22.00000 P	
6	IMAGE	Standard		Infinity	-			44.54058	

Figure 1.178. The prescription for the lens shown in figure 1.177.

image surfaces). Some items not shown in that plot are: (1) the average between the sagittal and tangential curves, and (2) the so-called *Petzval surface*, which is actually the sag profile for the focal position of the oblique rays when astigmatism is zero (for a review of what 'sag' is, see, for example, figure 1.114 in section 1.3.3). The first of these is intuitively comprehensible: in the presence of astigmatism (such as in figures 1.171 and 1.174), the average curve between the sagittal and tangential results in a sort of 'average field curvature' surface, which is rather physical in the sense that it is a surface on which a sensor would see the average focused spot across

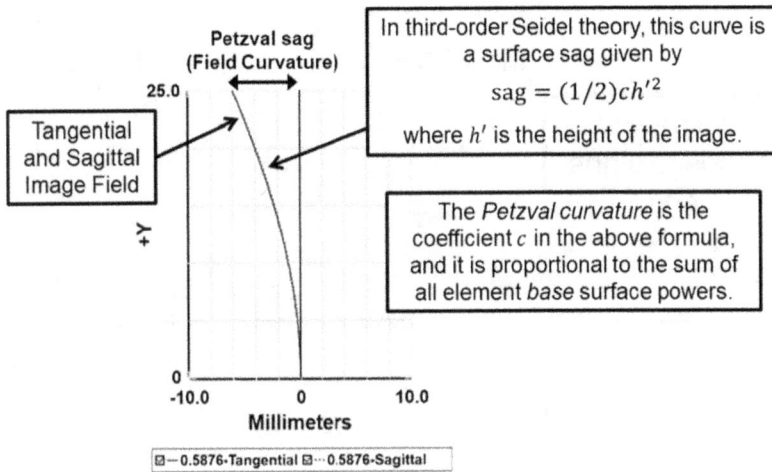

Figure 1.179. The OS Field Curvature plot for the lens in figure 1.177. The +Y axis units are degrees.

the image field. But the latter—the Petzval surface—is generally an *invisible* surface (it is there, but it is described by a mathematical expression).

To the third-order, the Petzval surface is a quadratic sag profile (i.e. a parabola) given by 1/2 times the *Petzval curvature* times the square of the image height (see the boxed description in figure 1.179) [6]. The Petzval curvature is proportional to the sum of the local *base* powers of all element surfaces in the optical system. Evidently, the reciprocal of the curvature is the radius of a sphere. Thus, the *Petzval field curvature* is governed by the sum of all element base surface powers. This sum is known as the *Petzval sum* or the *Petzval theorem* [5, 6]. When astigmatism is zero, the sagittal and tangential surfaces become the same surface, which is the Petzval surface, so that the Petzval surface becomes a physical image surface, or **image field**. Notice that I mentioned that the Petzval field curvature is governed by the *base surface powers* of optical elements. This means that here, we are referring to the curvature at the vertex of a surface (on-axis), not at any other location across an element's surface. For this reason, the Petzval field curvature is known as a *shape-independent aberration* in that aspheric surfaces play no role in controlling Petzval field curvature. However, aspheric surfaces do control the local tangential and sagittal powers across an element's surface, which affect astigmatism. So, astigmatism is a *shape-dependent aberration*.

Based on the discussion above, if we want the singlet in figure 1.177 to have a flat Petzval surface, we need the sum of its surface powers to equal zero. Therefore, a thick meniscus lens of equal radii could do the job, as one surface power would be the negative of the other, thereby canceling to yield zero Petzval curvature. The thickness of the lens would provide positive lens power to focus rays, while astigmatism would be dealt with by shifting the stop. An example of such a lens is shown in figure 1.180 and its field curvature plot is shown in figure 1.181. This

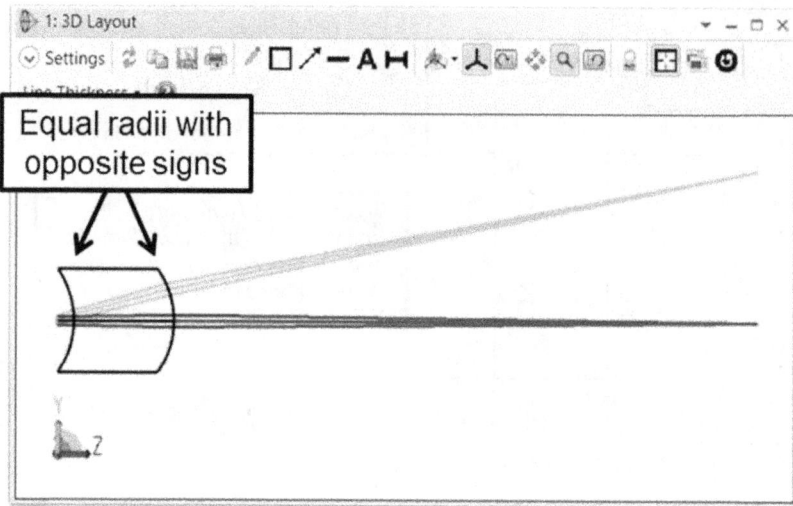

Figure 1.180. A thick meniscus lens with equal radii and zero Petzval curvature.

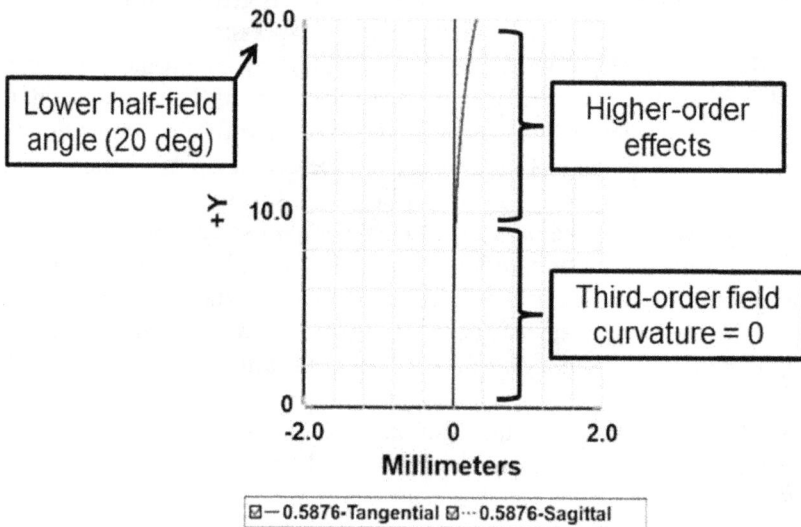

Figure 1.181. The Field Curvature plot for the lens in figure 1.180. The +Y axis units are degrees.

lens's left and right surface radii are −51.370 18 mm, its central thickness is 59.946 73 mm, and its material is N-BK7. The stop is 10.291 81 mm from the left vertex.

Note in figure 1.181 that the half-field angle had to be reduced from the 25° that was used in the previous examples to 20° in order to obtain a reasonably flat field over half the field. Even at this reduced field angle, higher-order field curvature is present, as indicated by the curve from 10° and up. This is an *extrinsic* aberration, which has been *induced* by residual third-order aberrations [5, 25]. If all third-order aberrations were zero (which is impossible for the current lens), then there would be no higher-order aberrations. Otherwise, the third-order aberrations are kept as low as possible, while the higher order is used to balance the residual third order. In an upcoming example, we will let higher-order field curvature balance the third-order Petzval such that the average field curvature is small.

Our next step is to learn (or at least, appreciate) that, in practical lens design, it is often the case that one uses multiple elements and surfaces to control aberrations. Recall that, in a paragraph just prior to figure 1.161, I mention (in bold letters) that a ray pencil is made to pass through localized areas across multiple surfaces of a lens system optimized in such a way that the tangential and sagittal rays are focused equally towards the image plane. Of course, this image plane should ideally be flat so that it is shared by well-focused axial rays (we generally assume that flat image sensors and screens are used rather than curved sensors, as the latter have not yet been made commercially available in large quantities and with many options). So, what else can be done to flatten the image field for the current lens?

We answer the above question by noting that earlier, we saw in figures 1.157–1.163 that a positive singlet has a local surface power *gradient* that increases from the axis towards the lens's edge. What if a second singlet placed to the right had the opposite transverse variation in local surface power? That is, what if we had a negative-powered singlet (such as a biconcave, plano-concave, or negative meniscus) singlet to the right of a positive-powered singlet? Would the second lens be able to cancel the first lens's field curvature? The answer is yes. In fact, it turns out that a negative thin singlet's local transverse negative power increases from the axial position to the edge of the lens. This transverse variation in local power scales with refractive index (this is true for both positive and negative elements). This is well suited for lens design, as we have found that we need to cancel out excessive positive power for oblique rays that are refracted through positive-powered elements (recall that in section 1.3.2, I discussed this point briefly to explain the improved astigmatism for the tube lens in figure 1.111). Therefore, a judicious selection of combinations of positive and negative-powered elements should help to balance the transverse power across the field such that we have a well-focused and flattened image. You may have heard of the so-called *Cooke triplet* lens, which comprises two positive singlets and a negative singlet between them. This combination can yield a reasonably flat image field. This design form has enough variables to control all of the primary third-order aberrations (spherical, coma, astigmatism, Petzval field curvature, image distortion) as well as the system EFL. As in the case of a two-lens system, a trivial case would be a second lens of precisely equal but opposite power to the first lens, thereby yielding effectively zero beam deviation. But this trivial case is not useful, as it just makes the two-lens combination into a window. In fact, it is known through aberration theory that several other possibilities exist. One possibility is for the first lens to be a positive thin lens of high index and the next lens to be a negative thin lens of low index; these two lenses are positioned next to each other. Another possibility is to maintain the same index for each singlet but to place the negative singlet close to the image plane. You have probably heard of this type of arrangement for the second lens—it is called a **field**

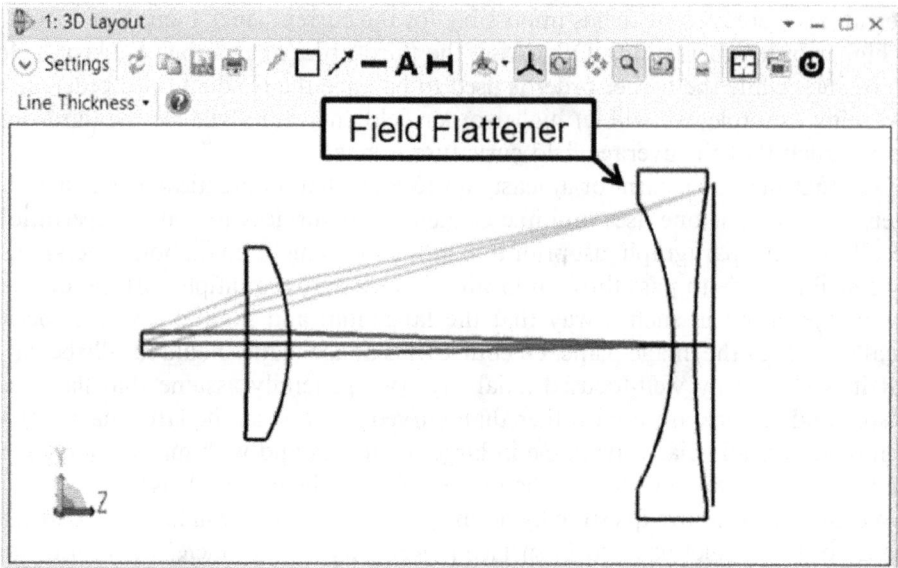

Figure 1.182. A positive singlet with a negative singlet (a field flattener).

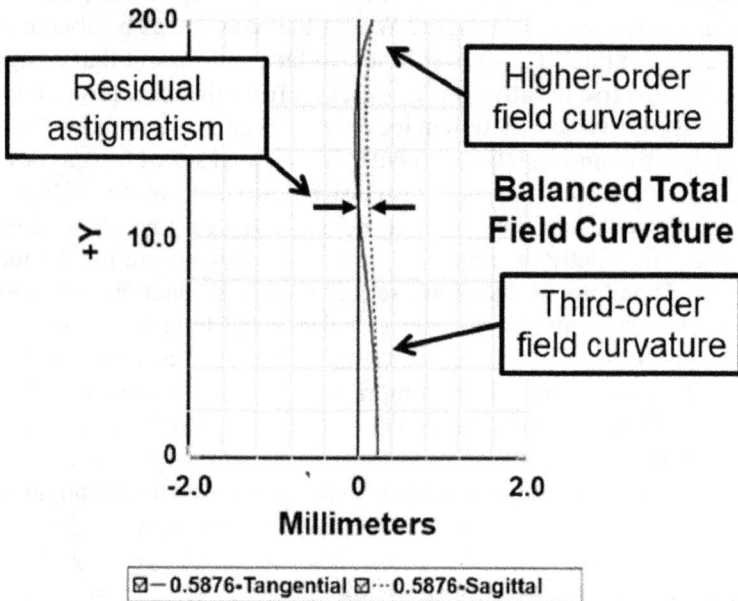

Figure 1.183. The Field Curvature plot for the lens system in figure 1.182. The $+Y$ axis units are degrees.

flattener. An example is shown in figure 1.182, whose field curvature plot, prescription, and merit function are shown in figures 1.183–1.185, respectively.

Note that the field curvature plot shown in figure 1.183 has a slight 'S' shape tò it. This is the slight balancing of third-order field curvature and astigmatism with the

Figure 1.184. The prescription for the lens system shown in figure 1.182.

#	Surface Type		Comment	Radius	Thickness	Material	Coating	Clear Semi-Dia	
0	OBJECT	Standard ▾		Infinity	Infinity			Infinity	
1	STOP	Standard ▾		Infinity	25.35885 V			2.50000	U
2	(aper)	Standard ▾	Lens 1	-768.93159 V	10.00000	N-BK7		22.00000	U
3	(aper)	Standard ▾		-47.65839 V	91.01529 V			22.00000	P
4	(aper)	Standard ▾	Lens 2	-67.85245 V	5.00000	N-BK7		33.00000	U
5	(aper)	Standard ▾		234.59401 V	4.44012 M			37.00000	U
6		Standard ▾		Infinity	-0.23506 V			37.07021	
7	IMAGE	Standard ▾		Infinity	-			36.90083	

Figure 1.184. The prescription for the lens system shown in figure 1.182.

Merit Function: 0.000940721222556651

#	Type	Ring	Wave	Hx	Hy	Target	Weight	Value	% Contrib
1	EFFL ▾		1			0.00000	0.00000	104.59633	0.00000
2	OPLT ▾	1				120.00000	1.00000	120.00000	0.00000
3	OPGT ▾	1				95.00000	1.00000	95.00000	0.00000
4	BLNK ▾	Using the difference of field curvatures between tan and sag rays:							
5	FCGT ▾			1 0.00000	1.00000	0.00000	0.00000	0.16218	0.00000
6	FCGS ▾			1 0.00000	1.00000	0.00000	0.00000	0.16219	0.00000
7	DIFF ▾	5		6		0.00000	1.00000	-7.81125E-06	1.37895E-03
8	BLNK ▾	Petzval curvature (1/radius):							
9	PETC ▾			1		0.00000	0.00000	-2.32238E-04	0.00000
10	BLNK ▾	RMS Spot							
11	RSCE ▾	3		1 0.00000	0.00000	0.00000	1.00000	1.40409E-03	44.55535
12	RSCE ▾	3		1 0.00000	1.00000	0.00000	1.00000	1.56628E-03	55.44327

Figure 1.185. The merit function for the lens system shown in figure 1.182.

higher order. If you look at the merit function in figure 1.185, no weight was placed on the PETC operand, which computes the third-order Petzval curvature. I purposely left it blank in order to allow the program to flatten the field using any order of the field curvature. However, note from this operand that the residual curvature is small (its reciprocal is roughly 4306 mm, a large radius), so the higher-order field curvature used for balancing is also small, as can be seen in the upper portion of the curve in figure 1.183. There is also residual astigmatism near the middle field point, as nothing was done to control it. Only the largest field point was targeted at zero astigmatism (for parabasal rays) using the difference between the FCGT and FCGS operands on rows 5–7 in the Merit Function Editor. Note also that this lens system its field reduced to ±20°. In theory, when astigmatism is present, the sagittal and tangential surfaces are separate from the Petzval surface, so they can

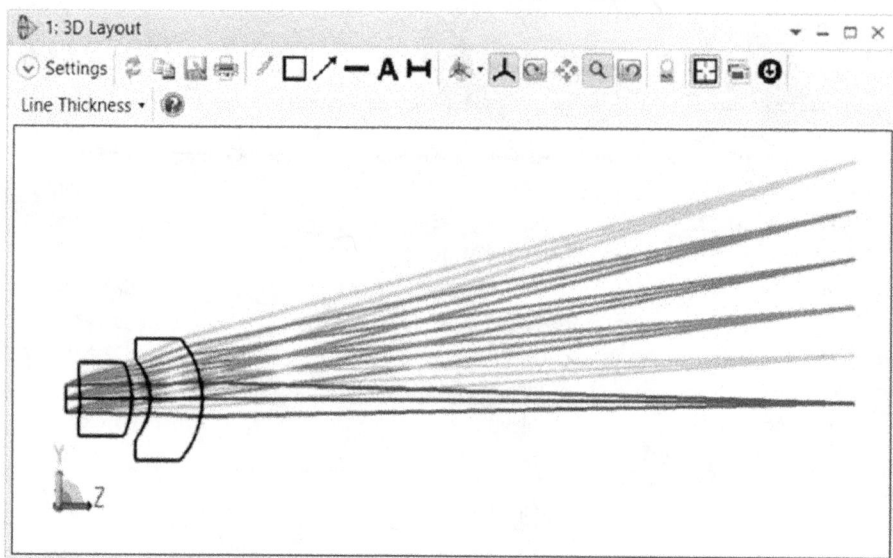

Figure 1.186. A two-element anastigmat optimized for 550 nm light.

each bend in a different direction from the Petzval surface. This is observed at the middle field in figure 1.183. This is another way the program has found to provide a flat field—that is, by letting the image plane be the average between the astigmatic curves. But at the maximum field in figure 1.183 the astigmatism is zero, yet the combined curve is bent towards the right, which is noticeably the opposite direction to the slightly left-bent curve near the axis. The only way this could happen is if there is slightly higher-order Petzval curvature. This is one way to tell whether there are higher-order aberrations at work.

It is possible to use a different arrangement of elements to minimize the astigmatism across a larger field. In particular, rather than using a field flattener, we can place two thick meniscus elements close to each other, the first of which has a positive power and the second of which has a negative power; both elements are made from the same material. Doing this at 550 nm, we end up with the low-aperture **anastigmat** shown in figure 1.186, whose field curvature/distortion plot, MTF, prescription, and merit function are provided in figures 1.187–1.190, respectively.

An anastigmat is a lens that is free of astigmatism, though nowadays it is also common to use this term for a lens that is corrected for both astigmatism and field curvature [5]. By now, it should have become quite apparent that the process of lens design is to involve a sufficient number of surfaces and elements to bend axial and oblique rays towards a common focal plane. Ray pencils from each field point (or angle) 'see' an effective optical system of locally powered surfaces along their path. This effective optical system should comprise positive and negative powers in order to obtain a flat focal plane across the image field. All anastigmats have this characteristic, though some have elements to the left and right of the stop, while

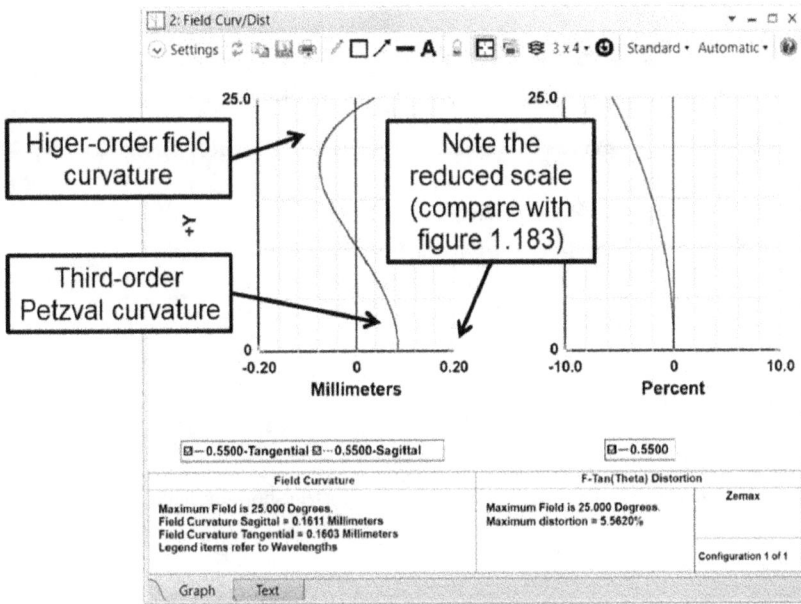

Figure 1.187. Field curvature (left) and distortion (right) for the lens system in figure 1.186. The units of the $+Y$ axis are degrees.

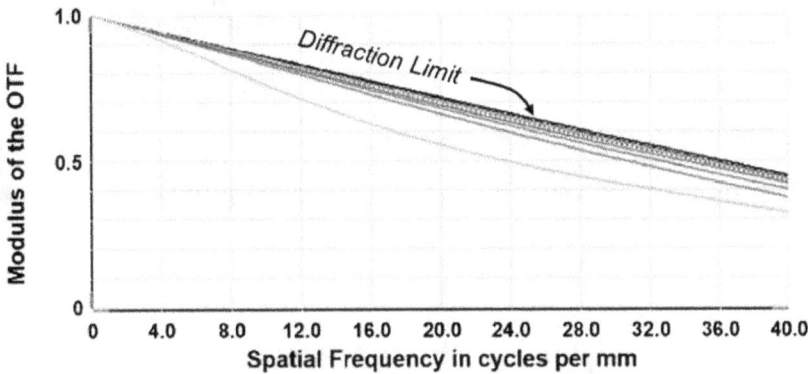

Figure 1.188. The MTF (at 550 nm) for the lens system in figure 1.186. Six fields are shown, namely 0°, 5°, 10°, 15°, 20°, and 25°.

others only have elements to the right (such as the current lens) or to the left (such as the so-called *landscape lens* that is often discussed in introductory texts on lens design [5]). Each of these types of arrangements for the stop position and elements have been determined by theory and experience to be beneficial for controlling certain dominant aberrations. Anastigmats of the form shown in figure 1.186 are mostly used for laser scan lens systems, in which the stop represents the location of a galvo-mirror (also called a 'scan mirror'), and each ray pencil is a purely geometric

Lens Data

Update: All Windows

Surface 0 Properties — Configuration 1/1

	Surface Type		Comment	Radius		Thickness		Material	Coating	Clear Semi-Dia	
0	OBJECT	Standard		Infinity		Infinity				Infinity	
1	STOP	Standard		Infinity		2.82037				2.50000	U
2		Standard	Lens 1	-30.67654	V	10.00000		N-BK7		3.71015	
3		Standard		-15.57473	V	3.42220	V			6.66749	
4		Standard	Lens 2	-11.02018	V	10.00000		N-BK7		7.26440	
5		Standard		-16.41407	V	123.32407	M			11.09464	
6		Standard		Infinity		-0.08618	V			44.02462	
7	IMAGE	Standard		Infinity		-				43.99812	

Figure 1.189. The prescription for the lens system shown in figure 1.186.

Merit Function Editor

Wizards and Operands — Merit Function: 0.00386802223231294

	Type		Wave			Target	Weight	Value	% Contrib
1	EFFL		1			100.00000	1.00000	100.00000	9.38986E-07
2	RSCE	3	0	0.00000	0.00000	0.00000	1.00000	2.42760E-03	3.22863
3	RSCE	3	0	0.00000	0.20000	0.00000	1.00000	3.35013E-03	6.14873
4	RSCE	3	0	0.00000	0.40000	0.00000	1.00000	4.61793E-03	11.68308
5	RSCE	3	0	0.00000	0.60000	0.00000	1.00000	4.90151E-03	13.16203
6	RSCE	3	0	0.00000	0.80000	0.00000	1.00000	4.77919E-03	12.51325
7	RSCE	3	0	0.00000	1.00000	0.00000	1.00000	9.57724E-03	50.25083
8	FCGS		1	0.00000	0.20000	0.00000	0.00000	0.06342	0.00000
9	FCGT		1	0.00000	0.20000	0.00000	0.00000	0.06305	0.00000
10	DIFF	8	9			0.00000	1.00000	3.73381E-04	0.07638
11	FCGS		1	0.00000	0.40000	0.00000	0.00000	5.46056E-03	0.00000
12	FCGT		1	0.00000	0.40000	0.00000	0.00000	4.20002E-03	0.00000
13	DIFF	11	12			0.00000	1.00000	1.26053E-03	0.87051
14	FCGS		1	0.00000	0.60000	0.00000	0.00000	-0.05651	0.00000
15	FCGT		1	0.00000	0.60000	0.00000	0.00000	-0.05728	0.00000
16	DIFF	14	15			0.00000	1.00000	7.73128E-04	0.32747
17	FCGS		1	0.00000	0.80000	0.00000	0.00000	-0.06974	0.00000
18	FCGT		1	0.00000	0.80000	0.00000	0.00000	-0.06802	0.00000
19	DIFF	17	18			0.00000	1.00000	-1.71795E-03	1.61689
20	FCGS		1	0.00000	1.00000	0.00000	0.00000	0.04121	0.00000
21	FCGT		1	0.00000	1.00000	0.00000	0.00000	0.04074	0.00000
22	DIFF	20	21			0.00000	1.00000	4.72298E-04	0.12221
23	MNCA	1	4			0.00000	0.00000	0.00000	0.00000
24	MNEA	1	4	1.00000	1	1.00000	0.20000	1.00000	0.00000

Figure 1.190. The merit function for the lens system shown in figure 1.186.

ray-based model of an unvignetted laser beam that reflects off the mirror at various rotational angles of the mirror [31, 32].

The lens in figure 1.186 can form the basis for a laser scan lens design provided that the stop is re-optimized to be farther to the left in order to provide sufficient space between the scan mirror and the first element. However, this is not all that this lens must satisfy. In addition, the lens's image distortion must satisfy a so-called *f-theta* condition. This is a condition whereby the image height must be equal to the product of the lens's EFL and the half-field angle (rather than the tangent of the half-field angle). The reason for this is that laser scan mirrors rotate at constant angular velocity, resulting in an acceleration of the focused laser spot across the image plane if the lens has zero image distortion. Assuming that the laser beam is modulated at some constant pulse rate during scanning, an accelerating spot at the image would cause uneven spots and lines in the image. But if the image height h' is given by

$$h' = f\theta, \tag{1.50}$$

where f is the lens's EFL, and θ is the half-field angle in radians (equivalently, it is the angle after reflection off the rotating scan mirror), then at any point in time t, a constant angular velocity $d\theta/dt$ translates into the linear transverse velocity dh'/dt. Since this imposes a criterion for the image height as a function of the field angle, and since the image height appears in the formula for percentage image distortion (see equation (1.8)), it imposes a distortion criterion for the image. Substituting equation (1.50) into (1.8) for the image height gives:

$$D\% = \left[\left(h' - h'_p \right)/h'_p \right] \times 100\%$$
$$= \left[\frac{f\theta - f \tan\theta}{f \tan\theta} \right] \times 100\% = \left[\frac{\theta}{\tan\theta} - 1 \right] \times 100\%. \tag{1.51}$$

All designed laser scan lenses suffer from the image distortion given by equation (1.51). This was why I displayed the image distortion plot in figure 1.187. Taking a quick glance at this plot and entering 25° (but converted to radians) into equation (1.51) indicates that the lens of figure 1.186 closely satisfies the f-theta condition, though not exactly, as this criterion was not included in the merit function during optimization. It may be possible to satisfy this criterion through further optimization. But if a high field angle and a lower f-number were needed, more elements might be required, such as those of the lens shown at the bottom of figure 1.153. Furthermore, a laser scan lens design must be checked for coherent beam propagation characteristics once the geometric ray-based optimization and modeling has been done. Most laser scan lenses are first designed through pure geometric ray tracing in which the diameter of the entrance pupil represents the $1/e^2$ diameter of a TEMoo Gaussian beam. The designer must ensure that the area of the scan mirror is larger than the diameter of the entrance pupil and that coherent beam propagation through the lens system results in acceptable beam quality at the focus.

Once a design has been completed (but more often, even during the design process), one has to check for the manufacturability of the lens. For example, the anastigmat in figure 1.186 has elements that are rather thick and of the meniscus form. Thick meniscus lenses are typically difficult to fabricate, especially during the *centering* or *edging* process (this is the process by which a lens's diameter is established by cutting with diamond tools such that the axis defined by the diameter is as collinear as possible with the axis defined by the two surfaces of the lens [88]). A rule in the lens design process is that one radius should either be significantly different from the other or at least satisfy the criterion $S = R_2 - R_1 - t$ as an operand in the merit function, where R_2 is the element's second radius, R_1 is the first radius, t is the element center thickness, and S is a figure of merit that you would need to agree upon with the manufacturing plant that is fabricating your lens [89]. Generally, the idea is simply to assure that $|R_2 - R_1| \neq t$ (as I was advised quite often by my lens processing/engineering colleagues at a manufacturing plant years ago). On examining the prescription in figure 1.189, we see that Lens 1 is easier to manufacture, as its radii are quite different. Fortunately, this problem can sometimes be overcome by either keeping all elements thin from the start of the design optimization (and not letting their central thicknesses be variables at all), or, if the elements have already become thick, by gradually thinning the elements (and not using their thicknesses as variables) and adding or subtracting elements during design optimization [89, 90].

In much of the discussion in the current section, we focused on a rather heuristic theory of the design of low-aperture lenses. This was done on purpose, with the aim at introducing the nature of lens design to newcomers in lens design in a highly graphical way such that complicated mathematics are avoided (note that algebra, calculus, and vectors are actually quite predominant in formal courses on lens design). Further, by keeping the system aperture low (or equivalently, having a large f-number for the lens), we were able to ignore most of the spherical aberration and coma (some coma creeps in at high obliquity), which become problematic at high apertures. Of course, we also ignored image distortion (which is present even at low apertures), but this is inconsequential for most of the discussion up until near the end, where the design of a two-element anastigmat with its stop towards the left naturally led to the topic of laser scan lenses, whose characteristic is to have a specific type of distortion in the image. Since many useful imaging systems must have their iris opened wide in order to produce a bright high-resolution image (due to the increased NA in image space), we should touch a little on high-aperture (low $F/\#$) lens design. We will see that at high apertures, spherical aberration is often minimized through bending element surface curvatures and *splitting* the powers of the elements (i.e. instead of having a single element do all of the ray bending, we divide its power amongst several elements and combine positive with negative elements). This is one reason why many high-resolution imaging lenses comprise multiple elements. Take, for example, laser scan lenses. A high-resolution laser scan lens with an EFL of 48 mm and $F/3$ (which is quite a high aperture for a scan lens) can have up to six elements [90]. Spherical aberration can also be corrected by using aspheric lenses. We saw this in section 1.3.8, where I briefly mentioned the use of spherical aberration compensator plates for improving the axial MTF of a microscope objective. On this note, if you are somewhat familiar with the typical lens design layout of a mobile phone lens, then you may note that they operate at roughly *F*/3 and below, they contain a number of highly aspheric elements, and they have the stop located at the left side of the lens system [91]. In this way, it may seem that a mobile phone lens somewhat resembles a laser scan lens. Yet, they are not used as such, even if they can be scaled up to similar sizes to those of laser scan lenses. As part of our learning journey in high-aperture imaging, it is instructive to understand why a scaled mobile phone lens would not work for laser scanning. This is the topic of the next section.

1.4.3 Why are mobile phone lenses not used as high-aperture laser scan lenses?

The short answer to the question in this section's title is that the highly aspheric surfaces of the elements in mobile phone lenses result in irregularly shaped image distortion curves, which do not satisfy the f-theta distortion criterion for laser scanning. These distortion curves remain when the lens is scaled to the typical size of a high-resolution scan lens. Recall from the previous section that ray bundles from each field angle 'see' an effective optical system through local areas at the surfaces they meet in their path, resulting in a gradient of effective focal lengths associated with the field. Figures 1.161 and 1.162 illustrate this effect, where the image formed by the oblique ray pencil is smaller than that produced by the axial ray pencil, due to the effectively smaller focal length associated with the oblique rays. This would occur with or without astigmatism.

Let us take an example from the first embodiment of US Patent 7643225B1 by Tsai [92]. Figure 1.191 shows the layout for this embodiment, which is a four-

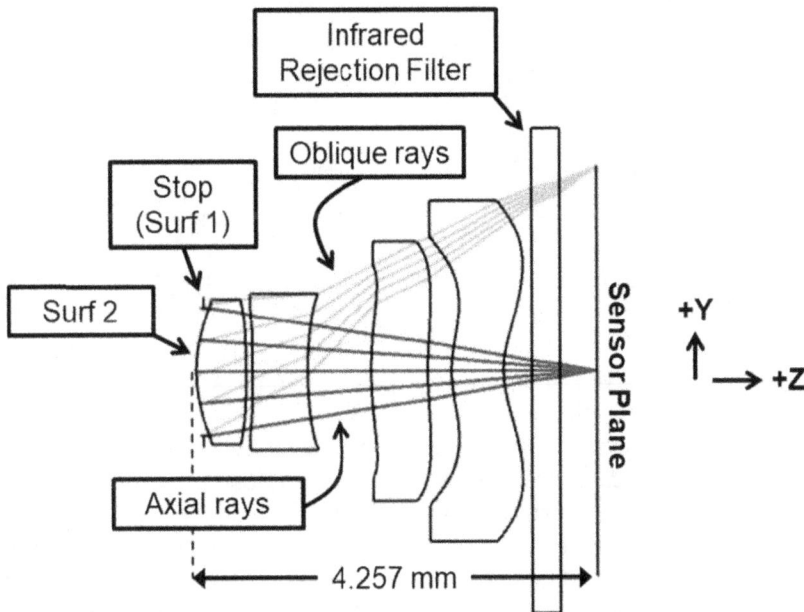

Figure 1.191. The layout of the first embodiment of US Patent 7643225B1 by Tsai [92].

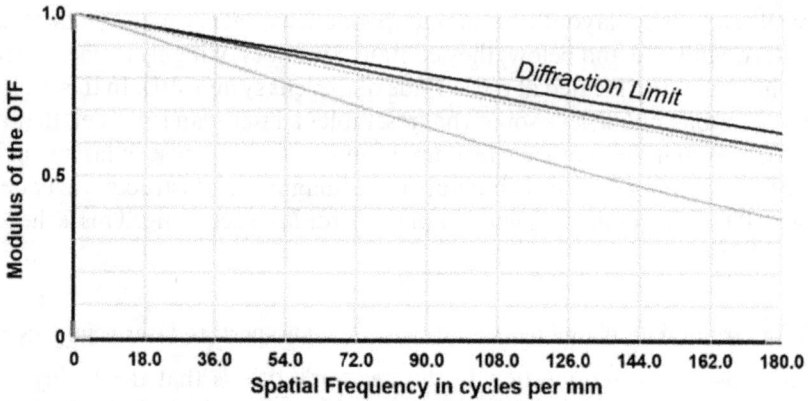

Figure 1.192. The polychromatic MTF (at F, d, C, weights = 1) for the lens in figure 1.191. Blue denotes the axial field, green denotes the oblique field.

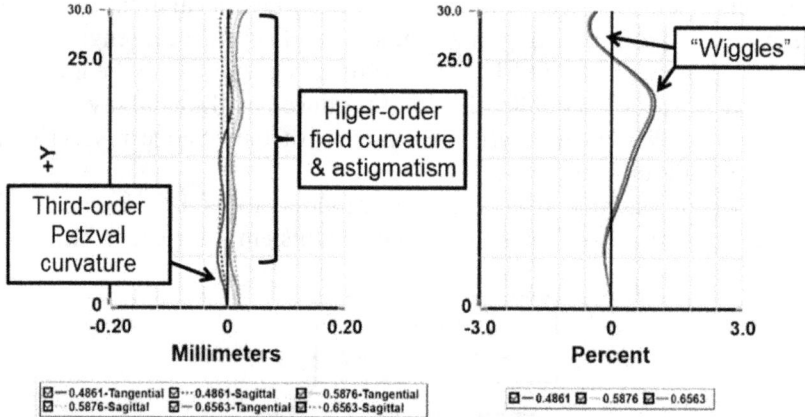

Figure 1.193. The field curvature (left) and image distortion (right) for the lens in figure 1.191 at the F, d, and C wavelengths. The +Y axis units are degrees.

element aspheric system for a mobile phone operating at $F/2.8$ with an EFL of 3.67 mm at 588 nm. The wavelengths are at the F (486.13 nm), d (587.56 nm), and C (656.27 nm) wavelengths, and all weights are set to 1. The maximum half-field for this lens is 33°, but I have limited it to 30°. The polychromatic MTF for this lens is displayed in figure 1.192 (the maximum spatial frequency has been chosen based on half the Nyquist frequency of a 12-megapixel image sensor with 1.4-micron pixel pitch, which was a typical sensor specification for a mobile phone lens in 2012 [91]). The field curvature and distortion plots are shown in figure 1.193.

We note from the MTF curves of figure 1.192 that this is a rather high-performance lens at $F/2.8$. The field curvature appears 'flat' (we will touch on this soon), and the image distortion possesses localized curvatures, indicating that there are various effective focal lengths (EFLs) associated with different field angles.

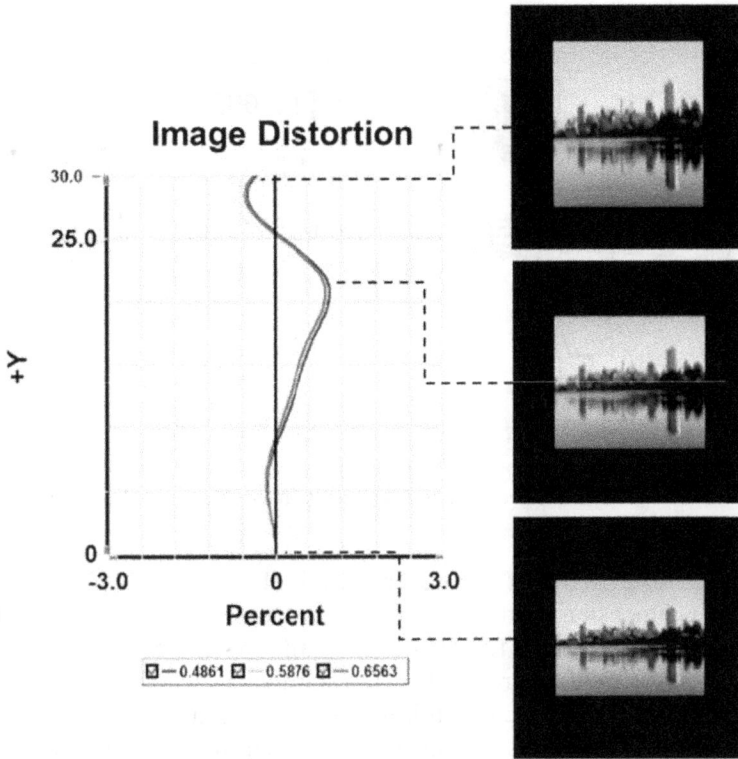

Figure 1.194. Simulated images at three locations across the image plane for the lens of figure 1.191. Each image spans a 1.5° field of view (±0.75°) from the three fields indicated on the distortion plot (this plot is the same as the one in figure 1.193).

These 'field wiggles' are caused by the highly aspheric surfaces of the elements for this lens. **Each ray bundle from an off-axis field passes through portions of this lens system, which are essentially localized areas of varying surface power for the sagittal and tangential rays, resulting in a transverse variation of EFL over the field.** As long as strong aspheres are present, it seems unlikely that any amount of re-optimization could change this distortion curve into an f-theta curve.

It is useful to visualize the effect of the different EFLs across the field for this lens (as was done in figures 1.161 and 1.162). There is an additional nonobvious nuance to explain regarding the image distortion curve of figure 1.193. Applying, once again, the Image Simulation feature in OS to the current lens system gives the three images of figure 1.194, each corresponding to the fields indicated in the distortion plot on the left.

In figure 1.194, the simulated images span a full field of view of 1.5° and sample the field at the locations indicated on the distortion plot. This means that the top image in this figure has a principal ray at 30°, the middle image has a principal ray at about 21°, and the bottom image is at the axial field point. The first thing to notice is that each image has a slightly different size, which is to be expected from the field

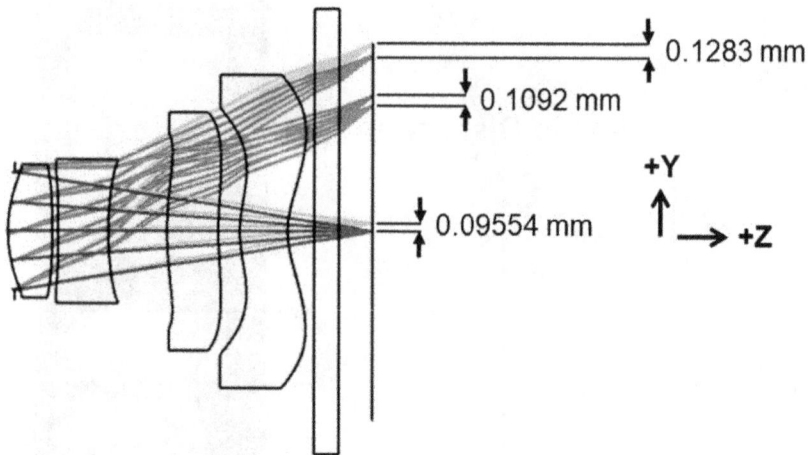

Figure 1.195. The image sizes obtained using real ray heights at the three image locations of figure 1.194.

wiggles on the distortion plot. However, why is the image at the maximum field *larger* in the y-dimension than the axial image? You might have expected it to be smaller, since the distortion at full field is less than zero (it is at roughly −0.4%). The reason for this apparent discrepancy is the manner in which image distortion is computed in the plot. In particular, it is given by equation (1.8). As described in the paragraph after that equation, the distortion is given by the difference between the height of the chief ray (given by exact ray tracing) and the height of the paraxial chief ray (i.e. the image height given by m times the object height, where m is the paraxial image magnification; or, if the object is at infinity, this height is given by the EFL times the tangent of the half-field angle, assuming zero image distortion). If, for the current lens, you compute this difference manually (using REAY and PARY operands in the merit function), you can see that indeed it is true that at the maximum field, the height of the real chief ray is less than the height of the paraxial chief ray. But if you were to trace real rays from the field, you would find that the local image size at full field is *larger* than that at the axis (figure 1.195). In other words, a real ray trace indicates that the local image at the farthest off-axis point at the image plane is slightly larger than the real image at the axis of the image plane. However, this does not mean that the distortion plot of figure 1.193 is necessarily wrong. It just provides a different set of information (for those who want to compare the difference in height between the real chief ray and the paraxial chief ray). In other circumstances, one might instead wish to use the CENY operand to estimate the RMS center of a ray bundle. This is also an option. Why not? It all depends on the objective of the analysis that you are performing.

There is another nuance to check in the distortion plot, which concerns its relationship to the relative illumination for this lens (actually, for all lenses). In particular, the wiggles of the distortion plot should sound an alarm to check the relative illumination, because the relative illumination is a function of a number of variables, one being image distortion and another being the instantaneous rate of

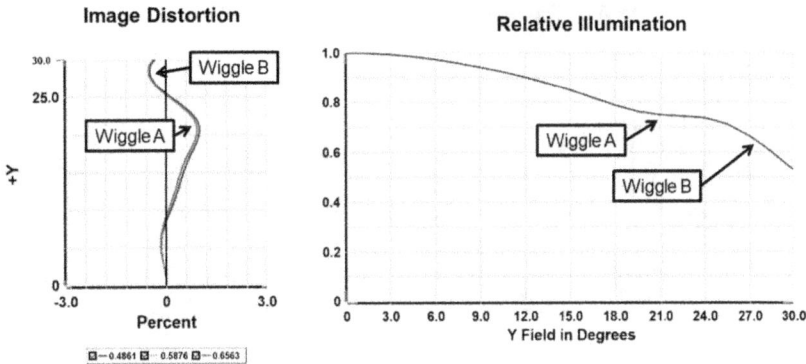

Figure 1.196. The wiggles on the image distortion plot and the relative illumination plot imply a direct relationship (not a statistical correlation) between these two curves.

change of distortion. Figure 1.196 provides an implied relationship (an equation expressing this relationship is provided in chapter 2).

Let us now return to the observation of the apparent flat field for the field curvature in this lens. In the previous section, I mentioned that when the astigmatism is zero, third-order (Petzval) field curvature is independent of element shape, and that this means it is not affected by aspheres. Here, in figure 1.193, we see an apparently flat *average* curve for each wavelength. In fact, when the PETZ operand is used in OS to compute the third-order Petzval radius for this mobile phone lens, the value obtained is roughly −21.84 mm (i.e. it is a sphere whose center of curvature is to the left of the image plane). This is undercorrected third-order Petzval field curvature. In fact, you can see it near the axial location of the field curvature plot. There is a noticeable curve bend to the left. Technically, the aspheres of this mobile phone lens are not correcting the field curvature. Stated more precisely, they are not correcting the third-order Petzval curvature (which requires adequate combinations of positive and negative base surface powers to cancel one another). If we take a point on the field curvature plot of this lens that is farthest displaced from the image plane, the displacement is roughly 0.015 mm. This means that there is a maximum residual 'field curvature' (more accurately, a defocus) at some point across the image of roughly 0.015 mm. If we take the percentage of the magnitude of the ratio of this to the lens's EFL, we obtain (0.015 mm/3.67 mm) × 100% ≈ 0.41%. So, the maximum residual defocus amount in the image is roughly 0.4% of this lens's EFL. Let us compare this value with that for the lens of figure 1.182, which has an EFL of 104 mm (see the EFFL operand on row 1 of figure 1.185). In examining the field curvature plot in figure 1.183, this spherical lens system's farthest displacement in field curvature is roughly 0.24 mm. Its percentage ratio to the lens EFL is (0.24 mm/ 104 mm) × 100% ≈ 0.23%, which is roughly half of the mobile phone lens's field curvature. Hence, it can be said that this mobile phone lens's maximum field curvature is worse than the two-element system of figure 1.182. Moreover, the Petzval radius of this lens is −4306 mm (the reciprocal of the curvature on row 9 of

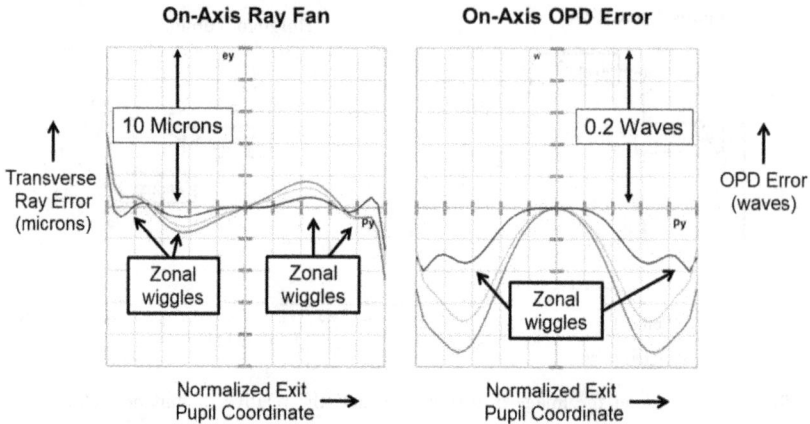

Figure 1.197. The on-axis transverse ray error (ray fan plot) and the wavefront OPD error for the lens in in figure 1.191. The blue, green, and red lines denote the F, d, and C wavelengths, respectively.

figure 1.183). This is 41 times this lens's EFL, while for the mobile phone lens in this section, its Petzval radius is only six times its EFL.

Of course, it is not an 'apples-to-apples' comparison, as the spherical lens system of figure 1.182 has a smaller field of view and a significantly smaller aperture. But the point is that third-order field curvature is not corrected by the use of aspheres. However, as highlighted in the previous section, the astigmatism is indeed correctable by aspheres, and if there are any residual third-order aberrations, then higher-order aberrations exist. In fact, if we use the ASTI operand in the Merit Function Editor to compute the third-order astigmatism (actually, in terms of the wavefront aberration, it is a fourth-order aberration), the value is roughly 28 waves (at the d wavelength)! The fourth-order spherical aberration wavefront error is -1.17 waves (which is a lot for this small lens), and the fourth-order coma wavefront error is 6.8 waves! Since the MTF is quite good (figure 1.192), this means that lower-order aberrations are being balanced by higher-order aberrations (mostly induced by the aspheres). You can see this in the ray fan and wavefront error (optical path difference (OPD)) plots, as shown in figure 1.197, which only displays the axial aberration. For the ray fan plot, one expects to see a cubic function if third-order spherical aberration is the only aberration present. But the curves are rather wiggly near the edges. These 'zonal wiggles' signify that higher-order spherical aberration is balancing the third-order aberration. In the OPD plot, one expects a fourth-power function of the exit pupil coordinate (so, since the ray fan plot for spherical aberration is the first derivative of the OPD plot, this is why I mentioned that we would expect a cubic curve for the left plot). Instead, we see wiggles here as well, resulting in a balanced cancellation between third-order spherical aberration and higher-order terms, yielding a total OPD error of less than 0.2 waves. Returning to the field curvature plot in figure 1.193, the presence of field wiggles indicates that higher-order aberrations are present to control the astigmatism curves such that the average of the tangential and sagittal curves lies roughly along the image plane.

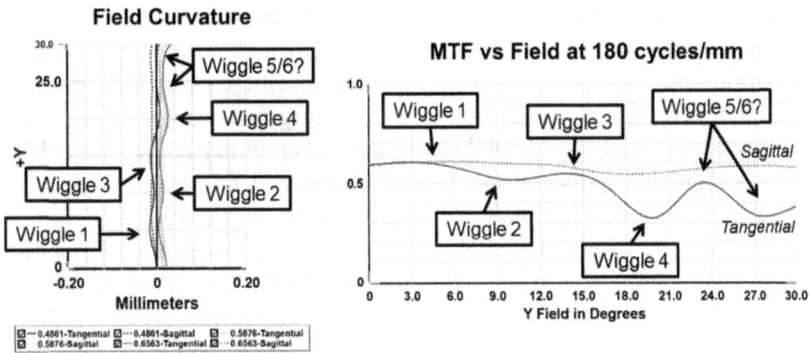

Figure 1.198. The effect of wiggles in the field curvature plot on wiggles in the MTF vs field plot at 180 cycles/mm for the lens in figure 1.191.

This is of course, effectively, a flat field on average, but it has residual defocus wiggles. You can see the direct effect of the defocus wiggles on the MTF plotted against field at 180 cycles/mm, as shown in figure 1.198. The image in the sagittal and tangential rays is not flat, but the wiggles average into flatness. Still, if this lens has been manufactured, tested, and mounted into commercial smartphones, then perhaps it could be an indication that this magnitude of wiggles is 'ok'.

There is, of course, one point that is often made about the aspheric surfaces of mobile phone lenses and the placement of the stop in front of the lens, which is that this configuration helps to achieve some form of telecentricity at the image for the chief ray in order to reduce vignetting at the microlenses of the image sensor. For a thick lens system, if the stop is located at one focal distance in front of the first principal plane, then image space telecentricity is attained. However, a simple check for this mobile phone lens shows that the first principal plane is 1.8933 mm towards the left of the stop, and the second principal plane is 3.6584 mm to the left of the image plane. Therefore, the stop is actually in between the two principal planes, which means that the telecentricity condition is not met by way of the first-order properties of the system (however, note that even for spherical lens systems with buried stops, it is possible to achieve image space telecentricity if either the stop is at the back focal plane of a next lens group to the rear of the stop or if negative distortion is introduced). Moreover, the exit pupil is 2.4092 mm to the left of the image plane, so the separation between the second principal plane and the exit pupil is 1.2492 mm, which is about 34% of the lens's EFL. This means that this lens's stop is not too far from its OC (see section 1.3.16 for a review of the OC concept). When the stop of a lens is at its OC, at low field angles (or if there is low image distortion), the principal ray exits the lens at the same field angle at which it enters. This mobile phone lens's chief ray is unvignetted, so it is also the principal ray. Since the stop for this mobile phone lens is near its OC, the chief ray tends to exit at an angle that is close to the field angle at which it entered the lens, at least for the low obliquity condition. At high obliquity, the chief ray's direction is influenced by the higher-order terms due to the aspheric surfaces. Thus, this lens's aspheres are working very hard against fundamental first-order lens properties. Note that nothing prevents the first-order properties from being applicable to highly aspheric systems, because rays near the optic axis are governed by the first-order (paraxial) properties. Also, consider the following: suppose this lens was originally comprised of spherical elements and the stop was at the OC, yielding equal chief ray angles in the object and image spaces (neglecting distortion). If the last element had been made aspheric to bend the chief ray down and make it telecentric to the image, then that asphere would have been the cause of new aberrations, which would then have had to be corrected by making the rest of the elements aspherical. Thus, there is the possibility that certain aspheres of a mobile phone lens may be correcting ray errors that at least one or more aspheres have created.

There is no doubt, however, that perhaps the most significant constraint for the design of a mobile phone lens is physical space—fitting all elements within a small volume. Must this constraint require all-aspheric elements? Steinich and Blahnik answer this question in an insightful study [91]. They found that shrinking (downscaling) a $F/2.8$ Biogon lens (with only spherical elements) yielded a resulting nominal performance that was not far from those of some mobile phone lenses (one of which is the lens being studied in this section). Recall from section 1.2.12 that downscaling a lens naturally reduces the aberrations. In fact, in the paper by Steinich and Blahnik, the downscaled Biogon lens outperformed the mobile phone lens in terms of image distortion and *longitudinal spherical aberration* (i.e. the variation of defocus with the pupil zone, also known as *zonal spherical aberration*). However, it is noted that the Biogon lens they used had eight elements, and the chief ray angle at the image was larger than that of mobile phone lenses. Should this number of elements be reduced in order to fit the Biogon into a small space, then the added aberrations might need to be corrected by making the remaining elements aspherical. This, I believe, should be the main reason for making any lens system aspherical.

Returning to the point of this section: despite the similarity in lens and stop configuration between laser scan lenses and mobile phone lenses and the fact that mobile phone lenses have high resolution capability (at $<F/3$), we find that an upscaled and re-optimized mobile phone lens that contains many highly aspheric surfaces is unlikely to be suitable for use as a high-resolution laser scan lens because of the irregularities in the image distortion, which cannot satisfy the f-theta criterion. These irregularities are caused by the highly aspheric surfaces of the elements. Therefore, it seems that a laser scan lens design form must remain all-spherical, such as the $F/3$ six-element design by Hopkins [93] and Murthy [90]. However, their designs were limited to a half-field angle of $\pm 9.55°$. Thus, there is often a trade-off between high aperture and field angle. For this reason, wide-angle lenses have low aperture, and high-aperture lenses have modest field angles, depending on the desired correction level of the aberrations and on what you consider to be sufficient image quality. To the third order, there are five main monochromatic aberration types: spherical aberration, coma, astigmatism, Petzval field curvature, and image distortion (though this last one can be considered not precisely an aberration in the sense that it does not cause a blurred image, but you get the idea). In the previous

section, at low aperture, we saw how to control the astigmatism and field curvature in the image using the concept of the transverse local power across element surfaces and stop-shifting. We ignored spherical aberration and coma (whilst letting distortion be always present). In the following section, we shall consider the rest.

1.4.4 Do not be afraid of 'aplanatism' (it is just a term for an optimized lens, except...)

At high aperture, a high-performance diffraction-limited lens system must be made **aplanatic**. An aplanatic imaging system is one that is free of spherical aberration and coma. In such systems, the image is also referred to as being aplanatic. At visible wavelengths, if a real image of the object must be formed, then for elements with only spherical surfaces, very near aplanatism over a small field can be achieved with just three to four surfaces (such as a cemented doublet or a thin air-spaced doublet). If aspheric surfaces are used, then for a small field, two surfaces suffice (see section 1.4.6). Actually, note that when real images are to be formed, a thin singlet can be bent to achieve approximate aplanatism, since the surfaces can be bent to have either zero spherical aberration or zero coma; in each case, the lens shape is such that the point of minimum spherical aberration is close to that of minimum coma [5, 6]. At infrared wavelengths, due to the refractive index of infrared materials, it is possible to attain both zero spherical and coma aberration using a highly meniscal shape for a singlet [6]. Returning to visible light, if only a virtual image is needed (or if rays are already converging into—or traversing out from—a surface), then a single spherical surface that either has the object at the center of curvature (such as the Bravais lens of the form in figure 1.139(a)) or one that has the object at a location that satisfies the *aplanatic points of a surface* can produce an aplanatic image [5].

Did the introduction above sound a bit intimidating? Experienced designers will have been familiar with everything said, but optical engineers who may not have had formal training in lens design probably found it rather scary. Many points were shared above and, all of a sudden, a new concept—the aplanatic points of a surface—was mentioned. But do not worry. Generally, during the process of optimization in an optical design program, through the mere application of appropriate operands to minimize RMS spot size (RSCE in OS) or RMS wavefront error (RWCE in OS), aplanatism tends to be achieved eventually, as this is the way to obtain a well-focused spot. This happens especially when there are many surfaces and elements in the lens system. More elements and surfaces mean more variables, which can help to minimize aberrations, especially if the initial lens sample from the literature is well chosen. If you have been following the examples in this book closely, you will have already experienced the process of designing (optimizing) a lens in section 1.2.12 without even knowing what aplanatism meant. So, in your work, you will ordinarily not be at all concerned by whether or not your lens is aplanatic and how it got there. However, there are benefits to knowing something about aplanatism, which are as follows:

1. Aplanatism is a fundamental concept, and knowing it will help one tell the difference between a real lens and a fake lens (a paraxial thin-lens model is a *fake ideal lens*).
2. In the study of aplanatism, one is led to understand that aplanatism is not wholly a necessary condition for **reducing the impact of aberrations on image quality** (in fact, one can sometimes balance several aberrations—making them *average out* towards a reasonable level—while having nonzero aplanatism). This is the reason for the implied exemption to aplanatism in the title of this section.
3. When we combine the knowledge that has been presented in the previous section with basic knowledge of the near-aplanatic conditions of lenses, we gain an understanding of how to select and use COTS components.
4. Aplanatism is intimately connected with the radiometric properties of optical systems. As such, an understanding of aplanatism helps in the design of illumination systems.

The fourth point in the list above is a topic addressed in chapter 2. In the current section, we will discuss the first three points. Let us begin with real lenses versus fake lenses. When an imaging system (either comprising lenses or mirrors) is aplanatic, there is a special behavior that characterizes rays focused onto the image plane, as illustrated in figure 1.199. In this figure, rays emerge from a spherical surface whose radius is centered on the axis of the image. At low apertures, this spherical surface tends towards a plane, which we ordinarily know as the second principal plane. But for an aplanatic imaging system at high apertures, this plane is a spherical surface, and we call it the **second principal surface** (hence, it is the non-paraxial generalization of the second principal plane under the conditions of a high aperture and a well-corrected imaging system). Moreover, the imaging system's exit pupil is also spherical and its curvature is centered at the axial position of the image (we applied this condition in figure 1.146 to derive some equations). However, note that the second principal surface shown in figure 1.199 applies only to the axial sagittal and tangential rays. The oblique rays emerge from a tilted spherical surface (not shown) that is centered at the height of the image.

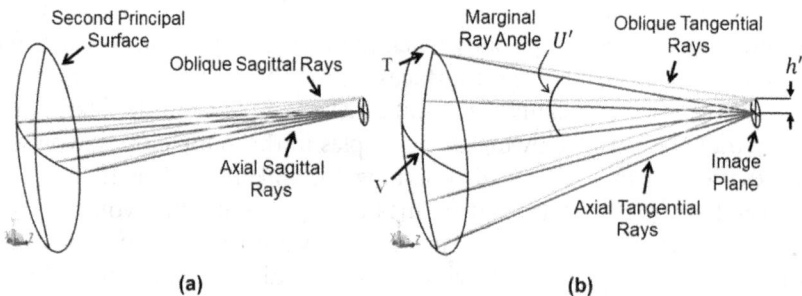

Figure 1.199. The characteristic geometry of focusing rays in an aplanatic imaging system. (a) Sagittal rays. (b) Tangential rays.

The systems depicted in figures 1.199(a) and (b) are the same, but the former displays only the sagittal rays, while the latter displays the tangential rays. The position labeled V locates the vertex of the second principal surface. The marginal ray angle U' is **in image space** (hence the use of the superscript prime) and is the ray emerging from the top of the exit pupil (here, it is also emerging from the point labeled T at the second principal surface). For an object at infinity, the distance between V and the center of the image is the EFL. In an aplanatic imaging system, the distance between T and the center of the image is also equal to the EFL. In a lens that is not corrected for spherical aberration and coma, this equality would not hold. For a *small object* at a finite distance, the magnification m of the image in an aplanatic imaging system is given by

$$m = \frac{h'}{h} = \frac{\sin U}{\sin U'} = \frac{u}{u'}.$$

(1.52)

Equation (1.52) is known as **Abbe's sine condition** [5], where U (not shown in figure 1.199) is the marginal ray at the object, and h (not shown in figure 1.199) is the height of the object. For this condition to hold, the spherical aberration must be zero, and the object height must be small (equivalently, the maximum field angle must be small). However, the marginal ray angle can be arbitrarily large. This means that the Abbe sine condition applies to coma near the axis of the image. The angles u and u' (also not displayed in figure 1.199) are the small angle approximations of U and U', respectively, for a low-aperture condition of the imaging system. In other words, when the imaging system is aplanatic, the ratio $\sin U/\sin U'$ is equal to u/u', the paraxial magnification of the image. **In contrast, the image magnification given by a paraxial thin-lens model is always determined by the ratio $\tan U/\tan U'$, and the first and second principal surfaces are coincident on a single plane.** Of course, at small marginal ray angles (i.e. at low apertures), since the tangent tends towards the sine, a paraxial thin-lens model tends towards an aplanatic imaging system. However, at large marginal ray angles, a paraxial thin-lens model (such as those we have been using in OS) is evidently not a model of a *real ideal lens* even though its rays always focus precisely into points. However, despite this, paraxial thin-lens models are highly useful for laying out a first-order imaging system. Moreover, as we shall see in chapter 2, paraxial thin-lens models can also be applied to lay out the first-order design for an illumination system, provided that certain corrections are applied to the magnitude of the flux density in the illumination.

If a lens is non-aplanatic, the geometry of sagittal and tangential rays focused towards the image deviates in a noticeable characteristic fashion from the conditions shown in figure 1.199. Take, for instance, the $F/1.88$ refractive singlet shown in figure 1.200 in a 3D layout. The object is at infinity, and the stop (not displayed) is at the left vertex and has a diameter of 40 mm. The lens's front and rear radii are 100 mm and −61.160 47 mm, respectively, and its central thickness is 10 mm. Its material is Schott N-BK7, and the wavelength used for the ray trace is the d wavelength (587.56 nm). This results in a lens EFL of 75 mm. The oblique rays are from a half-field angle of 2°. Based on the use of OS's SPHA and COMA operands

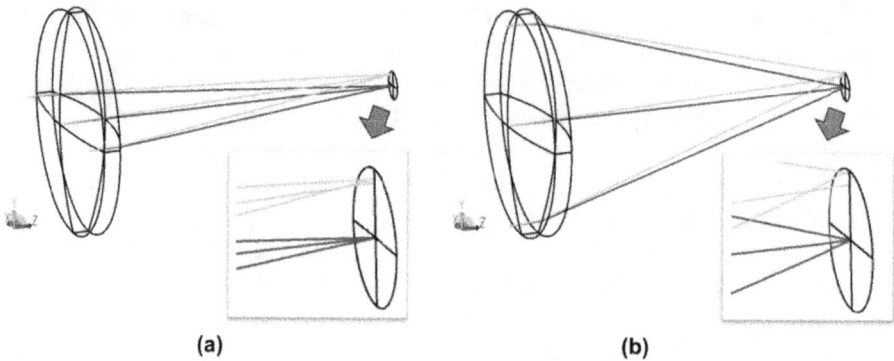

(a) **(b)**

Figure 1.200. The characteristic geometry of focusing rays for a non-aplanatic singlet. (a) Sagittal rays. (b) Tangential rays.

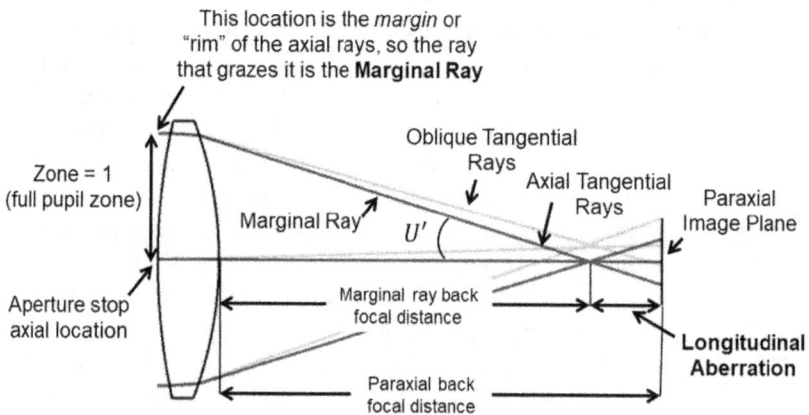

Figure 1.201. Longitudinal aberration and other classical definitions used in lens design as they are applied to the lens of figure 1.200, which is here displayed sideways.

for the third-order spherical aberration and coma in units of waves, we obtain SPHA ≈ 317 waves and COMA ≈ -57 waves. The image plane has been placed at the focal position of the marginal ray for the full pupil zone (this is why those rays appear to be focused well). However, the oblique sagittal and tangential rays have the appearance that is characteristic of non-aplanatic imaging in that they are not focused well, and the tangential rays exhibit more deviation than the sagittal rays. Moreover, due to the presence of **longitudinal aberration** (i.e. spherical aberration at the full pupil zone relative to the focus for axial paraxial rays), there is a defocus of the marginal rays when the image plane is placed at the paraxial focus, as shown in figure 1.201.

In classical lens design terms, the longitudinal aberration is taken by convention to be the spherical aberration at the full pupil zone. Thus, it is essentially the difference between the paraxial back focal distance and the focal distance of the marginal ray. However, in OS, the marginal ray is user defined for a specific zone of the entrance pupil. That is, the program lets you select which zone of the entrance pupil you are taking to be the marginal ray, and as such, in OS, a marginal ray is simply the rim ray for a specific zone. The full pupil zone is the normalized radial dimension (semi-diameter) of the entrance pupil. For the lens in figure 1.201, since the stop is at the left vertex of the lens, the stop is also the entrance pupil, so, when Zone = 1, it refers to the maximum height of the marginal ray at the stop. At intermediate zones of the pupil, axial rays can focus towards different locations at the axis of the system, creating **zonal spherical aberration**. However, in some classic texts [5], zonal spherical aberration normally refers to the difference in focal distance between the marginal ray at Zone = 0.7 and the paraxial ray. In any other zones, one would speak of *residual zonal spherical aberration* or, equivalently, *longitudinal aberration versus zone* (which is given by a plot by the name of Longitudinal Aberration in OS if one navigates to Analyze > Aberrations > Longitudinal Aberration).

Our next step is to explore bending this singlet of figure 1.201 to either make it as aplanatic as possible or to have it reach a state of balanced aberrations such that there is *reasonable* image quality, where 'reasonable' means anything that it needs to mean for some specific purpose (e.g. are we trying to resolve distant elephants or distant ants? Clearly, it depends on what we are imaging). In order to assess whether a lens is aplanatic, we need a measure of aplanatism. Of course, if a lens has zero coma and zero spherical aberration, then by definition, it is aplanatic. But a lens can also be partially aplanatic in the sense that the aplanatism is met only for rays passing through a single zone of the lens, and so in all other zones, the lens is non-aplanatic by definition. For instance, if only sagittal and tangential rays at the full pupil zone are focused perfectly, then aplanatism is achieved only for that zone. In all other zones, there can be residual zonal spherical aberration, so the lens is not 100% aplanatic over the full pupil. Yet, under such a condition, the lens can have quite good overall image quality within the limits of applicability of the Abbe sine condition (e.g. at low field angles but at high apertures). Think of this as you would for the correction of chromatic aberration. When the focal position for the blue and red wavelengths of a lens are equal, we have what is called a state of *primary axial color correction*. A lens with this level of chromatic aberration is an achromat, or an achromatic doublet. In achromats, the focal position of rays at the green wavelength remains uncorrected. This residual chromatic aberration is known as *secondary spectrum* (or *secondary color*, as mentioned in section 1.2.12). Yet, a lens that only has primary axial color correction can function fairly well when used for imaging over the full visible light spectrum, which makes achromats rather ubiquitous as COTS components in lens supplier catalogs.

A unitless quantity—attributed to A E Conrady—that essentially computes the partial aplanatic state of a lens is known as the *offense against the sine condition* (OSC), and, for an object at infinity, it may be expressed as [5]

$$\text{OSC} = \frac{F'}{f'}\left(1 - \frac{\delta s'}{Z'}\right) - 1, \tag{1.53}$$

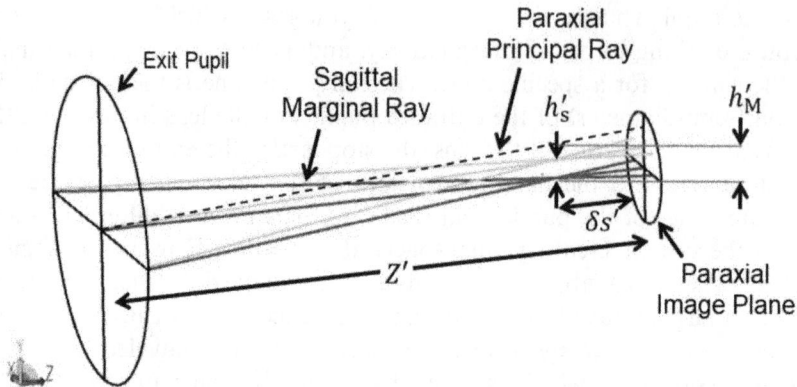

Figure 1.202. Definitions for some of the quantities in equation (1.53).

where F' is the EFL for the marginal ray *at the full pupil zone*, f' is the EFL given by first-order properties (i.e. it is the usual paraxial EFL, such as that given by the EFFL operand in OS), $\delta s'$ is the longitudinal aberration, and Z' is the distance between the exit pupil and the marginal image plane, as indicated in figure 1.202.

Equation (1.53) may be applied to any zone in the pupil, but it is ordinarily applied to the full pupil zone. In figure 1.202, the heights h'_s and h'_M are displayed because they are in fact the origin of the OSC, which is actually the ratio given by $(h'_M - h'_s)/h'_s = h'_M/h'_s - 1$. Here, h'_M is the height of the paraxial principal ray at the focal plane of the marginal sagittal ray, and h'_s is the **height of the image formed by the marginal sagittal ray**. The key point is that, when coma is zero at the full pupil zone, $h'_s = h'_M$, so that $F' = f'$. If, in addition, longitudinal aberration is zero, then $\delta s' = 0$, which, according to equation (1.53), yields zero OSC *for the full pupil zone*. For an object at a finite distance, the ratio F'/f' in equation (1.53) is replaced by M/m, where M is the magnification of the sagittal image and m is the paraxial magnification.

The OSC is computable in OS using the OSCD operand, which we will now apply to compare conditions for zero third-order spherical aberration and coma for a range of bend conditions of the lens in figure 1.200. Using the OS program, the lens EFL shall be maintained at 75 mm. Its surfaces shall be bent by adjusting the radius of the first surface from 30 mm to 100 mm and applying the Element Power solve to the second surface (this causes the second surface to vary from 117.536 47 mm to −61.160 47 mm). Using OS's Universal Plot feature (see section 1.2.17) to display the resulting third-order spherical aberration, coma, and the OSC, the results are shown in figures 1.203 and 1.204.

Based on figure 1.203, it can be seen that the minimum point for third-order spherical aberration and the zero point for coma are reached at nearly the same lens bend condition, which is for a radius of about 44 mm for the first surface and −300.894 79 mm for the second radius. In fact, it is known that the bend condition of a thin lens required to achieve minimum third-order spherical aberration is very close to the bend condition for minimum third-order coma [5, 6]. Apparently, our

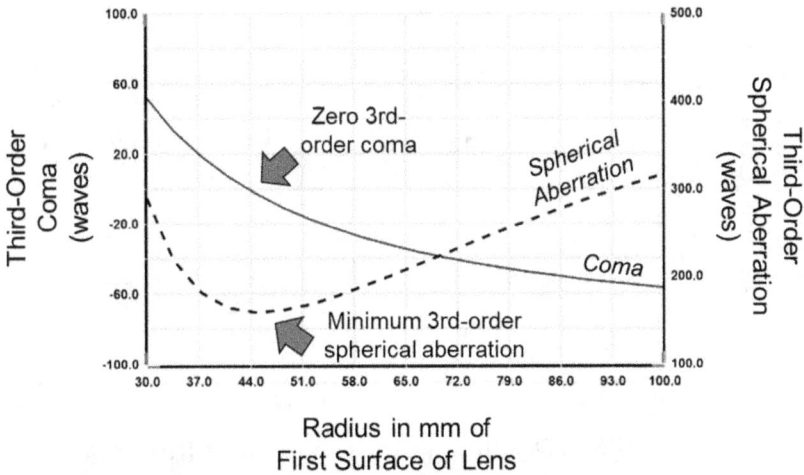

Figure 1.203. The variation of third-order spherical aberration and coma with bending of the surface radii of the lens of figure 1.200. Only the radius of the first surface is displayed.

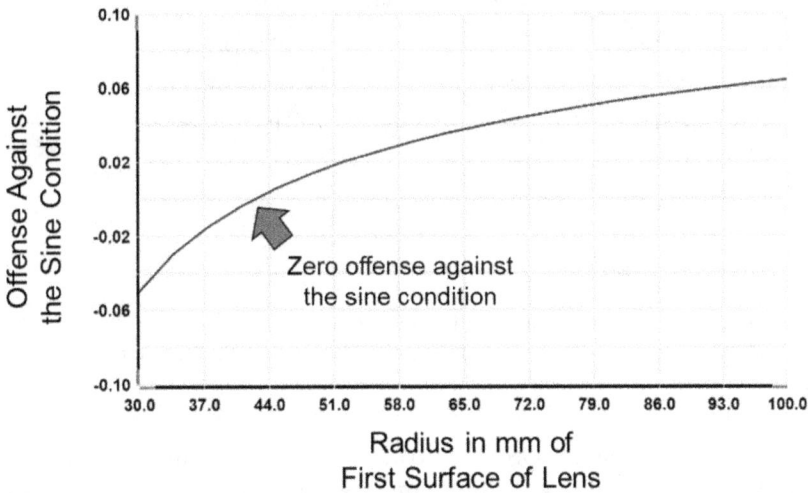

Figure 1.204. The variation of OSC with bending of the surface radii of the lens of figure 1.200. Only the radius of the first surface is displayed.

rather thick (but not overly thick) lens of figure 1.200 possesses this behavior, and so we see that certain lens design principles associated with thin lenses can be applied with reasonable accuracy to non-overly thick lenses. Indeed, there is much insight that can be gained from learning about the aberration properties of thin single elements [94]. What we have done here is to study classical knowledge graphically, using the tools made available to us in a modern optical design program (hence the title of this book).

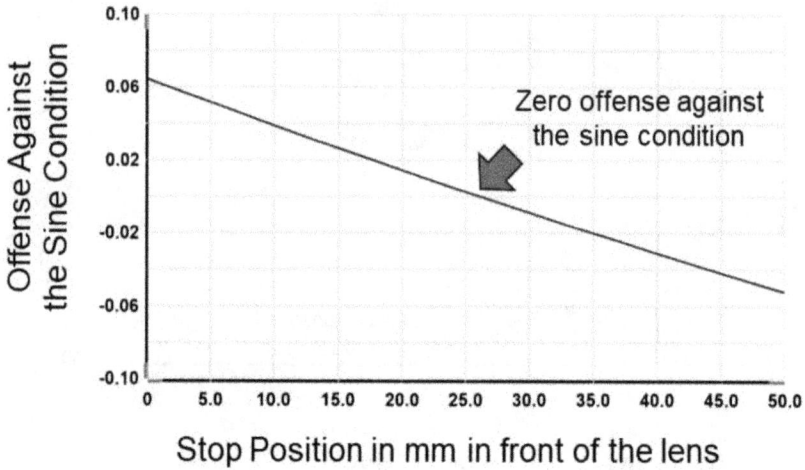

Figure 1.205. The variation of OSC with stop-shifting towards the left for the lens of figure 1.200. The third-order spherical aberration is fixed.

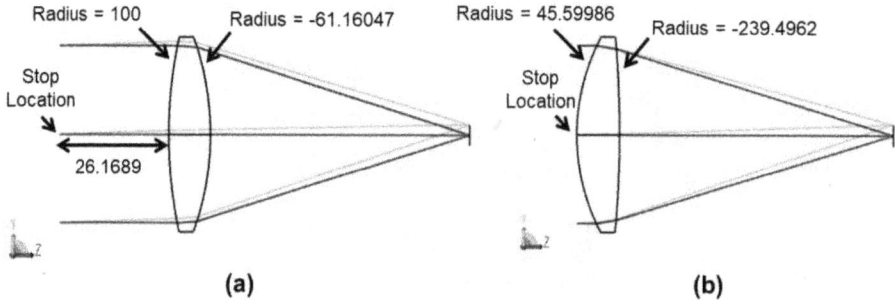

Figure 1.206. Two states of partial aplanatism illustrated in terms of satisfying the OSC for the singlet of figure 1.200. (a) Stop-shifted with finite spherical aberration. (b) A stop at the front vertex; the lens is bent for minimum third-order spherical aberration. Length units are in mm.

Figure 1.204 shows that the OSC is zero at a point close to (and again, not precisely equal to) the surface radii that yield minimum third-order spherical aberration and zero third-order coma. If one were to optimize the lens separately for OSC, spherical aberration, and coma, then one would find that the surface radii would be slightly different for each quantity. It is up to the designer to select the best radii for a given imaging condition, which is a state of balanced aberrations for the lens. Another example of a balanced state without full aplanatism is that in which the spherical aberration is left uncorrected and the stop is shifted to the left. The effect of stop-shift on the OSC is implied by the quantity Z' in equation (1.53). Beginning, again, with the state of the lens in figure 1.200, if we vary the stop from its current position at the left vertex to 50 mm to the left, we obtain the plot for the OSC in figure 1.205.

Based on the above analyses, two states of aplanatism for the lens are shown in figure 1.206. In figure 1.206(a), the lens maintains the third-order spherical aberration at roughly 317 waves, which has been used to balance the coma to yield zero OSC. In contrast, in figure 1.206(b), the lens has been bent to achieve a minimum spherical aberration of roughly 159 waves, with residual (but perhaps negligible) OSC. Due to the high residual spherical aberration, significant longitudinal aberration remains, which would become apparent if one were to place the image plane at the paraxial focus. In figure 1.206, we have placed the image plane at the focus of the tangential marginal ray at the full zone. Should the plane of focus be shifted, we would have an additional design variable, which would be to balance residual zone spherical aberration with defocus. Furthermore, if we could 'live' with an image with reduced brightness, we could lower the aperture (reduce the stop diameter). These are some of the options you have as a designer when you decide what is best for your system, based on requirements.

There is something else to be learned about the two states of aplanatism shown in figure 1.206. At the current refractive index, the shape of the lens in figure 1.206(a) is biconvex. Hence, when focusing rays from infinity at low fields and high aperture, you may use biconvex COTS elements (of similar index to that of the lens above) with a shifted stop location. On the other hand, as the lens in figure 1.206(b) is near plano-convex, then if plano-convex COTS elements (of similar index to that of the lens above) are available, then you may select these and place the stop at the front vertex. Further simple exercises along similar lines will reveal that at various object–image conjugates and refractive indexes, singlets of various bent shapes are preferred [6].

The two states shown in figure 1.206 have significant residual spherical aberration. As mentioned before, in most cases of custom lens design, multiple elements are used to correct aberrations. It is known that spherical aberration can be reduced by *splitting elements* such that multiple elements with reduced power are used to arrive at the total lens power [5, 6, 94]. This is done so as to either reduce the AOI at each surface of the elements (because spherical aberration is a function of the AOI for rays), or to let positive angles cancel negative angles, which can occur if there is a mix of positive and negative-powered surfaces. So, let us see what can be achieved by splitting, say, the singlet of figure 1.206(a) into two elements. To do this, the following procedure was applied to achieve the resulting design shown in figure 1.207, whose prescription and Merit Function Editor (at the final step) are shown in figures 1.208 and 1.209, respectively:

1. Double the surface radii of the original singlet such that its EFL is doubled. Reduce the central thickness by half (because a thick lens is not needed when radii are large).
2. Cut and paste the singlet to the right. Leave 1 mm of air space between them.
3. Assign variables to all surface radii but only optimize for minimum spherical aberration and system EFL. Remove radial variables when done.
4. Assign a variable to the stop location and optimize only for minimum OSC. Remove the variable when this is complete. Finally, place a variable at the back focus and optimize the RMS spot size for both fields. Remove the variable when this is complete.

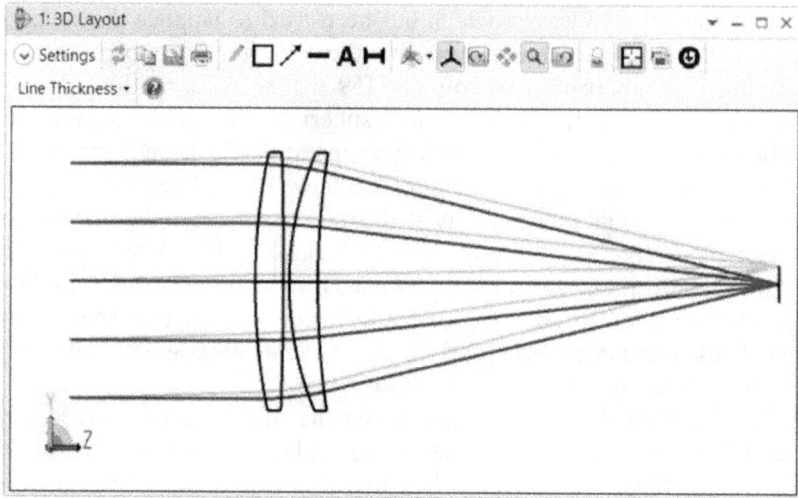

Figure 1.207. The layout in OS for a dual-element aplanat optimized for 587.56 nm, an object at infinity, and a ±2° field, based on the procedure outlined above.

Lens Data

Update: All Windows

Surface 0 Properties — Configuration 1/1

	Surface Type	Comment	Radius	Thickness	Material	Coating	Clear Semi-Dia	
0	OBJECT Standard		Infinity	Infinity			Infinity	
1	STOP Standard		Infinity	32.85825			20.00000	U
2	(aper) Standard	Element 1	105.46567	5.00000	N-BK7		22.00000	U
3	(aper) Standard		-688.08479	1.00000			22.00000	U
4	(aper) Standard	Element 2	55.59830	5.00000	N-BK7		22.00000	U
5	(aper) Standard		142.23410	81.20346			22.00000	U
6	IMAGE Standard		Infinity	-			3.15381	

Figure 1.208. The prescription for the lens system shown in figure 1.207.

Merit Function Editor

Wizards and Operands — Merit Function: 0.0372771846519219

	Type	Ring	Wave	Hx	Hy	Target	Weight	Value	% Contrib
1	EFFL		1			0.00000	0.00000	88.07485	0.00000
2	OSCD		1	1.00000		0.00000	0.00000	-6.31382E-12	0.00000
3	SPHA	0	1			0.00000	0.00000	19.66343	0.00000
4	COMA	0	1			0.00000	0.00000	0.36982	0.00000
5	RSCE	3	1	0.00000	0.00000	0.00000	1.00000	0.03680	48.73511
6	RSCE	3	1	0.00000	1.00000	0.00000	1.00000	0.03775	51.26489

Figure 1.209. The merit function for the lens system shown in figure 1.207.

Figure 1.209 shows that the EFL of 75 mm was not maintained, but the *f*-number is not very distant from the original *F*/1.9 condition (it is now at roughly *F*/2.2). Considering that *F*/1.9 is a 'tall order' for a singlet, the current lens system is in a much better state of correction than it was in its original single-element design. Therefore, I traded off some image brightness to obtain the current state of aberration correction. Moreover, I can safely call this lens a near-aplanat, due to the near-zero magnitude of the OSC (hence, I did not offend Abbe's sine condition by too much, which gives a rather pleasant feeling). This state of correction is a balanced one, in which the residual third-order spherical aberration (which is ≈ 20 waves, as shown on row 3 in the merit function) from step 3 was applied to balance the coma through stop-shifting such that the OSC was made zero.

The reader may have noticed that the design procedure above is vastly different from the 'scale-the-lens-and-optimize-everything' example given in section 1.2.12. In the current procedure, the optimization was performed step by step, so there was no point at which I optimized everything simultaneously. This can be done for simple systems, especially when the aberrations that need to be controlled are clear. For instance, in contrast to the low-aperture/high-field lenses we analyzed in the previous section, the current section's lens examples have high aperture and low field, which allowed us to ignore astigmatism, field curvature, and image distortion. This allowed us to focus on controlling only spherical aberration and coma. To understand what governs these two limiting conditions of low aperture/high field and high aperture/ low field, it is time to recall a helpful formula, known as the **wavefront aberration function** (sometimes also called the *wavefront aberration polynomial expansion*). But first, we need an illustration of what this is, as depicted roughly by the *Gap* labeled in figure 1.210.

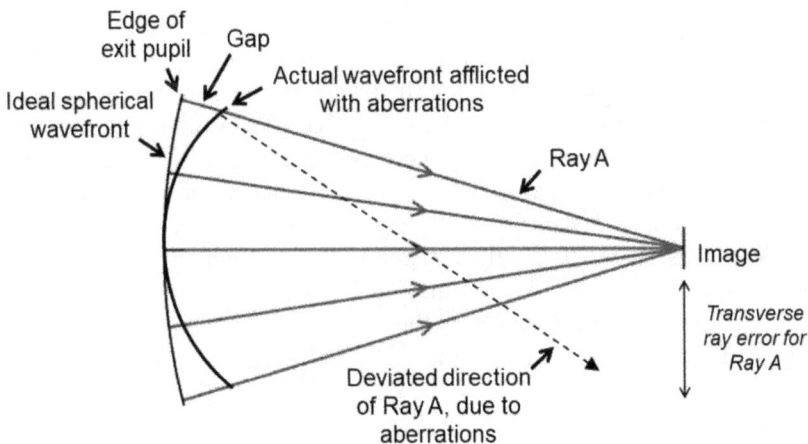

Figure 1.210. An aberrated wavefront emerges from the exit pupil of an optical system, resulting in a deviated direction of its rays, such as 'Ray A.'

In order to begin understanding figure 1.210 and the wavefront aberration function, you need to review everything you learned about rays and waves in your undergraduate optics class (or, if you did not go through formal training in this, you need to read about rays, waves, wavefronts, phase shifts, and the concept of **optical path length** and **optical path difference** [8]). If you can roughly recollect what these terms are, then you can probably follow the current quick review. Generally, we trace the paths of rays in lens design. But the theorem of Malus and Dupin tells us that rays are, in fact, just lines drawn orthogonally to wavefronts (e.g. [66]). Ray A in figure 1.210 is the line drawn orthogonally to the ideal spherical wavefront at its top edge. This ideal wavefront would focus towards the axial point in the image if it were not for the presence of aberrations. Hence, Ray A was meant to intersect the axial image point, but instead, due to aberrations, it deviates and intersects somewhere further down than the axial point, resulting in a transverse ray error in the vertical downward direction. This is a consequence of the aberrated wavefront, which has the result that Ray A's direction—being drawn orthogonally to the top edge of the aberrated wavefront—is influenced by the shape at the edge of the aberrated wavefront. Thus, there is a sort of 'gap' at the top edge between the spherical wavefront and the aberrated wavefront, and this gap is an illustration of *wavefront error*. Since the extent of the gap is a function of the radial dimension at the exit pupil (and also a function of field height), the gap's dimension can be expressed as a function of the exit pupil coordinate and the field height. This dimension is a difference in the optical path length (OPL) between the path that a ray would have taken from the object to the image in the ideal case, and the OPL that it actually took in the real-world case of being aberrated. This difference is the OPD for that ray. Hence, Ray A suffers an OPD at the top edge of the exit pupil. Other rays in other zones of the exit pupil suffer various other degrees of OPD.

We should talk a bit more about OPLs and OPDs. A ray is clearly not a wave, but in optics and optical design, it *has a wave in it* in that it is associated with points along wavefronts separated by the crests of the wave. We know this separation as the wavelength of the wave. You already know that a ray's wave-like property is observable when it strikes glass at a non-normal AOI, because it refracts. We know that this is a consequence of Snell's law, which can be seen as resulting from a change in wavelength for that ray (equivalently, it is a change in velocity by c/n, where c is the speed of light in vacuum, and n is the glass's refractive index). But when a ray from air strikes glass at normal incidence, you do not see the refraction effect as it passes into the glass, so it seems that there is no observable wave-like property for that ray. Yet, it is there. It is called a phase shift, in which the ray's wavefront literally shifts backwards by an amount equal to $kd(n - 1)$, the magnitude of the phase shift, where $k = 2\pi/\lambda$ (known as the wave number) and d is the thickness of the glass [8]. This phase shift can be thought of as k times the OPD between a ray traveling in the glass and a ray outside the glass. The ray in the glass may be thought of as traveling over an OPL equal to nd, while the ray outside the glass travels the OPL equal to d (there is no index outside the glass, and we assume that if the medium outside the glass is air, then its index is 1). Their OPD is the difference in their OPLs, which is $nd - d = d(n - 1)$. **Thus, when waves experience phase shifts (i.e. phase differences), their rays experience** *path shifts* **(i.e. OPDs).**

So, back at the top edge of the ideal spherical wavefront in figure 1.210, Ray A has traveled through many lenses and air gaps, resulting in a total OPL. If the lens system had had no aberrations, Ray A would have had a different OPL. It is the difference between these two OPLs that this ray has acquired by the time it arrives at the top edge of the exit pupil that we consider to be the wavefront error, which is the wavefront aberration function. Taking some of the notation from Welford [58], if this function is expressed to the fourth order in normalized rectangular coordinates at the exit pupil, then it may be written as

$$W\left(\rho_x, \rho_y, h_{\rm n}'\right) = \frac{1}{8}S_{\rm I}\left(\rho_x^2 + \rho_y^2\right)^2 + \frac{1}{2}S_{\rm II}\rho_y\left(\rho_x^2 + \rho_y^2\right)h_{\rm n}' + \frac{1}{2}S_{\rm III}\rho_y^2(h_{\rm n}')^2$$
$$+ \frac{1}{4}(S_{\rm III} + S_{\rm IV})\left(\rho_x^2 + \rho_y^2\right)(h_{\rm n}')^2 + \frac{1}{2}S_{\rm V}\rho_y(h_{\rm n}')^3. \tag{1.54}$$

In equation (1.54), the five *fourth-order wavefront error* terms are—in the order that they appear—identified as spherical aberration, coma, astigmatism, *astigmatic* field curvature (this is a term that I have somewhat concocted here, for reasons you will soon understand), and image distortion (from here on, we simply refer to this as *distortion*). The coefficients $S_{\rm I}$, $S_{\rm II}$, $S_{\rm III}$, $S_{\rm IV}$, and $S_{\rm V}$ are known as the Seidel aberration coefficients, and their values are determined through ray tracing from surface to surface. The quantities ρ_x and ρ_y are normalized coordinates at the exit pupil (so the maximum value for each is unity, or one). The quantity $h_{\rm n}'$ is the normalized height of the image. Equation (1.54) is not a formula in closed form. Rather, it is a polynomial expansion to the fourth order of the powers of the exit pupil and field, and, in view of the rotational symmetry of most optical systems about the optic axis, the terms need only to be even powers in the total of the pupil and field coordinates. Since, under the current expansion, coma is seen to be linear with field, it is known as *linear coma*. Hence, OSC is a measure of the amount of linear coma at a specific pupil zone. If further terms are included in the expansion of the wavefront error, there can be many more orders of aberrations, such as nonlinear coma (in which coma scales nonlinearly with the field) and oblique spherical aberration (which is a result of certain higher-order terms in astigmatism that vary with field). In this book, we will go no further with the theory of higher-order aberrations. You just need to be aware that many lens designs for high aperture and high field can involve such higher-ordered terms in the aberrations.

Equipped with equation (1.54), we are ready to take a different route towards understanding many of the results from the design examples of prior sections in this book as well as the material that follows momentarily and continues for the rest of the current section. First, returning to the point being made prior to our discussion of phase shifts and OPDs, I mentioned that at low aperture and high field, we may ignore most of the spherical aberration and coma, whilst tackling only astigmatism, field curvature, and distortion (but we did not constrain distortion and just left it as it is). At low aperture, ρ_x and ρ_y are small and, since they are normalized coordinates, when taken to a power greater than one, their values are even smaller. Thus, since spherical aberration scales according to $\left(\rho_x^2 + \rho_y^2\right)^2$, then at low aperture, it should make a considerably lower contribution than all of the other aberration terms. As for coma, since it is cubic with the y-coordinate in the pupil, it can be said to be somewhat low as well, relative to the other aberrations. However, coma scales linearly with the normalized image height, while astigmatism and astigmatic field curvature scale quadratically, so taken together, it can be said that coma is not necessarily negligible at low aperture and at high field (but in any case, we ignored it rather successfully in the examples of the previous section, which is fine). Distortion is cubic with image height and linear with the pupil, so in a sense, it is always there at high field and we really could not do anything about it in the previous section, especially with the few variables we had (but we will deal with it soon in this section). At high aperture and low field, the conditions are reversed, and we need to deal with the higher contributions due to $\left(\rho_x^2 + \rho_y^2\right)^2$ and $\rho_y\left(\rho_x^2 + \rho_y^2\right)$, which are the spherical aberration and coma.

Each of the terms in equation (1.54) can possess a positive or negative value, depending in part on the value of each of the five Seidel coefficients. The act of lens design optimization is to either minimize the values of the coefficients or to make them add and subtract each other so as to cancel one another (aberration balancing). On this note, defocus at the image plane adds a term to equation (1.54) that scales quadratically with the normalized pupil coordinate. For this reason, defocus can be used to counteract the Seidel terms (this is what is done whenever you see that I have placed a marginal ray solve at the last surface to set the paraxial focal distance and then added a final surface with a thickness variable). Incidentally, transverse ray errors in the x and y dimensions at the image plane are, to an approximation, given by

$$\varepsilon_x = \frac{1}{n'u'}\frac{\partial W}{\partial \rho_x}, \tag{1.55}$$

and

$$\varepsilon_y = \frac{1}{n'u'}\frac{\partial W}{\partial \rho_y}, \tag{1.56}$$

where W is the wavefront error from equation (1.54), n' is the index at the image side (of course, in this book, unless otherwise specified, the image is in air so that $n' = 1$), and u' is the angle (in radians) of the axial paraxial marginal ray at the image (see, for example, [6]). Thus, the ray fan plot of figure 1.197 is a plot of equation (1.55) for that mobile phone lens, and the OPD plot of figure 1.197 is a plot of equation (1.54) (but converted to units of waves by dividing the wavefront error by the primary wavelength set in the Settings menu for the plot). Note also that since the transverse ray errors are given by the first derivative of the wavefront error with respect to the exit pupil coordinates, this makes the transverse errors one order less than the powers of each aberration term. Thus, in terms of transverse ray errors, the spherical aberration term scales with the third power of the pupil. And since much of lens design involves tracing rays to obtain the transverse ray errors, in many cases, one speaks of third-order aberration theory, or third-order Seidel aberration coefficients, and so on. This is why I have been referring to the aberrations as third-order (it is the lens designer inside of me talking). Moreover, if the defocus wavefront term is included in equation (1.54), it would be taken to the power of one in terms of the transverse ray errors. Thus, to the first order in transverse ray errors, defocus is the only 'aberration' in the area of paraxial (first-order) optics. Of course, the name 'first order' that is given to paraxial optics is also a consequence of expanding the sines and cosines of ray AOIs and taking only the first-order terms.

Another insight to be gained from examining equation (1.54) is to note that the astigmatism and astigmatic field curvature terms can be shuffled algebraically to produce the expression:

$$\frac{1}{2}S_{\mathrm{III}}\rho_y^2(h_{\mathrm{n}}')^2 + \frac{1}{4}(S_{\mathrm{III}} + S_{\mathrm{IV}})\left(\rho_x^2 + \rho_y^2\right)(h_{\mathrm{n}}')^2$$
$$= \frac{1}{4}(3S_{\mathrm{III}} + S_{\mathrm{IV}})\rho_y^2(h_{\mathrm{n}}')^2 + \frac{1}{4}(S_{\mathrm{III}} + S_{\mathrm{IV}})\rho_x^2(h_{\mathrm{n}}')^2. \tag{1.57}$$

The right-hand side of this expression provides the explanation for the phenomenon seen in the field curvature plots for systems that remain uncorrected for astigmatism. That is, the first term is for the tangential rays (i.e. rays in the ρ_y coordinate that have the factor $(1/4)(3S_{\mathrm{III}} + S_{\mathrm{IV}})$ in their wavefront error), and the second term is for the sagittal rays (i.e. rays in the ρ_x coordinate that have the factor $(1/4)(S_{\mathrm{III}} + S_{\mathrm{IV}})$ in their wavefront error). When $3S_{\mathrm{III}} + S_{\mathrm{IV}} = 0$, the tangential rays no longer have the quadratic dependence with field, so they focus on a flat plane. **This was what you saw happen in** figure 1.171. On the other hand, when $S_{\mathrm{III}} + S_{\mathrm{IV}} = 0$, the sagittal rays lie on a flat plane. **This was what you saw happen in** figure 1.174. Moreover, it is known that S_{IV} is the Petzval sum that I discussed in section 1.4.2 (see figure 1.179), so this coefficient is related to the Petzval surface at the image. Equation (1.57) implies that, to the third order in transverse ray error, the focal location for tangential rays is three times further away from the Petzval surface than the focal location for sagittal rays. This is a well-known property of astigmatic oblique rays in lens design [5, 6].

There is a rule that applies in studying the wavefront error function of equation (1.54), which is that whenever you see a term that scales quadratically with the pupil coordinate, then that term represents either a convergent (if the term is positive) or divergent (if the term is negative) wavefront, which is a focus error relative to the (absent) defocus term that is associated with shifting the image plane axially. The origin of this rule is the first-order term in the Taylor series for the sag of a sphere, as seen in equation (1.24). Note that it is quadratic with the radial dimension. Thus, you can see that the two terms on the right-hand side of equation (1.57) represent rays focusing onto two different locations, due to the difference in the magnitude of their coefficients. One is for the tangential rays, while the other is for the sagittal rays. This is, of course, astigmatism. And, evidently, when $S_{\mathrm{III}} = 0$, both the tangential and sagittal terms share the same factor, which is $(1/4)S_{IV}(\rho_x^2 + \rho_y^2)(h_n')^2$. This is now a focusing wavefront with equal locations for the tangential and sagittal rays, and their focal positions lie on the Petzval surface defined by S_{IV}. **You saw this happen in** figure 1.179. Finally, due to the factor $(1/4)(S_{III} + S_{IV})$ in the second term on the left-hand side of equation (1.57), this term evidently represents either a focusing (if positive) or diverging (if negative) wavefront as a function of field (i.e. the quantity h_n'). Thus, this term simultaneously focuses both the sagittal and tangential rays towards different locations at different heights in the image. This is, indeed, field curvature, but there is also the presence of the first term on the left side of equation (1.57), which is for the tangential rays. Thus, when there is astigmatism, there is no physical surface that 'curves' across the field. Rather, there are only tangential rays and sagittal rays focusing onto their respective curved surfaces, and perhaps the average surface between them can be considered the effective surface (and if it has a curvature, then perhaps that is field curvature, if you wish). But the only instance in which one has a monotonically curved surface is when $S_{III} = 0$, which is the condition of zero third-order astigmatism. In this case, a physically monotonically curved surface is present, which is the Petzval surface defined by S_{IV}.

Some further insights related to the impact of aspheres on image quality can be gained from the expression in equation (1.54). Suppose that, at the top edge of the exit pupil in figure 1.210, the wavefront is meant to propagate in the positive z direction, described in complex form as $A(z, t) = A_o \exp[-i(kz - \omega t)]$, where $i = \sqrt{-1}$, A_o is the wave's peak amplitude, t is time, and ω is the angular frequency of the wave (given by $\omega = 2\pi\nu$, where ν is the number of cycles per unit time of the wave, which is the frequency). This is a heuristic description of the phenomenon, but let me show you where it takes us. If this wave experiences a phase shift from a glass slab of thickness d, then the amplitude is modified to become $A(z, t) = A_o \exp\{-i[kz - \omega t + kd(n - 1)]\}$. On the other hand, if the ray at this portion of the wavefront (i.e. Ray A) experiences a total OPD error of W, where W is given by equation (1.54), then the wave amplitude may be modified to become $A(z, t) = A_o \exp-[i(kz - \omega t + kW)]$. If highly aspheric surfaces are in the path of that ray, then higher-order OPDs are included in W. Suppose the added terms are lumped into the quantity W'; then, the wave's amplitude is $A(z, t) = A_o \exp-[i(kz - \omega t + kW + kW')]$. The addition of kW' can be either positive or negative, depending on the sum of all OPLs imposed on the ray by the aspheres. When $kW' < 0$, the aberration kW is corrected. This can be considered to be the physical optics perspective on the effect that aspheres have on the wavefront aberration function, which was the reason that I mentioned in the previous section that there is no reason why the current form of aberration theory cannot be applied to the use of highly aspheric surfaces in a lens system.

Based on the above and all the prior design examples, it should have become quite evident that lens design is an activity involving a variety of techniques and ways to understand what one is doing. These techniques can be graphical and they can be mathematical, and neither alone can be taken without the other. Often, a combination of techniques provides a clearer understanding of the physics behind designing and optimizing optical systems. On this note, we are ready to tackle distortion and, as you will see, also coma, and even an aspect of chromatic aberration called *lateral color*. To do this, we look again to equation (1.54) and see that the distortion term is proportional to the third power in the image height. There is a sign convention that if the height is below the optic axis, then it is negative. When tracing rays from surface to surface to compute the total Seidel coefficient for an aberration, we begin at the object and end at the image. If a lens is symmetrical about the stop (i.e. it has the same elements in front of and behind the stop), then since distortion is cubic with field, the aberration computation begins with one sign (say, negative if the object point begins below the optic axis) and it ends with the opposite sign at the image. The net effect is to cancel all of the surface-to-surface contributions between elements in front of and behind the stop. In fact, all aberrations that are odd powers of the field become zero for a symmetrical optical system with unit magnification (such as a 1:1 relay). These aberrations include lateral color (i.e. the transverse error among principal rays of different wavelengths at the image, or, equivalently, the chromatic variation of magnification) and third-order coma *at one zone of the pupil*. This is known as the *symmetrical principle* [5]. When the magnification is not unity, having elements about the stop can help to minimize such odd-powered aberrations with the field when the surface radii of the elements are optimized.

Applying the symmetrical principle given above to the lens in figure 1.207, let us simply replicate elements 1 and 2 and flip them to the front of the stop. We then let the object distance equal the image distance such that the lens system has unit magnification. Further, we reduce the stop diameter to 10 mm because we are purposely choosing a large half-field angle of 25° just to prove that the distortion is indeed zero. Doing this and ignoring, for now, the oblique field aberrations, yields the lens system shown in figure 1.211, whose prescription is shown in figure 1.212.

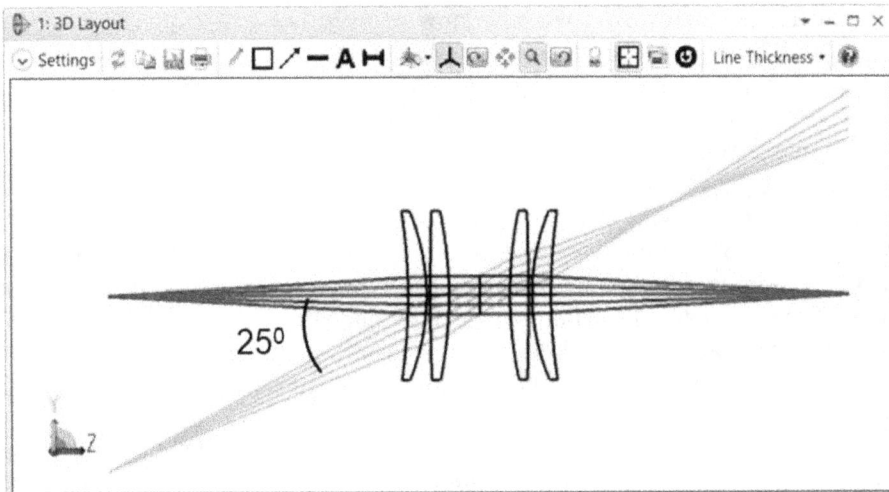

Figure 1.211. The layout for an unoptimized symmetrical lens system with unit magnification, based on replicating the elements from the lens in figure 1.207.

Lens Data

Update: All Windows

Surface 0 Properties — Configuration 1/1

	Surface Type		Comment	Radius	Thickness	Material	Coating	Clear Semi-Dia	
0	OBJECT	Standard		Infinity	81.20346			45.75809	
1	(aper)	Standard	1	-142.23410	5.00000	N-BK7		22.00000	U
2	(aper)	Standard		-55.59830	1.00000			22.00000	U
3	(aper)	Standard	2	688.08479	5.00000	N-BK7		22.00000	U
4	(aper)	Standard		-105.46567	8.21456			22.00000	U
5	STOP	Standard		Infinity	8.21456			5.00000	U
6	(aper)	Standard	3	105.46567	5.00000	N-BK7		22.00000	U
7	(aper)	Standard		-688.08479	1.00000			22.00000	U
8	(aper)	Standard	4	55.59830	5.00000	N-BK7		22.00000	U
9	(aper)	Standard		142.23410	81.20346			22.00000	U
10	IMAGE	Standard		Infinity	-			53.05489	

Figure 1.212. The prescription for the lens system in figure 1.211.

This system has an OSC of -6.2×10^{-5} at the full pupil zone, which is essentially zero at that zone. The distortion at the full field is 7.9×10^{-5} %, which is also essentially zero. Using the F, d, and C wavelengths, the lateral color (using the LACL operand) is zero. However, the curvature of the tangential field is quite apparent in figure 1.211, so this lens is rather useless for the current oblique field in terms of focus quality, even though the distortion and lateral color are zero.

The rather poor off-axis image quality in terms of the focus for the oblique rays is to be expected, as there are only two weak negative-powered surfaces (surfaces 1 and 9) in this lens. However, we have now evolved our lens from the use of the singlets and air-spaced doublets of section 1.4.2 to a lens system with four elements. Further, we have the stop between pairs of elements, thereby permitting good control of distortion. The form of this lens and its number of elements can permit a reasonable solution for a monochromatic system at perhaps half of the current field angle and also for a larger *f*-number, such as *F*/7. So, let us reduce the stop semi-diameter (the system aperture is set to float by stop size), reduce to half-field angle to 12.5°, maintain a single wavelength setting at 587.56 nm, remove the user-defined semi-diameters for each surface (so as to let the program decide what to do with them), ensure ray aiming is on, add merit function operands, and assign variables to the radii and thicknesses. The resulting starting layout is shown in figure 1.213 and its prescription and merit function are shown in figures 1.214 and 1.215, respectively.

Notice in figure 1.215 that although many operands have been included, only some have weights. This is often done (at least by me) to monitor the operands for cases in which I decide to switch optimization targets. The current list of operands is sufficient to 'see where the program takes the lens' through a first pass at

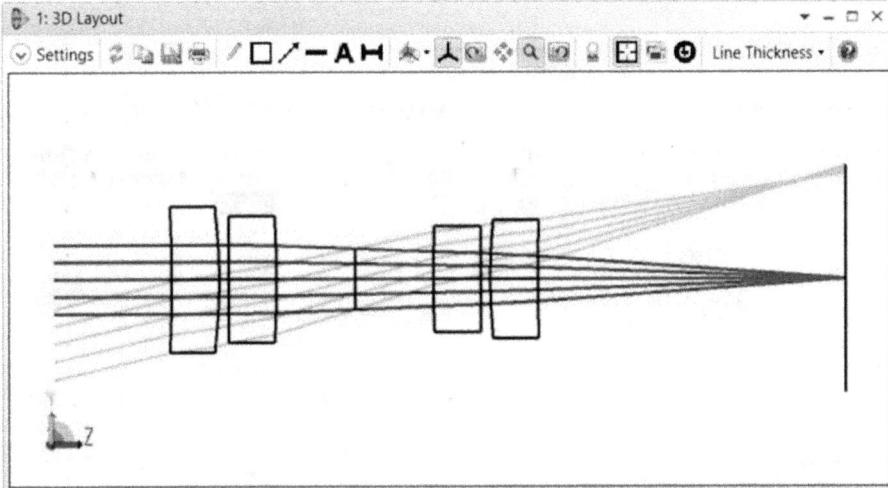

Figure 1.213. The modified layout prior to optimization, based on the lens shown in figure 1.212.

	Surface Type		Comment	Radius		Thickness		Material	Coating	Clear Semi-Dia
0	OBJECT	Standard ▾		Infinity		Infinity				Infinity
1		Standard ▾	Dummy	Infinity		12.50000				10.17776
2		Standard ▾	1	-142.23410	V	5.00000	V	N-BK7		7.44986
3		Standard ▾		-55.59830	V	1.00000	V			6.85158
4		Standard ▾	2	688.08479	V	5.00000	V	N-BK7		6.46721
5		Standard ▾		-105.46567	V	8.21456	V			5.62702
6	STOP	Standard ▾		Infinity		8.21456	V			3.10000 U
7		Standard ▾	3	105.46567	V	5.00000	V	N-BK7		4.83183
8		Standard ▾		-688.08479	V	1.00000	V			5.41194
9		Standard ▾	4	55.59830	V	5.00000	V	N-BK7		5.64507
10		Standard ▾		142.23410	V	32.47667	M			6.03425
11		Standard ▾		Infinity		0.00000	V			11.53145
12	IMAGE	Standard ▾		Infinity		-				

Marginal ray solve, height = 0, zone = 0

Figure 1.214. The prescription with variables for the lens shown in figure 1.213.

optimization, beginning by minimizing the RMS spot size (using the RSCE operands) for the axial and off-axis fields and also controlling the distortion, EFL, and the f-number, which is given by the computation performed using operands in rows 1 through 4. Note that I do not use the ISFN operand, which computes the *paraxial image space f-number*. I want the *real f*-number, given by

Merit Function Editor

Wizards and Operands — Merit Function: 0.290532592404586

	Type	Surf	Wave	Hx	Hy	Px	Py	Target	Weight	Value	% Contrib
1	EFFL		1					50.00000	1.00000	49.40174	88.33758
2	REAY	1	1	0.00000	0.00000	0.00000	1.00000	0.00000	0.00000	3.48377	0.00000
3	PROB	2		2.00000				0.00000	0.00000	6.96753	0.00000
4	DIVI	1	3					0.00000	0.00000	7.09028	0.00000
5	OPLT	4						7.20000	0.20000	7.20000	0.00000
6	OPGT	4						6.90000	0.20000	6.90000	0.00000
7	OSCD		1	1.00000				0.00000	0.00000	3.82632E-03	0.00000
8	SPHA	0	1					0.00000	0.00000	0.35112	0.00000
9	COMA	0	1					0.00000	0.00000	-4.97986	0.00000
10	DISG	1	1	0.00000	1.00000	0.00000	0.00000	0.00000	0.00000	0.21276	0.00000
11	OPLT	10						1.00000	0.20000	1.00000	0.00000
12	OPGT	10						-1.00000	0.20000	-1.00000	0.00000
13	PETZ		1					0.00000	0.00000	-67.03765	0.00000
14	FCGT		1	0.00000	1.00000			0.00000	0.00000	-4.71386	0.00000
15	FCGS		1	0.00000	1.00000			0.00000	0.00000	-2.19066	0.00000
16	DIFF	14	15					0.00000	0.00000	-2.52320	0.00000
17	RSCE	3	1	0.00000	0.00000			0.00000	1.00000	5.90673E-03	8.61120E-03
18	RSCE	3	1	0.00000	1.00000			0.00000	1.00000	0.21729	11.65381
19	RWCE	3	1	0.00000	0.00000			0.00000	0.00000	0.10491	0.00000
20	RWCE	3	1	0.00000	1.00000			0.00000	0.00000	5.17430	0.00000
21	MNEG	2	10	1.00000	1			1.00000	0.00000	1.00000	0.00000
22	MNCG	2	10					3.00000	0.20000	3.00000	0.00000
23	MXCG	2	10					15.00000	0.20000	15.00000	0.00000
24	MNCA	3	9					0.25000	0.20000	0.25000	0.00000
25	MNEA	3	9	1.00000	1			0.10000	0.20000	0.10000	0.00000
26	TTHI	2	9					0.00000	0.00000	38.42912	0.00000
27	OPLT	26						50.00000	0.20000	50.00000	0.00000

Figure 1.215. The starting merit function for the prescription shown in figure 1.214.

dividing EFL by the *real entrance pupil diameter*, which is twice the ray height at surface 1 (hence, the use of the *dummy* surface).

The optimization shall proceed as follows:

1. Use, as usual, the damped least-squares optimization algorithm by going to Optimize > Optimize!, selecting damped least squares, and clicking on 'Start.'
2. At the end, remove the weights from both of the RSCE operands of rows 17 and 18. Place them into the RWCE operands, but make each weight '2.'
3. Now, instead of using the local optimizer (which was done in step 1), go to Optimize > Hammer Current > Start.
4. At the end of the Hammer optimization execution, you are done.

At the end of the above procedure, upon tweaking the semi-diameters of the surfaces to make the lens look 'nice and ready,' we obtain the layout shown in figure 1.216, whose prescription, merit function, MTF, and field curvature/distortion plots are shown in figures 1.217–1.220, respectively. Notice in figure 1.216 that the diameters of the second, third, and fourth elements are oversized for the rays they serve. This is

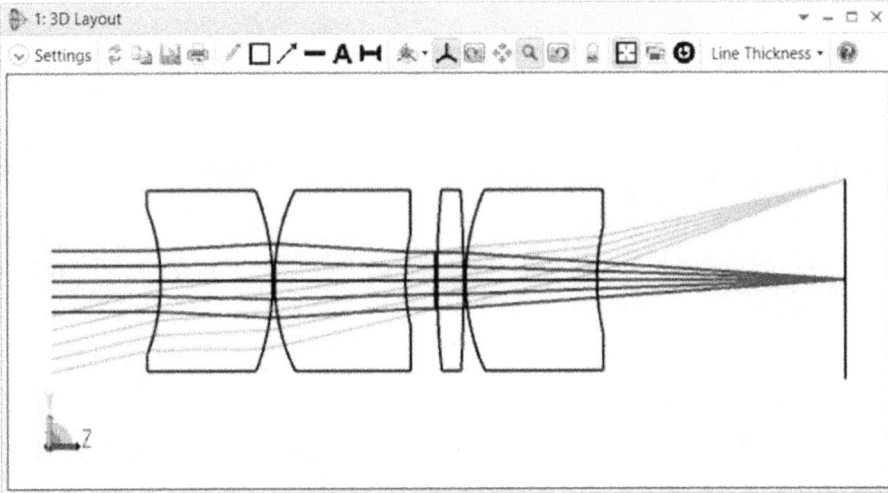

Figure 1.216. The optimized layout that began from figure 1.213.

	Surface Type		Comment	Radius	Thickness	Material	Coating	Clear Semi-Dia
0	OBJECT	Standard ▾		Infinity	Infinity			Infinity
1		Standard ▾	Dummy	Infinity	12.50000			10.11523
2	(aper)	Standard ▾	1	-21.92364 V	12.88368 V	N-BK7		8.00000 U
3	(aper)	Standard ▾		-26.78585 V	0.23983 V			10.00000 U
4	(aper)	Standard ▾	2	22.30652 V	15.00700 V	N-BK7		10.00000 U
5	(aper)	Standard ▾		21.60034 V	3.38514 V			5.00000 U
6	STOP	Standard ▾		Infinity	0.24921 V			3.10000 U
7	(aper)	Standard ▾	3	98.11102 V	2.99484 V	N-BK7		10.00000 U
8	(aper)	Standard ▾		-158.18931 V	0.24785 V			10.00000 U
9	(aper)	Standard ▾	4	24.23719 V	14.99999 V	N-BK7		10.00000 U
10	(aper)	Standard ▾		25.12202 V	28.66238 M			6.00000 U
11		Standard ▾		Infinity	-0.33282 V			11.07508
12	IMAGE	Standard ▾		Infinity	-			10.95578

Figure 1.217. The prescription for the lens shown in figure 1.216.

done in order to maintain good centering when the elements are assembled into a barrel that serves as the mechanical housing for the lens system. However, if the cost of the lens material outweighs the benefit of centering tolerances in assembly, then we ought to size down those elements (however, note that Schott N-BK7 is not an expensive glass).

Figure 1.218. The merit function at the end state, which is for the prescription in figure 1.217.

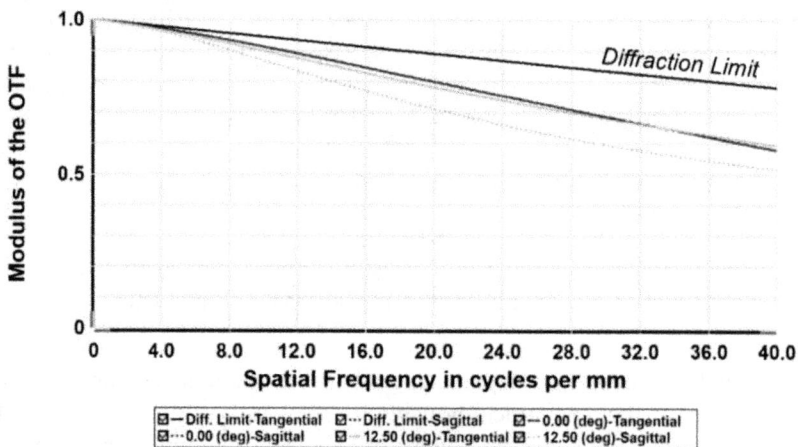

	Type	Surf	Wave	Hx	Hy	Px	Py		Target	Weight	Value	% Contrib
1	EFFL ▼		1						50.00000	1.00000	49.99720	0.01486
2	REAY ▼	1	1	0.00000	0.00000	0.00000	1.00000		0.00000	0.00000	3.39818	0.00000
3	PROB ▼	2		2.00000					0.00000	0.00000	6.79635	0.00000
4	DIVI ▼	1	3						0.00000	0.00000	7.35647	0.00000
5	OPLT ▼	4							7.20000	0.20000	7.35647	9.24562
6	OPGT ▼	4							6.90000	0.20000	6.90000	0.00000
7	OSCD ▼		1	1.00000					0.00000	0.00000	4.20343E-04	0.00000
8	SPHA ▼	0	1						0.00000	0.00000	1.12359	0.00000
9	COMA ▼	0	1						0.00000	0.00000	-0.49193	0.00000
10	DISG ▼	1	1	0.00000	1.00000	0.00000	0.00000		0.00000	0.00000	-0.44687	0.00000
11	OPLT ▼	10							1.00000	0.20000	1.00000	0.00000
12	OPGT ▼	10							-1.00000	0.20000	-1.00000	0.00000
13	PETZ ▼		1						0.00000	0.00000	-356.97158	0.00000
14	FCGT ▼		1	0.00000	1.00000				0.00000	0.00000	0.28105	0.00000
15	FCGS ▼		1	0.00000	1.00000				0.00000	0.00000	0.19105	0.00000
16	DIFF ▼	14	15						0.00000	0.00000	0.09000	0.00000
17	RSCE ▼	3	1	0.00000	0.00000				0.00000	0.00000	7.10442E-03	0.00000
18	RSCE ▼	3	1	0.00000	1.00000				0.00000	0.00000	9.78151E-03	0.00000
19	RWCE ▼	3	1	0.00000	0.00000				0.00000	2.00000	0.10032	38.00448
20	RWCE ▼	3	1	0.00000	1.00000				0.00000	2.00000	0.11805	52.62017
21	MNEG ▼	2	10	1.00000		1			1.00000	0.00000	1.00000	0.00000
22	MNCG ▼	2	10						3.00000	0.20000	2.99484	0.01004
23	MXCG ▼	2	10						15.00000	0.20000	15.00700	0.01848
24	MNCA ▼	3	9						0.25000	0.20000	0.23689	0.06485
25	MNEA ▼	3	9	1.00000		1			0.10000	0.20000	0.10000	0.00000
26	TTHI ▼	2	9						0.00000	0.00000	50.00755	0.00000
27	OPLT ▼	26							50.00000	0.20000	50.00755	0.02151

Figure 1.219. The MTF (at 587.56 nm) for the lens in figure 1.216.

Field Curvature

Distortion

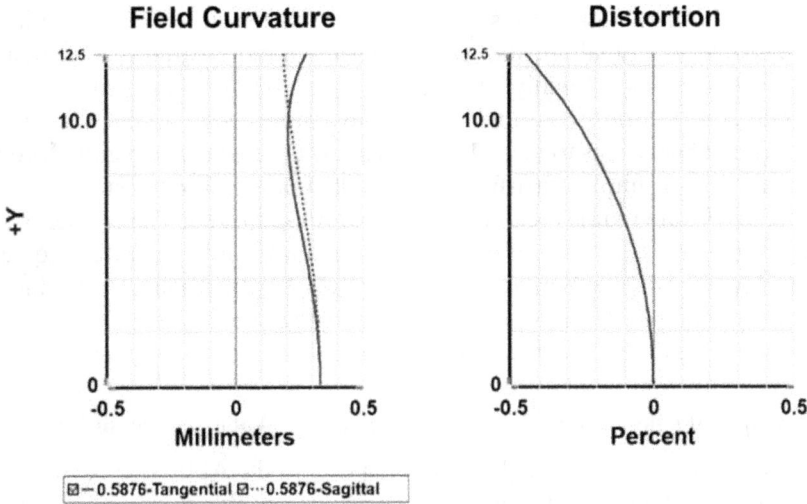

Figure 1.220. The OS field curvature and distortion plots for the lens in figure 1.213.

Although the current state of the lens has good image quality, it is not quite in a manufacture-friendly state. There are three thick meniscus lenses, which I had actually expected as the outcome, due to the operands used in the merit function (I did allow for a rather long maximum thickness of the elements on row 23 of the merit function). Therefore, the current state of the lens is generally not the end point if the current exercise were for a real design to be produced. All that is needed is either a number of further tweaks and optimization iterations, or to start over (usually, we save the starting design as a separate file so as to enable us to return to it). Sometimes, further tweaks to the current lens are possible. But at other times, the lens may be stuck in a *local minimum* of the design landscape, and so it may be better to start over. In the current exercise, I gave the lens a rather big nudge by starting from a state that was put together briskly without giving much consideration to manufacturability (the point of the exercise was just to demonstrate the symmetrical principle). This is sometimes a good thing, as Dilworth remarks [37], as the program's optimization algorithm is allowed to begin from a 'peak' on a mountain top (imagine many mountains and valleys, where the valleys represent various optimized states for the lens). Thereafter, the program may proceed towards a potentially unexpected result, which can be good or bad. In other instances, it may be better to start from a lens that is already close to what you want, and then tweak and optimize (as was done in section 1.2.12).

The current lens is a significant improvement over the simple singlets used in section 1.4.2, despite being limited to a single wavelength (this lens is achromatized in section 1.4.10). The principal idea was to provide an example of the evolution from using single-element components at the beginning of section 1.4 to the current design, in which multiple elements are combined to produce the desired image quality. In particular, we began with the notion of applying the local transverse power across a limited number of surfaces (two) to flatten either the tangential or the sagittal field. We then realized that combinations of more surfaces may be used to flatten the whole field. Finally, by applying the symmetrical principle and an appropriate stop location combined with many more lens variables, we now can control further aberrations to yield a decent result. Lens design can either proceed in this fashion (from a simple 'start-from-scratch' layout with limited surfaces) or from a pre-optimized state that is then just scaled and re-optimized. The former approach is obviously slower (but provides deep understanding), while the latter is quick (but can often involve trial and error along the way). This method of selecting a pre-optimized design and then re-scaling/re-optimizing would be quite suitable for conditions in which there are tight schedules and deadlines in product development.

1.4.5 The optical sine theorem is not the same as the Abbe sine condition

As figure 1.199 indicates, the angle U' that appears in the Abbe sine condition refers to the angle of the **tangential** marginal ray, which lies in the *meridional* plane (or the plane of the *meridian*, which is just another name for the Y–Z plane in OS). However, the use of this angle in the meridional plane is only valid when the sine condition is met, that is, when linear coma at the full zone is zero. Yet, in the sagittal plane, at low fields (small object and image heights), the magnification of the image formed by oblique sagittal rays is always valid at the full zone, regardless of the presence of coma and spherical aberration. Referring to figure 1.221, if the object height (not shown) is h_s, and the sagittal marginal ray angle in object space (not shown) is U_s, then the image magnification M_s at the full zone for the sagittal rays is [5, 58]:

$$M_s = \frac{h_s'}{h_s} = \frac{\sin U_s}{\sin U_s'}.$$
(1.58)

Equation (1.58) is known as the **optical sine theorem** (OST), written for the case of an air medium in object and image space. At low fields, it is always valid for the sagittal marginal ray, so it is completely unrelated to rays in the meridian. However, if the

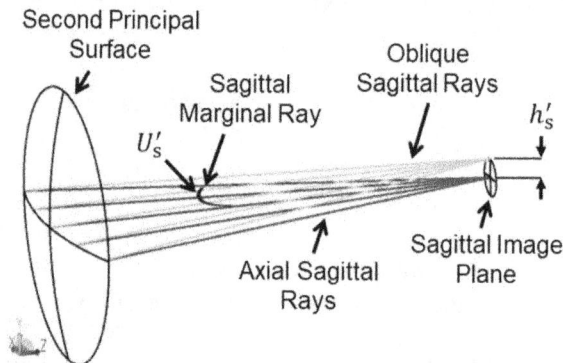

Figure 1.221. The geometry for sagittal rays in image space for equation (1.58).

lens is aplanatic at the full zone, then the relation is also equal to the ratio of the sines of the tangential marginal ray, as expressed by the Abbe sine condition of equation (1.52). In fact, the OSC is derived in part by applying the OST to determine h_s' in figure 1.202 [5, 58]. The point is to take note of the difference between the OST and Abbe's sine condition, which are two different relations that become equal when linear coma is zero at the full zone.

1.4.6 Analogous imaging systems: aspheric aplanatic singlets and Ritchey–Chrétien mirrors

In section 1.4.3, you saw that aspheric surfaces were used in a mobile phone lens design, which resulted in high axial MTF and high *average* MTF across the field. In section 1.4.4, I showed how spherical aberration is ordinarily corrected through the use of multiple elements, but that discussion was limited to the use of spherical surfaces. In a sense, one can say that the lens systems of sections 1.4.3 and 1.4.4 represent two limiting conditions: the former uses highly aspheric surfaces for all elements, while the latter uses no aspheric surfaces for any element. In mobile phone lenses, the asphere profile of lens surfaces usually contains many terms in addition to the sag formula of equation (1.23). But as you saw in equation (1.24), high-order terms naturally exist when equation (1.23) is expanded as a Taylor series, due to the presence of the conic constant K. The difference between these high orders and those used in mobile phone lenses is that the high-order terms of equation (1.24) have fixed values, defined solely by the conic constant. In section 1.4.4, we saw that third-order aberrations produce an OPD in the amount W, where W is the wavefront aberration function (or wavefront error) given by equation (1.54). I then mentioned that aspheric surfaces would introduce an additional OPD of W' such that the combined phase error in a wavefront converging towards the image plane is $kW + kW'$. Since a conic constant introduces high-order terms to its sag, we may regard W' as a consequence of OPDs added by all of the terms contributed by the conic. Hence, it is possible for the sum of all terms from one or more conic surfaces to combine with the aberrations such that $kW + kW' = 0$, and in some cases, $kW + kW' \approx 0$. Therefore, it should be possible for a lens (or a mirror) system to be made aplanatic if several surfaces are given a conic constant.

It turns out that specific values for the conic constant can make a system aplanatic when the field is sufficiently low. A good number of optical systems operate at low field, such as telescope and microscope objectives. Hence, such systems can make use of aspheres with just the conic constant alone. Let us begin with a non-aplanatic but **stigmatic** lens, as shown in figure 1.222. A stigmatic lens is only axially free of spherical aberration [95], which means that it does possess significant coma and other off-axis/oblique aberrations, so it cannot be used for off-axis fields. There are many solutions and forms for the stigmatic lens [95], but we will stick to the well-known form shown in figure 1.222, which has a hyperbola on the right-hand surface. The prescription and merit function for this stigmatic lens are provided in figures 1.223 and 1.224, respectively.

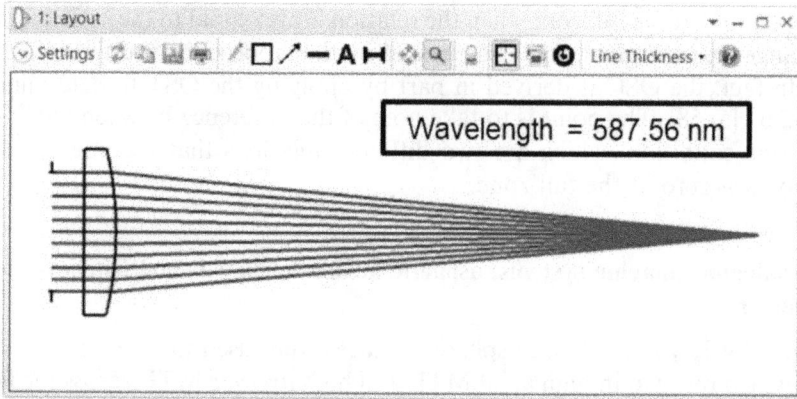

Figure 1.222. A stigmatic singlet for an object at infinity. The right-hand surface is a hyperbola.

	Surface Type	Comment	Radius	Thickness	Material	Coating	Clear Semi-Dia	Conic
0	OBJECT Standard		Infinity	Infinity			0.00000	0.00000
1	STOP Standard		Infinity	5.00000			9.00000 U	0.00000
2	(aper) Standard		Infinity	5.00000	N-BK7		12.50000 U	0.00000
3	(aper) Standard		-51.68000 X	100.00000 M			12.50000 U	-2.30068 V
4	IMAGE Standard		Infinity				1.75299E-09	0.00000

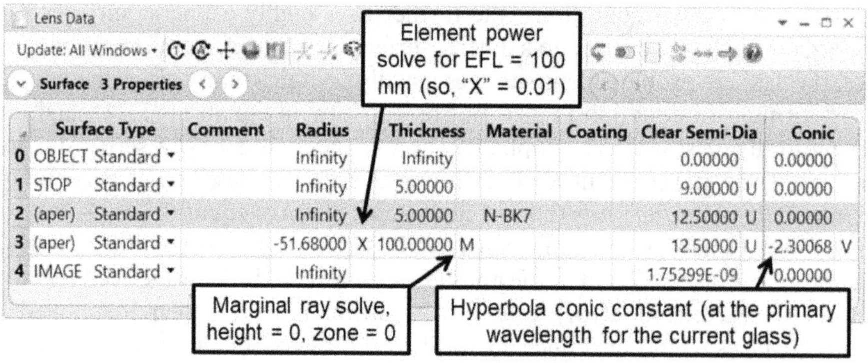

Figure 1.223. The prescription for the singlet in figure 1.222.

	Type	Surf	Wave	Hx	Hy	Px	Py	Target	Weight	Value	% Contrib
1	OPTH	4	1	0.00000	0.00000	0.00000	0.00000	0.00000	0.00000	112.58400	0.00000
2	OPTH	4	1	0.00000	0.00000	0.00000	0.70000	0.00000	0.00000	112.58400	0.00000
3	OPTH	4	1	0.00000	0.00000	0.00000	1.00000	0.00000	0.00000	112.58400	0.00000
4	DIFF	1	2					0.00000	1.00000	-2.84501E-11	5.55995
5	DIFF	1	3					0.00000	1.00000	-1.17254E-10	94.44005

Figure 1.224. The merit function for the singlet in figure 1.222.

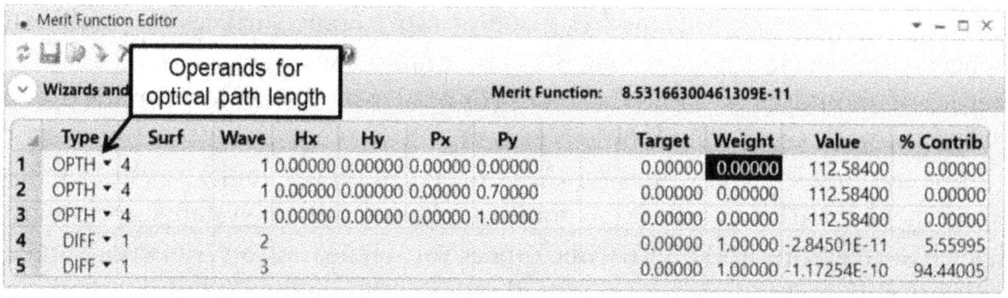

The current lens is perfectly stigmatic; in other words, it has a diffraction-limited MTF axially (and only axially). If you want to change its EFL, simply change the element power solve for the radius on surface 3. No further optimization is required (besides manually tweaking the lens's central thickness when the convex surface has a small radius), because a surface conic constant for a plano-convex singlet is only dependent on its refractive index. Notice in figure 1.223 that a variable has been assigned to the conic constant of surface 3. This was all that was needed to optimize the current singlet. In section 1.3.3, I mentioned that a hyperbola would have a conic constant of less than −1, so the value shown for this singlet in figure 1.223 satisfies this condition. I began this lens's optimization by making the conic constant −1, then let the program optimize it by way of the OPTH operands shown in figure 1.224. Suppose I did not know that the surface has to be a hyperbola, then I would have simply used trial and error in which the starting conic constant was set to either +1 or −1 (there are only two choices for sign, so the effort required for trial and error would not have been great). It is also possible to use the RSCE operand to minimize the axial RMS spot size instead of the OPTH operands. The reason I elected to use the OPTH operands was to highlight the connection between this lens design and the discussions provided earlier on aberrations, OPLs, and OPDs.

In the merit function shown in figure 1.224, the OPTH operand on row 1 computes the OPL for a paraxial ray (essentially, this would be the ray traveling along the optic axis in this case). The next two operands compute the OPL for the ray traversing the 0.7 and full entrance pupil zones, respectively. In order to make a spherical wave converge to a point on-axis, the wavefront error must be zero across the exit pupil, and in order for this to occur, the OPLs of all rays must be the same in all zones, but only three zones were needed for this lens (maybe even two would have been sufficient). This is the essence of lens design. **When a perfect image of a point source is formed, all rays from the point source traverse equal OPLs from source to image**, as a consequence of Fermat's principle, which was first highlighted in section 1.3.5. Hecht provides an illustrative example of applying Fermat's principle and arriving at the hyperbola as the necessary shape for a plano-convex lens to focus parallel rays to a perfect point [96]. For a mirror in air, the shape would be that of a parabola. Evidently, the determination of appropriate conic constants for lenses and mirrors does not require the use an optical design program. But here, we will make use of OS, as this is the point of this book and is not an uncommon approach even among experienced designers (see, for example, [30]).

For an axial point source at a finite distance, simply give the conic constant on surface 2 the same value as that on surface 3 and make the radius of surface 2 a variable. Then, re-run the optimization using the same merit function, but ensure that the 'Surf' column for each of the OPTH operands is set to the image surface. If you want to produce such a biconvex/bi-hyperbolic singlet, it may be hard, as the manufacturer may not be able to supply both surfaces as aspheres on a single element. In this case, just split the singlet into two plano-convex hyperbolic aspheres with an air gap between the plano sides. It will not change anything, but you may need to apply anti-reflection (AR) coatings to all four surfaces.

If oblique rays are involved, then a minimum of two conic surfaces are required. Suppose we include a half-field angle of 2° for the oblique rays of the current singlet and we want its EFL to be 50 mm at F/2.8. One solution is shown in figure 1.225, whose prescription and merit function are provided in figures 1.226 and 1.227, respectively. Note that in figure 1.226, conic constants are given to both surfaces of the lens, each of which has a value of less than −1, so both are hyperbolas. The

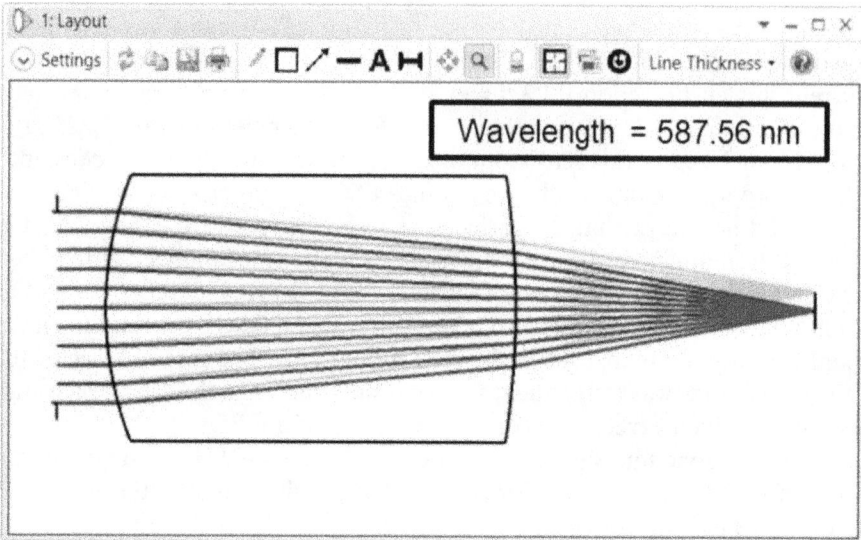

Figure 1.225. An aplanatic singlet at 2° half-field, *F*/2.8, and 50 mm EFL.

	Surface Type		Comment	Radius	Thickness	Material	Clear Semi-Dia	Conic
0	OBJECT	Standard		Infinity	Infinity		Infinity	0.00000
1	STOP	Standard		Infinity	5.00000		9.00000 U	0.00000
2	(aper)	Standard		32.61809 V	40.01382 V	N-BK7	12.50000 U	-0.64545 V
3	(aper)	Standard		-72.37500 X	29.10147 M		12.50000 U	-20.53425 V
4	IMAGE	Standard		Infinity			1.75530	0.0000

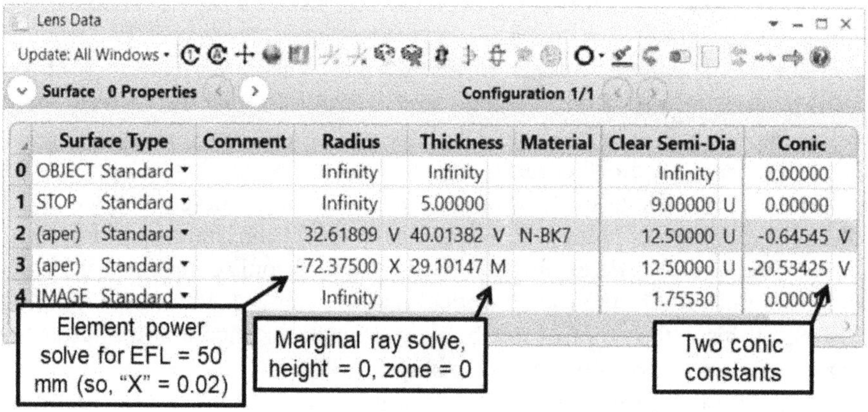

Figure 1.226. The prescription for the lens shown in figure 1.225.

	Type	Ring	Wave	Hx	Hy				Target	Weight	Value	% Contrib
1	EFFL		1						0.00000	0.00000	50.00000	0.00000
2	REAY	1	1	0.00000	0.00000	0.00000	1.00000		0.00000	0.00000	9.00000	0.00000
3	PROB	2		2.00000					0.00000	0.00000	18.00000	0.00000
4	DIVI	1	3						0.00000	0.00000	2.77778	0.00000
5	OSCD		1	1.00000					0.00000	0.00000	1.33615E-03	0.00000
6	SPHA	0	1						0.00000	0.00000	-0.48810	0.00000
7	COMA	0	1						0.00000	0.00000	-0.69136	0.00000
8	RWCE	5	1	0.00000	0.00000				0.00000	1.00000	0.11701	35.09816
9	RWCE	5	1	0.00000	1.00000				0.00000	1.00000	0.15900	64.80390
10	MNCG	2	3						5.00000	0.20000	5.00000	0.00000
11	MXCG	2	3						40.00000	0.20000	40.01382	0.09794

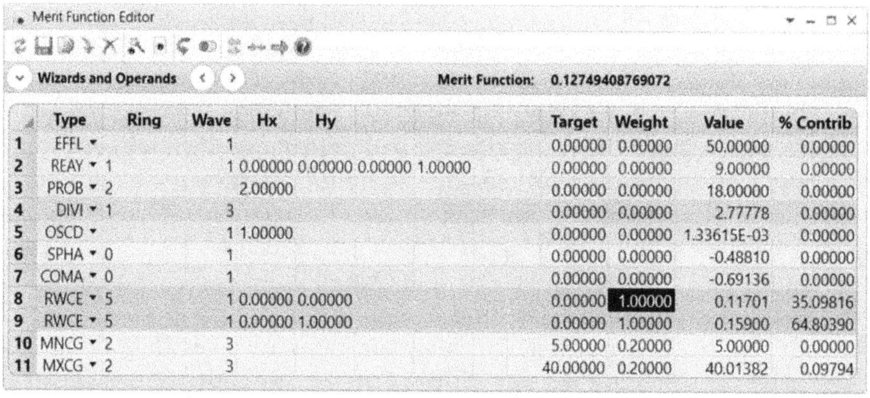

Figure 1.227. The merit function for the lens shown in figure 1.225.

central thickness was made variable and ended up at 40.013 82 mm, which implies that the program 'wants' a thick lens. Still, in its current state, the reader who enters the prescription into a program to view the MTF will see that the axial and off-axis MTFs are very nearly diffraction limited and that they have residual astigmatism and field curvature (which is expected).

In figure 1.227, note that the OPTH operands have been replaced by RWCE operands to minimize the RMS wavefront error for both axial and oblique fields. This is because, in the case of oblique rays, the wavefront from infinity is *tilted* at the half-field angle, which requires the stop to be tilted by the same amount in order to make it be orthogonal to the principal ray. It is possible to achieve this by setting up a multi-configuration system: one configuration for axial rays (with a stop normal to the optic axis), the other for oblique rays (with a tilted stop), and using only OPTH operands for optimization (see appendix A.3 for an example of this). In contrast, the RWCE operands automatically account for this by computing the difference in path lengths between a reference spherical wavefront (which is appropriately tilted) and the aberrated wavefront. During optimization, weights were only used for the axial and oblique wavefront error operands as well as for the min and max values for lens thicknesses (rows 18 and 19). The resulting OSC is small, and the residual third-order spherical and coma aberrations (at 587.56 nm) are left to balance the OSC and high-order aberrations. For this lens, if the oblique field is to be corrected further, then higher-order aspheric coefficients can be included for both surfaces.

Now, several paragraphs above, I mentioned that the parabola is the proper shape for a single reflective surface (in air) to focus rays from infinity into a perfect axial point, while the hyperbola is the proper shape for the convex side of a plano-convex lens to do the same. So, it seems that there is a kind of correspondence between stigmatic refractive surfaces and stigmatic reflective surfaces. If so, then there should be a correspondence between two-surface refractive aplanatic elements and two-surface reflective aplanatic systems. That is, by 'correspondence,' we roughly mean that for every aspheric aplanatic refractive singlet, there is an equivalent aspheric two-mirror system in air that has the same EFL and aplanatic condition. The only difference is that the former has glass between the surfaces, while the latter has air. However, if we continue using only conic constants without higher-order aspheric coefficients, there is a limitation to this correspondence in terms of the off-axis image quality of the mirror system. For instance, if, for now, we ignore the obscuring effects of a secondary mirror, using similar merit function operands (but replacing OPTH operands with RSCE), we can convert the lens in figure 1.225 into the fictitious near-equivalent two-mirror system shown in figure 1.228, whose prescription is provided in figure 1.229. This mirror system is not aplanatic (a reduction in field angle to 0.5° half-field and re-optimization would result in diffraction-limited MTF). However, it does have the advantage that, being only a mirror system in air, there is no chromatic aberration. Thus, the trade-off between using lenses and using mirrors is that lenses have dispersion but can operate at high field, while mirrors have no dispersion but are limited to low field.

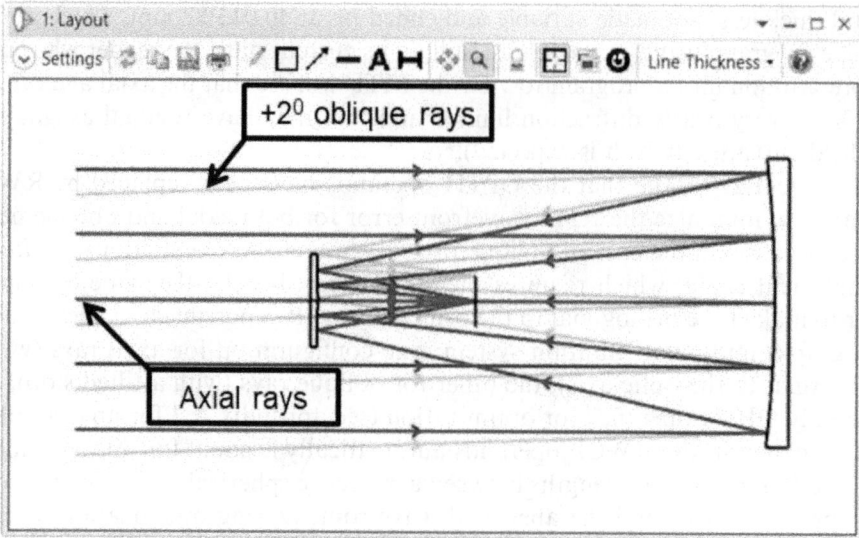

Figure 1.228. A fictitious two-mirror system approximately equivalent (non-aplanatic but reasonably corrected) to the lens in figure 1.225 at ±2° field, *F*/2.8, 50 mm EFL.

Figure 1.229. Prescription for the system in figure 1.228.

In its current state and limited to the sole use of conic constants, the off-axis image of the mirror system of figure 1.228 cannot be improved unless the field is reduced and the *f*-number is increased (as in the case of astronomical telescopes). Let us design a simple aplanatic reflective telescope for amateur astronomy with a primary mirror half-field of view of 0.2° and an *f*-number of *F*/8. The field of view is modest but will give excellent image quality. Of course, in this case, we must account for the obscuration of rays by the secondary mirror (surface 3 of the prescription in figure 1.229), and we must have a hole at the center of the primary mirror (surface 2) to let rays through to the image plane. First, we ignore both of these and get the correct first-order layout. To do this, consider the construction shown in figure 1.230.

By similar triangles, we have $y_2/(f_1 - d) = y_1/f_1$, where f_1 is the EFL of the primary mirror. Rearranging quantities and solving for y_2/y_1 gives:

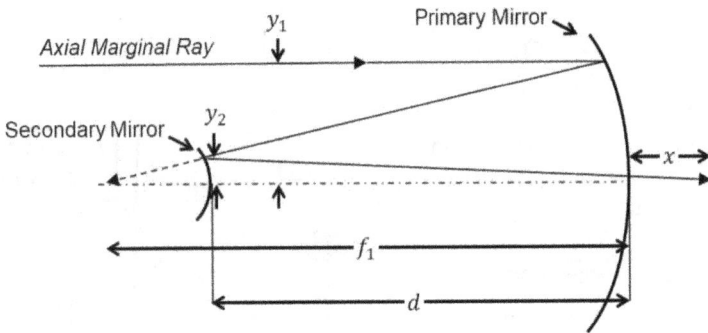

Figure 1.230. Some first-order quantities for a two-mirror reflective telescope with central obscuration from the secondary mirror.

$$\frac{y_2}{y_1} = 1 - \frac{d}{f_1}. \tag{1.59}$$

We also require that the final focal position produced by the secondary mirror be located at or to the right of the primary mirror. Let x in figure 1.230 be the distance to the right of the primary mirror for this focal position. Then, applying equation (1.20) for the secondary mirror yields:

$$\frac{1}{f_2} = \frac{-1}{f_1 - d} + \frac{1}{d + x}. \tag{1.60}$$

Combining equations (1.59) and (1.60) gives a simple relation for the case of $x = 0$:

$$\frac{f_1}{f_2} = \frac{2(y_2/y_1) - 1}{(y_2/y_1)[1 - (y_2/y_1)]}. \tag{1.61}$$

The quantity y_2/y_1 is known as the *obscuration ratio* for a telescope [13], and it is often lumped into a single variable in telescope design [97]. Let us suppose that we let this ratio be 0.25 (incidentally, the obscuration ratio for the Hubble space telescope is 0.13 [13]), and suppose the EFL of the primary mirror is arbitrarily set at 500 mm. Applying equation (1.61) to compute the focal length of the secondary mirror, we obtain $f_2 = -187.5$ mm, which implies that we have a convex reflective surface. Applying equation (1.60) to solve for d, we obtain $d = 375$ mm. Since d, f_1, and f_2 have been determined, this leaves the system EFL, which can be computed by applying equation (1.21). Doing this, we obtain $f = 1500$ mm, which is a reasonable magnitude (e.g. if we had an eyepiece with an EFL of 30 mm, then the magnification would be 50×, which is quite high). Since the EFL of a mirror surface is half its radius of curvature, the primary mirror's radius (accounting for the sign convention in OS) is −1000 mm, and the secondary mirror's radius is −375 mm. At *F*/8, this telescope's primary mirror (which is also the stop and therefore the entrance pupil) would have a diameter of 187.5 mm. Once we have entered all of the above first-order quantities into the Lens Data Editor, the only variables left to be optimized are

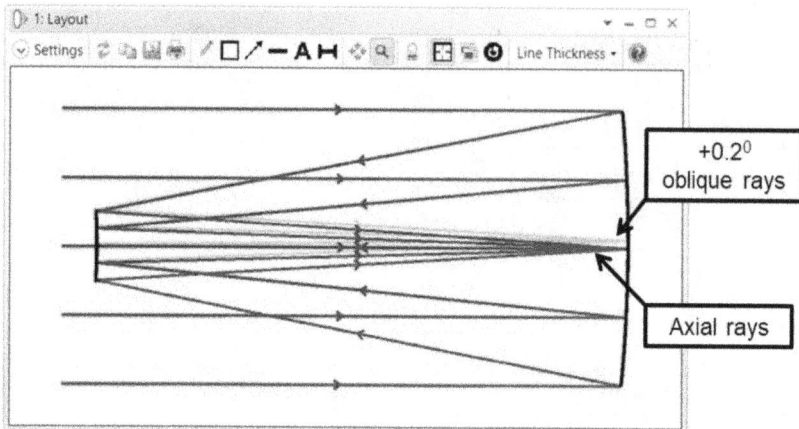

Figure 1.231. The layout of a simple Ritchey–Chrétien telescope.

	Surface Type	Comment	Radius	Thickness	Material	Clear Semi-Dia	Conic
0	OBJECT Standard ▾		Infinity	Infinity		Infinity	0.00000
1	Standard ▾	Dummy	Infinity	400.00000		95.14627	0.00000
2	STOP (aper) Standard ▾	Primary	-1000.00000	-375.00000	MIRROR	93.75000 U	-1.06621 V
3	Standard ▾	Secondary	-375.00000	375.00000 M	MIRROR	24.84898	-4.82488 V
4	IMAGE Standard ▾		Infinity	-		5.24125	0.00000

Figure 1.232. The prescription for the system shown in figure 1.231.

the conic constants for the primary and secondary mirrors. Let both conic constants begin at the value of -1, which makes them into hyperbolas. Applying only RWCE operands for the axial and oblique fields, we obtain the resulting layout for the so-called *Ritchey–Chrétien* telescope shown in figure 1.231, whose prescription, merit function, and MTF are shown in figures 1.232–1.234, respectively.

Of course, a real two-mirror telescope must have a central obscuration. We can achieve this for the primary mirror in OS by going to that surface and selecting Circular Aperture from the Surface Type. We then set the minimum radius to roughly 24.2 mm and the maximum radius to 93.8 mm. The result is that the MTF is reduced due to the increase in diffraction rings outside the central region of the PSF. In fact, for a telescope, we would actually be required to examine the PSF in addition to the MTF, because PSFs give information about how point-like sources (such as stars) appear, while the MTF provides information about the resolution of features on extended objects, such as the Sun, Moon, and large planets in our solar system. The layout with central obscuration is shown in figure 1.235. This telescope

Figure 1.233. The merit function for the system shown in figure 1.231.

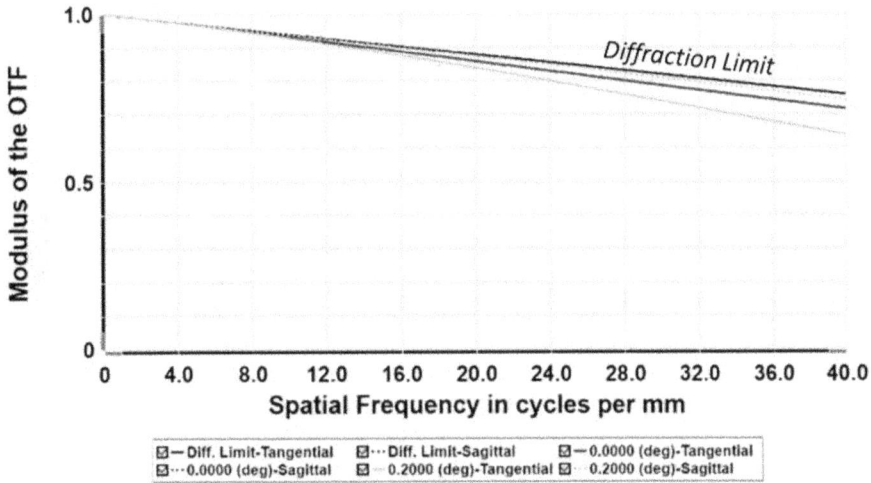

Figure 1.234. The MTF for the system shown in figure 1.231.

cannot be made any better for a larger field of view unless surface sag deviations using higher-order aspheric terms are included in addition to the conic constants for both mirrors. In particular, if we made the two surfaces into *even aspheres* and included the sixth- and eighth-order terms, then we would obtain a near-diffraction-limited MTF at double the current field.

The point of this section is to highlight the analogy between refractive aspheres and reflective aspheres, the latter being regarded simply as a 'reflecting lens' with air as its medium. It is therefore, a sort of *air lens*, except that the conventional definition of such a lens is refractive in the sense that it is a thick slab of glass with air in between doing the 'refraction' [5], as illustrated in figure 1.236; its prescription, merit function, and MTF (at wavelength of 587.56 nm) are provided in figures 1.237–1.239, respectively.

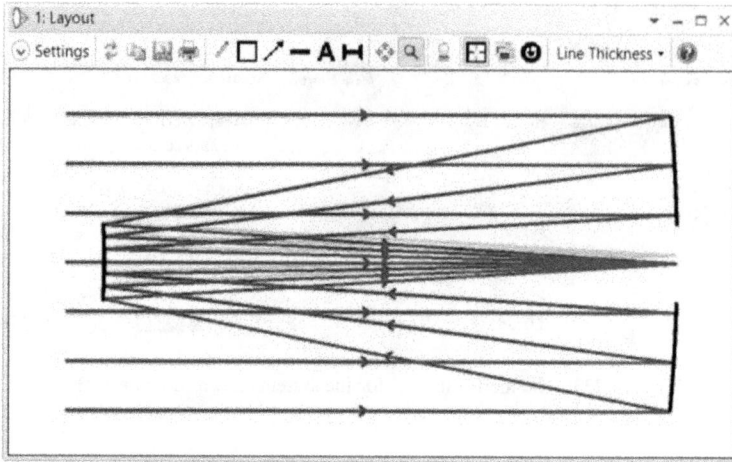

Figure 1.235. The layout of the system in figure 1.231 with a central obscuration.

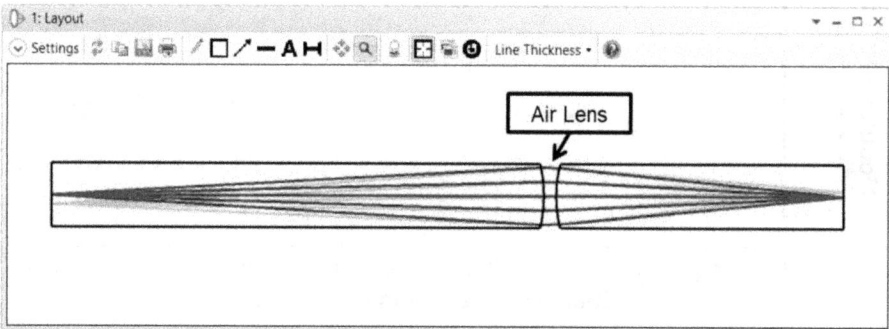

Figure 1.236. An air lens.

Figure 1.237. The prescription for the air lens shown in figure 1.236.

	Surface Type	Comment	Radius	Thickness	Material	Clear Semi-Dia	Conic
0	OBJECT Standard ▾	Glass 1	Infinity	200.00000	N-BK7	3.63061	0.00000
1	(aper) Standard ▾	Air Lens	-50.00000	5.00000		12.50000 U	-3.40402 V
2	STOP Standard ▾		Infinity	0.00000		11.00000 U	0.00000
3	(aper) Standard ▾	Glass 2	50.00000	115.70822 V	N-BK7	12.50000 U	2.11265 V
4	IMAGE Standard ▾	Glass 2	Infinity	-	N-BK7	2.15310	0.00000

Figure 1.238. The merit function for the air lens shown in figure 1.236.

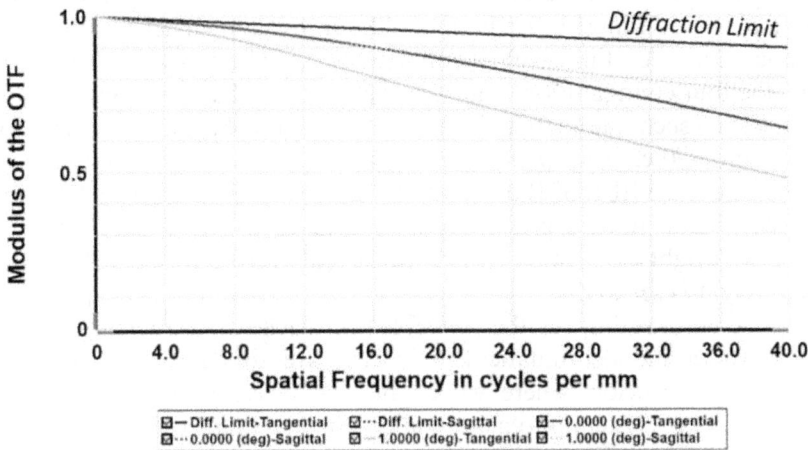

Figure 1.239. The MTF for the air lens shown in figure 1.236.

The air lens above can be improved in image quality if higher-order terms are added for the two aspheric convex glass surfaces (concave if regarded from the perspective of the air lens), which are both currently maintained as conic surfaces. Notice that the second air lens surface (surface 3) has a positive-valued conic constant, which means it is an *oblate ellipse*. If you are in water (such as when scuba diving), your mask or goggles is an air slab, and the apparent magnification you see is, to the first order, given by the index of the water. Since water's index is roughly 1.33, objects in the water appear 33% larger. If you are outside the water, then the air between the water's surface and the first air/eye interface (i.e. the cornea) is an air lens whose concave surface is the convex surface of your eye's cornea. The image, of course, is inside the fluid of your eye. This brings us to a final point in this section, which was also mentioned in section 1.4.2, which is that, in the OS program, when you have a medium between the image plane and the surface just prior to the image, the EFL in image space is not provided by the EFFL operand. Rather, you either need to compute it by multiplying the EFFL value by the index of the medium the image is in, or you need to refer to the output of the System Data feature by going to

Analyze > Reports > System Data. In the output list, scroll down till you see Effective Focal Length (in image space). The derivation of the EFL for a lens system immersed in a medium is discussed in appendix A.4.

1.4.7 Heuristic color correction theory

To an extent, the question of how to correct chromatic aberration for a lens system can be regarded as a question of whether or not an achromat (such as a doublet that has been corrected for its own chromatic aberration) can correct the dispersion introduced by other elements in the neighborhood of that achromat. In other words, given, say, an achromatic doublet, if a dispersive singlet is mounted at some distance either in front of or behind that doublet, can that doublet be made to correct the chromatic aberration introduced by the addition of that singlet to achieve a total achromatic state and still satisfy the system EFL? The answer is yes. Take, for example, the four configurations shown in figure 1.240. Each configuration is in an achromatic state such that marginal rays at 450 nm and 650 nm are focused to equal on-axis locations (i.e. the *primary axial color* is corrected). Further, each config-uration has the same EFL of 100 mm at 550 nm. Prescriptions for each are given in figures 1.241–1.244. In each prescription, a marginal ray solve has been set for the last thickness (height = 0, zone = 1). The wavelengths are set to 450, 550, and 650 nm and each weight is set to one.

The above exercise implies that possible approaches to correcting the primary axial color of a monochromatic lens system are to either add one or more achromatic doublets somewhere within the lens system or to convert specific elements in the lens system into achromatic doublets, followed by re-optimization. You could, for instance, pick a COTS achromatic doublet from a supplier's catalog, and, provided that the supplier has disclosed the names of the glasses, insert its prescription into your system and begin optimization, targeting the primary axial color at zero (an example is provided in section 1.4.10). The question of where to insert the doublet is decided through trial and error, but a reasonable starting point is to insert it into a space that has approximately collimated axial rays. However, note that the new doublet has a power, so you would need to reduce the power of one singlet or a pair of singlets that sandwich the new doublet.

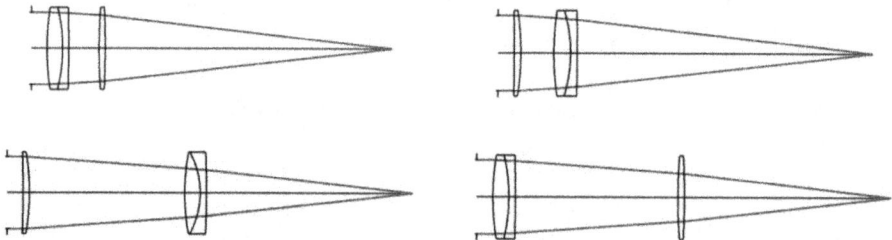

Figure 1.240. Four configurations of a lens system consisting of an achromatic doublet and a singlet. Each configuration is achromatized for the 450 and 650 nm rays.

Figure 1.241. The prescription for the top left configuration shown in figure 1.240.

Figure 1.242. The prescription for the top right configuration shown in figure 1.240.

But what if you need to select different glasses to improve the achromatic state? This is generally a matter of combining some knowledge in color correction theory with experience (trial and error) and automatic optimization of glass variables (index and dispersion). In automatic optimization, if the program allows it, you could even let the program vary the glass choices (e.g. see section 1.4.11). If you do not have time to do any of these, then give the hardcore detailed lens design work to a lens designer (they love it and live and breathe bending rays in their sleep). Of course, your boss has to set aside some company money for this.

Still, it is good to be familiar with basic color correction theory. Here, I share some basic theories and rules, beginning with the four configurations in figure 1.240.

Lens Data

Update: All Windows

Surface 0 Properties — Configuration 1/1

	Surface Type		Comment	Radius		Thickness		Material	Coating	Clear Semi-Dia	
0	OBJECT	Standard		Infinity		Infinity				0.00000	
1	STOP	Standard		Infinity		5.00000				11.00000	U
2	(aper)	Standard		-429.85482	V	2.00000		N-BK7		12.50000	U
3	(aper)	Standard		-72.31746	V	50.00000				12.50000	U
4	(aper)	Standard		65.51443	V	5.00000		N-BK7		12.50000	U
5	(aper)	Standard		-25.10380	V	2.00000		N-F2		12.50000	U
6	(aper)	Standard		-327.24229	V	65.37909	M			12.50000	U
7	IMAGE	Standard		Infinity		-				8.66719E-03	

Figure 1.243. The prescription for the bottom left configuration shown in figure 1.240.

Lens Data

Update: All Windows

Surface 0 Properties — Configuration 1/1

	Surface Type		Comment	Radius		Thickness		Material	Coating	Clear Semi-Dia	
0	OBJECT	Standard		Infinity		Infinity				0.00000	
1	STOP	Standard		Infinity		5.00000				11.00000	U
2	(aper)	Standard		74.12934	V	5.00000		N-BK7		12.50000	U
3	(aper)	Standard		-51.40735	V	2.00000		N-F2		12.50000	U
4	(aper)	Standard		-430.81855	V	50.00000				12.50000	U
5	(aper)	Standard		154.11584	V	2.00000		N-BK7		12.50000	U
6	(aper)	Standard		-235.37819	V	63.92472	M			12.50000	U
7	IMAGE	Standard		Infinity		-				8.80623E-03	

Figure 1.244. The prescription for the bottom right configuration shown in figure 1.240.

Before I optimized each configuration, I began with just the doublet alone. The glasses N-BK7 and N-F2 are common knowledge to lens designers (but I will explain why they were chosen in a moment). I minimized primary axial color for the doublet using the AXCL operand in OS at the full pupil zone (AXCL operands in OS compute the difference in axial focal positions between marginal rays at two wavelengths, for a specific zone). After that, I inserted the N-BK7 singlet behind the doublet (the top left configuration of figure 1.240) and fixed the air space between them, and I also fixed all element thicknesses. The only variables were applied to the curvatures. I then re-optimized the primary axial color by targeting the 450 nm and 550 nm rays (at the full pupil zone) at a zero difference in axial focal position. I also targeted the EFL at 100 mm and the optical path length difference between the

Merit Function Editor

Wizards and Operands — Merit Function: 4.69189920788669E-10

	Type	Wave1	Wave2	Zone				Target	Weight	Value	% Contrib
1	EFFL		2					100.00000	1.00000	100.00000	71.67434
2	OPTH	7	2	0.00000	0.00000	0.00000	0.00000	0.00000	0.00000	118.54907	0.00000
3	OPTH	7	2	0.00000	0.00000	0.00000	1.00000	0.00000	0.00000	118.54907	0.00000
4	DIFF	2	3					0.00000	1.00000	-3.73134E-10	21.08201
5	RSCE	3	2	0.00000	0.00000			0.00000	0.00000	4.98018E-04	0.00000
6	AXCL	1	3	1.00000				0.00000	1.00000	-2.18720E-10	7.24365

Figure 1.245. The merit function for the top right configuration shown in figure 1.240.

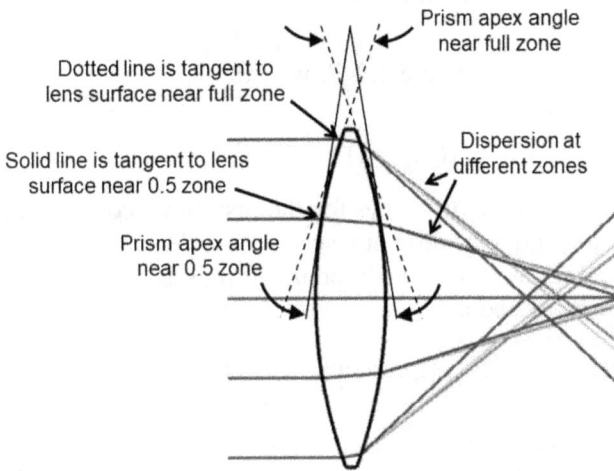

Figure 1.246. Different prism apex angles at different zones of a biconvex lens.

marginal ray (at the full zone) and axial ray at zero (using the OPTH operands). I did this for each configuration. Figure 1.245 shows the merit function for the top right configuration (the same merit function operands were used for all four configurations). The result is shown in figure 1.240. The resulting axial MTF (at 450, 550, 650 nm with all weights set to one) for all configurations exceeds 50% at 40 cycles/mm.

What fundamental mechanism lets the above method work for achromatizing a system of lenses? The idea was to take the view that a lens is like a prism with unique apex angles at specific zones, as illustrated in figure 1.246.

Based on figure 1.246, it is then possible to make an analogy between the equations for beam deviation performed by a prism and the equations for the focal power of a lens. To do this, recall that when a ray passes through a 'thin' prism (i.e. a prism with a small apex angle) in air, its direction upon exiting the prism suffers a *beam deviation* whose angle is proportional to $(n - 1)$, where n is the refractive index of the prism [29]. Figure 1.247 illustrates this situation.

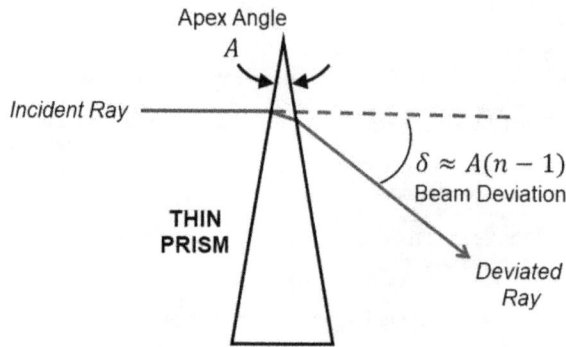

Figure 1.247. The definition of beam deviation performed by a thin prism.

Applying the formula in figure 1.247, we can express the total beam deviation from two thin prisms as

$$\delta_1 + \delta_2 = A_1(n_1 - 1) + A_2(n_2 - 1) = \delta_{\text{tot}}, \tag{1.62}$$

where the subscripts denote quantities for the first and second prisms, respectively, and δ_{tot} is the total beam deviation at a specific wavelength. Due to the dependence of index on wavelength, δ_{tot} can be different for rays at different wavelengths. This variation may be expressed as

$$\frac{\mathrm{d}}{\mathrm{d}\lambda}[A_1(n_1 - 1) + A_2(n_2 - 1)] \approx A_1\frac{\Delta n_1}{\Delta \lambda} + A_2\frac{\Delta n_2}{\Delta \lambda} = \frac{\varepsilon}{\Delta \lambda}, \tag{1.63}$$

where $\Delta n_1 = n_{1,\text{blue}} - n_{1,\text{red}}$ (i.e. the difference between the index at blue and at red for the first prism), and $\Delta n_2 = n_{2,\text{blue}} - n_{2,\text{red}}$. We can now either make the *difference* between δ_{tot} at the blue and δ_{tot} at the red zero (i.e. let $\varepsilon = 0$, which would cause the rays at each wavelength to emerge parallel to one another) or we can use a fixed value for ε (which would cause the rays at each wavelength to intersect at a point beyond the prism pair). The latter is generally the desired result when achromatizing a lens, because we do want rays at two wavelengths to intersect at the axis in the image (i.e. at the BFL of the lens). However, when it comes to glass selection, it is irrelevant whether or not the two rays at different wavelengths should have zero or nonzero ε. What matters is whether or not the selected glasses for specific elements can control the *sign* of ε. That is, **elements with the correct choices of glass for achromatization are capable of making $\varepsilon > 0$ or $\varepsilon < 0$ by varying their refractive powers, thereby enabling the blue and red rays to either intersect or diverge at a desired position in space.** This is the key point in color correction. **You can think of it as chromatic beam steering.** In the case of two prisms, this means that we want glasses such that different combinations of A_1, A_2, Δn_1, and Δn_2 can make $\varepsilon > 0$ or $\varepsilon < 0$. This can also be viewed in terms of an *inverse problem* as follows: for any desired choice of value for ε, there should exist a combination of realistic values for A_1, A_2, Δn_1, and Δn_2. To see this more explicitly, let us combine equation (1.62) with (1.63).

Upon performing some algebraic manipulations, the deviations for each prism may be expressed as

$$\delta_1 = \frac{\delta_{\text{tot}} + \varepsilon V_2}{1 - (V_2/V_1)}, \tag{1.64}$$

and

$$\delta_2 = \frac{\delta_{\text{tot}} - \varepsilon V_1}{1 - (V_1/V_2)}. \tag{1.65}$$

In the above, V_1 and V_2 are the following quantities:

$$V_1 = \frac{n_1 - 1}{n_{1,\text{Blue}} - n_{1,\text{Red}}}, \tag{1.66}$$

and

$$V_2 = \frac{n_2 - 1}{n_{2,\text{Blue}} - n_{2,\text{Red}}}. \tag{1.67}$$

In the above formulas, we regard n_1 and n_2 as the indexes at a wavelength between the blue and the red for the first and second prisms, respectively, and we of course recognize V_1 and V_2 as the Abbe dispersion values of the glasses, defined for appropriate indexes at specific wavelengths. Ordinarily, the Abbe dispersion values are written in text as 'Vd,' and in this case, the index in the numerator is that for the sodium d atomic wavelength (587.56 nm), while the index difference in the denominator is defined at the F (486.13 nm) and C (656.27 nm) atomic wavelengths for the blue and red, respectively. At visible wavelengths, the Abbe dispersion values range between 20 (high dispersion) and 90 (extremely low dispersion) and differences in beam deviations between rays at blue and red wavelengths are quite small, in the order of magnitude of $\pm 10^{-3}$ degrees (approximately zero *on average*). Hence, if we want to select glasses, it is simplest to analyze equations (1.64) and (1.65) by letting $\varepsilon \approx 0$.

Analogous equations exist for lenses. The sum of the refractive powers of a pair of thin lenses in close contact may be expressed as:

$$\Phi_1 + \Phi_2 = C_1(n_1 - 1) + C_2(n_2 - 1) = \Phi_{\text{tot}}, \tag{1.68}$$

where Φ_1 and Φ_2 are the reciprocals of the respective lens EFLs:

$$\Phi_1 = \frac{1}{f_1} = \left(\frac{1}{R_{1,1}} - \frac{1}{R_{1,2}} \right)(n_1 - 1), \tag{1.69}$$

and

$$\Phi_2 = \frac{1}{f_2} = \left(\frac{1}{R_{2,1}} - \frac{1}{R_{2,2}} \right)(n_2 - 1), \tag{1.70}$$

where $R_{1,1}$, $R_{1,2}$, $R_{2,1}$, and $R_{2,2}$ are the radii of curvature of the respective lenses. Accordingly, the variables C_1 and C_2 in equation (1.68) refer to the respective factors multiplying the refractive index quantities $(n_1 - 1)$ and $(n_2 - 1)$ in equations (1.69) and (1.70). Equation (1.68) is analogous to equation (1.62). By this analogy, if we apply similar mathematical operations to those performed in equation (1.63) to (1.68), we arrive at the expressions

$$\Phi_1 = \frac{\Phi_{\text{tot}} + \varepsilon V_2}{1 - (V_2/V_1)}, \tag{1.71}$$

and

$$\Phi_2 = \frac{\Phi_{\text{tot}} - \varepsilon V_1}{1 - (V_1/V_2)}, \tag{1.72}$$

where V_1 and V_2 are defined to be exactly the same as in equations (1.66) and (1.67). If the total lens system EFLs at the blue and red wavelengths are made equal, then $\varepsilon = 0$, and equations (1.71) and (1.72) degenerate to the well-known formulas for the powers of two thin lenses in close contact, and the total system EFL is achromatized for the blue and red rays [6]. Of course, real lenses have thickness, and we need the focal positions (not the EFLs) of the blue and red rays to intersect at the axial point in the image. Hence, ε is generally nonzero when a lens system has reached its achromatic state after design optimization. However, as mentioned earlier, the analysis of the glass choice is simplified by letting ε be in the *neighborhood* of zero (which is like saying that $d\Phi_{\text{tot}}/d\lambda$ should—in the language of variational calculus—be *stationary*). Doing this, we can see from equations (1.71) and (1.72) that an appropriate pair of glasses would cause V_1 and V_2 to be sufficiently different, leading to a common solution in which the first element is a positive-powered element with low dispersion (high magnitude for V_1) and the second element is a negative-powered element with higher dispersion than the first (i.e. $V_2 < V_1$). It turns out that the glasses N-BK7 (for lens 1) and N-F2 (for lens 2) from Schott are convenient choices (out of many others), as not only is N-BK7 glass commonly available but also their respective Abbe dispersions meet the requirement that $V_2 < V_1$. Further, we note that since the solution to eliminating primary axial color relies on element powers, we can maintain their powers and bend surface radii and shift the stop to minimize spherical aberration and coma (and therefore, the OSC). In multielement lens systems, we have even more variables available, so we generally do not bother so much with minimizing specific amounts of aberrations and instead simply minimize the RMS spot size and/or RMS wavefront errors.

Due to the equivalence between the prism equations and the lens equations above, let me show you how a prism pair is achromatized, followed by inserting a third prism that reintroduces dispersion (which is then corrected by the original prism pair). To that end, have a look at figure 1.248, whose prescription and merit function are provided in figures 1.249 and 1.250, respectively. On the left-hand side of figure 1.248, there are three input beams indicated (top, center, bottom). In order to appreciate the color correction that is occurring in the current prism pair example,

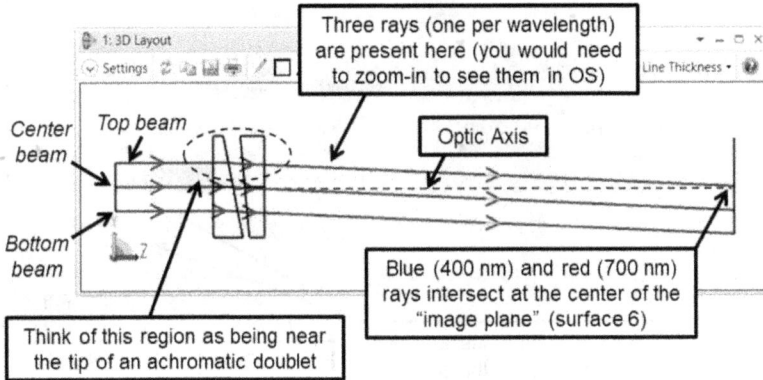

Figure 1.248. The layout of a prism pair model in OS.

Figure 1.249. The prescription in OS for the layout shown in figure 1.248.

you have to ignore the central and bottom beams entering from the stop. They are there only because I cannot ask the OS program to display just the top beam, which is the beam of interest in the current exercise. So, just ignore the central and bottom beams and look at what the top beam is doing within the dotted circle. Within this circle, the entering beam comprises three rays, each at a different wavelength (400 nm, 550 nm, and 700 nm). These three rays overlap precisely as they propagate from the top of the stop towards the first prism. But when they pass through the first prism, they begin to deviate due to dispersion. Upon emerging from the second prism, at the location indicated by the arrow in figure 1.248, the three rays in the top beam remain dispersed, but the ray vectors for the blue and red rays are already pointing towards the axial location at surface 6, due to the optimized state of the apex angles of the prism pair. As can be seen in rows 2 and 4 of the merit function (figure 1.250), their ray heights at surface 6 are essentially zero, while the ray height

Modern Classical Optical System Design

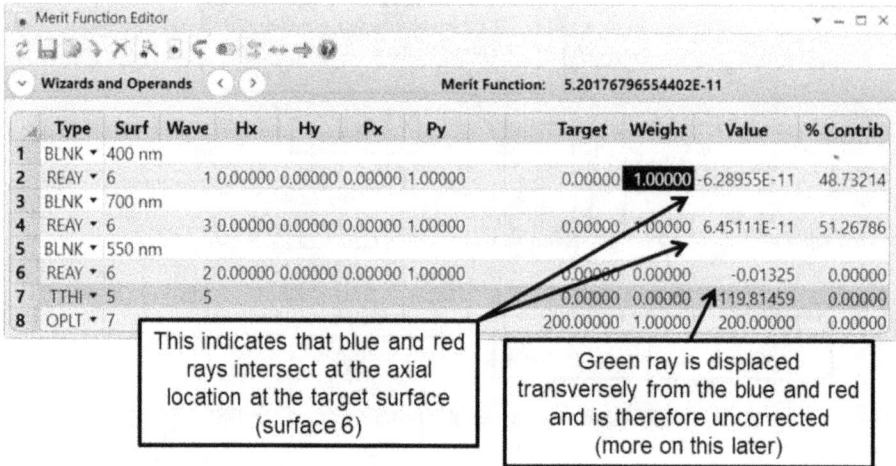

Figure 1.250. The merit function for the prescription shown in figure 1.249.

for the green ray is displaced below the axis. This is residual secondary color, and it has remained uncorrected by the current prism pair (hold that thought, as we will soon correct this). On row 8, the OPLT operand controls the maximum value for the thickness of surface 5. This is analogous to setting the proper BFL for a lens. For a prism pair, it is equivalent to setting a finite value for ε. Since it is cumbersome to determine an appropriate target for ε directly, I did so indirectly through constraining the thickness of surface 5 and also constraining the ray heights for the blue and red rays at the last surface. This effectively makes the top portion of the prism pair act like the top portion of an achromatic doublet.

In the merit function, if you use the RAID operand to compute the angles of incidence for the blue and red rays, then take the difference, you obtain the value for ε, which is very small (roughly 4.4×10^{-3} degrees), but not zero. If you remove the variable for the thickness of surface 5 and also remove all current weights (but leave variables for the X tilts of surfaces 3 and 4), then optimize to obtain $\varepsilon = 0$, you obtain the result that $\varepsilon = 1.139 \times 10^{-12}$ degrees, and the ray heights for the blue and red rays (given by the REAY operands on rows 2 and 4 in the merit function) become $-0.223\,89$ mm and $-0.214\,26$ mm, respectively. This is, of course, generally not regarded as an achromatic state. As mentioned earlier, we need a nonzero ε. However, this exercise provides an example of the validity and application of equations (1.64) and (1.65).

If we now add another prism of N-BK7 material in front of the current prism pair, we obtain a non-achromatized state due to the dispersion from the new prism. This is like having a multielement lens system where one or more elements are dispersive, and we need to have one or more achromats in the system to do work and achieve an overall achromatized state. But if we simply re-optimized, we would obtain the result shown in figures 1.251–1.253. This design exercise is closely equivalent to that performed for the systems in figures 1.240–1.245.

1-228

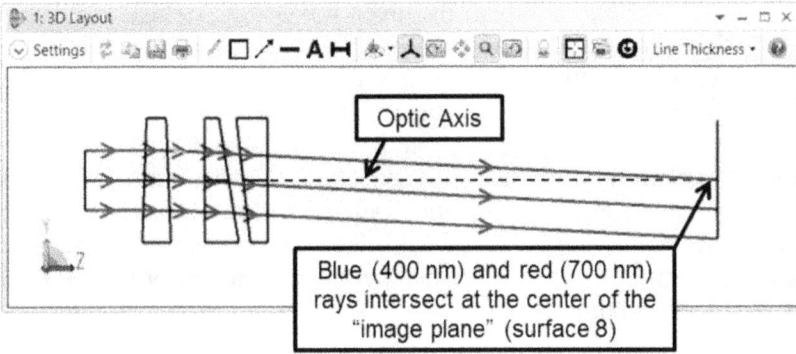

Figure 1.251. The layout for three prisms that are in an achromatic state.

Figure 1.252. The prescription for the layout in figure 1.251.

Figure 1.253. The merit function for the prescription in figure 1.252.

Now, suppose we want the ray at the green wavelength in this 'tri-prism' system to also intersect the blue and red rays at the axis of the last surface. In this case, the rays of all three wavelengths should be 'focused' by the top portions of the prism system at the axial location of the image plane. If we could achieve this, then it would be a state of *apochromatism*, and the prism system would become an *apochromat*. Apochromats correct secondary color aberration, while achromats correct just primary axial color aberration. Since we have learned that an achromat can correct the primary axial color of other singlets, we will guess that an apochromatic prism pair can correct the secondary color of other prisms. So, we begin by rewriting equation (1.63) as

$$\frac{d}{d\lambda'}[A_1(n_1 - 1) + A_2(n_2 - 1)] \approx A_1\frac{\Delta n_1'}{\Delta \lambda'} + A_2\frac{\Delta n_2'}{\Delta \lambda'} = \frac{\varepsilon'}{\Delta \lambda'}, \tag{1.73}$$

where $\Delta n_1' = n_1 - n_{1,\text{red}}$ and $\Delta n_2' = n_2 - n_{2,\text{red}}$ are now the differences between the indexes at the green and red wavelengths for the first and second prisms, respectively and ε' is the difference in total beam deviation between the green and red rays. If we perform the same type of mathematical operations that led from equation (1.63) to equations (1.64)–(1.67) (but now for the green and red rays), we obtain

$$\delta_1 = \frac{\delta_{\text{tot}} + \varepsilon' V_2'}{1 - (V_2'/V_1')}, \tag{1.74}$$

and

$$\delta_2 = \frac{\delta_{\text{tot}} - \varepsilon' V_1'}{1 - (V_1'/V_2')}. \tag{1.75}$$

And now, in the above, V_1' and V_2' are the following quantities:

$$V_1' = \frac{n_1 - 1}{n_1 - n_{1,\text{Red}}}, \tag{1.76}$$

and

$$V_2' = \frac{n_2 - 1}{n_2 - n_{2,\text{Red}}}. \tag{1.77}$$

Referring back to figure 1.248, if we want a prism pair to make the blue, green, and red rays intersect at the axis on the far right, then it needs to satisfy equations (1.64)–(1.67) as well as equations (1.74)–(1.77). Among these equations, δ_1, δ_2, and δ_{tot} must be equal. As before, we let $\varepsilon = \varepsilon' \approx 0$ so that the differences in ray color beam deviations fall in the neighborhood of zero. Doing this to the above equations, letting equation (1.64) equal (1.74), and also letting equation (1.65) equal (1.75), we find that we need to satisfy

$$\frac{1}{1 - (V_2/V_1)} \approx \frac{1}{1 - (V_2'/V_1')}, \tag{1.78}$$

and

$$\frac{1}{1 - (V_1/V_2)} \approx \frac{1}{1 - (V_1'/V_2')}. \tag{1.79}$$

Equations (1.78) and (1.79) are equivalent, so taking either of them alone yields

$$\frac{V_1}{V_1'} \approx \frac{V_2}{V_2'}. \tag{1.80}$$

Inserting equations (1.66), (1.67), (1.76), and (1.77) into equation (1.80), we obtain

$$\frac{n_1 - n_{1,\,\text{red}}}{n_{1,\,\text{blue}} - n_{1,\,\text{red}}} \approx \frac{n_2 - n_{2,\,\text{red}}}{n_{2,\,\text{blue}} - n_{2,\,\text{red}}}. \tag{1.81}$$

The quantities on the left and right sides of equation (1.81) are known as the *relative partial dispersions* (RPDs) of the glasses for the first and second elements (in this case, the first and second prisms), respectively. Equation (1.81) is a statement of the condition that the RPDs for the selected respective glasses must be roughly equal. They cannot be exactly equal, because we know we must not have $\varepsilon = \varepsilon' \approx 0$ in order to have rays at the three wavelengths intersect at the axial point in the image. The good news is that glasses that satisfy the approximation in equation (1.81) are available from glass suppliers (such as Schott, Ohara, Hoya, Chengdu, etc.). And there is further good news: we do not necessarily need to search for those special glasses in supplier glass maps that show a scatter plot of glasses with their RPD values on the vertical axis and the V_d numbers on the horizontal axis (though it is instructive to have a look by requesting them from glass suppliers). Most such plots only give data for the RPD vs V_d at specific wavelengths. What if we want to know the RPD at other wavelengths? We can use the OS program to search these special glasses at any wavelength through the use of automatic optimization. I originally briefly described this technique in a prior publication [98], followed by a specific example in a book [8]. Here, I have applied the same technique to the tri-prism system of figure 1.251, which resulted in the apochromatic prism system shown in figure 1.254, whose prescription and merit function are shown in figures 1.255 and 1.256, respectively. Note that in the merit function, the REAY values for each wavelength (rows 23, 25, and 27) are essentially zero and equal, indicating an apochromatic state.

The steps that I took to achieve the state of apochromatism for the above-mentioned prism system were as follows:

1. Remove all variables. Set the material solves on surfaces 4 and 6 to *Substitute*. This allowed OS to vary the glass choices from the current default supplier catalog (Schott). The original glasses for those surfaces were N-BK7 and N-F2, respectively (the same as in figure 1.252).
2. Remove all weights in the Merit Function Editor. Set weight to one only for the operands on rows 15 and 21. Rows 8 and 14 compute the RPDs for the glasses of surfaces 4 and 6. Row 15 divides their RPDs and sets a target of one. We know that we do not actually want their RPDs to be exactly equal,

Large apex angles indicative of high powers and curvatures if applied to lens surfaces

Top beam

Optic Axis

Blue (400 nm), green (550 nm) and red (700 nm) rays intersect at the center of the "image plane" (surface 8)

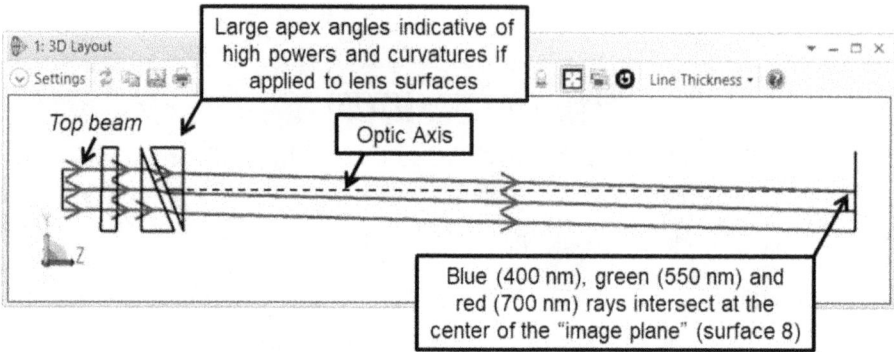

Figure 1.254. An apochromatic prism system at the specified axial location for the top beam.

	Surface Type		Radius	Thickness		Material	Clear Semi-Dia		Tilt About X		Tilt About Y
0	OBJECT	Standard ▾	Infinity	Infinity			0.00000				
1	STOP	Standard ▾	Infinity	12.50000			6.00000	U			
2	(aper)	Irregular ▾	Infinity	5.00000		N-BK7	12.50000	U	1.10724	V	0.00000
3	(aper)	Irregular ▾	Infinity	7.50000			12.50000	U	0.54588	V	0.00000
4	(aper)	Irregular ▾	Infinity	5.00000		SK14 S	12.50000	U	0.00000		0.00000
5	(aper)	Irregular ▾	Infinity	3.00000			12.50000	U	-22.69863	V	0.00000
6	(aper)	Irregular ▾	Infinity	5.00000		KZFSN2 S	12.50000	U	-22.74889	V	0.00000
7	(aper)	Irregular ▾	Infinity	208.66762	V		12.50000	U	0.00000		0.00000
8	IMAGE	Standard ▾	Infinity	-			11.94791				

Glass Substitute solves on surfaces 4 and 6

Figure 1.255. The prescription for the system shown in figure 1.254.

but we also do not know by how much they need to be different. Since OS only attempts to equalize their RPDs with appropriate glasses, we need not worry about setting the target to one. Row 21 constrains the Abbe dispersions to be different. This is necessary, as implied by equations (1.64)–(1.77). On row 25, I increased the maximum thickness of surface 7 to reduce the *stress* on the prism apex angles.

3. In the OS menu, go to Optimize > Hammer Current. Click 'Start' (do not use 'Automatic'). Let the program run for less than a minute and watch the merit function value decrease until it no longer changes, then manually terminate the optimization.

4. Notice that the glasses for surfaces 4 and 6 have changed (you may not get the same glasses as I did).

5. Leave the glass substitute solves alone (they will not be affected if subsequent optimizations are not performed using the Hammer optimization feature).

Assign variables to Tilt About X for surfaces 2, 3, 5, and 6. Remove the weights from rows 15 and 21. Set the weight to one on rows 23, 25, 27, and 29. Change the target on row 29 to 250.

6. Go to Optimize > Optimize! Click 'Start.' Exit when it ends.

Note that the glasses SK14 and KZFSN2 shown in figure 1.255 are actually obsolete glasses at Schott. In order to avoid letting OS use obsolete glasses when applying the Substitute Glass solve during Hammer optimization, we need to go to the 'Libraries' menu and click on 'Glass Substitution Template,' then place a tick mark on 'Use Glass Substitution Template' and ensure that no tick mark is present on 'Obsolete.' At any rate, the latest glasses for these two surfaces are called N-SK14 and N-KZFS2. If you have access to Schott's current glass catalog, you may check to see that indeed, the RPD values for N-SK14 and N-KZFS2 are very similar. We can therefore apply these two glasses to the top right configuration of the lens system of figure 1.240 (because this system is most similar to the tri-prism system of figure 1.254). However, there is a caveat. In the merit function of figure 1.256, I

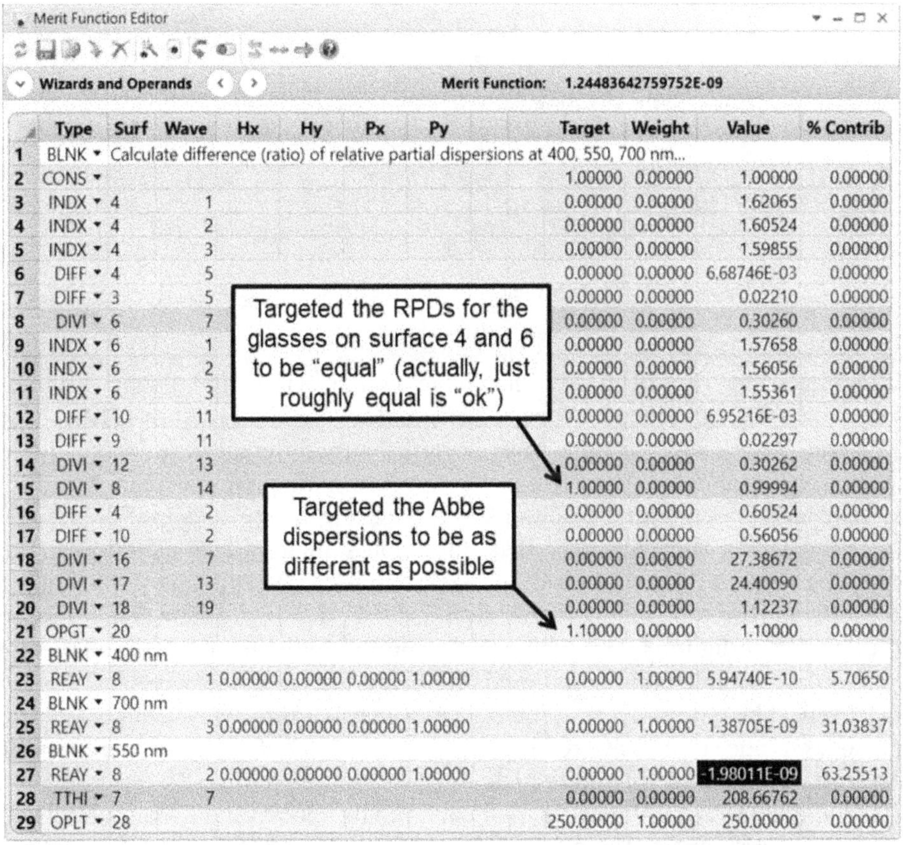

Figure 1.256. The merit function for the prescription in figure 1.255, displayed just subsequent to final optimizations performed *without* weights on rows 15 and 21 (the weights on these two rows were placed there in a prior optimization for glass substitutes).

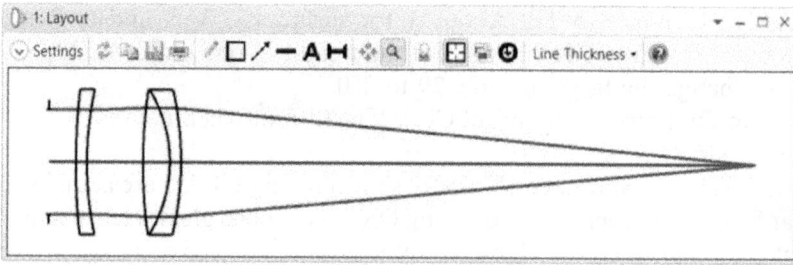

Figure 1.257. An apochromatic lens system with a dispersive front singlet.

	Surface Type	Comment	Radius	Thickness	Material	Coating	Clear Semi-Dia
0	OBJECT Standard ▾		Infinity	Infinity			0.00000
1	STOP Standard ▾		Infinity	5.00000			9.00000 U
2	(aper) Standard ▾		106.65365 V	2.00000	N-BK7		12.50000 U
3	(aper) Standard ▾		56.93593 V	10.00000			12.50000 U
4	(aper) Standard ▾		82.48339 V	5.00000	N-PK52A S		12.50000 U
5	(aper) Standard ▾		-22.13902 V	2.00000	N-KZFS2 S		12.50000 U
6	(aper) Standard ▾		-53.32180 V	102.82887 M			12.50000 U
7	IMAGE Standard ▾		Infinity	-			3.53567E-08

Figure 1.258. The prescription for the system shown in figure 1.257.

let OS select any pair of glasses that have similar RPDs and I placed a minor constraint on the difference in Abbe dispersion (see row 21). Equations (1.64)–(1.67) tell us that the Abbe dispersions must be sufficiently different in order for the lens powers to be minimal. If you look at the apochromatic prism pair in figure 1.254, they have high apex angles. If you think of these angles as the 'localized' slopes at the surfaces of lenses, you will note that this implies high surface curvatures (small radii). Therefore, a different pair of glasses is required. By applying steps similar to 1–6 above for the top right configuration in figure 1.240, I obtained the resulting layout shown in figure 1.257, whose prescription, merit function, *chromatic focal shift* plot (at the full pupil zone), and MTF are provided in figures 1.258–1.261, respectively. The plot in figure 1.260 indicates that the three wavelengths are at the same focus. Notice that I have reduced the aperture to $F/5.6$. Even so, the curvatures on surfaces 4 and 5 are still quite high.

The above exercises indicate that achromats/apochromats can correct the chromatic aberration in non-color-corrected systems. Therefore, color correction may be approached by way of inserting color-corrected lenses into uncorrected systems (or converting elements in uncorrected systems into achromats/

Figure 1.259. The merit function for the system shown in figure 1.257.

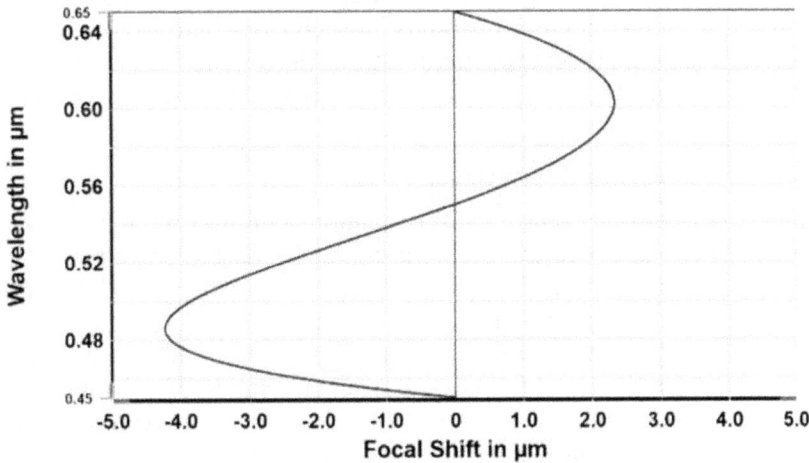

	Type	Surf	Wave	Hx	Hy	Px	Py	Target	Weight	Value	% Contrib
1	BLNK ▾	Calculate difference (ratio) of relative partial dispersions at 450, 550, 650 nm...									
2	CONS ▾							1.00000	0.00000	1.00000	0.00000
3	INDX ▾	4	1					0.00000	0.00000	1.50347	0.00000
4	INDX ▾	4	2					0.00000	0.00000	1.49830	0.00000
5	INDX ▾	4	3					0.00000	0.00000	1.49529	0.00000
6	DIFF ▾	4	5					0.00000	0.00000	3.01550E-03	0.00000
7	DIFF ▾	3	5					0.00000	0.00000	8.17945E-03	0.00000
8	DIVI ▾	6	7					0.00000	0.00000	0.36867	0.00000
9	INDX ▾	5	1					0.00000	0.00000	1.56935	0.00000
10	INDX ▾	5	2					0.00000	0.00000	1.56057	0.00000
11	INDX ▾	5	3					0.00000	0.00000	1.55545	0.00000
12	DIFF ▾	10	11					0.00000	0.00000	5.12003E-03	0.00000
13	DIFF ▾	9	11					0.00000	0.00000	0.01391	0.00000
14	DIVI ▾	12	13					0.00000	0.00000	0.36817	0.00000
15	DIVI ▾	8	14					1.00000	0.00000	1.00135	0.00000
16	DIFF ▾	4	2					0.00000	0.00000	0.49830	0.00000
17	DIFF ▾	10	2					0.00000	0.00000	0.56057	0.00000
18	DIVI ▾	16	7					0.00000	0.00000	60.92134	0.00000
19	DIVI ▾	17	13					0.00000	0.00000	40.30919	0.00000
20	DIVI ▾	18	19					0.00000	0.00000	1.51135	0.00000
21	OPGT ▾	20						1.50000	0.00000	1.50000	0.00000
22	EFFL ▾		2					100.00000	1.00000	100.00000	2.78299E-06
23	OPTH ▾	7	2	0.00000	0.00000	0.00000	0.00000	0.00000	0.00000	131.47857	0.00000
24	OPTH ▾	7	2	0.00000	0.00000	0.00000	1.00000	0.00000	0.00000	131.47857	0.00000
25	DIFF ▾	23	24					0.00000	1.00000	-1.38657E-08	0.10230
26	RSCE ▾	3	2	0.00000	0.00000			0.00000	0.00000	2.30445E-03	0.00000
27	AXCL ▾	1	3	1.00000				0.00000	2.00000	1.06167E-07	11.99547
28	AXCL ▾	2	3	1.00000				0.00000	2.00000	-2.87397E-07	87.90222

Figure 1.260. The chromatic focal shift (full zone) for the system shown in figure 1.257.

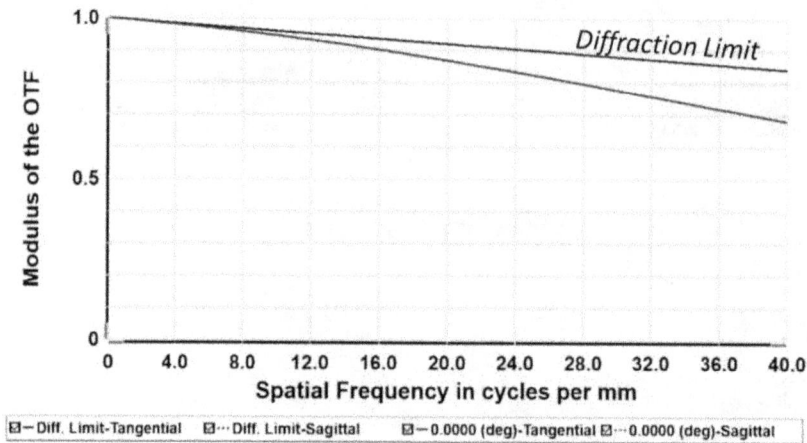

Figure 1.261. The axial MTF for the system shown in figure 1.257.

apochromats), followed by re-optimization. There is one other aberration: *lateral color* (chromatic variation of magnification). In the presence of lateral color, the principal ray at each wavelength has a different height at the defined image plane. Lateral color would be absent if every element were achromatized. However, even if each element is not achromatized, since lateral color is an odd aberration (like coma and distortion—see section 1.4.4), it is zero for a symmetrical optical system at unit magnification (i.e. when the object and image sizes are the same). If the magnification is not 1:1, then as long as some elements are achromatized, lateral color can be controlled by making the stop position and also all air spaces and thicknesses variables. These are the same types of variables as those used for coma and distortion. In most cases involving the design of a multielement lens system, you eventually arrive at good image quality through the use of RMS spot size, RMS wavefront error, and axial color operands; in such cases, lateral color tends to be well corrected even if no specific operand (such as the LACL operand in OS) is used to control this aberration. An example of this is provided in section 1.4.10.

1.4.8 Conrady's D-d method for achromatizing

In the previous section, we learned from equations (1.62)–(1.67) that at low prism apex angles, we can derive formulas for chromatic beam steering in prisms that are analogous to the primary axial color correction that is described by equations (1.68)–(1.72). The latter equations refer to the base (paraxial) powers of lenses. Despite the use of low-angle approximations in these formulas, we gained insight into achieving achromatic and even apochromatic states of prisms and lenses. Moreover, in figure 1.254, we observed the influence of color correction on prism apex angle: while attempting to steer the red, green, and blue rays through three prisms, we needed high apex angles which were analogous to high lens curvatures. In this sense, we were actually introduced to *spherochromatism*, which is the chromatic variation of spherical aberration. Prisms with high apex angles behave

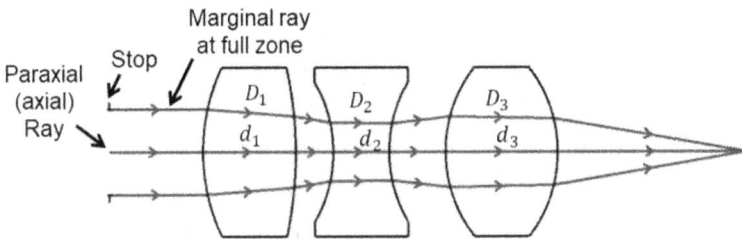

Figure 1.262. An illustration of he variables D_i and d_i related to equation (1.82).

like high-aperture lenses at the full zone, while prisms with small apex angles behave like lenses at the paraxial zone. Thus, for low-aperture lenses (say, lenses at apertures greater than $F/6$), spherochromatism is not much of an issue. But at high apertures, there is increased spherical aberration, and therefore lenses suffer from the influence of spherochromatism.

In classical lens design, A E Conrady developed a means to define a practical condition that partially corrects a lens for spherochromatism, which is called the D-d method. In this method, the equality given by

$$\sum_{i=1}^{N}(D_i - d_i)(n_{i,\,\text{blue}} - n_{i,\,\text{red}}) \approx 0 \qquad (1.82)$$

must be met [5], where D_i and d_i are lengths of the marginal and paraxial rays, respectively, for the ith *element* in the lens system, as illustrated in figure 1.262. Note that air spaces are not included, because $n_{i,\,\text{blue}} - n_{i,\,\text{red}} = 0$ in those spaces. This is why the values for $i = 1, \ldots, N$ are for the elements, not surfaces. If there are five lenses, then $i = 1, \ldots, 5$, and so on. Furthermore, the lengths D_i and d_i are traced for a ray at a wavelength somewhere between the extremes of the wavelength range of the light (e.g. between the blue and red). Usually, the rule is to use the wavelength of 'brightest light' (i.e. the peak in the effective spectrum of the light in the design; see section 1.2.12 for a review of what is meant by effective spectrum).

Does this seem complex? It is not really complex, especially if we are using a modern optical design program. In fact, when we use a program to perform ray tracing and apply the D-d method, we do not even need to compute the specific difference given by $D_i - d_i$ at each element. The origin of equation (1.82) is in extending Fermat's principle to consider rays at all wavelengths. Thus, in theory, the OPLs of all rays, though all zones, and at all wavelengths must be equal. In the days of classical lens design (when affordable powerful computers had not been made available), Conrady's D-d method simplified OPL calculations by considering rays in two zones (the full zone and the paraxial zone). The index difference $n_{i,\,\text{blue}} - n_{i,\,\text{red}}$ was computed at the extreme wavelengths (e.g. blue and red), but $D_i - d_i$ was traced for the ray that had the wavelength of the brightest light [5]. However, since most ray tracing programs can compute OPLs for any ray, in any zone, and at any wavelength, then all that is needed is to have the program equalize

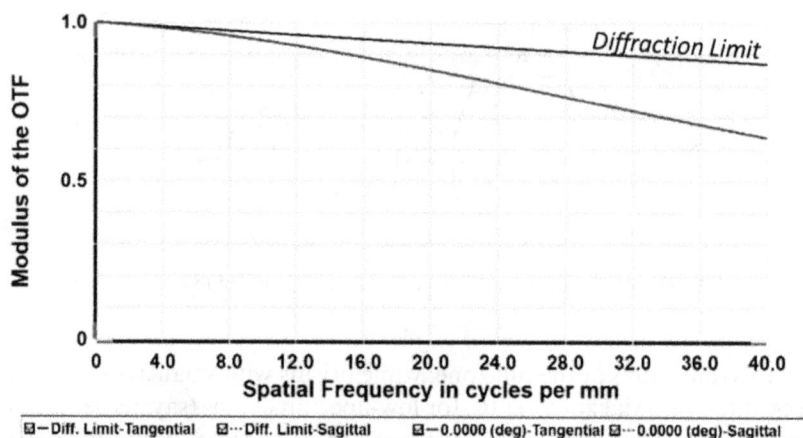

Figure 1.263. The axial MTF for the top right configuration of figure 1.240.

OPLs for a specified choice of rays for certain zones and at certain wavelengths. That said, not all lens design forms can be made to satisfy Fermat's principle for all rays, at all wavelengths, and through all zones (which would be a tall order). Even a sufficiently well-designed lens can possess residual aberrations. In practical lens design, what the D-d method provides is guidance to take the simplest case of considering the OPLs of two important rays: a marginal ray near or at the full zone, and a paraxial (usually, the axial) ray. In OS, we would use the OPTH operands to compute OPLs. Looking back at the merit function in figure 1.245, I used two OPTH operands, one to constrain the marginal ray at the full zone, and the other to control the paraxial ray (at zone = 0). Both were traced for the green wavelength (550 nm). I then used the AXCL operands to constrain the difference in axial focal positions for the blue and red marginal rays to be zero. This set of operands can be considered a case of indirectly applying the D-d *philosophy* in a practical manner. The resulting axial MTF and chromatic focal shift for this lens system are shown in figures 1.263 and 1.264, respectively. As can be seen, the result is quite good, and it is all that this lens system can handle. There is some residual spherochromatism, as can be seen in the variation of the focal positions for each wavelength across the zones of the pupil, as shown in the *longitudinal aberration* plot in figure 1.265. This plot shows the axial focal position on the horizontal axis versus the pupil zone on the vertical axis for each wavelength.

Through the technique of using the OPTH operand to control only the middle 'brightest' wavelength and using the AXCL operand to control the primary axial color aberration, we naturally do not obtain the same design result as a lens that is achromatized by explicit use of the D-d method. But this is acceptable because ultimately, what we want is a lens with good performance. **The real value of knowing about the D-d method is that it highlights the need to consider spherochromatism in a lens system.** It is too easy to forget this when one limits one's study of chromatic aberration to examining the paraxial formulas of equations (1.68)–(1.72). Note also

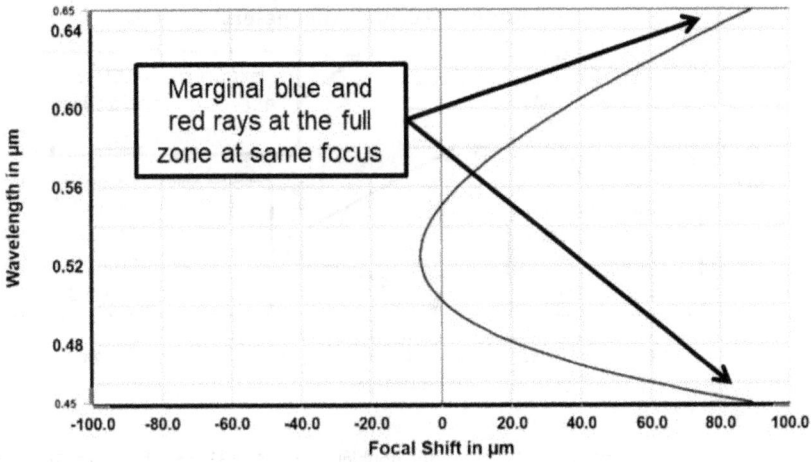

Figure 1.264. The chromatic focal shift for the top right configuration of figure 1.240.

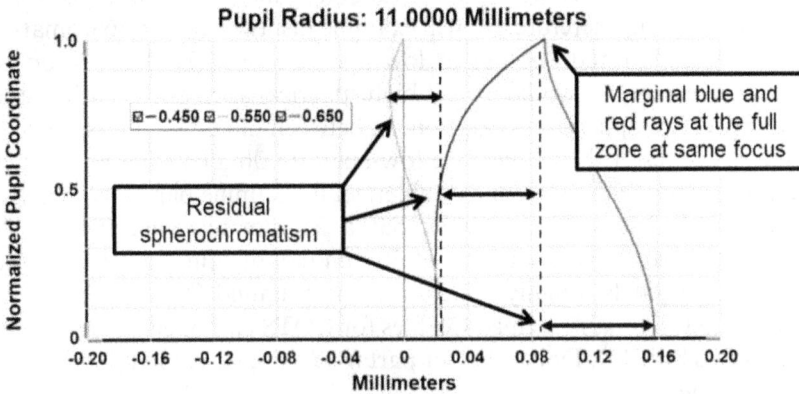

Figure 1.265. The longitudinal aberration for the top right configuration of figure 1.240.

that the D-d method is intimately related to transverse ray errors (which are in turn, related to axial focal errors by simple trigonometry) by way of applying equations (1.55) and (1.56) for appropriately chosen rays and wavelengths [5].

1.4.9 Do commercial off-the-shelf achromats satisfy D-d?

Based on the discussion in the previous section concerning the D-d method, the question posed in this section's title can be considered a question of whether or not spherochromatism has been considered in the design of COTS achromatic doublets. After all, we would ideally want rays at each wavelength to not suffer significantly different amounts of spherical aberration. We are not requiring that there is zero spherochromatism. Rather, what we prefer is a doublet achromat that has the marginal rays near the full zone at a common focus (or as close as possible) and good axial MTF, such as that displayed in figures 1.263–1.265.

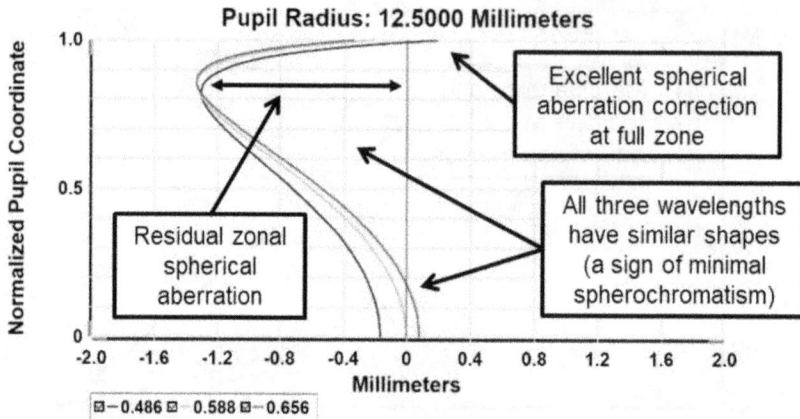

Figure 1.266. The longitudinal aberration for doublet part # 47-633 from Edmund Optics.

There are plenty of COTS achromats available, which makes it impossible for us to analyze all of them. Therefore, we will examine just two specific achromats, one at high aperture (F/1.25) and another at low aperture (F/16.7). The reason is quite clear: at high aperture, we would expect high spherical aberration. At low aperture, we would expect low spherical aberration. In between these two limits, we are likely to see an average result. Moreover, the low-aperture doublet shall be of interest to us, as you will soon see. For the first doublet, let us examine part # 47–633 from the supplier Edmund Optics. This doublet has a 25 mm diameter (with a 24 mm diameter clear aperture), and an EFL of 30 mm (so, dividing its EFL by the clear aperture diameter yields roughly F/1.25). At the Edmund Optics website, you can download publicly available Zemax OS files for COTS components labeled with the trademark TECHSPEC®. Doing this for part # 47–633, we see in figure 1.266 that it has excellent spherical aberration correction (defined as the difference in axial focal positions between zone = 0 and zone = 1) and very minimal spherochromatism (notice the similarity of the curves for each wavelength, which is a sign of satisfying D-d). Furthermore, the blue and red rays at the full zone and the paraxial zone are at roughly the same focal position (another sign of satisfying D-d). There is of course, residual zonal spherical aberration at about the 0.85 zone, but that is forgivable, because, after all, it is just a spherical doublet. Thus, there is also no point in examining this doublet's MTF, because at F/1.25, it is impossible for a spherical doublet to have high MTF. Instead, this COTS doublet may be considered sufficiently good for the purpose of light concentration (e.g. if you just want to focus light onto a photodiode for detection). In conclusion, this specific COTS achromat can be said to sufficiently satisfy D-d. At F/1.25, this is excellent.

By the way, I have not mentioned that spherical doublets have a sufficient number of variables to also be made aplanatic [5]. The doublet above is a good example. In particular, the OSC (at 587.56 nm) for this lens is minimized if you shift the stop to roughly 2 mm in front of the doublet. There is, of course, residual third-order

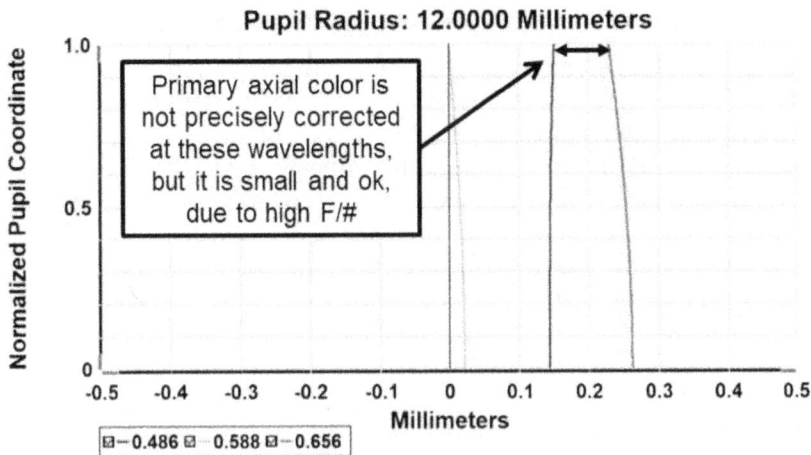

Figure 1.267. The longitudinal aberration for doublet part # 47-650 from Edmund Optics.

spherical aberration and coma, which is used for balancing the OSC as well as minimizing spot size.

Next, we turn our attention to part # 47–650 from the Edmund Optics catalog. This specific doublet is of interest to me because, in the next section, in one design example, I will insert this achromatic doublet into the four-element lens system of figure 1.216 (which was only designed at a monochromatic wavelength at 587.56 nm) to correct the lens system's chromatic aberration. This was the approach I recommended at the beginning of section 1.4.7. So, I will show you how it is done (plus I will also show a different method after that example). Part # 47–650 is an achromatic doublet with a 400 mm EFL and a 24 mm clear aperture diameter. This makes it roughly an $F/16.7$ lens, which is low aperture. Now, upon downloading the Zemax OS file for part # 47-650 from the Edmund Optics website and opening up the file, we see from the longitudinal aberration plot (figure 1.267, at the F, d, and C wavelengths) that there is negligible spherochromatism. This is no surprise, as this lens has a high $F/\#$, yielding low spherical aberration. If there is low spherical aberration, then there is low spherochromatism. At the displayed wavelengths of F, d, and C, the primary axial color is also not precisely corrected, but it is low nonetheless, due to the high $F/\#$ of the lens. In examining the chromatic focal shift plot, we see that rays at 486 nm and 636 nm are indeed at the same focus.

The basic conclusion here is that the specific COTS achromatic doublets examined here implicitly satisfy the D-d criterion for achromatization, even though no explicit D-d sum has been computed. As stated earlier, we are taking the practical approach that only the *philosophy* behind the D-d method is being considered, which is that marginal rays at the full zone for the extreme wavelengths are achromatized and that the spherical aberration of the middle wavelength is corrected. In addition, we require spherochromatism to be minimized. Finally, we ignore residual zonal spherical aberration.

1.4.10 Example: achromatizing a monochromatic four-element lens

Returning to the four-element lens system in figure 1.216, we would like to achromatize it across the visible spectrum. A reasonable selection of wavelengths is 450, 550, and 650 nm, and each is set to a weight of one. As a first example of a technique for achromatizing, we apply the principle that was discussed near the beginning of section 1.4.7, which is that an achromatic doublet can correct the primary axial color aberration from non-achromatic elements in a lens system. COTS lenses are usually designed for an object at infinity. So, the best place to insert a COTS doublet into an existing lens system for re-optimization would be in a space where axial rays are as collimated as possible. In examining the lens system of figure 1.216, a reasonable choice for insertion is the space just after the second element. Also, notice from the prescription of this lens system in figure 1.217 that all elements are 20 mm in diameter. Part # 47-650 from Edmund Optics has 25 mm diameter, which is a good match in terms of size. We also note that the second element of the system of figure 1.216 has an EFL close to 200 mm (it is about 212 mm at the d wavelength). So, this means that when we insert part # 47-650, we need to roughly double the EFL of the element before it, otherwise, there would be excessive lens power. Upon inserting part # 47-650 into the four-element lens system and removing the user-defined sizes from the elements (so that their diameters are automatically defined by OS in accordance with the entrance pupil diameter, which is defined by a 'Float By Stop Size' system aperture setting), we have the initial pre-optimized state shown in figure 1.268, whose prescription and merit function are shown in figures 1.269 and 1.270, respectively.

Note in the merit function that the AXCL operand has been added on row 17 (contrast this with this lens system's previous merit function shown in figure 1.218). Note also that the first optimization run is performed while minimizing the RMS spot size, using a weight of one for the RSCE operands on rows 18 and 19. This is a personal habit. I begin first optimizations with this operand, followed by switching to RMS wavefront error (which is why you see the RWCE operands on 'standby' on

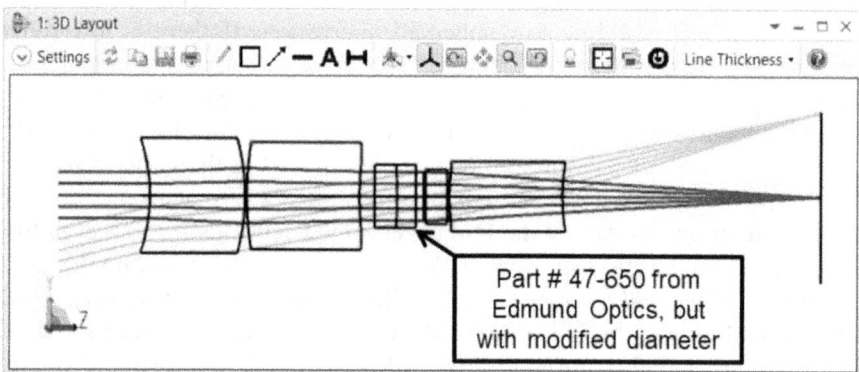

Figure 1.268. The layout of the system from figure 1.216 after inserting part # 47-650 from Edmund Optics and letting OS set the element diameters accordingly.

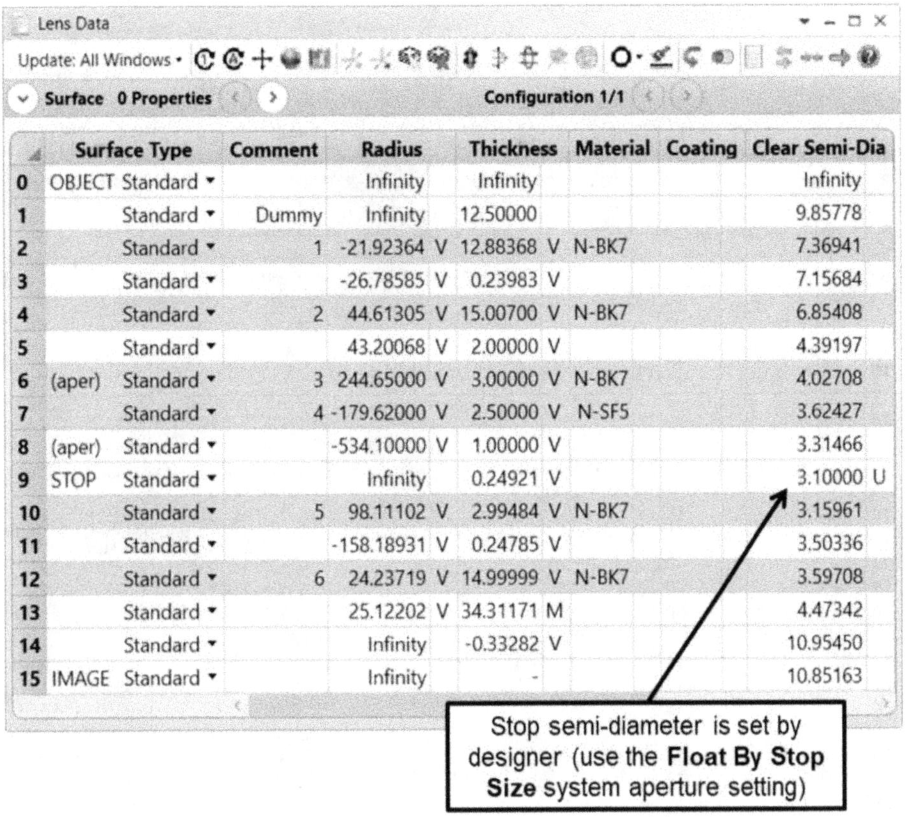

	Surface Type	Comment	Radius	Thickness	Material	Coating	Clear Semi-Dia
0	OBJECT Standard ▼		Infinity	Infinity			Infinity
1	Standard ▼	Dummy	Infinity	12.50000			9.85778
2	Standard ▼	1	-21.92364 V	12.88368 V	N-BK7		7.36941
3	Standard ▼		-26.78585 V	0.23983 V			7.15684
4	Standard ▼	2	44.61305 V	15.00700 V	N-BK7		6.85408
5	Standard ▼		43.20068 V	2.00000 V			4.39197
6	(aper) Standard ▼	3	244.65000 V	3.00000 V	N-BK7		4.02708
7	Standard ▼	4	-179.62000 V	2.50000 V	N-SF5		3.62427
8	(aper) Standard ▼		-534.10000 V	1.00000 V			3.31466
9	STOP Standard ▼		Infinity	0.24921 V			3.10000 U
10	Standard ▼	5	98.11102 V	2.99484 V	N-BK7		3.15961
11	Standard ▼		-158.18931 V	0.24785 V			3.50336
12	Standard ▼	6	24.23719 V	14.99999 V	N-BK7		3.59708
13	Standard ▼		25.12202 V	34.31171 M			4.47342
14	Standard ▼		Infinity	-0.33282 V			10.95450
15	IMAGE Standard ▼		Infinity	-			10.85163

Stop semi-diameter is set by designer (use the **Float By Stop Size** system aperture setting)

Figure 1.269. The prescription for the layout in figure 1.268.

rows 20 and 21). The optimization then proceeds in roughly the following steps (you should use additional trial and error if needed):

1. Run a single instance of damped least-squares local optimization.
2. When the above ends, exit and remove the weights from the RSCE operands. Set the weight to two for both RWCE operands.
3. Set proper element semi-diameters (I set the first element at 10 mm and the rest at 9 mm, with the exception of the concave surfaces).
4. Optimize using the Hammer feature (go to Optimize > Hammer Current, click 'Start').
5. When the optimization terminates, exit and end.

The result is the layout shown in figure 1.271, whose prescription, merit function, MTF, chromatic focal shift, and lateral color plots are shown in figures 1.272–1.276, respectively. Note that the lateral color is corrected even though I did not include any operands for it in the merit function. This was what I mentioned in the final paragraph of section 1.4.7.

	Type	Wave1	Wave2	Zone	Target	Weight	Value	% Contr
1	EFFL ▾		2		50.00000	1.00000	49.62715	27.171
2	REAY ▾ 1		2	0.00000 0.00000 0.00000 1.00000	0.00000	0.00000	2.94208	0.000
3	PROB ▾ 2			2.00000	0.00000	0.00000	5.88416	0.000
4	DIVI ▾ 1		3		0.00000	0.00000	8.43403	0.000
5	OPLT ▾ 4				7.20000	0.20000	8.43403	59.526
6	OPGT ▾ 4				6.90000	0.20000	6.90000	0.000
7	OSCD ▾		2	1.00000	0.00000	0.00000	-4.87615E-04	0.000
8	SPHA ▾ 0		2		0.00000	0.00000	0.17502	0.000
9	COMA ▾ 0		2		0.00000	0.00000	0.58833	0.000
10	DISG ▾ 1		2	0.00000 1.00000 0.00000 0.00000	0.00000	0.00000	-1.24716	0.000
11	OPLT ▾ 10				1.00000	0.20000	1.00000	0.000
12	OPGT ▾ 10				-1.00000	0.20000	-1.24716	2.387
13	PETZ ▾		2		0.00000	0.00000	-205.64779	0.000
14	FCGT ▾		2	0.00000 1.00000	0.00000	0.00000	-0.63309	0.000
15	FCGS ▾		2	0.00000 1.00000	0.00000	0.00000	-0.18127	0.000
16	DIFF ▾ 14		15		0.00000	0.00000	-0.45182	0.000
17	AXCL ▾ 1		3	0.70000	0.00000	0.00000	1.18183	0.000
18	RSCE ▾ 3		0	0.00000 0.00000	0.00000	1.00000	0.02163	0.091
19	RSCE ▾ 3		0	0.00000 1.00000	0.00000	1.00000	0.03603	0.253
20	RWCE ▾ 3		0	0.00000 0.00000	0.00000	0.00000	0.47147	0.000
21	RWCE ▾ 3		0	0.00000 1.00000	0.00000	0.00000	0.88173	0.000
22	MNEG ▾ 2		13	1.00000 1	1.00000	0.00000	1.00000	0.000
23	MNCG ▾ 2		13		3.00000	0.20000	2.49484	9.974
24	MXCG ▾ 2		13		15.00000	0.20000	15.00700	1.91302E
25	MNCA ▾ 3		12		0.25000	0.20000	0.23689	6.71338E
26	MNEA ▾ 3		12	1.00000 1	0.10000	0.20000	0.10000	0.000
27	TTHI ▾ 2		12		0.00000	0.00000	55.12241	0.000
28	OPLT ▾ 27				55.00000	0.20000	55.12241	0.589

Figure 1.270. The merit function for the layout in figure 1.268, which is the starting point for design optimization.

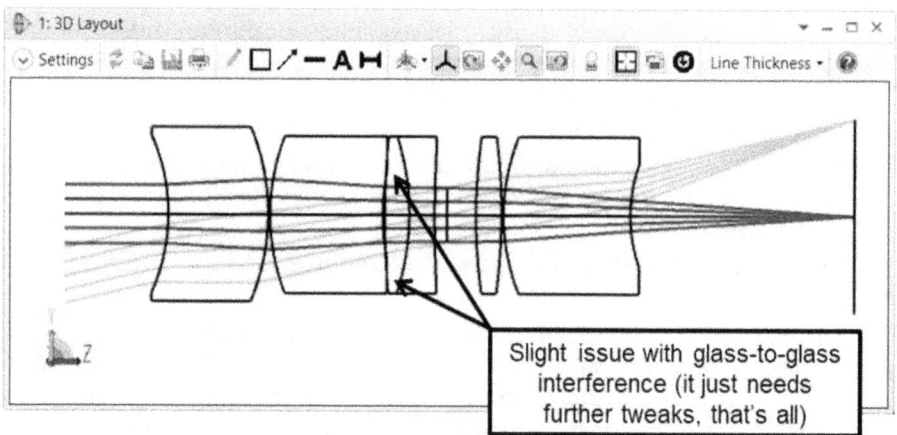

Figure 1.271. The layout of the optimized and achromatized six-element lens system.

Lens Data

Update: All Windows

Surface 0 Properties — Configuration 1/1

	Surface Type		Comment	Radius	Thickness	Material	Coating	Clear Semi-Dia
0	OBJECT	Standard ▾		Infinity	Infinity			Infinity
1		Standard ▾	Dummy	Infinity	12.50000			10.28975
2	(aper)	Standard ▾	1	-21.44015 V	11.99980 V	N-BK7		9.00000 U
3	(aper)	Standard ▾		-25.10805 V	0.23548 V			10.00000 U
4	(aper)	Standard ▾	2	23.55592 V	13.46022 V	N-BK7		9.00000 U
5	(aper)	Standard ▾		24.55111 V	0.25007 V			5.00000 U
6	(aper)	Standard ▾	3	93.85613 V	2.98777 V	N-BK7		9.00000 U
7	(aper)	Standard ▾	4	-27.58968 V	3.00003 V	N-SF5		9.00000 U
8	(aper)	Standard ▾		142.24240 V	1.49800 V			7.00000 U
9	STOP	Standard ▾		Infinity	3.37883 V			3.10000 U
10	(aper)	Standard ▾	5	64.63333 V	2.99962 V	N-BK7		9.00000 U
11	(aper)	Standard ▾		-66.96550 V	0.24975 V			9.00000 U
12	(aper)	Standard ▾	6	23.39301 V	14.95429 V	N-BK7		9.00000 U
13	(aper)	Standard ▾		17.45771 V	26.67719 M			6.00000 U
14		Standard ▾		Infinity	-0.24456 V			11.13564
15	IMAGE	Standard ▾		Infinity	-			11.04567

Figure 1.272. The prescription for the lens system shown in figure 1.271.

Note that in figure 1.271 there is some glass-to-glass interference between surfaces 5 and 6, but this is something that can be left for further tweaks. Other than that, we have a viable multielement lens to serve as an imaging objective for an object at infinity and a field of ±12.5°. But this is not the only possibility for an achromatized state of this lens. There are currently six elements (four plus a doublet). It is possible to begin from the original four-element lens system and convert the second element into a doublet. The idea is as follows:

1. The second element in the original four-element design is a reasonable choice for conversion into a doublet, because this is done in many double-Gauss lenses.
2. Further, the configuration at the top right of figure 1.240 has this geometry.
3. To convert the second element in the original four-element design into a doublet, just select two elements that have roughly the same refractive index (at the middle wavelength) but a sufficiently large difference in Abbe dispersion. Why? Because, first, if they have the same refractive index, then the rays that pass through the current second element of the four-element design will refract without significant deviation when I split it into a cemented doublet and let them pass through the positive-to-negative glass–glass interface in the doublet. Second, achromatization requires elements that

Merit Function Editor   ▾ – ☐ ×

⟳ 💾 📄 ✎ ✗ 🔧 • ⟲ 🔗 ⇌ ↔ ⇨ ❔

▾ **Wizards and Operands** (‹) (›)   **Merit Function:** 0.088991043430616

	Type	Wave1	Wave2	Zone		Target	Weight	Value	% Contrib
1	EFFL ▾		2			50.00000	1.00000	49.99553	0.03702
2	REAY ▾ 1		2	0.00000 0.00000 0.00000 1.00000		0.00000	0.00000	3.38110	0.00000
3	PROB ▾ 2			2.00000		0.00000	0.00000	6.76219	0.00000
4	DIVI ▾ 1		3			0.00000	0.00000	7.39339	0.00000
5	OPLT ▾ 4					7.20000	0.20000	7.39339	13.88971
6	OPGT ▾ 4					6.90000	0.20000	6.90000	0.00000
7	OSCD ▾		2	1.00000		0.00000	0.00000	3.08879E-04	0.00000
8	SPHA ▾ 0		2			0.00000	0.00000	1.02766	0.00000
9	COMA ▾ 0		2			0.00000	0.00000	-0.38140	0.00000
10	DISG ▾ 1		2	0.00000 1.00000 0.00000 0.00000		0.00000	0.00000	0.13753	0.00000
11	OPLT ▾ 10					1.00000	0.20000	1.00000	0.00000
12	OPGT ▾ 10					-1.00000	0.20000	-1.00000	0.00000
13	PETZ ▾		2			0.00000	0.00000	-449.79168	0.00000
14	FCGT ▾		2	0.00000 1.00000		0.00000	0.00000	0.22999	0.00000
15	FCGS ▾		2	0.00000 1.00000		0.00000	0.00000	0.13982	0.00000
16	DIFF ▾ 14		15			0.00000	0.00000	0.09017	0.00000
17	AXCL ▾ 1		3	0.70000		0.00000	0.00000	0.01685	0.00000
18	RSCE ▾ 3		0	0.00000 0.00000		0.00000	0.00000	6.24840E-03	0.00000
19	RSCE ▾ 3		0	0.00000 1.00000		0.00000	0.00000	8.74538E-03	0.00000
20	RWCE ▾ 3		0	0.00000 0.00000		0.00000	2.00000	0.09741	35.23683
21	RWCE ▾ 3		0	0.00000 1.00000		0.00000	2.00000	0.11675	50.62499
22	MNEG ▾ 2		13	1.00000	1	1.20000	0.00000	1.04604	0.00000
23	MNCG ▾ 2		13			3.00000	0.20000	2.98739	0.05908
24	MXCG ▾ 2		13			15.00000	0.20000	15.00000	0.00000
25	MNCA ▾ 3		12			0.25000	0.20000	0.23524	0.08094
26	MNEA ▾ 3		12	1.00000	1	0.10000	0.20000	0.10000	0.00000
27	TTHI ▾ 2		12			0.00000	0.00000	55.01387	0.00000
28	OPLT ▾ 27					55.00000	0.20000	55.01387	0.07142

Figure 1.273. The merit function for the lens system shown in figure 1.271.

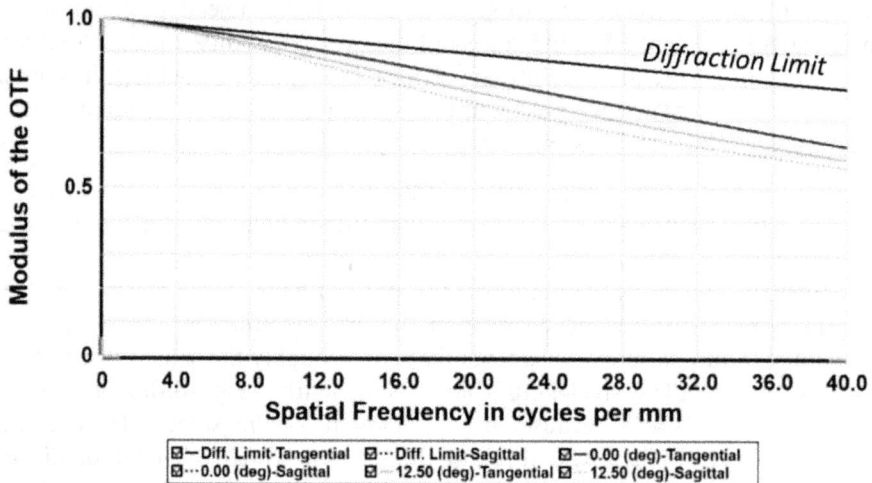

Figure 1.274. The MTF (polychromatic) for the lens system shown in figure 1.271.

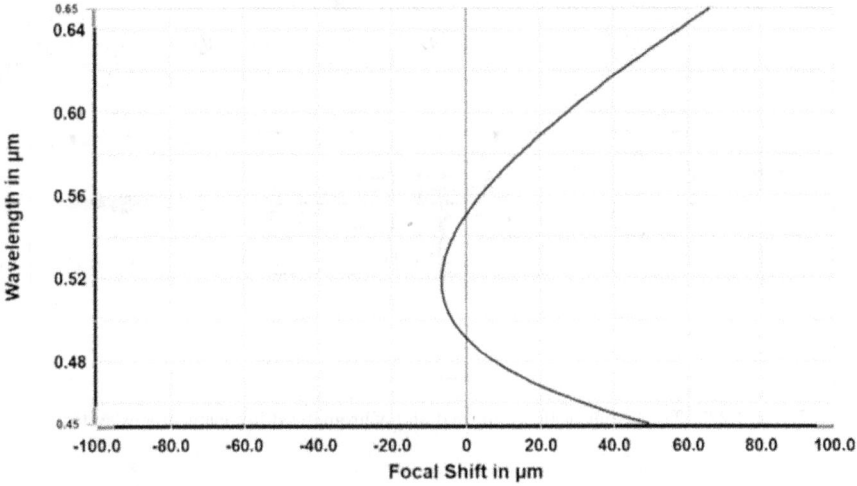

Figure 1.275. The chromatic focal shift for the lens system shown in figure 1.271.

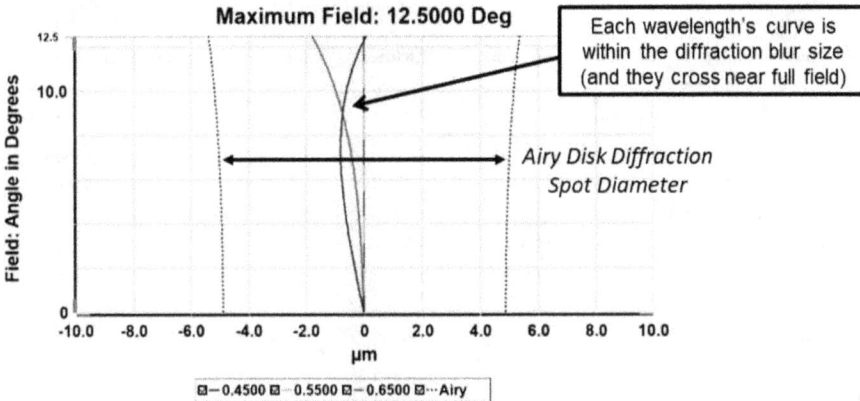

Figure 1.276. The lateral color for the lens system shown in figure 1.271.

have significantly different dispersions, according to equations (1.71) and (1.72).

4. I went to the Schott website to have a look at its (currently available) 'online interactive Abbe diagram' and decided that N-PSK53A and N-F2 had significant separation in their V_d values and approximately equal indexes. So, I split the second element of the original four-element design into a cemented doublet consisting of these two glass choices so that N-PSK53A was the positive element and N-F2 was the negative. I then manually adjusted the curvatures of their surfaces such that rays converged roughly towards the image plane (note that this pair of glasses has a higher index than that of N-BK7).

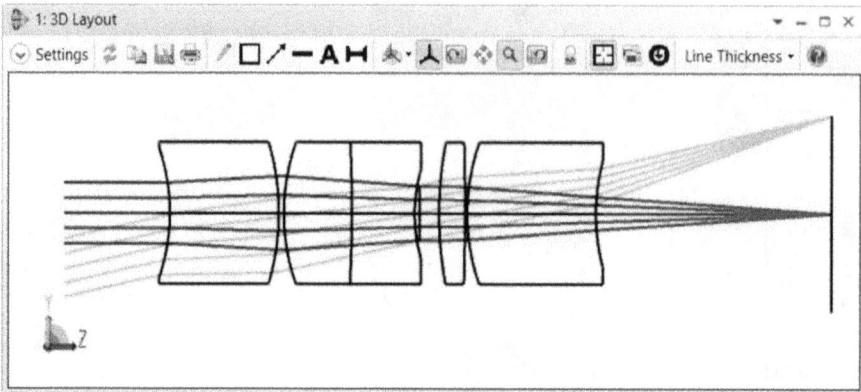

Figure 1.277. The layout of the optimized and achromatized five-element lens system.

	Surface Type		Comment	Radius	Thickness	Material	Coating	Clear Semi-Dia
0	OBJECT	Standard ▾		Infinity	Infinity			Infinity
1		Standard ▾	Dummy	Infinity	12.50000			9.53819
2	(aper)	Standard ▾	1	-21.35133 V	12.91089 V	N-BK7		7.30000 U
3	(aper)	Standard ▾		-26.41857 V	0.61560 V			8.00000 U
4	(aper)	Standard ▾	2	22.83697 V	7.96763 V	N-PSK53A		8.00000 U
5	(aper)	Standard ▾		-149.61407 V	7.39917 V	N-F2		8.00000 U
6	(aper)	Standard ▾		17.93434 V	0.61143 V			5.00000 U
7	STOP	Standard ▾		Infinity	2.23318 V			3.10000 U
8	(aper)	Standard ▾	3	42.22861 V	2.99993 V	N-BK7		8.00000 U
9	(aper)	Standard ▾		-286.36196 V	0.25575 V			8.00000 U
10	(aper)	Standard ▾	4	23.79286 V	15.00678 V	N-BK7		8.00000 U
11	(aper)	Standard ▾		23.04835 V	27.63591 M			6.00000 U
12		Standard ▾		Infinity	-0.16328 V			11.09597
13	IMAGE	Standard ▾		Infinity	-			11.03732

Figure 1.278. The prescription for the lens system shown in figure 1.277.

5. I then performed the same optimization steps described above; the resulting optimized layout is shown in figure 1.277 and its prescription, merit function, MTF, chromatic focal shift, and lateral color are displayed in figures 1.278–1.282, respectively. Note that the wavelengths are, as usual, 450, 550, 650 nm, and the weights are all one.

	Type	Wave1	Wave2	Zone	Target	Weight	Value	% Contrib
			Merit Function:	0.0653784208110922				
1	EFFL ▾		2		50.00000	1.00000	49.99942	1.17511E-03
2	REAY ▾ 1		2	0.00000 0.00000 0.00000 1.00000	0.00000	0.00000	3.43928	0.00000
3	PROB ▾ 2			2.00000	0.00000	0.00000	6.87856	0.00000
4	DIVI ▾ 1		3		0.00000	0.00000	7.26887	0.00000
5	OPLT ▾ 4				7.20000	0.20000	7.26887	3.26413
6	OPGT ▾ 4				6.90000	0.20000	6.90000	0.00000
7	OSCD ▾		2	1.00000	0.00000	0.00000	4.30073E-04	0.00000
8	SPHA ▾ 0		2		0.00000	0.00000	0.76572	0.00000
9	COMA ▾ 0		2		0.00000	0.00000	-0.55977	0.00000
10	DISG ▾ 1		2	0.00000 1.00000 0.00000 0.00000	0.00000	0.00000	-0.10062	0.00000
11	OPLT ▾ 10				1.00000	0.20000	1.00000	0.00000
12	OPGT ▾ 10				-1.00000	0.20000	-1.00000	0.00000
13	PETZ ▾		2		0.00000	0.00000	-925.04065	0.00000
14	FCGT ▾		2	0.00000 1.00000	0.00000	0.00000	0.11275	0.00000
15	FCGS ▾		2	0.00000 1.00000	0.00000	0.00000	0.09684	0.00000
16	DIFF ▾ 14		15		0.00000	0.00000	0.01591	0.00000
17	AXCL ▾ 1		3	0.70000	0.00000	0.00000	0.01745	0.00000
18	RSCE ▾ 3		0	0.00000 0.00000	0.00000	0.00000	4.93245E-03	0.00000
19	RSCE ▾ 3		0	0.00000 1.00000	0.00000	0.00000	5.25741E-03	0.00000
20	RWCE ▾ 3		0	0.00000 0.00000	0.00000	2.00000	0.07136	35.03951
21	RWCE ▾ 3		0	0.00000 1.00000	0.00000	2.00000	0.07054	34.23592
22	MNEG ▾ 2		11	1.00000 1	1.00000	0.00000	1.00000	0.00000
23	MNCG ▾ 2		11		3.00000	0.20000	2.99993	3.13097E-06
24	MXCG ▾ 2		11		15.00000	0.20000	15.00678	0.03160
25	MNCA ▾ 3		10		0.25000	0.20000	0.25000	0.00000
26	MNEA ▾ 3		10	1.00000 1	0.10000	0.20000	-0.09965	27.42757
27	TTHI ▾ 2		10		0.00000	0.00000	50.00037	0.00000
28	OPLT ▾ 27				50.00000	0.20000	50.00037	9.21706E-05

Figure 1.279. The merit function for the lens system shown in figure 1.277.

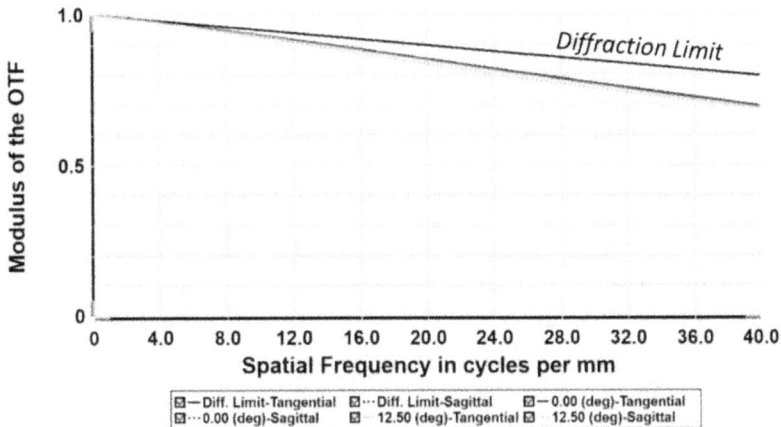

Figure 1.280. The MTF (polychromatic) for the lens system shown in figure 1.277.

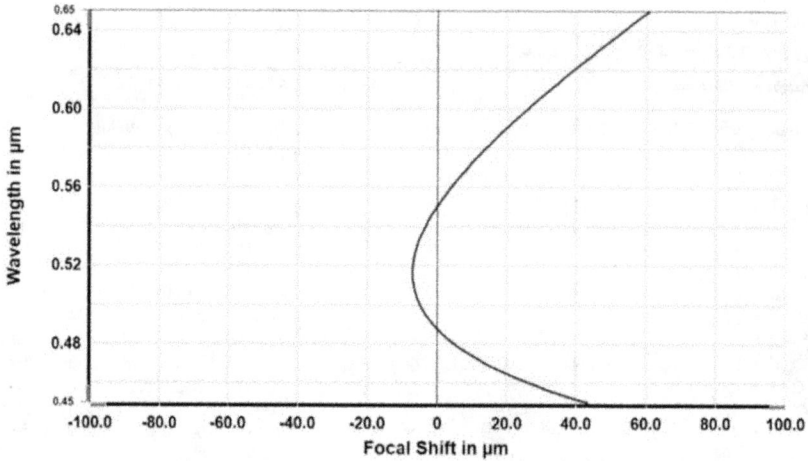

Figure 1.281. The chromatic focal shift for the lens system shown in figure 1.277.

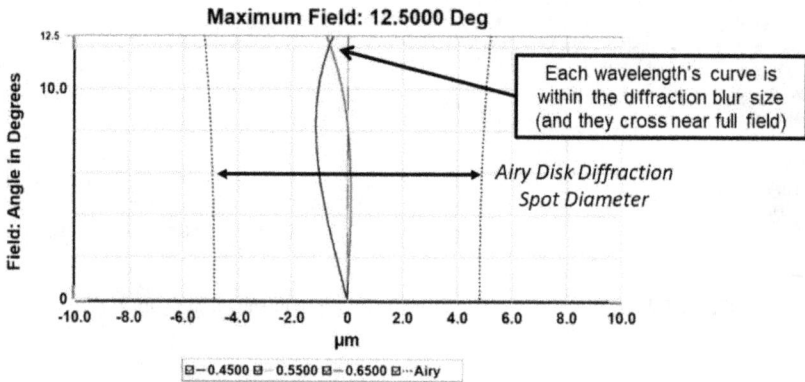

Figure 1.282. The lateral color for the lens system shown in figure 1.277.

Note that the MTF is quite good, and so are the chromatic focal shift and lateral color aberrations. The lens system size is also kept minimal, at a semi-diameter of 8 mm, and there are just five elements. However, N-PSK53A is not a cheap glass. According to the OS data file for this material (left-click on the Material cell for this element in the Lens Data Editor so that it is 'blackened,' then go to the menu and navigate to Libraries > Materials Catalog), its relative cost (i.e. relative to N-BK7) is 4.2199. So, it is roughly 4.2 times more costly than N-BK7. Everything has a cost/benefit ratio. In this case, the benefit could be high in terms of imaging performance. Ultimately, we would need to weight all pros and cons.

It should be noted the current lens is not the only possible achromatic design. It is possible to have a four-element dialyte lens (air-spaced achromat) with good MTF performance [5]. If you needed to open up the aperture, you could try this. You would begin by increasing the stop size incrementally as you changed the target values for the *f*-number. You would then re-optimize. As always, there would be much trial and error and iteration, supported by theory.

1.4.11 Example: an apochromatic microscope tube lens design

At its required spatial frequency (on the image side) of 10 cycles/mm, the COTS achromatic doublet of figure 1.105 has reasonable performance between 450 and 650 nm (weights = 1), but sometimes, biologists want to push the limit by viewing images over a broader wavelength range. Of course, it is possible to re-focus the microscope objective lens, but if this can be avoided, then all the better. It is possible to customize this tube lens for a wider band, such as 400–700 nm or more, by making it apochromatic in that range. We learned from section 1.4.7 that this requires special glasses, which can be determined through automatic optimization. The procedure is as follows:

1. Start with the COTS lens system of figure 1.105. Apply the glass substitute solves to each of the four materials (see section 1.4.7 for a review of what this is).
2. In the merit function, constrain the total system EFL to lie between 180 and 220 mm (in OS, use the OPLT and OPGT operands at a weight of 0.2 each).
3. In the merit function, insert two AXCL operands. Constrain the AXCL at Wave 1 and Wave 3 (pupil zone = 0.7) to be zero. Constrain the AXCL at Wave 2 and Wave 3 (pupil zone = 0.7) to be zero as well. Make each weight equal to ten.
4. Click on Optimize > Hammer Current > 'Start' (do not use 'Automatic'). Watch the merit function value decrease significantly. After a number of seconds (less than a minute), terminate the process manually. Check the new glasses (some may be obsolete if you have not placed appropriate tick marks in the Substitute Glass settings—see section 1.4.7).
5. Note that the above steps are different from the glass-finding procedure described in section 1.4.7. This time, instead of asking OS to search glasses using the criteria of equalizing the RPDs whilst obtaining different Abbe dispersions, we let OS select new glasses based on constraining the axial colors and system EFL. We are not even optimizing the radii! This approach works because axial color aberration is primarily a function of lens power, not curvature (unless spherochromatism is involved, which we are ignoring here). This approach is based on earlier work I have published [8, 98].
6. Next, remove the glass substitute solves. Now, assign variables to all surface radii and to the air space between the doublets. Do not assign any variables to element thickness. Include a marginal ray solve (height = 0, zone = 0) at the last surface. Include an extra surface beyond that, and add a variable for the thickness there (this is a defocus variable, as usual). Include merit function operands to constrain: (i) the EFL to 200 mm, (ii) the AXCL to zero for Waves 1 and 3 and Waves 2 and 3 (zone = 0.7), (iii) the minimum air space thickness to greater than some reasonable value, and (iv) the RMS spot size for both fields (axial and at a half-angle of 3.6°). Optimize using the Hammer feature. End.

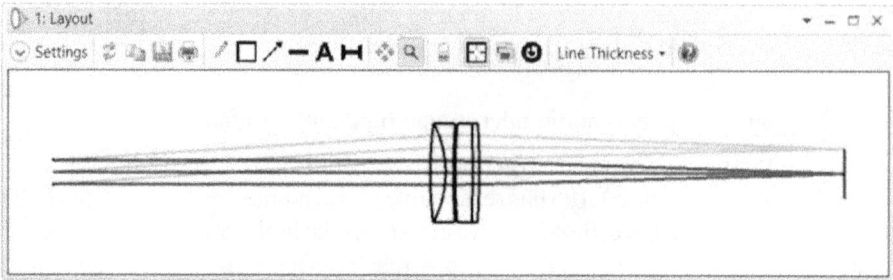

Figure 1.283. The layout of a four-element apochromatic microscope tube lens design.

	Surface Type		Comment	Radius	Thickness	Material	Coating	Clear Semi-Dia
0	OBJECT	Standard ▾		Infinity	Infinity			Infinity
1	STOP	Standard ▾		Infinity	200.00000			6.00000 U
2	(aper)	Standard ▾	1	202.92179 V	8.00000	N-PSK57		25.00000 U
3	(aper)	Standard ▾	2	-53.04308 V	4.00000	KZFS1		25.00000 U
4	(aper)	Standard ▾		346.58511 V	0.99998 V			25.00000 U
5	(aper)	Standard ▾	3	285.99936 V	9.00000	N-PSK57		25.00000 U
6	(aper)	Standard ▾	4	-5278.53754 V	3.50000	N-LASF46		25.00000 U
7	(aper)	Standard ▾		-388.94389 V	191.75442 M			25.00000 U
8		Standard ▾		Infinity	-0.09993 V			12.57485
9	IMAGE	Standard ▾		Infinity	-			12.57708

Figure 1.284. The prescription for the lens system shown in figure 1.283.

The resulting layout I obtained by applying the above steps is shown in figure 1.283, whose prescription, merit function, MTF (400, 550, and 700 nm, weights = 1), and chromatic focal shift are shown in figures 1.284–1.287, respectively. Note the classic 'S' curve indicative of an apochromatic state, with three crossings of the curve at the image plane location. If you were to use the above wavelength range in the COTS tube lens, the MTF would drop. Hence, the current customized version has indeed been optimized for a broader spectrum.

A final remark should be made concerning the current and prior optimizations. Generally, we would also need to control the image quality at a field point somewhere between the axial field and the full half-field. However, in this book, I have deliberately left out this third 'middle' field point, so as to let the program focus on optimizing for the most significant parameters that I am discussing (such as apochromatism in the current tube lens example or achromatism in the prior example). In some cases, leaving the middle field without a constraint produces acceptable results, such as those for the design examples in this book. But in the

Figure 1.285. The merit function for the system shown in figure 1.283.

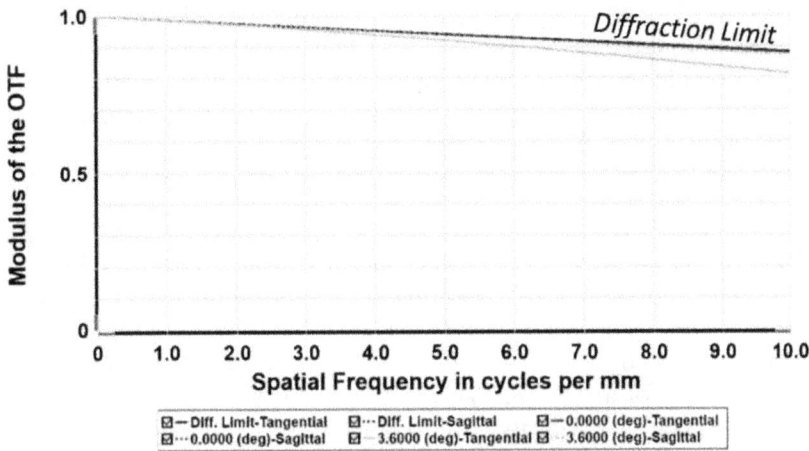

Figure 1.286. The MTF (400, 550, 700 nm, weights = 1) for the system shown in figure 1.283.

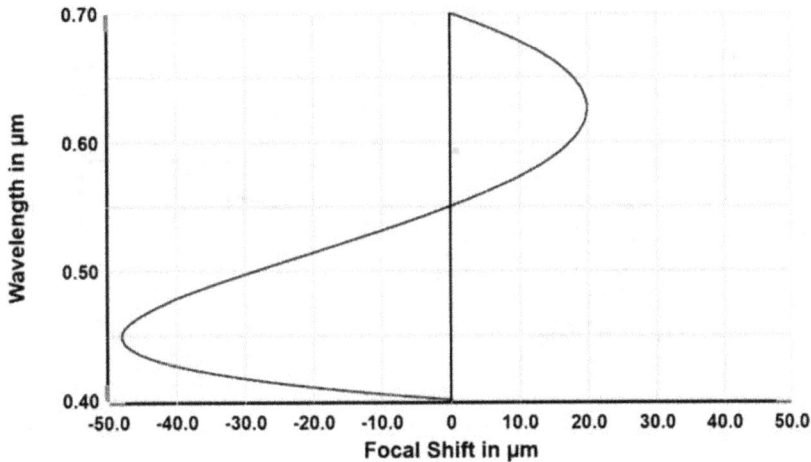

Figure 1.287. The chromatic focal shift for the system shown in figure 1.283.

general case, it is prudent to include multiple field samples in the merit function construction. You can begin with just three, such as the RSCE and/or RWCE operands shown in figures 1.68 and 1.71; then, iteratively involve more. In designs that use highly aspheric surfaces (such as mobile phones), you need significant field sampling, because there are many more *localized transverse effective optical systems* 'felt' by oblique rays as they propagate through those surfaces (see section 1.4.3).

1.4.12 Example: secondary color in a high-aperture double-Gauss lens

Previously, near the end of section 1.2.12, I showed a scaled-up version of a double-Gauss lens (see figure 1.77) that had an EFL of 100 mm EFL and a not particularly high polychromatic MTF at 450, 550, and 650 nm (each weighted at one). I mentioned that the issue with this lens is the residual secondary color. You can check this by simply removing the 550 nm wavelength setting, and you will see that the MTF improves. If, on the other hand, you retain the 550 nm wavelength but eliminate the blue and red (and then re-focus), you also obtain a high MTF. Finally, if you take a look at this lens's longitudinal aberration plot at all three wavelengths (see figure 1.288), you can see that the blue and red curves are aligned well, but the green is off. This is secondary color.

Armed with knowledge of color correction theory, we can now address this issue, as follows:

1. Save the lens system of figure 1.77 into a new file. Remove all user-defined apertures and simply let OS define the lens semi-diameters, except for the stop.
2. Observe from figure 1.258 that possible suitable choices for correcting the secondary color are N-PK52A and N-KZFS2 (this is an application of your 'prior experience'). Use them on the first doublet of the double-Gauss lens.
3. Observe from figure 1.284 that another possible glass pair is N-PSK57 (an obsolete glass) and N-LASF46. Select N-LASF46 for the negative element of

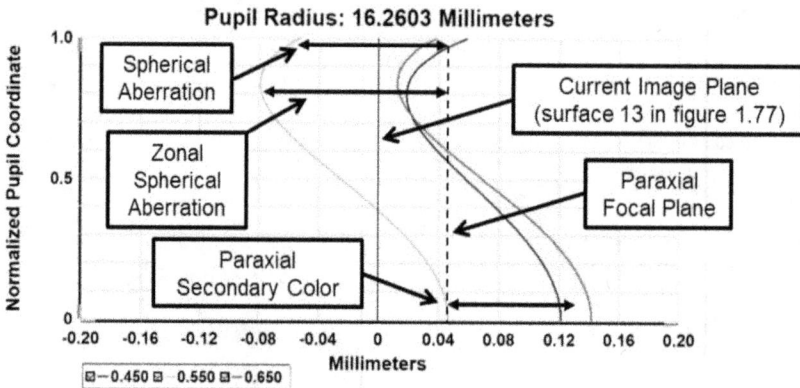

Figure 1.288. The longitudinal aberration (at 450, 550, and 650 nm, weight = 1 each) of the lens system shown in figure 1.77.

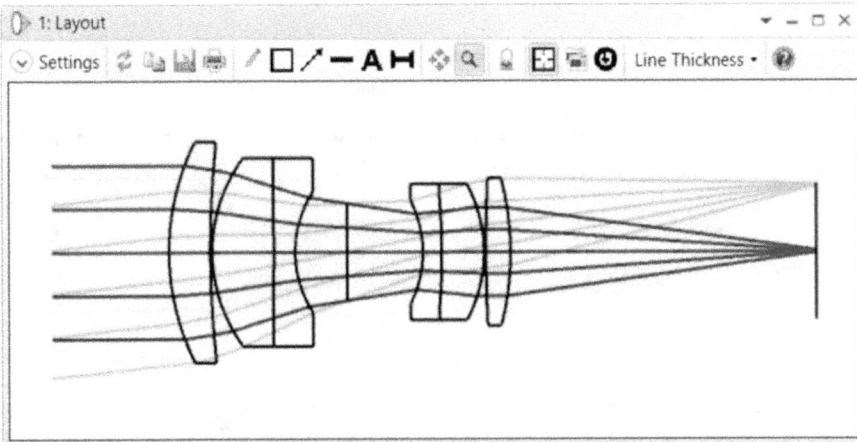

Figure 1.289. The starting layout of the double-Gauss system prior to optimization.

the second doublet in the double-Gauss lens, then try using N-PK52A for the positive element.

4. Make wild guesses for the front and last elements (no, really, you are allowed to do this). I used N-PSK53A (a low-dispersion glass) for the first and N-LAK12 for the last. I wanted a low-dispersion glass for the first, because I did not want the doublets to work too hard (but this approach is going *against* the convention—see, e.g. the advice from Smith [31]). I wanted a high-index glass for the last element, as this element seems to be doing significant refraction, so, I figured that it would help if its index were high.

5. The starting layout, prescription, and merit function construction I used are shown in figures 1.289–1.291, respectively.

Notice in figure 1.290 that all available variables have been assigned onto surfaces and thicknesses. The marginal ray solve for the thickness on surface 12 is at a height of zero and zone zero. As usual, there is a defocus variable assigned to the last surface prior to the image. In the merit function (figure 1.291), weights are assigned to the most important constraints for early optimization, such as the two RSCE operands used to minimize the RMS spot size for the axial and oblique fields (again, I am ignoring the middle field for this exercise).

The optimization proceeded roughly as follows:

1. I used the local damped least-squares optimization by going to Optimize > Optimize! and clicking 'Start.' I waited till the end, then exited.

2. I removed the weights from rows 7 and 8 (i.e. the RMS spot operands).

3. I assigned weights to the RWCE operands on rows 10 and 12 (the weights were set to two each). I left all the other operands alone.

4. I used the Hammer optimization feature by going to Optimize > Hammer Current and clicking on 'Automatic.' I waited till the end, then exited. The

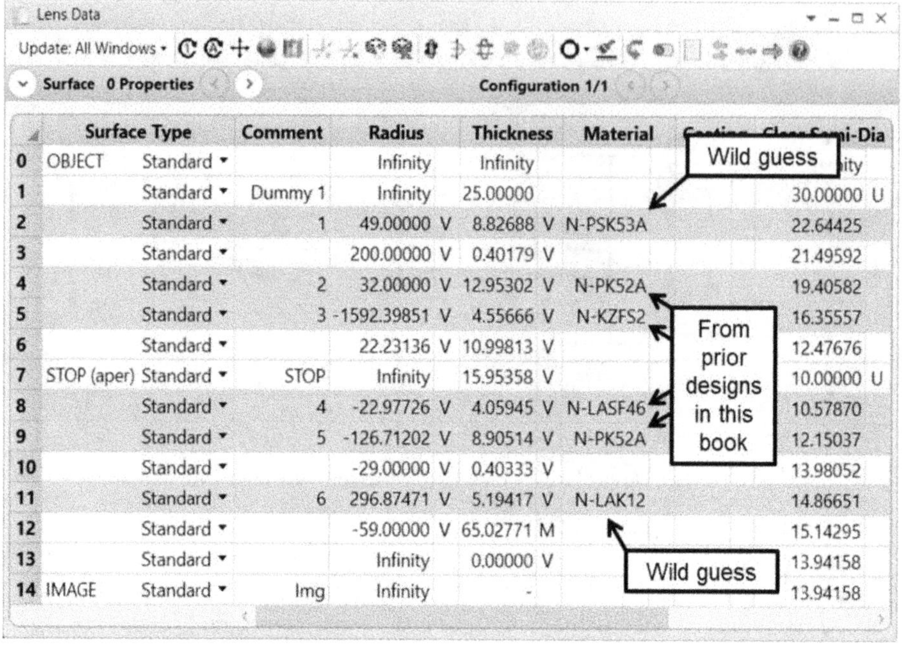

	Surface Type		Comment	Radius		Thickness		Material	i-Dia	
0	OBJECT	Standard		Infinity		Infinity			ity	
1		Standard	Dummy 1	Infinity		25.00000			30.00000	U
2		Standard	1	49.00000	V	8.82688	V	N-PSK53A	22.64425	
3		Standard		200.00000	V	0.40179	V		21.49592	
4		Standard	2	32.00000	V	12.95302	V	N-PK52A	19.40582	
5		Standard	3	-1592.39851	V	4.55666	V	N-KZFS2	16.35557	
6		Standard		22.23136	V	10.99813	V		12.47676	
7	STOP (aper)	Standard	STOP	Infinity		15.95358	V		10.00000	U
8		Standard	4	-22.97726	V	4.05945	V	N-LASF46	10.57870	
9		Standard	5	-126.71202	V	8.90514	V	N-PK52A	12.15037	
10		Standard		-29.00000	V	0.40333	V		13.98052	
11		Standard	6	296.87471	V	5.19417	V	N-LAK12	14.86651	
12		Standard		-59.00000	V	65.02771	M		15.14295	
13		Standard		Infinity		0.00000	V		13.94158	
14	IMAGE	Standard	Img	Infinity		-			13.94158	

Figure 1.290. The prescription of the starting system shown in figure 1.289.

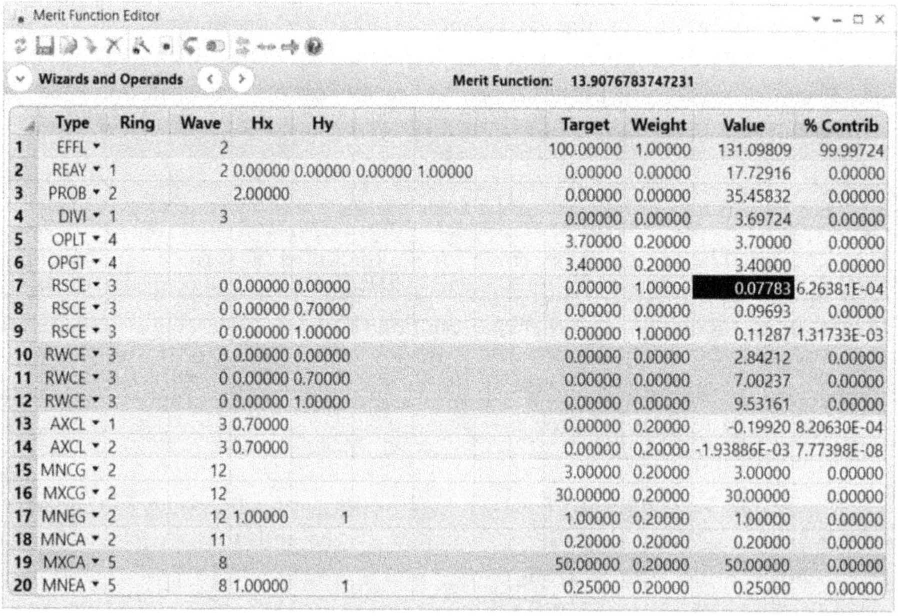

	Type	Ring	Wave	Hx	Hy			Target	Weight	Value	% Contrib
1	EFFL		2					100.00000	1.00000	131.09809	99.99724
2	REAY	1	2	0.00000	0.00000	0.00000	1.00000	0.00000	0.00000	17.72916	0.00000
3	PROB	2		2.00000				0.00000	0.00000	35.45832	0.00000
4	DIVI	1	3					0.00000	0.00000	3.69724	0.00000
5	OPLT	4						3.70000	0.20000	3.70000	0.00000
6	OPGT	4						3.40000	0.20000	3.40000	0.00000
7	RSCE	3	0	0.00000	0.00000			0.00000	1.00000	0.07783	6.26381E-04
8	RSCE	3	0	0.00000	0.70000			0.00000	0.00000	0.09693	0.00000
9	RSCE	3	0	0.00000	1.00000			0.00000	1.00000	0.11287	1.31733E-03
10	RWCE	3	0	0.00000	0.00000			0.00000	0.00000	2.84212	0.00000
11	RWCE	3	0	0.00000	0.70000			0.00000	0.00000	7.00237	0.00000
12	RWCE	3	0	0.00000	1.00000			0.00000	0.00000	9.53161	0.00000
13	AXCL	1	3	0.70000				0.00000	0.20000	-0.19920	8.20630E-04
14	AXCL	2	3	0.70000				0.00000	0.20000	-1.93886E-03	7.77398E-08
15	MNCG	2	12					3.00000	0.20000	3.00000	0.00000
16	MXCG	2	12					30.00000	0.20000	30.00000	0.00000
17	MNEG	2	12	1.00000	1			1.00000	0.20000	1.00000	0.00000
18	MNCA	2	11					0.20000	0.20000	0.20000	0.00000
19	MXCA	5	8					50.00000	0.20000	50.00000	0.00000
20	MNEA	5	8	1.00000	1			0.25000	0.20000	0.25000	0.00000

Figure 1.291. The merit function construction for the starting system shown in figure 1.289.

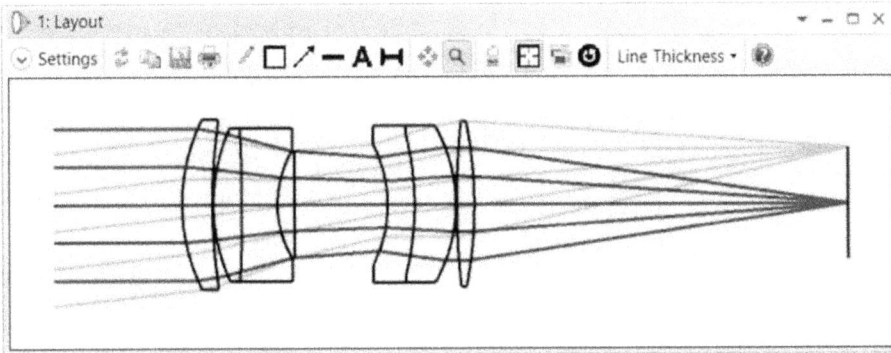

Figure 1.292. The layout of the optimized double-Gauss lens from figure 1.289.

Lens Data

Update: All Windows

Surface 0 Properties — Configuration 1/1

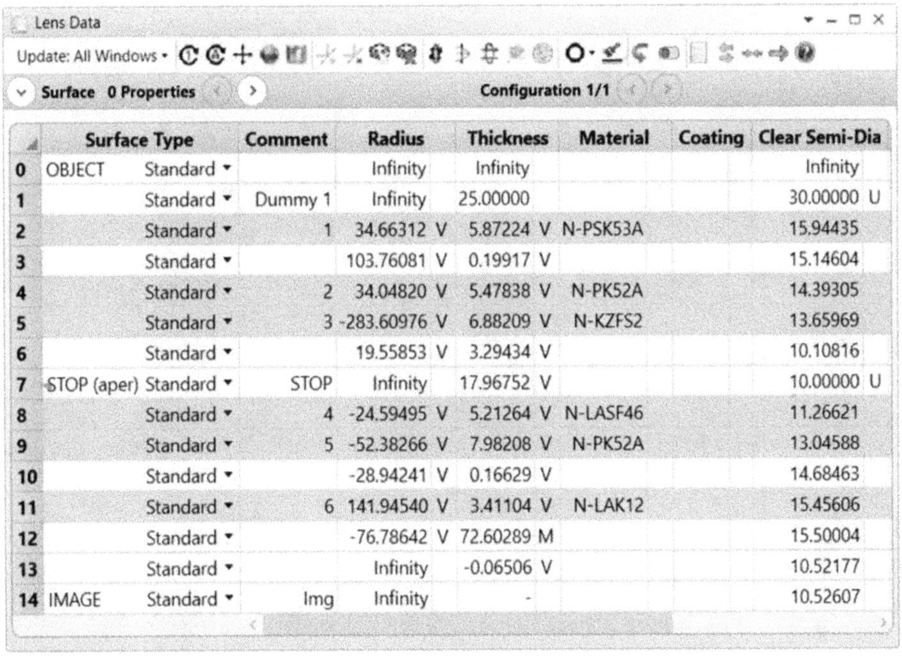

	Surface Type		Comment	Radius		Thickness		Material	Coating	Clear Semi-Dia	
0	OBJECT	Standard		Infinity		Infinity				Infinity	
1		Standard	Dummy 1	Infinity		25.00000				30.00000	U
2		Standard	1	34.66312	V	5.87224	V	N-PSK53A		15.94435	
3		Standard		103.76081	V	0.19917	V			15.14604	
4		Standard	2	34.04820	V	5.47838	V	N-PK52A		14.39305	
5		Standard	3	-283.60976	V	6.88209	V	N-KZFS2		13.65969	
6		Standard		19.55853	V	3.29434	V			10.10816	
7	STOP (aper)	Standard	STOP	Infinity		17.96752	V			10.00000	U
8		Standard	4	-24.59495	V	5.21264	V	N-LASF46		11.26621	
9		Standard	5	-52.38266	V	7.98208	V	N-PK52A		13.04588	
10		Standard		-28.94241	V	0.16629	V			14.68463	
11		Standard	6	141.94540	V	3.41104	V	N-LAK12		15.45606	
12		Standard		-76.78642	V	72.60289	M			15.50004	
13		Standard		Infinity		-0.06506	V			10.52177	
14	IMAGE	Standard	Img	Infinity		-				10.52607	

Figure 1.293. The prescription for the layout shown in figure 1.292.

resulting lens layout I obtained is shown in figure 1.292; its prescription, merit function, MTF, and longitudinal aberration are displayed in figures 1.293–1.296, respectively.

The current double-Gauss system (figures 1.292–1.296) clearly displays an overall improvement over its prior design (figures 1.77 and 1.288), despite not being a full apochromat. In an apochromat, the three curves at different wavelengths would

Figure 1.294. The merit function for the layout shown in figure 1.292.

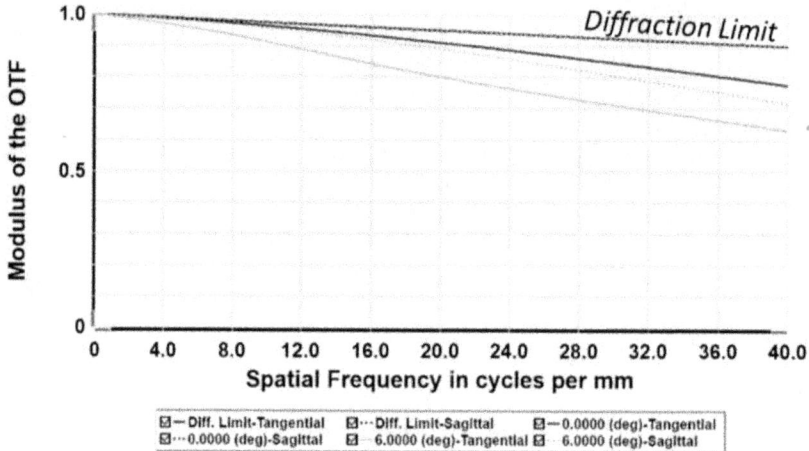

Figure 1.295. The MTF (at 450, 550, and 650 nm, weights = 1) for the layout shown in figure 1.292. Compare this with figure 1.77. The current plot displays an improvement in the MTF.

intersect at one or more zones in the longitudinal aberration plot. For the current lens system, if the spherical aberration can be properly corrected such that the marginal ray of the *green wavelength at the full zone* is made to focus at the paraxial focal plane, then it may be possible to achieve apochromatism. This would be a purposeful use of spherochromatism for this lens. In fact, it is known that residual

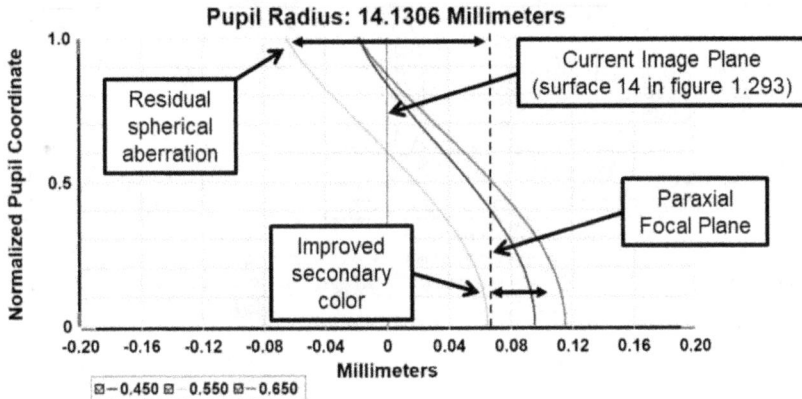

Figure 1.296. The longitudinal aberration for the layout shown in figure 1.292. Compare this with figure 1.288. The current plot displays an improvement in secondary color, but it is not apochromatic.

spherochromatism in a high-aperture double-Gauss lens may be controlled by adding a third surface to the pair of doublets to create triplets, resulting in a lens with an EFL of 100 mm at $F/3.8$ [5] (see also US Patent 2,823,583 by Altman and Kingslake).

1.5 Preparing drawings for optical fabrication

1.5.1 What an optical design drawing for production looks like

In section 1.2.13, we performed a brief tolerance analysis for a double-Gauss lens with an EFL of 50 mm whose prescription and layout are shown in figures 1.70 and 1.73, respectively. If that lens system were to be produced, then we would need to create lens production drawings for each of the elements, (including separate drawings for the two doublets). Figure 1.297 provides an example of a lens drawing for the first element of that double-Gauss lens system, based on the so-called ISO 10110 drawing standard.

While the ISO 10110 standard is not the only standard recognized by optical suppliers and manufacturers, it is used widely. A number of resources providing guidance on the standard are available, such as an open-access publication by OPTICA (formerly the Optical Society of America) edited by Kimmel and Parks [99], an open-access article by Lane [100], and a recent SPIE publication written by Herman, Aikens, and Youngworth [101], which provides detailed explanations on interpreting the standard. Further, OS has a tool that creates component drawings, such as that shown in figure 1.297. To generate the drawing in figure 1.297 using OS, go to Tolerance > ISO Element Drawing. A new window appears with a blank page. You need to enter data through the Settings dialog box by clicking on the 'Settings' button at the top left of that window. As usual, clicking on the '?' button at the top right of the window triggers the OS help file for this tool, which is actually quite helpful, as the help file (under The Tolerance Tab > Manufacturing Drawings and Data Group > ISO Element Drawing) provides a table associating the tolerance

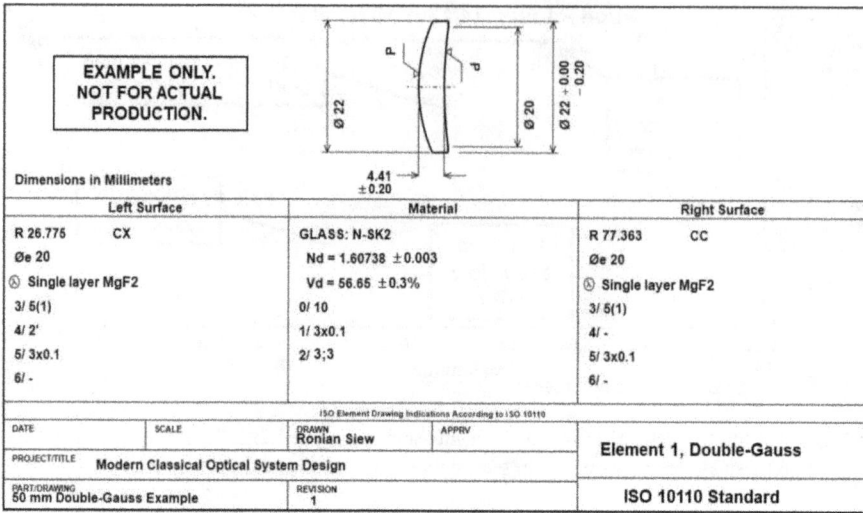

EXAMPLE ONLY.
NOT FOR ACTUAL
PRODUCTION.

Dimensions in Millimeters

4.41
±0.20

Left Surface		Material	Right Surface	
R 26.775	CX	GLASS: N-SK2	R 77.363	CC
Øe 20		Nd = 1.60738 ±0.003	Øe 20	
Ⓝ Single layer MgF2		Vd = 56.65 ±0.3%	Ⓝ Single layer MgF2	
3/ 5(1)		0/ 10	3/ 5(1)	
4/ 2'		1/ 3x0.1	4/ -	
5/ 3x0.1		2/ 3;3	5/ 3x0.1	
6/ -			6/ -	

ISO Element Drawing Indications According to ISO 10110

DATE		SCALE	DRAWN Ronian Siew	APPRV	
PROJECT/TITLE	Modern Classical Optical System Design				Element 1, Double-Gauss
PART/DRAWING 50 mm Double-Gauss Example			REVISION 1		ISO 10110 Standard

Figure 1.297. Drawing example for the first element of the lens system of figure 1.73.

operands used in OS with specific ISO specifications. In figure 1.297, many of the ISO 10110 specifications were entered directly using the OS tool, but some were edited for clarity after exporting the drawing into a JPEG file.

1.5.2 The relation between ISO 10110 specifications and tolerance operands

In OS, a number of practical tolerance operands are directly relatable to ISO 10110 specifications. The only exceptions are those related to specific material properties. This section provides a basic overview of the relationships between some of the specifications listed in the drawing of figure 1.297 and the tolerance operands used in figures 1.79 and 1.80. These relationships are summarized in table 1.1, followed by brief explanations of how each specification was determined (but note that they do not represent final decisions in a real design). For further in-depth descriptions and interpretations of these and other ISO 10110 specifications, the reader is directed to references [10, 99–101].

In figure 1.297, the central thickness of the element is specified as 4.41 mm ± 0.20 mm. The ±0.20 mm tolerance is directly taken from row 3 of the Tolerance Data Editor in figure 1.79. The element diameter tolerance is +0.00/−0.20 mm, which has a rather subtle relationship with the element decenter (TEDX, TEDY) operands at rows 45 and 51 in figure 1.80. Specified as such, the diameter tolerance allows the element's *semi-diameter* to reduce by −0.10 mm, thereby giving a ±0.10 mm decentration variability of the element inside a lens barrel, but the inner diameter of the lens barrel must also be given a tolerance by the opto-mechanical designer. If the opto-mechanical designer allows the barrel's inner diameter to grow by +0.10 mm (but with no allowance for reduction), then the element could decenter by an extra amount of the same magnitude. Thus, the element's decenter tolerance in

1-260

Table 1.1. ISO 10110 specifications in figure 1.297 and their corresponding OS tolerance operands in figures 1.79 and 1.80.

ISO 10110 specification	Meaning	Tolerance operand/s
4.41 ± 0.20	Thickness tolerance	TTHI
⌀22 +0.00 −0.20	Diameter tolerance	TEDX, TEDY (element decenter arising from diameter tolerances)
N_d, V_d	Index, Abbe dispersion tolerance	TIND, TABB
0/10	Material stress birefringence	None[1]
1/3 × 0.1	Material bubbles & inclusions	None[2]
2/3;3	Material inhomogeneity & striae	None[3]
3/5(1)	Surface form error & irregularity	TFRN, TIRR
4/2′	Surface tilt error	TSTX, TSTY
5/3 × 0.1	Surface imperfections	None[2]

Notes

[1] The ISO 10110 documentation for this specification provides guidance on how to specify.

[2] Usually considered cosmetic defects [10], these can theoretically be modeled in HNSC and NSC ray tracing by way of inserting small objects of varying indexes inside the glass, but this is rarely done.

[3] Some amount of modeling can be performed. See, for example, Elliot E and Deslis T 2021 *A function for generating random index inhomogeneity cases* Zemax Knowledgebase Article (online) https://support.zemax.com/hc/en-us/articles/1500005491041-A-function-for-generating-random-index-inhomogeneity-cases

the Tolerance Data Editor is a total of ±0.20 mm, which is the combined effect of its diameter tolerance and the housing's inner diameter tolerance. Implicit in the element drawing specifications is the assumption that sufficient discussion has taken place between the optical designer and the opto-mechanical designer. And by the way, these values are simply examples. In reality, a +0.10/−0.00 mm inner diameter tolerance for a lens barrel can be rather tight.

The index (N_d) and Abbe (V_d) tolerances in figure 1.297 (the TIND and TABB operands in figure 1.80) have been specified in accordance with Schott's 'Step Level' of '2' for these tolerances, which is usually stated in their glass catalog. This step level may or may not be typical, but since it corresponds to the index and Abbe tolerances used for the current lens system being discussed, I have used this step level. Generally, you would refer to the step levels (or any other type of level provided by a different glass supplier) prior to entering the index and Abbe tolerances into your tolerance analysis. The specification indicated by '0/' in the 'Material' box refers to *stress birefringence*. Think of this as a property of glass that has different indexes in orthogonal polarizations (birefringence) as a result of various processes applied to the glass at the glass supplier's factory. There is no direct tolerance operand for this specification. Fortunately, the ISO 10110 document for this specification provides some guidance on the values to be specified for various applications. In that document, a table lists 'Photographic Optics' as typically having a specification of 0/10, where 10 is a grade level for the birefringence. This is what you see in the drawing in figure 1.297.

The specification indicated by '1/' in figure 1.297 refers to the *bubbles and inclusions* in the glass. Bubbles are tiny gaseous voids in the glass, and inclusions are the opposite (i.e. tiny objects of various densities). Thus, in addition to being cosmetic defects, bubbles and inclusions can cause light to scatter. Nobody can tell you what is 'typical' here. We normally refer to prior design drawings in a factory for this (you can always ask a lens production plant for advice on what they have seen in specific drawings for lenses of similar glass, size, and application). From experience, I used 1/3 × 0.1, which means that three bubbles (or inclusions) are permissible, each with a maximum dimension of 0.1 mm (according ISO 10110, this is the square root of the area of the bubble/inclusion).

The specification indicated by '2/' in figure 1.297 refers to *inhomogeneity and striae* in the glass. Inhomogeneity is taken to mean the gradual variation in index over the volume of the glass, resulting in an induced OPD of Δn times the length a ray travels in the glass, where Δn is the change in index over that length in the glass. Striae are more localized inhomogeneities, in which the variation in index occurs over a significantly shorter distance. In ISO 10110, there are grade levels for inhomogeneity and striae. These are the two numbers you see in the specification indicated as 2/3;3. This means I have used grade level 3 for inhomogeneity and also for striae, respectively. In the list of inhomogeneity grade levels of the ISO 10110 documentation, level 3 corresponds to a maximum index change of $\pm 2 \times 10^{-6}$. My gut feel was that I would not want inhomogeneity to impart an OPD to a ray as large as the total residual wavefront error in my lens. I would like it to be as small as possible. Since the element in figure 1.297 has a thickness of 4.41 mm, a grade level 3 imparts roughly $(\pm 2 \times 10^{-6}) \times 4.41$ mm $\approx \pm 9$ nm. That seems acceptably small relative to a quarter of 550 nm (we normally want aberrations to be reduced to less than one quarter of the wavelength in the ray). As for striae, I simply picked a number in the middle of the list of grade levels (if you are unsure, the middle should be the average in terms of performance and cost). Lane [100] provides a useful guide, in which he states that, indeed, a specification of 2/3;3 is 'typical' and that 2/5;5 denotes high quality (note that 2/5;5 denotes the maximum grade levels for homogeneity and striae in the ISO 10110 documentation for this specification).

Back in figure 1.297, the specification indicated by '3/' for the left and right surfaces refers to the *surface form* tolerance, which is the tolerance for the radius of curvature of the surface and astigmatism in the surface. By this it is meant that the element's surface may not necessarily be polished into a perfect sphere. Rather, it could have different curvatures in the orthogonal axes (think of astigmatism of the human eye, in which the cornea or eye lens can have different curvatures in the x and y axes). The symbol 3/5(1) in the drawing specifies that the radius can be off by five fringes in *power* and one fringe in *irregularity*. The power specification is the '5' you see on rows 13 and 14 of figure 1.79 (using the TFRN operand), while the irregularity is the '1' you see on rows 23 and 24 in the same figure. You may note that the tolerance of the radius of curvature specification is given in terms of interferometric testing at a specified wavelength (you usually need to specify this in the drawing). By default, the test wavelength in OS is the primary wavelength in the Wavelength Data Editor, unless specified otherwise using a *tolerance control operand*.

The specification indicated by '4/' refers to the surface tilt error of the element. Sometimes, surface tilt is called 'wedge error' (and we will use this term synonymously with surface tilt in this book). An indication of '4/2' means that I have specified the left surface tilt (which is a physical wedge in the element) of no more than 2 arc minutes. This is the ±0.033 33° on row 33 in figure 1.79, which uses the TSTX and TSTY tolerance operands. An arc minute is one sixtieth of a degree, so 0.033 33° equals 2 arc minutes. In OS, this surface tilt is modeled by way of making the Surface Type 'Irregular' (see figure 1.57 in section 1.2.11). Now, note that while the left-hand surface is indicated with a surface tilt tolerance, the right-hand surface is not. The reason for this is that in the factory, an engineer or technician measures the surface tilt by sitting the element on one surface. So, one surface must be a reference surface (i.e. a 'datum'), and thus, in the drawing, the right-hand surface is a datum. For this reason, there are also no TSTX nor TSTY tolerance operands for surface 3 (the right-hand surface of the element in figure 1.297) in figure 1.79.

The specification indicated by '5/' in figure 1.297 refers to *surface imperfections* on the element surfaces. These are somewhat like the *scratch-dig* specifications of the United States's military (MIL) standard known as MIL-PRF-13830B, although there is no direct link to it. In ISO 10110, surface imperfections refer to many types of localized artifacts (scratches, pits, 'long' scratches, edge chips, and others) and many details for how they are specified are written in the references cited in this section. In figure 1.297, the specification 5/ 3 × 0.1 indicates that a maximum of three surface imperfections are allowed, each of which has a maximum dimension of 0.1 mm. I figured that if I can permit 1/3 × 0.1 for bubbles and inclusions, then it is reasonable to specify the same for surface imperfections. Some manufacturers require the rest of the 5/ values to be specified (i.e. there are 5/ indications for coating blemishes, long scratches, etc. which we ignore here).

You may have noted that there is no specification indicated for '6/' in figure 1.297. This is because '6/' indicates a specification for the laser damage threshold (for laser applications). Many online resources (e.g. white papers) are available that discuss the laser damage threshold, and the references cited in the current section provide further detail, which we will not spend time on here. You may also note that the element tilt operands TEDX and TEDY shown in figure 1.80 are not associated with any labeled specifications in the drawing of figure 1.297. Element tilts are the most difficult to model. They are a function of the opto-mechanical design for the lens mounts within the lens housing/barrel. The modeling of tilts requires conversations between the lens designer and the opto-mechanical designer (hence, all element tilt tolerance values that were set in figure 1.80 were best guesses, subject to further scrutiny). ISO 10110 component drawings may include many more specifications. Here, I have highlighted only the basic and most common ones.

1.5.3 Modeling the centering process of cemented elements

Production drawings often follow a sort of hierarchical structure (you will often hear terms such as 'high-level assembly,' 'assembly level,' 'subassembly level,' etc.). For instance, an opto-mechanical designer must create an assembly drawing that specifies the integrated elements as an imaging lens mounted into a housing (lens barrel). This is an example of a high-level assembly drawing. At a lower level than this (but higher than single elements), the doublets of the double-Gauss lens in figure 1.73 require their own separate 'next-level assembly drawing' in addition to drawings created for the individual elements comprising the doublets. This is because manufacturers need to cement the elements to make a doublet, and upon doing so, they need to know specific requirements such as the permissible final diameter of the combined doublet (because one element may be larger than the other) and the maximum centering error between the two elements (because each element of the doublet has a surface tilt tolerance specified by the '4/' symbol, as discussed above).

In consideration of the centering error of the doublet, manufacturers often attempt to compensate the combined wedge error tolerance of the cemented elements through an adjustment process in which they *slide* one element relative to the other about the cementing interface (prior to curing the cement) and qualify the centering error by passing a laser beam through the doublet (think of beam deviation due to two prisms) [102]. This adjustment is illustrated in figure 1.298.

In figure 1.298(a), the doublet has not yet been adjusted. The left-hand surface of the negative element has been given an exaggerated surface tilt error of 1.3° counter-clockwise about the minus X axis (this axis points out of the page), while the right-hand surface of the positive element has been given the same exaggerated surface tilt error but in the clockwise direction. The result is a deviated beam on the far-right

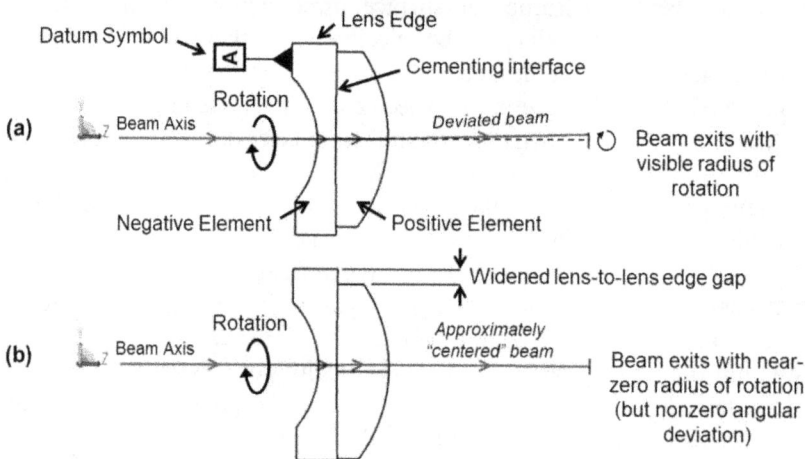

Figure 1.298. An example of doublet centering adjustment. (a) Prior to adjustment. (b) After adjustment. This doublet is the first from the left of the double-Gauss lens in figure 1.73.

side as the beam exits the doublet. The datum symbol with the 'A' inside is used in *geometric dimensioning and tolerancing* (GD&T) to indicate a reference surface [33]. In the case of figure 1.298(a), this datum indicates to the manufacturer that when cementing and centering adjustment is performed on the doublet, the surface of the element where the triangular base of datum A is placed can be used to position the negative element on a physical surface for mounting (thus, in this case, the centering is done with the doublet placed vertically and the laser beam entering from the bottom). Inspection is performed by rotating the doublet as a whole about the beam's axis (i.e. the axis of the beam before it enters the doublet, which is the $+Z$ axis in this figure) so that rotation of the beam's spot on a detector at the exit end of the doublet can be observed, as indicated by the circling dark arrow on the right. The idea is to minimize the radius of rotation of the beam's spot by way of the sliding process between the two elements in the doublet. The sliding is applied to the smaller (usually, positive) element. Due to this, manufacturers may also suggest specifying that one element in a doublet has a smaller diameter than the other, such that when the manufacturers apply the sliding effect, the smaller element is allowed to slide without having its edge stick out beyond the edge of the larger-diameter element, as illustrated in figure 1.298(b). Notice that, after adjustment, the smaller positive element on the right-hand side has been slid downwards, resulting in a widened lens-to-lens edge gap near the top. But since the positive element's diameter has been made smaller than the negative element's diameter, the shifted positive element does not protrude beyond the full diameter of the negative element, thus allowing the whole doublet to be inserted into a lens barrel.

There is a subtle point to be highlighted concerning the above factory adjustment for doublet centering. In particular, the process does not actually result in a truly centered lens system for the two elements unless the negative element is also tilted relative to the beam axis. The reason is as follows: the optic axis of an element is determined by joining the centers of curvature of its two surfaces with a line. Therefore, two elements are centered only when all four centers of curvature are collinear (i.e. they fall into a single axis). It follows that a truly centered lens system is one in which a beam entering the system exits without angular deviation and also without a radius of rotation. The only way to achieve this for the system in figure 1.298 is if, in addition to sliding the positive element about the cementing interface, the negative element is made to tilt as well (i.e. by tilting the entire doublet). This state of true centering can be modeled in OS, as illustrated in the layout in figure 1.299, whose prescription and merit function are provided in figures 1.300 and 1.301, respectively. In figure 1.300, certain unnecessary columns of the Lens Data Editor have been 'hidden' (you can right-click on any column at the top and select 'Hide Column'). Note that the Tilt About X values for surfaces 4 and 9 have been given the exaggerated surface tilt errors mentioned earlier. The Tilt About X on surface 3 provides the tilt adjustment for the negative element, while the Tilt About X on surface 7 provides the sliding adjustment for the positive element across the cemented interface. Note that this sliding adjustment is modeled by way of making the coordinate break on surface 6 'travel' a distance equal to the radius of surface 5, followed by the Tilt About X adjustment, which essentially rotates the

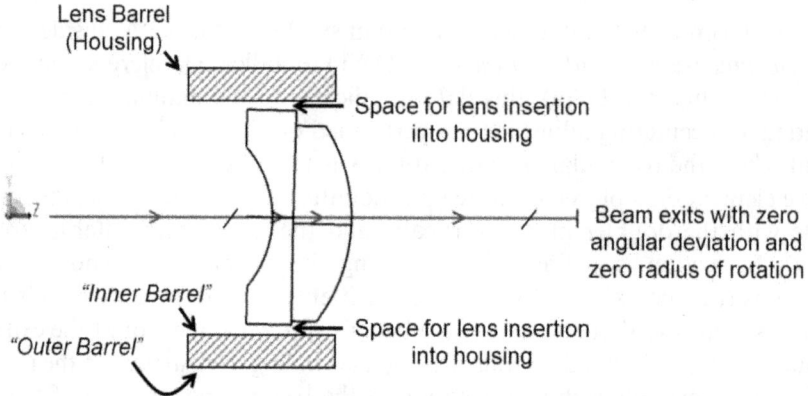

Figure 1.299. A truly centered version of the doublet of figure 1.298, achieved by way of tilting the entire doublet in an ideal lens barrel whose inner/outer axes are collinear.

	Surface Type	Comment	Radius	Thickness	Material	Semi-Diameter	Tilt About X	Tilt About Z	Order	Par
0	OBJECT Standard •		Infinity	Infinity		0.00000				
1	STOP Standard •		Infinity	25.00000		1.00000E-02 U				
2	Coordinate Break •	Rotate z (doub)	0.00000			0.00000	0.00000	0.00000	0	
3	Coordinate Break •	Tilt Neg El	0.00000			0.00000	1.27539 V	0.00000	0	
4	(aper) Irregular •	Element 2	-11.11600	2.28000	F5	7.50000 U	-1.30000	0.00000	0.00000	
5	(aper) Standard •		796.19900	0.00000		11.50000 U				
6	Coordinate Break •	Travel +	796.19900 P			0.00000	0.00000	0.00000	0	
7	Coordinate Break •	Rotate Pos El	-796.19900 T			0.00000	-0.04308 V	0.00000	1	
8	(aper) Standard •	Element 1	796.19900	6.48000	N-SK16	10.00000 U				
9	(aper) Irregular •		-17.90900	0.00000		10.00000 U	1.30000	0.00000	0.00000	
10	Coordinate Break •	Travel -	-6.48000 T			0.00000	0.00000	0.00000	0	
11	Coordinate Break •	Travel ++	796.19900 P			0.00000	0.00000	0.00000	0	
12	Coordinate Break •	Un-Rotate Pos El	-796.19900 P			0.00000	0.04308 P	0.00000	1	
13	Coordinate Break •	Return 1	6.48000 T			0.00000	0.00000	0.00000	0	
14	Coordinate Break •	Travel --	-8.76000 T			0.00000	0.00000	0.00000	0	
15	Coordinate Break •	Un-Tilt Neg El	8.76000 P			0.00000	-1.27539 P	0.00000	1	
16	Coordinate Break • Un-Rotate z (doub)		25.00000			0.00000	0.00000	0.00000 P	1	
17	IMAGE Standard •		Infinity	-		1.00000 U				

Figure 1.300. The prescription in OS for the doublet system shown in figure 1.299.

Merit Function: 4.90563271057711E-10

	Type	Surf	Wave	Hx	Hy	Px	Py	Target	Weight	Value	% Contrib
1	BLNK • Beam deviation (deg)...										
2	RAID •	17	1	0.00000	0.00000	0.00000	0.00000	0.00000	0.00000	2.61998E-13	0.00000
3	BLNK • In arc minutes...										
4	PROB •	2			60.00000			0.00000	1.00000	1.57199E-11	0.05134
5	BLNK • Minimize "deviation" (actually, minimize radius of rotation)...										
6	BLNK • ...at final plane (so, minimize global REAY height there)...										
7	REAY •	17	1	0.00000	0.00000	0.00000	0.00000	0.00000	1.00000	6.93583E-10	99.94866

Figure 1.301. The merit function for the doublet system shown in figure 1.299.

positive element about the center of curvature of surface 5. Note that two weights are used in figure 1.301: one on row 4 that targets the angular deviation of the exiting beam at zero and the other on row 7 that targets the radius of rotation of the exiting beam at zero on the observation plane (surface 17). Variables were assigned for the tilt of the negative element (Tilt About X on surface 3) as well as the sliding adjustment for the positive element (Tilt About X on surface 7). The result is a doublet in which all four centers of curvature are collinear, which now defines the doublet's optic axis. If you use the Footprint Diagram feature of OS (go to Analyze > Rays & Spots > Footprint Diagram), then you can model the rotation of the doublet by entering values into the Tilt About Z on surface 2 (but to see this effect, you would need to 'undo' the corrected state of the system by entering '0' into the Tilt About X on surface 3).

In figure 1.299, the shaded regions depict the inner and outer walls of a lens barrel that the doublet may be mounted into. If this lens barrel were an ideal barrel (i.e. one in which the mechanical axes of the inner and outer walls of the barrel were collinear), then a properly centered doublet in the lens barrel would be one in which the doublet's optic axis was collinear to the inner and outer walls of the barrel. In figure 1.299, I have labeled these inner and outer barrel walls as 'Inner Barrel' and 'Outer Barrel' for convenience. Clearly, this ideal situation is not always realized. Lens barrels must have a tolerance between the inner and outer barrel axes. Moreover, there has to be a gap between the edges of lens elements and the inner barrel in order to allow the elements to be inserted.

While the sliding action of the smaller (usually, positive) element in a doublet is routinely employed in the cementing and centering process in factories [102], there is no guarantee that this process will also involve tilting the larger (usually, negative) element. Consequently, a doublet often possesses some residual centering error relative to its edge. If a doublet is made to tilt in a lens barrel (as depicted in figure 1.299), then the tilt could be roughly aligned with either the inner or outer barrel's axis. These axes would not be in full alignment, due to the residual centering error of the doublet, but since the residual is usually small, the doublet's tilt inside a barrel could still be helpful. However, tilting a cemented pair of elements in a lens barrel amounts to a rather tedious alignment process during assembly. The mass production of many imaging lenses does not involve the active alignment of elements in the lens barrel (with the exception, perhaps, of high-performance microscope objectives). Instead, they are assembled through a 'drop-in' process whereby elements are simply inserted into their lens barrel—including the action of jiggling the barrel slightly in order to let elements fall into a roughly aligned state inside the barrel—and then locked and glued on the outside. In some cases, the opto-mechanical design may provide for mechanical spacers that sit on the element surfaces rather than their edges, resulting in a minor self-centering effect, while in some other cases, special designs can provide improved self-centering [103]. Otherwise, elements simply drop in between spacers in the lens housing, which results in a **statistical tolerance stack for centering errors, including possible subtractive interaction effects**. Under such conditions, it can be considered irrelevant whether or not the elements of doublets are precisely centered relative to each other and to the lens barrel's axes, because nothing is truly centered and one relies on the **statistical effect of cancellations among wedge errors (due to statistical interactions) to maintain the final image quality at a satisfactory level**. Sometimes, even a single element in the lens system is **purposefully decentered** in order to compensate for these tolerance stacks [33] (detailed examples of this are provided in section 3.7 and appendix C.2). As a result, a number of options exist for doublets:

1. Tighten the surface tilt tolerances of both elements such that no further adjustment for the elements is necessary (this was done for the tolerance operands of the doublets in figure 1.73). You would then need to specify to the factory that they should not perform the sliding adjustments for the positive element during cementing.

2. Tighten the surface tilt tolerance for the larger (usually, negative) element, but loosen the surface tilt tolerance for the smaller (usually, positive) element. This should result in very nearly centered axes among all centers of curvature of the doublet, because tightening the tolerance to wedge error for the negative element means that its two centers of curvature become nearly collinear to the entering beam's axis. Thereafter, sliding the positive element would align its axis with that of the negative element.

3. Specify that both tilting of the negative element and sliding of the positive element be done during cementing (but be prepared to absorb some added costs!)

In any case, when you perform Monte Carlo tolerancing analyses of optical systems, you may sometimes notice that a lens system's toleranced statistics indicate a rather robust optical system. You may therefore not always necessarily require that the elements are precisely centered with respect to each other. In fact, this is quite often the case in the mass production of hundreds of assemblies or more on a monthly basis, in which it becomes costly to align each element relative to the other and the lens housing during the assembly process. High-quality microscope objectives may not fall into this category, as their factories may have already been equipped with alignment techniques and processes, whose high costs may have already experienced their return on investment.

The situation may of course be somewhat different when setting up components on a lab bench, where there is often no lens barrel to mount elements into. In this case, singlets can be painstakingly aligned to each other using a laser or an autocollimator and a reference mirror. But doublets may still present a problem if no tilt adjustment is performed on the negative element in addition to the sliding/centering adjustment made to the positive element during cementing (unless, as stated earlier, the wedge error of the negative element is kept small). Consequently, even if the radius of rotation of a processed cemented doublet has been made minimal, the fact that the negative element may possess a significant wedge error means that the doublet as a whole may not necessarily be really centered, as mentioned earlier. In this case, if the doublet is to be mounted into a lens barrel and then onto a lab bench, then it may not be 'centerable' relative to the axes of the inner and outer barrels (see figure 1.299 for this terminology). Also, note that the outer barrel is the only physically accessible surface of a lens assembly and is therefore often used for mounting. If the inner barrel's axis is not well centered relative to the outer barrel's axis, then a mounted assembly on a lab bench will suffer from some slight misalignment. Further, under such conditions, the use of autocollimators (or lasers) to align this lens assembly relative to other lens assemblies on a lab bench may not necessarily be the best approach to achieving good final image quality. This is because when using an autocollimator, one normally regards 'good alignment' as a condition whereby either the axis of the autocollimator is centered relative to the axis of the lens assembly, or the axes of all lens assemblies are collinear with each other. But which axis is the axis of a lens assembly? There is no common optic axis when all elements in an assembly are dropped into a barrel without active centering on a per-component basis. They are also, therefore, not necessarily centered relative to the inner barrel nor to the outer barrel. Yet, an imaging lens assembly with no common axis can possess excellent image quality by virtue of being a potentially good sample out of a number of Monte Carlo samples.

1.5.4 Alternatives to design drawings: communicating with suppliers

By now, you may have noticed that the creation of optical production drawings can be rather time-consuming and is no easy task. It is not just a matter of placing numbers on dimensions and permissible variations. **Each specification in a drawing compels the manufacturer to take action to meet that specification, which involves laborious time and effort for interpretation, fabrication, and measurement.** Therefore, the total cost to you is both a function of the price you pay the manufacturer and the effort you put into specifying your optics in the drawing. Often, this effort can be underestimated. If insufficient effort is spent on ensuring the provision of meaningful specifications, then this can result in wasted time, as the drawings will be sent back. Further, if the specifications are overly simplistic, then the supplier could potentially accept them and fabricate a part that, although it meets the drawing specifications, may not work in your intended application. If your specifications are impossible to meet, then you may not be able to find a willing supplier. If a supplier accepts impossible specifications but nobody is aware of their absurdity, then nobody wins in the end.

   Hence, the task of designing lenses is a full-time job that includes the creation of meaningful lens drawings. If your job, like mine, mainly involves an 'optical system level' type of responsibility, in which components are selected, analyzed, and integrated to produce an entire instrument, then you may not be able to afford the time to optimize lens designs, perform tolerancing, and create lens drawings. Sometimes, you may have just enough time to perform optimization and tolerancing, leaving insufficient time for creating lens drawings. In this case, you may not need to. It is important to know that while the designer may be responsible for providing lens specifications in drawings, the designer may not necessarily have to be the one to create official production drawings, especially those used in the manufacturing plant. In some cases, you can just specify your components by way of your company's internal documentation procedures (i.e. without an ISO 10110 template, nor even with ISO 10110 specification symbols, such as those indicated by 0/, 1/, 2/, ...). Besides, lens manufacturers frequently have internal processes that are not disclosed to the public, which are documented in their own drawings. Therefore, you can sometimes specify your designs in a reasonable fashion and then simply start a productive conversation with potential manufacturers. However, do bear in mind that, optical design is generally performed concurrently with basic tolerancing analyses and communication with suppliers. This is an effective way to discover whether your optical system is on the right track towards being manufacturable.

References and further reading

[1] Siew R 2021 Basics of optical systems in real-time PCR instruments for virus detection *XLIV OSI Symp. on Frontiers in Optics and Photonics 2021*

[2] Creveling C M, Slutsky J L and Antis D 2003 *Design for Six Sigma in Technology and Product Development* (Upper Saddle River, NJ: Pearson Education) pp 112–7

[3] Bunch R M 2021 *Optical Systems Design Detection Essentials: Radiometry, Photometry, Colorimetry, Noise, and Measurements* ed R B Johnson (Bristol: IOP Publishing) pp 1–1–4 6-10–11

[4] Boyd R W 1983 *Radiometry and the Detection of Optical Radiation* (New York: Wiley) pp 82, 75–9, 99–103

[5] Kingslake R and Johnson R B 2010 *Lens Design Fundamentals* 2nd edn (Burlington, MA: Elsevier) pp 63–77, 101–14, 129, 132–3, 163–6, 174–81, 223–6, 245–8, 250–52, 255–68, 269–77, 289–92, 297–310, 301–4, 316–7, 323–30, 355–77, 459–62, 501–10

[6] Kidger M J 2002 *Fundamental Optical Design* (Bellingham, WA: SPIE) pp 10–14, 26–7, 63–5, 91–100, 132–5, 148–53, 159–62

[7] Shafer D 2017 *Highlights of My 51 Years in Optical Design* (www.slideshare.net/operacrazy/highlights-of-my-51-years-in-optical-design)

[8] Siew R 2017 *Perspectives on Modern Optics and Imaging: With Practical Examples Using Zemax® OpticStudio®* (Independently Published) pp 56–60, 181–9

[9] ANSYS, Inc 2023 https://www.ansys.com/en-gb

[10] Fischer R E and Biljana T-G 2000 *Optical System Design* (New York: McGraw-Hill) pp 5–6, 155–66, 315–85, 489–90

[11] Walker B H 2000 *Optical Design for Visual Systems* (Bellingham, WA: SPIE) pp 4–5, 8–9, 29–67

[12] Stewart J 1991 *Calculus* 2nd edn (Pacific Grove, CA: Brooks/Cole) pp 460–2

[13] Geary J M 2002 *Introduction to Lens Design: With Practical Zemax® Examples* (Richmond, VA: William Bell) pp 68–71, 123–4, 296–7, 370–1

[14] Mouroulis P 2008 Depth of field extension with spherical optics *Opt. Express* **16** 12995–3004

[15] Tang H and Kutulakos 2013 Utilizing optical aberrations for extended-depth-of-field panoramas *Computer Vision–ACCV 2012* (Lecture Notes in Computer Science vol 7727) ed K M Lee, Y Matsushita, J M Rehg and Z Hu (Berlin: Springer) pp 365–78

[16] Cossairt O and Nayar S 2010 Spectral focal sweep: extended depth of field from chromatic aberrations *2010 IEEE Int. Conf. on Computational Photography (Cambridge, MA, USA)* (Piscataway, NJ: IEEE) pp 1–8

[17] Fitzgerald N, Goncharov A V and Dainty C 2017 Extending the depth of field in a fixed focus lens using axial colour *Int. Optical Design Conf. 2017, Optical Design and Fabrication 2017 (Denver, Colorado, USA, 9–13 July 2017)* (Washington, DC: Optica Publishing Group) paper ITh4A.4

[18] Bakin D 2013 Extended Depth of Field Technology in Camera Systems *Smart Mini-Cameras* ed T V Galstian (Boca Raton, FL: CRC Press) pp 141–3

[19] Dowski E and Cathey W T 1995 Extended depth of field through wavefront coding *Appl. Opt.* **34** 1859–66

[20] Chi W and George N 2003 Computational imaging with the logarithmic asphere: theory *J. Opt. Soc. Am.* A **20** 2260–73

[21] Khare K, Butola M and Rajora S 2023 *Fourier Optics and Computational Imaging* 2nd edn (Switzerland: Springer) pp 249–59

[22] Isshiki M, Sinclair D C and Kaneko S 2006 Lens design: global optimization of both performance and tolerance sensitivity *Proc. SPIE* **6342** 63420N

[23] Moore K E 2019 Optimization for as-built performance *Proc. SPIE* **10925** 1092502

[24] Sasián J 2022 Lens desensitizing: theory and practice *Appl. Opt.* **61** A62–7

[25] Sasián J 2013 *Introduction to Aberrations in Optical Imaging Systems* (New York: Cambridge University Press) pp 4–9, 119–31, 187–90

[26] Rowlands D A 2020 *Physics of Digital Photography* ed R B Johnson (Bristol: IOP Publishing) 2nd edn pp 3-43–4, 5-21–22

[27] Ray S F 2002 *Applied Photographic Optics: Lenses and Optical Systems for Photography, Film, Video, Electronic and Digital Imaging* 3rd edn (Oxford: Focal Press)

[28] Boreman G D 2021 *Modulation Transfer Function in Optical and Electro-Optical Systems* 2nd edn (Bellingham: SPIE) pp 9–12, 85–95

[29] Pedrotti F L S.J. and Pedrotti L S 1993 *Introduction to Optics* 2nd edn (Englewood Cliffs, NJ: Prentice-Hall) pp 78, 118, 157–8

[30] Smith G H 1998 *Practical Computer-Aided Lens Design* (Richmond, VA: William Bell) pp 276–8, 339–60

[31] Smith W J 2005 *Modern Lens Design* 2nd edn (New York: McGraw-Hill) pp 42, 151–200, 319–29, 561–7

[32] Laikin M 2007 *Lens Design* 4th edn (Boca Raton, FL: CRC Press) pp 119–29, 245–52

[33] Schwertz K and Burge J H 2012 *Field Guide to Optomechanical Design and Analysis* (Bellingham, WA: SPIE) pp 59 110

[34] Goodman J W 1996 *Introduction to Fourier Optics* 2nd edn (New York: McGraw-Hill) pp 127–65, 172–218

[35] Siew R 2020 Variable focus machine vision lens without moving parts: freeform progressive eyeglasses for non-humans *Imaging and Applied Optics Congress: Imaging Systems and Applications 2020 (Washington, DC, 22–26 June, 2020)*

[36] Siew R 2020 Variable focus machine vision lens without moving parts: freeform progressive eyeglasses for non-humans *OSA Technical Digest* (Optica Publishing Group) paper ITh4E.4

[37] Dilworth D C 2020 *Lens Design: Automatic and Quasi-Autonomous Computational Methods and Techniques* 2nd edn (Bristol: IOP Publishing)

[38] Hazra L 2022 *Foundations of Optical System Analysis and Design* (Boca Raton, FL: CRC Press) pp 622–4

[39] Siew R 2002 Partial coherent imaging using the grating light valve *Proc. SPIE* **4773** 92–101

[40] Levi L 1969 Detector response and perfect-lens-MTF in polychromatic light *Appl. Opt.* **8** 607–16

[41] Barnden R 1974 Calculation of axial polychromatic optical transfer function *Opt. Acta* **21** 981–1003

[42] Vallmitjiana S, Davajas D, Barandalla J J and Moneo J R D E F 1983 A simple method of evaluating the polychromatic modulation transfer function for photographic systems *J. Opt.* **14** 25–8

[43] Saleh B E A and Teich M C 1991 *Fundamentals of Photonics* (New York: Wiley) p 650

[44] Palmer J M and Grant B G 2010 *The Art of Radiometry* (Bellingham, WA: SPIE) pp 53–7 195–197

[45] Lohmann A W 1989 Scaling laws for lens systems *Appl. Opt.* **28** 4996–8

[46] Brückner A 2014 Multiaperture cameras *Smart Mini-Cameras* ed T V Galstian (Boca Raton, FL: CRC Press) pp 239–98

[47] Siew R 2019 *Monte Carlo Simulation and Analysis in Modern Optical Tolerancing* (Bellingham, WA: SPIE)

[48] Geary J M 1993 *Introduction to Optical Testing* (Bellingham, WA: SPIE) pp 9–10

[49] Zhang Y and Gross H 2019 Systematic design of microscope objectives. Part II: Lens modules and design principles *Adv. Opt. Technol* **8** 349–84

[50] Johnson R B 2008 Correctly making panoramic imagery and the meaning of optical center *Proc. SPIE* **7060** 70600F

[51] Carlson W E 2017 *Computer Graphics and Computer Animation: A Retrospective Overview* (The Ohio State University) https://ohiostate.pressbooks.pub/graphicshistory/#main

[52] Zeitler K D 1987 Algorithms for ray tracing *Master of Mathematics Thesis* University of Waterloo, Waterloo, ON

[53] Kajiya J T 1986 The rendering equation *ACM SIGGRAPH Computer Graphics* **20** 143–50

[54] Muschaweck J and Rehn H 2022 *Designing Illumination Optics* (Bellingham, WA: SPIE) pp 94–5

[55] Vanderwerf D F 2010 *Applied Prismatic and Reflective Optics* (Bellingham, WA: SPIE) 33–60

[56] Tkaczyk T S 2010 *Field Guide to Microscopy* (Bellingham, WA: SPIE)

[57] Seward G H 2010 *Optical Design of Microscopes* (Bellingham, WA: SPIE)

[58] Welford W T 1986 *Aberrations of Optical Systems* (Boca Raton, FL: CRC Press) pp 86–7 139

[59] Alonso J, Goméz-Pedero J A and Quiroga J A 2019 *Modern Ophthalmic Optics* (Cambridge: Cambridge University Press)

[60] Siew R 2018 Progressive lens approach to variable focus without moving parts in electronic imaging systems *Inopticalsolutions Technical Note*

[61] Zhu J, Zhang B, Hou W, Bauer A, Rolland J P and Jin G 2019 Design of an oblique camera based on a field-dependent parameter *Appl. Opt.* **58** 5650–5

[62] Kingslake R 1983 *Optical System Design* (Orlando, FL: Academic) pp 52, 59–65, 72–3, 133–5, 248

[63] Cermax® Xenon arc lamp with elliptical reflector from Excelitas Technologies www.excelitas.com/product-category/cermax-ceramic-body-elliptical-xenon-lamps-and-modules (accessed 25 June 2023)

[64] Sasián J 2019 *Introduction to Lens Design* (Cambridge: Cambridge University Press) pp 161–2, 171–5

[65] Smith W 1997 *Practical Optical System Layout: And Use of Stock Lenses* (New York: McGraw-Hill)

[66] Walther A 1995 *The Ray and Wave Theory of Lenses* (New York: Cambridge University Press) pp 13–6, 29, 55–6

[67] Frolov D N 2021 *Microscope Design Volume 1: Principles* (Bellingham, WA: SPIE) p 55

[68] Born M and Wolf E 1980 *Principles of Optics* 6th edn (Oxford: Pergamon Press) pp 418–24, 522–6

[69] Hopkins H H 1950 The influence of the condenser on microscope resolution *Proc. Phys. Soc.* B **63** 737–44

[70] Zhang Y and Gross H 2019 Systematic design of microscope objectives. Part 1: system review and analysis *Adv. Opt. Technol* **8** 313–47

[71] Edmund Optics www.edmundoptics.com/knowledge-center/application-notes/optics/an-in-depth-look-at-spherical-aberration-compensation-plates/ (accessed 3 July 2023)

[72] Lackowicz J R 2006 *Principles of Fluorescence Spectroscopy* 3rd edn (New York: Springer)

[73] Kingslake R 1992 *Optics in Photography* (Bellingham, WA: SPIE) pp 78–83

[74] Falk D, Brill D and Stork D 1986 *Seeing The Light: Optics in Nature, Photography, Color, Vision, and Holography* (New York: Wiley) pp 133–41

[75] Arecchi A V, Messadi T and Koshel R J 2007 *Field Guide to Illumination* (Bellingham, WA: SPIE) p 59

[76] Sun C-C and Lee T-X 2022 *Optical Design for LED Solid-State Lighting* ed R B Johnson (Bristol: IOP Publishing)

[77] Delano E 1963 First-order design and the y, $\bar{y}$ diagram *Appl. Opt.* **2** 1251–6

[78] Bentley J and Olson C 2012 *Field Guide to Lens Design* (Bellingham, WA: SPIE) p 19

[79] Siew R 2004 Concept, design, and manufacture of a 3X rifle gun sight *Proc. SPIE* **5429** 281–7

[80] Blahnik V 2014 About the irradiance and apertures of camera lenses *Carl Zeiss AG Technical Article https://lenspire.zeiss.com/photo/en/article/overview-of-zeiss-camera-lenses-technical-articles* Accessed December 2023.

[81] Berek M 1930 *Grundlagen der Praktischen Optik* (Berlin: de Gruyter)

[82] Blahnik V and Schindelbeck O 2021 Smartphone imaging technology and its applications *Adv. Opt. Technol.* **10** 145–232

[83] Grant B G 2011 *Field Guide to Radiometry* (Bellingham, WA: SPIE) p 5

[84] Berkovic G and Shafir E 2012 Optical methods for distance and displacement measurements *Adv. Opt. Photonics* **4** 441–71

[85] Isa M A, Piano S and Leach R 2020 Laser triangulation *Advances in Optical Form and Coordinate Metrology* ed R Leach (Bristol: IOP Publishing) ch 3

[86] Forbes G 1988 Optical assessment for design: numerical ray tracing in the Gaussian pupil *J. Opt. Soc. Am.* A **5** 1943–56

[87] Johnson R B 1993 Balancing the astigmatic fields when all other aberrations are absent *Appl. Opt.* **32** 3494–6

[88] Williamson R 2011 *Field Guide to Optical Fabrication* (Bellingham, WA: SPIE) pp 6 32–4

[89] Smith W J 1988 The problem of the concentric meniscus element: a possible solution to the lens designer's dilemma *Opt. Eng.* **27** 121039

[90] Murthy E K 1992 Elimination of the thick meniscus element in high-resolution scanning lenses *Opt. Eng.* **31** 95–7

[91] Steinich T and Blahnik V 2012 Optical design of camera optics for mobile phones *Adv. Opt. Technol.* **1** 51–8

[92] Tsai T-H 2010 Optical lens system for taking image *U.S. Patent* 7643225B1

[93] Hopkins R E 1987 Optical system requirements for laser scanning systems *Opt. News* **13** 16

[94] Reidl M J 2017 *Single Optical Imaging Elements* (Bellingham, WA: SPIE)

[95] Gonzalez-Acuna R G and Chaparro-Romo H A 2020 *Stigmatic Optics* ed R B Johnson (Bristol: IOP Publishing)

[96] Hecht E 2002 *Optics* 4th edn (San Francisco, CA: Addison Wesley) pp 150–3

[97] Schroeder D J 1987 *Astronomical Optics* (San Diego, CA: Academic) pp 16–21

[98] Siew R 2016 Practical automated glass selection and the design of apochromats with large field of view *Appl. Opt.* **55** 9232–6

[99] OSA Standards Committee 1995 *Optics and Optical Instruments—Preparation of Drawings for Optical Elements and Systems: A User's Guide* ed R K Kimmel and R E Parks (Washington, DC: Optica Publishing Group)

[100] Lane J 2012 A practical tutorial for generating ISO 10110 drawings *Adv. Opt. Technol.* **1** 419–25

[101] Herman E, Aikens D M and Youngworth R N 2021 *Modern Optics Drawings: The ISO 10110 Companion* (Bellingham, WA: SPIE)

[102] Bliedtner J, Gräfe G and Hector R 2008 *Optical Technology* (New York: McGraw-Hill) pp 219–24, 318–9

[103] Lamontagne F and Desnoyers N 2019 New solutions in precision mounting *The 11th Int. Conf. on Optics-Photonics Design and Fabrication (ODF'18) (Hiroshima, Japan); Opt. Rev.* **26** 396–405

[104] Yang A, Gao X and Li M 2016 Design of apochromatic lens with large field of view and high definition for machine vision *Appl. Opt.* **55** 5977–85

IOP Publishing

Modern Classical Optical System Design
Fundamentals, techniques, tips, and tricks
Ronian Siew

Chapter 2

Illumination

This chapter brings us to the subject of designing optical systems for illumination. In contrast to imaging, the design of illumination systems generally does not require imaging. Nevertheless, imaging can be a useful technique in illumination design. Therefore, illumination design involves both imaging and nonimaging principles. For this reason, it was purposeful in this book to present the topic of imaging prior to that of illumination. Here, we shall combine radiometric principles with sequential and nonsequential computer-aided ray tracing approaches and apply them to the design of classical illumination configurations with rotational symmetry (e.g. critical and Köhler illumination, searchlights, light pipes, etc). In addition, the systems are all refractive and therefore do not involve mirrors. We also do not cover freeform designs and advanced techniques used in nonimaging optics that 'tailor' element shapes for prescribed illumination distributions. While such techniques are beneficial, many modern illumination systems still apply simpler classical methods and use rotationally symmetric configurations. You can therefore gain much practical skill in illumination design by acquiring knowledge of the ideas and techniques presented in this chapter.

2.1 The illumination problem

The problem in illumination is to shine light of sufficient *brightness* and *uniformity* across a screen. Many different configurations are possible, as depicted in figure 2.1.

If the illumination on the screen from just a single source possesses desirable characteristics, then that is all that is needed (figure 2.1(a)). If the illuminated area must have a finite extent, then perhaps an aperture between the source and screen may suffice (figure 2.1(b)). If higher brightness is desired, then perhaps a lens may be a suitable choice, provided that the surface of the source does not possess undesirable artifacts (figure 2.1(c)). If there are such artifacts on the source, then perhaps a simple defocus of the screen may solve the issue (figure 2.1(d)).

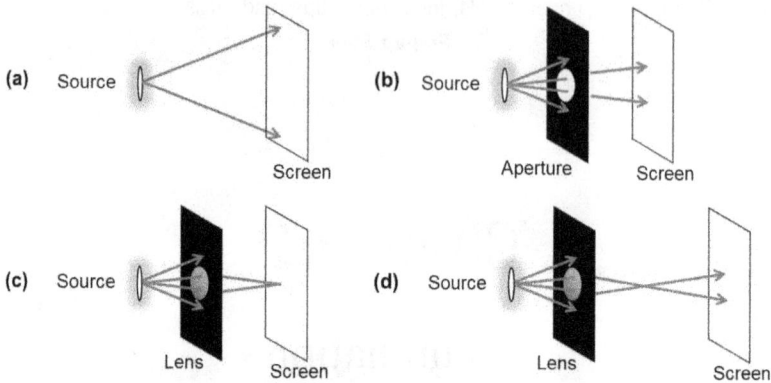

Figure 2.1. A sample of various possible configurations for illumination.

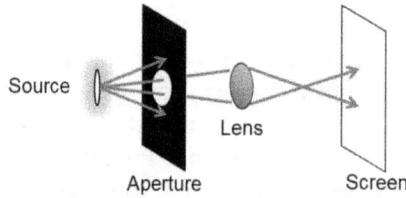

Figure 2.2. A lens shifted away from a source and an aperture can produce uniform illumination.

The configurations above are not the only types available. For instance, there is nothing to say that the lens in figure 2.1(d) cannot be shifted towards the right of the black opaque plane. If an aperture is retained on that black plane, a shifted lens can project an image of the aperture (rather than that of the source) onto the screen, as illustrated in figure 2.2. This can yield a reasonably uniform light distribution on the screen. If another lens were mounted at the aperture to focus the source into the shifted lens, then the illumination would be brightened. This is the configuration known as *Köhler illumination*, which you saw in sections 1.3.6, 1.3.7, and 1.3.10. But what exactly is meant by *brightness*? What exactly is meant by *uniformity*? These qualitative descriptions for illumination can be quantified in a practical manner, based on fundamental principles involving optical flux transfer. The branch of optics that provides these principles is known as *optical radiometry* (or simply, radiometry). To be precise, it is *photometry* that describes the *brightness* of sources as they are perceived by the human eye, while radiometry quantifies the flux content in the source and the light it emits. In this book, we will focus on radiometry. The next section provides a review of essential principles in radiometry for the purpose of tackling classical illumination problems. Note that although the surface to be illuminated in illumination problems can generally be of any shape or form (such as spherical, or some other 3D shape), in this book, we limit the illuminated surfaces to planes (i.e. a screen in this book is a plane). This is because the majority of illuminated targets in practical applications tend to be planar surfaces, such as microscope slides, liquid crystal displays (LCDs), and digital micromirror devices (DMDs).

2.2 Essential radiometry for illumination problems

2.2.1 What type of source is being modeled in sequential ray tracing?

You often see 'point sources' in ray tracing layouts. For the purposes of ray tracing and characterizing optical imaging systems, you may regard a point source as an idealized mathematical construct that is quasi-monochromatic and incoherent. But at the entrance pupil of an optical system, a point source can give rise to a degree of spatial coherence that permits us to integrate over amplitude and phase by virtue of the van Cittert–Zernike theorem [1–3]. For this reason, the Airy diffraction point spread function (PSF) is the limiting distribution at the image for a point source in the absence of aberrations. Still, if you see something that looks like the Airy pattern, this does not mean that point sources exist. From calculus, we know that in the limit that a source becomes infinitesimally small (or infinitely distant), the distribution of light at the image from that source tends towards a PSF. So, an image will look like the Airy pattern if the image is sufficiently small relative to the transverse extent of the Airy disk (more on this topic in section 2.2.6). According to the theory of *Fourier optics*, an image is regarded as being formed by convolving the imaging system's PSF with the idealized geometric image of that system. You came across this concept in chapter 1 (see also appendix A.1). According to Fourier optics, the PSF is the *impulse response* of the optical system to an idealized point source, and the MTF is proportional to the Fourier transform of the PSF. In imaging lens design, a qualitative assessment of image quality can be performed by way of inspecting the size (transverse spatial extent) of the PSF. When aberrations are more significant than diffraction effects, PSFs are described by spot diagrams. So, you can consider a spot diagram to be a sort of *geometric* PSF, which is rather visual and therefore quite useful.

However, in sequential (SC) ray tracing, rays from an idealized point source at the object plane do not behave like the rays we would expect from an actual point-like source, with the exception of conditions in which the source is at infinity (such as stars). In the layouts of OS used in SC mode, point sources at the object are purposefully made to illuminate the entrance pupil *uniformly*. For layouts with the stop in front, if you look at ray heights across the entrance pupil, they are equally spaced. In real life, small point-like sources generally do not illuminate a screen uniformly. Thus, a point source in SC ray tracing is not only a mathematical construct, but the manner in which it illuminates the entrance pupil of an optical imaging system is somewhat fictionalized (but that is acceptable, as you will see). When you examine spot diagrams, the Settings menu provides selections that introduce additional fictional ray *patterns* into the entrance pupil. The options are Square (sometimes called Rectangular), Hexapolar, and Dithered. Each setting provides different spacings for rays in the entrance pupil, which result in different *ray densities* (the number of rays per unit area—see section 2.3) across the spot diagram. Each pattern lets the lens designer perform qualitative assessments of the impact of aberrations on the image of an ideal point source. Thus, they are not meant to provide quantitative information about the radiometric properties of PSFs (however, it can be said that the dithered ray pattern—being *pseudorandomized* in the pupil—removes systematic artifacts from spot diagrams and therefore results in more realistic ray densities in a geometric PSF). As Kingslake and Johnson put it: 'It should be recognized that there is no best pattern' [4]. Interestingly, a recent study by Díaz and Navarro suggests that, on the basis that accurate wavefront aberration data should be obtained from spot diagrams, then a so-called *spiral* pattern for rays in the pupil provides the best choice [5]. Fortunately, the overall shapes of spots in spot diagrams do not change with each choice of ray pattern in the pupil. That is, if there is coma in the lens, then the spot diagram always appears like coma, regardless of the chosen square, hexapolar, or dithered pupil ray pattern settings. Also, the quantitative assessment of transverse ray aberrations (i.e. ray intercept plots) fortunately does not require any specific ray pattern. If you have gone through all of the design examples in chapter 1, you will have seen that none of them required any consideration of the radiometric properties of rays. Thus, the fictionalized ray patterns in the pupil from point sources are generally inconsequential to image quality analyses performed in SC ray tracing.

Yet sometimes, SC ray tracing can model approximate illumination distributions at the image by *apodizing* the entrance pupil. This involves setting a transmittance function in the entrance pupil such that there is a transverse variation of ray density across the pupil. When this is done, spot diagrams tend to possess approximately realistic ray distributions. Further, certain settings in the GIA and GBI analysis tools of the SC ray tracing mode in OS can be used to estimate illumination distributions, as you have seen in section 1.2.10. For these reasons, SC ray tracing can be applied to illumination design. In fact, I often begin with SC ray tracing when I conceptualize and design illumination systems, as you will see throughout chapter 2.

2.2.2 What is different about sources in nonsequential ray tracing?

In nonsequential (NSC) ray tracing, you can model a real source's properties by setting various parameters for NSC source objects, such as their flux, *radiant intensity* (see section 2.2.4), and others. Sources in NSC ray tracing can be modeled as having size; in this case, we call such sources *extended sources* or *extended objects* if the object is the source. NSC sources may also be modeled as points with no area (idealized mathematical constructs). The former is clearly more realistic, but the latter is a useful tool for performing alignments between and among NSC objects in the system as well as a means to compare results between your NSC and SC models.

2.2.3 Flux, radiance, and étendue in illumination design

In section 1.1.3, you were introduced to equation (1.1), which describes an element of flux $d\phi$ in a cone of rays from a small *Lambertian* source—a term that we will formally define in this section. By a 'small' source we mean one of elemental area dA. For any optical system with a circular entrance pupil subtending a half-angle u

from a small Lambertian source, equation (1.1) says that the amount of flux $d\phi$ passing through the optical system and onto a receiver at the end is $L\pi dA \sin^2 u$. If there is loss due to reflection and/or absorption by anything between the source and receiver, then a transmittance factor τ can be multiplied by the transferred flux.

Note that the radiance L in equation (1.1) is constant, which is not the general case for arbitrary sources. A source of constant radiance is referred to as being *Lambertian*, while a source whose radiance is a function of ray direction is considered *non-Lambertian*. Think of radiance as a *radiation density function*. Its role in optical radiometry is analogous to the role that quantities such as mass density (mass per unit volume) serve in other branches of physics. For instance, if mass is m and volume is V, then mass density μ may be defined as $\mu = dm/dV$. If μ is constant (e.g. in a homogeneous gas, liquid, or solid), then the total mass is simply mass times volume. If μ has continuous variation in space, then the total mass in that space is $\int \mu dV = \iiint \mu(x, y, z) dx dy dz$. Similarly, the radiance of a source can have continuous variation. In particular, it can vary over directions in space to which rays are emitted, and it can vary across the surface area of the source from which the rays are emitted. Integrating the radiance over a source area and angular space yields the total flux. To see this, we express radiance as [6]

$$L = \frac{d^2\phi}{dA_{proj}d\Omega},\qquad(2.1)$$

where $d^2\phi$ is an element of emitted flux, dA_{proj} is an elemental *projected surface area of the source*, and $d\Omega$ is an elemental *solid angle* in the direction of an emitted ray bundle, as illustrated in figure 2.3. In this figure, the axis of the source is orthogonal to the source's surface, and dS is an elemental area of illumination that is orthogonal to the axis of the solid angle defined by dS/r^2. The angle u is subtended by the semi-diameter of dS. The unit of solid angle is the *steradian* (sr).

We can now rearrange equation (2.1) to be

$$d^2\phi = L dA_{proj}d\Omega.\qquad(2.2)$$

For a centered optical system, $u = 0$, so that the axis of the solid angle is collinear to the axis of the source, and the optical system's entrance pupil has an elemental area dS subtending the source. Under this condition, $dA_{proj} = dA$ when a ray bundle is captured by an elemental pupil area dS. However, for a large pupil, the factor $\cos u$

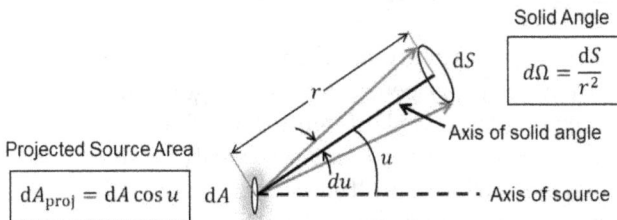

Figure 2.3. The ray geometry defining equation (2.1).

must always be accounted for. Keeping this in mind, we now note that rays from an elemental source arrive at the entrance pupil resembling a cone of rays of solid angle given by

$$\Omega = 4\pi \sin^2(u/2), \tag{2.3}$$

where u is the marginal ray at the full zone [6]. Differentiating equation (2.3), we obtain

$$d\Omega = 2\pi \sin u du. \tag{2.4}$$

Substituting equation (2.4) into (2.2) and noting that $dA_{\text{proj}} = dA \cos u$ gives

$$d^2\phi = 2\pi L dA \sin u \cos u du. \tag{2.5}$$

For a Lambertian source, L is constant. If this source is small, then integrating both sides of equation (2.5) yields

$$d\phi = \pi L dA \int_0^u 2 \sin u \cos u du = \pi L dA \int_0^{\sin u} 2 \sin u d(\sin u)$$

$$= 2\pi L dA \left(\frac{\sin^2 u}{2} \Big|_0^u \right) = \pi L dA \sin^2 u, \tag{2.6}$$

which is, of course, seen to be the same as equation (1.1). In fact, this is the origin of equation (1.1). Because many rough surfaces can, to a first approximation, be assumed to have roughly Lambertian scattering properties, experienced optical designers often use equation (2.6) to quickly assess the amount of flux collected by an optical system from a small scattering surface. This would also apply to small LEDs with an output distribution of rays that is roughly Lambertian. You would not need to model this in any program.

If a source is not Lambertian, then the radiance in equation (2.5) is a function of u. In this case, equation (2.5) may be expressed as

$$d^2\phi = 2\pi dA_{\text{proj}} L(u) \sin u du, \tag{2.7}$$

which produces no analytic solution unless $L(u)$ is expressible in closed form and the integral over u can be determined without recourse to numerical integration (however, see section 2.2.4). Of course, the NSC mode of OS can model non-Lambertian sources, and it can do it in a number of ways, which we discuss in section 2.2.4.

In SC ray tracing, if you are forming the image of an extended object (i.e. a non-point source) and you see a ray pencil in a layout going from a point at the object towards a small area of a lens, think of figure 2.3. Of course, the local surface of the lens at which the ray pencil strikes is not necessarily parallel to dS. In general, as rays refract through lens surfaces, dS 'follows' the central ray in the ray pencil and is always made orthogonal to that ray. Therefore, the axis of the solid angle in figure 2.3 is the central ray in a ray pencil. Upon refraction at an interface between two media of index n and n', the **radiance theorem** [6] states that

$$\frac{L}{n} = \frac{L'}{n'}, \tag{2.8}$$

where L' is the radiance in the medium of index n'. Equation (2.8) is a statement of the conservation of *radiance over index*, and it is proven all the time in texts on radiometry. The proofs often involve **a ray pencil refracting from a medium into another medium**. Thus, equation (2.8) tells us that if a **ray pencil** begins in air at the source and ends in air at a screen, then the radiance at the screen is the same as that at the source. Any losses can, as usual, be accounted for by multiplication by a transmittance factor, τ. This is clearly also true in imaging. That is, if a ray pencil begins in air at the object and ends in air at the image, then the radiance at the image is the same as that at the object. This means that **even when a source with rotationally symmetric emission is not Lambertian and its radiance is therefore a function of u, its radiance within a localized ray pencil at angle u is conserved all the way from source to screen**. This applies to ray pencils at any angle u. This leads us to the concept of **étendue** and **sub-étendue**, which are quantities that you will often deal with in illumination.

In section 1.1.3, you were introduced to the term étendue by way of equations (1.1) and (1.2). In the current section, I showed you how equation (1.1) came about through the derivation that led to equation (2.6). In this formula, the quantity $\pi \mathrm{d}A \sin^2 u$ is the étendue in a cone of rays (of half-angle u and from a small Lambertian source) passing through the unvignetted pupils of an optical system with a circular stop. If the source is not Lambertian, then we revert to using equation (2.7), where $2\pi \mathrm{d}A \sin u \mathrm{d}u$ is an element of the étendue. In this book, we will refer to any element of étendue as a **sub-étendue**. In cross-referencing equations (2.5) and (2.7), we note that **the sub-étendue in a ray pencil is always conserved, regardless of whether or not the source is Lambertian or non-Lambertian**. This is because $L(u)$ is conserved along the central ray in a ray pencil, as described above.

You may have noticed that the sub-étendue given by $2\pi \mathrm{d}A \sin u \mathrm{d}u$ contains all quantities in an element of flux other than radiance. Therefore, **an element of flux is always given by radiance times a sub-étendue**. If we have a Lambertian source with rotationally symmetric emission, integrating the sub-étendue over the angle u and the area $\mathrm{d}A$ yields the system's **total étendue** Therefore, for Lambertian sources, the total flux is given by the product of the radiance and the total étendue. You can also see this by expressing equation (2.2) as

$$\mathrm{d}^2\phi = L \mathrm{d}A_{\mathrm{proj}} \mathrm{d}\Omega = L \mathrm{d}^2\varepsilon, \tag{2.9}$$

where $\mathrm{d}^2\varepsilon$ is the sub-étendue. For Lambertian sources, integrating both sides of equation (2.9) shows that the total flux is given by the product of its radiance and total étendue. If the source is non-Lambertian and emits non-rotationally symmetric radiation, then we assume that L is a function of all the quantities in equation (2.1); there is no general way to separate radiance from étendue (i.e. there is no such thing as 'total radiance' and 'total étendue' for non-Lambertian sources). **But elemental fluxes within the ray pencils of non-Lambertian sources are always given by the radiance in that ray pencil times the sub-étendue of that ray pencil, and both of these quantities are always conserved.**

The reason that all of this is useful is because the design of illumination systems involves transferring flux within a ray pencil towards a screen, and the manner in which the ray pencil is *guided* on its way influences the uniformity and brightness in the illumination at the screen. The radiant intensity of a source influences our decision on how to guide rays towards screens to obtain a desired result in the illumination. The spread of flux in a ray pencil is governed by its sub-étendue, and this in turn influences how a ray pencil is guided. Sometimes, it is as simple as letting rays pass through apertures (such as in figure 2.1(b)) or using a lens to focus them (such as in figure 2.1(c)). At other times, it can involve dividing up a full beam into little ray pencils, such as in the case of using so-called fly's eye arrays, which is discussed in sections 2.15 and 2.16. Moreover, aberrations *break up the total étendue into many sub-étendues*, which enables the design of illumination systems that can shape coherent and incoherent beams into highly uniform distributions, a topic that will be discussed in further detail in section 2.5. Thus, the concept of étendue and sub-étendue is important in illumination design.

As a final remark before we continue to the next section, you may notice that in section 1.4, a distinction is made between the angles u and U at the object side and the angles u' and U' at the image side. In that section, it was important to distinguish between paraxial angles (given by the smaller letters, u and u') and marginal ray angles (given by the larger letters, U and U'). In the current section, we will often use small letters to symbolize ray angles, but I will highlight the use of the larger letters where appropriate. Sometimes, radiometric quantities are intertwined with imaging quantities, so I will try to point out specific distinctions where needed.

2.2.4 From radiance to radiant intensity: modeling sources

In many cases, the quantity $\mathrm{d}AL(u)$ in equation (2.7) may be modeled by expressing it as

$$\mathrm{d}A_{\mathrm{proj}}L(u) \approx I_o(\cos u)^m, \tag{2.10}$$

where I_o is the **radiant intensity** (flux per unit solid angle, defined by $\mathrm{d}\phi/\mathrm{d}\Omega$), and m is any value within the bounds $0 < m < \infty$. OS lets you set the values for m, but they are capped at a maximum of 100. Inserting equation (2.10) into (2.7) results in an integrable formula:

$$\mathrm{d}\phi \approx 2\pi I_o(\cos u)^m \sin u\,\mathrm{d}u = 2\pi I_o(\cos u)^m[-\mathrm{d}(\cos u)]. \tag{2.11}$$

Integrating both sides of equation (2.11) yields

$$\phi \approx 2\pi I_o\left[-\frac{\cos^{(m+1)} u}{m+1} \Bigg|_0^{|u} \right] = \frac{2\pi I_o[1 - (\cos u)^{m+1}]}{m+1}. \tag{2.12}$$

Note that setting $m = 1$ in equation (2.12) yields equation (2.6). Thus, where applicable, equation (2.10) may be applied to model emission from sources with rotationally symmetric angular ray outputs. This is useful if, for example, an LED's emission is non-Lambertian and has a narrow angular distribution. In this case, a suitable choice for m can be used to model the distribution, and so equation (2.12) may be applied to estimate the flux within a cone of rays subtended by u for such sources. But in some other cases, equation (2.10) may not adequately model the ray emission from an actual source. If a datasheet is provided for the source, then you can often find a plot of the radiant intensity data showing the flux per unit solid angle on the vertical axis and angle on the horizontal axis. The NSC mode of OS lets you enter the radiant intensity data of non-Lambertian sources into appropriate cells of the NSC Editor when the *source radial* NSC object is used for the source object (see section 1.2.15).

Incidentally, in equation (2.10), when $m = 1$, the resulting quantity given by $I_o \cos u$ is indeed the dependence of radiant intensity on u for a Lambertian source, but in this case, the dependence on u is not because of the function $L(u)$. Recall that L is always constant for a Lambertian source. Since radiant intensity is $d\phi/d\Omega$, returning to equation (2.2), in the case of a Lambertian source, we can write

$$\frac{d^2\phi}{d\Omega} = d\left(\frac{d\phi}{d\Omega}\right) = dI = LdA_{\text{proj}} = LdA \cos u = dI_o \cos u, \qquad (2.13)$$

where I is the radiant intensity and $dI_o = LdA$, which is an element of radiant intensity at $u = 0$. Equation (2.13) can now be reconciled with the model given by equation (2.10). The approximation in equation (2.10) is due to regarding I_o as being roughly equal to the product of radiance with the source's area. According to equation (2.13), the source's area should be an elemental area. But for small sources, integrating over the area in equation (2.13) yields $\int I dA = \int LdA \cos u \approx LA \cos u = I_o \cos u$. Thus, for small Lambertian sources and sufficiently distant screens to receive the illumination, the radiant intensity at the screen is roughly proportional to $\cos u$.

2.2.5 The concept of source spread functions and the irradiance of images

A source spread function (SSF) results from reversing the rays in an optical imaging system such that they travel from a point at the image towards the object plane [7]. Thus, the SSF of an optical system is the PSF in a reversed ray trace, as illustrated in figure 2.4. The significance of this is that the transverse size of a SSF represents the minimum size for an object to be considered as an extended source. This point was highlighted in 1966 by Gilmore [8], although he did not use the term SSF.

Different field positions at the object are associated with different SSFs, just as there can be many point sources from the image side that can be reverse traced towards the object. At the axial position labeled O in figure 2.4, an object whose size is at least equal to the size of its SSF would be an extended source centered at the lens's axis such that, from the perspective of an axial point Q in the image, rays from across the source's surface would fully fill the pupils of the imaging system. Under this condition, if the object is a Lambertian source, then the amount of flux per unit

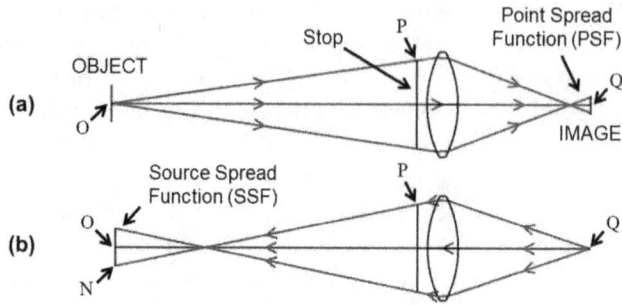

Figure 2.4. The definition of the source spread function (SSF) of an optical system. (a) A forward trace yields the PSF. (b) A reversed trace yields the SSF.

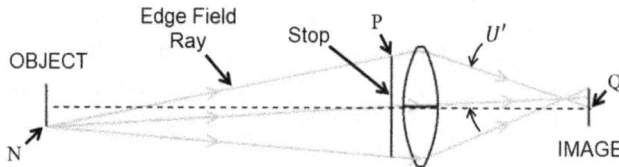

Figure 2.5. The edge field ray yields axial irradiance given by equation 2.14.

area (i.e. the **irradiance**, generally defined as $d\phi/dA$) at the axial point in the image may be determined by applying equation (2.6), in which dividing both sides of the equation by an elemental area dA' in the image yields

$$\frac{d\phi}{dA'} = E = \pi L \sin^2 U',\qquad(2.14)$$

where E is the irradiance, and the angle U' is given by an *apparent marginal ray from the image side* defined by the ray segment NPQ in figure 2.4, which is subtended by the full zone at the exit pupil. Equivalently, this ray is the edge field ray indicated by the segment NPQ in figure 2.5. This result is true regardless of the presence of aberrations [7].

When the size of the object is smaller than the SSF that is associated with the object's position in the field, equation (2.14) no longer holds for the image marginal ray, and the angle U' must be reduced accordingly to some other angle u' that is associated with a different edge field ray (at lower field) that intersects the axis at the image. However, if the optical system is aberration-free near the axis, such as when it is at least aplanatic, then equation (2.14) always holds (neglecting diffraction), where the angle U' is once again subtended by the exit pupil of the imaging system at the full zone. Another way to see this is to apply the Abbe sine condition of equation (1.52) to equation (2.6) for the total flux received by the entrance pupil and the same total flux (by the conservation of flux) emerging from the exit pupil. For example, applying equation (2.6), the total flux received by the entrance pupil is $\pi L dA \sin^2 U$, where U is the marginal ray at the full zone of the entrance pupil. If we let dA' be an element in the image, then the total flux received at the axial point in the image is $\pi L dA' \sin^2 U'$. By flux conservation, we equate these two quantities so that $dA'/dA = \sin^2 u/\sin^2 u'$, which is essentially the square of the Abbe sine condition. Accordingly, due to flux conservation, we can divide $\pi L dA' \sin^2 U'$ by dA' to obtain the irradiance at the axial point in the image, yielding equation (2.14). In section 1.4.4, I mentioned that aplanatism is intimately connected with the radiometric properties of optical systems. This is one example of this connection. Also, near the end of section 1.3.15, I mentioned that the irradiance in the image is proportional to $1/(2F/\#')^2$, where $F/\#'$ is the working f-number defined by equation (1.42). Here, you can see that letting u in equation (1.42) be equal to U' in equation (2.14) provides this connection. Further, if the stop of a lens system is noncircular, the working f-number may be replaced by an *effective f-number* [9].

When the object is an extended Lambertian source, the irradiance profile in the image is given by the relative illumination of a lens system, regardless of the presence of aberrations (i.e. when the object's size is greater than its SSF). Thus, if the image of an extended source can be used for illumination in a practical application, then an optical imaging system is a form of illumination design, and its relative illumination provides the irradiance distribution for the illumination. Irradiance is therefore the primary metric used to judge the quality of uniformity of illumination, and its magnitude describes the 'brightness' impact on the observer (to be precise, the corresponding term in photometry for the irradiance 'felt' by the human eye is *illuminance*, which includes the eye's sensitivity). The higher the irradiance on a plane of illumination, the brighter the illumination appears. More importantly, in applications involving an electronic sensor or any non-human target at the image or plane of illumination, the higher the irradiance, the greater the amount of absorbed or scattered radiation. Here, we have finally given specific definitions for what is meant by *brightness* and *uniformity* in illumination design (still, there is more to say about brightness in the next section).

If no imaging is performed, then the illumination on a screen can possess specific irradiance distributions, depending on the nature of the source. If the source is a small Lambertian emitter, then applying equation (2.1) and re-arranging variables allows us to write:

$$\frac{d^2\phi}{d\Omega} = d\left(\frac{d\phi}{dA}\right)\frac{r^2}{\cos u} = dE\frac{r^2}{\cos u} = L dA_{\text{proj}} = L dA \cos u, \quad (2.15)$$

where r is the dimension shown in figure 2.3. Solving for the elemental irradiance gives

$$dE = \frac{L dA \cos^2 u}{r^2}. \quad (2.16)$$

Finally, since, for an axial distance z between the source and screen, $\cos u = z/r$, substituting this into equation (2.16) yields

$$dE = \frac{L dA \cos^4 u}{z^2}. \quad (2.17)$$

Equation (2.17) is the so-called cosine fourth-power 'law' for a small source relative to a large screen subtending a half-angle u from the source. It is only valid under this condition or when the screen is significantly far from the source, such that $z \gg dA$. In this limit, the elemental area dA of the source may be replaced by its surface area, and $dE \approx E$. As is implied by equation (2.17), the irradiance due to a small Lambertian source at a screen can vary quite significantly when u is large, a situation depicted in figure 2.1(a). This can be said to be the source of all problems in illumination, which stems from étendue, the tendency for radiation to spread. The only exception is when the radiation is a coherent Gaussian beam at the waist, where rays are essentially parallel, yielding practically zero étendue in a loose sense. But there is no known precise correspondence between the radiometry of incoherent beams and coherent beams [3], so care should be taken to make such analogies.

2.2.6 If a source radiates and nobody is there to see it, does it shine?

In the previous section, I mentioned that brightness is measured by a source's irradiance at a screen. However, traditionally, it is *radiance* that is often considered to be *brightness* (or, to put it in photometric terms, *luminance*). Yet, a different perspective can be provided, which we now discuss. First, there is the philosophical question of whether or not a source can be said to be bright if no detector is present to receive the radiation. Radiance, on its own, is simply the amount of flux *emitted* per unit projected source area, per unit solid angle in the ray direction. One needs a sensor to receive this beam. If there is a single source, then the sensor receives equal radiance from the source no matter which direction it points towards, with an exception when its surface is parallel to the rays from the source. This is a consequence of the radiance theorem mentioned in the previous section. So, radiance is independent of direction, and it is also independent of distance. Thus, in this sense, radiance is a rather robust quantity and appears to be a suitable measure of the radiation *output*. But is radiation output the same as 'brightness'? Hold that thought.

If there are multiple sources and a single sensor, then there is no way this sensor can discern between the sources present, unless the sensor is fitted into a box with an aperture, like a pinhole, and the sensor now comprises pixels, such as those of a charge-coupled device (CCD) or a complementary metal–oxide–semiconductor (CMOS) image sensor. If the image sensor is highly sensitive, then a pinhole may allow sufficient flux to pass through from each source. If the pinhole images of the sources do not overlap at the image sensor, then each source is distinguishable. The radiance is invariant through the pinhole, so each source's radiance may be measured. Therefore, the 'brightness' of each source can be quantified. And yet, if the sensor is not sufficiently sensitive, then the pinhole must open up, resulting in overlapping projected sources at the image sensor. But if a lens is inserted into the opened pinhole, then the images are well focused and separated, thereby yielding sufficiently high irradiance at the image sensor. The higher the radiance from a source, the higher the irradiance of its image.

Let us now consider different boxes of image sensors with different lens sizes. According to equation (2.14), the axial irradiance in the image at the image sensor is proportional not just to radiance, but also to the angle U' subtended by the lens's semi-diameter (to be precise, its exit pupil). This means that smaller lens pupils yield smaller irradiance, while larger lens pupils yield larger irradiance. It can be argued that since this does not change the radiance (assuming lossless transmission through lenses, or assuming that lens transmittance τ has been included through some means of calibration), then radiance is still a measure of a source's brightness. This can lead to some confusion. When I was in graduate school, a teaching assistant told me that, due to the above argument, a source's 'brightness' is invariant with distance, which got me lost in deep thought. This would mean that if a distant star possessed the same radiance as our Sun, then that distant star should burn my eye as much as the Sun does when I look at it. However, I then reasoned that the irradiance in my eye from the star should be lower than that from the Sun. But equation (2.14) does not describe this reduction. Do you see it? There is no variable in equation (2.14) that accounts for source distance, and it is not entirely wrong. If you work out the geometry of rays from source to image in a lens system, when a source's distance is increased, although its flux through the entrance pupil is reduced, its image size is proportionately reduced, yielding a constant image irradiance. Thus, it would seem that not only is radiance invariant with distance, but so is the irradiance in the image! So, now what to do?

The solution to the above apparent paradox is to account for the wave character of image formation, which involves diffraction and, in general, aberrations [7, 10]. In particular, if the PSF is shift-invariant, as you know, the image is the convolution of the PSF with the expected ideal geometric image. Under this condition, the irradiance distribution in the image may be expressed as

$$E(x, y) = \frac{E}{V} \int_{-\infty}^{+\infty} \int_{-\infty}^{+\infty} P(x - x', y - y')G(x', y')\mathrm{d}x'\mathrm{d}y', \qquad (2.18)$$

where E is given by equation (2.14), $P(x - x', y - y')$ denotes the PSF of the system being shifted, V is the volume of the PSF (i.e. it is the integral of P across an infinite plane), and $G(x', y')$ is the normalized geometric image profile that defines the boundaries for integration over the 'dummy' space variables x' and y' [7, 10]. That is, $G(x', y') = 1$ everywhere in the bounds of integration, and $G(x', y') = 0$ outside. Note that $E(x, y)$ is not just a relative irradiance distribution but the absolute magnitude of the irradiance. For instance, in the limit in which the PSF's size (defined, perhaps, by its full width at half maximum (FWHM)) is significantly smaller than the bounds defined by $G(x', y')$, then the integral in equation (2.18) is equal to the volume of the PSF, effectively canceling V in the denominator and yielding equation (2.14). This is the limiting condition that is observed at the image plane when imaging extended sources. In this condition, it is indeed the case that the image irradiance is invariant with source distance, but up to a point. Remember that the geometric image reduces when the source's distance is increased. Beyond a certain source distance, there is the other limiting condition that the geometric image $G(x', y')$ reduces until its size becomes comparable with the PSF's size. As the geometric image size reduces even further, the PSF's size dominates in equation (2.18), resulting in a negligible effect of $P(x', y')$ in the integral. In this condition, the

PSF may be taken out of the integral, so that $\int_{-\infty}^{+\infty}\int_{-\infty}^{+\infty} G(x', y')\mathrm{d}x'\mathrm{d}y' = A'$, the area of the image. Multiplying this area by E, the irradiance, yields flux. This is the flux in the image, which is also the flux ϕ through the entrance pupil of the imaging system. Equation (2.18) then becomes

$$E(x, y) = \frac{\phi}{V}P(x, y). \tag{2.19}$$

Equation (2.19) describes what a distant star would look like in your eye, or on the image sensor whose lens is imaging that star. When no aberrations are present, by cross-referencing equation (2.19) with a derivation by Born and Wolf for the far-field diffraction PSF distribution of a distant quasi-monochromatic point source being imaged by a lens [1], it is revealed that the quantity V in equation (2.19) is given by $(\lambda^2 f^2)/(\pi r^2)$, where f is the focal length of the lens, and λ is the central wavelength of the source. Thus, for distant objects (or small objects), there is a distribution of image irradiance that would not be given by equation (2.14). Rather, it is given by equation (2.19), and it tells us that, as the object gets farther away, its flux ϕ through the entrance pupil is lesser, thereby reducing its image irradiance. For such conditions, nobody would say that this object's 'brightness' does not change with distance, nor would any individual say that placing a light bulb significantly far away would make it just as bright as it would appear when it is next to that individual. Since the irradiance in equation (2.18) is consistent with the classical radiometric formula of equation (2.14) as well as the diffraction PSF formula of equation (2.19), it can be said that the descriptor for the brightness of a source (assuming it is being observed either by a human or a sensor) is irradiance, not radiance. However, I do have an alternative descriptor for radiance: *potential brightness*. Clearly, the higher the radiance of a source, the higher is the expected irradiance of its image in any sensor.

How is this all relevant to illumination design? Equation (2.18) is consistent with the concept of SSFs [7]. For image-forming illumination systems, in which the size of the object is given by a superposition of SSFs across the field, the image's irradiance profile is given by the imaging system's relative illumination plot, and the magnitude of the irradiance may be scaled accordingly by E in equation (2.14) at the axis. Of course, an NSC ray trace can eliminate the need to perform such calculations, but it is still instructive to know the physics behind the ray tracing. At the opposite limiting condition, in which the object's size is significantly less than its SSF, it is not an extended source, so its image's irradiance distribution is nearly like a PSF. In this case, the system's relative illumination plot is no longer a suitable descriptor for the illumination in the image, but the PSF itself—which has a light distribution across the image plane—can be said to be an illumination distribution at that plane. In fact, it is possible to engineer the PSF to have a very flat irradiance profile, which is what we want in many cases of illumination. In this book, this method of illumination is called *point spread function illumination*, and it is discussed in section 2.18.

2.2.7 Why is the full width at half maximum often the width of a distribution?

When the illumination system is an imaging system, the width of the illumination is governed by the width of the image. If the image is sharp, then the boundary of the illumination is distinct. But illumination distributions do not always require sharp boundaries (think of a cheap handheld LED flashlight, or the gradual roll-off of the relative illumination in a lens system). Thus, the images of sources used for illumination purposes can be or appear grossly defocused as long as the region of interest (ROI) lies within a sufficiently bright and uniform area in the illumination. However, even defocused (or highly aberrated) images may still require some definition of the acceptable width of the illumination.

In the previous section, we talked at length about convolution in imaging. I mentioned briefly that the width of an irradiance distribution may be given by its FWHM. You may have often also seen this in the course of your work or college courses. Of course, you can define the width of a distribution by a full dimension that does not necessarily correspond to the half point of the maximum irradiance. But it is rather interesting to note that if the PSF of an imaging system is rotationally symmetric, and if its size is significantly smaller than the area of convolution, then it turns out that the FWHM of the image resulting from the convolution is very nearly the actual width of the area of convolution. Since the area of convolution is the area of the ideal geometric image, this means that under the condition that the (rotationally symmetric) PSF width is much less than the width of the expected sharp image, the resulting convolution of the PSF with the image yields a width that is essentially the width of the ideal image.

To see how this comes about, note that the convolution of the PSF with the ideal image is proportional to the volume of the PSF. This is implied by equation (2.18). The convolution operation is a computation of the volume of the PSF bounded by $G(x', y')$, and dividing the result by V yields a fraction of the PSF volume. When $G(x', y')$ spans an infinite surface area, then everywhere in that area, the irradiance is multiplied by $V/V = 1$, a full volume of the PSF divided by that volume. Thus, everywhere in that area receives the maximum irradiance. It follows that locations measuring half of that maximum irradiance must be receiving half the volume of the PSF. For a rotationally symmetric PSF distribution, the convolution yields half of the volume of the PSF when the peak of the PSF is at a boundary between $G(x', y') = 1$ and $G(x', y') = 0$, because this condition divides the PSF into two halves. Thus, the boundary (full width) of an ideal image $G(x', y')$ that spans an area significantly larger than that of the PSF is defined by half the volume of the PSF, which is proportional to half the maximum irradiance. This condition does not apply under more general conditions, such as when $G(x', y')$ has curves and shapes with localized surface areas that may be comparable to the size of the PSF. And for those conditions, we would just need to redefine what is meant by a width.

2.2.8 Is the image of a Lambertian source a Lambertian source?

Due to the invariance of radiance over index (and therefore, radiance itself when source and illumination spaces are in air), there can be a tendency to say that the image of a Lambertian source has Lambertian emission (because the radiance in the image is the same as at the source). But this is rather inaccurate. The radiance in a ray pencil at and beyond the image (supposing we have an intermediate image) is the same as that of the source, but the image generally does not possess Lambertian emission, especially when there is high aberration near its edges. However, when an optical imaging system has high image quality and is *telecentric* in image space (i.e. the principal ray is parallel to the optic axis at the image), then the ray 'emission' from the image can appear to have Lambertian character in the sense that when the image plane is shifted sufficiently far to the right (assuming left-to-right ray travel), then the irradiance profile within the bounds of all rays is describable to a reasonable extent by equation (2.17). This is good to know, as it could be used as another configuration for illumination.

2.2.9 Is chromatic aberration important in illumination?

Yes and no. If imaging is the means by which we illuminate, and if it is intended that the image possess structural properties across the object (think of a projector in a classroom or theater), then chromatic aberration is just as important as it should be in lens design [11]. However, if the source is white and uniform in irradiance, then any presence of chromatic aberration would only be observable at the boundary of the illumination, where perhaps lateral color may show up. In all other areas within the ROI of the illumination, colors mix together and are therefore *self-homogenized*.

In some other applications, it could be desirable to perform *color mixing*, such as when an optical system requires white light projection from red, green, and blue color sources. Such mixing can be performed in several ways, such as using beam splitters as beam combiners, using the Köhler configuration [12], and mixing elements such as light pipes and lenslet arrays [13] (see, for example, section 2.5).

2.2.10 The radiometry of LEDs and the use of source files in nonsequential ray tracing

Many of the concepts that have been presented for radiometry essentials are applicable to most types of incoherent sources, such as halogen lamps, arc sources, and LEDs. But since LEDs have become rather dominant in many modern applications (and because we simply do not have space in this book to discuss every type of source), the current section will focus on providing a basic overview of the radiometric characteristics of LEDs so that you know what optical specifications to look out for in a LED datasheet. The idea is to pick out the most significant information necessary for the purpose of designing an illumination system that uses LEDs as sources.

In many practical cases and as far as nominal optical properties are concerned, the first range of LED properties of interest in an LED's datasheet are: (1) Its total radiometric flux (and available *flux bins*), (2) its spectrum (and available *wavelength bins*), (3) its typical radiant intensity, (4) emitter dimensions, and (5) the presence of any structural properties across the surface/s of the emitter/s (sometimes, there is more than one emitter on an LED surface). By 'emitter,' I mean the physical area that is emitting the rays. These five properties enable you to model the LED's output in an optical design program. Of course, there are other additional important properties related to variation in the LED's output, such as the variation of spectrum with current and the variation of flux with junction temperature (just think of the junction as somewhere inside the emitter, where all of those electron–hole pairs are generated, giving rise to emission). There is also the LED's long-term behavior, such as the variation of flux and spectrum over time. Of course, you do need to examine these, but without the first five properties listed above, optical design would not even be possible. So, we will focus on those.

The first is radiometric flux, which is often problematic, because most LED datasheets provide this in the form of the photometric term, *luminous flux*, in units of lumens (lm). The relation between luminous flux ϕ_ν and radiometric flux ϕ is

$$\phi_\nu = 683 \int_{-\infty}^{+\infty} \phi(\lambda)V(\lambda)\mathrm{d}\lambda, \tag{2.20}$$

where $V(\lambda)$ is the normalized *spectral luminous efficacy* (SLE, whose peak is 1 at 555 nm) and the factor 683 is associated with the SLE such that 683 times $V(\lambda)$ equals the magnitude of the SLE, in units of lm W^{-1} [6]. Sometimes, $\phi(\lambda)$ is called *the spectral radiometric flux* of the source. In this book, we will just call it spectral flux. The standard SLE curve is also known as the bright-adapted *photopic* response of the human eye [6]. Sometimes, the 'E' in SLE is interchanged with 'efficiency,' while at other times, the term *spectral luminous efficiency* specifically refers to the dark-adapted *scotopic* response of the human eye [6]. You just need to verify which curve is meant by those you communicate with.

If an LED's datasheet specifies luminous flux, and one needs the radiometric flux value in an application (as I often do), then one needs to extract the LED's spectral flux from its specified luminous flux. To do this, one first needs to obtain a standard curve for $V(\lambda)$. Although the actual standard must be purchased from the website of the International Commission on Illumination (CIE), many resources provide this curve (sometimes, it is even provided in the LED's datasheet), including textbooks and guides on radiometry, such as those by Boyd [6], Bunch [14], McCluney [15], Grant [16], and Wolfe [17]. Each text is an excellent resource, and you must include at least one (if not all) on your desk. The normalized version of the LED's spectrum $\phi(\lambda)$ is usually provided, in which the magnitude of the spectrum is unitless (i.e. it is a curve with a relative amplitude on the vertical axis with some peak value, usually one or 100). The trick to determining the total flux is as follows. First, let ϕ_o be the total radiometric flux, which is to be determined. Let $S(\lambda)$ be the spectral flux given in the datasheet in arbitrary units. Then, express the spectral flux as

$$\phi(\lambda) = \phi_o \frac{S(\lambda)}{\int_{-\infty}^{+\infty} S(\lambda)\mathrm{d}\lambda}. \tag{2.21}$$

In equation (2.21), $S(\lambda)$ may be plotted using the LED's datasheet, and $\int_{-\infty}^{+\infty} S(\lambda)\mathrm{d}\lambda$ may be computed in a spreadsheet or some math program. Equations (2.20) and

(2.21) may then be used together. The idea is to make iterative adjustments to the value of ϕ_o until ϕ_v equals the luminous flux value provided in the datasheet, thereby revealing that the final value for ϕ_o is the radiometric flux of the LED. Thereafter, $\phi(\lambda)$ possesses the actual magnitude of the LED's spectral flux in units of W per unit wavelength (did you notice that equation (2.21) is synonymous with equation (1.10)?). You now have flux and spectral flux. But you must be careful about which *bin* you are drawing your flux and spectra from. Binning refers to the manufacturer's way of sorting LEDs into different categories of performance. It is like taking an LED of a specific flux and spectrum and placing it into a bucket (a bin) at the factory (they probably really do have buckets/bins for this). This is the only way for them to be profitable, because there is no realistic way for manufacturers to control the flux and spectra so tightly that only a single flux value and spectrum is produced.

The next properties to consider for optical modeling are the specified radiant intensity and the dimensions of the emitter. Sometimes, datasheets just use the term 'intensity' (or *angular intensity*) without the term 'radiant.' Intensity plots are often normalized to some value (it need not be one, though it usually is), and they are displayed as either a polar 'directivity' plot (figure 2.6(a)) or a regular intensity vs angle plot (figure 2.6(b)). In these example plots (which were produced in OS's NSC mode using 20×10^6 rays) the LED is assumed to have Lambertian emission, and the emitter's size is 1 mm × 1 mm. Further, a *polar detector* object is used and it is located 1000 mm from the LED (figure 2.7). The reason for this distance is to position the detector as far as possible from the LED. It does not actually need to be this distant, because according to the CIE 127:2007 standard, intensity may be

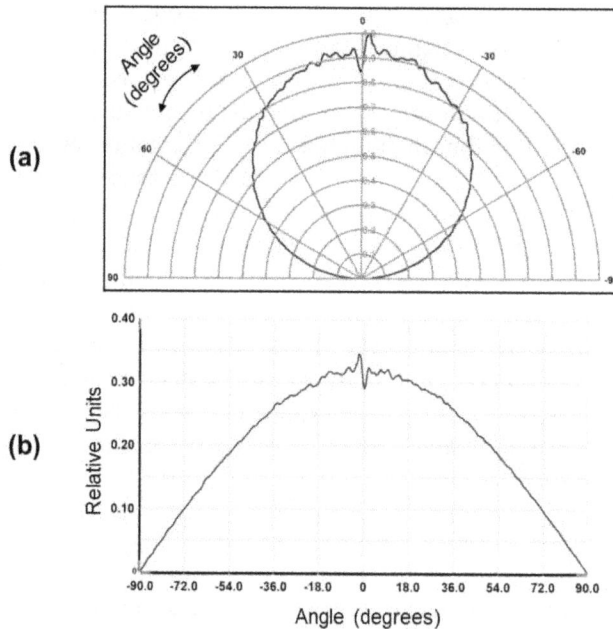

(a)

(b)

Figure 2.6. Typical LED intensity plot formats. (a) A polar plot. (b) An angular plot.

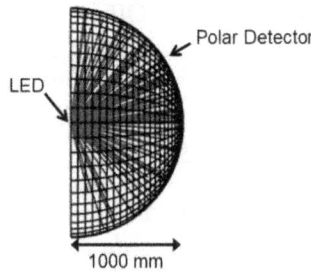

Figure 2.7. The polar detector in OS NSC layout used to produce the plots in figure 2.6.

measured at 316 mm from the LED [18]. Still, I tend to use polar intensity plots far away because of the approximation that is assumed in equation (2.10). Note also that the CIE 127:2007 standard uses the term *averaged LED intensity*, which means that the sensor used for measurement, being 1 cm^2 in area, averages the flux it receives in the solid angle.

The flux per unit solid angle in equation (2.10) can only describe a very small point-like source. However, the CIE 127:2007 standard also has a closer measurement distance condition of 100 mm, and as far as I can tell, LED datasheets do not necessarily specify which distance is used. Therefore, if your application requires strict adherence to certain requirements, you must contact the LED manufacturer to gain clarity. Otherwise, LED modeling is often a combination of approximate modeling and iteration based on empirical tests (which is actually not a bad approach, as nothing beats a real experiment).

Sometimes, neither ϕ_ν nor ϕ_o is provided by the LED manufacturer, but the *peak intensity* is provided, which is the intensity at the 0° angle in figure 2.6. However, the peak intensity may be given either in radiometric flux per unit sr, or in photometric terms, which is the *luminous intensity*, with units of candles per steradian (cd sr^{-1}). At other times, either ϕ_ν or ϕ_o is provided, and the intensity data is normalized. Either way, all of the necessary formulas and techniques have been provided for you to make the appropriate connections and arrive at the proper units of interest. For instance, suppose the datasheet's intensity plot shows that the point at which intensity drops to half of its peak is roughly 60°, and the radiation falls off monotonically at angles up to 90°, as in the example shown in figure 2.6. This indicates that the LED is near-Lambertian, because equation (2.13) tells us that, for small Lambertian sources, the intensity is proportional to $\cos u$. At $u = 60°$, this cosine factor is 0.5. As such, according to equation (2.6), for a small Lambertian emitter, the total emitted flux over a full 90° half-angle is roughly $\pi L \mathrm{d}A$, where $\mathrm{d}A$ is the emitter's surface area. Thus, if the emitted flux is given in radiometric units, then $\phi_o \approx \pi L \mathrm{d}A$. If it is in photometric units, then $\phi_\nu \approx \pi L_\nu \mathrm{d}A$, where L_ν is the photometric radiance called *luminance*. In these two formulas, the peak intensity is given by $I_o \approx L \mathrm{d}A$ if the intensity is radiometric or by $I_{o\nu} \approx L_\nu \mathrm{d}A$ if it is photometric. On the other hand, if the intensity profile is non-Lambertian, then you may attempt to apply the cosine-powered factor $(\cos u)^m$ in equation (2.10) to fit the curve on the datasheet, followed by using equation (2.12) to obtain the flux. In these formulas, again, I_o relates to the radiometric case. Otherwise, it is replaced by I_{ov} for the photometric case.

The last of the five properties to consider mentioned earlier was the presence of any structural characteristics on the LED emitter's surface. For example, are there any wires running across the emitter? Are there any visible opaque structures on the emitter? Is the emitter inside a dome lens? The presence of wires and opaque structures indicates to you that you cannot to use imaging as a means to illuminate a screen, as this would let those structures appear and therefore not result in a uniform distribution at the screen. The presence of a dome lens (such as the Bravais lens illustrated in figure 1.139(a)) indicates that magnification is present, so when you model the emitter as an NSC source object in OS, you must give the object a larger size in accordance with the magnification. Of course, you may need to assume a value for the index of the dome lens. This is guesswork (or you can contact the supplier). Given the uncertainties and variability involved in LED flux bins, current, junction temperature, and spectra, the uncertainty in the guess of your guesswork in LED modeling is usually inconsequential. The practical thing to do is to allow for empirical studies and proofs of concept in the development of your illumination system when you plan your project with your team members.

Quite often, there are so-called *source files* provided by the LED manufacturer in various formats appropriate for an optical design program. They contain actual source emission properties measured using a device called a goniophotometer [19], and the measured properties are in the form of data that can be read and used by the optical design program. Thus, NSC rays traced using these source files are emitted according to measured radiant intensities, yielding fairly accurate models for those rays. In OS, several file types are supported, including DAT files (with the .DAT file extension) and SDS files (with the .SDS file extension). Recently, *Source IESNA* files (with the .IES file extension) have also been supported, in accordance with standards provided by the Illumination Engineering Society (IES). However, the reader ought to be cautioned that such files represent measurements performed on a single sample and with a finite number of traceable rays (sometimes as few as only a million) [19]. Further, when placing the NSC IESNA source object at a specific location in the NSC Editor, it is not always clear what part of the LED that position refers to. (For example, is it the emitter's physical surface? Is it the dome lens of the LED?) In situations in which the source is to be located at the focus of a lens or mirror, this ambiguity can be problematic. Some good guidance has been provided by Muschaweck [20], and the reader is encouraged to consult this paper and also open up conversations with the supplier of these source files. We will use an LED source file in an example in section 2.13.

Before closing this section, some points should be highlighted concerning the *near-field problem in LED illumination*. Recall from above that the intensity plots provided in LED datasheets may be assumed to be made at a distance of 100 mm or 316 mm from the LED. These are not exactly long distances, but the described measurement process appears to some extent to be indicative of the far-field assumption, as the LED drawing example shows a small source. On the other hand, the term *averaged intensity* may be somewhat indicative of accounting for near-field effects, because the detector's size is 1 cm^2 (quite big, relative to the 100 and 316 mm distances), and the averaging effect can be considered to involve the source's size as well. However, no guidance is provided on what is considered a maximum LED size. Finally, goniophotometer measurements are sometimes performed in the far field and, more recently, performed in the near field (with provisions for also obtaining far-field results) [21, 22], but LED manufacturers may not state which method is used or which LED distances are considered valid when using the source data files generated from these measurements. Researchers and engineers continue to work towards overcoming certain challenges in near-field goniophotometric measurements of LEDs [21–23], and techniques are being developed for modeling near-field radiation patterns from LEDs [24]. Meanwhile, optical designers need to sometimes rely on source files, and in some optical systems, the elements used for collecting and directing the rays from LEDs are rather close to the LEDs. However, from my experience, there has not yet been significant concern here. I have modeled LED emitter sizes with diameters (or lengths) from ~1 mm to ~3 mm and larger mainly through the use of radiant intensity models based on either Lambertian emission (when the LED datasheet indicates that the LED is near-Lambertian) or non-Lambertian emission described by equation (2.10); the distance from the LED to the first optical element (usually a lens) can be as short as 10 mm. Yet the modeled irradiance distributions at the final screen or ROI are often close to the actual measurements done in a laboratory during proof-of-concept testing. If or when there are slight variations from the model, then slight adjustments are made to the prototype, and the measured data becomes feedback to the model for improvement.

2.2.11 There is no free étendue

At the beginning of this chapter, I highlighted that the problem in illumination consists of shining light onto a screen to achieve a desired brightness (irradiance) level and uniformity. You may wonder why I did not include the *efficiency* of the illumination system, or the efficiency of the *luminaire* (a luminaire is an optical system comprising a source and some optical element that shines light onto a surface). Ideally, one would wish to have an efficient flux collector in a luminaire such that if, for instance, 1 W of electrical power is fed into the luminaire, then 1 W of that flux should be converted into light, yielding 100% efficiency. Indeed, this would be wonderful. However, have you heard of the so-called *project management triangle* [25], or the *triple constraints in project management* [26]? It concerns the balancing of three constraints: time, quality, and cost. But quite often, you may hear engineers and managers express that they may in fact be not wholly balanceable, resulting in only satisfying two of the three constraints. Analogously, in illumination optical design, we have three constraints: irradiance, uniformity, and efficiency. In my career, which has included developing luminaires for scientific instrumentation (such as real-time PCR and microscopy, which require high irradiance and uniformity), I have often prioritized irradiance and uniformity over and above the efficiency of the luminaire.

Let me explain why. Each time I attempt to maximize efficiency with uniformity, I have to 'fight' with étendue. And each time that happens, étendue wins. Étendue is as fundamental as is flux, and it is costly. It is because of étendue—the incessant need for light to spread—that the illumination on a screen is nonuniform (e.g. think of equation (2.17)), unless optical elements impose a burden on a spreading beam to shape its form. But these elements are often only successful at shaping uniformity when not all of the flux from the beam is used. This is the key point in illumination. There is always a trade-off between irradiance and uniformity, and the thing standing in between them is efficiency. Certain types of luminaires called *hybrid optics* or *pseudo-collimators* are very efficient; they achieve a flux collection efficiency of more than 80% and transfer it all onto a screen [27]. But quite often, the center of the ROI has higher irradiance than the edge. This condition is not really efficient, because photons are wasted at the center, while the edge is always begging for more. In many applications involving the use of scientific/analytical instruments that require illumination, you have to let efficiency go if you want high uniformity. If you want higher flux, you need to spend more money on the extra energy, for example by paying for high-powered LEDs and good thermal management hardware. This is not the same as home lighting, in which nobody pays much attention to uniformity.

It is only when a source has *zero étendue* that its rays can easily be shaped to have complete uniformity. What is a zero étendue source? It does not really exist, but certain sources get close (physics allows mathematical constructs to be reached by way of a limiting condition). Small incoherent sources are examples of near-zero étendue sources but at the expense of having low flux. In the limit of zero size, therefore, they would have zero flux. In classical radiometry, there can be no idealized point sources, for otherwise, they would have infinite flux in an infinitesimal area (zero), according to equation (2.1). Accordingly, Necodemus (whose name appears in every published text on radiometry due to his many contributions in this field) has stated, 'Thus, here and in actuality, there is no such thing as a point source, with all of the power traveling along rays which intersect at a single mathematical point' [28]. Two idealized types of zero-étendue sources are illustrated in figure 2.8. Neither can be said to truly exist, but the one on the right—parallel rays—is very nearly closely approximated by the Gaussian laser beam at its *waist*. But such beams cause other problems, such as speckle interference. Furthermore, when a Gaussian laser beam is made to spread and is shaped into a highly uniform irradiance profile (which we will discuss in an example in section 2.18), then its irradiance drops significantly. Much of the high irradiance of a laser's beam is the result of

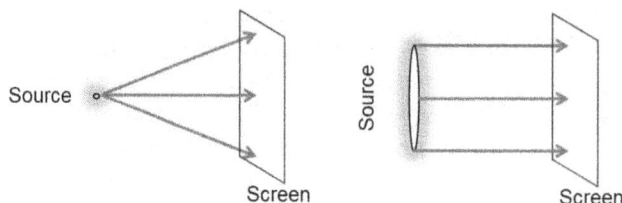

Figure 2.8. Two idealized zero-étendue sources. Left: zero size. Right: zero angle.

concentrating flux within a small area. Once that area is enlarged, you lose that irradiance. You can regain some irradiance by purchasing a high-powered laser, but that comes with an increase in cost. Thus, there is no free étendue.

2.3 The concept of ray density in illumination design

Figure 2.9 shows three examples of ray patterns at the entrance pupil produced by an idealized point source in OS's SC mode. In this example, let us consider the entrance pupil to be a screen. The first (figure 2.9(a)) is what SC ray tracing does by default, in which rays strike the pupil with uniform transverse spacing. In figure 2.9(b), a *cosine-cubed* apodization is applied to the pupil, in which OS models the object as an idealized point source with irradiance at the pupil that varies as the cube of the cosine of the half-angle u with respect to the optic axis. An idealized point source has this characteristic irradiance at a plane because it is considered to be a source with *isotropic* emission in that its radiant intensity is constant with angle u. So, for such a source at an axial distance z from a plane, $I_o = $ constant $= d\phi/d\Omega = [d\phi/(dA \cos u)]r^2 = Er^2/\cos u$, where r is, as usual, a ray connecting the source to a point on the vertical at the plane of illumination (such as in figure 2.3). But $\cos u = z/r^2$, so that $I_o = Ez^2/\cos^3 u$. Thus, the irradiance is $E = I_o \cos^3 u/z^2$, which is the cosine-cubed dependence of the irradiance with u. In figure 2.9(c), a Gaussian (bell-curve) apodization is applied to the pupil, which means that the irradiance profile at the entrance pupil is given a Gaussian transmittance function. The amount of spread in this Gaussian function depends on the *apodization factor* given to it in the program (go to System Explorer > Aperture > Apodization Type).

In illumination design, you can often get a feel for the uniformity of irradiance at a screen by simply observing how many rays strike an area on the screen. Simply put, the number of rays per unit area (i.e. the **ray density**) on a plane of illumination is proportional to the irradiance. For instance, notice in figures 2.9(b) and (c) that the axial area on the entrance pupil (our 'screen' in this example) has a greater concentration of rays than the edge. Compare these to figure 2.9(a), where the rays are equally spaced across the screen. The relative irradiance distribution at the

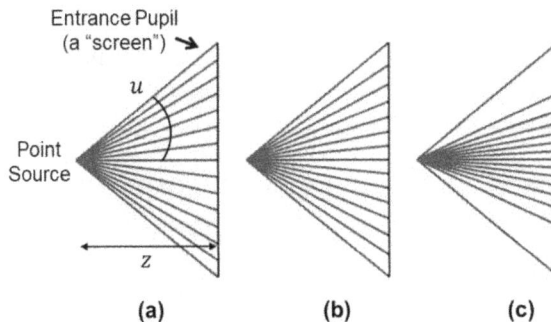

Figure 2.9. Three systems with different ray densities on a 'screen.' (a) Uniform ray density. (b) Cosine-cubed ray density. (c) Gaussian ray density.

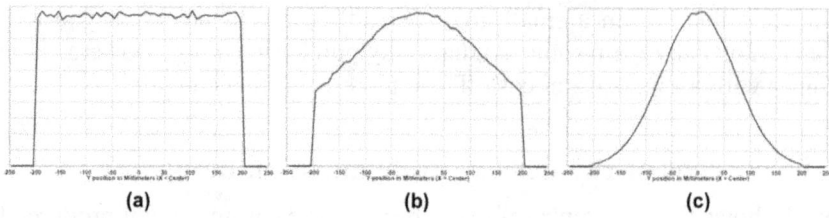

Figure 2.10. The irradiance at the respective screens of figure 2.9. (a) Uniform ray density. (b) Cosine-cubed ray density. (c) Gaussian ray density.

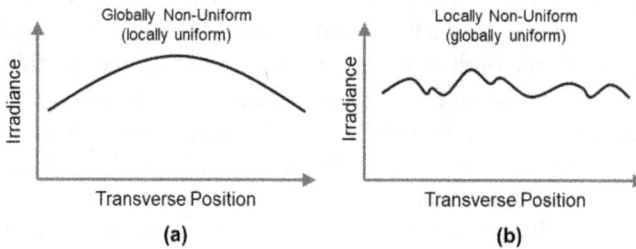

Figure 2.11. Two types of (non)uniformities in irradiance distributions. (a) Globally nonuniform but locally uniform. (b) Locally nonuniform but globally uniform.

entrance pupil for each of these conditions can be analyzed using the GIA feature (see section 1.2.10 for a review of this analysis feature), as displayed in figure 2.10.

The fact that ray density is proportional to irradiance means that when you perform illumination analysis in NSC mode, any presence of consistently equal ray densities at various locations across an illuminated plane implies uniform irradiance across the plane. By 'consistent,' it is meant that when you perform a NSC ray trace, due to the statistical way in which rays are emitted by the source, there may be no obvious indication that equal ray densities are present across the plane of illumination even if the density is actually uniform. You need to trace many rays (usually millions) and do it several times. If, at each ray trace, you see similar (consistent) ray densities at sampled locations across the illuminated plane, then chances are that there is uniform illumination of that plane. If, on the other hand, you are analyzing your design in SC mode, then what is needed is proper entrance pupil ray patterns, which may be achieved through apodization, as highlighted here and in section 2.2.1.

2.4 The concepts of global and local uniformity

Figure 2.11 illustrates what is meant in this book by global and local uniformity. Global uniformity refers to the *envelope* of the illumination (think of this as the average profile across an irradiance distribution), while local uniformity is the *modulation* in the envelope. Thus, a distribution can be globally nonuniform but have no localized variations (figure 2.11(a)). In contrast, when a distribution has 'hot spots' and 'cold spots' (localized regions of high and low irradiance) but is otherwise 'flat' on average, then it can be said to possess local nonuniformities and a globally uniform profile (figure 2.11(b)). Think of the image of a tungsten light bulb as an example of a locally nonuniform profile. On the other hand, think of a lens system's relative illumination plot with a gradual fall in irradiance at the edge as an example of a globally nonuniform profile.

2.5 The concepts of étendue division and superposition

In many cases of illumination design, it is useful to think of a total étendue as being a *global étendue* comprised of many different *local sub-étendues*. For instance, a cone of rays from a source can be said to possess global étendue that spreads and become globally nonuniform at a screen, but dividing it into pieces of the beam (**étendue division**) and letting each sub-étendue spread and overlap the others (**superposition**) can improve the uniformity, as illustrated in the layout for a simple lenslet array system in figure 2.12. In this rather simplified example, a globally nonuniform collimated beam from an idealized point source is divided by a lenslet array into individual beams, each with a sub-étendue. Lenslet arrays are also called *fly's eye arrays* (inspired by the eyes of flies). Think of a fly's eye array as an étendue divider. When a large lens is mounted at the back focal plane of the fly's eye array, it introduces a *global focal plane* at which superimposition of the divided sub-étendues produces a globally uniform distribution. In theory, the same configuration may also be used for color mixing (figure 2.13). The use of small extended sources in fly's eye array systems is discussed in section 2.15.

Étendue division and superposition are also responsible for homogenizing local nonuniformities across the source in the Köhler configuration. In this case, any absence of light in the dark areas (which can be considered the absence—and

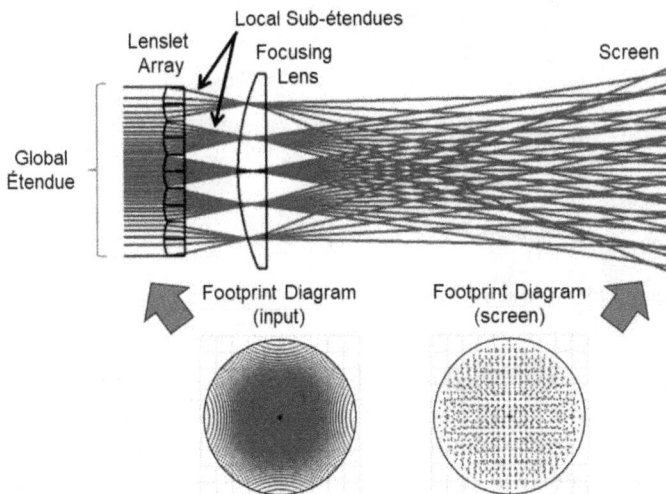

Figure 2.12. An example of étendue division using a lens array and superposition by a focusing lens to homogenize a globally nonuniform input beam.

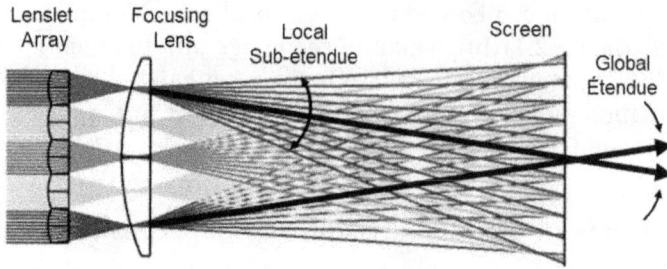

Figure 2.13. The use of the lens array from figure 2.12 for color mixing.

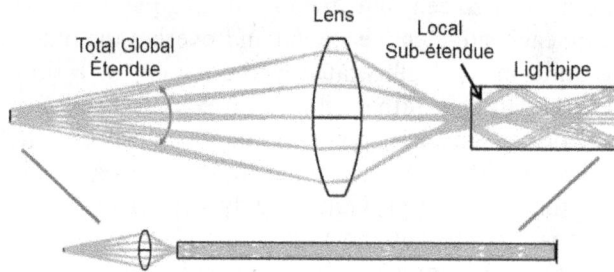

Figure 2.14. An example of 'global' étendue being broken up by the light pipe of figure 1.96 into 'local' sub-étendues, which combine (by superposition) to homogenize at the end.

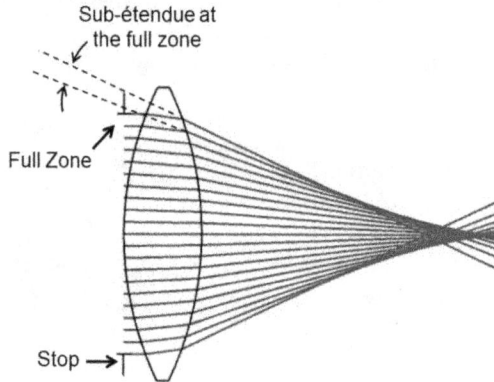

Figure 2.15. An example of étendue being broken up by spherical aberration.

therefore division—of sub-étendue) of the source is compensated for by an overlap (superposition) of rays in those areas at the screen. Sometimes, I refer to étendue division as breaking up the étendue. Light pipes break up global étendue into sub-étendues that homogenize at the exit (figure 2.14). Optical aberrations can also be said to influence étendue by breaking it up into sub-étendues. For example, figure 2.15 shows how spherical aberration divides a globally low-étendue beam

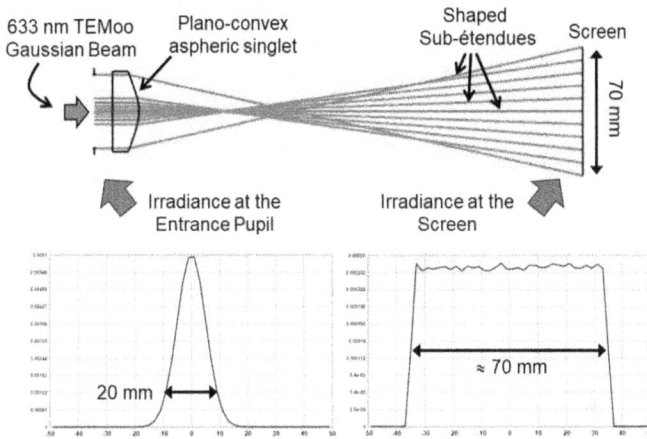

Figure 2.16. An example of dividing a collimated Gaussian laser beam into sub-étendues to shape the full beam and flatten its global uniformity.

from a distant source into sub-étendues, their sum being equal to the total étendue. Equivalently, the localized surface powers of the lens re-shape incident local sub-étendues into different sub-étendues. Each zone in the entrance pupil is associated with its own local sub-étendue and with a unique focal length. For a Lambertian source, due to flux conservation, the total global étendue is constant, so it must be divided into sub-étendues that sum to the total étendue. Alternatively, if the incident beam is an idealized zero-étendue collimated beam (with infinite radiance), then each sub-étendue has zero étendue (and infinite radiance), since the sub-étendues have zero areas at their local focal points. In all cases, flux is always conserved.

Étendue division is also the mechanism by which a collimated Gaussian laser beam can be shaped by an aspheric element to have improved global uniformity. In this case, the aspheric surface of the element introduces a high degree of spherical aberration. By the same principle as that illustrated in figure 2.15, the spherical aberration re-shapes local sub-étendues into new and different sub-étendues that are projected with new local surface areas at the screen. A hint of the physics of this mechanism is illustrated in figure 2.16, where a quick optimization was performed. Examples of such designs are provided in section 2.18.

2.6 'First-order' illumination design

2.6.1 Illumination using paraxial thin lens models

Your first encounter with an illumination system using paraxial thin lenses was in figure 1.137 of section 1.3.10. There, **appropriate marginal and chief ray solves** were used to form a Köhler configuration from two paraxial thin lenses, in which the source was projected by the first lens onto the second lens, and the second lens projected the condenser's surface onto a screen. Figure 2.17 shows the irradiance on the screen for the system in figure 1.137, using the GIA feature in OS's SC mode at

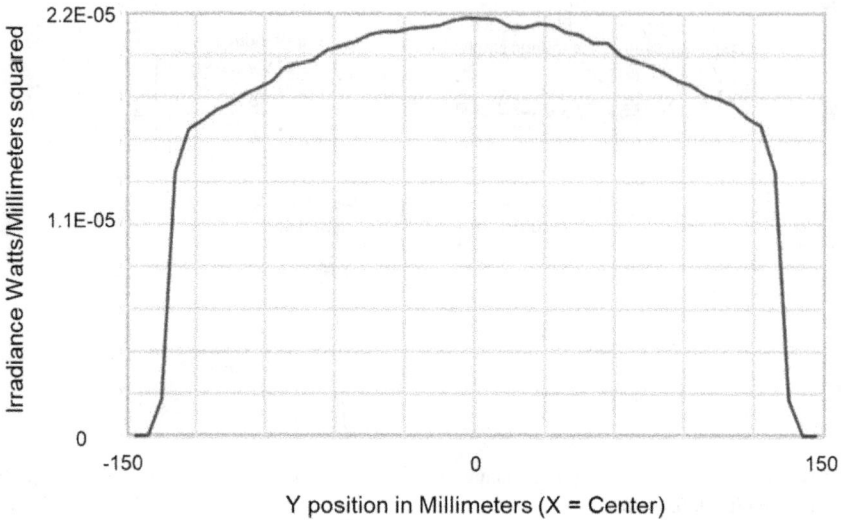

Figure 2.17. The irradiance profile at the screen for the system in figure 1.137 (section 1.3.10).

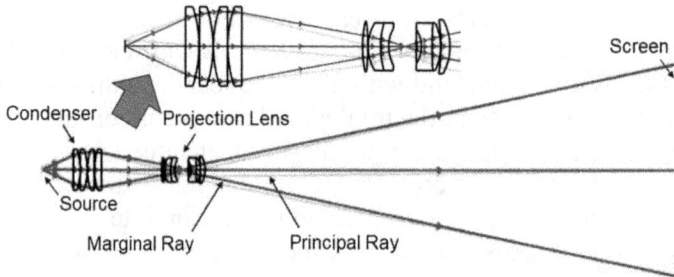

Figure 2.18. The real lens version of the paraxial thin lens Köhler system of figure 1.137.

550 nm. The plot assumes a Lambertian disk source 14 mm in diameter. If you were to replace the source with a different .IMA file (such as the letter F or an array of alphabets), the irradiance profile would retain its excellent local uniformity (due to étendue division and superposition), but the global dome-shaped nonuniformity would remain. This is because the illumination at the screen is the image of the condenser, where the irradiance profile across its surface is given by the source. For example, if there is a small Lambertian source, then the illumination at the condenser is given by equation (2.17). However, global uniformity is achievable in a Köhler system if either an unrealistically large source is used or if the condenser is aplanatic (e.g. see section 2.10).

Would you believe that the profile in figure 2.17 belongs to a pair of paraxial thin lens models? For comparison, figure 2.18 shows an approximately equivalent system using real lenses, whose irradiance profile (at 550 nm) and prescription are provided in figures 2.19 and 2.20, respectively. The equivalence is approximate for two reasons. First, the real lens system's projection lens is slightly too small to receive the

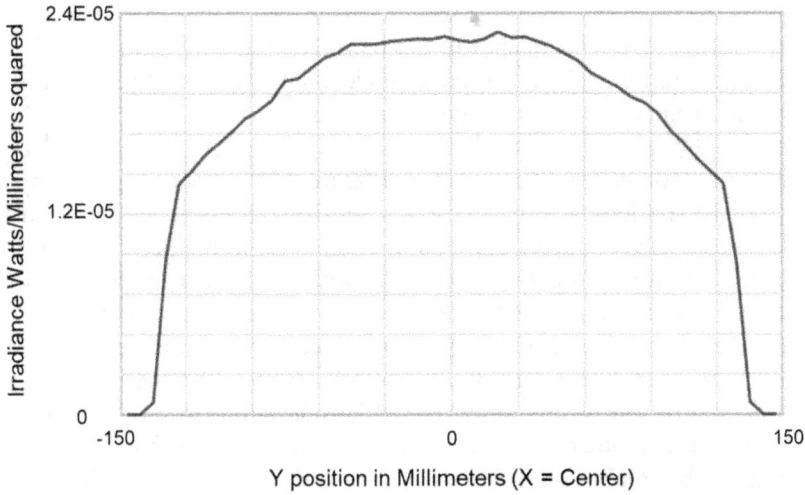

Figure 2.19. The irradiance profile at the screen for the system in figure 2.18.

Figure 2.20. The prescription for the system in figure 2.18.

full field from the principal ray at the source, so the source in the real lens system has been given half the diameter of that in the paraxial system of figure 1.137. Second, the irradiance magnitude for the paraxial thin lens system is not strictly at the correct level expected from radiometric calculations. This point is addressed in the next section.

The condenser for the real lens system in figure 2.18 was put together by way of selecting four COTS lenses with part # 48-248 from the supplier Edmund Optics, which are labeled as such under the Comment column in the prescription provided in figure 2.20. The stop is located between lenses, at surface 5, because this reduces the vignetting for oblique rays (note that Ray Aiming is on). Surfaces 10–20 comprise the projection lens, which is the reversed form of the double-Gauss lens from figure 1.292 in section 1.4.12. It was the most convenient choice for the current example and is certainly not the only type of lens design form available for use as a projection lens in a Köhler illumination configuration. Note that surface 5 in the prescription has been labeled a 'PUPIL,' which is the stop of the projection lens, as this is the main point about Köhler illumination, in which the source should be projected by the condenser into the *entrance pupil* of the projection lens. Accordingly, this would also position the source's image at the projection lens's stop. However, a close examination of the layout in figure 2.18 shows that the source has not been accurately projected onto the stop of the projection lens. This is a trivial matter for two reasons. First, it does not have a significant impact on the irradiance profile. Second, the current real lens system has not been extensively optimized. Generally, a Köhler system's condenser and projection lens may be designed separately (which was the approach in the current example) and can be further optimized when integrated with each other.

2.6.2 How to correct the radiance problem (because paraxial thin lenses are fake lenses)

Paraxial thin lenses help us to conceptualize layouts in imaging and in illumination, so we will often use them to explain a general design layout. However, the magnitude of irradiance that paraxial thin lens models produces is inconsistent with radiometry. For an elemental object and an elemental image, the radiance through a paraxial thin lens is given by

$$L \cos^4 u = L' \cos^4 u', \tag{2.22}$$

where L and L' are the radiances before and after the paraxial thin lens, respectively, and u and u' are the half-angles labeled in figure 2.21, respectively [29] (see also appendix B.1). These half-angles are not the marginal ray angles of the paraxial thin lens. Rather, they subtend the middle ray in a ray pencil's solid angle from the axial positions in the object and image.

Due to equation (2.22), the axial image irradiance of a paraxial thin lens is modified. For instance, it can be shown (see appendix B.1) that when equation (2.22) is accounted for in calculating the flux transferred to the image, equation (2.14) is modified to be

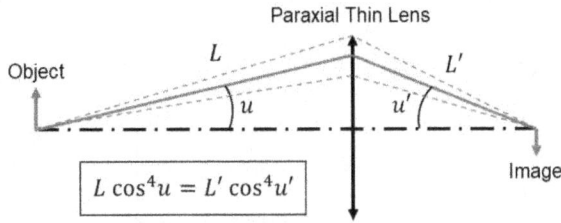

Figure 2.21. The radiance through a paraxial thin lens in air.

$$E_{\mathrm{p}} = \frac{\pi L \sin^2 U}{m_{\mathrm{p}}^2}, \tag{2.23}$$

where U is the object-space marginal ray angle, L is the source's radiance (assuming it is Lambertian), E_{p} is the axial irradiance in the image pertaining to a paraxial thin lens rather than that due to a real lens, and m_{p} is specifically the paraxial magnification of the image defined by

$$m_{\mathrm{p}} = \frac{\tan U}{\tan U'}, \tag{2.24}$$

where U and U' are the marginal rays (with arbitrary magnitude) in object and image space, respectively. Note that we are now purposely ensuring that the definition in equation (2.24) is specific to a paraxial thin lens and should generally be distinguished from the magnification m stated in in section 1.4. Substituting equation (2.24) into (2.23) yields

$$E_{\mathrm{p}} = \pi L \cos^2 U \tan^2 U'. \tag{2.25}$$

When paraxial thin lens models in OS are used in NSC mode (they are available as a Paraxial Lens NSC object), the axial irradiance in a focused image is given by equation (2.25). Note that only in the limit of small angles for U and U' is this formula equal to equation (2.14). Accordingly, if paraxial thin lens objects are used to lay out an illumination system in OS's NSC mode, the axial irradiance must be corrected. This can be done by simply dividing the measured axial irradiance in an NSC detector at the image plane by the value computed by equation (2.25) and multiplying the result by the value computed by equation (2.14). In other words, for the images formed by paraxial lens objects in NSC mode, multiply the axial irradiance by E/E_{p}, where E is given by equation (2.14) and E_{p} is given by equation (2.25).

2.6.3 Relative illumination is called *critical illumination* in illumination design

Figure 2.22 shows a paraxial thin lens that has an EFL of 25 mm EFL in OS's SC mode. The relative illumination plot for this lens is shown in figure 2.23(a), which indicates a roll-off of the illumination towards the edge. But figure 2.23(b) shows the

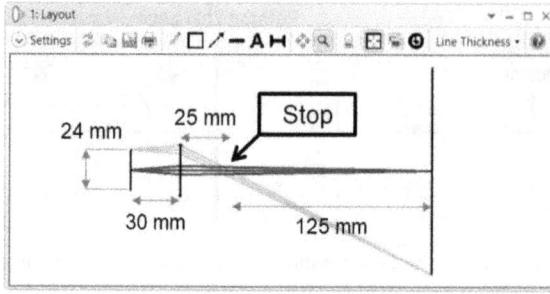

Figure 2.22. The OS layout of a paraxial thin lens in SC mode.

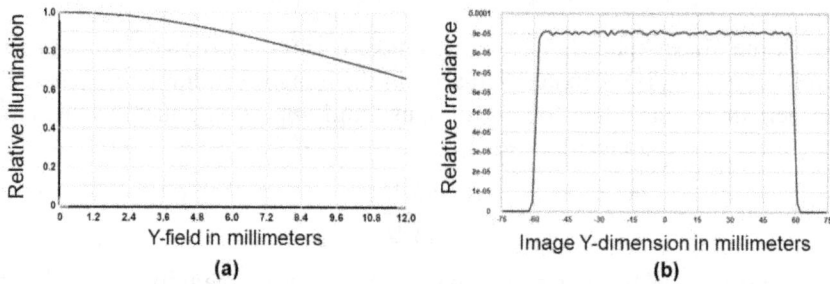

Figure 2.23. An extended scene analysis for the layout in figure 2.22. (a) Relative illumination. (b) GIA irradiance plot across the image.

GIA plot of the irradiance profile for the same lens, which is completely uniform. What is going on?

As will be shown in the next section, when there is no vignetting of oblique rays, the relative illumination in the image is describable by two formulas. One of them involves image distortion, the rate of change of image distortion, the aberration of the entrance pupil, and the angle of the principal ray in object space. The layout in figure 2.22 is telecentric in object space, as the stop is located one focal length to the right of the thin lens, resulting in a principal ray angle of zero. Under this condition, when a *real* lens has no distortion in the field and also possesses no aberrations related to the entrance pupil (as pupils are images of the stop, they can generally be aberrated), the irradiance profile at the image is flat. It so happens that the paraxial thin lens model in OS has no distortion and no pupil aberrations. Therefore, its irradiance profile is correctly given by the plot in figure 2.23(b). If the system were modeled in NSC mode, the irradiance profile that would be observed at a detector would indeed be given by the plot in figure 2.23(b). In fact, the GIA plot is essentially an NSC plot for rays that pass through the entrance pupil. Under the Source option in the Settings menu of the GIA plot, we can select 'Lambertian.' This is the setting used for the plot in figure 2.23(b). Hence, the GIA plot correctly projects Lambertian emission into the entrance pupil. As for the reason why the relative illumination in figure 2.23(a) shows a drop-off, this shall be made clear in the next two sections.

In illumination design, the technique of projecting a source directly onto a screen is called **critical illumination**. The technique is a good one as long as the source has no local nonuniformities, and as long as the image's relative illumination can be made globally uniform. The layout in figure 2.22, for example, projects a magnified image of a locally uniform Lambertian disk source 24 mm in diameter onto a screen. This paraxial thin lens possesses no image distortion and no pupil aberrations, yielding uniform relative illumination, as highlighted above. Therefore, you would expect that if a real lens were made to be telecentric in object space and to possess no distortion in the field, and if it also had no pupil aberrations, then the irradiance profile should also be uniform. This would indeed be the case, but achieving a completely zero-distortion field and eliminating pupil aberrations are not always possible for general cases of imaging. Still, all is not lost. As the next section will show, the relative illumination can be made uniform even when distortion and pupil aberrations are present. The idea is to balance all of the variables in such a way as to create a flat illumination profile.

2.7 How to design for uniform relative illumination

When the object of an imaging system is a flat Lambertian source at a finite distance, it can be shown that at low aperture and for an unvignetted system, the relative illumination RI in the image may be expressed as

$$\text{RI} = \frac{(A/A_o)\cos^4 \bar{u}}{(1 + D)[1 + D + h(\mathrm{d}D/\mathrm{d}h)]}, \tag{2.26}$$

where A and A_o are the oblique and axial entrance pupil areas, respectively, $\bar{u}$ is the principal ray angle of arbitrary magnitude (i.e. here, it is not considered a paraxial ray angle) in object space, D is the image distortion in fractional terms (i.e. it is given by equation (1.8) but without multiplying by 100% so that if the distortion is 1%, then $D = 0.01$), h is the object height, and $\mathrm{d}D/\mathrm{d}h$ has been called the **differential distortion** in the image, which is the instantaneous rate of change of the image distortion [30]. Figure 2.24 illustrates the ray geometry use to define these quantities. If h' is the image height and h'_p is the paraxial image height, then

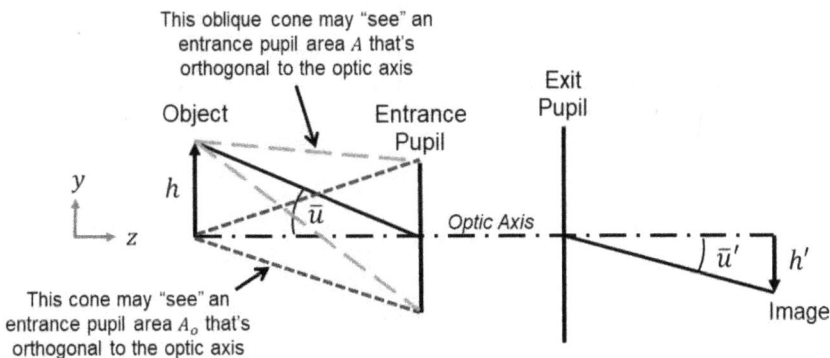

Figure 2.24. The geometry of the rays defining the quantities in equation (2.26).

$$h\frac{dD}{dh} = h'_{\mathrm{p}}\frac{dD}{dh'_{\mathrm{p}}}. \tag{2.27}$$

Therefore, for an object at infinity, the right-hand side of equation (2.27) may be used in equation (2.26) rather than $h(dD/dh)$. Alternatively, it can be shown [31] that for an object at infinity, equation (2.26) may be recast as

$$\mathrm{RI} = \frac{(A/A_o)\cos^4 \bar{u}}{(1 + D)[1 + D + \sin \bar{u} \cos \bar{u}(dD/d\bar{u})]}. \tag{2.28}$$

You may notice that the principal ray angle $\bar{u}'$ in image space does not appear in equations (2.26) and (2.28). Actually, by flux conservation through the entrance and exit pupils, it can be shown [29] that there is a corresponding formula in terms of the exit pupil given by

$$\mathrm{RI} = (A'/A'_o)\cos^4 \bar{u}', \tag{2.29}$$

where A' and A'_o are the exit pupil areas, respectively. In 1948, Reiss [32] defined two formulas for the illumination across the image—one from the point of view of the entrance pupil and the other from the point of view of the exit pupil. Of these two, the entrance pupil formula provided by Reiss was expressed differently from (but is algebraically equivalent to) equation (2.28). Kingslake [33] calls these two formulations the 'object-space formula' and the 'image-space formula.' As a result of flux conservation through the pupils, both formulas are the same and they compute the relative illumination in the image in the limit of low aperture. More recently, Reshidko, Sasián, and Johnson [34–36] have made further contributions by expanding the cosine fourth-power dependence in both pupils into a power series and relating the relative illumination to aberrations in the pupils.

In section 1.4.3, it was highlighted that the wiggles displayed in the relative illumination of a mobile phone lens can be attributed in part to image distortion (figure 1.196). Here, equations (2.26)–(2.28) show the explicit relationship. As an example of the application of the formulas presented here, consider the layout shown in figure 2.25, whose relative illumination, distortion, prescription, and merit

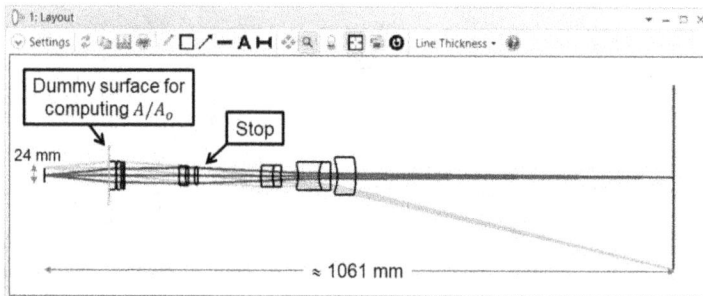

Figure 2.25. The layout of a projection lens system with uniform relative illumination.

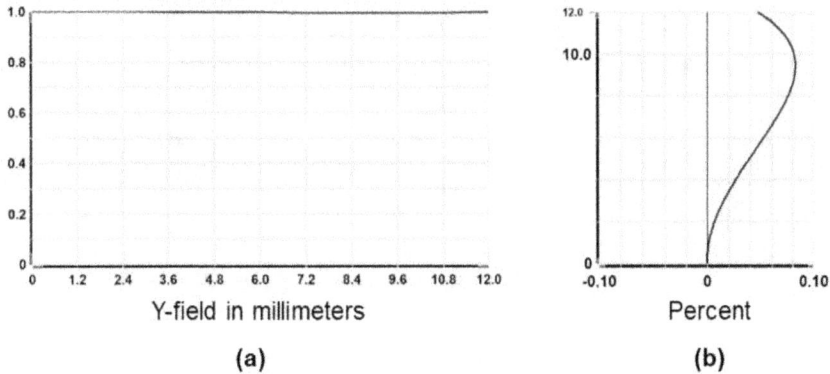

Figure 2.26. The relative illumination (a) and distortion (b) for the lens system of figure 2.25.

	Surface Type	Comment	Radius	Thickness	Material	Clear Semi-Dia	Mech Semi-Dia
0	OBJECT Standard ▾		Infinity	109.75695 V		12.00000	12.00000
1	Standard ▾	Dummy	Infinity	0.00000		50.00000	50.00000
2	Standard ▾		-1.37492E+04 V	12.14921 V	N-SK4	22.13356	22.73655
3	Standard ▾		-152.27334 V	0.13617 V		22.73655	22.73655
4	Standard ▾		199.94680 V	8.11299 V	N-SK4	22.73808	22.73808
5	Standard ▾		-203.71389 V	4.14558 V		22.49736	22.73808
6	Standard ▾		-146.30220 V	1.73050 V	SF1	21.97362	21.97362
7	Standard ▾		-403.14305 V	92.08889 V		21.92858	21.97362
8	Standard ▾		280.70791 V	10.99791 V	SF1	15.65269	15.65269
9	Standard ▾		64.49651 V	4.90500 V	N-SK4	14.88322	15.65269
10	Standard ▾		1046.80666 V	11.14548 V		14.71643	15.65269
11	Standard ▾		122.64267 V	5.80078 V	N-SK4	13.84287	13.84287
12	Standard ▾		752.69384 V	10.51610 V		13.38593	13.84287
13 STOP	Standard ▾	Stop	Infinity	96.09073 V		12.00000 U	12.00000
14	Standard ▾		131.73431 V	19.75685 V	N-SK4	17.05890	17.05890
15	Standard ▾		61.60971 V	13.78707 V	SF1	16.68940	17.05890
16	Standard ▾		150.83085 V	28.78033 V		16.26903	17.05890
17	Standard ▾		-62.99022 V	36.00001 V	N-SK4	16.77571	22.96161
18	Standard ▾		55.06927 V	20.00272 V	SF1	21.75277	22.96161
19	Standard ▾		-517.91619 V	11.41788 V		22.96161	22.96161
20	Standard ▾		-65.07818 V	30.58905 V	N-SK4	23.75556	30.57951
21	Standard ▾		-159.16678 V	533.34652 V		30.57951	30.57951
22 IMAGE	Standard ▾		Infinity	-		150.08365	150.08365

Figure 2.27. The prescription for the lens system shown in figure 2.25.

function are provided in figures 2.26–2.29, respectively (the plots are monochromatic at 550 nm).

The lens system of figure 2.25 is based on an optimized state for a projection lens from a prior study by this author [30], which was then re-optimized using the merit

	Type	Op#1	Op#2					Target	Weight	Value	% Contrib
		Merit Function Editor									
		Wizards and Operands	<	>		Merit Function:	0.00548199699711687				
1	BLNK ▾	Principal ray angle in object space									
2	RAID ▾	1	1	0.00000	1.00000	0.00000	0.00000	0.00000	1.00000	6.71791E-05	2.88794E-03
3	BLNK ▾	cos^4(U-bar)									
4	COSI ▾	2	1					0.00000	0.00000	1.00000	0.00000
5	PROD ▾	4	4					0.00000	0.00000	1.00000	0.00000
6	PROD ▾	5	5					0.00000	0.00000	1.00000	0.00000
7	BLNK ▾	"Ao", "Relative" entrance pupil area (axial)									
8	REAY ▾	1	1	0.00000	0.00000	0.00000	1.00000	0.00000	0.00000	10.23602	0.00000
9	REAX ▾	1	1	0.00000	0.00000	1.00000	0.00000	0.00000	0.00000	10.23602	0.00000
10	PROD ▾	8	9					0.00000	0.00000	104.77615	0.00000
11	BLNK ▾	"A", "Relative" entrance pupil aeaa (oblique)									
12	REAY ▾	1	1	0.00000	1.00000	0.00000	1.00000	0.00000	0.00000	22.13520	0.00000
13	REAY ▾	1	1	0.00000	1.00000	0.00000	-1.00000	0.00000	0.00000	1.76561	0.00000
14	DIFF ▾	12	13					0.00000	0.00000	20.36959	0.00000
15	DIVB ▾	14	2.00000					0.00000	0.00000	10.18479	0.00000
16	REAX ▾	1	1	0.00000	1.00000	1.00000	0.00000	0.00000	0.00000	10.21937	0.00000
17	PROD ▾	15	16					0.00000	0.00000	104.08220	0.00000
18	BLNK ▾	A/Ao									
19	DIVI ▾	17	10					0.00000	0.00000	0.99338	0.00000
20	BLNK ▾	D (at full field)									
21	DISG ▾	1	1	0.00000	1.00000	0.00000	0.00000	0.00000	0.00000	0.04706	0.00000
22	DIVB ▾	21	100.00000					0.00000	0.00000	4.70591E-04	0.00000
23	BLNK ▾	(1+D)									
24	CONS ▾							1.00000	0.00000	1.00000	0.00000
25	SUMM ▾	24	22					0.00000	0.00000	1.00047	0.00000
26	BLNK ▾	(dD/dh)									
27	DISG ▾	1	1	0.00000	1.00010	0.00000	0.00000	0.00000	0.00000	0.04702	0.00000
28	DIVB ▾	27	100.00000					0.00000	0.00000	4.70197E-04	0.00000
29	DISG ▾	1	1	0.00000	0.99990	0.00000	0.00000	0.00000	0.00000	0.04710	0.00000
30	DIVB ▾	29	100.00000					0.00000	0.00000	4.70985E-04	0.00000

Figure 2.28. The merit function (rows 1–30) for the lens system shown in figure 2.25. Rows continue in figure 2.29.

function in figure 2.29, resulting in the prescription in figure 2.28. Variables have been left in the prescription so that you can see the variables used for optimization. The lens's EFL is 51.86 mm at 550 nm. Notice that the relative illumination is uniform, and the distortion is maintained low but positive and nonzero. However, the differential distortion is not considered low near the full field, and thus contributes to the relative illumination, as I will now explain. In the merit function (figures 2.28 and 2.29), titles for specific computed quantities are labeled in white rows, and the final quantity is the last row for that title. For example, row 2 computes the principal ray angle $\bar{u}$ in object space, which is appropriately labeled by the title in row 1. However, row 6 computes $\cos^4 \bar{u}$, whose title is provided in row 3, and so on.

For optimization, high weights were assigned only to rows 2 and 44, whose quantities are $\bar{u}$ and $(A/A_o)/\{(1 + D)[1 + D + h(\mathrm{d}D/\mathrm{d}h)]\}$, respectively, from equation (2.26). The idea was that if $\bar{u}$ were made zero, then the lens system would be telecentric in object space and so there would be no contribution from $\cos^4 \bar{u}$. What remains are the rest of the quantities in the formula. Further, since distortion aberration can be of any order higher than the third order (which is indeed the case, as shown in figure 2.26(b)), and since A/A_o can also have any value, it was better to target the entire ratio in equation (2.26) at one. Note that A and A_o in the merit

Merit Function Editor

Wizards and Operands   Merit Function: 0.00548199699711687

#	Type	Op#1	Op#2					Target	Weight	Value	% Contrib
31	DIFF ▾	28	30					0.00000	0.00000	-7.88031E-07	0.00000
32	REAY ▾	0	1	0.00000	1.00010	0.00000	0.00000	0.00000	0.00000	12.00120	0.00000
33	REAY ▾	0	1	0.00000	0.99990	0.00000	0.00000	0.00000	0.00000	11.99880	0.00000
34	DIFF ▾	32	33					0.00000	0.00000	2.40000E-03	0.00000
35	DIVI ▾	31	34					0.00000	0.00000	-3.28346E-04	0.00000
36	BLNK ▾	h x (dD/dh)									
37	REAY ▾	0	1	0.00000	1.00000	0.00000	0.00000	0.00000	0.00000	12.00000	0.00000
38	PROD ▾	37	35					0.00000	0.00000	-3.94016E-03	0.00000
39	BLNK ▾	[1 +D+ h(dD/dh)]									
40	SUMM ▾	25	38					0.00000	0.00000	0.99653	0.00000
41	BLNK ▾	(1+D)[1 +D+ h(dD/dh)]									
42	PROD ▾	24	40					0.00000	0.00000	0.99653	0.00000
43	BLNK ▾	A/Ao divide by (1+D)[1 +D+ h(dD/dh)]									
44	DIVI ▾	19	42					1.00000	1.00000	0.99684	6.40862
45	BLNK ▾	Relative illum using formula									
46	PROD ▾	6	44					0.00000	0.00000	0.99684	0.00000
47	BLNK ▾	Relative illum using OpticStudio									
48	RELI ▾	3	1	2	0			0.00000	0.00000	0.99662	0.00000
49	BLNK ▾	Principal ray AOI at image									
50	RAID ▾	22	1	0.00000	1.00000	0.00000	0.00000	0.00000	0.00000	12.99976	0.00000
51	OPGT ▾	50						13.00000	0.20000	12.99976	7.39350E-03
52	BLNK ▾	Lens constraints									
53	MNCG ▾	2	21					1.00000	0.20000	1.00000	0.00000
54	MNEG ▾	2	21	1.00000	1			0.20000	0.20000	0.20000	0.00000
55	MXCG ▾	2	21					36.00000	0.20000	36.00001	3.77012E-06
56	MNCA ▾	3	20					0.10000	0.20000	0.10000	0.00000
57	MNEA ▾	3	20	1.00000	1			0.10000	0.20000	0.10000	0.00000
58	RSCE ▾	3	1	0.00000	0.00000			0.00000	1.00000	7.23390E-03	33.48606
59	RSCE ▾	3	1	0.00000	1.00000			0.00000	1.00000	9.69080E-03	60.09504

Figure 2.29. The merit function (rows 31–59) for the lens system shown in figure 2.25.

function are not the values of the axial and oblique entrance pupil areas, respectively. However, for the purpose of computing relative illumination, only A/A_o is required, which is computable by placing a dummy surface anywhere between the object and the first element. Surface 1 in the prescription was used for this purpose. From these quantities, row 46 calculates the resulting relative illumination using equation (2.26), while row 48 uses the RELI operand to obtain the relative illumination that is computed by OS's algorithm (we will discuss this computation in the next section). We can, of course, simply target RELI at one, but here, we want to see how equation (2.26) is applied. Moreover, this formula reveals all of the essential physics involved in relative illumination, even though it is limited to low aperture. Rows 53–57 constrain the lens thicknesses and edges, and the final two rows help to focus the image.

As can be seen in the merit function (figure 2.29), the relative illumination computed in row 46 is quite close to that computed in row 48. Notice that, due to non-negligible *negative* differential distortion, the quantity $h(dD/dh)$ in row 38 is subtracting D from $(1 + D)$ in row 40. But the combined effects in the computation $(A/A_o)/\{(1 + D)[1 + D + h(dD/dh)]\}$ make the relative illumination equal to 0.996 84 at the full field. OS's algorithm yields 0.996 62. These values are reflected in the relative illumination plot in figure 2.26(a). If A/A_o were made to be exactly equal to one, then it would have been difficult to balance the distortion with differential distortion. Note that dD/dh is not necessarily zero even if D is zero at some field point. Rather, it would only be zero everywhere in the field if $D = 0$ throughout the field, which is a difficult condition to meet (unless you are using a paraxial thin lens!)

Some further points are worth discussing. Notice that on row 51 of the merit function, I targeted the principal ray angle in image space at a value greater than 13° (an arbitrary angle). This was because I wanted to show that it is possible to achieve flat relative illumination even when this ray angle is nonzero. It is a common misconception that the relative illumination is uniform only when the principal ray is telecentric in image space. This is far from the truth, as has been illustrated here. If you want an image-space telecentric system and also want the relative illumination to be uniform, then you must also ensure that the exit pupil areas for the axial and oblique rays are equal, as implied by equation (2.29). Alternatively (and equivalently), you may apply the formulation in object space by way of equations (2.26) or (2.28). The advantage of the object-space formulation is that it reveals the explicit contributions of image distortion and differential distortion to the relative illumination. This lets you know right away if you are over-constraining your design, since you may have specific operands for image distortion in your merit function. If you are performing custom lens design, this also highlights the importance of checking your customer's specifications for distortion and relative illumination, to be sure that there are no contradicting requirements.

There is another point to highlight, which is that in certain applications, a target plane may require illumination by a separate subsystem, which is then imaged into a camera by a lens. In this case, if the illumination is symmetric and globally nonuniform, it is possible to compensate for this in the imaging lens if the relative illumination on the imaging side is designed to have the opposite irradiance profile. The final illumination in the image of the target plane is then given by the product of the illumination from the separate subsystem and the relative illumination in the image. As the current section has shown, you would apply the equations for relative illumination and decide which variables to adjust and/or target.

We now return to the problem of the relative illumination roll-off in figure 2.23(a). In OS, the computation of relative illumination is done through the image-space formulation, but it does not use equation (2.29). However, at low aperture, it is roughly given by this formula (you will learn more about OS's method of computation in the next section). Applying equation (2.29) for a paraxial thin lens, since its surface is both the entrance and the exit pupil, there is no aberration of the pupils. Thus, $A'/A_o' = 1$ and we are left with $\cos^4 \bar{u}'$. If you model this paraxial thin lens system in OS and use the RAID operand to compute the AOI for the principal ray at the image, you get roughly 25.64°. Take the cosine of this angle to the fourth power, and you get 0.66. This is the magnitude of the relative illumination at the full field in figure 2.23(a), which should actually not apply to paraxial thin lenses.

As shown in section 2.6.2, paraxial thin lenses are fictitious lenses in that they do not satisfy the radiance theorem. Rather, for a paraxial thin lens, the radiance L' at the image side is given by equation (2.22). Another way to express this radiance

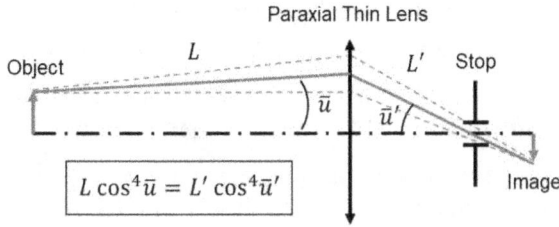

Figure 2.30. An alternate geometry for radiance through a paraxial thin lens.

formula is by way of the geometry shown in figure 2.30. There, L' is now governed by the angles $\bar{u}$ and $\bar{u}'$, so that

$$L \cos^4 \bar{u} = L' \cos^4 \bar{u}'. \tag{2.30}$$

For an oblique ray pencil, if we regard the thin lens's surface as a small source with area dA, we can apply equation (2.17) and say that the irradiance at an off-axis point in the image is roughly $(L'dA \cos^4 \bar{u}')/z^2$. But due to equation (2.30), $L' = L \cos^4 \bar{u}/\cos^4 \bar{u}'$. So, substituting this into the former yields an off-axis irradiance given by

$$E_{\text{pob}} = \frac{LdA \cos^4 \bar{u}}{z^2}, \tag{2.31}$$

where the subscript 'pob' denotes that this is the oblique (off-axis) image irradiance for a paraxial thin lens (at low aperture). On the axis, the irradiance is LdA/z^2, which is essentially the same as equation (2.25), for which $\cos^2 U \approx 1$ at low aperture, so that $\tan^2 U' \approx dA/z^2$. Hence, for a paraxial thin lens, the relative illumination in terms of the image-space formulation is

$$RI = \frac{E_{pob}}{E_p} = \cos^4 \bar{u}. \tag{2.32}$$

Equation (2.32) is consistent with the argument made earlier in terms of the object-space formulation. That is, the relative illumination for a paraxial thin lens whose principal ray is telecentric in object space has $\cos^4 \bar{u} = 1$, because $\bar{u} = 0$, resulting in a flat relative illumination plot (i.e. figure 2.23(b)). Moreover, equation (2.32) indicates that in fact, the relative illumination for a paraxial thin lens model at an arbitrary angle $\bar{u}$ is always given by equation (2.32). Indeed, in OS, when using the GIA plot to display the relative illumination for a paraxial thin lens (e.g. as done in figure 2.23(b)), varying the stop location relative to a paraxial thin lens controls the irradiance profile in a manner describable by equation (2.32). That said, note that the irradiance profile in figure 2.17 is not the relative illumination for the second thin lens in a paraxial Köhler system. Rather, that profile resulted from imaging the irradiance profile on the first thin lens, given by a small disk Lambertian source. Comparing this profile with the profile in figure 2.19 from a real and approximately equivalent Köhler system, it appears that using the GIA feature in SC mode is an effective way to perform a 'first-order' analysis of illumination.

2.8 Relative illumination in direction cosine space

If no ray aberrations are present, it can be shown [9, 29] that the axial image irradiance given by equation (2.14) may be computed numerically by tracing *circumferential rays* (sometimes called 'rim rays') that sample the edge of the exit pupil, expressed as

$$E = \pi L \frac{1}{N} \sum_{i=1}^{N} \sin^2 U_i'. \tag{2.33}$$

The quantity $\pi \frac{1}{N} \sum_{i=1}^{N} \sin^2 U_i'$ in equation (2.33) is known as the *projected solid angle* for the axial circumferential rays. There can also be oblique circumferential rays, and in this case, we can label the axial and oblique projected solid angles as Ω_o and Ω, respectively. Following the construction shown in figure 2.31, we may express these quantities as

$$\Omega_o = \pi \frac{1}{N} \sum_{i=1}^{N} \sin^2 U_i', \tag{2.34}$$

and

$$\Omega = \pi \cos \bar{u}' \frac{1}{N} \sum_{i=1}^{N} \sin^2 Q_i', \tag{2.35}$$

where $\cos \bar{u}'$ is needed in equation (2.35) because of the obliquity of that solid angle. In figure 2.31, only four rim rays for each projected solid angle are shown, so $N = 4$ in that illustration. However, many more rim rays can be included if the stop (and therefore, also the exit pupil) is noncircular and not rotationally symmetric about the optic axis. Assuming the same number of rim rays is used for axial and oblique rim rays, the relative illumination is the ratio of the two projected solid angles:

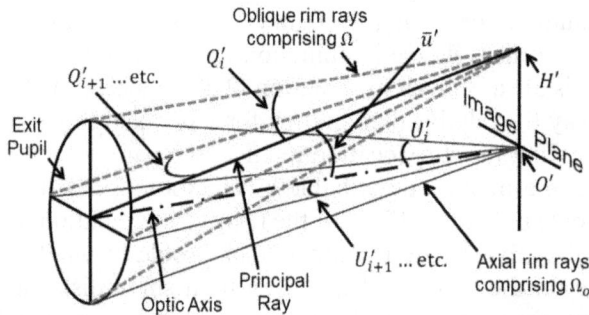

Figure 2.31. A construction used to define the axial and oblique projected solid angles.

$$RI = \frac{\Omega}{\Omega_o} = \cos \bar{u}' \frac{\sum_{i=1}^{N} \sin^2 Q_i'}{\sum_{i=1}^{N} \sin^2 U_i'}. \tag{2.36}$$

If, in equation (2.36), we take angles defined by $\alpha_i' = 90° - U_i'$ and $\beta_i' = 90° - Q_i'$ instead of U_i' and Q_i', then the relative illumination can be expressed as

$$RI = \cos \bar{u}' \frac{\sum_{i=1}^{N} \cos^2 \beta_i'}{\sum_{i=1}^{N} \cos^2 \alpha_i'}. \tag{2.37}$$

The quantities $\cos \alpha_i'$ and $\cos \beta_i'$ can be considered *direction cosines* for the rim rays, though not in rectangular coordinates. For a circular exit pupil, the quantities $\cos \alpha_i'$ are equal in magnitude, and they trace out a circular area of radius $\cos^2 \alpha_i'$ at the image plane. Thus, on that plane, there is a *direction cosine space* of vectors, each vector of which points radially away from the axial position (indicated by O' in figure 2.31) at the image and each of which possesses the magnitude $\cos \alpha_i'$. For oblique rim rays passing through a circular exit pupil, the quantities $\cos \beta_i'$ trace out an oval on a plane orthogonal to the principal ray and centered at the image field point that they focus onto, indicated by H' in figure 2.31. Rimmer [37] showed that this oval can be considered approximately an ellipse because, from the position H' at the image, when one looks towards the exit pupil along the principal ray, one sees a flattened exit pupil in the vertical (so, it is a *prolate ellipse*). Rimmer expressed that the relative illumination can therefore be estimated quite accurately by taking the ratio of the area of an ellipse (for the oblique rim rays) to the area of a circle (for the axial rim rays), where both areas are computed in a direction cosine space. OS's help file on relative illumination states that the program's algorithm is based on Rimmer's paper. When ray aberrations are present, then transverse ray errors ε_x and ε_y are used to correct the direction cosines [37] (see section 1.4.4 for a review of the nature of ε_x and ε_y).

Let us perform some relative illumination calculations for a high-aperture double-Gauss lens, using the various methods discussed so far. Figure 2.32 shows a layout based on a 1973 patent by Nakagawa [38], whose prescription, relative illumination,

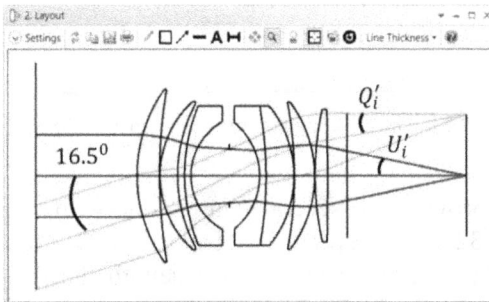

Figure 2.32. The OS layout of US Patent 3743387 by Nakagawa [38].

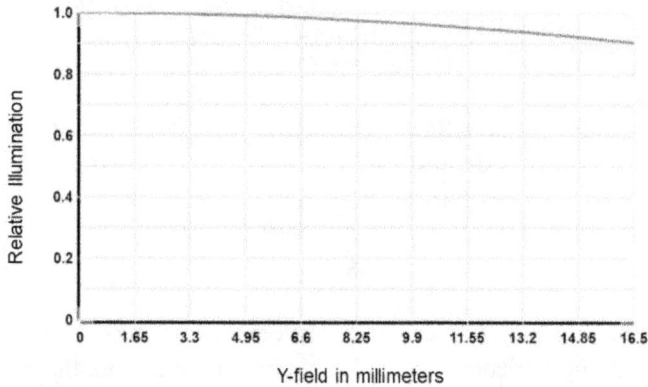

Figure 2.33. The relative illumination for the lens shown in figure 2.32.

	Surface Type		Comment	Radius	Thickness	Material	Clear Semi-Dia	Mech Semi-Dia
0	OBJECT	Standard ▾		Infinity	Infinity		Infinity	Infinity
1		Standard ▾	Dummy 1	Infinity	0.50000		0.54750	0.54750
2	(aper)	Standard ▾	1	0.71650	0.11030	N-LASF44	0.41652 U	0.41652
3	(aper)	Standard ▾		2.51380	2.10000E-03		0.41321 U	0.41652
4	(aper)	Standard ▾	2	0.47020	0.08900	N-LASF43	0.35753 U	0.35753
5	(aper)	Standard ▾		0.71940	0.03470		0.34780 U	0.35753
6	(aper)	Standard ▾	3	0.86630	0.03200	N-SF14	0.33670 U	0.33670
7	(aper)	Standard ▾		0.29570	0.19130		0.26041 U	0.33670
8	STOP	Standard ▾		Infinity	0.15000		0.13000 U	0.13000
9	(aper)	Standard ▾	4	-0.32670	0.03200	N-SF10	0.25808 U	0.33601
10	(aper)	Standard ▾	5	-1.77810	0.13880	N-LAF34	0.31650 U	0.33601
11	(aper)	Standard ▾		-0.51930	1.80000E-03		0.33601 U	0.33601
12	(aper)	Standard ▾	6	-1.83940	0.10140	N-LAK14	0.35323 U	0.35558
13	(aper)	Standard ▾		-0.57410	2.10000E-03		0.35558 U	0.35558
14	(aper)	Standard ▾	7	1.46550	0.06230	N-LAK14	0.32000 U	0.32000
15	(aper)	Standard ▾		-16.67680	0.10000		0.32000 U	0.32000
16		Standard ▾	Dummy 2	Infinity	0.00000		0.30036	0.30036
17		Standard ▾		Infinity	0.59284 M		0.30138 U	0.30138
18		Standard ▾	Focus	Infinity	-9.81898E-04		0.29615	0.29615
19	IMAGE	Standard ▾		Infinity	-		0.29577	0.29577

Figure 2.34. The prescription for the lens shown in figure 2.32.

and merit function (provided in several figures, due to its significant length) are provided in figures 2.32–2.39, respectively. The lens's EFL is 1 mm at 587.56 nm, and all ray traces and plots are done at this wavelength. The half-field angle is 16.5°. Without doubt, it is going to be a little difficult for you to read the merit function provided in figures 2.35–2.39, but if you spend time looking through it, you can see

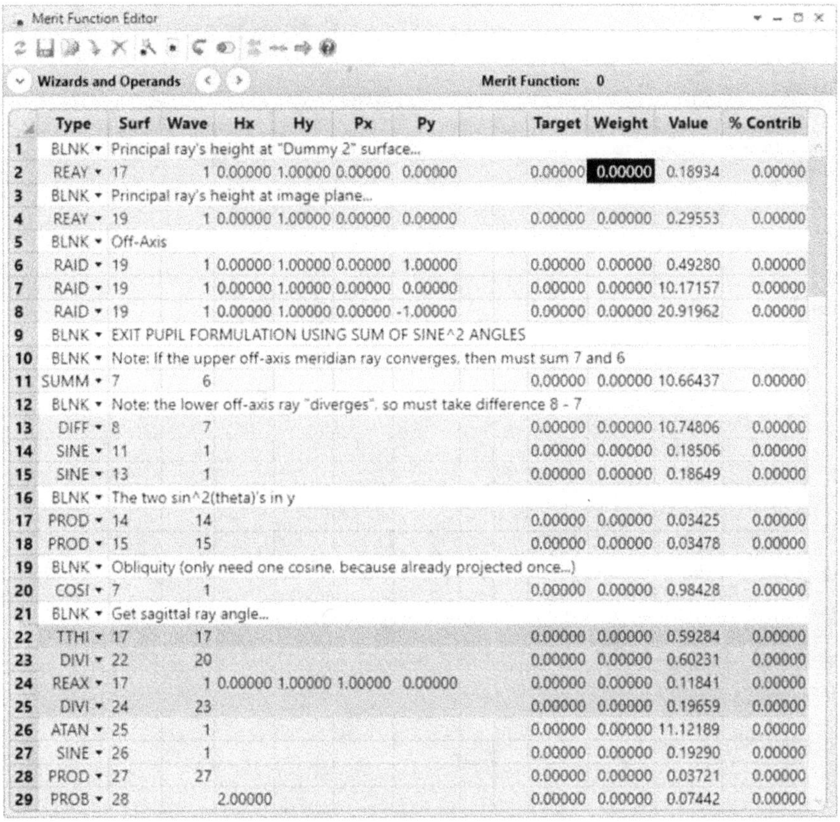

Merit Function Editor

Wizards and Operands — Merit Function: 0

	Type	Surf	Wave	Hx	Hy	Px	Py	Target	Weight	Value	% Contrib
1	BLNK	Principal ray's height at "Dummy 2" surface...									
2	REAY	17	1	0.00000	1.00000	0.00000	0.00000	0.00000	0.00000	0.18934	0.00000
3	BLNK	Principal ray's height at image plane...									
4	REAY	19	1	0.00000	1.00000	0.00000	0.00000	0.00000	0.00000	0.29553	0.00000
5	BLNK	Off-Axis									
6	RAID	19	1	0.00000	1.00000	0.00000	1.00000	0.00000	0.00000	0.49280	0.00000
7	RAID	19	1	0.00000	1.00000	0.00000	0.00000	0.00000	0.00000	10.17157	0.00000
8	RAID	19	1	0.00000	1.00000	0.00000	-1.00000	0.00000	0.00000	20.91962	0.00000
9	BLNK	EXIT PUPIL FORMULATION USING SUM OF SINE^2 ANGLES									
10	BLNK	Note: If the upper off-axis meridian ray converges, then must sum 7 and 6									
11	SUMM	7	6					0.00000	0.00000	10.66437	0.00000
12	BLNK	Note: the lower off-axis ray "diverges", so must take difference 8 - 7									
13	DIFF	8	7					0.00000	0.00000	10.74806	0.00000
14	SINE	11	1					0.00000	0.00000	0.18506	0.00000
15	SINE	13	1					0.00000	0.00000	0.18649	0.00000
16	BLNK	The two sin^2(theta)'s in y									
17	PROD	14	14					0.00000	0.00000	0.03425	0.00000
18	PROD	15	15					0.00000	0.00000	0.03478	0.00000
19	BLNK	Obliquity (only need one cosine, because already projected once...)									
20	COSI	7	1					0.00000	0.00000	0.98428	0.00000
21	BLNK	Get sagittal ray angle...									
22	TTHI	17	17					0.00000	0.00000	0.59284	0.00000
23	DIVI	22	20					0.00000	0.00000	0.60231	0.00000
24	REAX	17	1	0.00000	1.00000	1.00000	0.00000	0.00000	0.00000	0.11841	0.00000
25	DIVI	24	23					0.00000	0.00000	0.19659	0.00000
26	ATAN	25	1					0.00000	0.00000	11.12189	0.00000
27	SINE	26	1					0.00000	0.00000	0.19290	0.00000
28	PROD	27	27					0.00000	0.00000	0.03721	0.00000
29	PROB	28		2.00000				0.00000	0.00000	0.07442	0.00000

Figure 2.35. The merit function for the lens shown in figure 2.32. (Continued in the next figure.)

how the different approaches to relative illumination compare. In the computations provided, I have used four rim rays (top, bottom, left, right), though only three are actually needed, due to symmetry in the sagittal plane. Row 41 computes the relative illumination according to equation (2.36), using all of the operands above that row. Note that its value is close to that computed by OS using Rimmer's method in row 43 (using the RELI operand). One difference is that I did not apply corrections to ray angles due to residual aberrations, which is done in OS for the RELI operand. Another possible difference is in the way OS samples rays in the pupil. Next, row 64 computes the relative illumination according to equation (2.29), using all of the operands above it. Note that its value is close to OS's value shown in row 43. Finally, row 129 computes the relative illumination according to equation (2.28) and obtains a value close to OS's value shown in row 131.

As a final remark, note that in all cases of relative illumination discussed, the condition is that of a flat Lambertian source object, and the observation plane is the image plane. The calculation techniques shown are invalid when there is a switch of conjugates, as in Köhler illumination. Thus, for the system in figure 2.18, you cannot

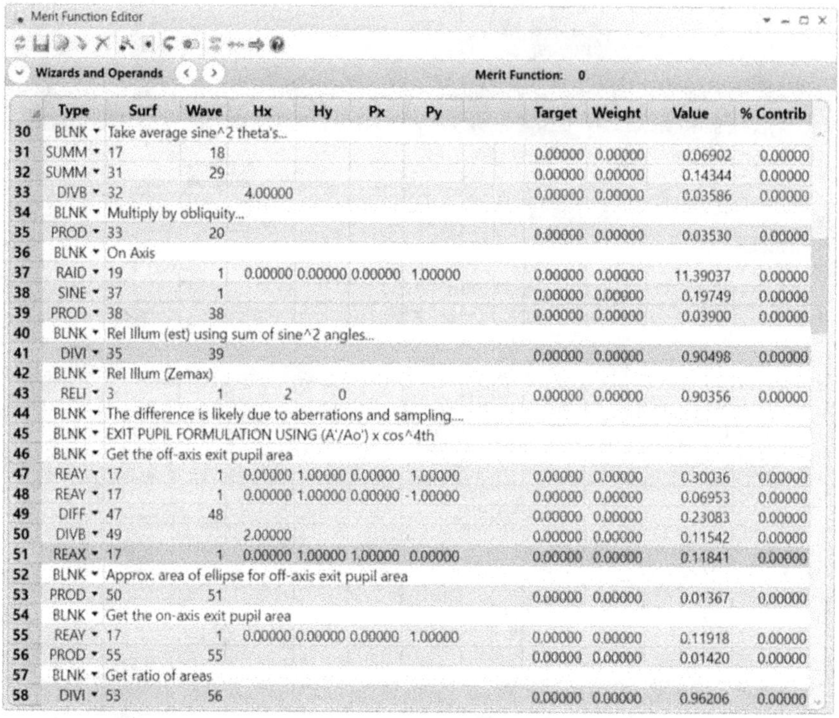

Merit Function Editor

Wizards and Operands — Merit Function: 0

	Type	Surf	Wave	Hx	Hy	Px	Py	Target	Weight	Value	% Contrib
30	BLNK	Take average sine^2 theta's...									
31	SUMM	17	18					0.00000	0.00000	0.06902	0.00000
32	SUMM	31	29					0.00000	0.00000	0.14344	0.00000
33	DIVB	32		4.00000				0.00000	0.00000	0.03586	0.00000
34	BLNK	Multiply by obliquity...									
35	PROD	33	20					0.00000	0.00000	0.03530	0.00000
36	BLNK	On Axis									
37	RAID	19	1	0.00000	0.00000	0.00000	1.00000	0.00000	0.00000	11.39037	0.00000
38	SINE	37	1					0.00000	0.00000	0.19749	0.00000
39	PROD	38	38					0.00000	0.00000	0.03900	0.00000
40	BLNK	Rel Illum (est) using sum of sine^2 angles...									
41	DIVI	35	39					0.00000	0.00000	0.90498	0.00000
42	BLNK	Rel Illum (Zemax)									
43	RELI	3	1	2	0			0.00000	0.00000	0.90356	0.00000
44	BLNK	The difference is likely due to aberrations and sampling....									
45	BLNK	EXIT PUPIL FORMULATION USING (A'/Ao') x cos^4th									
46	BLNK	Get the off-axis exit pupil area									
47	REAY	17	1	0.00000	1.00000	0.00000	1.00000	0.00000	0.00000	0.30036	0.00000
48	REAY	17	1	0.00000	1.00000	0.00000	-1.00000	0.00000	0.00000	0.06953	0.00000
49	DIFF	47	48					0.00000	0.00000	0.23083	0.00000
50	DIVB	49		2.00000				0.00000	0.00000	0.11542	0.00000
51	REAX	17	1	0.00000	1.00000	1.00000	0.00000	0.00000	0.00000	0.11841	0.00000
52	BLNK	Approx. area of ellipse for off-axis exit pupil area									
53	PROD	50	51					0.00000	0.00000	0.01367	0.00000
54	BLNK	Get the on-axis exit pupil area									
55	REAY	17	1	0.00000	0.00000	0.00000	1.00000	0.00000	0.00000	0.11918	0.00000
56	PROD	55	55					0.00000	0.00000	0.01420	0.00000
57	BLNK	Get ratio of areas									
58	DIVI	53	56					0.00000	0.00000	0.96206	0.00000

Figure 2.36. The merit function (continued from figure 2.35) for the lens shown in figure 2.32.

display the relative illumination plot in OS. Instead, you must use the GIA plot (or convert everything to NSC).

2.9 The phase space viewpoint of relative illumination

You may sometimes hear the term *phase space*, especially when it comes to topics related to illumination design, étendue, and nonimaging optics. Although we will not be applying the concept of phase space very much in this book, I will provide a brief explanation of what it is and try to put it in some practical context by way of describing how to think of the relative illumination problem in terms of phase space.

If you look at equation (2.11), you may notice that the elemental étendue is associated with the quantity $\mathrm{d}A_{\mathrm{proj}}\mathrm{d}\Omega$. Based on figure 2.3, we note that $\mathrm{d}A_{\mathrm{proj}} = \mathrm{d}A\cos u$ and $\mathrm{d}\Omega = \mathrm{d}S\cos u/r^2$. Thus, the elemental étendue may be expressed as

$$\mathrm{d}^2\mathcal{E} = \frac{\mathrm{d}A\cos u\mathrm{d}S}{r^2}. \tag{2.38}$$

If we let $\mathrm{d}A = \mathrm{d}x\mathrm{d}y$ and $\mathrm{d}S = \mathrm{d}x_\mathrm{S}\mathrm{d}y_\mathrm{S}$, then equation (2.38) may be written as

$$\mathrm{d}^4\mathcal{E} = \frac{\mathrm{d}x\mathrm{d}y\mathrm{d}x_\mathrm{S}\mathrm{d}y_\mathrm{S}\cos u}{r^2}. \tag{2.39}$$

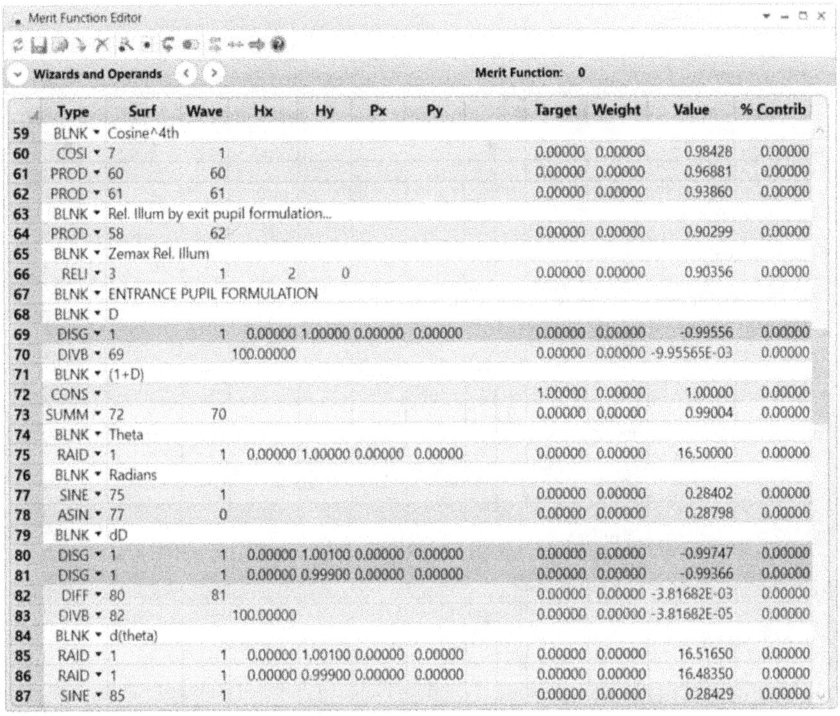

Merit Function Editor

Wizards and Operands — Merit Function: 0

	Type	Surf	Wave	Hx	Hy	Px	Py	Target	Weight	Value	% Contrib
59	BLNK	Cosine^4th									
60	COSI	7	1					0.00000	0.00000	0.98428	0.00000
61	PROD	60	60					0.00000	0.00000	0.96881	0.00000
62	PROD	61	61					0.00000	0.00000	0.93860	0.00000
63	BLNK	Rel. Illum by exit pupil formulation...									
64	PROD	58	62					0.00000	0.00000	0.90299	0.00000
65	BLNK	Zemax Rel. Illum									
66	RELI	3	1	2	0			0.00000	0.00000	0.90356	0.00000
67	BLNK	ENTRANCE PUPIL FORMULATION									
68	BLNK	D									
69	DISG	1	1	0.00000	1.00000	0.00000	0.00000	0.00000	0.00000	-0.99556	0.00000
70	DIVB	69		100.00000				0.00000	0.00000	-9.95565E-03	0.00000
71	BLNK	(1+D)									
72	CONS							1.00000	0.00000	1.00000	0.00000
73	SUMM	72	70					0.00000	0.00000	0.99004	0.00000
74	BLNK	Theta									
75	RAID	1	1	0.00000	1.00000	0.00000	0.00000	0.00000	0.00000	16.50000	0.00000
76	BLNK	Radians									
77	SINE	75	1					0.00000	0.00000	0.28402	0.00000
78	ASIN	77	0					0.00000	0.00000	0.28798	0.00000
79	BLNK	dD									
80	DISG	1	1	0.00000	1.00100	0.00000	0.00000	0.00000	0.00000	-0.99747	0.00000
81	DISG	1	1	0.00000	0.99900	0.00000	0.00000	0.00000	0.00000	-0.99366	0.00000
82	DIFF	80	81					0.00000	0.00000	-3.81682E-03	0.00000
83	DIVB	82		100.00000				0.00000	0.00000	-3.81682E-05	0.00000
84	BLNK	d(theta)									
85	RAID	1	1	0.00000	1.00100	0.00000	0.00000	0.00000	0.00000	16.51650	0.00000
86	RAID	1	1	0.00000	0.99900	0.00000	0.00000	0.00000	0.00000	16.48350	0.00000
87	SINE	85	1					0.00000	0.00000	0.28429	0.00000

Figure 2.37. The merit function (continued from figure 2.36) for the lens shown in figure 2.32.

Based on figure 2.40, let $\vec{v}$ be a unit vector pointing in the direction of the middle ray in the solid angle. If u_x is a rotation of unit vector $\vec{v}$ about the x-axis and u_y is a rotation of the same vector about the y-axis, then its resultant angle u with respect to the z-axis may be determined through a dot product given by $\vec{v}$ and its component along the z-axis:

$$\vec{v} \cdot \ \vec{v}_x = \| \vec{v} \| \| \ \vec{v}_x \| \cos u = \cos u_x \cos u \tag{2.40}$$

$$\begin{aligned} &= \left(\cos u_x \sin u_y, \ \sin u_x, \ \cos u_x \cos u_y\right) \cdot (0, \ 0, \ \cos u_x) \\ &= \cos^2 u_x \cos u_y. \end{aligned} \tag{2.41}$$

In equation (2.40), we made use of the fact that $\| \vec{v} \| = 1$ (it is a unit vector), and $\| \ \vec{v}_x \| = \cos u_x$. Then in equation (2.41), we let $\vec{v} = (\cos u_x \sin u_y, \sin u_x, \cos u_x \cos u_y)$ and $\vec{v}_x = (0,0, \cos u_x)$. Thus, based on the results in equations (2.40) and (2.41), we obtain

$$\cos u = \cos u_x \cos u_y. \tag{2.42}$$

Substituting equation (2.42) into (2.39), we obtain

	Type	Surf	Wave	Hx	Hy	Px	Py		Target	Weight	Value	% Contrib
88	SINE ▼	86	1						0.00000	0.00000	0.28374	0.00000
89	ASIN ▼	87	0						0.00000	0.00000	0.28827	0.00000
90	ASIN ▼	88	0						0.00000	0.00000	0.28769	0.00000
91	DIFF ▼	89	90						0.00000	0.00000	5.75959E-04	0.00000
92	BLNK ▼	dD/d(theta)										
93	DIVI ▼	83	91						0.00000	0.00000	-0.06627	0.00000
94	BLNK ▼	sin(theta)										
95	SINE ▼	75	1						0.00000	0.00000	0.28402	0.00000
96	BLNK ▼	cos(theta)										
97	COSI ▼	75	1						0.00000	0.00000	0.95882	0.00000
98	BLNK ▼	sin(theta) x cos(theta)										
99	PROD ▼	95	97						0.00000	0.00000	0.27232	0.00000
100	BLNK ▼	sin x cos x dD/d(theta)										
101	PROD ▼	99	93						0.00000	0.00000	-0.01805	0.00000
102	BLNK ▼	[1+D + sin x cos x dD/d(theta)]										
103	SUMM ▼	73	101						0.00000	0.00000	0.97200	0.00000
104	BLNK ▼	(1+D) x above										
105	PROD ▼	73	103						0.00000	0.00000	0.96232	0.00000
106	BLNK ▼	Entrance Pupil Off Axis semi-dia x										
107	REAX ▼	1	1	0.00000	1.00000	1.00000	0.00000		0.00000	0.00000	0.20044	0.00000
108	BLNK ▼	Entrance Pupil Off Axis semi-dia y										
109	REAY ▼	1	1	0.00000	1.00000	0.00000	1.00000		0.00000	0.00000	-0.14153	0.00000
110	REAY ▼	1	1	0.00000	1.00000	0.00000	-1.00000		0.00000	0.00000	-0.54750	0.00000
111	DIFF ▼	110	109						0.00000	0.00000	-0.40597	0.00000
112	DIVB ▼	111		2.00000					0.00000	0.00000	-0.20298	0.00000
113	ABSO ▼	112							0.00000	0.00000	0.20298	0.00000
114	BLNK ▼	Entrance Pupil Off Axis Area										
115	PROD ▼	107	113						0.00000	0.00000	0.04069	0.00000
116	BLNK ▼	Entrance Pupil On Axis Area										

Figure 2.38. The merit function (continued from figure 2.37) for the lens shown in figure 2.32.

	Type	Op#					Target	Weight	Value	% Contrib
117	REAY ▼	1	1	0.00000 0.00000 0.00000	1.00000		0.00000	0.00000	0.19899	0.00000
118	REAY ▼	1	1	0.00000 0.00000 0.00000	0.00000		0.00000	0.00000	0.00000	0.00000
119	PROD ▼	117	117				0.00000	0.00000	0.03960	0.00000
120	BLNK ▼	A'/A								
121	DIVI ▼	115	119				0.00000	0.00000	1.02756	0.00000
122	BLNK ▼	Cosine^4th (Chief Ray in Object Space)								
123	COSI ▼	75	1				0.00000	0.00000	0.95882	0.00000
124	PROD ▼	123	123				0.00000	0.00000	0.91934	0.00000
125	PROD ▼	124	124				0.00000	0.00000	0.84518	0.00000
126	BLNK ▼	A/A' x cos^4th								
127	PROD ▼	121	125				0.00000	0.00000	0.86847	0.00000
128	BLNK ▼	Rel Illum (est). by Entrance Pupil Formula								
129	DIVI ▼	127	105				0.00000	0.00000	0.90248	0.00000
130	BLNK ▼	Rel Illum (Zemax)								
131	RELI ▼	3	1	2	0		0.00000	0.00000	0.90356	0.00000

Figure 2.39. The merit function (continued from figure 2.38) for the lens shown in figure 2.32.

$$d^4\mathcal{E} = \frac{dx\,dy\,dx_S\,dy_S\cos u_x \cos u_y}{r^2}. \tag{2.43}$$

Now, let $dx_S/r = du_x$ and $dy_S/r = du_y$. Substituting these into equation (2.43) gives

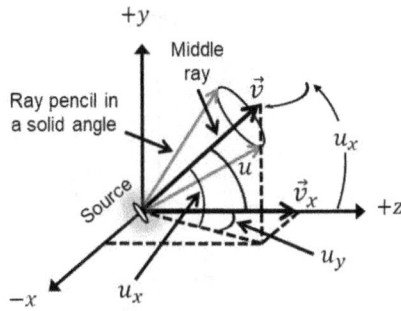

Figure 2.40. The rotated middle ray with unit vector $\vec{v}$.

$$\mathrm{d}^4\mathcal{E} = \mathrm{d}x\mathrm{d}y\, \cos u_x \mathrm{d}u_x \cos u_y \mathrm{d}u_y. \tag{2.44}$$

But note that we can write $\cos u_x \mathrm{d}u_x = \mathrm{d}(\sin u_x)$ and $\cos u_y \mathrm{d}u_y = \mathrm{d}(\sin u_y)$. If we then let $\theta_x = 90° - u_x$ and $\theta_y = 90° - u_y$, then $\mathrm{d}(\sin u_x) = -\mathrm{d}(\cos \theta_x)$, and $\mathrm{d}(\sin u_y) = -\mathrm{d}(\cos \theta_y)$. But $\mathrm{d}(\cos \theta_x)$ and $\mathrm{d}(\cos \theta_y)$ are the differentials of the direction cosines $\cos \theta_x$ and $\cos \theta_y$, respectively. So, if we let $l = \cos \theta_x$ and $m = \cos \theta_y$, then equation (2.44) may be expressed as

$$\mathrm{d}^4\mathcal{E} = \mathrm{d}x\mathrm{d}y\mathrm{d}l\mathrm{d}m. \tag{2.45}$$

In equation (2.45), the elemental étendue is seen to be an elemental *volume* in a *space* of four dimensions, x, y, l, and m [39] (did you notice the change of 'power' from two to four in the differential for étendue starting at equation (2.39)?) The *phase* in the space of x, y, l, and m does not, in a strict sense, actually refer to the quantities l and m, because phase usually refers to angles rather than cosines and sines. As Winston *et al* [39] have highlighted, $\mathrm{d}m$, for instance, is taken as a 'measure of the angle' associated with the direction cosine for m, so it is a measure of the phase for that dimension. Another way to think of this is to recall the term *phasors*. That is, we recall that when we want to add two scalar wave equations, we can add them graphically by plotting the cosine and sine components of the wave amplitudes on the x and y axes, respectively, which is considered a *phasor diagram*. Since l and m are cosines in orthogonal axes, they too can be plotted on a graph in a *phase space diagram*. However, the areas $\mathrm{d}x\mathrm{d}l$ and $\mathrm{d}y\mathrm{d}m$ can also be considered. Due to flux conservation, since $\mathrm{d}^4\mathcal{E}$ must be conserved within a ray pencil, the volume $\mathrm{d}x\mathrm{d}l\mathrm{d}y\mathrm{d}m$ is conserved (like an incompressible fluid), which means that plots of the areas of $\mathrm{d}x\mathrm{d}l$ and $\mathrm{d}y\mathrm{d}m$ at any elemental area intersecting the path of propagation of a ray pencil must be maintained. Since, for instance, $\int \mathrm{d}y\mathrm{d}m$ is the total area, we can plot ray positions in y versus ray directions in m at various planes of interest to study the *phase space area* given by $\int \mathrm{d}y\mathrm{d}m$ at those planes.

Let us take an example. Figure 2.41 depicts the propagation of rays from object to image in an imaging system using the usual Cartesian space (figure 2.41(a)) and using phase space diagrams (figure 2.41(b)). In phase space, ray positions in y at

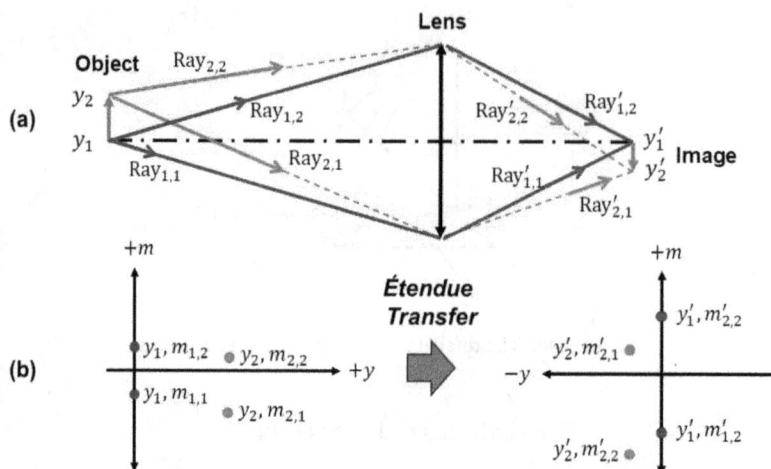

Figure 2.41. Cartesian space (a) and phase space diagrams (b) for an imaging system.

the object are plotted against ray directions m. The subscripts of the y and m values associate them with the rays. For example, y_1, m_1 is the phase space point in the object plane for Ray$_1$, and so on. Points at the image are labeled with superscript primes. At the image, we would expect a plot of y values to be either lengthened (if the image is magnified) or shortened (if the image is demagnified), and we would expect the corresponding m values to either have a smaller spread (if the image is magnified) or a larger spread (if the image is demagnified). This is étendue conservation (equivalently, it is precisely the Lagrange invariant if the ray angles and field heights are small). In the case depicted in figure 2.41, since the image is demagnified, there is a compression in the y-length but a widened spread for the m values. Connecting the four phase space points at the object yields a boundary covering an area of phase space at the object, while connecting the other four points at the image yields a boundary for a phase space area in the image. Both areas must be equal, but their shapes can change. Many other detailed insights about analyzing such shapes of phase space diagrams are provided by Muschaweck and Rehn [19].

In the context of relative illumination, if you first return to equation (2.11), you may note that the irradiance is obtained by dividing both sides of this equation by an elemental area in the image. Therefore, for Lambertian sources, image irradiance may be thought of as L times an elemental étendue in image space divided by an elemental area in the image. Since the phase space volume in a ray pencil is the étendue in that ray pencil, then on the image side, dividing a phase space volume by an elemental area yields a quantity proportional to image irradiance (it is not equal to the irradiance unless it is multiplied by radiance). Assuming symmetric phase space areas in the x and y dimensions at the object and image planes, a 1D version of irradiance in phase space at the image is given by radiance times the phase space areas **divided by length for each phase space area**.

Let us call this the *phase space irradiance*, which is essentially the phase space area per unit length. Dividing each phase space irradiance across the image plane by the axial phase space irradiance cancels the radiance, yielding the relative illumination in the image. Thus, in the case of relative illumination, one would first plot several phase space areas at fixed line segments in, say, the y-axis, in the object plane. By 'line segment,' we mean a line connecting, for instance y_1 with y_2. A subsequent phase space area at the object would be divided by another line segment, such as that from y_3 to y_4, and so on. At the object plane, each line segment may be given equal lengths. Now, each phase space area at the object 'maps' to its corresponding phase space area in the image plane, but the lengths of the line segments at the image may not be equal to each other, due to distortion. If the lens system has nonuniform relative illumination, **the ratios given by the phase space area divided by the segment length for each phase space area at the image** are unequal. Thus, to achieve uniform relative illumination, a design must equalize these ratios (so, perhaps a separate *phase space irradiance diagram* given by phase space area divided by segment length may be plotted).

2.10 Aplanatism and the relative illumination in the pupil

If the entrance and exit pupils of an optical imaging system are images of the stop, then there should be a relative illumination profile at the pupils, depending on the illumination at the stop. Sasián [40] showed that when a lens is aplanatic for an object at infinity, and if the source is Lambertian, the irradiance profile at the exit pupil is (globally) uniform. One way to think about this is to consider the image-space formulation for relative illumination given by equation (2.29). In this formula, for a Lambertian source, the image has uniform relative illumination if two conditions are met. First, the principal ray angle $\bar{u}'$ must be telecentric at the image (i.e. $\bar{u}' = 0$). Second, the exit pupil must not vary in size for the axial and oblique rays. Suppose we think of the exit pupil as the image plane (of the stop, of course) and suppose we consider the stop to approximate a Lambertian source object. Then in this case, if a lens has no spherical aberration, then $\bar{u}' = 0$ at the exit pupil. Further, it is known that the exit pupil size in an imaging lens is a function of the coma of the pupil (*pupil coma*), and that pupil coma is a function of image distortion [40]. So, if the exit pupil is the image, then its distortion impacts A'/A_o' in equation (2.29). But it is also known that the distortion in the pupil is a function of the coma in the lens [41]. It follows that if a lens has no coma, then when the exit pupil is the image plane, the quantity A'/A_o' is one. This means that if an imaging lens has no spherical aberration and no coma (i.e. if it is aplanatic), then the relative illumination at the exit pupil should be uniform, as Sasián points out [40].

The above fact can be applied to illumination design, as it means that if a Lambertian source is located at the image plane of an aplanatic lens, then using this lens in reverse would result in a globally uniform irradiance profile at its exit pupil (which is the lens's entrance pupil when the lens is used in its original orientation). Thus, if an additional lens projects the exit pupil of the first lens towards a screen, we obtain uniform illumination on that screen (provided that the additional lens's

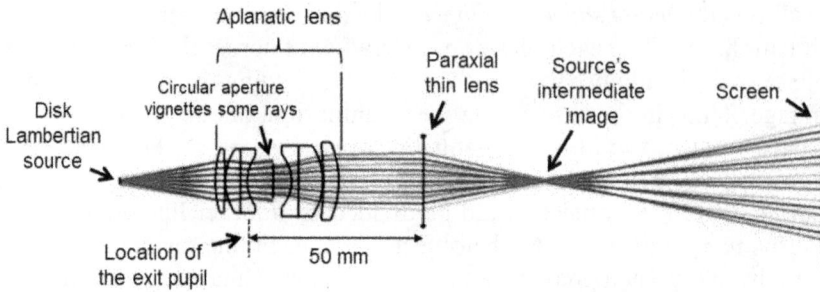

Figure 2.42. A layout that produces uniform illumination using an aplanatic lens.

	Surface Type		Comment	Radius	Thickness	Material	Coating	Clear Semi-Dia
0	OBJECT	Standard ▾		Infinity	28.37078			1.00000
1	STOP	Standard ▾	Stop	Infinity	0.00000			5.30000 U
2	(aper)	Standard ▾		31.54955	2.77643	N-SK16		9.00000 U
3	(aper)	Standard ▾		-146.73154	0.18444			9.00000 U
4	(aper)	Standard ▾		16.13621	4.41213	N-SK16		9.00000 U
5	(aper)	Standard ▾		Infinity	2.00000	F5		9.00000 U
6	(aper)	Standard ▾		11.43730	7.79647			7.00000 U
7	(aper)	Standard ▾		Infinity	5.42083			5.00000 U
8	(aper)	Standard ▾		-11.14702	2.05129	F5		7.50000 U
9	(aper)	Standard ▾		Infinity	6.95514	N-SK16		10.00000 U
10	(aper)	Standard ▾		-17.90062	0.19915			10.00000 U
11	(aper)	Standard ▾		-79.22306	4.36008	N-SK2		10.00000 U
12	(aper)	Standard ▾		-27.11146	24.72489			11.00000 U
13		Paraxial ▾			116.66537			12.50000 U
14	IMAGE	Standard ▾		Infinity				23.18149

Figure 2.43. The prescription for the system shown in figure 2.42. Note that surface 1 is not displayed in figure 2.42 (it is hidden in order to avoid clutter).

intrinsic relative illumination is uniform). Let us take an example. Figure 2.42 shows an SC layout using the 35 mm EFL lens from figure 1.35 in reverse, combined with a paraxial thin lens that has an EFL of 35 mm. The prescription for this system is provided in figure 2.43, and the GIA plot for the irradiance profile at the final plane is given in figure 2.44. Note the highly uniform global profile, with some Monte Carlo ray trace noise modulating the globally flat average profile. The settings for the GIA plot are provided in figure 2.45. The object is a Lambertian disk source 2 mm in diameter, but the uniform irradiance profile is maintained if you select any other .IMA file for the source (go to the File drop down menu in the Settings for the GIA plot). In other words, if the file is a letter 'F,' the irradiance profile is

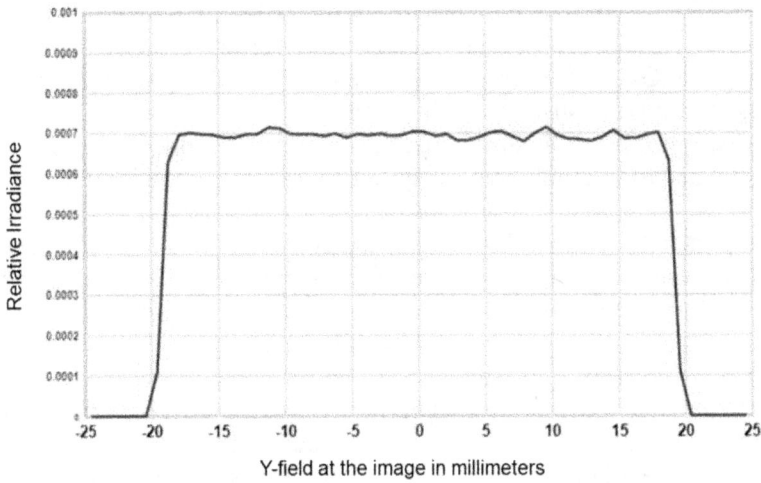

Figure 2.44. The GIA plot in OS for the system shown in figure 2.42.

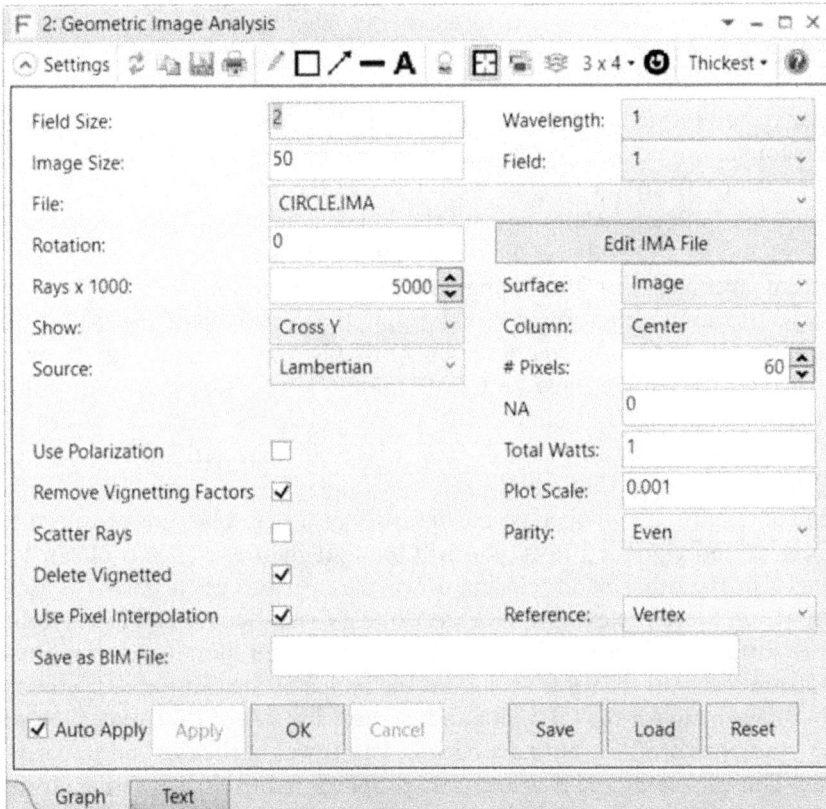

Figure 2.45. The settings for the GIA plot shown in figure 2.44.

unchanged. This is because the current system is like a Köhler configuration in which the object-image conjugate is not the source-screen conjugate. In this case, the aplanatic double-Gauss lens serves as the condenser and the paraxial thin lens serves as the projector. It is not strictly a Köhler configuration in the sense that the source is not projected onto the paraxial thin lens. However, the basic principle is the same. All that is needed for a Köhler type of system is that the source is not projected onto the screen.

As mentioned above, an aplanatic lens has the characteristic that its exit pupil is uniform when the source is Lambertian. So, the paraxial lens is projecting the exit pupil onto the screen (surface 14 in the prescription). Having first determined the location of the exit pupil of the double-Gauss lens, the paraxial thin lens is purposely positioned 50 mm from the exit pupil. Now comes a tricky part. The thickness value of 116.665 37 mm on surface 13 is determined by using a chief ray solve at height = 0, which places the last surface at the focus of the unvignetted chief ray (so, it is actually the principal ray), but this action must be performed by setting surface 7 as the stop and turning on ray aiming. This results in proper SC ray tracing for unvignetted rays through the double-Gauss and paraxial thin lens system. However, when using the GIA feature to plot the irradiance, **you must set the first surface as the stop and remove ray aiming, because this enables the Lambertian source setting in the GIA plot to properly sample the entrance pupil with Lambertian ray densities**.

You may be tempted to replace the paraxial thin lens with a real lens, such as a simple biconvex singlet, but the result would not be uniform. The reason for this is that aberrations from the singlet would cause the irradiance profile to change (think of what happens at and beyond the intermediate image of the source when an uncorrected lens introduces high aberration). This is one of the drawbacks of the current approach for producing uniform illumination. If an aplanatic lens produces uniform illumination at its exit pupil, then subsequent lenses that serve to project this exit pupil towards a screen must also be aplanatic or at least well corrected (and the source must be Lambertian). Of course, a paraxial thin lens is not aplanatic and does not conform to the principles of radiometry, but it does perform ideal point-to-point imaging (so it is 'well corrected'). Thus, for convenience, I used a paraxial thin lens.

One final remark: if you open up the diameter of the circular aperture on surface 7 until there are no vignetted rays, you will see that the irradiance profile's width is expanded. On the other hand, closing down the aperture on surface 7 reduces the profile's width. Since the exit pupil of the double-Gauss lens is conjugate to its stop (which is surface 7 when the double-Gauss lens is used alone), this confirms that what you are seeing at the screen is indeed the image of the exit pupil. If you now go to the Settings menu for the GIA plot and increase the size of the source, you will see that the irradiance profile eventually becomes narrower at the top and trapezoidal at the sides. Thus, the source size is inversely proportional to the useable uniform area at the screen. Recall that, as discussed in section 2.2.11, there is no free étendue. This is one example. When the source's size is increased, this increases the total source étendue. In real life, this could take the form of, for example, switching to a larger source with more flux (e.g. the chip-on-board (COB) type of white LED with a phosphor coating and multiple emitters underneath has a large diameter and high flux). But an increase in the flux for the current system results in a reduction in the useable area of uniform illumination (this is the trade-off between flux and uniformity). To compensate, you would need to increase the size of the double-Gauss lens and the paraxial thin lens (i.e. scale up the system size), but this becomes a costly system (clearly, larger optics leads to higher cost). Hence, again, there is no free étendue.

2.11 Regions of uniformity in collimated light: the searchlight optical layout

When the surface of an extended disk Lambertian source is locally uniform and is located at the back focal plane of a lens, there are two regions of globally uniform irradiance in a space of collimated rays. Such a configuration is known as a *searchlight* [6, 27, 42, 43]. Of course, the rays possess angular divergence for the oblique paths, defined by the size of the source (étendue is at work again). Figure 2.46 shows four layouts for a searchlight, each of which uses different elements for collimation. Shaded areas (Region I and Region II) indicate regions of globally uniform irradiance. Each layout's irradiance profile at the screen is shown in a respective plot in figures 2.47(a)–(d). In figure 2.46(a), a paraxial thin lens becomes the basis for a concept design. In figure 2.46(b), a plano-convex lens (part # 32–484 from Edmund Optics) replaces the paraxial thin lens. In figure 2.46(c), the plano-convex singlet is replaced by a Fresnel lens model (5 grooves/mm, index = 1.5) by way of using OS's hybrid NSC (HNSC) mode. In figure 2.46(d), the same Fresnel lens is modeled in pure NSC mode. The *total track* (source to screen distance) is fixed at 300 mm in each layout, which results in slightly different source diameters in three of the layouts. In order from (a) to (d), they are 10, 9.63, 11.75, and 11.75 mm, respectively. The reason for these differences is that each lens model has slightly different characteristics. For instance, the plano-convex lens has a central thickness, while the paraxial thin lens does not. The Fresnel lens model's grooves are thin

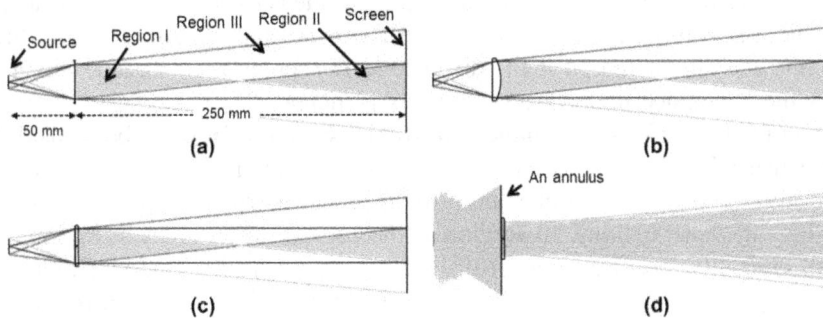

Figure 2.46. Searchlight optical layouts using: (a) a paraxial thin lens, (b) a plano-convex lens, (c) a Fresnel lens in OS's HNSC mode, (d) a Fresnel lens in pure NSC mode.

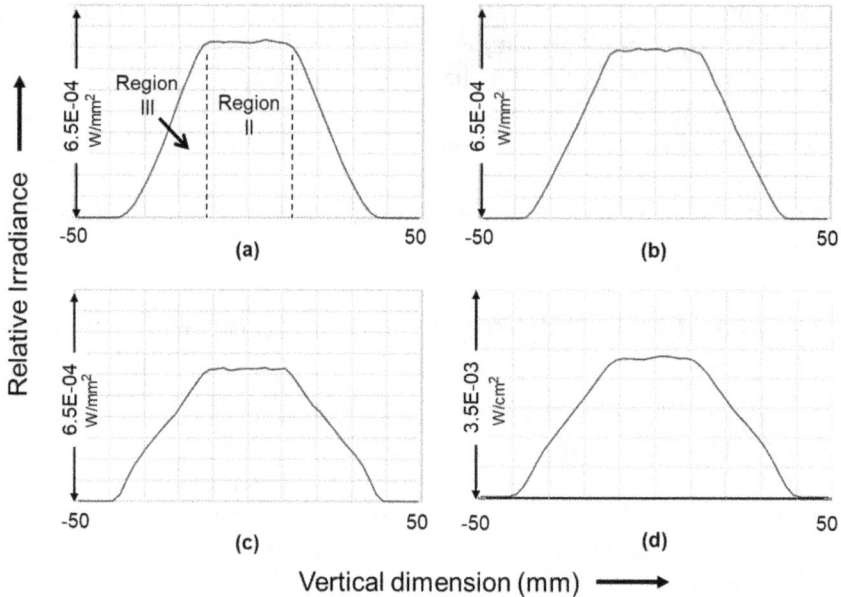

Figure 2.47. Irradiance at the screen for the layouts in figure 2.46: (a) paraxial thin lens, (b) plano-convex lens, (c) Fresnel lens (hybrid NSC mode), (d) Fresnel lens (pure NSC mode).

prisms, so they do not bend rays as regular lenses do. And so on. Due to these differences, each layout's rays had to be collimated using slightly different dimensions for the back focal length, and the width of the irradiance profile had to be equalized by adjusting the source's size.

The irradiance profiles in figures 2.47(a)–(c) are plotted in SC mode using OS's GIA feature, as usual, while the profile in figure 2.47(d) is plotted in pure NSC mode using a rectangular NSC object detector. It can be seen from the layouts and irradiance plots that the region of flat global uniformity at the screen lies within a narrow area given by the intersection of Region II with the screen. At a fixed screen distance, this area of uniformity can be widened if the source's size is increased, but its flux must also increase if the irradiance magnitude is to be maintained (there is no free étendue). Alternatively, if the source's size is maintained, then the area of uniformity may be increased if the distance to the screen is increased, which again reduces the irradiance unless the source flux is increased.

The globally uniform irradiance in Regions I and II may be considered a consequence of étendue division and superposition in the sense that the source's surface may be divided into elemental areas, each of which contributes elemental irradiances in those regions. In Region I, the superposition is given by integrating over all elemental contributions from the source, while in Region II, the super-position is given by integrating over all elemental contributions from the exit pupil of the lens. Outside Regions I and II, not all elemental areas of the source contribute rays, and the number of contributions decreases as the position of observation slides

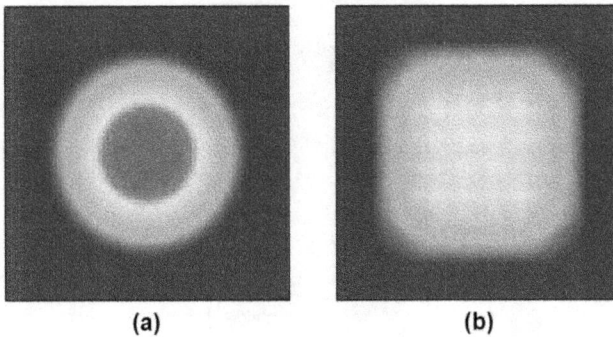

(a)   **(b)**

Figure 2.48. False-color irradiance plots at the screen for the layout in figure 2.46(a). (a) Using a disk Lambertian source. (b) Using a grid of 6×6 line sources.

farther out of the beams in the transverse direction. This results in the slopes on the sides of the irradiance profiles in figures 2.47(a)–(d), which are in Region III.

The arrangements of sources and lenses shown in figures 2.46(a)–(d) can reasonably be considered those of a *classical searchlight* configuration. Many modern searchlights (which are essentially high-powered flashlights) have additional features and nonimaging characteristics (a search through the internet will link to examples from various suppliers). A drawback of the classical searchlight configuration is that it requires the source's surface to be locally uniform, otherwise, patterns can become visible on the screen. For example, figure 2.48 shows a comparison between two false-color irradiance distributions at the screen for the layout in figure 2.46(a). Each is produced by a different source. The source for figure 2.48(a) is a disk Lambertian source 10 mm in diameter, while in figure 2.48(b), the source is a grid of 6×6 lines with Lambertian emission. The pattern is visible on the screen. This can be problematic when, for instance, a large COB type of white LED (the kind that has an array of emitters under a phosphor coating) is used as the source. In this case, the array of emitters can become quite visible in collimated space when a searchlight configuration is used for illumination.

A rather interesting phenomenon occurs for the irradiance along the axis from the lens to the screen. In Region I, the axial irradiance is constant (for an aplanatic lens), because the source's angular size is constant up to the intersection of Regions I and II, where the apparent size of the source becomes limited by the lens diameter [6] (see appendix B.2 for the derivation). Beyond that intersection, in Region II, the axial irradiance is given by equation (2.14), where U' in this case is the angle subtended by the lens's semi-diameter. Thus, in Region II, the axial irradiance drops as one moves farther from the lens. A simple way to observe these beam characteristics in a relative sense is to use OS's pure NSC mode and place a long rectangular detector so that it lies flat along the axis, as shown in figure 2.49 for the case of the plano-convex lens. Of course, the magnitude of the irradiance arriving at the detector is not the axial irradiance (due to the orthogonal orientation of its surface to the rays striking it) but the irradiance at the middle column of pixels on the detector does show the expected trend.

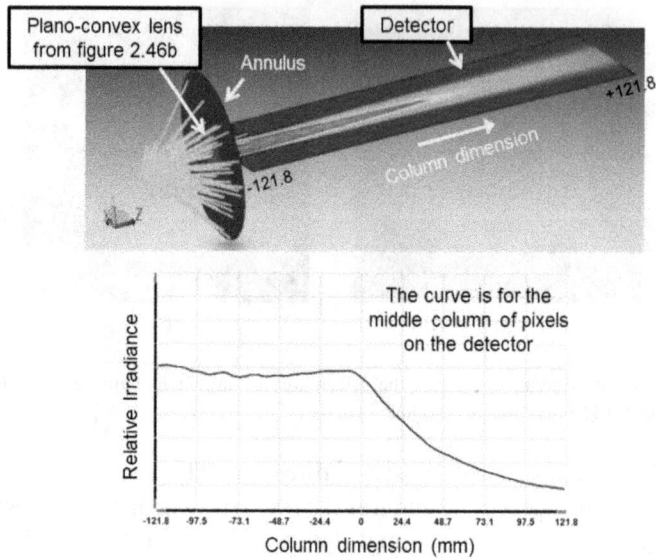

Figure 2.49. The axial irradiance in collimated space for the lens shown in figure 2.46(b).

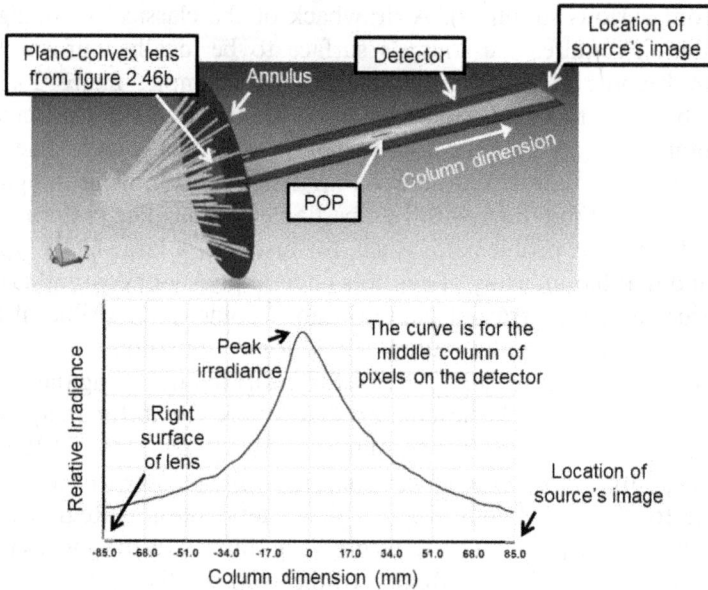

Figure 2.50. The relative irradiance along the axis when a real image of the source is formed by the searchlight's lens, resulting in a POP between the lens and the image.

If the lens is shifted away from the source such that a real image of the source is formed between the lens and the screen, then a position of peak irradiance (POP) can be found somewhere between the lens and the image [7, 29, 44], as shown in figure 2.50 (see appendix B.2). By applying the construction for the *edge field ray*

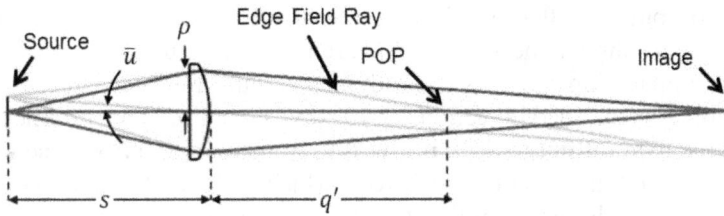

Figure 2.51. The construction relating to equation (2.46).

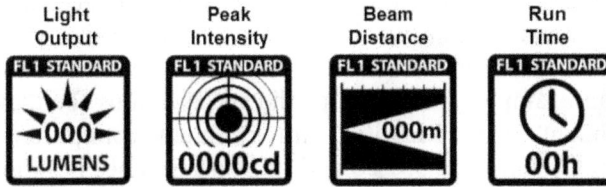

Figure 2.52. Some symbols from the ANSI/PLATO FL1-2019 standard. Reproduced from [46], with the permission of PLATO.

shown in figure 2.51, the POP's location may be estimated by way of a first-order equation [29, 44], expressed as

$$\frac{1}{q'} \approx \frac{1}{f} - \frac{1}{s} + \frac{\tan \bar{u}}{\rho},$$ (2.46)

where f is the EFL of the lens [29, 44]. For example, in figure 2.50, $f \approx 50$ mm, $s \approx 66.4$ mm, $\rho = 12.5$ mm, and $\bar{u} \approx 4.6^0$, giving $q' \approx 87$ mm.

2.12 The specification of flashlights and searchlights based on the ANSI FL1 Standard

Just as the IES has provided guidance and standards on lighting specifications and testing, the American National Standards Institute (ANSI)—in cooperation with other societies, such as the National Electrical Manufacturers Association (NEMA) and the Portable Lights American Trade Organization (PLATO)—has provided guidance and standards for the specification of certain lighting practices. The current revised standard is known as the ANSI/PLATO FL1-2019 standard [45]. These standards are available at the organizations' respective websites, but searching on the internet provides links to some general information about these standards.

It could be useful to be familiar with specification standards for flashlights and searchlights, because sometimes, a simple flashlight could be used as a light source for your illumination application, especially during the prototype development stages. In the ANSI/PLATO FL1 2019 standard, some useful optical specifications are *light output, peak beam intensity, beam distance*, and *runtime*, whose symbols are shown in figure 2.52.

The light output is the flashlight's luminous flux (think of equation (2.20)). This is reasonable, as flashlights and searchlights are meant for human vision. But if your illumination application requires the radiometric flux, then the procedure for the conversion of units outlined in section 2.2.10 may be applied. The runtime is the time it takes for the light output to drop below 10% of the initial output. The peak beam intensity is specified in units of *candela* (cd), which is the unit for luminous flux per unit solid angle (i.e. luminous intensity). The peak beam intensity is therefore the quantity I_{ov} discussed in section 2.2.10. The beam distance is the distance at which the *illuminance* (luminous flux per unit area, in lm m^{-2}) is equal to that produced by a full moon on a clear night. According to a table of illuminance values provided by Boyd [6], the illuminance from a full moon *at the zenith* (i.e. when the moon is directly above you) is 0.27 lm m^{-2}.

Let us take an example. According to the specifications for the Solitaire LED Keychain Flashlight from Maglite® (https://maglite.com/products/solitaire-led-flashlight), its peak intensity is 539 cd and its beam distance is 46 m. This distance should therefore be the distance at which the LED's illumination of a surface yields an illuminance of 0.27 lm m^{-2}. Recall from section 2.2.4 that the intensity I is $d\phi/d\Omega$, so that the luminous intensity I_v is $d\phi_v/d\Omega$. When the plane of illumination is orthogonal to the source's axis, the peak luminous intensity I_{ov} (see section 2.2.10) is $d\phi_v/d\Omega = (d\phi_v/dS)z^2 = E_v z^2$, where E_v is the illuminance at the plane and z is distance between the source and the plane. Thus,

$$E_v = \frac{I_{ov}}{z^2}. \tag{2.47}$$

For the Maglite® Solitaire LED mentioned above, $z = 46$ m and $I_{ov} = 539$ cd. Inserting these values into equation (2.47), we obtain $E_v = 0.25$ lm m^{-2}, which is close to the 0.27 value stated above.

2.13 Searchlights, critical illumination, and Köhler illumination: a comparison at equal flux and track length

Figures 2.53(a)–(c) show paraxial thin lens models for three techniques of illumination. Their corresponding irradiances at the screen are provided in figures 2.54(a)–(c):

Given three common classical illumination techniques (searchlight, critical, and Köhler), it makes sense to compare them so that their pros and cons can be considered when selecting a specific method of illumination for some desired application. A fair approach to the comparison is to maintain equal flux, source size, and track length for each. This is done in figures 2.53(a)–(c), where a locally uniform disk Lambertian source is assumed that emits 1 W of flux into a 25 mm diameter stop located 50 mm from the source in each system. Thus, in figures 2.53(b) and (c), $s = 50$ mm.

The collimating lens in figure 2.53(a) has, of course, an EFL of 50 mm. It is the same lens as that used in figure 2.46(a), except that the source's diameter has been increased to 14 mm, which results in a wider distribution of globally uniform

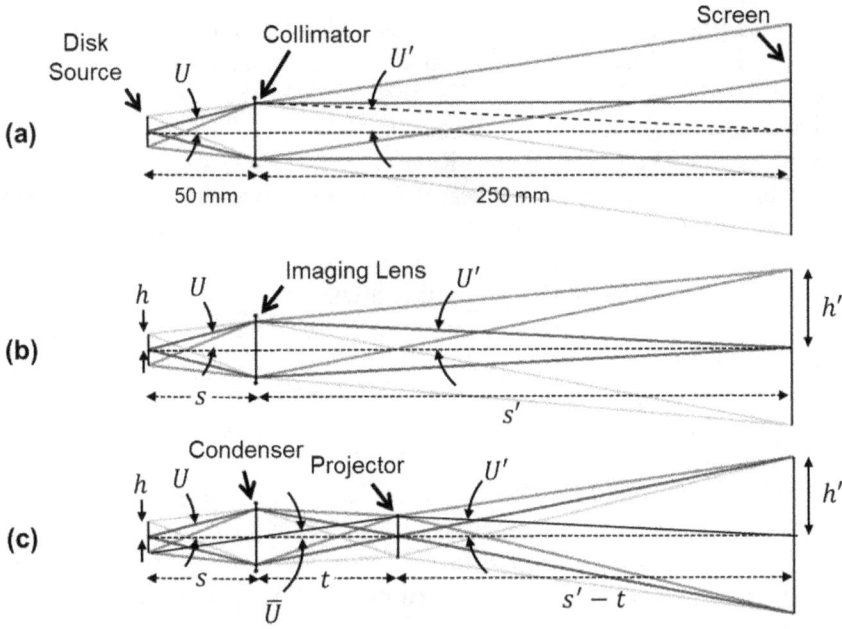

Figure 2.53. Layouts at equal flux, source size, and track length. (a) Searchlight. (b) Critical illumination. (c) Köhler illumination.

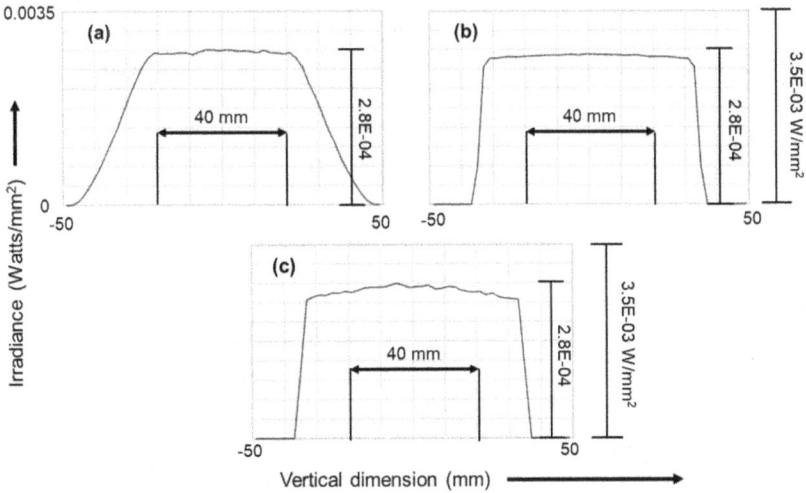

Figure 2.54. The irradiance on the screen for the layouts in figure 2.53. (a) Searchlight. (b) Critical illumination. (c) Köhler illumination.

irradiance so as to cover a minimum circular region of interest (ROI) at the screen 25 mm in diameter. As can be seen, the area of high global uniformity falls within a 40 mm diameter, allowing an ample 7.5 mm of room on both sides of the ROI in consideration of any tolerance stack due to the positioning of the source, lens, etc.

When I worked on figures 2.53(a)–(c), the searchlight system was the first one laid out, because it is known—*a priori*—that it has the smallest ROI when all three layouts have equal track length. To see this, connect the principal ray from the maximum source height all the way to the screen, which it strikes at a point between the edge of Region II and the edge of Region III (see, for example, figure 2.46(a)). Since Region II is the ROI for the searchlight, and since the height at which the principal ray strikes the screen would be the height of the source's image if the lens's EFL were set up for critical illumination, it follows that the searchlight always has a smaller ROI than the critical illumination configuration.

In figure 2.53(b), the imaging lens has to focus the source onto the screen whilst maintaining its distance of 50 mm from the source in order to maintain equal flux through the stop. Thus, if we let f_{cr} be the EFL for the imaging lens (the subscript denoting that it is for critical illumination), then applying equation (1.20) for $s = 50$ mm and $s' = 250$ mm gives $f_{cr} = 41.6667$ mm. Of course, in OS, this EFL is determined through optimization. I performed the optimization by setting a marginal ray solve (height = 0, zone = 0) for the imaging lens's distance to the screen and setting the TTHI operand for that distance to target 250 mm. Then, as I set f_{cr} to be a variable, the optimization quickly determined that f_{cr} was 41.6667 mm.

The image height h' is constrained to be the same in figures 2.53(b) and (c), which is done to create a fair comparison of irradiance and global uniformity. According to first-order optics, $h'/s' = h/s$. Since $s' = 250$ mm, $h = 7$ mm, and $s = 50$ mm, this makes $h' = 35$ mm. Notice in figures 2.54(b) and (c) that the FWHM of the plots is indeed, close to 35 mm (which is interesting—see section 2.2.7). In order to constrain the Köhler layout to have the same image height, we may apply a first-order relationship for the sizes of the condenser, projector, and h' given by

$$\frac{1}{R} = \frac{1}{\rho} + \frac{1}{h'}, \qquad (2.48)$$

where R is the semi-diameter of the projector and ρ is the semi-diameter of the condenser. To derive this formula, note from figure 2.53(c) that $(s' - t)/t = h'/\rho$. Solving for t, we obtain

$$t = \frac{s'}{1 + h'/\rho}. \qquad (2.49)$$

Now, note also from the same figure that $t/s = R/h$. If you substitute equation (2.49) into this formula, you obtain equation (2.48) after performing some algebraic manipulation. Since the stop diameter is 25 mm, $\rho = 12.5$ mm. Applying $h' = 35$ mm and $s' = 250$ mm to equation (2.49), we obtain $t = 65.789\,47$ mm. The EFL f_{kop} of the projector in figure 2.53(c) is determined through $1/f_{kop} = 1/t + 1/(s' - t)$, yielding $f_{kop} = 48.476\,45$ mm. The EFL f_{koc} of the condenser in figure 2.53(c) is determined through $1/f_{koc} = 1/s + 1/t$, yielding $f_{koc} = 28.409\,09$ mm. Of course, again, these can all be determined through optimization in the optical design program, but the computations here are instructive.

We note from comparing the irradiance profiles in figure 2.54 that under the current conditions, the searchlight layout yields the best global uniformity within the desired minimum ROI diameter of 25 mm. However, critical illumination provides the widest globally uniform illumination when the relative illumination is appropriately 'tuned' (recall section 2.7). Köhler illumination presents the least global uniformity under the present conditions, because its irradiance profile is the profile at the condenser (however, see section 2.10). Also, it so happens that under the present fair conditions for comparison, both critical and Köhler illumination yield the same axial irradiance on the screen. We can understand why this is so by taking the ratio $R/(s' - t)$ in figure 2.53(c), which is the tangent of the angle at the screen subtended by the semi-diameter of the projector. Applying equations (2.48) and (2.49), this ratio is

$$\frac{R}{s' - t} = \left(\frac{h'\rho}{\rho + h'} \right) \left[s' - \frac{s'}{1 + (h'/\rho)} \right]^{-1} = \frac{\rho}{s'}. \tag{2.50}$$

The result in equation (2.50) is precisely the tangent of the angle given by the imaging lens's semi-diameter and screen distance in figure 2.53(b). In figures 2.53(b) and (c), $\tan U' = \rho/s'$, which means that the axial irradiance at the screen in both the critical and Köhler illumination configurations is the same in the case of real lenses, which is given by $\pi L \sin^2 U'$ (see appendix B.3). For paraxial thin lenses, we apply equation (2.25) but replace U in the cosine by $\overline{U}$ so that E_p is $\pi L \cos^2 \overline{U} \tan^2 U'$, where L is the source's radiance (see appendix B.3 for the derivation). How far is E_p from E in the Köhler layout of figure 2.53(c)? In that figure, $\rho = 12.5$ mm and $s' = 250$ mm, so that $\tan U' = 12.5/250 = 0.05$. This means that $U' = 2.8624°$. At the source side, $\overline{U} = \text{atan}(h/s) = \text{atan}(7/50) = 7.9696°$. Thus, $E_p = \pi L \cos^2(7.9696°)\tan^2(2.8624°) \approx 0.0025\pi L$ watts/area. For a real lens, $E = \pi L \sin^2(2.8624°) = 0.002\,493\,7565 \approx 0.0025\pi L$ watts/area. Therefore, under the current conditions, E and E_p are virtually the same. How does E_p for the Köhler layout (figure 2.53(c)) compare with E_p for the critical layout (figure 2.53(b))? For critical illumination in a paraxial layout, we apply equation (2.25), for which $U = \text{atan}(\rho/s) = \text{atan}(12.5/50) = 14.0362°$, so that $E_p = \pi L \cos^2(14.0362°)\tan^2(2.8624°) \approx 0.0024\pi L$. Indeed, the peak of the profile in figure 2.54(b) is slightly less than that in figure 2.54(c). As for the searchlight, its axial irradiance is characterized by the type of plot shown in figure 2.49 (see also appendix B.2 for a slightly lengthier discussion).

You may notice that I did not bother determining L in the above calculations. This is because, in the cases of both the paraxial thin lens model and a real lens, L is the same (i.e. it is the radiance in object space). Nevertheless, it is instructive to know how to compute it so that E_p may be determined and compared with the axial irradiance shown in figure 2.54. Moreover, at the end of section 1.2.18, I mentioned that I will show you how to obtain absolute irradiance quantities in the GIA plot by determining the appropriate flux value to enter into the Settings menu for the GIA plot. Now is a good time to talk about it, as it is related to determining L. Figure 2.55 shows the Settings menu for the GIA plot related to figure 2.54(a).

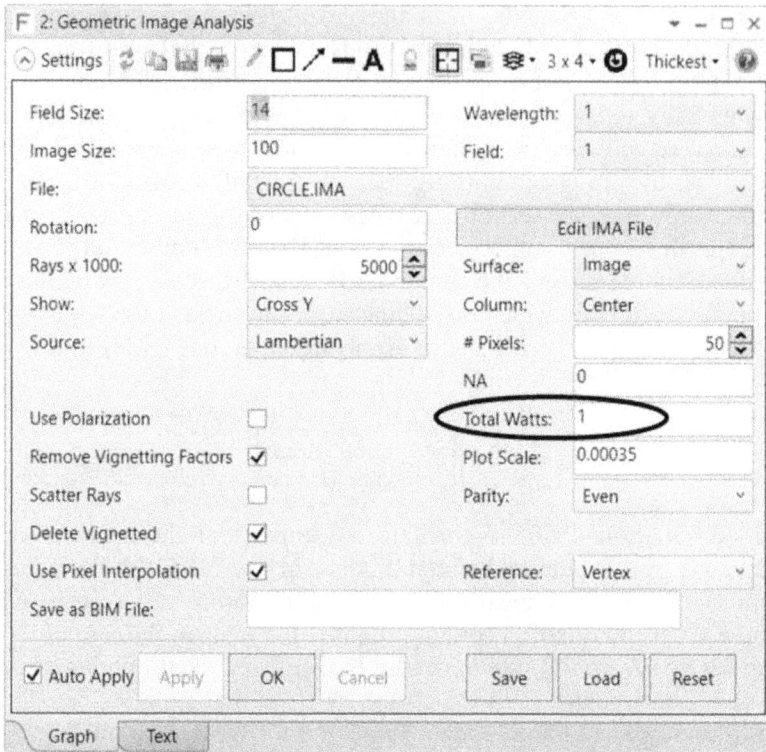

Figure 2.55. The Settings menu for the GIA plot related to figure 2.54(a).

Notice that the flux value entered is 1 W. As mentioned in section 1.2.18, this is the flux you must provide to the GIA plot, assuming that it is known *a priori* how much flux enters the entrance pupil from the source. In section 2.2.10, I showed that for a small Lambertian source of area dA, $\phi_o \approx \pi L dA$, where ϕ_o is the total flux emitted into a hemisphere by the source. In section 2.2.3, I showed that out of this total flux, the amount emitted into (or captured within) a cone of half-angle u is $\pi L dA \sin^2 u$ (see equation (2.6)). Thus, $\sin^2 u$ is always the fraction of flux into a half-angle u from a small Lambertian source of radiance L. In the layout of figure 2.53(a), the stop (which is at the collimating lens) is the entrance pupil, so 1 W is the flux entering it. If we make the approximation that the disk Lambertian source in that figure is of area dA, then 1 W is the flux given by $\pi L dA \sin^2 U$, where U is the angle shown in the figure, which has been determined to be $14.0362°$. Since the source's height is 7 mm, its area is 49π mm^2. Therefore, $L = 1/[49\pi^2 \sin^2 (14.0362°)] \approx 0.035\ 15$ W mm^{-2} sr^{-1}. Applying this result to E_p determined above for the Köhler layout, we obtain $E_p \approx (0.0025)(\pi)(0.035\ 15) \approx 2.76 \times 10^{-4}$ W mm^{-2}, which is essentially the axial irradiance in figures 2.54(a)–(c). Not bad, considering that the source in those figures is not an elemental source. This is why equation (2.6) is so useful to remember.

Table 2.1. A summary of the properties of the searchlight, critical, and Köhler illumination techniques.

Technique	Global uniformity	Local uniformity	Solutions and options
Searchlight	High (within the desired region of interest)	Modest (depends on source's surface)	Use engineered diffusers[1] to improve local uniformity
Critical	Moderate (depends on relative illumination)	Low (in fact, undesirable)	Defocus screen, or use engineered diffusers[1] to improve local uniformity
Köhler	Low (unless source emits uniformly at first lens)	High (the best of the three methods here)	Apply aplanatism to improve global uniformity (see section 2.10)

Notes[1] See, for example, diffusers produced by Luminit® (www.luminitco.com), HOLO/OR (www.holoor.co.il), and RPC Photonics (www.rpcphotonics.com).

Despite the shortcomings of the present configuration of Köhler illumination in terms of its global uniformity, we once again note that it has the highest local uniformity out of the three classical techniques when the source is locally nonuniform. In this case, it is obvious that critical illumination possesses the worst case of local uniformity, followed by the searchlight. These properties for the three classical illumination configurations can now be summarized in table 2.1, which highlights their pros and cons.

Let us use a *real* source and check our three layouts. It so happens that the LED manufacturer Luminus (www.luminus.com) supplies a number of white (broad visible spectrum) COB array LEDs with *light-emitting surface* (LES) diameters of 14.5 mm, which is rather convenient, as it is close to the 14 mm diameter source used in figures 2.53 and 2.54. Moreover, many of their COB LEDs have near-Lambertian emission. One example is part # CIM-14, which is part of their line of 'Standard Gen 6 COB Array' products. On their website, a source file of the 'SDF' (spectral color format) file type is available for download. Upon download, the file has a rather long filename, 'Luminus CIM-14-30-90-36-TC60 FP230221050049B 朗明 RevA 450mA #1-20230312_5000000Rays_ZEMAX,' and the end of the filename indicates that it has five million rays. When you use SDF source files, you must place them into the following Zemax directory: Documents > Zemax > Objects > Sources > Source Files. If you have done this, set up a pure NSC system model with the NSC Editor prescription and Multi-Configuration Editor shown in figures 2.56 and 2.57, respectively. This produces the layouts shown in figure 2.58(a) (3D side view) and figure 2.58(b) (solid model diagonal view).

In figure 2.56, because my computer mouse's cursor was on the cell in black (object 1, comment column), the rest of the column labels were automatically made to refer to the properties of object 1. Those readers who have gone through the examples in section 1.2.15 may already be familiar with this characteristic of OS's pure NSC user interface. In order to fit the NSC Editor into a figure on a page in this book, some of the columns have been 'hidden' from view such that only relevant

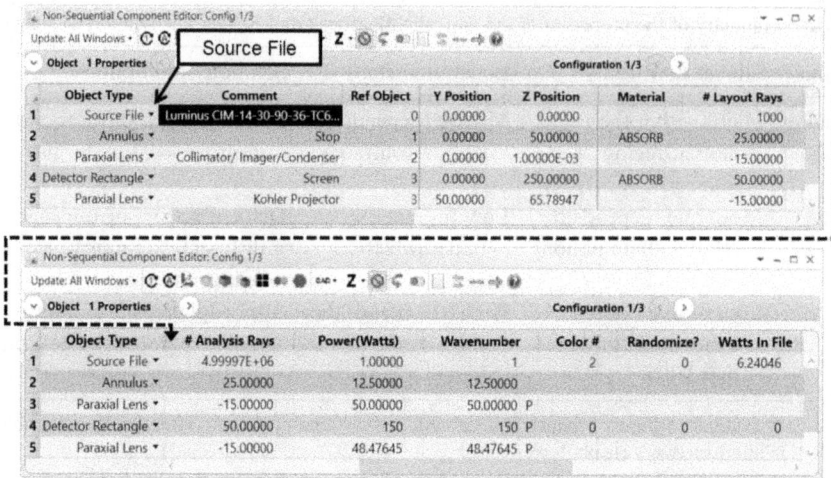

Figure 2.56. The NSC Editor prescription in OS for the layout shown in figure 2.58.

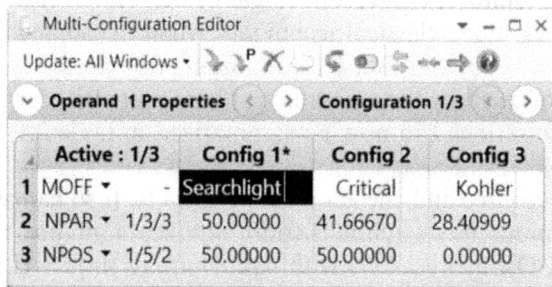

Figure 2.57. The Multi-Configuration Editor settings for the layout shown in figure 2.58.

columns are displayed. The top prescription only displays the columns up to the # Layout Rays column, so the rest of the columns towards the right that could not be displayed are continued at the bottom, starting with # Analysis Rays. Note that there are pickups in columns for objects 4 and 5. They are there merely to obtain the same values for the pixel number and focal length for those objects, respectively.

In figure 2.57, three configurations have been set, and they are labeled in the blank MOFF row. On the second row, the NPAR operand sets the focal length parameter for object 3 (the paraxial lens model), such that the lens attains the appropriate EFL that was calculated earlier for the searchlight, critical, and Köhler layouts. The NPOS operand on row 3 sets the Y position of object 5 (i.e. the Köhler projector lens) to be at 50 mm for configurations 1 and 2. But for configuration 3, this lens is inserted at its appropriate location, which is why the NPOS setting is 0 mm for that configuration. Note the * symbol on Config 1, indicating that the prescription in figure 2.56 is that for the first configuration. This is why you see this lens being vertically shifted out of the way in figure 2.58. Thus, the Multi-

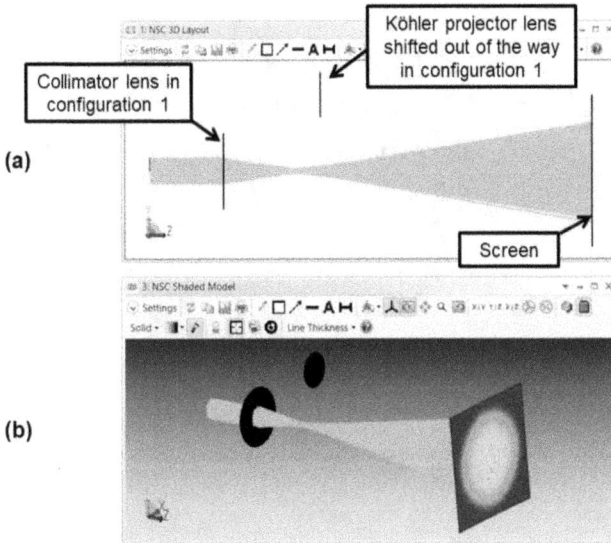

Figure 2.58. The NSC mode layouts for the prescriptions shown in figures 2.56 and 2.57.

Configuration Editor lets us set up three different configurations in a single OS file, which is rather convenient.

The rays displayed in figure 2.58(a) appear to be collimated when leaving the source, for unknown reasons. A thousand of these rays are being displayed using the # Layout Rays setting for object 1. The SDF source file comes with its own flux and wavelength values, so we need to worry about what to set for those, but I have opted to display the rays in green in the layouts. The irradiance distribution shown in figure 2.58(b) was the last ray trace result. This 3D solid model layout with the displayed irradiance profile in false color is set up using the settings shown in figure 2.59. The actual false-color detector display is shown in figure 2.60(a), and the one-dimensional middle column profile is displayed in figure 2.60(b). Note the presence of ray trace noise. Note also that since this source file possesses its own flux value, we shall not be concerned with the irradiance magnitude for the plots, only their profiles.

If we now double-click on the Config 2 cell in the Multi-Configuration Editor, the system switches to the critical illumination layout, with the resulting irradiance profiles shown in figure 2.61. Look! The COB array emitters are visible! This is obviously the drawback of critical illumination. It may seem that these emitters are not visible in the irradiance profile for the searchlight configuration, but this is because there are not enough rays in the source file to resolve them, so you mostly see ray trace noise. Hold that thought—I will soon show you a trick for determining whether the array emitters would indeed be visible in the searchlight configuration. First, let us go to the Köhler configuration, whose irradiance profile is shown in figure 2.62. This configuration actually has excellent local uniformity, but the ray trace noise is excessive. This is the drawback of using ray source files, even though

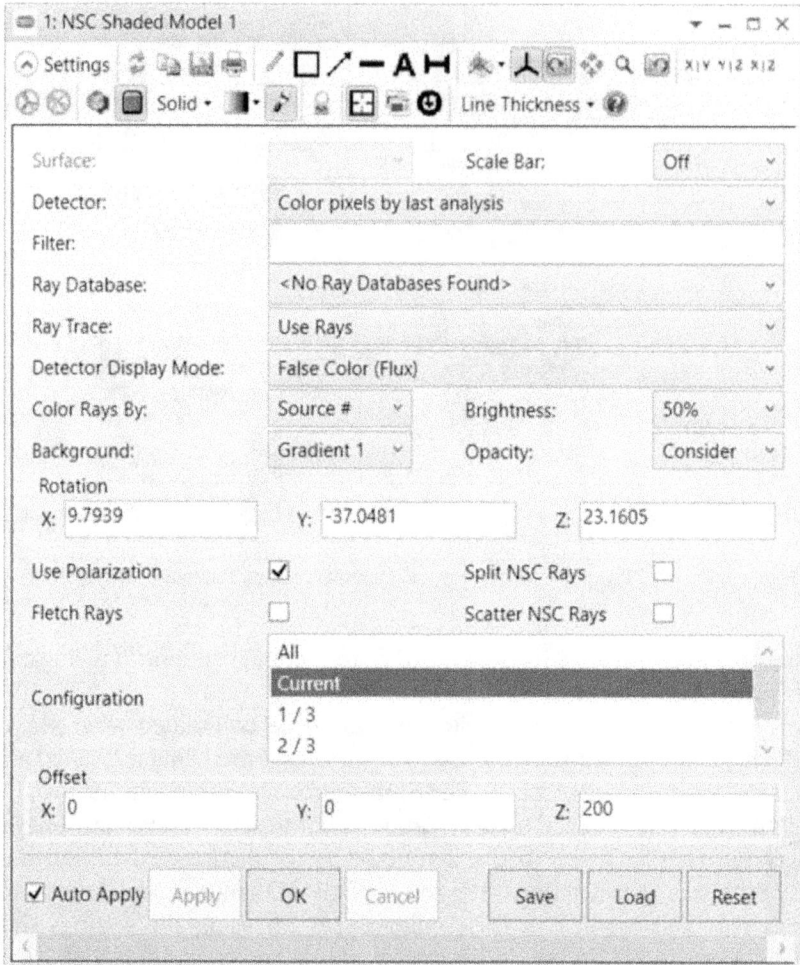

Figure 2.59. The settings used to display the layout shown in figure 2.58(b).

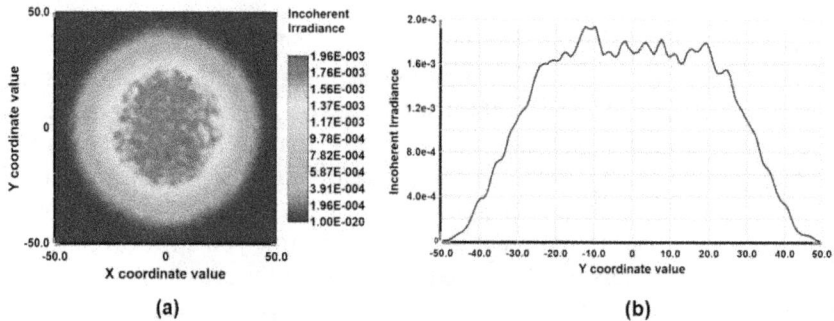

(a)

(b)

Figure 2.60. Irradiance profiles for the layout in figure 2.58 (note that ray trace noise is present). (a) False color. (b) Middle column profile.

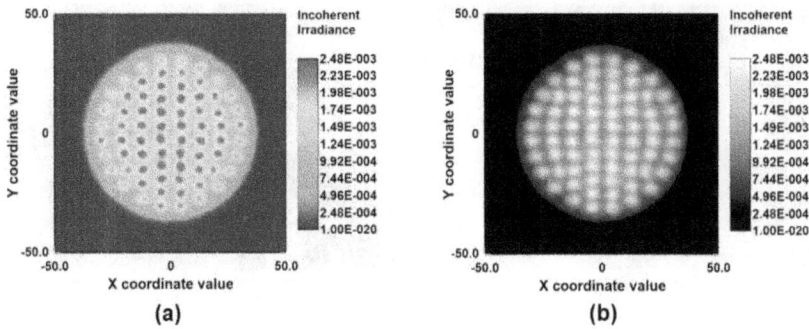

Figure 2.61. The irradiance profile for the critical illumination configuration. (a) False color. (b) Inverted gray scale.

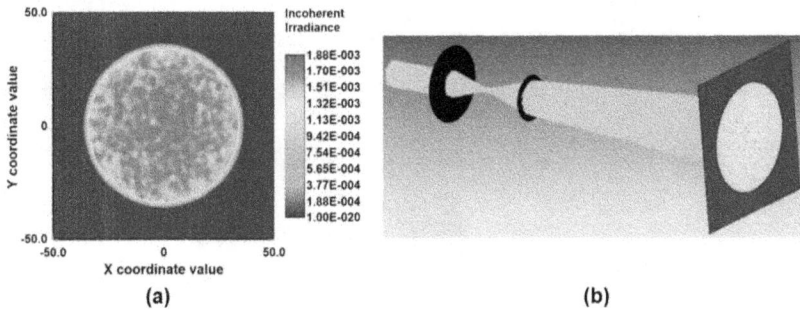

Figure 2.62. The irradiance profile for the Köhler illumination configuration (note that ray trace noise is present). (a) False color. (b) 3D layout with profile.

they contain actual ray data (which enabled us to use the critical illumination configuration to check for any presence of visible array emitters).

Now, it is time for the trick. You can export the image shown in figure 2.61(b) into a bitmap file, crop it, then place it into the following directory: Documents > Zemax > IMAFiles. We then return to the SC model for the searchlight system (i.e. figure 2.53(a)) and this time, we use the geometric bitmap image (GBI) feature (see sections 1.3.13 and 1.3.14 for a review of this analysis feature in OS) to display the irradiance profile on the screen. The advantage of this analysis feature is that many more rays can be traced. Profiles produced using 125 million rays are shown in figure 2.63. In figure 2.63(a), the lens is given an EFL of 41.6667 mm so that it can operate in critical illumination in order to check the bitmap image file for the presence of the COB array emitters, which are evidently visible. Having confirmed this, the lens was switched back to searchlight mode, resulting in the irradiance profile shown in figure 2.63, where the array emitters are still difficult to see, but they are present if you look closely. In figure 2.63(c), the distance from the lens to the screen was increased to 500 mm, and the screen size was doubled. The emitters are not quite visible. The settings used to display the GBI profile of figure 2.63(c) are shown in figure 2.64.

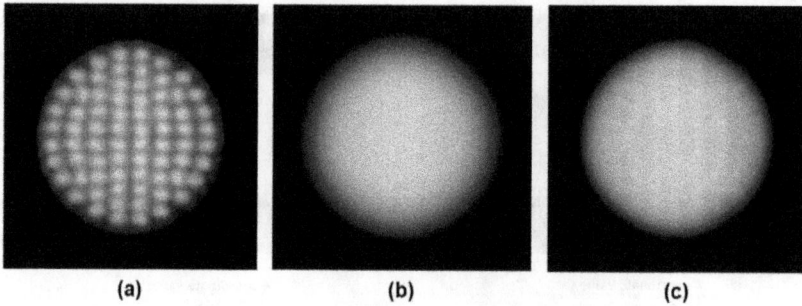

Figure 2.63. Irradiance profiles produced using GBI in SC mode. (a) Critical illumination. (b) Searchlight configuration. (c) Searchlight with a 500 mm distance to the screen.

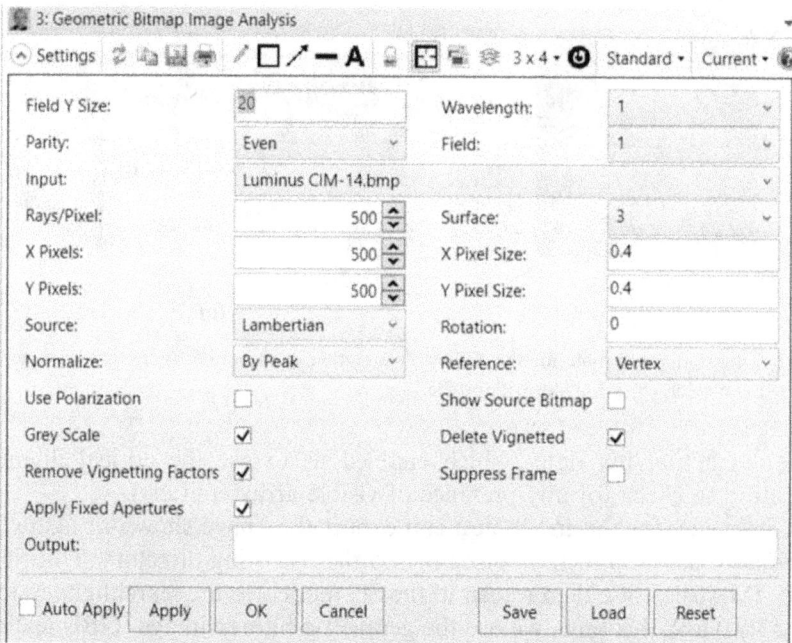

Figure 2.64. The GBI settings for the plot shown in figure 2.63(c).

In figure 2.64, the wavelength setting of 1 refers to 550 nm, and the field is the axial location at the object. Using pixel size of 0.4 mm with 500 pixels in X and Y yields a screen size of 200 mm × 200 mm, which is double the size used in the GIA plots of figure 2.54. The highly visible array emitters in figure 2.63(c) are, of course, not images of the emitters. Rather, they are present because of overlapping bright and dark regions in collimated space. I have personally observed these patterns for a searchlight setup in real life, but using a different COB array LED from a different manufacturer. All COB array LEDs possess this characteristic, which is why the use of the Köhler configuration is so vital in illumination design.

2.14 How to lay out light pipes for uniform illumination

Although it is often a simple matter to insert light pipe models into the NSC layout of an optical system (see, for example, section 1.2.18), it is not always obvious to designers *how* a light pipe is actually used, because the homogenizing effect occurs only at the exit surface of the pipe. Further, as a consequence of global étendue conservation, the exiting rays diverge. Therefore, even if it occurred to a designer to project the exit surface of a light pipe towards a screen, the size of the projection lens could become rather large. Thus, a practical way to use a light pipe is to add a relay lens to first focus all exit rays onto an intermediate plane, as shown conceptually by the simple system in figure 2.65 for the case of a rectangular light pipe (re-used from figure 1.96 but in pure NSC mode). The image of the entrance surface of the pipe forms on that plane, along with multiples of this image caused by the reflections in the light pipe. Next, the entrance pupil of a projection lens (or simply, the 'projector') should be conjugate to the focal plane of the relay. The projector then projects the exit surface of the pipe onto a screen 125 mm from the projector. The irradiance profile is shown in figure 2.66. Because the intermediate plane is the projector's entrance pupil, the focused rays from the relay end up at the stop of the projector. Thus, the projector's stop is conjugate to the relay's focal plane, while the exit surface of the light pipe is conjugate to the screen [47, 48].

Although the irradiance profile is not globally fully flat, you may note that had it not been for the light pipe, the irradiance on the screen direct from the source would have had high global nonuniformity. This is because the source model used is a small 2 mm diameter disk with a radiant intensity proportional to the cosine to the 80th power, which is a very narrow angular distribution. We can calculate this without simulation. What is the half-angle divergence for such a distribution? It is the angle

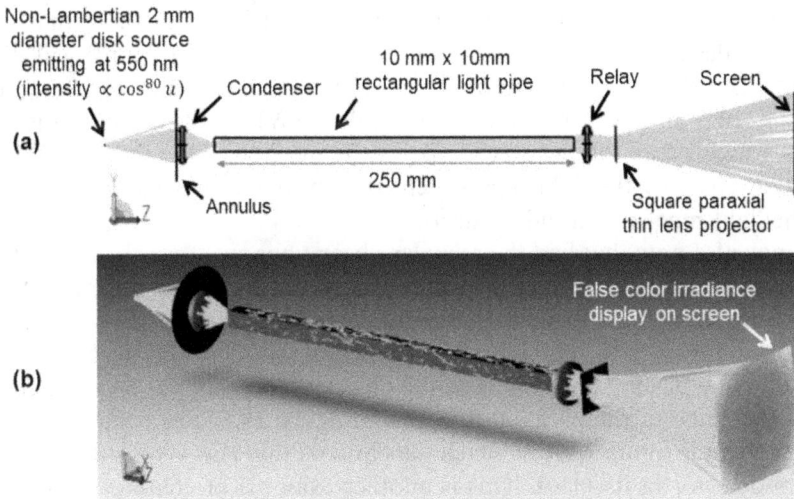

Figure 2.65. A layout for a light pipe system. (a) Side layout. (b) 3D shaded layout.

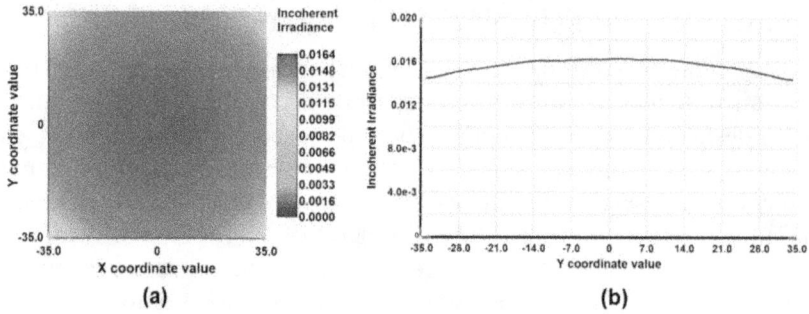

Figure 2.66. The irradiance profile on the screen of figure 2.65. (a) False color. (b) 1D profile for the middle column.

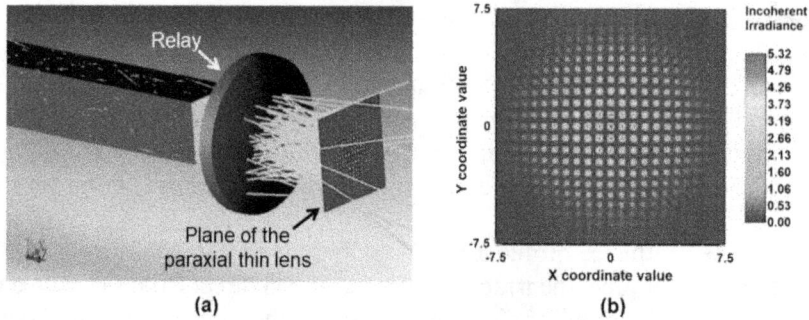

Figure 2.67. The view looking into a rectangular light pipe. (a) The layout with the detector. (b) The irradiance profile at the detector.

for which $\cos^{80} u = 0.5$. Solving this for u yields $u \approx 7.5°$. From equation (2.47), the irradiance on a screen is proportional to the radiant intensity (note that the $1/z^2$ factor just scales the irradiance magnitude). So, if rays from the source strike a plane 125 mm away, then the half-width of the irradiance distribution is 125 mm times $\tan(7.5°) \approx 16.5$ mm. So, the irradiance profile's FWHM on the screen is 33 mm, which is roughly half of the screen size in figure 2.66. Yet, the FWHM of the distribution given by the light pipe system is seen to extend wider than the screen. Thus, the light pipe system did a fair job.

You may also note that had the relay lens been your eye, then the focused exiting rays would end up at the surface of your retina. You would then see multiples of the entrance surface, similar to those seen through a kaleidoscope. In fact, those who have looked through a light pipe can relate to this description. We can easily simulate this by placing an NSC detector object at the plane of focus of the relay, which is the plane of the paraxial thin lens in figure 2.65. The resulting irradiance profile is shown in figure 2.67, in which the paraxial thin lens is replaced by a 15 mm × 15 mm detector in its place. This is another example of étendue division. Their superposition occurs at the screen, as shown in figures 2.65 and 2.66.

2.15 How to lay out fly's eye arrays for uniform illumination

Figure 2.68 shows a 2D layout and a 3D layout of a fly's eye array system that were modeled in OS's pure NSC mode. The irradiance profile on the screen is shown in figure 2.69, and the system's prescription is provided in figure 2.70. I will help you understand how this system is laid out by going about it in reverse. That is, I will begin from the end. I will describe certain characteristics of the 'final product,' and show you what you need to achieve, which sets the direction for concept and design. I will then describe my approach and explain the basic principle of the fly's eye array (i.e. how it works, why it works).

In figure 2.68(a), there are three annuli, each of which serves as an aperture (objects 2, 5, and 8 in the prescription), which are needed for different purposes. The first annulus (object 2) is like the first aperture prior to the first collimator element (object 3) in the collimator's lens barrel. You may notice that objects 3 and 4 are labeled 'backwards' (i.e. Element 2, then 1). The reason is because they were first designed in reverse in SC mode, from infinity space towards the source (I will describe this step later). So, in the SC file, Element 1 was the first lens. The second annulus (object 5) was the aperture stop in the SC file used to design the collimator lens pair (does this bring back a memory? Think of figure 1.207). The lenslet arrays are identical, but the second array is the reverse of the first. Each has 5×5 plano-convex lenslets which are identical to those used in figures 2.12 and 2.13). The reason that we have a pair of arrays will be made clear soon (it has to do with the fact that the source is not a point source). The third annulus (object 8) is needed to block any

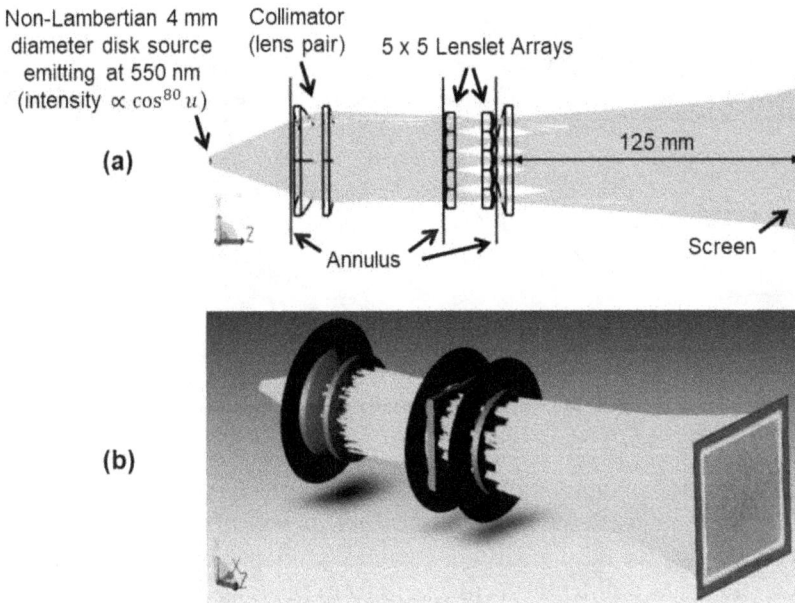

Figure 2.68. A fly's eye array system. (a) Two-dimensional layout. (b) Three-dimensional shaded model with traced rays on a screen.

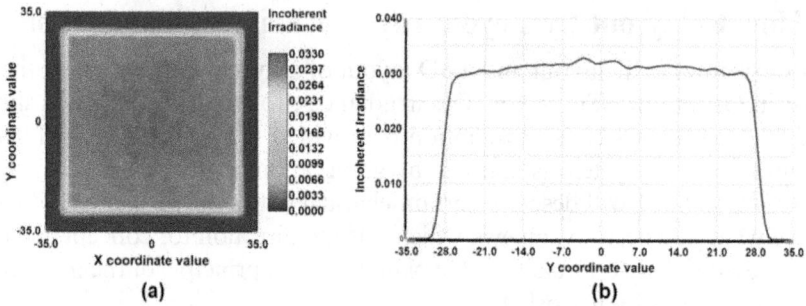

Figure 2.69. The irradiance profile on the screen for the system shown in figure 2.68. (a) False color. (b) For the middle column on the screen.

Non-Sequential Component Editor

Update: All Windows

Object 1 Properties — Configuration 1/1

	Object Type	Comment	Ref Object	Z Position	Tilt About X	Material	# Layout Rays	# Analysis Rays	Power(Watts)
1	Source Ellipse		0	0.00000	0.00000		3000	2E+06	1.00000
2	Annulus	Aperture 1	0	34.90000	0.00000	ABSORB	35.00000	35.00000	20.00000
3	Standard Lens	Element 2	0	40.12636	0.00000	N-SF11	-53.75965	0.00000	20.00000
4	Standard Lens	Element 1	3	9.00000	0.00000	N-SF11	-1579.23116	0.00000	23.00000
5	Annulus	Aperture 2	0	100.88478	0.00000	ABSORB	35.00000	35.00000 P	20.00000
6	Lenslet Array 2	Fly's eye 1	0	100.88578	0.00000	N-BK7	4.00000	4.00000	5.00000
7	Lenslet Array 2	Fly's eye 2	6	22.00000	180.00000	N-BK7	4.00000	4.00000	5.00000
8	Annulus	Aperture 3	0	123.88478	0.00000	ABSORB	35.00000	35.00000 P	20.00000
9	Standard Lens	Global focuser	0	123.88578	0.00000	N-BK7	70.00000	0.00000	23.00000
10	Null Object	Ref	9	7.00000	0.00000				
11	Detector Rectangle	Screen	10	125.00000	0.00000	ABSORB	35.00000	35.00000	100

Non-Sequential Component Editor

Update: All Windows

Object 1 Properties — Configuration 1/1

	Object Type	Wavenumber	Color #	X Half Width	Y Half Width	Source Distance	Cosine Exponent	Gauss Gx	Ga
1	Source Ellipse	1	2	2.00000	2.00000	0.00000	80.00000	0.00000	0
2	Annulus	20.00000 P							
3	Standard Lens	23.00000	8.00000	-34.03516	0.00000	23.00000	23.00000		
4	Standard Lens	23.00000	6.00000	-76.09827	0.00000	23.00000	23.00000		
5	Annulus	20.00000 P							
6	Lenslet Array 2			10.00000	0.00000	0.00000	0.00000	5	
7	Lenslet Array 2			10.00000	0.00000	0.00000	0.00000	5	
8	Annulus	20.00000 P							
9	Standard Lens	23.00000	7.00000	0.00000	0.00000	23.00000	23.00000		
10	Null Object								
11	Detector Rectangle	100 P	0	0	2	0	0.00000	0	

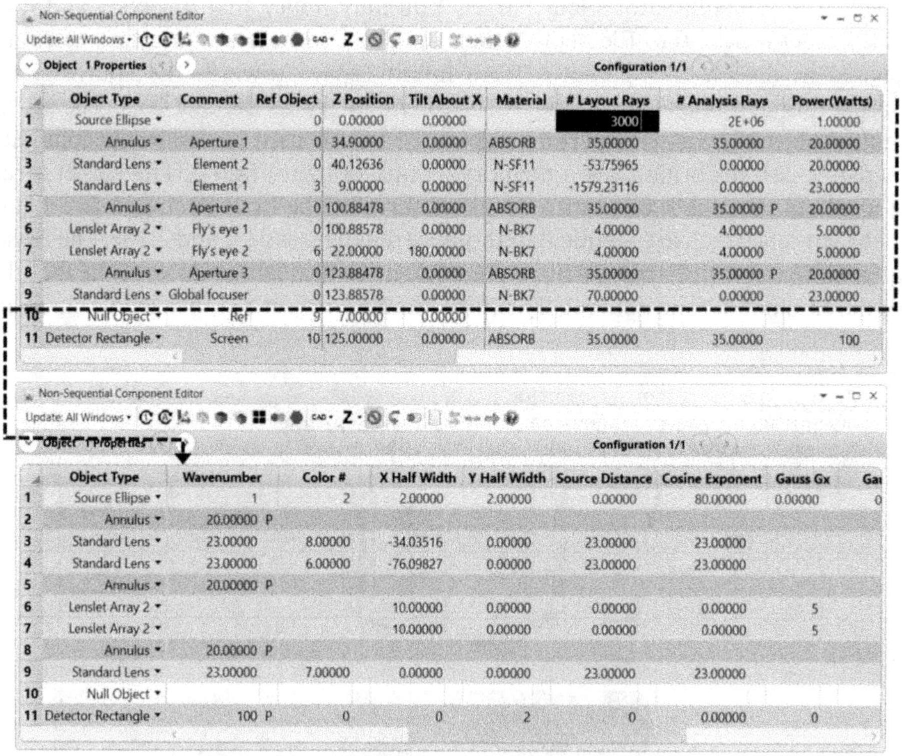

Figure 2.70. The prescription for the model in figure 2.68.

stray rays that miss the lenslet arrays for any reason. The plano-convex lens to the right of the second lenslet array is the *global focuser* (think of figure 2.13). Its purpose is to superpose the divided sub-étendues produced by the lenslet arrays onto the screen, but it has a different focal length from the focusing lens in figure 2.13 in order to widen the distribution on the screen.

As always, the blackened cell for object 1 under '# Layout Rays' is there because it was the last cell that my computer mouse's cursor was placed in. Due to this, the labels for each column (e.g. # Layout Rays, # Analysis Rays, etc.) are only meant to apply to that object. When the mouse cursor is moved to the cells of other objects, those labels change (and you know what they are when you use the OS program).

The irradiance profile shown in figure 2.69(a) is square because the lenslets are square. Notice that the footprint diagrams shown in figure 2.12 are circles, because circular lenslets were used (in SC mode) in that figure. Generally, the shape of the distribution's perimeter is defined by the shape of the lenslets. Thus, rectangular lenslets produce rectangular distributions, and so on. This is due to the fundamental mechanism of étendue division and superposition used in fly's eye array designs. Imagine that at the plane of the second annulus (object 5), the 5×5 array of lenslets divides the irradiance distribution into 25 small areas that sample the distribution—much like a low-resolution image sensor. The idea is to then enlarge these areas and overlap them (via superposition) at the screen. The power (and spacing) of the lenslet pair defines the distribution's width. When the power is increased (and the spacing is reduced accordingly), the distribution widens (or you can also shift the screen farther away).

If the source had been an idealized point source, then the rays reaching the first lenslet array would have been axially collimated, and there would have been no need for a second lenslet array (like in figures 2.12 and 2.13). But sources, such as small LEDs, always have a size. In the current example, I used a 4 mm diameter disk source (an arbitrary size). Its radiant intensity and wavelength of emission (550 nm) are the same as those used in the light pipe example of the previous section. In order to understand how the use of a finite-sized source impacts the irradiance distribution on the screen, we model the fly's eye system in SC mode, beginning with the single lenslet array system from figure 2.12, which is now shown in figure 2.71, where fields at ±2.3° have been included. The rays in the layout are color coded by field. The off-axis fields represent the collimation of light from an extended disk source with a

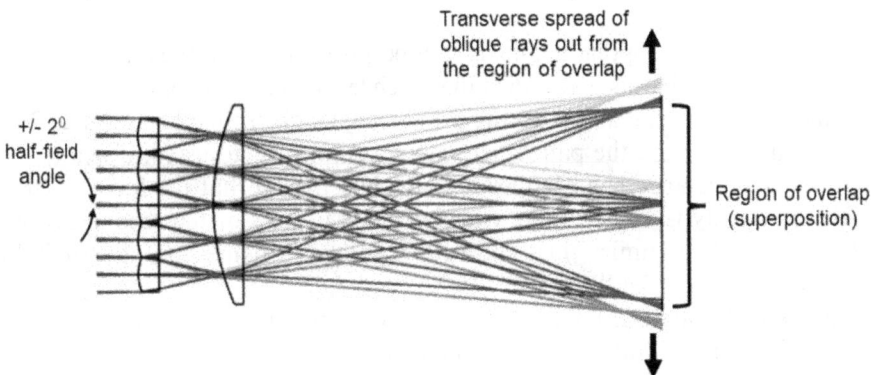

Figure 2.71. The effect of finite source size on the irradiance profile at the screen.

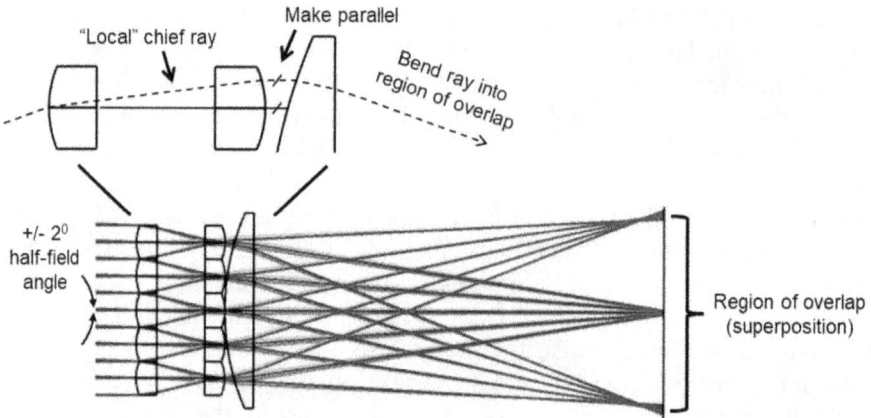

Figure 2.72. The bending of the local chief ray by a second lenslet array.

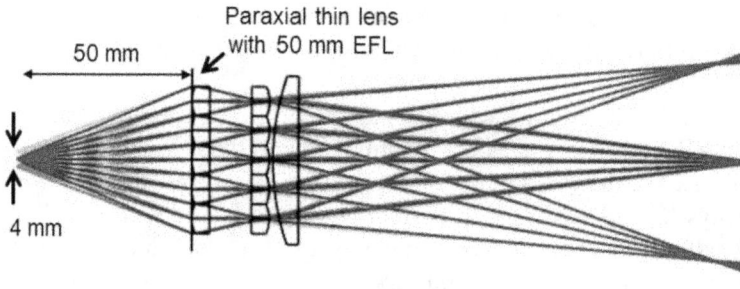

Figure 2.73. Using a paraxial thin lens to collimate rays from an extended source.

4 mm diameter using a 50 mm EFL lens (thus, an arctan of 2 mm over 50 mm equals roughly 2.3°). Notice the transverse spread of rays at the screen. This means that globally uniform irradiance only occurs in the region of overlap. Outside this region, the edge of the distribution tapers down, resulting in a trapezoidal one-dimensional profile.

The way to bring the spreading fields to superposition within the region of overlap is to mount a second lenslet array such that each lenslet in the second array bends the 'local chief ray' from the lenslet in the first array, as illustrated in figure 2.72:

The distance between the pair of lenslet arrays may be adjusted, if necessary. In figure 2.72, this distance has been arbitrarily chosen such that the focus of all rays from the first array is approximately at the right surfaces of the second array. No effort was made to optimize this (and yet, as shown in figure 2.69, the global and local uniformity is not bad).

In figures 2.71 and 2.72, the fields were set by angle. In real life, we would have a finite-sized source at a finite distance from a collimating lens. Since I wanted to have a 4 mm disk source and a lens with an EFL of 50 mm, I began with the simplest approach, i.e. using a paraxial thin lens model, as shown in figure 2.73:

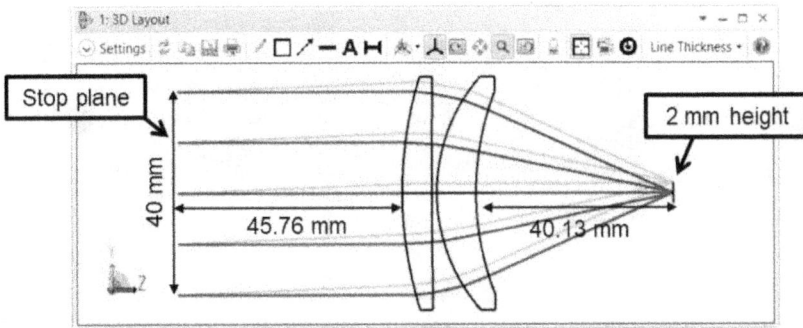

Figure 2.74. A collimating lens design layout for a 4 mm disk source.

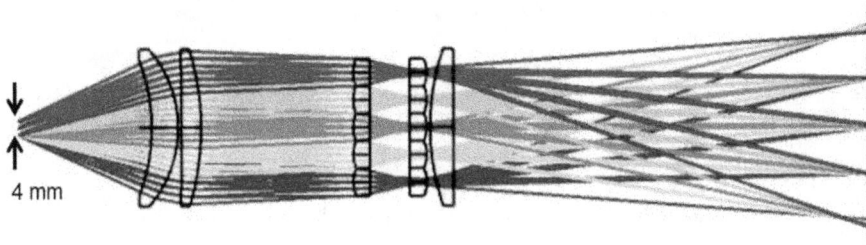

Figure 2.75. An SC layout for a fly's eye array system with an extended source.

If I now want to use a real lens for collimation, then it must mimic the condition for rays at the left surfaces of the first lenslet array. This is required so that rays traveling beyond that array more or less follow the paths depicted in figure 2.73. One way to accomplish this is to make use of the partial aplanatic double lens from figure 1.207, because we note that when reversed, its stop can be made coplanar with the left surface of the first lenslet array. However, we must re-optimize the aplanat to give it an EFL of 50 mm for a source 4 mm in diameter. This was done by switching to higher-index glasses (Schott N-SF11) and reusing the merit function from figure 1.209 (but with the EFL targeted at 50 mm), resulting in the layout shown in figure 2.74. The paraxial thin lens model in figure 2.73 was then replaced by the lens in figure 2.74 (which was, of course, reversed), resulting in the SC ray tracing model shown in figure 2.75, whose prescription and Multi-Configuration Editor settings are shown in figures 2.76 and 2.77, respectively (indeed, I used a multi-configuration setup to model the lenslet arrays).

In figure 2.75, note that rays are now color coded by configuration in order to illustrate the action of rays from lenslets superposing at the screen. In figure 2.77, the operand PRAM is used on rows 2 and 3 to set the Y heights for the lenslets. In particular, PRAM 6/2 on row 2 of the Multi-Configuration Editor sets parameter 2 (Y Decenter) on surface 6 (the coordinate break) in the Lens Data Editor (figure 2.76), while PRAM 10/2 on row 3 in the editor sets parameter 2 on surface 13. Thus, the trick here is that there is only a single lenslet per array (see surfaces 7

2-75

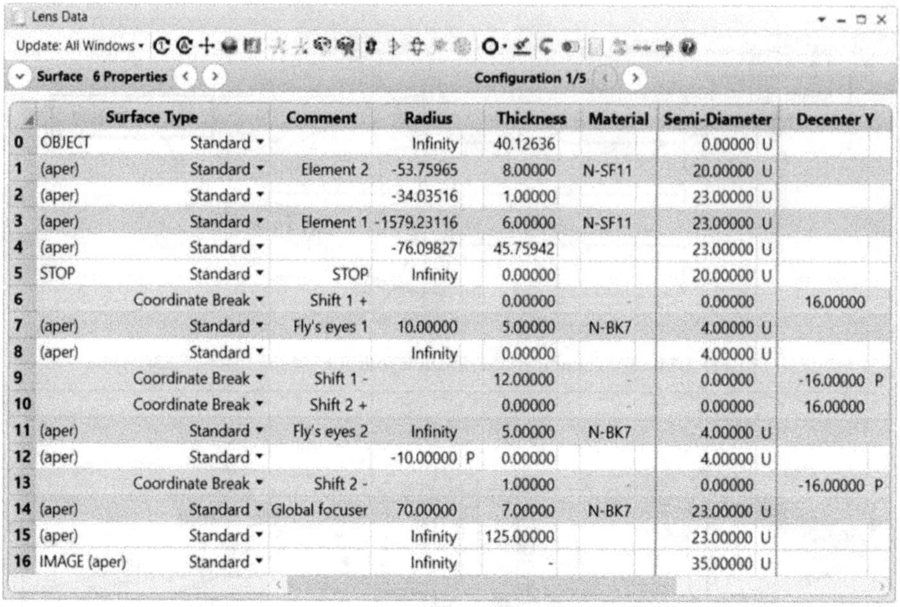

Figure 2.76. The prescription for the system shown in figure 2.75.

	Surface Type	Comment	Radius	Thickness	Material	Semi-Diameter	Decenter Y
0 OBJECT	Standard ▾		Infinity	40.12636		0.00000 U	
1 (aper)	Standard ▾	Element 2	-53.75965	8.00000	N-SF11	20.00000 U	
2 (aper)	Standard ▾		-34.03516	1.00000		23.00000 U	
3 (aper)	Standard ▾	Element 1	-1579.23116	6.00000	N-SF11	23.00000 U	
4 (aper)	Standard ▾		-76.09827	45.75942		23.00000 U	
5 STOP	Standard ▾	STOP	Infinity	0.00000		20.00000 U	
6	Coordinate Break ▾	Shift 1 +		0.00000		0.00000	16.00000
7 (aper)	Standard ▾	Fly's eyes 1	10.00000	5.00000	N-BK7	4.00000 U	
8 (aper)	Standard ▾		Infinity	0.00000		4.00000 U	
9	Coordinate Break ▾	Shift 1 -		12.00000		0.00000	-16.00000 P
10	Coordinate Break ▾	Shift 2 +		0.00000		0.00000	16.00000
11 (aper)	Standard ▾	Fly's eyes 2	Infinity	5.00000	N-BK7	4.00000 U	
12 (aper)	Standard ▾		-10.00000 P	0.00000		4.00000 U	
13	Coordinate Break ▾	Shift 2 -		1.00000		0.00000	-16.00000 P
14 (aper)	Standard ▾	Global focuser	70.00000	7.00000	N-BK7	23.00000 U	
15 (aper)	Standard ▾		Infinity	125.00000		23.00000 U	
16 IMAGE (aper)	Standard ▾		Infinity	-		35.00000 U	

Figure 2.77. The SC Multi-Configuration Editor settings for the system shown in figure 2.75.

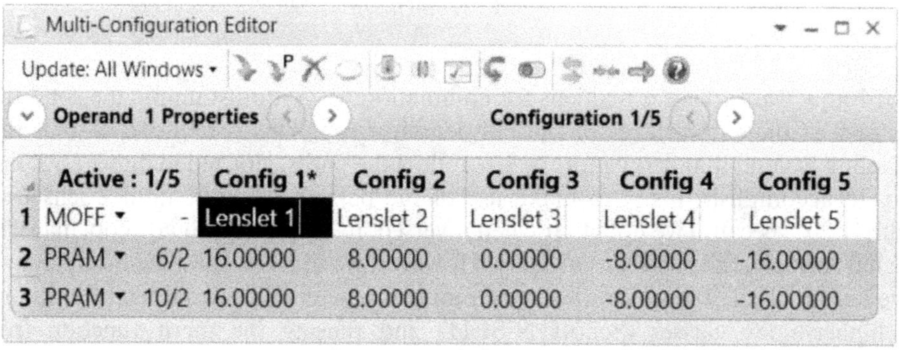

	Active : 1/5	Config 1*	Config 2	Config 3	Config 4	Config 5
1 MOFF ▾	-	Lenslet 1	Lenslet 2	Lenslet 3	Lenslet 4	Lenslet 5
2 PRAM ▾	6/2	16.00000	8.00000	0.00000	-8.00000	-16.00000
3 PRAM ▾	10/2	16.00000	8.00000	0.00000	-8.00000	-16.00000

and 11 in figure 2.76), but five positions in the Y dimension have been set by the Multi-Configuration Editor; therefore, when the layout in figure 2.75 is set to display all configurations simultaneously, the systems appear to have five lenslets in the vertical. This is a rather convenient trick used in SC ray tracing to model the arrangement of complex components, such as fly's eye arrays in the current example. This SC model was then used as a reference in building the pure NSC model in figure 2.68.

Beginning each illumination design in SC mode is helpful, as it allows one to check the paths of rays. When you model an optical system, you ask yourself what you are specifically trying to achieve with the current model. If you have a different question to answer, then you may need to build a separate model. In figure 2.75, I knew that I could not realistically model the irradiance at the screen, due to the multi-configuration approach used for the lenslet array. If I had wanted to, I could have dispensed with the multi-configuration setup and simply inserted an NSC surface to create a HNSC environment, where I would have inserted a pair of NSC lenslet arrays (like the ones used in objects 6 and 7 in figure 2.70). In this way, I would have been able to use OS's GIA feature as usual (or the GBI feature) to simulate irradiance at the screen. However, I elected not to do so in this case, as I wanted to switch between using HNSC and SC models in this book for various systems (as you have probably noticed) in order to illustrate how various systems can be modeled in different ways.

2.16 Fly's eye arrays that have negative-focal-length lenslets

It is not unreasonable to consider having lenslets with negative focal length. Figure 2.78 shows an SC layout for a theoretically plausible system, in which rays have been color coded to show their superposition at the screen. This layout has been modified from the system in figure 2.75. Notice that we now only have a single lenslet array. In this modified layout, each lenslet is a plano-concave lens with a −6 mm radius of curvature on the left side and a 3 mm central thickness. Further, the source's size has been reduced to a 2 mm diameter, as this system has only a single lenslet array and therefore cannot handle extended sources too well.

The pure NSC version of this layout is shown in figure 2.79 and its irradiance profile at the screen is shown in figure 2.80. Certainly, this system does not possess high global uniformity compared to the positive-focal-length version, but it is still broader than the source's irradiance (recall the calculation described for the source near the end of section 2.14, as it is the same source). Moreover, it is a pedagogically appealing system, as it proves the concepts of étendue division (using negative powered lenslets this time) and superposition.

Are there negative-focal-length lenslet arrays? As of the writing of this book, these are not available in mass production glass quantities, as far as I know. However, LCD arrays with positive and negative focal lengths have been explored [49], and negative-focal-length microlens arrays produced by photolithographic methods have been investigated [50]. Thus, we may yet have some use for such arrays in the future.

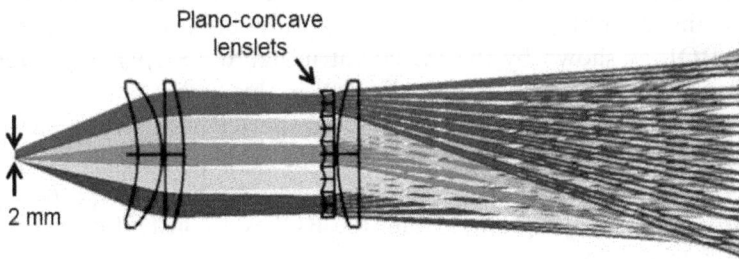

Figure 2.78. An SC layout for a fly's eye system with negative-focal-length lenslets.

Figure 2.79. An NSC layout for a fly's eye system with negative-focal-length lenslets.

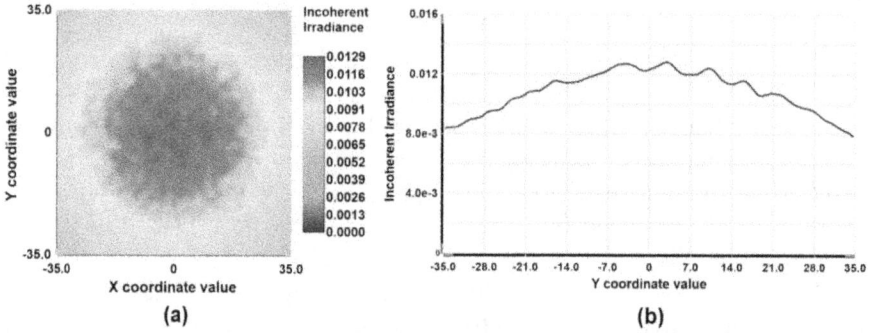

Figure 2.80. The irradiance profile on the screen for the system shown in figure 2.79.

2.17 Uniform oblique illumination

In section 2.10, we saw that the exit pupil of an aplanatic lens is globally uniform if the source is Lambertian and if the lens is optimized for an object at infinity space. Because such a lens system has minimal spherical aberration, its axial rays can remain collimated for a very long distance. Hence, if a small locally uniform Lambertian source is placed at the back focal plane of such a lens system, a surface can be illuminated obliquely and retain globally uniform irradiance within a reasonable ROI, as shown by the SC layout in figure 2.81, whose prescription is provided in figure 2.82. As illustrated, the ROI in this layout falls roughly within a 30 mm length in the Y dimension (it would be shorter in the X dimension, as the shape of the ROI is elliptical). The reason that the profile is globally uniform within the ROI on an inclined plane is because the rays have low divergence; hence, the only factor influencing the profile within the ROI is a constant cosine factor that multiplies the irradiance.

The lens system in figure 2.81 was derived from the COTS objective in figure 1.129 by scaling to an EFL of 50 mm and then re-optimizing (at visible wavelengths of

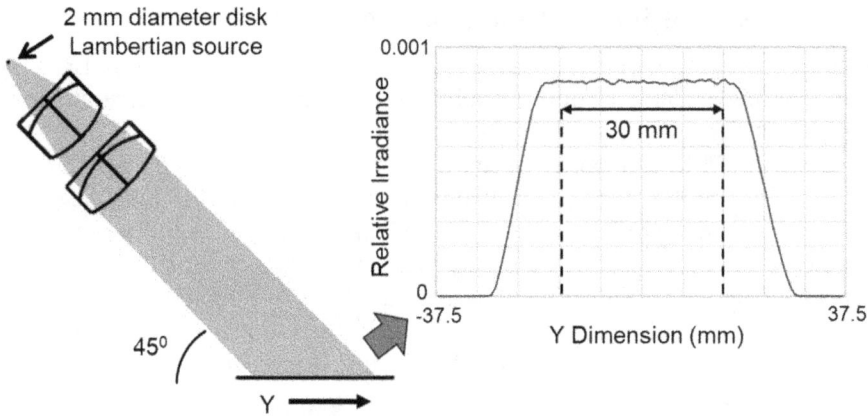

Figure 2.81. Oblique illumination produced by an aplanatic lens system. (a) SC layout. (b) Irradiance profile (using the GIA feature in OS).

	Surface Type		Radius	Thickness	Material	Semi-Diameter	Tilt About X
0	OBJECT	Standard ▾	Infinity	23.65257		1.00000	
1	STOP (aper)	Standard ▾	Infinity	0.00000		8.00000 U	
2	(aper)	Standard ▾	-616.76461	3.00034	N-SF10	17.00000 U	
3	(aper)	Standard ▾	37.90804	22.15175	N-BAF10	17.00000 U	
4	(aper)	Standard ▾	-44.24656	10.46579		17.00000 U	
5	(aper)	Standard ▾	476.49296	2.92877	N-SF10	21.00000 U	
6	(aper)	Standard ▾	37.81856	17.77187	N-BAF10	21.00000 U	
7	(aper)	Standard ▾	-72.09145	125.00000		21.00000 U	
8		Coordinate Break ▾		0.00000		0.00000	45.00000
9	IMAGE	Standard ▾	Infinity	-		37.50000 U	

Figure 2.82. The prescription for the system shown in figure 2.81.

450, 550, and 650 nm, each with a weight of one). The idea was to make it useable with small white LED chips. If the size of the LED is too large, then the full irradiance profile in the inclined direction (i.e. in the Y-dimension in figure 2.81) is trapezoidal, due to non-overlapping (non-superposed) areas of oblique rays with the axial rays. Of course, if a larger ROI requires illumination, then the lens system must scale accordingly, unless multiple luminaires surrounding the ROI are used. On the other hand, if the obliquity is not too large and if it is required that a single luminaire of a fixed size should be used, then one way to achieve this is to create a condition of spreading rays and apply a tilt about a location between the lens system and the plane of illumination [51].

2.18 Point spread function illumination

In section 2.11, we saw that the region of globally uniform illumination (Region II) for the searchlight configuration is enabled by locating a large source at the back focal plane of a lens. When the source is small, Region II vanishes, and the result is a large geometric PSF that possesses some form of distribution. If the distribution happens to be desirable, then it is a form of illumination. In this book, we shall call this *point spread function illumination* or PSF illumination. It turns out that it is possible to shape the PSF into a globally uniform irradiance distribution by way of controlling aberrations. You saw this in section 2.10, where a flat profile was enabled by minimizing spherical aberration and coma. In the following sections, we will do *almost* the opposite. That is, by introducing just *spherical aberration alone*, it is possible to shape a PSF into a top-hat distribution.

2.18.1 The coherent case: Gaussian to top-hat laser beam shaping

Just as a focused laser spot is a PSF (think of a laser scan lens design), so is a projected expanded laser spot. In some cases, the Gaussian profile of an expanded laser spot may be undesirable, and a requirement may instead exist for it to be globally uniform. The conversion of a transverse electromagnetic zero-zero (TEMoo) laser Gaussian beam into a flat-top or 'top-hat' irradiance distribution is well known and well documented [52–59]. As Rhodes and Shealy [58] have described, geometrical ray tracing can be applied to design lenses for shaping Gaussian beams. Kim [59], in particular, has described a simple design process for optimization in OS (neglecting diffraction and speckle), which is the method we will discuss here.

Figure 2.83 illustrates what needs to be done to design a lens to shape a Gaussian beam into a top-hat distribution:

In figure 2.83, ρ_{max} is defined to be the maximum semi-diameter of the entrance pupil. The idea is to apply conservation of flux between the beam incident at the entrance pupil of the beam-shaping lens (BSL) and the same beam when it is incident on the screen. For the purpose of pedagogy, we will neglect back-reflections from the plano side of the BSL towards the laser (let us assume that there is an ideal AR coating on both surfaces of the lens). It is known [60] that the irradiance profile for a Gaussian beam is given by

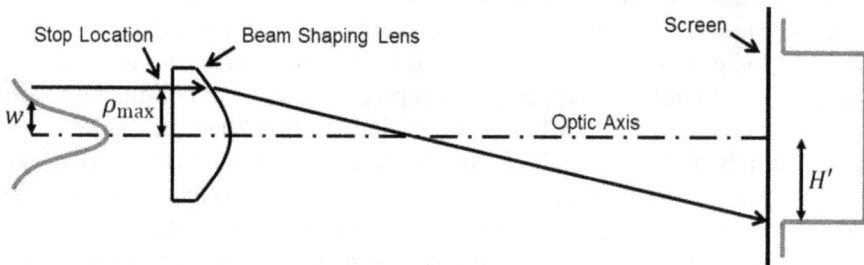

Figure 2.83. The construction used to derive a formula for laser beam shaping.

$$E_G(x, y) = E_{Gp} \exp\left[-2\left(\frac{x^2 + y^2}{w^2}\right)\right], \tag{2.51}$$

where E_{Gp} is the peak irradiance and w is the $1/e^2$ half-width of the beam. Due to symmetry about the optic axis, we can let $x^2 + y^2 = \rho^2$, so that

$$E_G(\rho) = E_{Gp} \exp\left[-2\left(\frac{\rho^2}{w^2}\right)\right], \tag{2.52}$$

where ρ is the radial dimension at the stop (i.e. the entrance pupil) of the BSL such that $0 < \rho < \rho_{max}$. Within a radius ρ, the flux $\phi(\rho)$ in the beam may be determined through integration in circular coordinates:

$$\phi(\rho) = 2\pi E_{Gp} \int_0^\rho \exp\left[-2\left(\frac{\rho^2}{w^2}\right)\right]\rho\,d\rho. \tag{2.53}$$

Let $u = 2\rho^2/w^2$, so that $du = 4\rho d\rho/w^2$. Substituting this into equation (2.53) gives

$$\phi(\rho) = \frac{E_{Gp}\pi w^2}{2} \int_0^{2\rho^2/w^2} \exp(-u)\,du = \frac{E_{Gp}\pi w^2}{2}\left[1 - \exp\left(\frac{-2\rho^2}{w^2}\right)\right]. \tag{2.54}$$

In equation (2.54), since, in the limit $\rho \to \infty$, $\phi = E_{Gp}\pi w^2/2$, then the total flux ϕ_{tot} in the beam is $E_{Gp}\pi w^2/2$ (note that, in OS, we will take $\rho_{max} \approx \infty$ relative to the half-width w of the Gaussian beam). Thus, we have

$$\phi(\rho) = \phi_{tot}\left[1 - \exp\left(\frac{-2\rho^2}{w^2}\right)\right]. \tag{2.55}$$

Equation (2.55) is well known and is a standard textbook formula (see, for example, [61]). At the screen, if the irradiance profile is turned into a top-hat distribution, then the irradiance E' at the screen is constant within the top-hat region, and its total flux is therefore

$$\phi_{tot} = E'\pi H'^2. \tag{2.56}$$

This also means that, at the screen, the flux within a radius $h' < H'$ is

$$\phi(h') = E'\pi h'^2. \tag{2.57}$$

Here comes the key ingredient. We let a ray at ρ travel towards the screen at h', thereby requiring that the flux in $\phi(\rho)$ be made equal to $\phi(h')$. Rays can therefore be said to 'map' from the entrance pupil of the BSL towards the screen. So, letting equation (2.55) be equal to equation (2.57) gives

$$\phi_{tot}\left[1 - \exp\left(\frac{-2\rho^2}{w^2}\right)\right] = E'\pi h'^2. \tag{2.58}$$

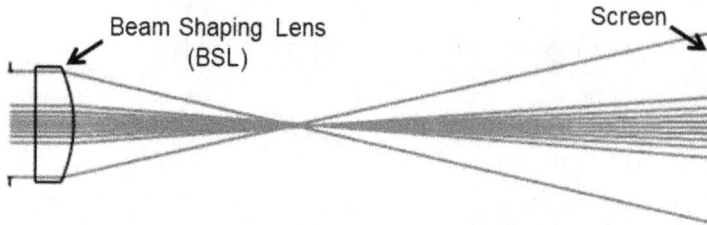

Figure 2.84. A starting layout in OS SC ray tracing for a BSL design.

Figure 2.85. The prescription for the layout shown in figure 2.84.

Next, we replace E' in equation (2.58) with that given by equation (2.56) to yield

$$1 - \exp\left(\frac{-2\rho^2}{w^2}\right) = \frac{h'^2}{H'^2}. \tag{2.59}$$

Finally, solving for h', we obtain

$$h' = \pm H'\sqrt{1 - \exp\left(\frac{-2\rho^2}{w^2}\right)}. \tag{2.60}$$

Equation (2.60) tells us that, for a Gaussian beam whose half-width is w, if we want a top-hat distribution at the screen to possess a maximum half-width of H', then a ray at some height ρ in the entrance pupil must reach a height of h'. In OS, we can therefore create REAY operands **in an SC ray tracing layout** that trace rays beginning at the normalized radial coordinates ρ/ρ_{max} (recall that the entrance pupil coordinates Px and Py in OS are normalized such that the maximum radial height is Py = 1 and Px = 1) and ending at heights h'. However, we note from figure 2.83 that our construction requires the ray to cross the optic axis. Therefore, we must remember to set ray heights at the screen for which $h' < 0$ (i.e. take the solution for which $H' < 0$).

Figure 2.84 is a starting layout for a BSL design whose prescription and merit function are provided in figures 2.85 and 2.86, respectively. There is only a single

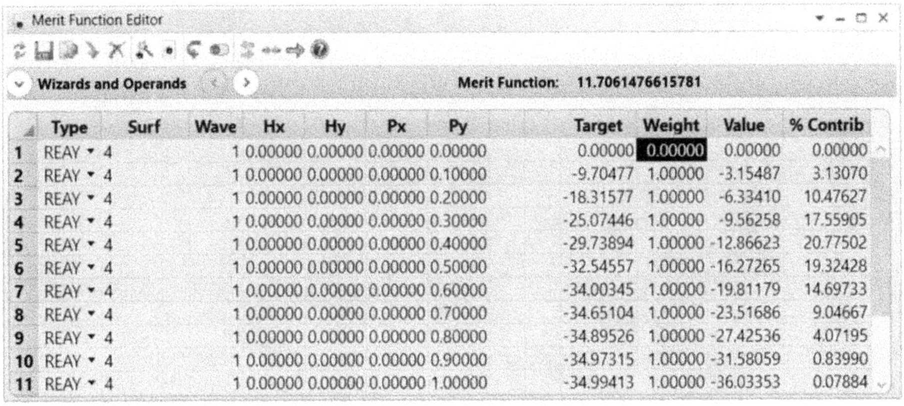

	Type	Surf	Wave	Hx	Hy	Px	Py	Target	Weight	Value	% Contrib
1	REAY	4	1	0.00000	0.00000	0.00000	0.00000	0.00000	0.00000	0.00000	0.00000
2	REAY	4	1	0.00000	0.00000	0.00000	0.10000	-9.70477	1.00000	-3.15487	3.13070
3	REAY	4	1	0.00000	0.00000	0.00000	0.20000	-18.31577	1.00000	-6.33410	10.47627
4	REAY	4	1	0.00000	0.00000	0.00000	0.30000	-25.07446	1.00000	-9.56258	17.55905
5	REAY	4	1	0.00000	0.00000	0.00000	0.40000	-29.73894	1.00000	-12.86623	20.77502
6	REAY	4	1	0.00000	0.00000	0.00000	0.50000	-32.54557	1.00000	-16.27265	19.32428
7	REAY	4	1	0.00000	0.00000	0.00000	0.60000	-34.00345	1.00000	-19.81179	14.69733
8	REAY	4	1	0.00000	0.00000	0.00000	0.70000	-34.65104	1.00000	-23.51686	9.04667
9	REAY	4	1	0.00000	0.00000	0.00000	0.80000	-34.89526	1.00000	-27.42536	4.07195
10	REAY	4	1	0.00000	0.00000	0.00000	0.90000	-34.97315	1.00000	-31.58059	0.83990
11	REAY	4	1	0.00000	0.00000	0.00000	1.00000	-34.99413	1.00000	-36.03353	0.07884

Figure 2.86. The merit function for the layout shown in figure 2.84.

field at 0° (i.e. just axial rays), the wavelength is 633 nm (assume that it is for a HeNe laser), and there is a Gaussian apodization factor of four set for the beam. This setting as well as all of the necessary properties for this starting layout shall be explained in the paragraphs to follow.

In OS, when Gaussian apodization is applied, the *amplitude* transmittance is given by

$$A(\rho) = \exp\left(-G\rho_N^2\right), \tag{2.61}$$

where G is defined as the apodization factor for the Gaussian transmittance and ρ_N is the normalized pupil's radial coordinate (e.g. it is Py in the REAY operand). In other words, $\rho_N = \rho/\rho_{max}$. Since equation (2.61) is an amplitude, it is the square root of equation (2.52). By cross-referencing these two equations, we note that $G\rho_N^2 = \rho^2/w^2$. Since $\rho_N = \rho/\rho_{max}$, $G = \rho_{max}^2/w^2$. This lets us set an appropriate value for G, given a preferred choice of values for ρ_{max} and w. In the starting layout, I have arbitrarily decided to let $w = 10$ mm and $\rho_{max} = 20$ mm. In the real world, this is not very realistic, as diffraction due to the stop can cause serious problems. But we will neglect this issue for the purpose of pedagogy. Based on these settings, $G = 20^2/10^2 = 4$. Thus, this has been used as the apodization factor.

Since, in the previous sections, I was using a screen height of 35 mm, I arbitrarily set $H' = -35$ mm (note the negative sign, as the pupil's ray must cross the optic axis, as illustrated in figure 2.83). Using all the values computed, I then entered equation (2.60) into a spreadsheet to compute h' for the range from $\rho = 0$ mm to $\rho = \rho_{max} = 20$ mm; this means a range from Py = 0 mm to Py = 1 mm in steps of 0.1 mm, yielding target values for $h' = 0, -9.704\,77\ldots-34.994\,13$ mm, which is what you see in the merit function shown in figure 2.86. No weight is assigned to Py = 0 because of its triviality. The next thing to do is to generate a GIA plot to check the starting beam's irradiance profile at the screen. The plot and its settings are shown in figures 2.87 and 2.88, respectively.

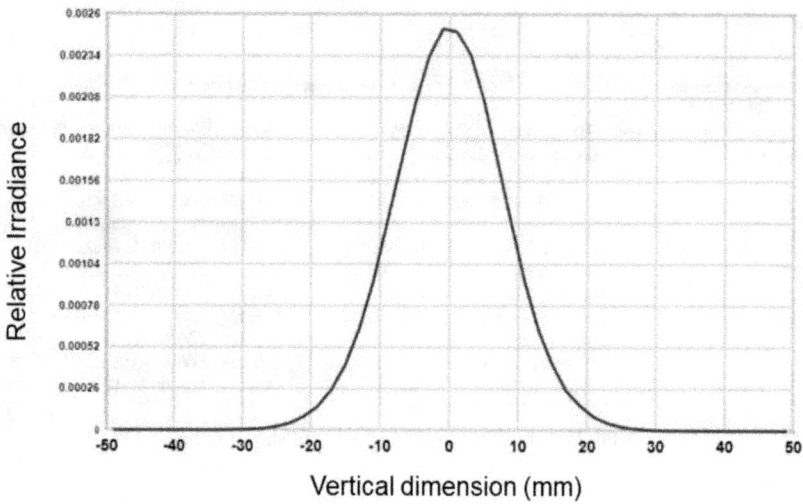

Figure 2.87. A GIA plot of the beam's starting profile on the screen.

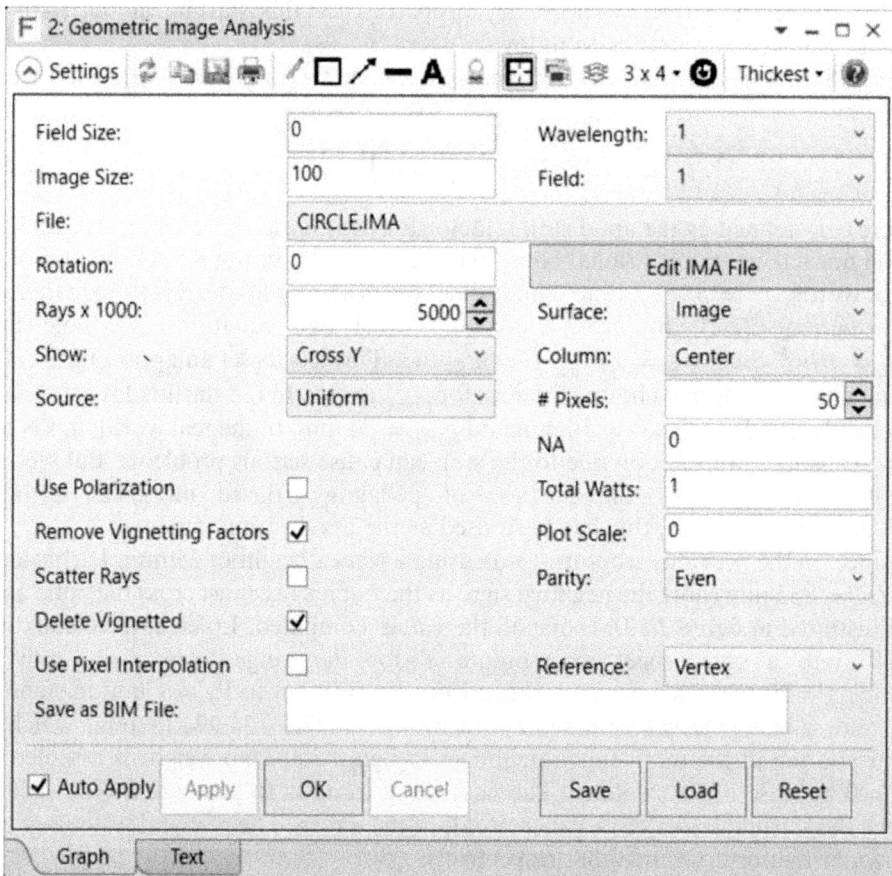

Figure 2.88. The settings used for the plot in figure 2.87.

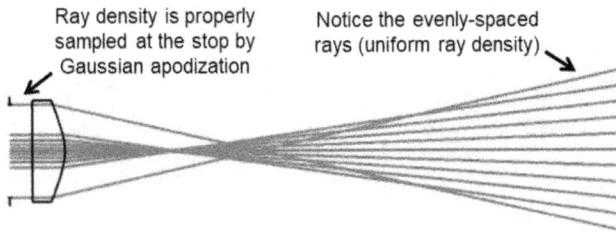

Figure 2.89. The optimized layout for the BSL.

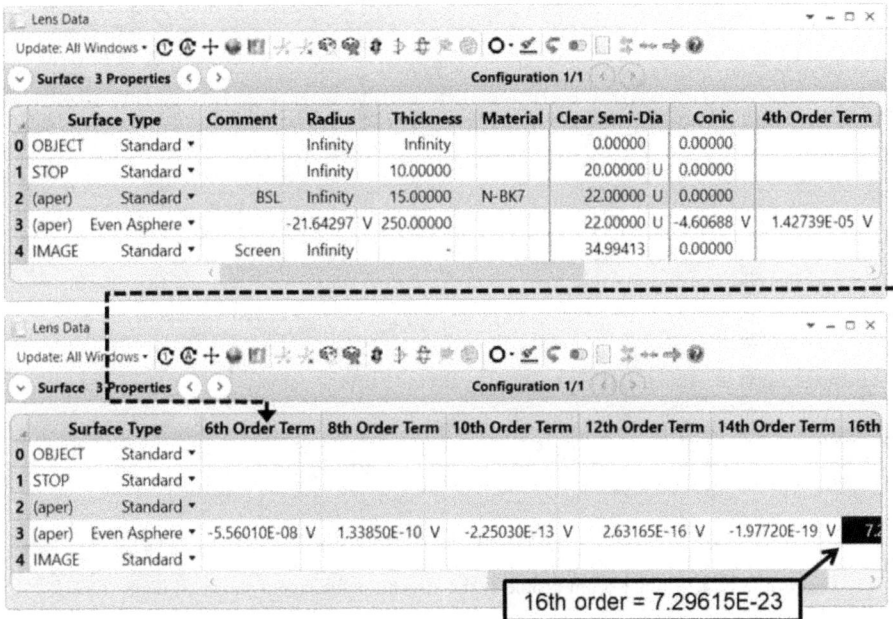

	Surface Type	Comment	Radius	Thickness	Material	Clear Semi-Dia	Conic	4th Order Term
0 OBJECT	Standard ▾		Infinity	Infinity		0.00000	0.00000	
1 STOP	Standard ▾		Infinity	10.00000		20.00000 U	0.00000	
2 (aper)	Standard ▾	BSL	Infinity	15.00000	N-BK7	22.00000 U	0.00000	
3 (aper)	Even Asphere ▾		-21.64297 V	250.00000		22.00000 U	-4.60688 V	1.42739E-05 V
4 IMAGE	Standard ▾	Screen	Infinity	-		34.99413	0.00000	

	Surface Type	6th Order Term	8th Order Term	10th Order Term	12th Order Term	14th Order Term	16th
0 OBJECT	Standard ▾						
1 STOP	Standard ▾						
2 (aper)	Standard ▾						
3 (aper)	Even Asphere ▾	-5.56010E-08 V	1.33850E-10 V	-2.25030E-13 V	2.63165E-16 V	-1.97720E-19 V	7.2
4 IMAGE	Standard ▾						

16th order = 7.29615E-23

Figure 2.90. The prescription for the layout shown in figure 2.89.

The result I obtained by performing a single local damped least-squares optimization is provided in figures 2.89–2.92. The settings for the GIA plot in figure 2.92 were the same as those used for figure 2.88. Increasing the number of rays (and also increasing the # Pixels) yields a smoother and sharper profile.

The mechanism responsible for producing the top-hat profile in this design is the introduction of spherical aberration. How do we know this? Because spherical aberration is the only aberration that can be present axially. As you can tell, we interestingly did not need to compute the magnitude of the spherical aberration needed to any order. We simply relied on the conservation of flux between the pupil and the screen. You may notice that the even asphere made use of coefficients from the fourth order to the sixteenth, even though I mentioned in section 1.3.3 that if a conic constant were used, then we would generally at least try to avoid the use of the fourth-order coefficient. I explained that, due to equation (1.24), the conic constant

	Type	Surf	Wave	Hx	Hy	Px	Py	Target	Weight	Value	% Contrib
1	REAY ▾ 4		1	0.00000	0.00000	0.00000	0.00000	0.00000	0.00000	0.00000	0.00000
2	REAY ▾ 4		1	0.00000	0.00000	0.00000	0.10000	-9.70477	1.00000	-9.70478	27.42377
3	REAY ▾ 4		1	0.00000	0.00000	0.00000	0.20000	-18.31577	1.00000	-18.31576	47.29937
4	REAY ▾ 4		1	0.00000	0.00000	0.00000	0.30000	-25.07446	1.00000	-25.07447	20.64337
5	REAY ▾ 4		1	0.00000	0.00000	0.00000	0.40000	-29.73894	1.00000	-29.73893	0.89295
6	REAY ▾ 4		1	0.00000	0.00000	0.00000	0.50000	-32.54557	1.00000	-32.54557	1.37872
7	REAY ▾ 4		1	0.00000	0.00000	0.00000	0.60000	-34.00345	1.00000	-34.00346	1.80803
8	REAY ▾ 4		1	0.00000	0.00000	0.00000	0.70000	-34.65104	1.00000	-34.65104	0.50470
9	REAY ▾ 4		1	0.00000	0.00000	0.00000	0.80000	-34.89526	1.00000	-34.89526	0.04774
10	REAY ▾ 4		1	0.00000	0.00000	0.00000	0.90000	-34.97315	1.00000	-34.97315	1.35458E-03

Merit Function: 6.59651953846957E-06

Figure 2.91. The merit function for the layout shown in figure 2.89.

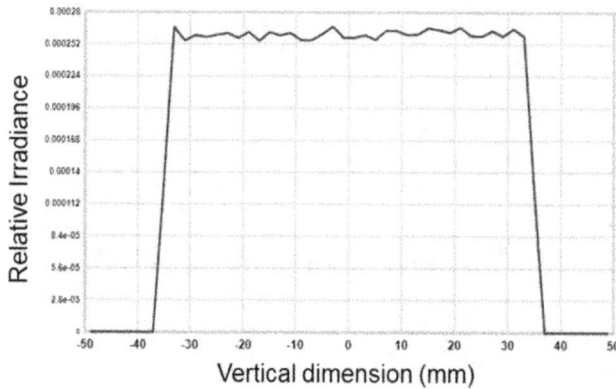

Figure 2.92. The top-hat beam profile at the screen for the layout shown in figure 2.89.

introduces its own set of higher-order terms. However, there is some justification for the use of all available orders in the even asphere of the current example. The idea is that in the current design, we do not know what the sag of the aspheric surface should be in order to produce the desired top-hat profile, but it is known *a priori* that a conic constant of K is needed, because this is what is introduced onto a surface to control the spherical aberration. So, the inclusion of a specific amount of K onto the convex surface can be considered a starting point that guides the design optimization path (it is like a design form for an asphere). By virtue of equation (1.24), higher-order coefficients on that surface are introduced when K is present, but we want specific perturbations that make the surface sag just right to flatten the irradiance at the screen. For this reason, it is beneficial to place variables on all orders of the coefficients, with the exception of the second order, as that is clearly redundant, considering that the base radius is wholly controlled by that surface's radius of curvature. The result is the design you see here, and it was this layout that was used in the illustration shown in figure 2.16.

2.18.2 The incoherent case: LED Lambertian to top-hat beam shaping

By way of applying aberration theory, Sasián [62] provided an example of a singlet asphere design that produces globally uniform irradiance at a distant screen for the case of a small Lambertian source. Alternatively, it is possible to apply the method described in the previous section [63, 64]. For a Lambertian emitter, recall that the flux in a cone of angle u is given by equation (2.6). In section 2.2.10, I showed that the quantity πLA in equation (2.6) is the total flux entering a hemisphere from a small Lambertian source. Hence, if ϕ_{tot} is the total flux, then applying the construction shown in figure 2.93, the flux reaching an entrance pupil of semi-diameter ρ from a small Lambertian LED emitter of area A is

$$\phi(\rho) = \pi LA \sin^2 u = \phi_{tot}\frac{\rho^2}{\rho^2 + z^2} = \frac{\phi_{tot}}{1 + (z^2/\rho^2)}. \tag{2.62}$$

If a ray at height ρ reaches the screen at height h', then once again, the flux is given by equation (2.57). Hence, equating the right-hand sides of equations (2.62) and (2.57) gives

$$\frac{\phi_{tot}}{1 + (z^2/\rho^2)} = E'\pi h'^2. \tag{2.63}$$

The total flux passing through the stop is given by letting $\rho = \rho_{max}$ in equation (2.62), yielding

$$\phi(\rho_{max}) = \frac{\phi_{tot}}{1 + (z^2/\rho_{max}^2)}. \tag{2.64}$$

Accordingly, this flux is equal to the right-hand side of equation (2.56). So, combining these gives

$$\frac{\phi_{tot}}{1 + (z^2/\rho_{max}^2)} = E'\pi H'^2. \tag{2.65}$$

Finally, combining equations (2.63) and (2.65) and solving for h' yields

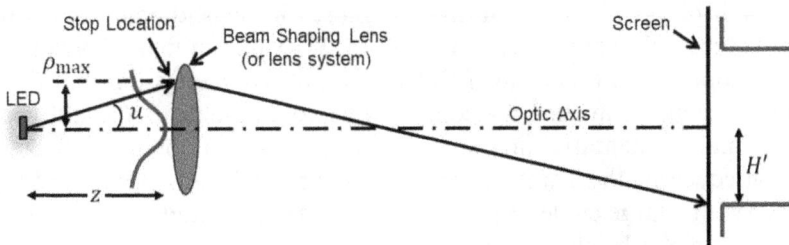

Figure 2.93. The top-hat beam profile at the screen for the layout shown in figure 2.89.

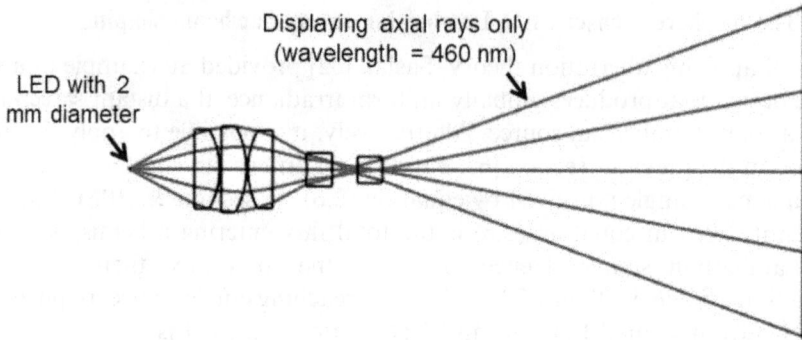

Figure 2.94. The starting layout for a five-element lens system.

$$h' = \pm H' \sqrt{\frac{1 + (z^2/\rho_{max}^2)}{1 + (z^2/\rho^2)}}. \qquad (2.66)$$

Once again, in OS, we can set REAY merit function operands to trace rays at pupil heights and target values for h', having decided on the values to be used for H', ρ_{max}, and now also z. It has been shown [63–65] that the BSL in figure 2.93 can comprise a number of spherical elements. The reason that this works is because spherical lenses—by their nature—possess spherical aberration. In fact, Shafer [55, 56], Shimmel and Wyrowski [58], and Dilworth [66] have described the use of spherical lenses to shape a Gaussian beam into a top-hat distribution. In a prior work [64], I used five spherical elements to shape the rays, which shall be the approach used here. Figure 2.94 shows a starting layout with five elements, whose prescription, merit function, and irradiance profile (again, using the GIA plot feature in OS) are shown in figures 2.95–2.97, respectively. The settings used for the GIA plot are shown in figure 2.98. The design wavelength is 460 nm.

In figure 2.96, the merit function now includes operands (on rows 12–16) to constrain element thicknesses and air gaps. If you have been following my design exercises in this book, you may have noticed that I often set weights for such constraints to be about 20% of the main weights, such as for the REAY operands on rows 1–11. To produce the target h' values in these rows, I settled on using 70, 20, and 40 mm for H', ρ_{max}, and z, respectively. Using these, I then applied equation (2.66) to compute those h' targets using a spreadsheet, at the Py intervals shown. Figure 2.94 displays only the axial rays (i.e. those for the field point 0), but the GIA setting for object full diameter is 2 mm. So, the GIA plot in figure 2.97 displays the irradiance from a 2 mm diameter LED with Lambertian emission, but the REAY operands in the merit function are computed based on the assumption that we have an infinitesimal (elemental) source size, which is the assumption that led to equation (2.66). Consequently, the irradiance profile in figure 2.97 is essentially a convolution of the PSF with a large circle of finite size, defined by the angle subtended in object space by a source of height 1 mm.

Lens Data

Update: All Windows

Surface 0 Properties — Configuration 1/1

	Surface Type		Comment	Radius		Thickness		Material	Clear Semi-Dia	
0	OBJECT	Standard		Infinity		40.00000			1.00000	
1	STOP	Standard		Infinity		0.00000			20.00000	U
2		Standard	1	Infinity		15.00000	V	N-SK11	20.00000	
3		Standard		-45.00000	V	1.00000	V		22.73358	
4		Standard	2	Infinity	V	15.00000	V	N-SK11	23.91658	
5		Standard		-50.00000	V	1.00000	V		24.79348	
6		Standard	3	45.00000	V	15.00000	V	N-SK11	22.48533	
7		Standard		Infinity	V	20.00000	V		19.70472	
8		Standard	4	-100.00000	V	15.00000	V	N-SK11	9.42826	
9		Standard		200.00000	V	15.00000	V		5.29189	
10		Standard	5	Infinity	V	15.00000	V	N-SK11	3.46376	
11		Standard		-50.00000	V	250.00000			7.45881	
12	IMAGE	Standard	Screen	Infinity		-			95.92300	

Figure 2.95. The prescription for the layout shown in figure 2.94.

Merit Function Editor

Wizards and Operands — Merit Function: 12.5265663385228

	Type	Surf	Wave	Hx	Hy	Px	Py	Target	Weight	Value	% Contrib
1	REAY	12	1	0.00000	0.00000	0.00000	0.00000	0.00000	0.00000	0.00000	0.00000
2	REAY	12	1	0.00000	0.00000	0.00000	0.10000	-7.81647	1.00000	-9.06874	0.09085
3	REAY	12	1	0.00000	0.00000	0.00000	0.20000	-15.57480	1.00000	-18.17936	0.39302
4	REAY	12	1	0.00000	0.00000	0.00000	0.30000	-23.21895	1.00000	-27.36873	0.99768
5	REAY	12	1	0.00000	0.00000	0.00000	0.40000	-30.69703	1.00000	-36.66403	2.06279
6	REAY	12	1	0.00000	0.00000	0.00000	0.50000	-37.96283	1.00000	-46.07883	3.81616
7	REAY	12	1	0.00000	0.00000	0.00000	0.60000	-44.97706	1.00000	-55.61081	6.55113
8	REAY	12	1	0.00000	0.00000	0.00000	0.70000	-51.70802	1.00000	-65.24240	10.61255
9	REAY	12	1	0.00000	0.00000	0.00000	0.80000	-58.13184	1.00000	-74.94658	16.38037
10	REAY	12	1	0.00000	0.00000	0.00000	0.90000	-64.23223	1.00000	-84.70116	24.27357
11	REAY	12	1	0.00000	0.00000	0.00000	1.00000	-70.00000	1.00000	-94.51628	34.82188
12	MNCG	2	11					2.50000	0.20000	2.50000	0.00000
13	MXCG	2	11					16.00000	0.20000	16.00000	0.00000
14	MNEG	2	11	1.00000	1			1.00000	0.20000	1.00000	0.00000
15	MNCA	3	10					0.20000	0.20000	0.20000	0.00000
16	MNEA	3	10	1.00000	1			0.10000	0.20000	0.10000	0.00000

Figure 2.96. The merit function for the layout shown in figure 2.94.

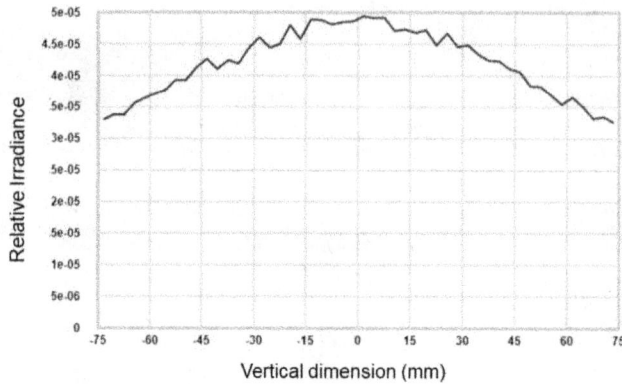

Figure 2.97. The irradiance profile at the screen of the layout shown in figure 2.94.

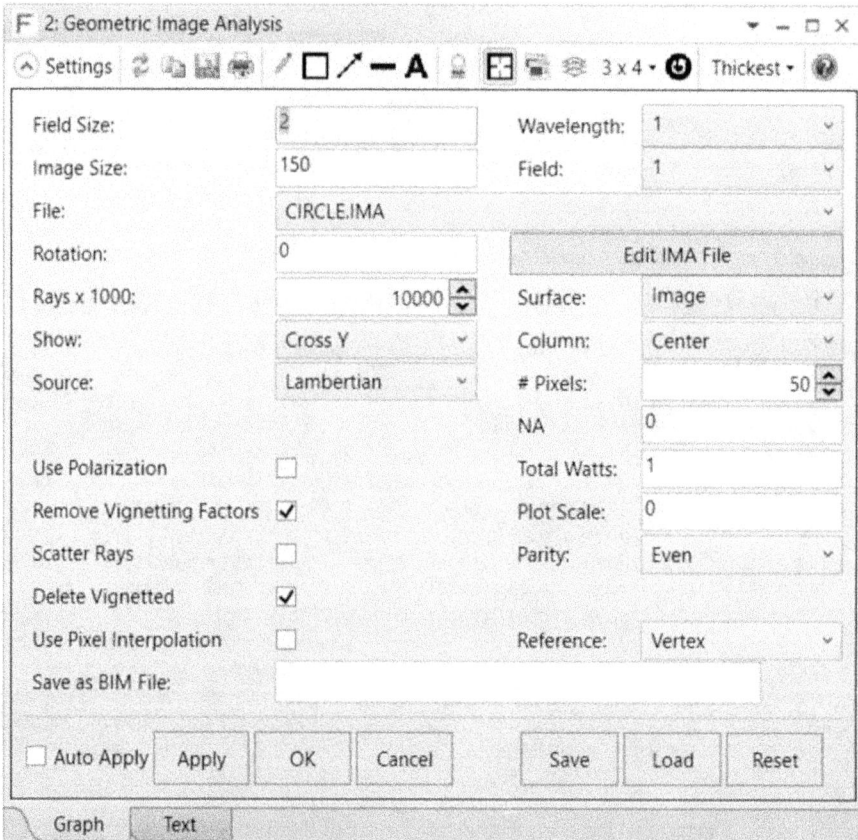

Figure 2.98. The settings for the GIA plot shown in figure 2.97.

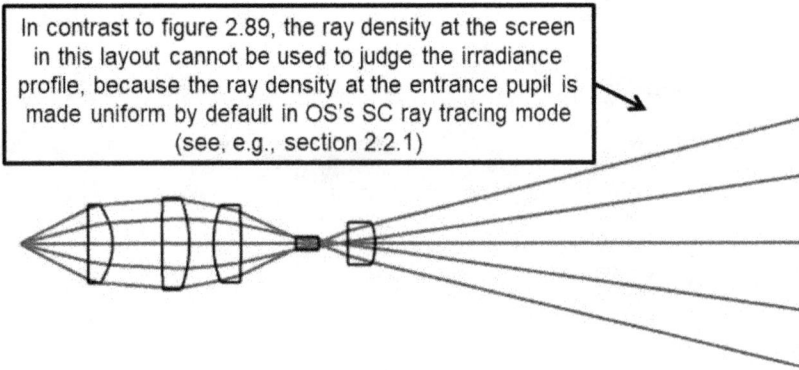

In contrast to figure 2.89, the ray density at the screen in this layout cannot be used to judge the irradiance profile, because the ray density at the entrance pupil is made uniform by default in OS's SC ray tracing mode (see, e.g., section 2.2.1)

Figure 2.99. The optimized layout that began with the system shown in figure 2.94.

	Surface Type		Comment	Radius	Thickness	Material	Clear Semi-Dia
0	OBJECT	Standard ▾		Infinity	40.00000		1.00000
1	STOP	Standard ▾		Infinity	0.00000		20.00000 U
2		Standard ▾	1	Infinity	13.87297 V	N-SK11	20.00000
3		Standard ▾		-36.94497 V	28.63350 V		22.03628
4		Standard ▾	2	518.55797 V	16.00000 V	N-SK11	25.28515
5		Standard ▾		-70.83404 V	13.94289 V		25.69145
6		Standard ▾	3	58.47041 V	16.00000 V	N-SK11	21.78843
7		Standard ▾		-294.63345 V	31.89283 V		18.97424
8		Standard ▾	4	-35.22605 V	13.11902 V	N-SK11	3.85776
9		Standard ▾		-72.99120 V	16.93144 V		1.41599
10		Standard ▾	5	-124.57824 V	15.82615 V	N-SK11	8.12315
11		Standard ▾		-35.97075 V	250.00000		11.93444
12	IMAGE	Standard ▾	Screen	Infinity	-		70.92509

Figure 2.100. The prescription for the system shown in figure 2.99.

After a single local damped least-squares optimization, the result is the layout shown in figure 2.99, whose prescription, merit function, and irradiance profile are shown in figures 2.100–2.102, respectively. The settings for the irradiance plot are the same as those used for figure 2.98.

2.19 A summary of the approaches used in illumination

Table 2.2 provides a summary of the various classical illumination design approaches that we have discussed in this book, with perhaps a *modern classical*

	Type	Surf	Wave	Hx	Hy	Px	Py	Target	Weight	Value	% Contrib
1	REAY ▾ 12		1	0.00000	0.00000	0.00000	0.00000	0.00000	0.00000	0.00000	0.00000
2	REAY ▾ 12		1	0.00000	0.00000	0.00000	0.10000	-7.81647	1.00000	-7.81373	10.07957
3	REAY ▾ 12		1	0.00000	0.00000	0.00000	0.20000	-15.57480	1.00000	-15.57151	14.46663
4	REAY ▾ 12		1	0.00000	0.00000	0.00000	0.30000	-23.21895	1.00000	-23.21788	1.52812
5	REAY ▾ 12		1	0.00000	0.00000	0.00000	0.40000	-30.69703	1.00000	-30.69922	6.42823
6	REAY ▾ 12		1	0.00000	0.00000	0.00000	0.50000	-37.96283	1.00000	-37.96631	16.22304
7	REAY ▾ 12		1	0.00000	0.00000	0.00000	0.60000	-44.97706	1.00000	-44.97807	1.36932
8	REAY ▾ 12		1	0.00000	0.00000	0.00000	0.70000	-51.70802	1.00000	-51.70491	12.93986
9	REAY ▾ 12		1	0.00000	0.00000	0.00000	0.80000	-58.13184	1.00000	-58.12890	11.54834
10	REAY ▾ 12		1	0.00000	0.00000	0.00000	0.90000	-64.23223	1.00000	-64.23642	23.44904
11	REAY ▾ 12		1	0.00000	0.00000	0.00000	1.00000	-70.00000	1.00000	-69.99879	1.96784
12	MNCG ▾ 2	11						2.50000	0.20000	2.50000	0.00000
13	MXCG ▾ 2	11						16.00000	0.20000	16.00000	2.76514E-06
14	MNEG ▾ 2	11	1.00000		1			1.00000	0.20000	1.00000	0.00000
15	MNCA ▾ 3	10						0.20000	0.20000	0.20000	0.00000
16	MNEA ▾ 3	10	1.00000		1			0.10000	0.20000	0.10000	0.00000

Figure 2.101. The merit function for the system shown in figure 2.99.

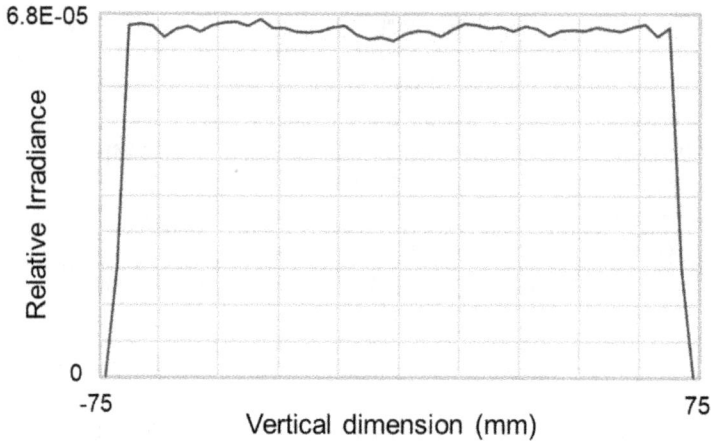

Figure 2.102. The irradiance profile at the screen for the system shown in figure 2.99.

'twist' in terms of how they are applied using various tools and features of a modern optical design program (such as OS's SC ray tracing operands and GIA/GBI plotting features). In addition, the method of PSF illumination is surely a modern classical and practical approach that does not require the application of complicated mathematics. This table should be helpful to you as a quick reference, especially since the methods discussed have been provided in miscellaneous sections of the book.

Table 2.2. A summary of the illumination design approaches discussed in this book.

Technique	The basis of the technique	The type of uniformity achievable
Searchlight	Étendue superposition (sections 2.5 and 2.11–2.13)	Global (diffusers may help if the source is locally nonuniform)
Critical/ relative illumination	SSF (sections 2.2.5 and 2.7–2.9)	Global (diffusers may help if the source is locally nonuniform)
Köhler	Étendue division and superposition (sections 1.3.7, 1.3.10, 2.5, and 2.13)	Local (global uniformity is also possible if the condenser is aplanatic)
Light pipes	Étendue division and superposition (sections 1.2.18, 2.5, and 2.14)	Global and local
Fly's eye arrays	Étendue division and superposition (sections 2.5 and 2.16)	Global and local
PSF illumination	Étendue division and superposition (sections 2.5 and 2.18)	Global and local

In addition to the techniques listed in table 2.2, there are, of course, many more forms of illumination designs that offer various benefits. Of these, the most common modern designs involve so-called *hybrid optics* (mentioned briefly in section 2.2.11) that combine reflection, refraction, and faceted surfaces (an example of étendue division) to produce the desired illumination profile at a surface (which may not necessarily be flat) [19, 27, 47]. Hybrid optics are what you often see in a modern LED flashlight, although you can also call them *LED collimators* [19]. The design of such luminaires is beyond the scope of this book and my experience, but perhaps I can at least highlight one of their important characteristics, which is to think of the starting point in their design as a nonimaging concentrator in reverse [27]. For example, a *compound parabolic concentrator* (CPC) is a reflective concentrator, but when it is used in reverse (by placing a Lambertian source at the smaller end), globally and locally uniform irradiance is produced at a screen. However, many modern designs are variants of the CPC approach, often involving both reflective and refractive surfaces. Examples of some manufacturers of hybrid optics are Carclo Optics (https://www.carclo-optics.com/), Fraen (https://www.fraen.com/), Khatod (https://www.khatod.com/en/), and LEDiL (https://www.ledil.com/).

2.20 Tips on optimization and tolerancing in nonsequential ray tracing

Experienced designers often have preferred ways to accomplish a design task, and illumination is certainly no different. In my work on illumination, which involves the use of luminaires in analytical instruments for fluorescence excitation and general *bright-field* illumination, I have often mainly relied on using SC ray tracing for both the design optimization and analysis of illumination systems. The use of NSC ray tracing is often a final step in checking the results. Even in tolerancing, as I will show, while the final quality of the illumination must be inspected by way of NSC ray tracing, the act of tolerancing (such as performing Monte Carlo simulations) can be performed completely through SC ray tracing. Moreover, I want to show you design tips that are not necessarily limited to the OS program. However, I will begin this section with a simple example of optimization in NSC ray tracing (using, of course, the OS program). Then, I will explain how to perform tolerancing of illumination systems in SC mode, which I believe may be applied to any program.

In OS, when in pure NSC ray tracing mode, optimization is performed by the usual way of inserting relevant merit function operands (meant only for NSC objects and NSC detectors) into the Merit Function Editor, but there is an initial crucial step, which is that the first row in your merit function must use the NSDD operand and the second row must use the NSTR operand. The reason for this is because the NSDD operand on the first row is used to clear your detectors of any last ray trace result and the NSTR operand performs the ray trace. Think of this as equivalent to the two steps you would normally take when you perform manual NSC ray traces. You would generally have a detector viewer on screen to see what is on your detector, which you would first need to clear prior to a new ray trace. So, this is what the two operands NSDD and NSTR are meant to do in the Merit Function Editor. Thereafter, you can insert other NSC-related operands in subsequent rows of the Merit Function Editor, which normally include two types of NSC operands: (1) detector operands (such as any additional NSDD operands used to *read* data from pixels on the NSC object detector) and (2) operands for constraints on NSC object data (such as constraints on object size, position, and other object properties).

Let us take an example. Suppose that, for the fly's eye system shown in figure 2.68 (whose prescription in the NSC Editor is shown in figure 2.70), we now desire a larger region of interest (ROI) for illumination such that the distribution of rays has a wider fill in the 70 mm × 70 mm area detector, as illustrated in figure 2.103. In section 2.15, I mentioned that the ROI can be widened if the lenslets (for the pair of arrays) have higher power and their gap is adjusted accordingly (so that the local chief ray is bent to become parallel with the optic axis). I also mentioned that, alternatively, one may shift the screen farther away. Since the latter approach is

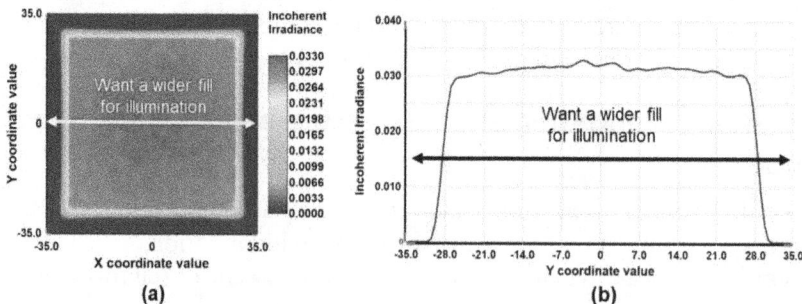

Figure 2.103. A new desired ROI for the fly's eye system of figure 2.68.

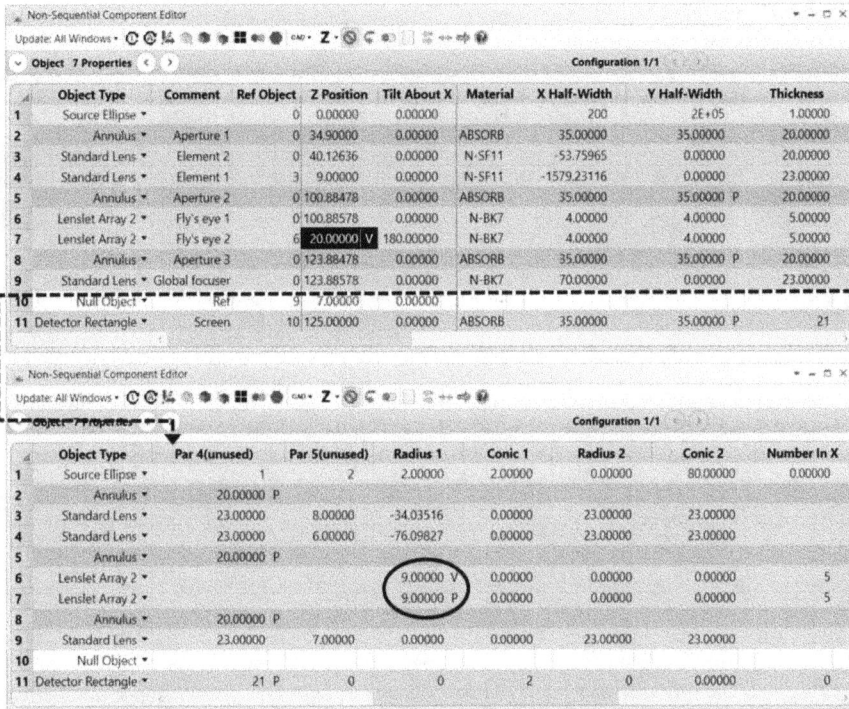

Figure 2.104. A prescription for the initial state of a fly's eye system prior to optimization.

trivial, it is more interesting to try the former, which means that variables should be set on the radii of the convex surfaces of the lenslets in both arrays. In addition, we should set their gap to be a variable.

At this point, a number of tricks are involved. As usual, I will first show you the setup for the starting point in the optimization, and then I will explain the steps. Figure 2.104 shows the NSC Editor for the initial setup, in which variables have been placed on the lenslet curvatures (object 6, column 'Radius 1') and their separation (object 6, column 'Z Position'). Note that there is a Pickup Solve for the radius on the second lenslet array, which is made to mirror the radii from the first array (they are symmetrical arrays). But note that the radius has been pre-adjusted to be at 9 mm (it was 10 mm previously, in figure 2.70) and the separation is 20 mm (it was 22 mm previously, in figure 2.70). Moreover, the numbers of X and Y pixels in the detector have been made smaller (object 11, see the column under 'Thickness,' which would have been labeled '# X Pixels' if my computer mouse's cursor were placed there). Note that these numbers are now set at 21×21 pixels. Finally, note that the number of analysis rays has been reduced to 200 000. These will all be explained soon. Next, figure 2.105 shows the merit function for the system in figure 2.104, which has been set up ready for optimization. Notice the use of the NSDD and NSTR operands on rows 1 and 2, followed by two NSDD operands and a computation of the ratio of the first (row 3) to the second (row 4). The value of 11

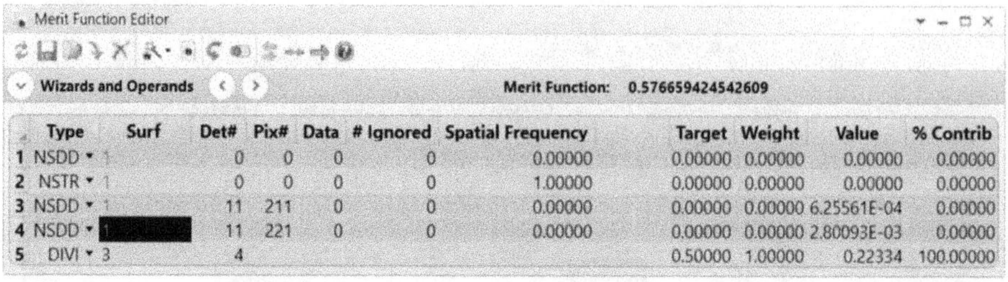

Figure 2.105. The merit function for the system shown in figure 2.104.

Figure 2.106. The numbering format for pixels on detectors in OS's NSC mode. This example shows the numbering for a rectangular detector with 9×9 pixels. Reproduced from [67], with the permission of ANSYS, Inc.

has been entered into the column 'Det#' for the two NSDD operands, because they refer to object 11 (the detector, which is the screen in the layout of the fly's eye system). On the other hand, '0' has been entered as the Det# for the NSDD on row 1, which is done for no reason other than personal habit (a '0' means that all detectors present are cleared prior to performing the NSTR ray trace in the merit function). Under the column labeled 'Pix#,' the NSDD operand on row 3 has the value 211, while the next has the value 221. These values refer to the locations of specific pixels on the rectangular detector (object 11). In OS, pixels on a rectangular detector are numbered from bottom left to bottom right, then the numbering moves up one row and begins again from left to right, as indicated in figure 2.106 for the case of a rectangular detector with 9×9 pixels (this figure is provided in the OS help file for NSC detectors). In the case of the rectangular detector in the current NSC model, I used 21×21 pixels, and I wanted to obtain the flux in the pixel located at the center of the detector as well as the flux located at the middle left side. It turns out that the central pixel number is 211, and the middle left-side pixel number is 221. This is equivalent to the following: say I had used a 9×9 array of pixels (see, for example, figure 2.106), then the middle pixel would have been 41 and the middle left-side number would have been 37. The DIVI operand on row 5 in the merit function takes the ratio of the flux in the middle left-side pixel to the flux in the central pixel, yielding the approximate fraction of the 'irradiance' at the center. Since I am using single pixels, I do not need to divide the flux by area. The flux in the pixel already samples the local irradiance in that area.

Continuing with our discussion of the merit function in figure 2.105, you may have noticed that the target value in the DIVI operand is 0.5. This is to *guide* the optimization to widen the irradiance distribution until the total width of the detector is roughly the FWHM for the irradiance profile. The result will not be exact, due to the low sampling of pixels and traced rays, but it will get close. The value '0' has been entered into the column labeled 'Data' for the NSDD detectors, because this is the value which indicates that flux is the desired quantity in those pixels. You can ignore the '# Ignored' and 'Spatial Frequency' columns.

At this point, you may begin to see why I entered a rather low pixel number in the detector. First, it is easier to determine the locations of pixels (I manually counted rows and columns rather than deriving a counting formula for the array). Second, I wanted to integrate the flux over a larger surface area on the detector, which helps to average out the ray trace noise, especially since I have entered only 200 000 rays to be traced. If I had used millions of rays, the optimization time would have been significantly lengthened, as you can imagine. So, this is a trick in NSC optimization —use fewer rays, use fewer pixels, perform averaging to get approximate results, then make a final assessment using a higher pixel count and more rays.

I mentioned that the radii of curvature of all lenslets have been tweaked slightly and so has the separation between them. The reason is because, in my experience in NSC optimization, the efficiency of the optimization process is significantly increased if the desired value in the merit function (i.e. the numerical value under the 'Value' column for the operand with the weight) is set close to the desired target value. Of course, this does not mean that you should manually tweak the variables for optimization until the merit function value attains the exact target value, for that would defeat the whole purpose of optimization. However, it is beneficial to manually tweak the variables to get a nonzero value for the desired merit function value. If I had not made this tweak to the radii and separation for the lenslet arrays, then the merit function value would have been zero. This *confuses* the algorithm in NSC optimization and can make it go completely off track. So, I manually tweaked the radius and separation until the value became nonzero and positive. You see that I have left it at the value of 0.223 34. Think of this as you would when you tweak and set up a lens design in SC mode prior to optimization (of course, this does not apply to programs and algorithms that let you start optimization from plane parallel glasses). The concept here is to let your mind and hands do some initial coarse adjustments to the variables, then let the program find the exact solution. So, thereafter, I performed a single local damped least-squares optimization, which took about four minutes to complete, and the results are shown in figures 2.107–2.109.

You can see in the prescription (figure 2.107) that new values have been determined through optimization for the separation and radii. In figure 2.108, the target value of 0.5 has been attained. Then, to obtain the plots in figure 2.109, I saved the optimized state into a separate file and renamed it, followed by changing the detector's pixels back to the original 100×100 and increasing the number of

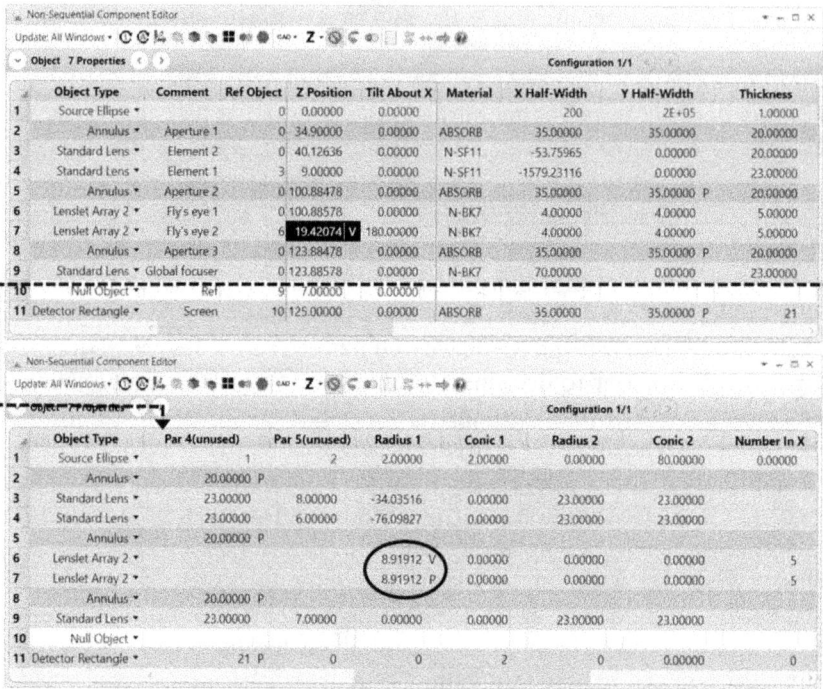

Figure 2.107. The prescription for the optimized state of the fly's eye system.

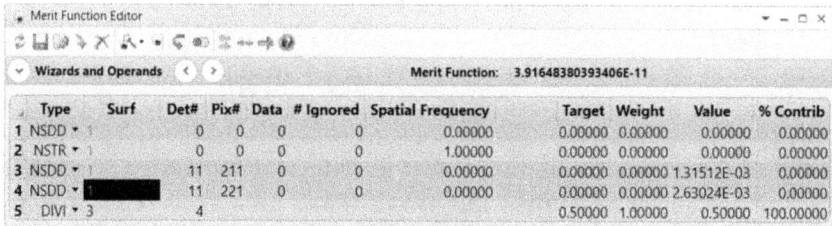

Figure 2.108. The merit function for the optimized state of the fly's eye system.

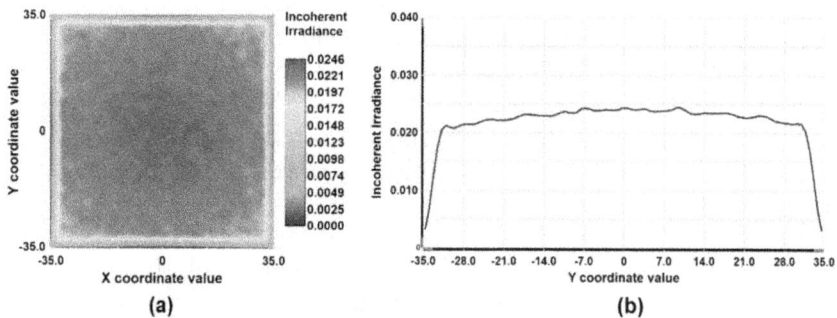

Figure 2.109. The irradiance profile for the optimized state of the fly's eye system.

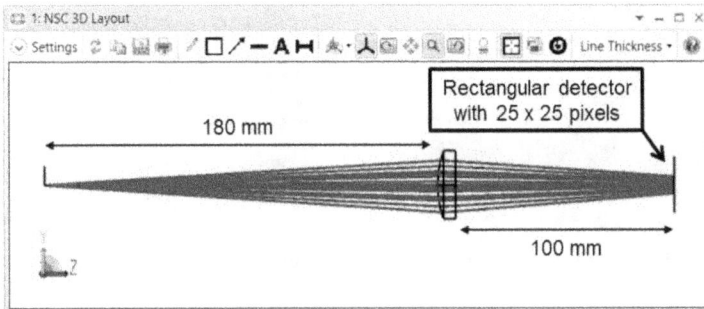

Figure 2.110. An NSC layout for a plano-convex lens and a point source.

analysis rays to 2E+06. The result shows that, indeed, the ROI has been broadened, but note that the FWHM is not quite at the full width of the detector. As mentioned above, the low pixel sampling results in an approximate optimized state. In reality, a bit of trial and error is involved (e.g. one can set the target value in row 5 of the merit function to something higher than 0.5).

It should become apparent that optimization in NSC ray tracing involves a combination of manual tweaks and applying suitable merit function operand construction. Sometimes, it can even involve new algorithms (every experienced illumination engineer using a preferred program has developed personal bags of tricks and algorithms for design and optimization). In the current example, I showed one of the simplest merit function constructions by far. Let me show you another simple one. Suppose you had to optimize the radius of a plano-convex lens in pure NSC mode to focus axial rays to a sharp spot. In SC ray tracing, we are *spoiled* for choices of pre-developed codes for operands, such as RMS spot size, RMS wavefront error, MTF, and so on. In NSC mode, setting the 'Pix#' value to '-9' selects an algorithm that computes the RMS spot size, and setting it to '−14' or '−15' computes the geometric MTF (you can also enable the *optimization wizard* on the merit function toolbar). But let me show you a simple alternative. When a lens images a point source in NSC mode, the desired result is to achieve the maximum flux at the smallest possible number of pixels on the detector, so that the rest of the pixels should be dark. This is a condition of *maximum* std dev of pixel fluxes on the detector. In OS's NSC ray tracing mode, setting 'Pix#' to the value of '−4' lets the program compute the std dev of flux values across all pixels on the detector. However, the merit function is designed to *minimize* everything that is set as targets. We cannot just use std dev alone as the operand, because we want it *maximized*. The solution is to use a second operand, RECI, to compute the reciprocal of the std dev. We would then target the reciprocal for minimization (using, say, '0' as the target value). Clearly, if the program minimizes 1/(std dev), then it maximizes the std dev. Figure 2.110 shows a layout for a simple plano-convex lens and a point source in pure NSC mode. The source's rays are set to be emitted within a cone of 4° half-angle. In the layout, only 25 random rays in this cone are displayed, while 2000 analysis rays are entered for optimization (an arbitrary number). The wavelength is

Figure 2.111. The merit function for the system shown in figure 2.110.

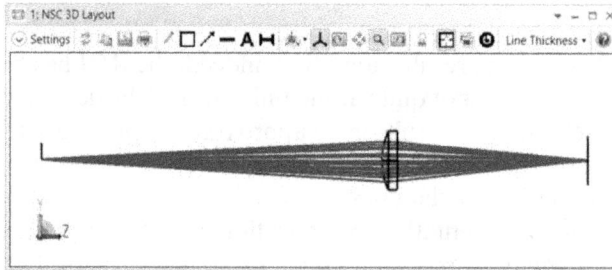

Figure 2.112. The optimized layout for the lens shown in figure 2.110.

Figure 2.113. The merit function of the optimized state in figure 2.112.

550 nm (but the rays are displayed in blue). The lens material is N-BK7 and its radius on the convex side is 45 mm. The lens's central thickness is 8 mm and its diameter is 30 mm. A variable has been assigned to the lens's radius. Figure 2.111 shows the merit function for this system in its initial state, i.e. prior to optimization. Note the target of '0' for the reciprocal of the std dev, given by using Data# = −4 for the NSDD operand on row 3. Upon performing a single local damped least-squares optimization, the result is the layout shown in figure 2.112, and the final merit function is shown in figure 2.113. The lens's radius is now 35.818 87 mm.

When the same lens is optimized in SC mode, using the RSCE operand to minimize the RMS spot size, the radius on the convex surface is determined to be 35.921 91 mm and the RMS spot radius value is roughly 0.2127 mm. If the radius is changed to 35.818 87 mm, then the RMS spot radius becomes 0.2172 mm.

These values are quite close, signifying that the method above worked for optimizing the focus in pure NSC mode.

Returning to the fly's eye system of figure 2.68, suppose we wish to perform a tolerancing analysis. Clearly, in pure NSC mode, this could take a very long time. So, the question to ask is: 'What quantities in SC ray tracing mode can be used to describe the quality of illumination of an illumination system?' Recalling the topic of discussion in section 2.2.5, if the PSF of an optical imaging system is large compared to the size of the ideal image, then the PSF has a significant impact on the irradiance profile of the image. The transverse boundary defined by the ideal image sets the limits of integration in the convolution integral. Therefore, a comparison made between the PSF's size (defined perhaps by its FWHM) and the size of the ideal image is equivalent to comparing the transverse extent of the PSF with a ROI (in this case, the ROI is the boundary defined by the ideal image). Analogously, for an illumination system, the ROI of illumination defines a boundary for the superposition of rays at a screen. In some cases, those rays may be considered to comprise a PSF. For instance, in the method of PSF illumination, it is clear that the illumination on the screen is, in fact, a PSF. Therefore, tolerancing PSF illumination systems is just a matter of using the RMS spot size as the figure of merit in the merit function (in other words, using it to judge the quality of illumination). To obtain an illumination profile, one simply performs a Monte Carlo simulation and examines the illumination profile for the Monte Carlo sample file of the worst RMS spot size (or one can save all files and inspect all illumination profiles). In OS, one can use the GIA or GBI feature, but one can also convert any of the saved Monte Carlo files into NSC files and then perform quality inspection through NSC ray tracing.

Even if an illumination system is not based on the PSF method, the irradiance profile may still be considered to be the result of superposing PSFs (a consequence of the concept of étendue division and superposition—see, for example, section 2.5). For example, consider the system shown in figure 2.114, which is the layout of the fly's eye system from figure 2.75 (the sequential model of the system in figure 2.68).

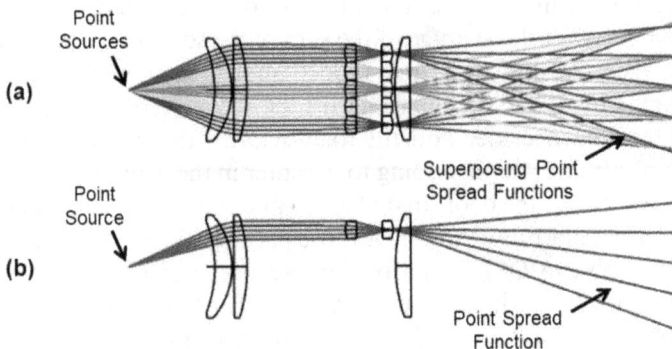

Figure 2.114. The layout from figure 2.75, illustrating PSF superposition. (a) Rays passing through all lenslets superpose at the screen. (b) Isolated rays from just the top lenslet.

Here, only rays from the axial field point are displayed for each configuration. As before, the rays being displayed are colored by configuration. Figure 2.114(a) shows rays passing through all lenslets, while figure 2.114(b) shows only the rays passing through the upper lenslet. This is meant to show that what we have is a condition of superposed PSFs.

There is also a subtle difference between the layouts of figures 2.114 and 2.75, which is that in figure 2.114, the left surface of the lenslets on the first array has been made into a stop surface. If you look at the prescription in figure 2.76, this means that instead of placing the stop on surface 5, I have placed the stop on surface 7, so that the stop's semi-diameter is 4 mm. Also, ray aiming is 'on,' so that for each configuration, rays are properly aimed through each lenslet. This setup in the prescription allows the SC model to have a point source from the object travel unvignetted through the system and end up on the screen as a large PSF. The boundary for the superposition of the PSFs is defined by the screen's ROI, but the size of each PSF is comparable to the ROI. Therefore, it is the sum of all PSFs that defines the illumination profile. In this case, we may use a sum of RMS spot sizes to define the quality of illumination. However, it turns out that this sum alone is insufficient to qualify the illumination, because taking a sum of RMS spot sizes is equivalent to averaging the nonuniformities across the ROI. Moreover, a sum of RMS spot sizes is more of a measure of the transverse size of the illumination profile. While this can be a parameter of interest, we must also have a measure of nonuniformity. In this case, the std dev of RMS spot sizes can be that measure. We do not need to derive a formula to relate the std dev of ROI spot sizes to a metric for the actual irradiance profile's nonuniformity (for example, by taking a profile's max/min ratio and relating this to the std dev of RMS spot sizes). Rather, one may perform a Monte Carlo simulation and save the file with the worst std dev of RMS spot size, then inspect the irradiance profile.

Let us take a simple example. Returning to the SC model for the fly's eye system in figure 2.75, we construct a merit function to compute the std dev of the RMS spot size for just the axial field in each configuration, as shown in figure 2.115. Row 24 provides the resulting std dev, which is 17.991 57 mm. Note that RMS spot 'size' refers to the average length of the radial dimension of the spot, so that the std dev here is the square root of the average of the sum of the squares of the spot radii. If we now return to the prescription for this model (shown in figure 2.76), we can add a pair of coordinate breaks to decenter and tilt the second lenslet array. This shall be our way of modeling decenter and tilt tolerances to the mounting of the second array. For simplicity, we are only going to decenter in the Y dimension and tilt about the X axis. So, we create the coordinate breaks and then insert tolerancing operands into the Tolerance Data Editor to create those perturbations (see section 1.2.13 for a review of tolerancing in OS). In particular, we use two TPAR operands: one to decenter the second array within ± 1 mm and the other to tilt the second array within $\pm 2°$. We then perform a Monte Carlo simulation to produce 1000 Monte Carlo samples of the fly's eye system, each of which possesses random values for the decenter and tilts, using a uniform distribution. At the end of *this* Monte Carlo run, OS outputs a window with two tabs (see figure 2.116, right side). The left tab is

Figure 2.115. The merit function used to compute the std dev of the RMS spot size in the SC fly's eye array system of figure 2.75.

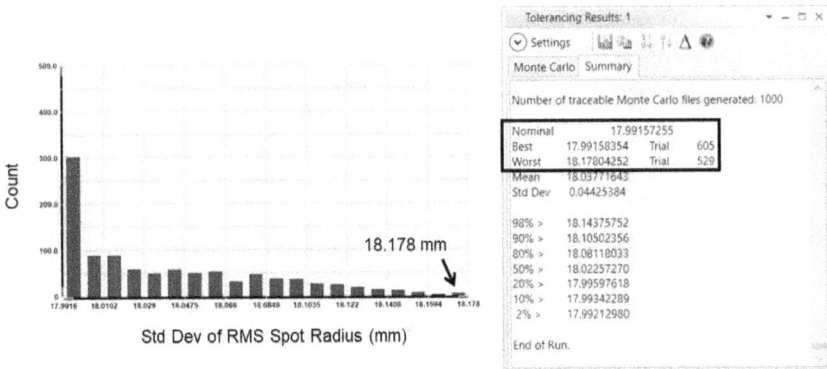

Figure 2.116. A histogram of the Monte Carlo run (left) and the tolerance results summary window (right), which indicates the best and worst std devs of the RMS spot sizes.

labeled 'Monte Carlo,' and the right tab is labeled 'Summary.' I have enclosed the most most important results in the right tab in a rectangle. The 'Best' Monte Carlo sample produced a value of 17.991 58 mm for the std dev of the RMS spot size, which is expected, because the perturbations are performed in a uniform distribution about the nominal state of the system, whose nominal value for the std dev of RMS spot size is 17.991 57 mm (see figure 2.115, bottom circled value). Close enough. The 'Worst' value for the std dev of RMS spot size is 18.178 mm, which is also the value observed in the histogram.

In figure 2.116, towards the right-hand side of the Best and Worst values, the Monte Carlo file numbers associated with those values are provided: Trial # 605 for the Best, and Trial # 529 for the Worst. They are called 'Trials' because a Monte Carlo run simulates the act of trying an experiment or a dice roll, i.e. relying on chance. Clicking on the 'Monte Carlo' tab in the output window for Tolerancing Results shows a table of statistics (figure 2.117). Scrolling down, we can locate Trial # 529 (labeled 'MC_529' for 'Monte Carlo file # 529') and also Trial # 605, which are the Worst and Best trials. The next two columns list the random tolerance values (under a uniform distribution) that were given to those trials. This means that we can open up the NSC version of the fly's eye model and enter those perturbed values into the Y position and tilt about X for the second array. Ordinarily, we would save the Monte Carlo trials and open them up for inspection. But for the current simple exercise, we will not be doing that.

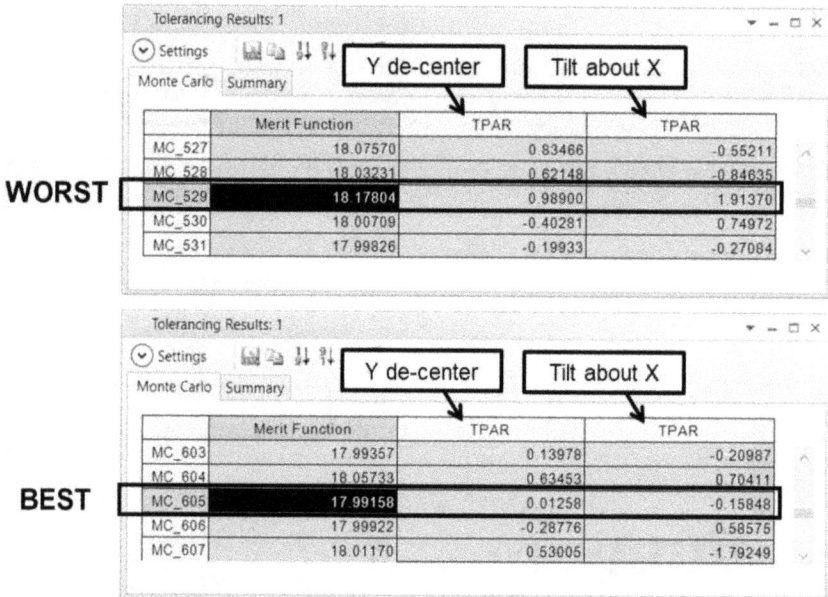

Figure 2.117. The worst and best std devs of the RMS spot sizes in mm (Merit Function column). The corresponding *Y decenter* and *tilt about X* tolerances applied are shown in the next two columns. The Monte Carlo file number is listed at the left-most column.

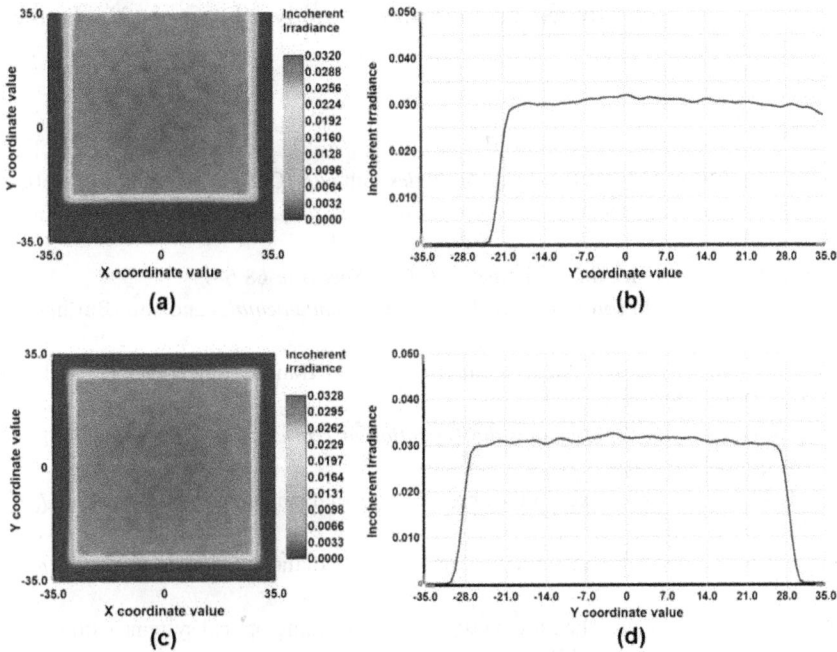

Figure 2.118. Irradiance profiles for the NSC layout of the fly's eye system. (a) and (b): The worst-case std dev of RMS spot size. (c) and (d): The best-case std dev of RMS spot size. Note that (b) and (d) plot the profile along the middle column on the detector.

So, we then open up the NSC model and enter the perturbations given by the values for *Y decenter* (in mm) and *tilt about X* (in °) for the Worst and Best cases into the NSC model. The result is shown in figure 2.118. You can see that in figure 2.118(a), the irradiance profile is shifted. In figure 2.118(b), the middle column of pixels is plotted, showing the 1D profile in that dimension. Figures 2.118(c) and (d) show the best-case result, which looks good, as expected. This means that the worst-case std dev of the RMS spot size successfully *detected* or *quantified* an undesirable illumination profile, which is that of a shifted profile relative to the ROI. And the best-case std dev of the RMS spot sizes correlated well with a good state for the irradiance profile. We were able to make these correlations and detections without recourse to any tolerancing done in pure NSC mode because of our understanding of how PSFs impact the irradiance profiles of illumination systems in the limit that the PSF is large compared to the ROI size. Otherwise, as you can imagine, performing tolerancing 'live' in pure NSC mode would have taken an entire night to trace the Monte Carlo rays.

The approach described is applicable to many classical illumination configurations. For instance, in Köhler illumination, if we trace rays from the source to the screen in SC mode, we have another case in which the PSF is comparable in size to the ROI (see, for example, figures 1.137 and 2.18). Therefore, the tolerancing of Köhler illumination designs may be performed by the method described here, and the resulting worst-case Monte Carlo file may be analyzed to assess the final quality of the illumination.

References

[1] Born M and Wolf E 1980 *Principles of Optics* 6th edn (Oxford: Pergamon) pp 508–34

[2] Wolf E 2007 *Introduction to the Theory of Coherence and Polarization of Light* (Cambridge: Cambridge University Press)

[3] Wolf E 1978 Coherence and radiometry *J. Opt. Soc. Am.* **68** 6–17

[4] Kingslake R and Johnson R B 2010 *Lens Design Fundamentals* 2nd edn (Burlington, MA: Elsevier) pp 245–8

[5] Díaz J A and Navarro R 2023 Geometrical optical transfer function and pupil sampling patterns *Optik* **279** 170746

[6] Boyd R W 1983 *Radiometry and the Detection of Optical Radiation* (New York: Wiley) pp 16–7, 86–9, 95–103

[7] Siew R 2022 Image irradiance in the absence of aplanatism and isoplanatism *Eur. J. Phys.* **43** 035304

[8] Gilmore H F 1966 The determination of image irradiance in optical systems *Appl. Opt* **5** 1812–7

[9] Siew R H 2005 *f*/No. and the radiometry of image forming optical systems with non-circular aperture stops *Proc. SPIE* **5867** 586701

[10] Siew R 2008 Corrections to classical radiometry and the brightness of stars *Eur. J. Phys.* **29** 1105

[11] Rehn H 2019 *Beleuchtungsinduzierte Chromatische Abbildungsfehler DGaO Proceedings* (www.dgao-proceedings.de/archiv/120_chronologisch_d.php)

[12] Chaves J 2016 *Introduction to Nonimaging Optics* 2nd edn (Boca Raton, FL: CRC Press) pp 461–520

[13] Muschaweck J and Rehn H 2019 Illumination design patterns for homogenization and color mixing *Adv. Opt. Technol.* **8** 13–32

[14] Bunch R M 2021 *Optical Systems Design Detection Essentials: Radiometry, Photometry, Colorimetry, Noise, and Measurements* ed R B Johnson (Bristol: IOP Publishing) pp 6–35 9

[15] McCluney R 1994 *Introduction to Radiometry and Photometry* (Norwood, MA: Artech House) pp 33–47

[16] Grant G B 2011 *Field Guide to Radiometry* (Bellingham, WA: SPIE) pp 85–6

[17] Wolfe W 1998 *Introduction to Radiometry* (Bellingham, WA: SPIE) pp 79–83

[18] Goodman T *et al* 2007 *Measurement of LEDs CIE Technical Report* 127:2007

[19] Muschaweck J and Rehn H 2022 *Designing Illumination Optics* (Bellingham, WA: SPIE) pp 21–3, 38–42, 51–2

[20] Muschaweck J 2023 Virtual focus of a ray data source *Opt. Eng.* **62** 070401

[21] López M, Bredemeier K, Schmidt F and Sperling A 2010 Near-field goniophotometry: a metrological challenge *Simposio de Metrologia (Mexico, Oct. 27–29 2010)* SM2010-S3B-2

[22] López M, Bredemeier K, Rohrbeck N, Véron C, Schmidt F and Sperling A 2012 LED near-field goniophotometer at PTB *Metrologia* **49** S141

[23] Audenaert J, Acuña P C R, Hanselaer P and Leloup F B 2015 Practical limitations of near-field goniophotometer measurements imposed by a dynamic range mismatch *Opt. Express* **23** 2240–51

[24] Moreno I and Viveros-Méndez P X 2021 Modelling the irradiation pattern of LEDs at short distances *Opt. Express* **29** 6845–953

[25] Rudder A and Main K 2023 What is the project management triangle? *Forbes* (www.forbes.com/advisor/business/project-management-triangle/)

[26] Dobson M S 2004 *The Triple Constraints in Project Management* (Oakland, CA: Berrett-Koehler)

[27] Arecchi A V, Messadi T and Koshel R J 2007 *Field Guide to Illumination* (Bellingham, WA: SPIE) pp 63, 76–81

[28] Necodemus F E 1963 Radiance *Am. J. Phys.* **31** 368–77

[29] Siew R 2017 *Perspectives on Modern Optics and Imaging: With Practical Examples using Zemax® OpticStudio®* (Independently Published) pp 84–6, 105–12, 112–22

[30] Siew R 2017 Relative illumination and image distortion *Opt. Eng.* **56** 049701

[31] Siew R 2020 Relative illumination and image distortion part 2: how to 'sense' the entrance pupil *Inopticalsolutions Technical Note (Revision 3)*

[32] Reiss M 1948 Notes on the Cos^4 law of illumination *J. Opt. Soc. Am.* **38** 980–6

[33] Kingslake R 1965 Illumination in Optical Images *Applied Optics and Optical Engineering Vol. II: The Detection of Light and Infrared Radiation* (New York: Academic) pp 214–7

[34] Reshidko D and Sasián J 2016 The role of aberrations in the relative illumination of a lens system *Proc. SPIE* **9948** 994806

[35] Reshidko D and Sasián J 2016 Role of aberrations in the relative illumination of a lens system *Opt. Eng.* **55** 115105

[36] Johnson T P and Sasián J 2020 Image distortion, pupil coma, and relative illumination *Apt. Opt.* **59** G19–23

[37] Rimmer M P 1986 Relative illumination calculations *Proc. SPIE* **0655** 99–104

[38] Nakagawa J 1973 Large aperture six component photographic lens *U.S. Patent* 3743387

[39] Winston R, Miñano J C and Benítez P 2005 *Nonimaging Optics* (Burlington, MA: Elsevier) pp 18–21, 45–7

[40] Sasián J 2019 *Introduction to Lens Design* (Cambridge: Cambridge University Press) pp 60–1, 158–60

[41] Kidger M J 2002 *Fundamental Optical Design* (Bellingham, WA: SPIE) pp 135–6

[42] Hulbert E O 1946 Optics of searchlight illumination *J. Opt. Soc. Am.* **36** 483–91

[43] Stewart S M and Johnson R B 2017 *Blackbody Radiation: A History of Thermal Radiation Computational Aids and Numerical Methods* (Boca Raton, FL: CRC Press) pp 150–1

[44] Siew R 2016 Axial nonimaging characteristics of imaging lenses: discussion *J. Opt. Soc. Am. A* **33** 970–7

[45] ANSI/PLATO FL 1-2019 Standard (www.plato-usa.org/about/standard) accessed 8 November 2023

[46] PLATO 2023 https://www.plato-usa.org/about/standard

[47] Koshel R J 2013 *Illumination Engineering* (Hoboken, NJ: Wiley) pp 220–5, 282–9

[48] Pan J-W and Wang H-H 2013 High contrast ratio prism design in a mini projector *Appl. Opt.* **52** 8347–54

[49] Feng W, Liu Z and Ye M 2022 Liquid crystal lens array with positive and negative focal lengths *Opt. Express* **30** 28941–53

[50] Lin C-H and Chen K-H 2021 Depth-sensing technology using a negative microlens array *Int. J. Optomechatronics* **15** 170–81

[51] Siew R, Wei S Y and Hung J-S 2021 Optical system, and method of illuminating a sample plane *U.S. Patent* 11567007

[52] Frieden B R 1965 Lossless conversion of a plane laser wave to a plane wave of uniform irradiance *Appl. Opt.* **4** 1400–3

[53] Dickey F M 2014 *Laser Beam Shaping Theory and Techniques* 2nd edn (Boca Raton, FL: CRC Press)

[54] Dickey F M, Holswade S C and Shealy D L 2006 *Laser Beam Shaping Applications* (Boca Raton, FL: CRC Press)

[55] Shafer D 1982 Gaussian to flat-top intensity distributing lens *Opt. Laser Technol.* **14** 159–60

[56] Shafer D 1997 Gaussian to flat-top in diffraction far-field *Appl. Opt.* **36** 9092

[57] Shimmel H and Wyrowski F 2007 Designing beam shaping systems basing on spherical catalog lenses *Proc. SPIE* **6663** 66630C

[58] Rhodes P W and Shealy D L 1980 Refractive optical systems for irradiance redistribution of collimated radiation: their design and analysis *Appl. Opt.* **19** 3545–53

[59] Kim N-H 2021 (updated by Wilczynski A) How to design a Gaussian to top hat beam shaper Zemax Knowledgebase Article (https://support.zemax.com/h.c./en-us/articles/1500005489161-How-to-design-a-Gaussian-to-Top-Hat-beam-shaper)

[60] Milonni P W and Eberly J H 1988 *Lasers* (New York: Wiley) pp 484–90

[61] Pedrotti F L S J and Pedrotti L S 1993 *Introduction to Optics* 2nd edn (Englewood Cliffs, NJ: Prentice-Hall) pp 473–4

[62] Sasián J 2020 Formulae for the geometrical propagation of a beam of light *Appl. Opt.* **59** G24–32

[63] Siew R 2020 How to use spherical optics to create uniform top hat illumination Zemax Envision 2020 (Oct. 7, 2020)

[64] Siew R 2020 Uniform top hat illumination for extended sources using only spherical lenses *Inopticalsolutions White Paper*

[65] Gao G, Lin L and Huang Y 2005 Using spherical aberrations of a singlet lens to get a uniform LED illumination *Proc. SPIE* **5638** 551–60

[66] Dilworth D 2020 *Lens Design: Automatic and Quasi-Autonomous Computational Methods and Techniques* 2nd edn (Bristol: IOP Publishing) ch 15

[67] ANSYS, Inc. 2023 https://www.ansys.com/en-gb

Chapter 3

Optical system product development

The previous two chapters aimed to provide you with basic imaging and illumination design skills, which are essential for a designer of optical systems. The current chapter will equip you with further knowledge to survive in the world of product development. In this world, we often need to know more than how to bend rays in a program. Certain optical systems are so complex that they require us to think hard about the relationships between optical parameters and system parameters, the latter being quantities that are most important to the end application, which may not be obvious to a designer of a subsystem. On this note, many experienced designers of optical systems rightfully highlight the need to look at optical design from a *systems perspective* [1–3]. We will get to know more about what is meant by this. To that end, the current chapter begins with storytelling, where I share some personal experiences from my journey into product development—the challenges, failures, self-doubt, perseverance, and to my delight, some surprising successes along the way. This story may be helpful to early-career designers, and it is meant to demonstrate that having a broad *systems perspective* on our personal and professional development can be beneficial in achieving success in product development. Beyond these personal experiences, we discuss basic statistical principles that will help you make some sense of the role of variability in tolerancing complex optical systems. It is highlighted that the Monte Carlo method is the preferred choice for performing tolerancing analysis on such systems and that it may even be applied to justify an alignment philosophy. We then touch on some specific nuances of optical systems that one often experiences in product development (glare, stray light, drift effects, etc). The chapter progresses towards closure by presenting two conceptual case studies involving complexity in optical systems. The purpose of these is to help put some of the concepts discussed in this chapter into a practical context, so that you get an idea of how they may be applied in turning an optical design into an optical system product.

doi:10.1088/978-0-7503-6059-3ch3

3.1 Lights at the ends of tunnels (not light pipes, but a personal story)

3.1.1 Gratification and enlightenment

It can be truly gratifying to see your design turned into an actual product that is sold and used by others. The product could be as 'simple' as an imaging lens, or it could be a complex instrument comprising optics, mechanics, electronics, firmware, and software. In my career, I have felt both fortunate and unfortunate that my first job involved a very complex system. It was an instrument (called a *prepress system*) almost *as big as a small car*, comprising everything you could imagine, and its purpose was to scan a modulated high-power near-infrared beam onto specially coated aluminum plates that are used by bigger machines for printing and publishing. I felt fortunate in that it was my first exposure to complex instruments, which taught me a lot about integrating optics with other functions. But I felt unfortunate that I did not know how to model and design optical systems using a program. I often felt helpless when there were engineering problems that I thought could be solved if only I knew how to use a ray tracing program. I did many analyses *by hand* and relied on spreadsheets and a mathematics program. Of course, you could say that those laborious analyses done on paper, spreadsheets, and math programs were helpful lessons, which certainly gave me some deep analytical skills that I probably still possess today. But I want to highlight that not being able to design and model using an advanced optical design program can put one at a huge disadvantage, especially today. Just look at the designs and simulations presented in chapters 1 and 2 in this book. They could not have been produced without using an optical design program.

In my first job, not only did I feel helpless about not being able to use an optical design program, but I was also not very mature in the way I handled projects. I was quite disorganized and was unable to set priorities. While I knew what the project's objectives were, I did not know how to be effective in getting there. It was only in my second job that I learned from my boss the importance of having a weekly plan just for work. It sounds so straightforward, but it was not obvious to me then. Once I began to create a weekly work plan, I became more effective and was able to chart out my path. While my weekly work plan helped me prioritize my day job tasks, I kept a separate list of 'things I still need to understand,' a list that I had kept since I was little. But as a working adult, my list started to include items such as learning how to use an optical design program, and looking for research papers and textbook chapters that could help me achieve my work tasks. Remember the topics discussed in section 1.1? Many of them came from my use of a weekly plan with separate lists of priorities.

3.1.2 Challenges in academic life

School and the college experience was another tough journey. I had a difficult time in class. It was a matter of personal style in learning: I was not good at taking notes; I was better at studying from textbooks, which let me learn material at my own pace, ask myself questions, and work out solutions in different ways. Yet, neither books nor lectures told me everything. Sometimes, a professor would share special tricks and techniques not found in books. In other cases, some books provided insights not found in lectures. It would have been best to be able to understand material through both methods—lectures and books. So, I decided to knock on doors, bothering professors and graduate students, asking all of my questions. I would occasionally go to classes and, in instances where I became excited about a topic, I would also offer some further points, or ask further questions, but these were rare moments.

In university, I began to realize that my academic journey had two parts: (1) the part that society imposed on me, which was to pass exams and obtain a degree; (2) the part that involved things I really wanted to do, such as going into a lab to try a new idea that nobody else cared about or digging up textbooks to read unspecified chapters for nothing but personal gratification. Today, I am glad I did these things. Back then, I did not expect to get credit for such extracurricular activities. But through them, I have gained skills and knowledge that have become useful today, such as in writing a book on *Modern Classical Optical System Design*. As mentioned in section 1.1.7, it is helpful to divide work/life into two parts: one part is for doing what employers need us to do; the other part is just for ourselves.

3.1.3 Transition to the real world and product development

Do you know what is strange? I never thought that I would be involved in product development. Back in high school and college, I dreamt of being a professor. I studied for exams by going into an empty classroom or lecture hall, and, with my notes and textbooks closed, pretended to deliver lectures on topics I needed to know. I fantasized about having my own lab and my own office with a window facing a river, where I could daydream about the next big theory and write a paper on it. When I was eleven and throughout school years, I even wrote 'inventions' and 'theories' on scrolls of paper, thinking that they were publication-worthy research.

Then I realized how difficult it was to possess *real* knowledge. In my second year as an undergraduate, all of what I thought I knew about Newton's laws was replaced by the *Lagrangian* and the *Hamiltonian*. Suddenly, I realized I knew nothing. *Hilbert space* was no space I had heard of, and I did not understand it very well. Everything was very abstract. Quantum mechanics was not just about reading popular books on Schrödinger's cat, nor was it just about applying Bragg's formula to determine the angle of diffraction of x-ray quanta from crystals. No, it was about a differential equation that seemed unsolvable except in the simplest of cases, such as that of a particle in a 'box,' except that it was not literally a box. Rather, it was a *potential*, and I did not understand what that meant. I was just a naïve immature student.

I realized that I had to prioritize my knowledge. While I was not particularly good at classical and quantum mechanics at the time, I knew I was not bad at geometrical optics, physical optics, and radiometry. I additionally found that I could support my understanding of these core subjects with sound knowledge in calculus, complex numbers, trigonometry, and of course, algebra—mathematical tools that I had enjoyed applying to solving problems other than my homework (with the exception of statistical theory, which I did not possess knowledge of at the time).

I then figured that someday, I would perhaps try to master other subjects at my own pace and perform research in my personal time, which would not require me to study for exams and be pressured to go to class. But to do all that, I needed to find work to support myself. Fortunately, during my years as a master's degree student, I was presented with an opportunity to be part of what was called a co-operative (co-op) program at the Institute of Optics in the University of Rochester. This was a program where they let you interview with companies who would be willing to pay a student for a year to work for them, followed by the chance to return to college to complete the final semester for the master's degree. It was the ideal opportunity for me, as I thought it would give me a year off from a rather stressful academic life (and I did not even know that work life would eventually be equally stressful).

So, I went for the program and spent a year in Boston. I enjoyed it. It was liberating. I figured that this was my path. I could work, get paid, then do research at home. I was also fortunate to have found work in the same company that had hired me for the co-op program (it seemed that they did not mind having me back, so I must have done a fair job). Then I realized something else when I started *real work*. Other than my lack of skill in using an optical design program (and my inability to prioritize my tasks), I realized how challenging it was to do research after long work hours. At the end of each day (night) at the office, I just wanted to sleep, play sports, and go out with friends to a movie. While I occasionally went to the local university library to read (at the time, I was living close to the MIT campus in Cambridge, Massachusetts), and while I had plenty of ideas, I could not come up with any research worthy of scientific publication. In retrospect, I do not think that time and energy were significant factors in my lack of research progress. Quite simply, I found it difficult to follow the material in research papers, and I did not possess sufficient knowledge to identify real problems to solve that could lead to research. I do believe that it was plainly a matter of immaturity, lack of experience in doing real research, lack of some guidance from an experienced research advisor, and insufficient exposure to known problems in the area of optical system design. Basically, I just did not know what I was doing.

3.1.4 The light at the end

After four productive years at work (but unproductive years in terms of identifying a problem of significance worthy of scientific publication), two things happened. First, I was going to leave the company because I was moving to Singapore. Second, *the* elusive research problem that I was hoping would inspire me to write a publication-worthy paper had actually been right in front of me all that time. In the job I was in, as part of my responsibilities, I was tasked to understand the physics of a new spatial light modulator, invented by David Bloom, called the *Grating Light Valve*. I knew how it worked according to conventional theory. But I had a gut feeling that a different theory could be applied. I eventually figured it out and, with the encouragement of my boss at the time (who was one of three conference chairs of the *Optical Scanning* topical area at SPIE), I wrote and presented my first conference paper [4].

But this was just the beginning. Work life was not without struggle. After moving to Singapore, while I did enjoy successes at work (including publishing a couple more papers), I also faced career setbacks. Not long after those setbacks, my wife and I moved to Vancouver, Canada. Then, I fell into deep depression in the years 2014–5. I never thought it could happen to me. But it did. I was no longer passionate about optics and was a bit lost. After some further struggles, I made up my mind to buy an expensive book in a specific area of optics that remained a challenge for me. I figured that since it was expensive (a few hundred dollars), it would motivate me to make good use of it. And it worked. I read it over and over, even when I did not understand certain topics. That led me to some new ideas. I began writing papers again. In the course of about a year, I published four peer-refereed papers, and I was beginning to feel better. This activity led me to write a book, then another, and another. It also led me to make new friends and acquaintances through social media, such as Linkedin. I also got to know Prof. R. Barry Johnson, who, even if he may not know it, I consider as my mentor today.

In retrospect, while I do feel that I put in significant effort to achieve my goals, those achievements were not all possible without influence and support from others. Looking back, if it were not for the co-op program at the University of Rochester, I would not have found my first job. If it were not for the support of the management team in my first job, I would not have had the opportunity to write the SPIE paper on my idea. And these were only part of the whole program in my life. If I traced back further, it would obviously start with my parents supporting me and my education. While many events in my life were challenging, many were also rather fortuitous, yet their significances were not obvious until now. Today, I am grateful for the opportunity to write this book with support from Barry Johnson and IOP Publishing's editorial team. The light that I am seeing now—after a long road and deep tunnel of challenges I had faced in academia and at work—is not just the result of my effort, but it has also been due to the support of my parents, my brothers, my wife, teachers, professors, colleagues, and even *some* managers and bosses.

3.1.5 The systems perspective on optical system product development

The act of tracing the roots of our successes to the support we received from people is similar to tracing the roots of successes for an optical system product. It is often not possible for a product to be successful by working alone. Taking a systems perspective means that one has to consider things from different points of view. For instance, the output from a complex instrument is a function of many variables. In order to know what those variables are, we need to listen to the views and facts expressed by team members from different functions and disciplines, whose work may involve those variables. In the next section, I will give a specific example of an output from a complex instrument that can be expressed by formulas involving variables from optics, mechanics, electronics, software, and even chemistry.

3.2 An example of a complex optical system: virus detection using real-time quantitative PCR instruments

3.2.1 What is the minimum viable product for such a device?

The minimum viable product (MVP) for a real-time quantitative PCR (qPCR) instrument is *partially* depicted in figure 1.2. The rest of it consists of the information displayed on the 'computer & data display subsystem.' The basic idea in using a real-time qPCR instrument to detect a virus is as follows: suppose that there is a medical patient who is suspected of being infected with a specific virus. In order to detect that virus, its genetic signature (DNA) must be known, which is achieved by the method known as *gene sequencing*. Suppose this has been done. Then, in a medical facility, a sample of some form (such as saliva or blood) is extracted from the patient, and some chemical process is performed to isolate the virus's RNA (ribonucleic acid, which is essentially single-stranded DNA) and convert it into its DNA form. At this stage, it remains unknown whether or not that conversion has been successful, because it is unknown whether the virus is present. Nobody knows anything until the chemically processed sample is inserted into a real-time qPCR instrument.

To insert the sample into the instrument, the sample is placed into a tray with 'wells' (see the bottom of figure 1.2), often called a *microtiter plate*, or simply a plate. Each well on the plate is a test tube, but it is often called a well in molecular biology. In some cases, such as when it is affordable to do so, the suspected viral sample is placed into more than a single well, such as maybe a population of 20 or more wells. This is like replicating your experiment 20 or more times in order to obtain a statistically significant result with some confidence. Also, in some cases, a second population of wells is filled with a known concentration of the same viral sample. This is depicted in figure 3.1(a), where two shaded populations of wells are shown, one in black and the other in patterned lines. Further, a population of wells is filled

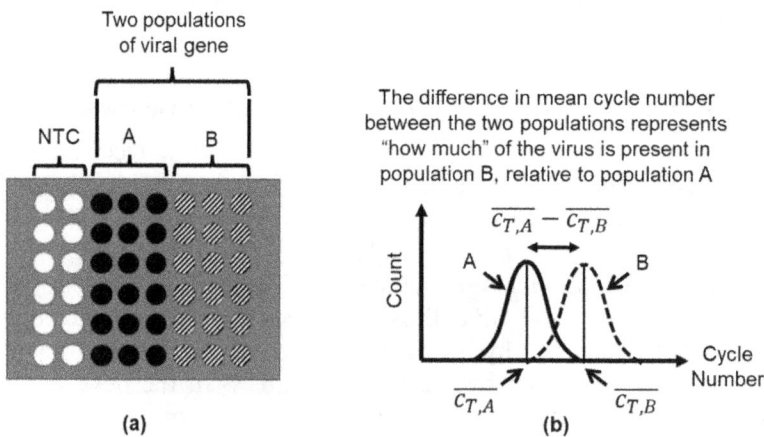

Figure 3.1. A real-time PCR assay. (a) A plate of wells filled with NTCs and two different populations of the same viral sample. (b) A possible outcome of the PCR assay.

with *buffer solution*, which you may think of as plain water with perhaps some treatment. These are called the *no-template-control* (NTC) wells. Think of them as your experimental control, which does not contain any virus. In theory, nothing should happen in the NTC wells, so if something is detected in any one of them, then it signals that the test is invalid and should perhaps be repeated.

In biology and chemistry, an experiment is often called an *assay*. So, a PCR test is an assay that has been carefully developed to detect the presence or absence of the virus in a sample from a patient. It does not actually look for the physical virus. Rather, it tests for the presence of the viral gene of interest, which is its DNA (converted from its RNA). Again, while some process is used to convert the virus's extracted RNA to DNA, nobody knows whether that conversion has been successful prior to performing the PCR assay in a real-time qPCR instrument. **The virus's DNA, if present, undergoes a process of being multiplied exponentially in the wells when the PCR instrument is activated**. This multiplication process of the target viral DNA is often called *PCR amplification*. It is essentially a process which clones the target viral DNA—if that DNA is present. Inside each of the shaded wells is a *template* (called a *primer*) of the suspected virus's DNA, among other chemical reagents. Only if the suspected virus's DNA is present are the templates paired (like a jigsaw puzzle) with the genetic sequences in the virus's DNA. During PCR amplification, the qPCR instrument undergoes a cyclic process of heating and cooling of the plate of wells. This is called *thermocycling*. During each cycle, elevated temperatures split up DNA into halves, so that if the suspected virus's DNA is present, templates 'join' with the viral DNA molecules. At lowered temperatures, a special enzyme in the wells connects and 'seals' the split-up DNA molecules back together. Thus, if there were one viral DNA molecule at the beginning of a thermocycle, then there would be two after that first thermocycle. On the second cycle, two DNA molecules would split and be joined with templates to form four DNA molecules. And so on. This is an exponential doubling [5] in cloned DNA molecules of the sought virus, cycle by cycle, which may be expressed as

$$C_i = C_o 2^i, \tag{3.1}$$

where C_o is the initial concentration (expressed as the quantity per unit volume) of the target viral DNA molecules, and $i = 1, 2, 3...$ thermocycles [6]. Equation (3.1) is the ideal case, in which the chemistry involved in the PCR process is said to be 100% efficient. We will return to this point later. Again, note that this cyclic process of amplification (cloning) of the viral DNA happens only if the viral DNA is present. Otherwise, the templates cannot pair with any split DNA, and there is no amplification.

Returning to figure 3.1(a), suppose that the population of wells filled with the suspected viral gene sample is labeled B, and a population of known concentration of that virus's DNA is present in the wells labeled A. So, population A has a known concentration of the known virus's DNA, and population B has an **unknown concentration** of the suspected presence of a specific virus's DNA (whose DNA sequence is known—otherwise, we would not know what viral DNA to insert into population A for reference). The first two columns of wells are filled with blank samples (buffer), so they are labeled as NTC wells. The qPCR instrument usually runs the PCR assay (including thermocycling) for more than 40 cycles, which takes roughly an hour. Suppose that the target viral DNA were actually present in population B. During the thermocycling process, fluorescent light would then be emitted from special *probes*, which are dye molecules (often called *fluorophores*) that are made to specifically detect the viral gene of interest. However, due to a process called *fluorescence resonance energy transfer* (FRET) [7], these dyes only light up when the viral gene of interest is present (I am simplifying the actual processes quite drastically here, just to highlight the main points). Since, ordinarily, only a single probe is used to detect a single viral DNA molecule, the intensity of fluorescent light emitted by the probes is proportional to the number of detected viral DNA molecules. So, if population B has the suspected virus's DNA molecules, then that entire population of wells lights up and gets brighter at each thermocycle. **Thus, the presence of the virus from an infected patient is detected by way of the intensity of the fluorescent light from a population of wells containing that virus's extracted/ converted DNA.** The fluorescence is, of course, activated by way of illuminating the wells using the type of optical system shown in figure 1.2, which is called a *fluorescence detection system*. The emitted fluorescence is captured by a sensor, such as that depicted at the top portion of figure 1.2, and the detected light is fed through electronics to a computer display. It turns out that the emitted fluorescence (which may be expressed in either units of flux per unit solid angle, or even in irradiance at some surface, such as that of a sensor) is approximately proportional to the concentration of amplified DNA, which we may express as

$$S_i = \alpha C_i = \alpha C_o 2^i, \tag{3.2}$$

where α is a proportionality factor and S_i may be regarded as some generic quantity related to the detected signal fluorescence (e.g. it could be a current signal), depending on what we insert into α for the type of light-to-signal conversion being considered.

The origin of the proportionality relation in equation (3.2) is in the first-order term of the expansion of the Beer–Lambert law for a beam propagating through absorbing media. This law states that $I(z) = I_o \exp(-\epsilon Cz)$, where I is, say, the radiant intensity of the beam in the absorbing medium, z is the dimension in which the beam is traveling, I_o is the intensity at $z = 0$, C is the concentration of the *active ingredient* (such as the fluorophores) responsible for light absorption in the medium, and ϵ is called the *extinction coefficient* of the absorbing ingredient [7]. Since the absorbed intensity upon propagation is $I_o - I(z)$, which gives rise to fluorescence, then the fluorescence is proportional to $I_o[1 - \exp(-\epsilon Cz)]$, which, upon expansion into a Taylor series, yields $I_o\epsilon Cz$ for its first-order term. In cross-referencing this with equation (3.2), the quantity αC_o may be identified as being proportional to $I_o\epsilon Cz$. So, if $C = C_o$, then $\alpha = I_o\gamma\epsilon z$, where γ is a proportionality factor, but this is not the only possibility for α. If there are any other elements or optical components in the path of the beam in or out of the absorbing medium (such as filters with transmittance), then additional factors may be included in α. And so on.

To quantify the amount of the virus thought to be present in the sample from the patient, one speaks of the concentration of that virus's DNA per well in the population of wells with that sample (e.g. population B in figure 3.1(a)). The higher the concentration, the higher the *magnitude* of infection, relative to some known concentration of that virus used as a sort of 'baseline' reference (sometimes, this is called the *positive control* [8, 9]). Thus, we need the amplification result of the population of wells with known concentration, which is population A in the current example. Suppose that population A happens to have a higher concentration C_A of the viral DNA of interest. Then the fluorescent light from population A would reach a specific *signal threshold* of fluorescence intensity at a thermocycle number lower than the cycle number for population B. Let S_T be the signal threshold (which is arbitrary, as its exact value is inconsequential). Let $c_{T, A}$ be the *cycle threshold* of population A. According to equation (3.2), we then have, for population A,

$$S_T = \alpha C_A 2^{c_{T, A}}. \tag{3.3}$$

Similarly, let C_B be the concentration of viral DNA from population B and let $c_{T, B}$ be its cycle threshold. Then, for population B,

$$S_T = \alpha C_B 2^{c_{T, B}}. \tag{3.4}$$

Dividing equation (3.3) by (3.4) and solving for C_B gives

$$C_B = C_A 2^{(c_{T, A} - c_{T, B})}. \tag{3.5}$$

Thus, in theory, for a known concentration C_A, the factor $2^{(c_{T, A} - c_{T, B})}$ is simply a multiplier that enables us to determine concentration C_B. This can be said to be the result on a per-well basis in the population of wells for A and B on the plate. Due to a combination of statistical errors and systematic variability, there is ordinarily a population of cycle threshold numbers for each population, as each well's amplified concentration may vary. These distributions of cycle numbers for both populations may be plotted as a histogram pair, as depicted in figure 3.1(b). Each distribution has its own mean cycle threshold number, so that the difference $c_{T, A} - c_{T, B}$ in equation (3.5) is usually a difference in the mean cycle threshold numbers for the relative quantities of each DNA concentration. Consequently, equation (3.5) should be expressed as

$$\overline{C_B} = \overline{C_A} 2^{(\overline{c_{T, A}} - \overline{c_{T, B}})} \pm \Delta C_B, \tag{3.6}$$

where a bar over a variable denotes an average and ΔC_B is the uncertainty in the concentration of the target viral DNA.

So far, so good. But many other factors often need to be considered in the estimate for $\overline{C_B}$. For instance, equation (3.1) is based on the assumption that the PCR amplification process is 100% efficient, which means that each thermocycle exactly doubles the viral DNA. But in some cases, the PCR amplification process is not as efficient as would be expected, resulting in less than a doubling effect, so that the factor of two in the equation is replaced by a *PCR efficiency*, say, E_{ef}, whose value is between 0 and 1 [6, 10, 11]. Thus, equation (3.1) would then be expressed as

$$C_i \approx C_o(1 + E_{\text{ef}})^i. \tag{3.7}$$

And this is not all. As E_{ef} can vary from cycle to cycle, the amplification should, in theory, be a product given by

$$C_i = C_o \prod_{j=1}^{i} (1 + E_{\text{ef}})^j. \tag{3.8}$$

Returning to the question posed in the title of this section, the MVP for a real-time qPCR instrument can be said to be divided into a collection of 'sub-MVPs,' the sum of which gives the complete product. The hardware system depicted in figure 1.2, for example, may be considered the hardware sub-MVP, while equation (3.5) may be considered the chemistry's sub-MVP. The end users of a real-time qPCR instrument are usually molecular biologists or trained medical professionals working in a test laboratory, and it can be said that they are generally most interested in the quantity $\overline{C_B}$. However, they may also be interested in the uncertainty ΔC_B, because otherwise, if ΔC_B is large compared to $\overline{C_B}$, then the instrument (or reagents) may be considered unreliable. Therefore, it can be said that ΔC_B is a significant MVP for a real-time qPCR instrument. An optical designer may not know it, but the optical system plays a significant role in this quantity. And yet, this is not all. The mechanical engineer must realize that ΔC_B is also influenced by variation in E_{ef}, which is a function of the design of the *thermocycler* (i.e. the heating and cooling 'engine' underneath the plate of wells) [12, 13]. In the next section, we will briefly talk about what the designers of real-time qPCR systems need to do to make ΔC_B as small as possible. We will, of course, not attempt a deep dive into the design specifics. The point is to highlight the significance of taking a systems perspective in the design and development of complex optical instruments.

3.2.2 Why you cannot be 'just' a lens designer when designing real-time qPCR instruments

The optical system designer involved in the development of a real-time qPCR instrument has to begin by temporarily leaving the optical design program alone. One has to think hard about what influences ΔC_B, for otherwise, what is the optical design for? In order to understand which optical variables affect ΔC_B, we first look back at equation (3.2) and recall that $\alpha = I_o \gamma \epsilon z$. Substituting this into equation (3.2) yields

$$S_i = I_o \gamma \epsilon z C_o 2^i. \tag{3.9}$$

Further, for reference, let us also make the definition:

$$S_o = I_o \gamma \epsilon z C_o. \tag{3.10}$$

So, we have

$$S_i = S_o 2^i. \tag{3.11}$$

We are, of course, currently assuming that the PCR amplification efficiency is 100%. Now, since S_i is a signal (say, a current or voltage signal from the sensor), there can always be fluctuations in the sensor's response in addition to natural fluctuations in the fluorescent light, especially at low light levels (due to Poisson noise). Thus, there is, say, a noise amount σ_r in the signal, so that equation (3.8) should be written as

$$S_i = S_o 2^i \pm \sigma_r, \tag{3.12}$$

where the subscript r in σ_r denotes that the std dev arises from random signal noise. Further, returning to figure 1.2, if the illumination of the wells has residual irradiance nonuniformity from well to well, then the emitted fluorescence possesses this systematic nonuniformity. This nonuniformity should, in theory, not be quantified in terms of a std dev, but for simplicity, let us suppose that σ_s denotes the std dev of the systematic signal variation arising from the illumination nonuniformity. Equation (3.9) may then be roughly expressed as

$$S_i = S_o 2^i \pm \sigma_r \pm \sigma_s. \tag{3.13}$$

In cross-referencing between equations (3.13) and (3.6) and tracing equation (3.6) back to equations (3.3)—(3.5), it becomes apparent that σ_r and σ_s are involved in ΔC_B. This means that the optical system designer must minimize σ_r and σ_s. There is more to this. One of the MVP requirements would normally involve the total time required for thermocycling. Generally, one prefers to minimize this time, which affects the amount of time available for optical detection. In some systems, this detection takes place by way of imaging the plate of wells using a lens and image sensor. Thus, camera integration time is involved, so that one seeks to image the wells in the shortest possible integration time. But the shorter the integration time, the lower the signal. Therefore, the optical system designer must maximize S_i, which equivalently implies maximizing I_o in equation (3.10). Quite evidently, even when imaging is involved in the development of a real-time qPCR instrument, the development process is not just about lens design and aberrations.

Further to the above, recall I mentioned that the design of the thermocycler impacts the PCR efficiency E_{ef} [12, 13]. In addition, if the thermocycler has variation in temperature from well to well, then there is variation of E_{ef} from well to well. This can result in an additional spread of values for the detected fluorescence signal S_i from cycle to cycle. Thus, if we let σ_{ef} be a measure of that spread, then this quantity increases the variation on the right-hand side of equation (3.13). Consequently, this adds further uncertainty to ΔC_B. It is also widely known among developers of real-time qPCR instruments that E_{ef} may also be a function of the reaction efficiency and the interactions among molecules of the reagents used for PCR, which are in turn influenced by thermocycler temperature. Hence, there are intimate relationships among all of the variables listed so far: σ_r and σ_S (optical) and E_{ef} (mechanical and chemical). But electronics and software are also involved, such as circuitry for the sensor and software algorithms to process data in the presence of noise. Therefore, a team of optical, mechanical, electronic, and software engineers as well as chemists (and biologists) has to be involved in the development of a real-time qPCR instrument. It is quite a team effort, and listening to each other's points from a systems perspective was the reason that I am able to express the formulas involving variables other than optical variables in this section.

3.3 Statistical principles for optical system product development

3.3.1 The 'expectation value' is not the value you should expect to get

Figures 3.2(a) and (b) depict two probability distributions of random values x and their means $\bar{x}$, but the skewed distribution is characterized by an additional value, x_p. As you may know, the mean is the average, and the average is also often called the *expectation value* (or *expected value*). But the expectation value for the distribution in figure 3.2(a) is not the value that you get *most of the time* when you take samples from that distribution. Rather, you get values close to x_p, because that is the value under the peak of the curve, which makes it the **most frequently occurring value** in the distribution. Another way of saying this is that in the case of the distribution in figure 3.2(a), x_p is the **most probable value**, while $\bar{x}$ is the expectation value. Therefore, the expectation value is not the value you should *expect* to get from sampling a probability distribution. In fact, if we let $E[X] = \bar{x}$, then as Ross [14] states, 'Thus, even though we call $E[X]$ the *expectation* of X, it should not be interpreted as the value that we *expect* X to have but rather as the average value of X in a large number of repetitions of the experiment.'

Of course, in the case of the Gaussian or *normal* distribution in figure 3.2(b), $\bar{x}$ is the most frequently occurring value, so that when you draw samples from that distribution, values of x are often close to $\bar{x}$. Interestingly, if you were to take, say, N samples of x from the skewed distribution and compute $\bar{x}$, and you repeated this many times until you had enough $\bar{x}$ values to plot a histogram of means, then $\bar{x}$ would be normally distributed and the width of the distribution would be given by $\sigma/\sqrt{N}$, where σ is the std dev of x (provided that $N \gg 1$). In fact, you would get this result for any distribution of random numbers, and it is known as the **central limit theorem** (CLT), found in many standard texts on probability and statistics (e.g. [14]).

The above principles of statistical theory are helpful in optical system design. For example, certain LEDs have radiant intensity profiles that do not have a central peak. Rather, they may have the curve shown in figure 3.3(a), which is often called a 'batwing' intensity profile. In such cases, when you perform NSC ray traces, as you saw in chapter 2, using a finite number of Monte Carlo rays does not generally yield the smooth curve given in figure 3.3(a). If you repeat the ray trace a number of times

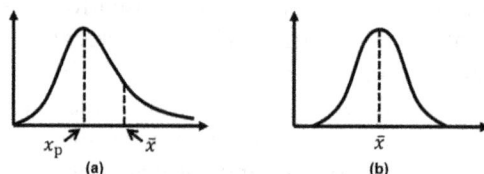

Figure 3.2. A skewed probability distribution (a) and a Gaussian distribution (b).

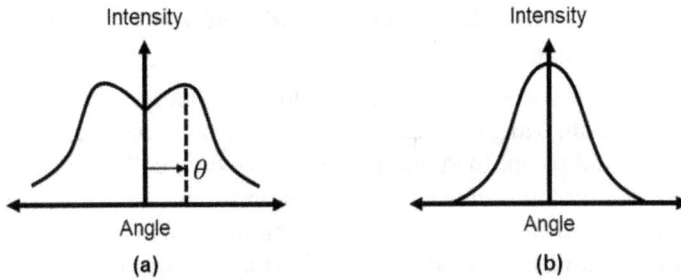

Figure 3.3. A 'batwing' LED intensity (a) and a narrow 'spot' intensity (b).

using the same number of rays each time, then you find that the most frequently occurring locations of pixels on a screen with high ray density are in an annulus at an angle of θ, but the average location (centroid) of the pixels (weighted by the intensity distribution) is at the center. Moreover, if a number of detector outputs are summed and divided by the total number of plots, then the averaged irradiance profile is indeed given by the curve in figure 3.3(a). If the magnitude of the irradiance profile is not important to you (because only the profile is important), then summing is also averaging (the denominator in the average is just a scaling factor). So, to make assessments of illumination uniformity, you may either capture many ray traces and take the average, or you may look for the most frequently occurring areas of high and low ray densities. If, on average, all areas have roughly equal occurrences of rays, then it is likely that the illumination is uniform. Similarly, in the case of an LED with a narrow 'spot' radiant intensity profile (figure 3.3(b)), the most frequently occurring location of rays is at the center, and the average over many ray traces tends towards the spot profile.

If you need to compare the total flux in the irradiance profile with the total flux in another profile, you may find it cumbersome that, due to the Monte Carlo ray trace noise, fluctuations in the flux values make the comparison seem impossible. In this case, you may increase the number of rays traced, which is equivalent to taking the average flux from a number of ray traces that trace fewer rays. Either case is an application of the CLT. For instance, if you had the time, you could perform a number of NSC ray traces for both profiles and plot histograms of each profile's *average total flux*. Note that this is not a histogram of pixel fluxes. Rather, it is a histogram of the sums or averages of the pixel fluxes across many ray traces. You would likely find that the histograms of mean fluxes for both profiles are normally distributed by virtue of the CLT. If you can distinguish between the peaks of the two normal distributions, then you can determine whether or not the fluxes are the same or different, *statistically speaking*. I am avoiding the term *statistical significance*, which applies the notion of statistical hypothesis testing and so-called *p-values*. These terms and their meaning are found in standard texts on probability and statistics, which are not always useful and necessary. In fact, professional statisticians and academicians in universities still debate this issue [15].

3.3.2 The difference between the standard deviation of a function and the standard deviation of random values

In many cases of optical system product development, detection is involved, which may be performed using various types of sensors such as photodiodes, image sensors (CCDs, CMOSs), and so on. In an image sensor, the data collected across pixels can vary either due to random events (such as noise in the pixels from dark current) or simply due to the presence of an image formed on the array of pixels. The images of objects and sources have spatial variation, which may be considered some systematic continuous function $f(x, y)$ across the image sensor, sampled by the pixels. You can compute a std dev of pixel intensities across the image, but you would need to consider whether it has any significance [16], because it is tempting to discretize $f(x, y)$ into $f_i = f(x_i, y_i)$ and compute a std dev for it, which may be expressed as

$$\hat{\sigma} = \sqrt{\frac{\sum_{i=1}^{N}(f_i - \bar{f})^2}{N - 1}}, \tag{3.14}$$

where $\hat{\sigma}$ is known as the *sample std dev* of σ, the *population std dev* (given by $\sqrt{(N - 1)/N}$ times $\hat{\sigma}$ [14, 16]), and $\bar{f}$ is the average over all sampled values of f_i. For $N \gg 1$, $\hat{\sigma} \to \sigma$. The square of $\hat{\sigma}$ is known as **sample variance**, and the square of σ is simply the **variance**.

Equation (3.14) is well known and can be used to compute the std dev of any sample of numbers, but if f_i is a discretized sample of a systematic function, then there remains a question as to what $\tilde{\sigma}$ physically means. In the case of the focusing exercise described in figures 2.110–2.113, the std dev of pixel intensities was a convenient means to quantify the systematic spread in pixel intensities, but it was not considered 'noise.' In fact, a well-focused image has high spatial variation, so that it has a maximal spread of pixel intensities, but it is not a 'noisy' image. If the image is meant for illumination and has high local uniformity but low global uniformity, then a std dev of pixel intensities may be employed to quantify that spread, but the Monte Carlo ray trace noise in the pixels must somehow be dealt with. On the other hand, if the maximum and minimum *average* pixel intensities are known, then taking their difference (or ratio) would be a simple and reasonable measure of uniformity, so that there would be no need to compute a std dev of pixel intensities across the image. If more information is required to characterize the image, then perhaps other measures may be employed. The point is that the std dev of a set of measured values is generally reserved for *random variables*. As we will see in section 3.4.2, the use and misuse of the std dev to quantify variation can significantly influence key decisions in optical system product development.

3.3.3 Statistical principles related to error analysis, error bars, sensitivity analysis, and optical tolerancing analysis

In product development, you will encounter many instances in which you take measurements. You have to acquire the habit of taking more than a single measurement, otherwise your data will be rather unreliable. For instance, suppose you want to compare the fluxes produced by two LEDs. For the first LED, you obtain 1 W. For the second LED, you obtain 1.1 W. How does one tell whether these two LEDs are emitting equal or different fluxes? It is impossible to tell, because each value represents only a single instance of measurement. However, if a number of measurements are performed for each, and if the results show that the first LED has a flux of 0.9 ± 0.2 W and the second LED has a flux of 1.1 ± 0.2 W, then since the first LED can have a flux as high as 1.1 W and the second LED can have flux as low as 0.9 W, then this would imply a likelihood that both LEDs have the same average flux. Let us put this into statistical language.

In section 3.3.1, I mentioned that while the std dev of N random values of x (for which $N \gg 1$) is σ, the std dev of the mean $\bar{x}$ of those random values is $\sigma/\sqrt{N}$. In the context of *error analysis* and taking measurements of quantities with random errors, σ may be regarded as the error (or equivalently, the *uncertainty*) of a single measurement, while $\sigma/\sqrt{N}$ is the error in the mean of those measurements [17]. You can think of this as a consequence of how errors sum (i.e. *error propagation*), which is that we take the square root of the sum of their squares. That is, assuming equal errors in each measurement,

$$\frac{x_1 + x_2 + \ldots + x_N}{N} \pm \frac{\sqrt{\sigma_1^2 + \sigma_2^2 + \ldots + \sigma_N^2}}{N} = \bar{x} \pm \frac{\sqrt{N\sigma^2}}{N} = \bar{x} \pm \frac{\sigma}{\sqrt{N}}. \qquad (3.15)$$

The result above is a common practical approximation for the computation of the total error or uncertainty of measurements. Each measurement of x is an *estimate* of its true value, and $\bar{x}$ is the *best estimate* of that true value. The std dev σ is an *error bar* associated with each measurement, and $\sigma/\sqrt{N}$ is the uncertainty in the best estimate. Thus, measurements are best estimates, plus or minus uncertainty, a phrase taken from Taylor [17] (which was quoted in section 1.1.11). And therefore, if we regard data as a form of reliable *truth* about something, then truth is a best estimate, plus or minus an uncertainty in it.

Returning to the example of the LED fluxes, suppose that three flux measurements are made for each LED and the std dev of the flux is roughly 0.4 W for each set of measurements. This then means that each time a single flux measurement is made, the uncertainty is ±0.4 W. But if an average is taken of three measurements, then the uncertainty in the average flux is ±0.4/$\sqrt{3} \approx 0.2$ W. Thus, if 0.9 W is the average flux for the first LED, then it may be reported as 0.9 ± 0.2 W, and if 1.1 W is the average flux for the second LED, then it may be reported as 1.1 ± 0.2 W.

In the context of optical tolerancing analysis, variations in measured or specified optical component variables (such as radii, central thickness, etc.) may be regarded as having some magnitude $\pm n\sigma$, where n is any number (it does not have to be an integer). However, in OS, n can only take integer values from two to a maximum of ten (the default setting is $n = 2$). This means that when you specify min and max values for a tolerance operand, the min is $-n\sigma$ and the max is $+n\sigma$. No other values are allowed beyond these, and the tolerances are then said to be *truncated* at the min and max values.

When *sensitivity analysis* is considered in a tolerancing run in OS, the concept applies the definition that each change in tolerance operand in the amounts $+n\sigma$ and $-n\sigma$ gives rise to *independent* changes in some optical parameter (RMS spot size, MTF, etc.) by amounts $+\Delta$ and $-\Delta$. The changes $\pm\Delta$ are then said to be the sensitivities of the optical parameter to $\pm n\sigma$. The idea is that a large magnitude of Δ_i indicates high sensitivity to influences from the tolerance $n\sigma_i$. This is generally a good approximation of perturbations due to tolerances (we will talk a little about higher-order mixed effects called **interactions** in the next section). In OS, if $+\Delta$ is a change in a parameter as a result of a change $+n\sigma$ in a tolerance operand, and if $-\Delta$ is the change in the parameter for a change $-n\sigma$ in the operand, then the total parameter change for N operands in the positive direction is assumed to be $\sum_{i=1}^{N}(+\Delta_i)^2$, while the total parameter change in the negative direction is $\sum_{i=1}^{N}(-\Delta_i)^2$. OS then estimates that the maximum squared change in the parameter is $\Delta_{max}^2 = \left[\sum_{i=1}^{N}(+\Delta_i)^2 + \sum_{i=1}^{N}(-\Delta_i)^2\right]/2$. Taking the square root yields the maximum estimated change. This is called the *root sum square* (RSS) method of tolerance stack estimation. This means that if the nominal EFL of a lens is, say, 100 mm, and if the maximum estimated RSS change in the EFL is Δ_{max}, then under the influence of all tolerances, the EFL could be as large as 100 mm $+ \Delta_{max}$ or as small as 100 mm $-\Delta_{max}$ —at least, according to the RSS rule. But as we will see in section 3.3.4, **the actual min and max changes in the optical parameter due to tolerances are not necessarily given by the maximum RSS change, due to higher-order mixed effects in the tolerances.** Moreover, when *compensators* are used (such as focus adjustment), there is a **negative covariance term** in the RSS stack, so that the maximum change may be reduced.

3.3.4 In optical tolerancing analysis, a merit function is a *function* of random variables

We know that optical parameters (such as the EFL) are functions of optical variables (such as radii, central thicknesses, etc). It therefore follows that if q is any general optical parameter that is a function of the variables $x_i = x_1, x_2..., x_N$, then q may be written as a Taylor series in x_i. In multivariable calculus, a Taylor series of a multivariable function to the second order is known as the *quadratic approximation* of that function [18]. For simplicity, suppose an optical parameter q is a function of two variables, x and y. Its quadratic approximation is then

$$q(x, y) = q_o + \frac{\partial q}{\partial x}(x - x_o) + \frac{\partial q}{\partial y}(y - y_o) + \frac{\partial^2 q}{\partial x \partial y}(x - x_o)(y - y_o)$$

$$+ \frac{\partial^2 q}{\partial x^2}(x - x_o)^2 + \frac{\partial^2 q}{\partial y^2}(y - y_o)^2, \tag{3.16}$$

where $q_o = q(x_o, y_o)$. Many insights into optical tolerancing analysis may be gained from studying equation (3.16). Let us begin by asking: what is the change in the optical parameter (relative to its nominal value of q_o) that results from a change in its variables? Since the changes in its variables are given by $(x - x_o)$ and $(y - y_o)$, the consequential change in q is given by $\Delta = q(x, y) - q_o$. So, we obtain

$$\Delta = \frac{\partial q}{\partial x}(x - x_o) + \frac{\partial q}{\partial y}(y - y_o) + \frac{\partial^2 q}{\partial x \partial y}(x - x_o)(y - y_o)$$

$$+ \frac{\partial^2 q}{\partial x^2}(x - x_o)^2 + \frac{\partial^2 q}{\partial y^2}(y - y_o)^2. \tag{3.17}$$

Equation (3.17) is known to experienced optical designers as the quadratic approximation to a perturbation in an optical parameter when performing advanced optical tolerancing analyses [19–24]. They apply this series expansion to tolerancing because the series includes an important term that quantifies the influence that one variable may have on another. For instance, in comparing equation (3.17) with the RSS changes (Δ) described in the previous section, it becomes apparent that an RSS change for a parameter does not involve the cross-term $\frac{\partial^2 q}{\partial x \partial y}(x - x_o)(y - y_o)$. It is this term that quantifies the influences between variables. Let us make this observation more explicit. A change in the amount $n\sigma$ for *one* variable, say, in x, could result in

$$\Delta_x = n\sigma_x = \frac{\partial q}{\partial x}(x - x_o) + \frac{\partial^2 q}{\partial x^2}(x - x_o)^2 + ... + \frac{\partial^N q}{\partial x^N}(x - x_o)^N, \tag{3.18}$$

which may involve many orders of the expansion series for that variable. Also, a change in another variable, say, in y could result in

$$\Delta_y = n\sigma_y = \frac{\partial q}{\partial y}(y - y_o) + \frac{\partial^2 q}{\partial y^2}(y - y_o)^2 + ... + \frac{\partial^N q}{\partial y^N}(y - y_o)^N. \tag{3.19}$$

For simplicity, let us ignore the positive and negative changes in the RSS and consider only changes in one direction, say, the positive direction. Then, the squared RSS estimate in the total parameter changes can be expressed as

$$\Delta_{\max}^2 = \Delta_x^2 + \Delta_y^2. \tag{3.20}$$

The mixed partial derivatives are missing from equation (3.20), i.e. the cross-terms given by $\frac{\partial^2 q}{\partial x \partial y}(x - x_o)(y - y_o) + ... +$ higher-order mixed terms. Such cross-terms in the series expansion are called **interactions** in the language of statistical theory (you will often hear it called that in statistical design of experiments (DOE) and the analysis of variances (ANOVA)). In section 1.2.13, I mentioned that sensitivity analyses do not account for the interactions between and among variables. Here, you can finally see what I mean. In cases of optical tolerancing in which the variations $\pm n\sigma$ are small, the interactions may possibly be neglected. Otherwise, such interactions may be present, and unfortunately, they are not always computed in optical design programs. However, I also did mention in section 1.2.13 that Monte Carlo simulations can account for such interactions. Let us now discuss what is meant by this.

If q were the EFL of a singlet, then to the second order, x and y could be its radii of curvature. Terms can be added for the central thickness, refractive index, wedge error, and so on. In a tolerancing analysis, all variables would be given random values, which should be such, because in reality (under manufacturing conditions), all variables are subject to tolerances in production. Therefore, **in a tolerancing analysis, an optical parameter is a function of random variables**. In statistical theory, since random variables are governed by probability distributions, a function of random variables also has specific probability distributions, and they need not be normal [24, 25]. They are also associated with variances. **Since an optical parameter can be a quantity that is computed in the merit function of an optical design program, merit function values become functions of random variables in a tolerancing analysis.** The quantity in the merit function could be a lens's EFL, or it could be an RMS spot size, and so on. During a Monte Carlo simulation in OS, every trial generates a random lens sample and many combinations of random changes $\pm n\sigma$ are applied to the variables, resulting in a total random change $\pm\Delta$ in the merit function. For N Monte Carlo trials, the variance σ_{MF}^2 of the merit function value can therefore be expressed as

$$n^2\sigma_{MF}^2 \approx \frac{1}{N}\sum_{i=1}^{N}\Delta_i^2. \tag{3.21}$$

The reason for the presence of the n^2 factor in equation (3.21) is that, in OS, the changes Δ_i in the parameter result from $n\sigma$ changes in the variables (the reason for the approximation is that the changes in the variables are truncated). What variations can be expected to influence the quantity σ_{MF}^2 in equation (3.21)? These would be everything contained in Δ_i^2. Suppose that, for a function of two variables, perturbations in variables were made to vary about their nominal values in a *fair* symmetric fashion, such as when the distributions of random variables are either normal or uniform. In the case of equation (3.17), we would then obtain $\bar{x} = x_o$ and $\bar{y} = y_o$. Thus, $\bar{q} = q_o$. We can therefore express equation (3.17) as

$$\Delta^2 = \left[\frac{\partial q}{\partial x}(x - \bar{x}) + \frac{\partial q}{\partial y}(y - \bar{y}) + \frac{\partial^2 q}{\partial x \partial y}(x - \bar{x})(y - \bar{y}) \right.$$

$$\left. + \frac{\partial^2 q}{\partial x^2}(x - \bar{x})^2 + \frac{\partial^2 q}{\partial y^2}(y - \bar{y})^2 \right]^2. \tag{3.22}$$

Inserting equation (3.22) into (3.21), squaring, and then summing, we find that if we have independent random variables, all terms involving factors of $(x - \bar{x})$ and $(y - \bar{y})$ vanish [16, 17], leaving the result

$$n^2\sigma_{MF}^2 \approx \frac{1}{N}\sum_{i=1}^{N}\left[\frac{\partial q}{\partial x}(x_i - \bar{x}) \right]^2 + \frac{1}{N}\sum_{i=1}^{N}\left[\frac{\partial q}{\partial y}(y_i - \bar{y}) \right]^2$$

$$+\frac{1}{N}\sum_{i=1}^{N}\left[\frac{\partial^2 q}{\partial x \partial y}(x_i - \bar{x})(y_i - \bar{y})\right]^2 + \frac{1}{N}\sum_{i=1}^{N}\left[\frac{\partial^2 q}{\partial x^2}(x_i - \bar{x})^2\right]^2$$

$$+\frac{1}{N}\sum_{i=1}^{N}\left[\frac{\partial^2 q}{\partial y^2}(y_i - \bar{y})^2\right]^2. \tag{3.23}$$

By virtue of equation (3.23), we can see that in a Monte Carlo simulation, the cross-term is always involved in the variance of an optical parameter. Even though this formula has been derived for only two variables, by induction, there is no reason why a function of multiple random variables would not include the interaction term in its variance. In comparing equation (3.23) with equation (3.20), we find that they differ by the extra term given by interaction in equation (3.23). Thus, taking the difference between equations (3.23) and (3.20) yields the interaction term. This means that if, in OS, we wish to know the magnitude of the interactions contribution to an optical parameter being toleranced, then provided that a sufficiently large number of Monte Carlo trials are generated, then one may compute $n^2\sigma_{MF}^2$ and the RSS change Δ_{max}^2, then take the difference given by $n^2\sigma_{MF}^2 - \Delta_{max}^2$. Alternatively, one can first take the difference between the maximum parameter value and the nominal value from the Monte Carlo run, which would be an estimate of $n\sigma_{MF}$. Then, take the difference between the square of that result and Δ_{max}^2. Of course, this only provides the total contributions from the interactions rather than individual contributions. However, it does at least give some analytical indication of the interaction contributions. Also, evidently, note that $n^2\sigma_{MF}^2$ is the quantity that is most closely associated with the *worst-case* Monte Carlo trial.

Typically, certain adjustments are possible for a manufactured product, which help the product to attain a level of performance that is better than that of its *as-built* version. These adjustments are called **compensators**. Focus adjustment is an obvious compensator, but sometimes, there may be other forms of compensators, such as tilting a lens (as mentioned in section 1.2.11), or decentering an element (as mentioned in section 1.5.3), which *cancels* the tilt and decentering tolerance stack from other perturbed elements so that the final image quality is improved. Due to compensators, neither the estimated maximum RSS change nor the worst-case Monte Carlo parameter change are necessarily indicators of the worst-case performance of a parameter being considered for a manufactured product. Overall, we expect better worst-case performance after compensators have been applied. This means that, statistically speaking, equations (3.20) and (3.23) should include a *negative term* that represents the effect of the compensator. This negative term is the **covariance** of the variables in the total variance [24].

To understand covariance, one notes that when a compensator is used, it does its work by *going against the direction of change for one or more variables* that are influencing the parameter in an unfavorable manner, so that the compensator effectively cancels the effect/s of those variables. Thus, in this way, a compensator— which is itself a variable, but one that is being actively adjusted by an engineer/ technician—is *negatively correlated* with one or more *offending tolerances*. When this happens, the partial derivatives of the correlated variables have opposite signs. For example, suppose there are only two variables, so that the maximum change in the optical parameter (to the second order) is given by equation (3.23). Now, suppose that variable x is a compensator for variable y. Equation (3.23) then becomes

$$
\begin{aligned}
n^2 \sigma_{MF}^2 \approx \frac{1}{N} \sum_{i=1}^{N} & \left[\frac{\partial q}{\partial x}(x_i - \bar{x}) \right]^2 + \frac{1}{N} \sum_{i=1}^{N} \left[\frac{\partial q}{\partial y}(y_i - \bar{y}) \right]^2 \\
& + \frac{1}{N} \sum_{i=1}^{N} \left[\frac{\partial^2 q}{\partial x \partial y}(x_i - \bar{x})(y_i - \bar{y}) \right]^2 + \frac{1}{N} \sum_{i=1}^{N} \left[\frac{\partial^2 q}{\partial x^2}(x_i - \bar{x})^2 \right]^2 \quad (3.24) \\
& + \frac{1}{N} \sum_{i=1}^{N} \left[\frac{\partial^2 q}{\partial y^2}(y_i - \bar{y})^2 \right]^2 - \frac{2}{N} \sum_{i=1}^{N} \left[\frac{\partial q}{\partial x}(x_i - \bar{x}) \frac{\partial q}{\partial y}(y_i - \bar{y}) \right].
\end{aligned}
$$

Notice the extra negative nonzero term. This is a covariance term [17]. Suppose that somehow, the compensation can be made precisely equal and opposite, to the first order. In this case, $(\partial q/\partial x)(x_i - \bar{x}) = -(\partial q/\partial y)(y_i - \bar{y})$, and the result is an exact cancellation of the negative covariance term with the first two terms, leaving only the quadratic and interaction terms. If the effects of the higher-order quadratic terms can be *sensed* by the engineer (or test system), then perhaps the compensation may even be adjusted to obtain precise cancellation to the second order, leaving only the interaction term, if any interaction exists.

Before we end this section, there is one other point to highlight. In particular, notice that each term on the right-hand side of equation (3.23) may regarded as n^2 times a variance term. If we cancel both sides by n^2, then the right-hand side is simply a sum of variances, yielding a total variance on the left-hand side. It turns out that this is an expression of a kind of *reproductive property* for independent random variables whose distributions are normal, Poisson, or chi-squared, which is that the total variance is given by the sum of the variances of the individual independent random variables [26]. Thus, in cases in which the distributions of certain random values are normal-like, or Poisson-like, or chi-squared-like, then the sum of their variances can be said to be roughly equal to the total variance. This is helpful in noise analysis. You may have seen datasheets from manufacturers of photodiodes and image sensors stating that if a sensor's dark current noise is characterized by the variance σ_D^2 and if the shot noise from the signal is σ_S^2, then the total noise is $\sigma_D^2 + \sigma_S^2$. If there are more noise sources, then simply sum their variances. It turns out that signal shot noise has a Poisson distribution [27], and so if the dark current noise has a similar distribution, then it is justifiable to sum their variances. In closing, we also note that the individual variance terms on the right-hand side of equation (3.23) may be thought of as the *components of variance* of a total variance, with the caveat that specific algebraic *partitioning methods* be applied in order to associate them with statistical *estimators* of variance components in the ANOVA technique for analyzing variability [28].

3.3.5 The std dev of a merit function has a std dev

In statistical theory, by right, the computation of any quantity (such as the sample std dev) from data must possess a variance [29], and therefore, a std dev. If a std dev is computed from data that is normally distributed, then it can be shown [16, 17] that the std dev of that std dev is given by

$$\hat{\delta} = \frac{\hat{\sigma}}{\sqrt{2(N-1)}}. \tag{3.25}$$

Thus, $\hat{\delta}$ is the error in the estimate of error, or the std dev of the sample std dev. It is a measure of the variation of variation. Of course, for $N \gg 1$, this error is negligible.

Since we have been regarding an optical parameter in a merit function as a function of random variables in a tolerance analysis, then a std dev computed from the merit function has a std dev. If the distribution of merit function values is normal, then the std dev of its std dev is given by equation (3.25). When tolerances are small, relative to the nominal (or average) of the parameter being toleranced, then merit function values tend to be normal. What this means is that if you performed a first Monte Carlo run with, say, 100 trials, then in this first run, let us assume you obtained a std dev of a merit function value (e.g. it could be the std dev of the EFL of a lens system for 100 trials in a Monte Carlo run). In the next Monte Carlo run (another 100 trials), you obtained another std dev for the merit function. Equation (3.25) tells us that the first and second std devs may not be the same, and the uncertainty in the computed std dev is in the amount given by equation (3.25) for which $N = 100$. The uncertainty is negligible, in this case. We are spoiled by the ease of letting a program run many trials of Monte Carlo samples. But if you were taking measurements in the lab, you would normally take a few at a time. In such cases, it would be wise to keep equation (3.25) in mind.

3.3.6 The different ways in which engineers and statisticians solve problems

Statisticians view the world from a perspective that can seem puzzling to engineers. For many of us, practical engineering and design often involve determinism. We apply formulas, such as the formula for the EFL of a lens system, to solve problems. These formulas offer predictive power that is so good that in fact, we may easily forget that variability and randomness in events can alter the values computed using the equations. In contrast, statisticians often take the opposite view, which is that all events are possibly random. However, statisticians consider that if random events have repeatable histories called *probability distributions*, then it is possible to determine useful information from the events. Such information could be, for example, the mean and std dev of the events.

An example of a random event with repeatable history is the outcome of tossing a coin or dice, which is governed by the binomial distribution [17]. In optical system design, the tolerances of components are assumed to have random values with repeatable history. During the design optimization phase of the optical system, the nominal values of the optical parameters (such as the EFL of a lens system) are considered deterministic in the sense that they are known functions of their variables, but they possess randomness when tolerances are imposed on the variables. This makes the optical parameter into a function of random variables, as discussed in section 3.3.4. Thus, in optical system design, we are constantly dealing with functions of random variables when we perform Monte Carlo simulations for tolerancing analyses. Under such conditions, these functions possess repeatable distributions that may be visualized through plotting their histograms (e.g. figures 1.83 and 2.116). If one is interested in finding out how such distributions arise, one needs only to consult some textbooks and guides on probability and statistics (e.g. [25, 26, 30]).

In statistical theory, all data are regarded as being part of an abstract *parent population* (or simply, 'the population') of random values governed by specific probability distributions. In that pool of values, a *true value* exists for that piece of sought data, but it is never known. For example, if you were taking measurements of the flux emitted by an LED, then the measured flux values are just a sample of data collected from a parent population of flux values, all 'waiting' to be further sampled. In that population, the LED's true flux exists, but it is unknown and not equal to any of the flux measurements you made. The best that can be done to obtain that elusive true flux value is to *estimate* it by taking many more samples of data from that population. A quantity computed from those samples is called a *statistic* [28] (so now you know that this term is not just the title of a subject!), and when a statistic is used to estimate the true value in the population, it is called an *estimator*. When the estimator is a single value, it is called a *point estimator*. Thus, the mean in a sample of values from a population is a point estimator of the true value in that population. If X is the true value of the LED's flux, and if you computed the mean $\bar{x}$ from a sample of flux measurements, then $\bar{x}$ is regarded as one of many means surrounding X. In fact, according to the CLT, $\bar{x}$ is normally distributed about X. If you compute a std dev of flux values, then provided that the sample size N is large, then the std dev you compute is roughly σ, the std dev of the population, so that the std dev of $\bar{x}$ is $\sigma/\sqrt{N}$. The larger the sample size, the higher the precision (smaller $\sigma/\sqrt{N}$), so that in the limit of infinite sample size (reaching, therefore, the size of the population), $\bar{x} \to X$. The true value X is also called the *population mean*. Under these conditions, $\sigma/\sqrt{N}$ is a *confidence interval* in the estimate of X.

In the language of error analysis, the confidence interval is known as the *error bar* [16, 17, 31] or simply the uncertainty interval [32]. Interestingly, the term error bar is not used often in standard texts on probability and statistics. This indicates something of significance in terms of how a team of skilled people work together and communicate in any endeavor, be it a project in product development, or arguing about climate change. If there is disagreement, could it be that both parties are correct, but using different terms? As another example, the expression $q = f(x, y)$ is known as a scalar function of two variables in multivariable calculus, but in *six-sigma* training camps, it is called the *response surface method* to 'model system performance' [33]. So, if engineers and six-sigma trainers argue, could it be that both parties have had different experiences in education and must now determine the facts that are common to both? Are both parties right without even knowing it? Are they using different (but consistent) formalisms? A famous story is told of four physicists, Richard Feynman, Julian Schwinger, Shinichiro Tomonaga, and Freeman Dyson, who, in the 1940s, worked on problems in the field of quantum electrodynamics (QED) [34]. According to the story, Feynman, Schwinger, and Tomonaga each worked out generalized solutions to problems in QED, but each of them went about it differently. While Tomonaga worked independently, Feynman and Schwinger decided to compare their results, which were consistently the same. Their approach was quite effective—neither cared what method was used, only that they both got the same results. Later, Freeman Dyson showed that all three physicists' approaches were consistent with one another, thereby providing further support for their independent work. Feynman, Schwinger, and Tomonaga went on to share the 1965 Nobel Prize in Physics for their work in QED.

Back to statistics. Just as the sample mean $\bar{x}$ is a point estimator of the population mean X, the sample variance $\hat{\sigma}^2$ is a point estimator of the population variance, σ^2. In statistical theory, if the average of a point estimator of a parameter in the population is found to be equal to the parameter in the population, then the estimator is called an *unbiased estimator*. It turns out that the presence of the denominator $n - 1$ in equation (3.14) for the sample std dev makes its square (i.e. the sample variance) an unbiased estimator of σ^2 [24, 26, 28]. It can also be shown that $\bar{x}$ is an unbiased estimator of X [35]. However, the sample std dev is not an unbiased estimator of the population std dev [36]. Unbiased estimators are often seen to be accurate quantities for determining the *true values* of parameters. Yet, it is said that sometimes, biased estimators are preferred [36]. Apparently, this is because it may be possible to determine an estimator (a biased one) whose values computed from the data have a smaller spread about the parameter than those from an unbiased estimator, even though the average value of the biased estimator may be farther from the parameter that it is estimating, compared to the average of an unbiased estimator [36]. Such is the fascinating world of probability and statistics!

In closing, it should be noted that the above statistical concepts and methods of analysis fall into the category of *parametric statistics*, in which probability distributions are known [14, 28]. When distributions are thought to exist but remain unknown, then statisticians call their methods of analysis *non-parametric statistics*. Finally, when partial knowledge is available about the distributions of the statistics, the methods of analysis are called the *Bayesian* approach [14, 28]. In product development, parametric statistics are often applied by system engineers and members of the quality and leadership team (such as project managers) for analysis, because statistical estimation is the means by which one attempts to arrive at certain conclusions about product quality [33, 37]. However, combinations of parametric and non-parametric statistics as well as Monte Carlo methods are applied in work involving reliability engineering [38, 39]. In this book, much of what we have discussed in terms of error analysis, sensitivity analysis, and Monte Carlo simulations has mostly involved the methods of parametric statistics. However, we never truly know the distributions of optical tolerances in a factory. Therefore, in a Monte Carlo optical tolerance analysis, one can only assume certain distributions for the tolerances. In my approach, the best that can be done has often been to assume a *fair* distribution for the tolerances, which is the uniform distribution [24].

3.4 The concept of the signal-to-noise ratio

3.4.1 What exactly is the signal, and what do you mean by 'noise'?

Figure 3.4 shows two signals. Imagine that figure 3.4(a) is the output from a digital oscilloscope that is connected to a photodiode, and the photodiode is accepting a fraction of optical flux from a source. For figure 3.4(b), imagine that a low-quality lens (e.g. one with a somewhat low MTF) has focused the image of the source onto a CCD or CMOS image sensor, and the profile displayed is the image across a horizontal row of pixels over some integrated period of time. The solid horizontal line in figure 3.4(a) represents the average of the signal over time, while the solid curve in figure 3.4(b) represents the 'true' image of the source. What are the signal-to-noise ratios (SNRs) of the two figures?

The question posed above is rather ambiguous. Someone has to say what is meant by the signal and the noise in these plots. However, for the plot in figure 3.4(a), it is common to regard the horizontal solid line as the signal and the std dev over time as the noise, their ratio being the SNR. As for figure 3.4(b), suppose that in an application, we want the sum of the pixel intensities covering the entirety of the image to be the signal. The situation could be, for instance, that we want to use an image sensor rather than a large-area photodiode to capture the flux from the source. Further, perhaps we do not require the summed pixel intensities to be calibrated in flux values, such as watts. Suppose it is only required that the accumulated photoelectron signal in a pixel is proportional to the flux. In this case, it is sensible to use the sum of the pixel intensities to represent the signal. But what is the noise in that image? Many different definitions are possible. But if the application is to simply represent the flux in the image by way of the sum of pixel intensities, then the question to ask is: 'What is the noise in the sum of pixel intensities?' We address this question in the next section.

3.4.2 How does the signal-to-noise ratio scale with the size of a region of interest?

If the application in the previous section for the signal in figure 3.4(b) is to integrate as much flux as possible from a source by forming the image of the source on an

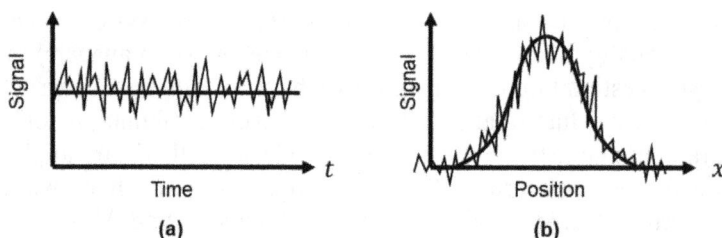

Figure 3.4. Two signals. (a) Voltage vs time for a photodiode connected to a digital oscilloscope. (b) Integrated photoelectrons vs position on an image sensor.

image sensor, then the SNR in the total signal is describable by asking the question: 'What happens if we repeat the measurement several times?' We will find that the sum of the pixel intensities changes a little each time we take the measurement. This incurs an error bar for each measurement, given by the std dev across all measurements, each of which is the sum of the pixel intensities in the image. So, the SNR is calculated by dividing the average across all measurements by the std dev across all measurements, namely:

$$ \mathrm{SNR} = \frac{(1/N)\sum_{i=1}^{N} P_i}{\hat{\sigma}}, \tag{3.26} $$

where P_i is the ith sum of the pixel intensities in the measurements

$$ P_i = \sum_{j=1}^{M} p_{i,j} \tag{3.27} $$

and p_j is the intensity of the jth pixel on the image sensor. By 'pixel intensity,' here, we specifically refer to the number of photoelectrons integrated by a pixel in an integration period, the period being a fixed quantity used for each integration of the image. The std dev $\hat{\sigma}$ is given by

$$ \hat{\sigma} = \sqrt{\frac{1}{N-1}\sum_{i=1}^{N}(P_i - \overline{P})^2}, \tag{3.28} $$

where $\overline{P}$ is the mean sum of the pixel intensities given by the numerator on the right-hand side of equation (3.26). Defined in this manner, the SNR of equation (3.26) is the SNR of a single flux measurement, P_i. If, on the other hand, one asks for the SNR of $\overline{P}$, then the noise $\hat{\sigma}$ is divided by $\sqrt{N}$, resulting in an increased SNR. This is why taking the mean across a number of measurements is better than a single measurement.

Let us apply the above SNR definition to a specific example. If we return to equations (3.3) and (3.4) for the virus detection example discussed in section 3.2, suppose that each signal in a well is obtained by imaging a well (which is now 'the source'), so that the signal threshold S_T is a sum of pixel intensities. If the image of a well looks like the profile in figure 3.4(b), then the noise in the signal is given by $\hat{\sigma}$ in equation (3.28). In this case, equations (3.3) and (3.4) may be expressed as

$$ S_T \pm \hat{\sigma} = S_T\left(1 \pm \frac{\hat{\sigma}}{S_T}\right) = S_T\left(1 \pm \frac{1}{\mathrm{SNR}}\right) = \alpha C_A 2^{(c_{T,\,A} \pm \Delta)}, \tag{3.29} $$

and

$$ S_T \pm \hat{\sigma} = S_T\left(1 \pm \frac{\hat{\sigma}}{S_T}\right) = S_T\left(1 \pm \frac{1}{\mathrm{SNR}}\right) = \alpha C_B 2^{(c_{T,\,B} \pm \Delta)}, \tag{3.30} $$

where Δ is now the error (uncertainty) in the cycle thresholds. Thus, we see a direct link between the SNR for the sum of pixel intensities and the uncertainty in determining the important $c_{T,\,A}$ and $c_{T,\,B}$ values required in equation (3.6) for the relative quantification of the viral DNA concentration. The noise $\hat{\sigma}$ in this case corresponds to the random component σ_r of variation in equation (3.13).

Now, if the lens system used to image the source has high MTF, then the profile in figure 3.4(b) will be like a top-hat distribution, so that one should sum up all pixel intensities under the top-hat profile whilst not including any pixels outside the profile. This is because each pixel in itself possesses noise in the sensor, such as dark current noise. The region under the top-hat profile is therefore the region of interest (ROI) in the image. But not all lenses are required to have high MTF (especially if you want to save on cost). So, the image of a source can be of the form shown in figure 3.4(b), and in this case, a question may arise concerning how large the ROI area should be when summing pixel intensities. On one hand, if the ROI is too large, then more dark current is integrated from pixels that are not receiving much light. On the other hand, the signal from just one pixel (say, a pixel under the peak of the curve in figure 3.4(b)) cannot possibly have a high SNR, because not much signal can be obtained from just a single pixel. Moreover, that pixel's dark current noise may be significant relative to the signal in that pixel (and that signal in the pixel also possesses its own intrinsic shot noise). Thus, it becomes a lab exercise to determine an optimum ROI size by simply capturing images and computing the SNR vs the ROI size.

We can gain some insight into the above problem of SNR vs ROI size in an analytic way by assuming that we have a Gaussian profile for the image given by

$$s(\rho) = s_o \exp\left(-2\frac{\rho^2}{w^2}\right), \tag{3.31}$$

where $s(\rho)$ is the mean number of photoelectrons per unit area captured at the image sensor, s_o is the mean peak, ρ is the radial size, and w is the half-width at the $1/e^2$ level of the Gaussian profile. For simplicity, we are neglecting the use of the 'bar' symbol above the quantity to denote that it is the mean. The total mean of the signal photoelectrons captured in an ROI of radial size ρ is obtained by the method of integration shown in equation (2.54), yielding

$$S(\rho) = S_{\text{tot}}\left[1 - \exp\left(-2\frac{\rho^2}{w^2}\right)\right], \tag{3.32}$$

where S_{tot} is the total number of mean photoelectrons captured when $\rho \to \infty$. The intrinsic photoelectron shot noise in this signal is governed by a Poisson distribution [27], so that the variance in the noise is equal to the signal mean. Thus, in terms of the std dev, the shot noise σ_S in the signal is simply the square root of equation (3.32):

$$\sigma_S = \sqrt{S(\rho)} = \sqrt{S_{\text{tot}}\left[1 - \exp\left(-2\frac{\rho^2}{w^2}\right)\right]}. \tag{3.33}$$

If the intrinsic sensor noise (such as dark current noise, read noise, and any other noise) in a single pixel is σ_p, and a pixel's surface area is A_p, then in a circular area

$\pi\rho^2$, the pixel's noise in terms of its variance is $(\pi\rho^2/A_p)\sigma_p^2$. The total noise N_{tot} in the ROI defined by radial size ρ is therefore

$$N_{tot} = \sqrt{\sigma_S^2 + (\pi\rho^2/A_p)\sigma_p^2} = \sqrt{S_{tot}\left[1 - \exp\left(-2\frac{\rho^2}{w^2}\right)\right] + (\pi\rho^2/A_p)\sigma_p^2}. \qquad (3.34)$$

The SNR in the ROI is therefore the ratio given by dividing equation (3.32) by (3.34), yielding

$$SNR = \frac{S_{tot}\left[1 - \exp\left(-2\frac{\rho^2}{w^2}\right)\right]}{\sqrt{S_{tot}\left[1 - \exp\left(-2\frac{\rho^2}{w^2}\right)\right] + (\pi\rho^2/A_p)\sigma_p^2}}. \qquad (3.35)$$

In the limits $\rho \rightarrow \infty$ and $\rho \rightarrow 0$, the SNR in equation (3.35) is seen to approach zero. Therefore, there must be a maximum SNR somewhere between these limits. Figure 3.5 shows a plot of equations (3.35) and (3.31) for $S_{tot}= 2E+04$ electrons, $w = 1.5$ mm, $A_p = 10^{-4}$ mm^2 (i.e. a ten micron pixel size), and $\sigma_p = 1$ electron. For these values, the SNR is seen to peak at roughly $\rho \approx 1.3$ mm, which is slightly less than the $1/e^2$ half-width of the image.

The reader may find it odd that equation (3.35) was arrived at using the pixels in a single image, whereas equation (3.26) suggests that the SNR should be determined by taking N images. In fact, they give the same results. If we consider that pixel intensities are **ergodic** (i.e. taking an average over an *ensemble* of images at a point in time is equal to taking an average over time [27, 40, 41]), then taking the mean of the totals of pixel intensities across N images is roughly the same as multiplying the numerator in equation (3.35) by N/N. The same applies to the denominator, thus canceling all N values.

Section 3.3.2 mentioned that a misuse of the concept of std dev can influence key decisions in optical system product development. In one experience, a team had

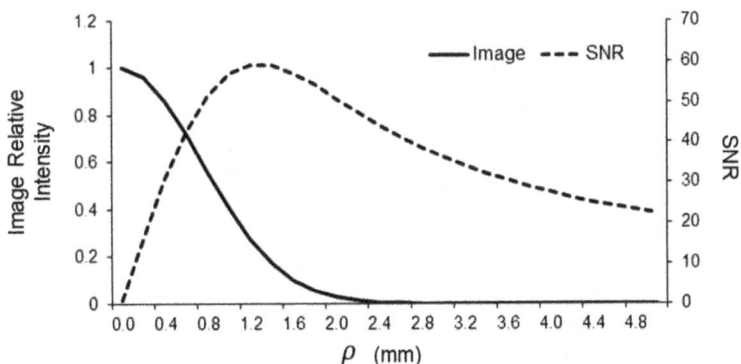

Figure 3.5. The SNR (given by equation (3.35)) and the normalized image profile (given by equation (3.31)) vs the ROI given by the radial size ρ.

decided that the SNR associated with an image of the form similar to that shown in figure 3.4(b) should be calculated by dividing the mean pixel intensities in the ROI by the std dev of those pixel intensities. But due to the profile of the image, they found (unsurprisingly, to readers of this book) that the inclusion of more pixels across the ROI area reduced the mean and increased the std dev, yielding 'low' SNR for an ROI that covered all of the image's pixels. Believing that it was the non-flatness of the profile in the image that caused this apparent reduction in the SNR, they decided to use only a small group of the brightest pixels at the center of the profile (i.e. near and at the peak of the profile). This flawed judgment was due to miscomprehending the difference between systematic variation (i.e. the profile of the image) and random noise (i.e. the fluctuations in the pixel intensities due to signal shot noise and dark current). Had they realized that in their application, the sum of pixel intensities was the relevant quantity of interest, and had they performed the analysis shown above, they would have realized that there is an ROI size that maximizes the SNR of the sum of pixel intensities in the ROI.

3.4.3 How does the signal-to-noise ratio scale with integration time?

Given any ROI, any flux projected onto a sensor, and any sensor, integration over time always increases the SNR, even when it increases the intrinsic noise of the signal. In this condition, one is said to be *signal shot noise limited* (i.e. detection of the signal is limited by its own intrinsic shot noise). To see this, take any mean photon flux per unit time $\overline{F}$ at some wavelength, any sensor noise given by a mean dark current i_D (whose noise current, assuming it has a Poissonian distribution, is $\sqrt{i_D}$), as well as some sort of *readout noise* σ_R at any instance in time (common in CCDs). At integration time t and sensor quantum efficiency η, the total integrated signal is $\overline{F}\eta t$. Since the photon flux's shot noise is $\sqrt{\overline{F}}$, the total noise at integration time t is $\sqrt{\overline{F}\eta t + i_D t + \sigma_R^2}$ (the readout noise does not scale with integration time in a CCD [42]). The SNR is therefore

$$\text{SNR} = \frac{\overline{F}\eta t}{\sqrt{\overline{F}\eta t + i_D t + \sigma_R^2}}. \tag{3.36}$$

In the limit $(\overline{F}t + i_D t) \ll \sigma_R^2$, the SNR is limited by the readout noise. However, in the limit $(\overline{F}t + i_D t) \gg \sigma_R^2$, equation (3.36) becomes

$$\text{SNR} = \frac{\overline{F}\eta t}{\sqrt{\overline{F}\eta t + i_D t}} = \frac{\overline{F}\eta \sqrt{t}}{\sqrt{\overline{F}\eta + i_D}}. \tag{3.37}$$

Thus, further increases in integration time result in the SNR scaling with $\sqrt{t}$. Finally, if the sensor can somehow be cooled, the dark current may be reduced until $\overline{F}\eta \gg i_D$. In this limit, $\text{SNR} = \sqrt{\overline{F}\eta t}$, so that the SNR becomes completely signal shot noise limited.

3.4.4 Does camera 'gain' increase the signal-to-noise ratio?

The answer to the question above depends on what is meant by camera *gain*. In some CMOS image sensors, there is the option in the user interface (such as software that you can install in a personal computer connected to the camera) to adjust a quantity called the *gain* of the camera, depending on the camera's manufacturer. Can this quantity amplify the signal without amplifying any internal noise from the sensor? I think that this would be very unlikely. Once optical flux strikes a sensor, its conversion into current or voltage *must* involve existing dark current or some form of circuit noise. Even photomultipliers (which are well-established sensors) are subject to an underlying dark current and fluctuations in the amplification process of the dynodes [27]. If gain in a CMOS camera could amplify signals, it would be a game changer (which it isn't).

Then what is the advantage of having a gain function for CMOS cameras? A reasonable application for it occurs when a real-time observer of the camera output wishes to see bright-looking images at short exposures (i.e. at high frame rates), especially when observing events in motion. Evidently, events that are in motion would be blurred if long integration times were used for the camera. Reducing the integration time would reduce the motion blur, but because this would also reduce the apparent brightness of the scene, the camera's gain can compensate for this, but the reduced SNR (due to the reduced integration time) remains. However, one may also define the 'noise' in such cases to involve motion blur. In this case, a new type of definition for SNR (such as, perhaps, 'motion SNR'?) may be employed, which would indicate that this new type of SNR *increases* with the camera gain.

3.4.5 What is 'charge conversion efficiency'?

Today, everyone has used an image sensor, because they hold one in their hands all the time, snapping digital photographs and taking selfies. Engineers, however, often perform analyses of digital images captured from image sensors, and when they do, they often use software such as ImageJ [43], which is convenient, as it is free to download and use. When you open a digital image in such software, you may notice that the 'intensities' of the pixels are associated with digital values from zero to a maximum given by the *bit depth* of the camera. In context, if one says that the camera is an eight-bit camera, then its bit depth is eight bits, which means that the pixel intensity can range between 0 and 2^8 (i.e. 0, 1, 2, ..., 255).

In previous sections, I have been referring to pixel intensities in terms of the number of photoelectrons, because they are the physical entities that accumulate in a sensor pixel that is exposed to light. But electronics in the image sensor eventually convert these accumulated photoelectron numbers for each pixel into a digital number (DN) that is proportional to the number of photoelectrons in the pixel. Sometimes, DN values are referred to as analog-to-digital-converted (ADC) numbers. In this book, we shall simply call them DN values. Since DN values correspond to the number of accumulated photoelectrons in a pixel, there is a *conversion gain* (Cg) that multiples a DN, yielding a number of photoelectrons [44–46]. On various camera suppliers' websites, Cg is sometimes called the *charge conversion efficiency* (CCE). The unit for

Cg and/or CCE is the number of electrons per DN (e⁻/DN). Thus, to convert the number of photoelectrons to DN, divide it by the CCE. For example, to convert the quantity $\bar{F}\eta t$ in equation (3.36) to DN, divide $\bar{F}\eta t$ by the CCE for the image sensor being used.

3.5 The concept of the limit of detection

3.5.1 The noise in the background after subtracting the background

Figure 3.6 shows two 'signals' in the absence of an actual signal. In figure 3.6(a), imagine that a photodiode is meant to capture nothing, but some average *background signal* is observed that is slightly greater than the zero-signal level. In figure 3.6(b), the mean background level has been subtracted, leaving only the noise. Sometimes, even when no signal is present, some ambient light or some dark current may be present in the sensor. These 'signals' may be considered the background, and they represent an unwanted nuisance on top of the wanted signal when the signal is also present. Once the mean background level is subtracted (an action often called *baselining*), the background noise remains, and the level of this residual noise becomes the **limit of detection** (LOD) for the sensor, which sets the *noise floor* for detecting signals.

The reason that the residual background noise may be called a noise floor is because, roughly speaking, a signal must be *above that floor*, otherwise, it is *lost* in the noise. Sometimes, a *detection system* may comprise the sensor, the optics for signal collection, and some electronics whose output is fed into a computer. In this case, one speaks of the LOD of the detection system. A real-time qPCR instrument, for example, is a fluorescence detection system. As such, it is characterized by an LOD. Since the level of background noise may be quantified by a std dev $\hat{\sigma}_B$ (or some factor M times $\hat{\sigma}_B$), the LOD may be defined in terms of the quantity $M\hat{\sigma}_B$, where M is any number (it does not have to be an integer).

3.5.2 The limit of detection is like an 'apparent nuisance signal'

Suppose that, when background is not present, a mean photon flux $\bar{F}$ arrives at a sensor of a detection system, whose expected output mean signal at integration time t and quantum efficiency η is $\bar{S} = \bar{F}\eta t$. If background signals are present and if the

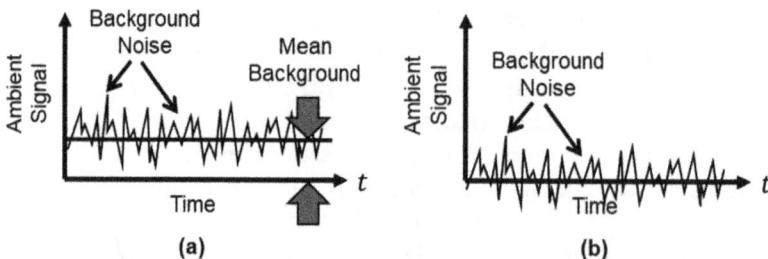

Figure 3.6. (a) An ambient 'background' signal with mean level and noise. (b) After subtraction of the mean background, noise remains.

level of the background noise is $M\hat{\sigma}_B$, then one can say that there is an 'apparent nuisance signal' $\overline{B}$ due to the noise level given by $M\hat{\sigma}_B = \overline{B}\eta t$. If we combine this with $\overline{S} = \overline{F}\eta t$, eliminate ηt, and solve for $\overline{B}$, we obtain $\overline{B} = M\hat{\sigma}_B\overline{F}/\overline{S}$. We may then call $\overline{B}$ the detection system's LOD:

$$\text{LOD} = \frac{M\hat{\sigma}_B\overline{F}}{\overline{S}}. \tag{3.38}$$

Equation (3.38) is helpful in some applications, such as in virus detection using a real-time qPCR instrument. In this application, the viral concentration in a patient is sought. If we assume that the viral concentration in a patient is proportional to the concentration of its extracted DNA (converted from its RNA) in a sample of blood or saliva (or something else) provided by the patient, then one may quantify viral content by way of equations (3.5) and (3.6). But background noise (and other noise sources) in the detection process can hinder the detection of the viral DNA concentration in those equations. Therefore, one often seeks to determine the minimum detectable concentration of the DNA sample in the presence of the noise. Indeed, as Klymus *et al* have stated: 'LOD can be defined as the lowest concentration of target analyte that can be detected with a defined level of confidence' [47]. Since the 'level of confidence' may be considered to be $M\hat{\sigma}_B$, we can express the statement above in an approximate formula as follows. In real-time qPCR applications, it is common to relate the mean fluorescence signal $\overline{S}$ with a known concentration C of the fluorophore used for the PCR amplification assay. In this case, applying equation (3.2) gives $\overline{S} = \alpha C$, so that the ratio $\overline{S}/C$ replaces the ratio $\overline{S}/F$ in equation (3.38), yielding

$$\text{LOD} \approx \frac{M\hat{\sigma}_B C}{\overline{S}}. \tag{3.39}$$

Equation (3.39) is considered approximate not only because $\hat{\sigma}_B$ is an *estimate* of the background noise but also because $\hat{\sigma}_B$ can involve other noises. In fact, many researchers have developed further techniques and formulas to include additional noise sources and levels in the LOD of a qPCR instrument [48–50]. Also, when the LOD is defined in the manner shown in equation (3.39), it may also be considered a measure of the signal-to-background noise ratio (so, it is not a *complete* SNR in the sense that the ratio does not involve the signal's shot noise). To see this, divide both sides of equation (3.39) by C, then take the reciprocal. You obtain the relation $C/\text{LOD} = \overline{S}/M\hat{\sigma}_B$, for which $\overline{S}/M\hat{\sigma}_B$ is the ratio of the mean signal to the background noise level defined at a *noise confidence interval* of $\pm M\hat{\sigma}_B$. Finallly, the LOD in equation (3.39) may be considered to be a *noise equivalent concentration*. This is somewhat like the *noise equivalent power* (NEP) discussed in texts on radiometry (e.g. [27]), defined as the input power of a signal that yields an output SNR of one. In the case of equation (3.39), the noise equivalent concentration is the sample concentration that equals the noise level in the instrument's detection system.

3.5.3 The relationship between the limit of detection and the signal-to-noise ratio

If we return to equation (3.36), we would note that while it contains background noise from dark current and readout noise, it does not include shot noise from any ambient background radiation present. If we let σ_A^2 be the variance of the noise due to ambient background radiation, then equation (3.36) may be expressed as

$$\text{SNR} = \frac{\overline{F}\eta t}{\sqrt{\overline{F}\eta t + i_D t + \sigma_A^2 t + \sigma_R^2}}. \tag{3.40}$$

Notice that since ambient background light can scale with the integration time, and since its variance is equal to its mean, the integration time t has been included for that factor. Equation (3.40) lets us keep track of what other noise components may be included or excluded in the denominator under various conditions, but it is also instructive to express it in a more compact form by grouping all background noise components into the quantity $\hat{\sigma}_B^2 = i_D t + \sigma_A^2 t + \sigma_R^2$. If we do this, we may also combine it with equation (3.38) and $\overline{S} = \overline{F}\eta t$ to express the SNR as

$$\text{SNR} = \frac{\overline{S}}{\sqrt{\overline{S} + \left[\frac{(\text{LOD})\overline{S}}{M\overline{F}}\right]^2}} = \frac{1}{\sqrt{\frac{1}{\overline{S}} + \left[\frac{\text{LOD}}{M\overline{F}}\right]^2}}. \tag{3.41}$$

If we take the square of the reciprocal of equation (3.41), we obtain

$$\frac{1}{(\text{SNR})^2} = \frac{1}{\overline{S}} + \left[\frac{\text{LOD}}{M\overline{F}}\right]^2. \tag{3.42}$$

Equation (3.42) is useful in the sense that the right-hand side is seen to be the sum of the reciprocals of the squares of the SNRs. The term $1/\overline{S}$ is one over the square of the signal's own SNR (the ratio of the mean signal to its own shot noise), and the term $[\text{LOD}/(M\overline{F})]^2$ is one over the square of the ratio of the input signal to the background noise. Equivalently, the right-hand side of equation (3.42) may be seen as noise propagation in terms of the sum of the squares of coefficients of variation (because the coefficient of variation is the std dev divided by the mean). When a virus is detected using real-time qPCR, the quantity $\overline{F}$ is the input fluorescence signal from fluorophores being amplified as the viral DNA is amplified, and $\overline{S}$ is the output sensor signal from the fluorescence. Thus, in this case, the term $[\text{LOD}/(M\overline{F})]^2$ says that, in a real-time PCR assay, the fluorescence signal $\overline{F}$ must reach an LOD beyond which the signal becomes dominant over and above the background noise. And under this condition, the signal integrity is only limited by its own shot noise, which is the limiting condition for any detection system. This is the ingenuity of real-time PCR chemistry—the ability to amplify the signal without amplifying the background.

3.6 Remarks concerning the tolerancing of complex optical systems in product development

In section 3.3.4, we saw that the RSS technique of tolerance stack estimation for the worst-case total change in an optical parameter does not include interaction effects. Even if an RSS tolerance stack includes mixed partial derivatives, it may only involve second-order terms, due to the complexity of the mathematics used to compute derivatives numerically. In sections 3.4 and 3.5, we saw that if the output from an instrument (such as detectability of a virus) depends on a detection system's SNR, many noise components are involved, so that one must somehow relate the performance of any optical components or subsystems to the final instrument's output. In section 3.4.2, an example was provided in which the SNR of a signal depended on size of the ROI in an image signal. Therefore, in such cases, if there is any variation in the image profile, there can be variation in the optimum ROI size, which can affect the final instrument's SNR. Finally, in section 1.2.9, we also discussed reasons for which the MTF of an imaging system is not the only way to judge imaging performance. As appendix A.1 shows, optical crosstalk can sometimes provide a more direct assessment of the impact of an image profile on end performance. If an imaging lens is forming images of wells on an image sensor in a real-time qPCR instrument for virus detection, and if the image profiles are of the form shown in figure 3.4(b), then optical crosstalk can cause light from an image of a well with signal to leak into the image of a NTC well (see figure 3.1(a)). Moreover, the spillover of light from the image of a well containing a positive control into the image of a well containing a sample can result in false positives in the case of a sample that has been taken from a patient who is not actually infected by the virus of interest. Hence, in this case, optical crosstalk is directly relevant to the application.

The examples provided above are meant to demonstrate that complex optical systems require tolerancing analyses that are beyond trivial computations involving RSS tolerance stacks and limiting one's scope of design analysis to one's own subsystem. If, in some application, images of the sort illustrated in figure 3.4(b) are desired in order to obtain the total signal in a ROI, then prior to prototyping, a software algorithm developer in a team may wish to receive some possible worst-case samples of image profiles so that the developer can try to apply work-in-progress algorithms to determine ROI sizes and SNRs. RSS tolerance stacks in an optical tolerance analysis would not provide such information, but Monte Carlo analyses would. In the OS program, as we have seen in sections 1.2.13 and 2.20, worst-case trials can provide samples of any figure of merit. Moreover, all trials can be saved as OS files, which may be opened for analysis. In this case, in the example of imaging system that forms images with profiles such as that shown in figure 3.4(b), one can open up a worst-case file to output such an image from OS in any of the available formats (such as jpeg, bitmap, or tiff) and supply them as image samples to the software developer. In this way, from a systems perspective, the *inputs* to an optical simulation for tolerancing analysis can become *outputs* to a larger team in a project involving product development, as discussed by Bunch [1].

Monte Carlo simulation is therefore seen to be a powerful technique for performing optical tolerancing analyses [24]. Due to the nature of a Monte Carlo simulation, in which all variables are given random perturbations within min and max tolerances, all interactions are included, up to any order of a Taylor series, within the bounds of the min and max of the tolerances. The Monte Carlo method is not limited to its use in predicting manufacturing yields in large volume production. On the contrary, when there is just a single instrument to be produced (say, as a prototype), **the Monte Carlo method allows one to perturb all variables in order to determine the worst-case combination**. Indeed, as stated by Shonkwiler and Mendivil [51], 'Yet, in many applications, Monte Carlo is unsurpassed. It still enjoys almost exclusive dominion over its original application, simulating complex interactions in any area where quantitative models are possible.' In the following section, we shall apply the Monte Carlo tolerancing method to reveal nonobvious properties of optical systems that can influence decisions about the way in which optical assemblies should or should not be aligned to achieve a desired result. To do this, we will take a problem related to the integration of two multi-element lens systems as a simple example.

3.7 Monte Carlo tolerancing as a means to justify an alignment philosophy

Suppose an optical system comprises a number of components and assemblies that are to be mounted in some way for system integration. When a tolerance analysis indicates that certain tolerances for mounting and integration cannot be met by way of production specifications, one often identifies compensators and adjustments to set the system right. If required, an alignment procedure may also be developed for mounting and integration. Since a Monte Carlo simulation can provide the worst-case scenario (with interactions) under production tolerances, it can also reveal the worst-case perturbations to be fixed by the compensators [52]. In OS, Monte Carlo trials are files that can be saved and opened for analysis. Therefore, the worst-case file may be used to simulate compensation and alignment procedures, targeting the offending tolerances that cannot be addressed by factory production. Alternatively, OS provides a means for us to write simple *tolerance scripts* that instruct the program to perform specific actions (which could involve alignments) on components and assemblies during a live Monte Carlo simulation; the output is a report of the desired measures of quality (such as a list of merit function operand values for MTF, RMS spot size, etc).

As an example, let us suppose we have the dual-assembly system shown in figure 3.7, whose prescription and merit function are provided in figures 3.8 and 3.9, respectively. The wavelength of operation is 550 nm (so the stated EFLs are at this wavelength). The objective is a scaled-down version of the lens from figure 1.70, but reversed. The COTS tube lens from figure 1.105 is reused. Due to the ratio of their EFLs, the system projects an image at a magnification of 5.3×. By design, the integration of these two lenses in their nominal states requires that the exit pupil of the objective (when used in reverse, such as shown in figure 3.7) is at the same

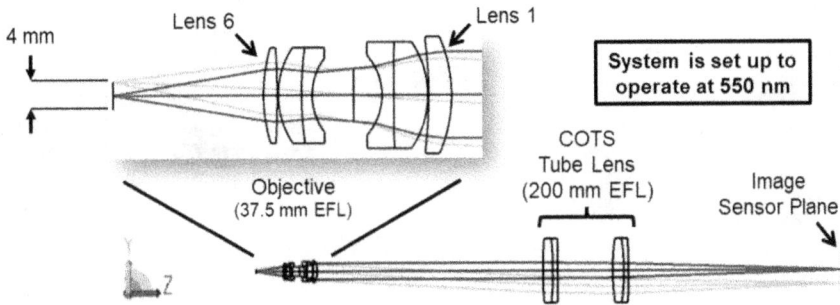

Figure 3.7. Two imaging assemblies integrated to project an image at a magnification of 5.3×.

	Surface Type		Comment	Radius	Thickness	Material	Semi-Diameter	Decenter X	Decenter Y
0	OBJECT	Standard ▾		Infinity	21.72101		2.00000		
1		Coordinate Break ▾	COMP +/-		0.00000		0.00000	0.00000	0.00000
2	(aper)	Standard ▾	Lens 6	23.54175	1.94781	N-SK16	6.75000 U		
3	(aper)	Standard ▾		-111.32802	0.15125		6.75000 U		
4		Coordinate Break ▾	POSITION		-2.09906 T		0.00000	0.00000	0.00000
5		Coordinate Break ▾	COMP -/+		2.09906 P		0.00000	0.00000 P	0.00000 P
6	(aper)	Standard ▾	Lens 5	12.09429	3.33943	N-SK16	6.75000 U		
7	(aper)	Standard ▾	Lens 4	47.51701	1.52229	F5	6.75000 U		
8	(aper)	Standard ▾		8.61647	5.98259		5.25000 U		
9	STOP (aper)	Standard ▾	STOP	Infinity	4.12430		3.75000 U		
10	(aper)	Standard ▾	Lens 3	-8.33676	1.70875	F5	5.62500 U		
11	(aper)	Standard ▾	Lens 2	597.14944	4.85738	N-SK16	7.50000 U		
12	(aper)	Standard ▾		-13.43212	0.15067		7.50000 U		
13	(aper)	Standard ▾	Lens 1	-58.02206	3.31008	N-SK2	7.50000 U		
14	(aper)	Standard ▾		-20.08110	181.28000		8.25000 U		
15		Standard ▾	Dummy	Infinity	0.00000		27.00000 U		
16	(aper)	Standard ▾	45-271	305.74000	8.00000	N-BK7	25.00000 U		
17	(aper)	Standard ▾	45-271	-223.20000	4.00000	N-SF5	25.00000 U		
18	(aper)	Standard ▾		-663.82000	43.86832		25.00000 U		
19	(aper)	Standard ▾	45-181	173.11000	9.00000	N-BAK4	25.00000 U		
20	(aper)	Standard ▾	45-181	-164.03000	3.50000	N-SF10	25.00000 U		
21	(aper)	Standard ▾		-709.83000	172.82490		25.00000 U		
22		Coordinate Break ▾	Cam Shift		0.00000		0.00000	0.00000	0.00000
23		Standard ▾		Infinity	0.00000		10.65148		
24	IMAGE (aper)	Standard ▾	Camera	Infinity		-	15.00000 U		

Figure 3.8. The prescription for the system shown in figure 3.7.

location as the entrance pupil of the COTS tube lens. This is called *pupil matching*. For this reason, the prescription in figure 3.8 shows that the thickness of surface 14 is not the 200 mm distance that is in the prescription for the tube lens that is shown in figure 1.106. When analyzed alone, the objective's exit pupil is found to be roughly 18.72 mm to the left of Lens 1's right vertex, so the 200 mm pupil distance of the COTS lens has to be taken up by 18.72 mm. The difference is therefore the thickness shown for surface 14.

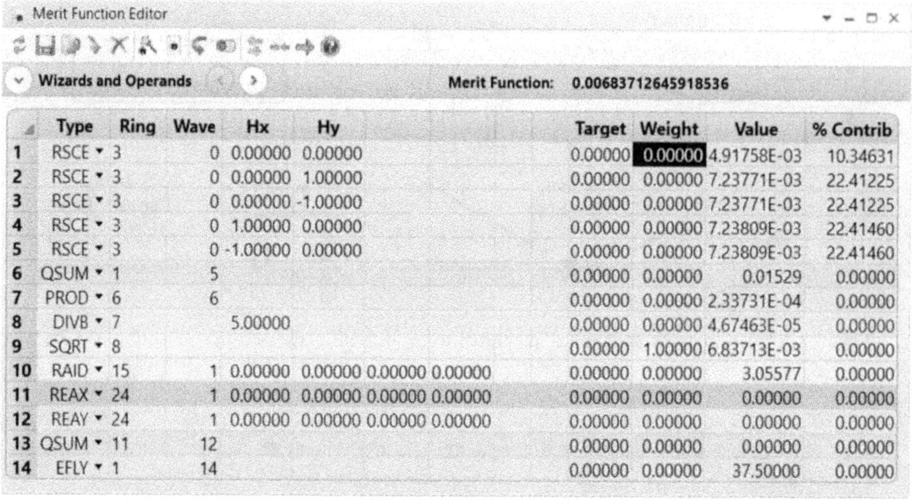

	Type	Ring	Wave	Hx	Hy			Target	Weight	Value	% Contrib
1	RSCE	3	0	0.00000	0.00000			0.00000	0.00000	4.91758E-03	10.34631
2	RSCE	3	0	0.00000	1.00000			0.00000	0.00000	7.23771E-03	22.41225
3	RSCE	3	0	0.00000	-1.00000			0.00000	0.00000	7.23771E-03	22.41225
4	RSCE	3	0	1.00000	0.00000			0.00000	0.00000	7.23809E-03	22.41460
5	RSCE	3	0	-1.00000	0.00000			0.00000	0.00000	7.23809E-03	22.41460
6	QSUM	1	5					0.00000	0.00000	0.01529	0.00000
7	PROD	6	6					0.00000	0.00000	2.33731E-04	0.00000
8	DIVB	7		5.00000				0.00000	0.00000	4.67463E-05	0.00000
9	SQRT	8						0.00000	1.00000	6.83713E-03	0.00000
10	RAID	15	1	0.00000	0.00000	0.00000	0.00000	0.00000	0.00000	3.05577	0.00000
11	REAX	24	1	0.00000	0.00000	0.00000	0.00000	0.00000	0.00000	0.00000	0.00000
12	REAY	24	1	0.00000	0.00000	0.00000	0.00000	0.00000	0.00000	0.00000	0.00000
13	QSUM	11	12					0.00000	0.00000	0.00000	0.00000
14	EFLY	1	14					0.00000	0.00000	37.50000	0.00000

Figure 3.9. The merit function for the system shown in figure 3.7.

In the merit function (figure 3.9), operands have been listed for the reporting values that would be reported by a tolerance script (to be discussed soon). Only operand 9 has been given a weight, as we will be letting the program optimize for the average RMS spot radial size across the five fields that are indicated on rows 1–5. Of course, it would have been the same to place weights on those five operands, but as will be revealed soon, we want the tolerance script to report values specifically for operands 9 (average RMS spot size), 10 (beam deviation), 13 (image shift), and 14 (EFL of the objective), rather than 1, 2, 3, 4, 5, 10, 13, and 14. So, it is convenient to manually compute the averaged RMS spot size on row 9 in the merit function.

Figures 3.10–3.12 list tolerance operands relevant to the prescription *for only the objective* in figure 3.8. The reason for tolerancing only the objective shall be made clear soon. Note that rows 46–49 are present to set the tolerance on parameters 1–4 (Decenter *X*, Decenter *Y*, Tilt About *X*, Tilt About *Y*) on surface 1 (Lens 6 on the objective), which is a coordinate break in the prescription. This not only serves to set those tolerances, but a tolerance script shall make use of the x and y decenters for Lens 6 to serve as a compensator to optimize the RMS spot size during Monte Carlo simulation. Recall that in section 1.5.3, I mentioned that an element's decenter may sometimes be used as a compensator [53]. Here, we shall apply this technique. As we will see, its effectiveness as a compensator in production has a significant influence on the alignment philosophy for this dual-assembly system.

The scenario that we are studying in the current example is as follows. We have two assemblies, each of which is a lens system with a known design and tolerances. The objective is a customized lens design, whilst the tube lens is a *customized integration* of COTS doublets. The combination resembles a digital microscope system, but is used under monochromatic conditions as a simple substitute for an expensive high-performance commercial microscope. When the objective is used on

Tolerance Data Editor

Operand 2 Properties

	Type	Surf	Adjust	Nominal	Min	Max
1	STAT	1	3			
2	TTHI	2	2	1.94781	-0.15000	0.15000
3	TTHI	3	3	0.15125	-0.15000	0.15000
4	TTHI	6	6	3.33943	-0.15000	0.15000
5	TTHI	7	7	1.52229	-0.15000	0.15000
6	TTHI	8	8	5.98259	-0.15000	0.15000
7	TTHI	9	9	4.12430	-0.15000	0.15000
8	TTHI	10	10	1.70875	-0.15000	0.15000
9	TTHI	11	11	4.85738	-0.15000	0.15000
10	TTHI	12	12	0.15067	-0.15000	0.15000
11	TTHI	13	13	3.31008	-0.15000	0.15000
12	TFRN	2		0.00000	-5.00000	5.00000
13	TFRN	3		0.00000	-5.00000	5.00000
14	TFRN	6		0.00000	-5.00000	5.00000
15	TFRN	7		0.00000	-5.00000	5.00000
16	TFRN	8		0.00000	-5.00000	5.00000
17	TFRN	10		0.00000	-5.00000	5.00000
18	TFRN	11		0.00000	-5.00000	5.00000
19	TFRN	12		0.00000	-5.00000	5.00000
20	TFRN	13		0.00000	-5.00000	5.00000
21	TFRN	14		0.00000	-5.00000	5.00000
22	TIRR	2		0.00000	-1.00000	1.00000
23	TIRR	3		0.00000	-1.00000	1.00000
24	TIRR	6		0.00000	-1.00000	1.00000
25	TIRR	7		0.00000	-1.00000	1.00000

Figure 3.10. The tolerance operands for the system shown in figure 3.7 (continued in figures 3.11 and 3.12).

its own, it is intended to be used with an object at infinity, but as rays are reversible, we have decided to reuse it as a kind of simple microscope objective. As its image will be at infinity, it is natural to pair it with a tube lens. Since the COTS tube lens in figure 1.105 is available, it is a natural choice. When the objective is analyzed on its own, tracing rays from infinity to the image plane (which is the object plane in figure 3.7) shown that the decentering of Lens 6 in the x–y transverse dimension is a suitable compensator for improving image quality that has been degraded due to stacked tolerances. Therefore, this objective's entire assembly (which includes its housing) contains a last element (Lens 6) that is always purposefully decentered and glued, achieving improved image quality even when all other elements in the objective have a tolerance stack. A customer who purchases and uses this objective

	Type	Surf	Adjust	Nominal	Min	Max
26	TIRR ▾	8		0.00000	-1.00000	1.00000
27	TIRR ▾	10		0.00000	-1.00000	1.00000
28	TIRR ▾	11		0.00000	-1.00000	1.00000
29	TIRR ▾	12		0.00000	-1.00000	1.00000
30	TIRR ▾	13		0.00000	-1.00000	1.00000
31	TIRR ▾	14		0.00000	-1.00000	1.00000
32	TSTX ▾	3		0.00000	-0.03333	0.03333
33	TSTX ▾	6		0.00000	-0.03333	0.03333
34	TSTX ▾	8		0.00000	-0.03333	0.03333
35	TSTX ▾	10		0.00000	-0.03333	0.03333
36	TSTX ▾	12		0.00000	-0.03333	0.03333
37	TSTX ▾	14		0.00000	-0.03333	0.03333
38	TSTY ▾	3		0.00000	-0.03333	0.03333
39	TSTY ▾	6		0.00000	-0.03333	0.03333
40	TSTY ▾	8		0.00000	-0.03333	0.03333
41	TSTY ▾	10		0.00000	-0.03333	0.03333
42	TSTY ▾	12		0.00000	-0.03333	0.03333
43	TSTY ▾	14		0.00000	-0.03333	0.03333
44	TEDX ▾	2	8	0.00000	-0.30000	0.30000
45	TEDY ▾	2	8	0.00000	-0.30000	0.30000
46	TPAR ▾	1	1	0.00000	-0.30000	0.30000
47	TPAR ▾	1	2	0.00000	-0.30000	0.30000
48	TPAR ▾	1	3	0.00000	-0.20000	0.20000
49	TPAR ▾	1	4	0.00000	-0.20000	0.20000
50	TEDX ▾	6	8	0.00000	-0.30000	0.30000

Figure 3.11. The tolerance operands for the system shown in figure 3.7 (continued from figure 3.10).

as a camera lens will never notice this decentered element, because it is at the end of the lens system and always 'hidden' inside a camera with a suitable image sensor.

Earlier, I mentioned that tolerance operands have only been included to tolerance the objective. The reason for this is that we shall regard the objective as being a standalone 'existing product' that has been aligned for its own image quality. When we repurpose it for the current integrated microscope layout, we expect there to be a decentered Lens 6 in that objective. When imaging assemblies are integrated, questions may arise concerning how to align them. The use of alignment tools (such as lasers and autocollimators) requires the axes of each assembly to be aligned with respect to each other [54, 55]. But we know in this case that the objective does not possess an axis, due to the decentered last element. So, we now want to know the consequence of having this type of objective assembled with a nominally centered tube lens assembly. We will later find that it is not difficult to make certain

Tolerance Data Editor

▼ — □ X

⌄ Operand 2 Properties ‹ ›

Type	Surf	Adjust	Nominal	Min	Max
51 TEDX ▾	10	12	0.00000	-0.30000	0.30000
52 TEDX ▾	13	14	0.00000	-0.30000	0.30000
53 TEDY ▾	6	8	0.00000	-0.30000	0.30000
54 TEDY ▾	10	12	0.00000	-0.30000	0.30000
55 TEDY ▾	13	14	0.00000	-0.30000	0.30000
56 TETX ▾	6	8	0.00000	-0.20000	0.20000
57 TETX ▾	10	12	0.00000	-0.20000	0.20000
58 TETX ▾	13	14	0.00000	-0.20000	0.20000
59 TETY ▾	6	8	0.00000	-0.20000	0.20000
60 TETY ▾	10	12	0.00000	-0.20000	0.20000
61 TETY ▾	13	14	0.00000	-0.20000	0.20000
62 TIND ▾	2		1.62041	-1.00000E-03	1.00000E-03
63 TIND ▾	6		1.62041	-1.00000E-03	1.00000E-03
64 TIND ▾	7		1.60342	-1.00000E-03	1.00000E-03
65 TIND ▾	10		1.60342	-1.00000E-03	1.00000E-03
66 TIND ▾	11		1.62041	-1.00000E-03	1.00000E-03
67 TIND ▾	13		1.60738	-1.00000E-03	1.00000E-03
68 TABB ▾	2		60.32365	0.28325	-0.28325
69 TABB ▾	6		60.32365	0.30162	-0.30162
70 TABB ▾	7		38.02992	0.19015	-0.19015
71 TABB ▾	10		38.02992	0.19015	-0.19015
72 TABB ▾	11		60.32365	0.30162	-0.30162
73 TABB ▾	13		56.65011	0.30162	-0.30162

Figure 3.12. The olerance operands for the system shown in figure 3.7 (continued from figure 3.11).

conclusions about the types of additional effects that the tube lens's tolerances would have had on the image if they had been added (and, if needed, this can always be verified). Since each assembly has its own housing, the idea here is to determine whether any active alignment is required between the two assemblies when they are joined. When a Monte Carlo tolerancing analysis is performed for the complete system of two assemblies but perturbs only the objective (to mimic its production tolerances *as well as to include the decentering adjustment to Lens 6*), certain characteristics of the final image are revealed. To that end, a tolerance script in OS has been written, which is shown in figure 3.13. In this script, all lines beginning with the exclamation symbol '!' are for comments. I have included them as a means to help you understand the purpose of each script command.

Before we proceed to the tolerance analysis, we should highlight some of the assumptions the analysis makes about the way in which the assemblies are mounted. The **global coordinate** of the system is surface 0, the object plane at the far left of the layout. A biological sample is located on this plane. Recall that when the objective is

```
MCOSD_Script.TSC                                                    ▾ ▬ ▢ ✕
🖫🖫📇🖶⮌ℭ ▣ ≣⚲
 1│! This tolerancing script is for the file "Figure 3.7+ Dual-Assembly System".
 2 ! Clear all existing compensators, if any.
 3 CLEARCOMP
 4
 5 ! First, define the focus compensator.
 6 COMP 0 0
 7
 8 ! Optimize the focus for the current Monte Carlo file.
 9 OPTIMIZE 5
10
11 ! Next, clear the focus compensator.
12 CLEARCOMP
13
14 ! Now, define new compensators for adjusting the last element.
15 CPAR 1 1
16 CPAR 1 2
17
18 ! Now, optimize again to achieve higher image quality.
19 OPTIMIZE 5
20
21 ! Evaluate the final merit function.
22 GETMERIT
23
24 ! Report merit function operand values.
25 REPORT "AVG RMS Spot: " 9
26 REPORT "Beam Deviation: " 10
27 REPORT "Image Shift: " 13
28 REPORT "EFL: " 14
```

Figure 3.13. Tolerance script to be used on the system shown in figure 3.7 during a Monte Carlo simulation.

used on its own, it is an imaging lens used with a camera and its image sensor, but due to the decentered last element, the image is shifted. To address this, the image shift is known and calibrated at the factory, and we shall assume that this objective is therefore always assembled into a housing whose camera mounting interface has a mechanical axis centered to that of most commercial camera mounts. A conceptual schematic for this idea is illustrated in figure 3.14. In this way, the image is always centered in the camera (within some acceptable tolerance), regardless of the decentered last element. Due to this, we know that if the axis of the object in figure 3.7 is nominally centered to the objective, then the image at the end of the tube lens must surely be decentered. So, we expect this result and therefore include a merit function operand (row 13 in figure 3.9) to compute this shift. This is also reported by the tolerance script (line 27 in figure 3.13). In fact, note that the tolerance script reports lines 9, 10, 13, and 14, labeled accordingly with the operands in the merit function. The beam deviation (line 26 in figure 3.13) is computed by making the axial ray intersect a dummy surface (surface 15 in figure 3.9, but not drawn in the layout). This measures the beam's incident angle on that surface, which is also equal to the deviation of the beam relative to the axis of the object plane (the global coordinate). The reason that this quantity is of interest is to check for any correlation between the axial beam's deviation angle and the image quality (RMS spot size) for the entire sample of Monte Carlo trials. A correlation would indicate that optical

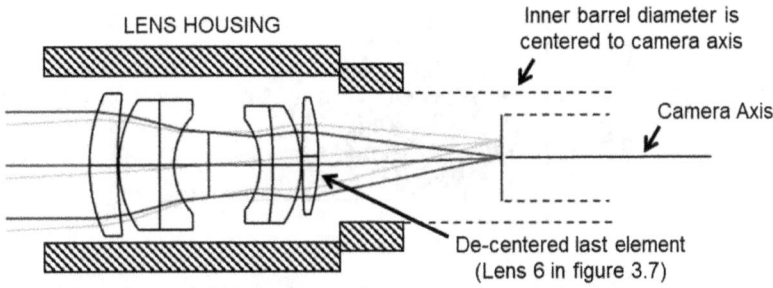

Figure 3.14. A conceptual schematic depicting the centering of the objective's housing relative to a camera's image sensor, within some level of tolerance.

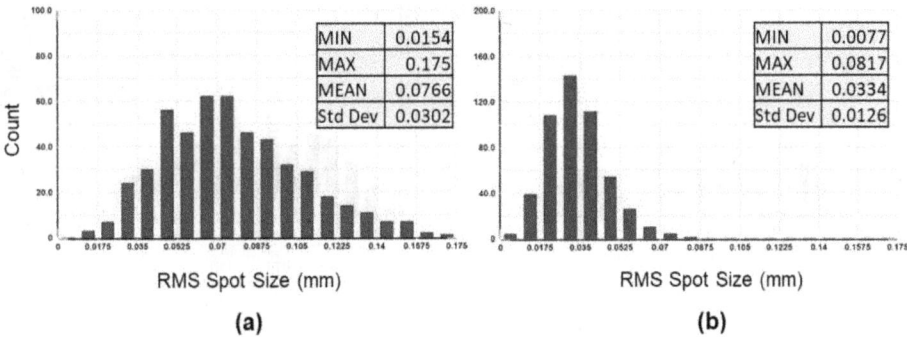

Figure 3.15. A histogram of the average RMS spot size. (a) With focus compensation only. (b) With focus compensation and lateral decenter compensation performed by Lens 6 in the objective.

alignment may be performed using a laser or an autocollimator. Otherwise, neither of these would be beneficial.

The Monte Carlo simulation is performed in the usual way, as described in earlier sections of this book (e.g. section 1.2.13), with the exception that under the 'Criterion' option (see bottom left in figure 1.81), one selects 'User Script,' and then, under the list of 'Script' choices, for the current simulation, one selects the name of the tolerance script 'MCOSD_Script.TSC' (this file is made available as supplementary material at https://doi.org/10.1088/978-0-7503-6059-3, and it must first be placed into the directory Documents > Zemax > Tolerance). For comparison, a first Monte Carlo run is performed with only focus compensation (applied to the thickness on surface 0, which is the object distance). The second run applies the tolerance script shown in figure 3.13, which first focuses the image and then performs the X and Y decenter for Lens 6 in the objective. Each run has 500 trials, and their RMS spot size results (averaged across five fields, see operand 9 in figure 3.9) are plotted in the histograms shown in figure 3.15.

In both histograms of figure 3.15, the maximum value in the horizontal axis is 0.175 mm. Notice the compression of the distribution in figure 3.15(b), due to the use of the additional decenter compensator on Lens 6 of the objective. Was there a

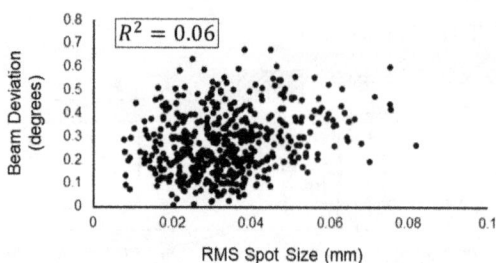

Figure 3.16. A scatter plot of the beam deviation and RMS spot size for 500 Monte Carlo trials, based on the tolerances and script in figures 3.10–3.13, respectively.

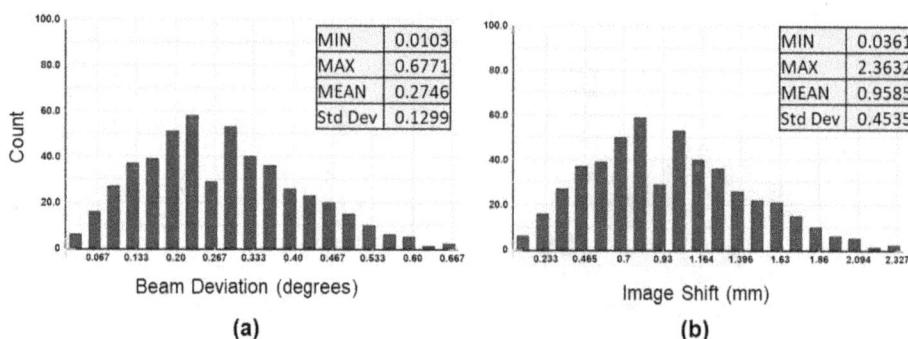

Figure 3.17. Histograms of the beam deviation (a) and the image shift (b).

correlation between RMS spot size and beam deviation out of the objective? Figure 3.16 shows a scatter plot of the two, whose R^2 value is 0.06, indicating virtually no *linear* correlation, but also no obvious relationship whatsoever. This is indeed expected, due to the decentering adjustment of Lens 6 in the objective to improve the image quality. Note again that by *beam deviation*, we mean the angle of incidence of the axial ray originating from the object and ending on surface 15, which is a surface that is parallel to the object plane. Hence, this angle should be zero if an autocollimator is aligned to the object and to an objective that has no centering error. But the objective here has nonzero centering, due to the decentering of Lens 6.

Histograms for the beam deviation, the transverse shift of the final image formed by the tube lens, and the objective's EFL are shown in figures 3.17(a) and (b) and 3.18, respectively. Figures 3.17(a) and (b) reveal that if all tolerances from the tube lens are held at zero, the image shifts transversely by up to roughly 2.36 mm, due to the beam deviation from the objective towards the tube lens (a scatter plot will—to no surprise—reveal that they are perfectly linearly correlated). This means that an additional transverse alignment is necessary for the camera if the image is to be centered relative to the image sensor. **Thus, the Monte Carlo simulation has so far revealed the need for one additional alignment procedure**.

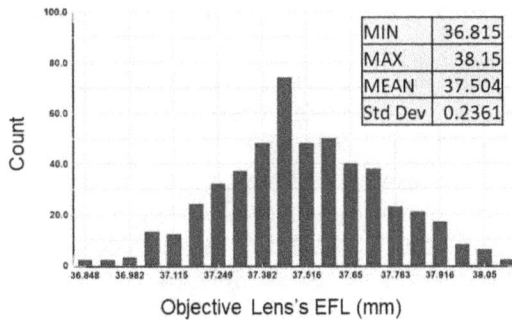

Figure 3.18. A histogram of the EFL of the objective lens.

The EFL of the objective lens is analyzed because one may wish to know whether the magnification of the system requires adjustment. Holding the tube lens's EFL at 200 mm, since the objective's EFL is seen to vary between a min of 36.82 mm and a max of 38.15 mm, the magnification varies between 5.24 × and 5.43 ×. Additional tolerances from the tube lens assembly would increase this variation unless we were to apply an air gap adjustment between the two doublets (see section 1.3.2). However, if this were to be considered, then further Monte Carlo simulations would have to be performed to determine the effect of this air gap adjustment on other parameters, such as the image quality.

Additional conclusions can be made about the alignment philosophy. It is not difficult to see that if transverse decenter tolerances were included in the entire tube lens assembly, the image shift would be inconsequential if the camera were attached to the assembly. Therefore, with the provision that the camera be attached to the tube lens assembly, the tolerances on that assembly's x–y positioning may be relaxed, hence eliminating the need for that adjustment, with the additional provision that oblique rays from the objective do not become vignetted. But the tilt of the tube lens assembly has not yet been accounted for. Also, the decenter tolerances for the individual doublets and their air space have not been included. If the objective lens is to continue to be regarded as a standalone product, then it means that the tube lens must also function as a well-assembled assembly. This means that the tube lens must be toleranced first, on its own, and tolerance scripts should be written to align it for best image quality. Subsequently, a full tolerance analysis that involves tolerance scripts that mimic factory adjustments of both independent assemblies should be applied, followed by an analysis of the final image quality and other characteristics (such as image shift). Alternatively, a company that owns these assemblies could consider that, for this specific dual-assembly to become a good product, perhaps the objective lens and tube lens may be toleranced together *but without adjusting Lens 6 of the objective*. This approach could open up the possibility of identifying new compensators that might offer further benefits to the integrated system. If this path were to be considered, then further discussions with the opto-mechanical designer would be helpful, as there might perhaps be ways to tighten the tolerances of the elements by way of applying various mounting techniques [53, 56, 57].

At this point, one might ask: 'But if we were purchasing and integrating all COTS assemblies, how would we know whether there is a special compensating element in a multi-element lens system that has been decentered or even tilted?' The quick answer is that, indeed, you would not know whether such compensators have been used in COTS assemblies, and so you must either ask the supplier (who is unlikely to tell you the answer) or you must characterize these assemblies. To perform the latter, you may consider first aligning the assembly by centering it to the axis of an autocollimator and checking the image quality using a gold-standard resolution target. Does zero centering error (beam deviation) for this lens result in optimum image quality? If not, then the entire lens assembly may be checked further by tilting it to see if this improves the image quality. If there is improvement, then it is possible that compensating features (including natural statistical interactions) are present (see appendix C.2 for an example in which no compensating element is present and the image quality does not correlate with centering error).

3.8 Some nuances of optical systems in product development

3.8.1 When a lens images an intermediate transmissive or reflective surface

We saw in sections 1.3.5–1.3.10 that any plane in the path of rays from a first lens can be the object conjugate to an image plane for a second lens, and as long as no vignetting occurs at the second lens, all rays are captured in the final image. An example of this can be seen in the classical Köhler projector shown in figure 1.137, where the plane of the condenser is imaged onto a screen. There, the condenser bends rays into the projection lens, thereby avoiding any vignetting at the projection lens. It also so happens that the projection lens is made to 'fit' the size of the image of the source formed by the condenser, resulting in a full transfer of all rays captured by the condenser. If the EFL of the projection lens is maintained and the screen location is shifted to the right, then a different plane is imaged at the screen. If a first and second lens system is set up to be telecentric in object and intermediate image spaces, and if a third lens projects the intermediate image towards a screen, then its diameter may be maintained at a reasonable size, as illustrated in figure 3.19. In this case, if the third lens is to be designed on its own (but it is intended to be used with the double-telecentric first/second lens pair), then its system aperture must be set to receive rays from the telecentric lens pair in an appropriate manner (e.g. in OS,

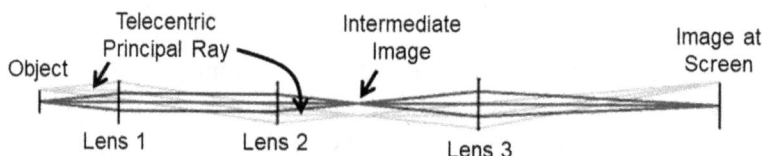

Figure 3.19. An object-image telecentric lens pair with a third lens relay.

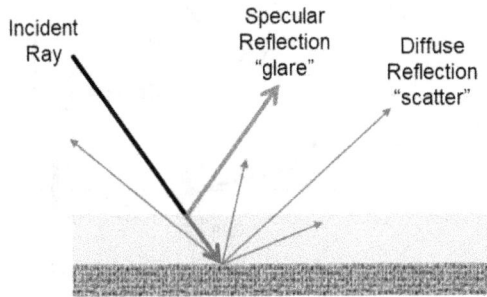

Figure 3.20. Sources of rays from a 'glossy' surface (such as a magazine cover).

instead of setting *Float by Stop Size*, set the system aperture to *Telecentric* with an appropriate object space NA).

If the intermediate plane is a mirror, the concepts above still apply, as a mirror is just a reflective form of the transmissive surface. However, the reflective intermediate plane is sometimes a physical surface, such as perhaps the polished surface of a tabletop, or even a glossy magazine cover. The situation could be that one illuminates the cover of a magazine, or a smooth tabletop, or some other shiny surface for the purpose of taking a photograph of the surface. In such cases, there are two sources of rays at the illuminated surface, one from specular reflection (glare), and the other from diffuse reflection, as illustrated in figure 3.20.

Glare may be considered to be a form of stray light. In figure 3.20, the glossy–shiny portion that gives rise to the glare can be thought of as an upper layer of homogeneous material (e.g. on a smooth tabletop, it may be the polish or varnish). Usually, this 'layer' is a material capable of 'filling in' spaces between texture or particles (or fibers in paper), which are, of course, the source of the diffuse reflection. It is the diffuse portion of the surface that one would normally be trying to image. Evidently, the glare results in an image of the source when both specular rays and diffuse rays are captured by a lens and camera. To get rid of the glare, two linear polarizers (one at the source, the other at the lens) at orthogonal orientations may be used, but this obviously cuts down the image brightness significantly. Alternatively, the glare may be avoided altogether if a source is located beyond a *glare radius* [58], as illustrated in figure 3.21. The idea is to reverse trace the chief ray (the principal ray, if there is no vignetting) from the lens towards a height above the lens's axis. This height R is the glare radius. Since the angle θ is the half-FOV of the lens, anything lying outside it cannot be seen by the lens, unless the FOV is a *soft boundary* (e.g. if it defines a point in the field at which the relative illumination is 50%). Just make sure that the source is mounted sufficiently far from R!

3.8.2 The eternal challenge of stray light analysis and control

Another cause of stray light is implied by the flood of rays from an ambient scene, as shown in figure 1.3(a). As mentioned in section 1.1.1, the stray light is not obvious because sequential ray tracing layouts only show the rays that make it through the

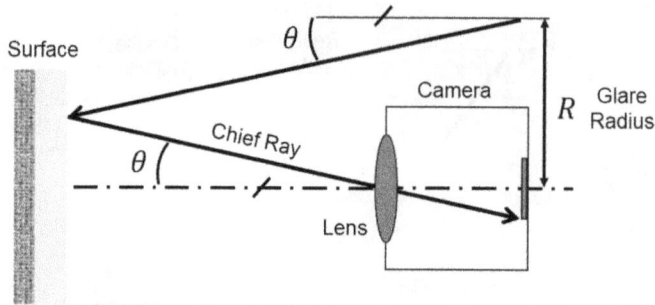

Figure 3.21. Mounting a source beyond a *glare radius* to avoid glare.

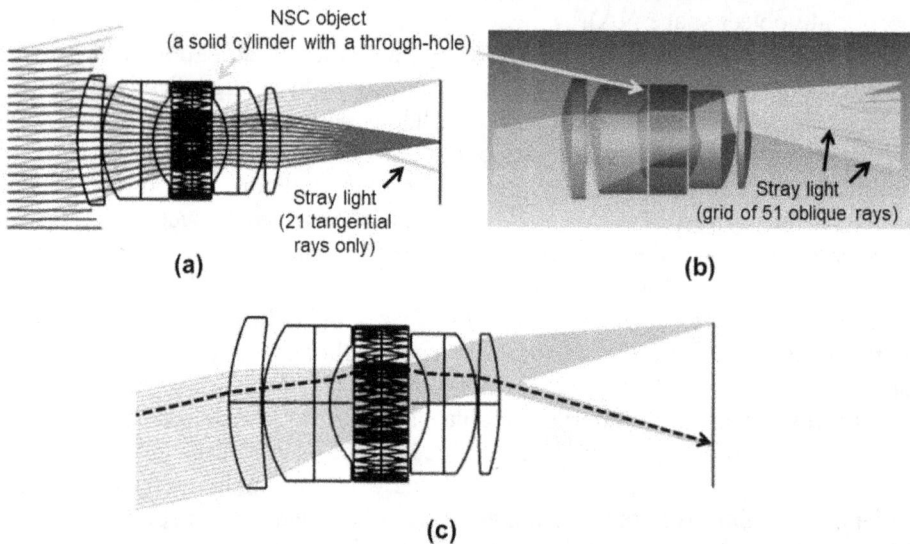

Figure 3.22. Stray reflected rays from a reflective spacer cylinder used as the stop. (a) An HNSC layout that displays only 21 rays per field in the tangential plane. (b) An HNSC shaded layout displaying a grid of 51 rays for the oblique field. (c) Layout with only the reflected stray ray.

stop. As an exaggerated example, we can apply the hybrid NSC (HNSC) feature to the layout in figure 1.3(a), in which the stop is made into a specular reflective NSC cylinder object such that rays reflect off the internal cylinder of the stop, as shown in figure 3.22. Of course, in reality, the stop should be the smallest aperture in the cylinder. Here, however, we have purposely exaggerated the spacer model to illustrate the consequence of out-of-field rays reflecting off the spacer.

If we display only the tangential (Y–Z plane) rays, the magnitude of the reflected stray light is not made apparent, especially if the number of rays traced is not large (figure 3.22(a)). If the ray number is increased and a grid of those rays is used, the extent of the reflected stray light is made apparent (figure 3.22(b)). Finally, if one deletes all the vignetted oblique rays, the source of the reflected stray light is revealed (figure 3.22(c)). The presence of the stray rays suggests that they are always present if the actual object you are using to create the stop in reality is anything close to what is being modeled here. Hence, the more absorptive the material, the less the stray light. Or, if the internal surface of the cylinder used for the stop is black and threaded (if it is in the form of vanes in baffles, for example), then the stray light is further reduced. Vanes in baffles are useless if their surfaces are diffuse. The idea of using vanes in baffles is to let rays undergo multiple reflections such that, on each reflection, rays suffer losses through partial transmission into an absorptive layer. Think of the two-layer model in figure 3.22, in which the lower layer is blackened material.

Figure 3.22 also indicates that there is some advantage in performing ray trace analysis in a mixed HNSC mode. Sometimes, it may not be necessary to apply scatter properties to all surfaces and perform an entire NSC ray trace analysis (including ray tracing from the image plane to the source). Of course, this ultimately depends on the needs of the application, such as how crucial it is to model the scattering properties of surfaces and whether the magnitude of scatter is important. However, if you feel that you need to model the scattering properties of surfaces, you may want to think about how many surfaces you need to apply these properties to: every surface? Including glass? It would take you a long time to perform this task. A more practical approach is to specify certain finishes on the mechanical parts, then build a real-life prototype of the system in order to obtain measurements of the performance in the presence of any stray light, as such measurements may be used as the reference design (often called the *baseline* design). From the baseline, one may scale the results of future ray trace predictions.

In some cases, it may be necessary to obtain *bidirectional reflectance distribution functions* (BRDFs) to model the scattering properties of surfaces [59]. Think of the BRDF as a generalized measure of reflectance in that it also accounts for scatter. The BRDF is defined by the output radiance in some direction (given by rotations about X and Y in cartesian space—hence, the term 'bidirectional') divided by an input irradiance in a specified direction [59]. In the limit of a Lambertian surface, the BRDF is numerically equal to the fraction of scattered light divided by π [59, 60]. Heuristically, this is because, for a Lambertian surface, the incident irradiance may be thought of as $\pi L \sin^2(90°)$, and the output radiation is just L times the fraction of total scattered light. Thus, if the fraction of total scattered light is ρ, then the BRDF is simply $\rho L/[\pi L \sin^2(90°)] = \rho/\pi$. Today, some researchers are finding ways to simplify scatter analysis by limiting the number of BRDFs required to model wide cases of scattering properties, which they hope will make scatter analyses more efficient [61]. An in-depth discussion of the BRDF is beyond the scope of this book, so the reader is directed to the references cited above.

3.8.3 Ghosts and 'narcissists' (I mean narcissus) effects

A *ghost image* is simply the secondary, tertiary, or multiply produced images of the object, as illustrated in figure 3.23 (refraction has been neglected, for simplicity). Of course, suitable AR coatings minimize the intensity of multiply reflected rays. On the

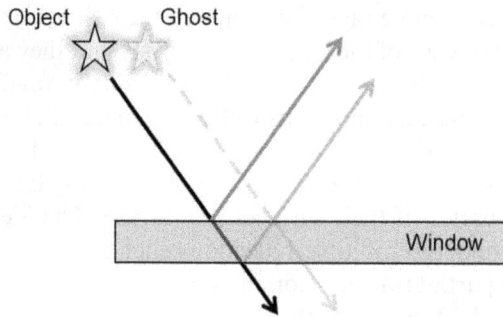

Figure 3.23. Ghost images of a source produced by multiple reflections.

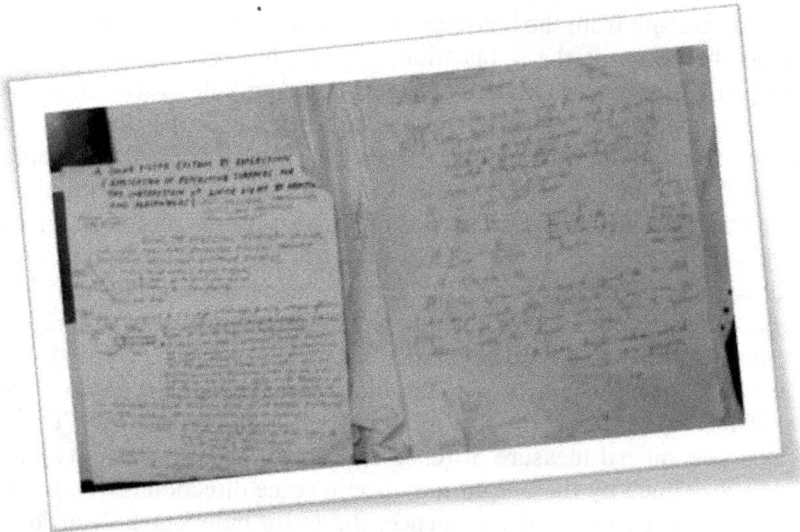

Figure 3.24. The author's (1995) illegible high school concept scribbles for producing a solar filter using ray reflections between a glass slab and mirror. Courtesy of Ronian Siew.

other hand, multiple reflections from non-AR coated glass slabs may be used as neutral density filters, provided that the surfaces of the glasses are as flat as possible. An example is provided in figures 3.24 and 3.25, where in a personal high school experiment, the author attempted to create a solar filter using a glass slab and a mirror with a wedge angle in between.

Of course, in the author's experiment, the glass and mirror surfaces were not sufficiently flat for high-quality imaging. The glass was from a bookcase in my room, and the mirror was from my parents' bathroom. The wedge was created by using a large office binder clip, so that there was some cantilevering between the glass and mirror, possibly resulting in additional distortion of the surfaces of the glass and mirror. Nevertheless, the concept demonstrates the neutral density filtering effect achievable using glass slabs.

Figure 3.25. The author's high school experimental setup (left), and the result (right). Courtesy of Ronian Siew.

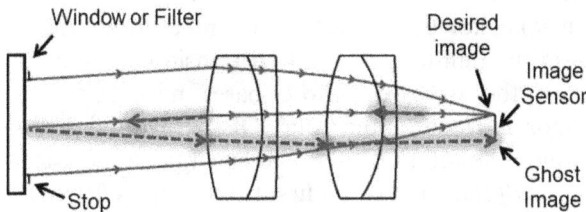

Figure 3.26. A ghost image of the object formed by retro-reflection from a near-telecentric principal ray at the window of an image sensor.

To find ghost images, a rule is that whenever rays are collimated (either partially or fully) and pass through windows and filters, look out for ghosting in the final image plane, such as that illustrated in figure 3.26 for the case of a filter located near the stop of an imaging lens that is almost telecentric in image space. At the sensor of such a lens, the principal ray may bounce and reverse trace towards the other side of the sensor, resulting in a secondary image (a ghost) of the object.

In some cases, when the image sensor itself is visible in the image, this is a ghost reflection of rays from the image sensor off one or more surfaces of lens elements and windows in the lens system. This phenomenon—known as the *narcissus* effect—is more common in cooled infrared imaging systems, due to the lowered temperature of the image sensor (which is actively cooled by some means, such as Dewars, thermoelectric (Peltier) coolers, or a *Joule–Thomson* cooler), relative to the *hotter* image from the object [62, 63]. It is sometimes considered to be a special case of *veiling glare*, which is a general term used in visible-light optical systems for stray light originating from any reflections off surfaces, whose light is not image-forming at the image [64]. Thus, veiling glare reduces image contrast. The stray light shown in figure 3.22 may therefore be considered a form of veiling glare. Since veiling glare can originate from non-ghost-forming reflections, in order to minimize it, one should not only blacken lens mounts and the internal surfaces of lens housings but also minimize dust and scratches on optical surfaces that may scatter rays.

3.8.4 The spectral 'blueshifts' of thin-film filters and beam splitters

Thin-film filters and dichroic beam splitters have a characteristic nuance of shifts to shorter wavelengths in their transmission spectra as a function of increasing angles of incidence (AOI) [65–67]. To understand why, we need only to derive the formula for interference between two reflected rays, using the construction in figure 3.27. The interference is

$$I_{\text{tot}} = I_1 + I_2 + 2\sqrt{I_1 I_2} \cdot \cos\left(\frac{2\pi}{\lambda}\text{OPD}\right), \tag{3.43}$$

where OPD is the optical path difference between the two rays, λ is the wavelength *outside* the thin film, and I_{tot} is the total intensity (the time-averaged square of the electric field amplitude). From Snell's law, $\sin\theta = n\sin\theta'$, where θ' (not shown) is the refracted angle of the incident ray in the thin film and n is the index of the thin film. In the time it takes for the second ray to travel inside the thin film until just prior to its exit from the thin film, its total OPL inside the thin film is n times twice the hypothenuse of the triangles whose bases make up length s. Since the hypothenuse's length is $t/\cos\theta'$, the total OPL traversed by the second ray is $2nt/\cos\theta'$. Meanwhile, the total OPL for the first ray, relative to the second ray, is $s(\sin\theta)$. But $\tan\theta' = s/(2t)$, so that the first ray's OPL is $2t\tan\theta'\sin\theta$. Therefore, the OPD between the rays is

$$\text{OPD} = \frac{2nt}{\cos\theta'} - 2nt\tan\theta'\sin\theta. \tag{3.44}$$

Applying Snell's law and $\tan\theta' = \sin\theta'/\cos\theta'$ to equation (3.44), we obtain

$$\text{OPD} = \frac{2nt}{\cos\theta'} - 2nt\frac{\sin^2\theta}{n\cos\theta'} = \frac{2nt}{\cos\theta'}\left[1 - \frac{\sin^2\theta}{n^2}\right]. \tag{3.45}$$

Now, applying $\cos\theta' = \sqrt{1 - \sin^2\theta'}$ and also Snell's law, equation (3.45) becomes

$$\text{OPD} = \frac{2nt}{\sqrt{1 - \frac{\sin^2\theta}{n^2}}}\left[1 - \frac{\sin^2\theta}{n^2}\right] = 2nt\sqrt{1 - \frac{\sin^2\theta}{n^2}}. \tag{3.46}$$

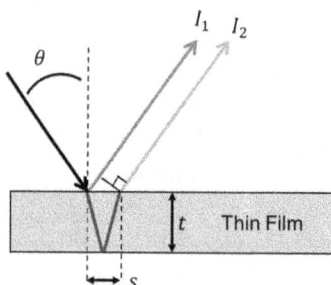

Figure 3.27. Two 'rays' (paths orthogonal to reflected plane waves) interfering upon reflection from the surfaces of a thin film.

Substituting equation (3.46) into (3.43), we obtain

$$I_{\text{tot}} = I_1 + I_2 + 2\sqrt{I_1 I_2} \cdot \cos\left(\frac{4\pi n t}{\lambda}\sqrt{1 - \frac{\sin^2\theta}{n^2}}\right). \tag{3.47}$$

At $\theta = 0$, the cosine factor in equation (3.47) becomes equal to $\cos(4\pi n t/\lambda)$, so that I_{tot} possesses some fixed value. At any other AOI greater than zero, this value of I_{tot} is maintained if there is a different wavelength λ' for which the cosine factor is made to equal its value when $\theta = 0$. This means that we must have

$$I_1 + I_2 + 2\sqrt{I_1 I_2} \cdot \cos\left(\frac{4\pi n t}{\lambda}\right) = I_1 + I_2 + 2\sqrt{I_1 I_2} \cdot \cos\left(\frac{4\pi n t}{\lambda'}\sqrt{1 - \frac{\sin^2\theta}{n^2}}\right). \tag{3.48}$$

Therefore, equating the cosine factors on both sides of equation (3.48) and solving for λ' gives us the relation

$$\lambda' = \lambda\sqrt{1 - \frac{\sin^2\theta}{n^2}}. \tag{3.49}$$

Equation (3.49) says that, in a single-layer thin film on a substrate, a *reflectance* at wavelength λ is shifted to a *smaller* value λ' when $\theta > 0$. For this reason, the shift is often called a *blueshift* [68]. Since the same OPD geometry applies to transmitted rays, equation (3.49) also holds true for the thin film's transmittance. There is no similar formula that can be written in closed form for a stack of multilayer thin films on a substrate, but their behavior is the same as if an *effective refractive index* n_E is associated with the stack of thin films [65–67]. How do filter suppliers obtain this n_E? They generally simulate the reflected or transmitted spectrum for the filter at various AOIs, then fit a curve for λ' vs θ at the best value for n_E that allows equation (3.49) to match the curve. In this way, we can insert n_E into equation (3.49) and write the expression

$$\lambda' = \lambda\sqrt{1 - \frac{\sin^2\theta}{n_E^2}}. \tag{3.50}$$

The shift in spectrum is therefore given by the difference $\lambda' - \lambda$, and it is seen to scale with wavelength. When you are selecting thin-film filters and beam splitters for an application, you need to consider the impact of rays in your system that may become incident on the filter at non-normal AOIs. So, you need to request their effective refractive indexes. Some suppliers provide these values in their online catalog. Others provide them upon request. Note that equation (3.50) only applies to collimated beams passing through tilted filters. If a beam has a cone angle and its central ray is normally incident on the filter, then to a good approximation, you can take the average between the λ' for which $\theta = 0$ and the λ' for which the cone's half-angle is θ [65]. Alternatively, if you have the spectral transmittance $\tau(\lambda)$, then assuming a simplified case of 100% peak transmittance and 0% out-of-band transmittance (for the purpose of visualizing just the spectral shift effect), you may account for the cone angle by first modeling $\tau(\lambda, \theta)$ by way of a formula, such as perhaps a *super-Gaussian* function expressed as

$$\tau(\lambda, \theta) = \exp\left\{-\left[\frac{\lambda(\theta)}{w}\right]^N\right\}, \tag{3.51}$$

where N is any even integer greater than zero, w is a suitable half-width, and $\lambda(\theta) = \lambda - \lambda_c(\theta)$, where, in applying equation (3.50), $\lambda_c(\theta)$ is given by

$$\lambda_c(\theta) = \lambda_c\sqrt{1 - \frac{\sin^2\theta}{n_E^2}}, \tag{3.52}$$

where λ_c is the central wavelength of the spectral transmittance. If we want a half-width at which the transmittance is 50% of its peak, then let $\tau = 0.5$ and $\lambda(0) = $ HWHM in equation (3.51), where HWHM is the *half-width at half-maximum*, which is the half-width at 50% of the peak transmittance. Applying the above in equation (3.51) and solving for w, we obtain

$$w = \frac{\text{HWHM}}{(-\ln 0.5)^{1/N}}. \tag{3.53}$$

Substituting equation (3.53) into (3.51) yields

$$\tau(\lambda, \theta) = \exp\left\{-\left[\frac{\lambda(\theta)}{\text{HWHM}}\right]^N [-\ln(0.5)]\right\}. \tag{3.54}$$

On its own, equation (3.54) may be used to model the spectral transmittance subject to shifts due to θ for collimated rays passing through the filter. But if there is a cone angle to the rays and the central ray is normally incident to the filter, then we can multiply the right-hand side of equation (3.54) with the right-hand side of equation (2.5) and then integrate to obtain the total spectral transmittance $\tau(\lambda)_{\text{tot}}$ as follows:

$$\tau(\lambda)_{\text{tot}} = \frac{\int_0^\theta 2\pi L dA \tau(\lambda, \theta) \sin\theta \cos\theta d\theta}{\pi L dA \sin^2\theta}$$

$$= \frac{\int_0^\theta \tau(\lambda, \theta) \sin(2\theta) d\theta}{\sin^2\theta}, \tag{3.55}$$

where θ is the *half-angle* of the full cone of rays incident on the filter (whose plane is normal to the central ray in the cone), as depicted in figure 3.28:

As examples, figure 3.29 shows plots of equation (3.54) for the case of a collimated beam at AOI = 20 degrees and at 0 degrees incident to a bandpass filter, for which $n_E = 1.7$, HWHM = 15 nm, $N = 20$, and $\lambda_c = 550$ nm. Figure 3.30 shows plots of equation (3.55) for the case of a cone at a half-angle of $\theta = 20$ degrees and at zero degrees.

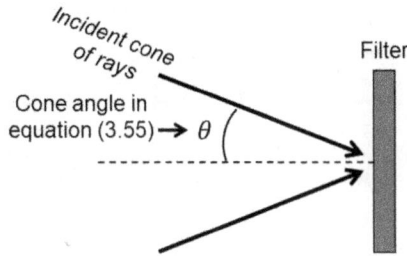

Figure 3.28. A cone of rays incident on a filter.

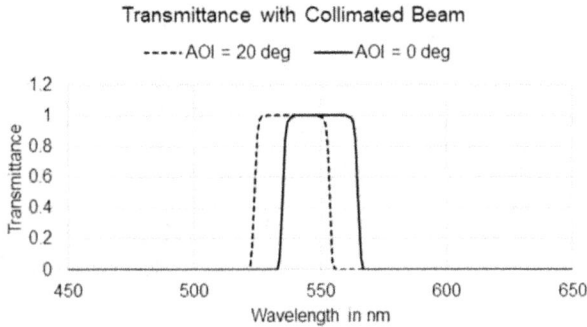

Figure 3.29. A plot of equation (3.54) for two different AOIs.

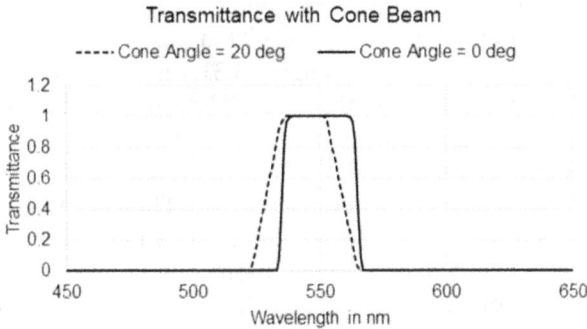

Figure 3.30. A plot of equation (3.55) for two different cone angles θ.

Both plots in figures 3.29 and 3.30 were constructed using Microsoft's Excel® spreadsheet. If you are wondering how to calculate the integral in equation (3.55) using a spreadsheet (supplied as the file named *MCOSD Blue-Shift Transmittance v1.1* at this book's website), one creates wavelengths in a column, while incremental cone angles are created in a row. The idea is to perform the numerical integration $\int_0^\theta \tau(\lambda, \theta) \sin(2\theta) d\theta = \sum_{i=1}^n \tau(\lambda, \theta_i) \sin(2\theta_i) \Delta\theta$ across a row and then finally plot

	A	B	C	D	E	F	G	H	I	J	K	L	M	N	O
2	-	(deg)	(rad)	(nm)	(nm)	(nm)	-	(deg)	(rad)	(deg)	(rad)	-			
3	Eff. Index	AOI	AOI	Lambda c	Lambda c of theta	HWHM	N	Cone Angle	Cone Angle	Delta Angle	Delta Angle	sin^(Cone Angle)			
4	1.7	20	0.349066	550	538.75392	15	20	20	0.349066	2	0.034907	0.128149			
5															
6															
7		Collimated													
8	Lambda	T		Deg	0	2	4	6	8	10	12	14	16	18	20
9	300	0		Rad	0	0.034907	0.069813	0.10472	0.139626	0.174533	0.20944	0.244346	0.279253	0.314159	0.349066
10	301	0		Rad	0	0.034907	0.069813	0.10472	0.139626	0.174533	0.20944	0.244346	0.279253	0.314159	0.349066
11	302	0		Rad	0										
12	303	0		Rad	0										
13	304	0		Rad	0										
14	305	0		Rad	0										
15	306	0		Rad	0										
16	307	0		Rad	0										
17	308	0		Rad	0										
18	309	0		Rad	0										
19	310	0		Rad	0										
20	311	0		Rad	0										
21	312	0		Rad	0										
22	313	0		Rad	0										
23	314	0		Rad	0										
24	315	0		Rad	0										
25	316	0		Rad	0										
26	317	0		Rad	0										
27	318	0		Rad	0	0.034907	0.069813	0.10472	0.139626	0.174533	0.20944	0.244346	0.279253	0.314159	0.349066

Transmittance with Collimated Beam

----- AOI = 20 deg ——— AOI = 0 deg

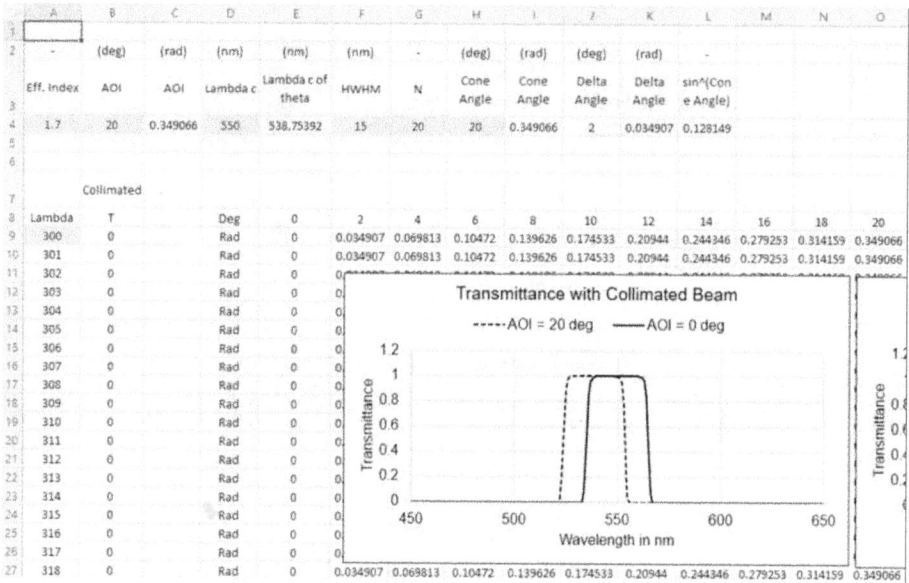

Figure 3.31. The Microsoft Excel® worksheet used to plot figures 3.29 and 3.30.

$\tau(\lambda)_{tot}$ in a last column. For example, figure 3.31 shows a partial screenshot from the worksheet I used, where all of the constants are inputs on row 4. Cell E4 computes the shifted central wavelength using equation (3.52). Cells in yellow (by personal habit) are inputs, while all white cells with numerical values contain computations. Evidently, the plot that is wholly visible in figure 3.31 is for the case of a collimated beam and was what I used to display figure 3.29 in this book, while the plot used for figure 3.30 has been cut off from view in figure 3.31, due to the limited space in which to display the entire spreadsheet on this page. The values at cell B9 and below are used to plot figure 3.29. As for figure 3.30, notice that cells F8 to O8 have values 2, 4, 6...20. These are incremental angles $\theta_i = i\Delta\theta$ for the cone in steps of $\Delta\theta = 2$ degrees, ending at the half-cone angle of $10\Delta\theta = 20$ degrees. This means that I only used ten incremental angle steps (i.e. $n = 10$ in the summation above) to perform the integration in equation (3.55). This integration is done in a column all the way over to a far-right portion of the spreadsheet (not shown here), at column AX. The reason is that, first, columns F to O are used for listing the incremental angles (in radians). In columns Q to Z (not shown), I listed the values for the sine in equation (3.55) computed as $\sin(2i\Delta\theta)$. In columns AB to AK, I listed the values for $\tau(\lambda, i\Delta\theta)$. In columns AM to AV, I listed the products $\tau(\lambda, i\Delta\theta)\sin(2i\Delta\theta)$. Finally, in column AX, I computed $\sum_{i=1}^{10}\tau(\lambda, i\Delta\theta)\sin(2i\Delta\theta)\Delta\theta$. Thus, in cells AX9 and below, I plotted figure 3.30. Note, however, that the value for $\sin^2\theta$ in the denominator of equation (3.55) was also computed using numerical integration, given by $\sum_{i=1}^{10}\sin(2i\Delta\theta)\Delta\theta$. The reason for this is that in numerical integration, an approximation is made by virtue of summing up discrete values for $\tau(\lambda, i\Delta\theta)\sin(2i\Delta\theta)$. As such, the

$$\lambda := 450, 451 .. 650 \qquad ne := 1.7 \qquad HWHM := 15 \qquad N := 20 \qquad \Theta(deg) := deg \cdot \frac{\pi}{180}$$

$$\lambda c := 550 \qquad \lambda cc(\theta) := \lambda c \cdot \sqrt{1 - \left(\frac{\sin(\theta)}{ne}\right)^2}$$

$$\tau(\lambda, \theta) := e^{-\left(\frac{\lambda - \lambda cc(\theta)}{HWHM}\right)^N \cdot (-\ln(0.5))}$$

$$\lambda cc(\Theta(20)) = 538.754$$

$$T(\lambda, CA) := \frac{\int_0^{\Theta(CA)} \tau(\lambda, \theta) \cdot \sin(2 \cdot \theta) \, d\theta}{(\sin(\Theta(CA)))^2}$$

$\tau(\lambda, 0)$

$\tau(\lambda, \Theta(20))$

$T(\lambda, 1)$

$T(\lambda, 20)$

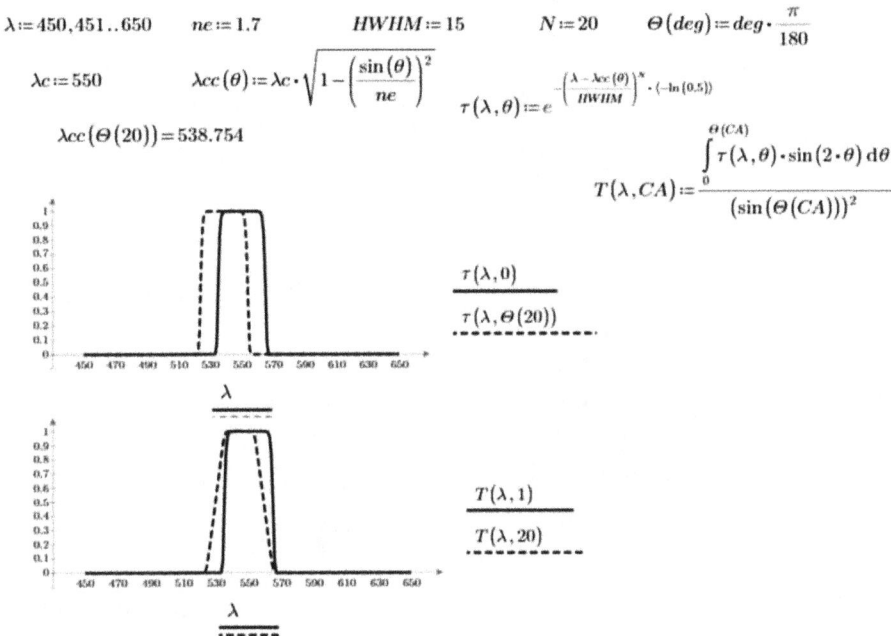

Figure 3.32. Plots of figures 3.29 and 3.30 obtained using PTC Mathcad® Prime (version 9).

denominator in equation (3.55) must also be discretized into a sum, for otherwise, the ratio would give a peak value greater than unity.

How accurate is the simple Riemann type of numerical integration performed using Excel®? It is quite good. Figure 3.32 shows the same plots as produced using PTC Mathcad® Prime (version 9, obtained through a 30-day free trial). Notice the similarities between the Mathcad plots and those in figures 3.29 and 3.30.

If either a collimated beam or a cone beam is incident on a dichroic beam splitter, then equation (3.52) may be modified to

$$\lambda_c(\theta) = \lambda_{cref} \sqrt{1 - \frac{\sin^2(45^0 \pm \theta)}{n_E^2}}, \qquad (3.56)$$

where λ_{cref} may be regarded as a *reference wavelength* such that when $\theta = 0$, λ_c is the wavelength of interest anywhere in the spectral transmittance of the dichroic beam splitter (e.g. it could be the cut-on wavelength for the spectrum). Thus, λ_{cref} would first need to be determined, based on a choice for λ_c, so that $\lambda_c = \lambda_{cref} \sqrt{1 - \sin^2(45^0)/n_E^2}$. Thereafter, equation (3.56) could once again be applied to equation (3.54) with the provision that one uses the definition that $\lambda(\theta) = \lambda - \lambda_c(\theta)$.

3.8.5 Optical density and the transmittance of stacked filters

Suppose we have two filters, as shown in figure 3.33. What is the total transmittance τ of an incident **incoherent** beam with intensity I_o? Due to possible multiple

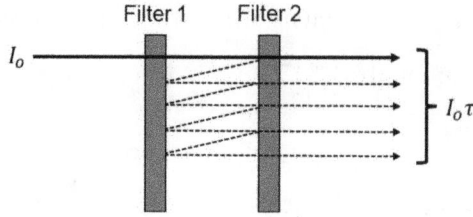

Figure 3.33. The transmittance of two stacked filters.

reflections in an air space between the two stacked filters, the total transmittance is obtained by summing all reflected and transmitted beams. Let τ_1 and τ_2 be the transmittances of the first and second filters, respectively, then assuming no absorption in the space between the filters, we obtain

$$\tau = \tau_1\tau_2 + \tau_1\tau_2(1 - \tau_1)(1 - \tau_2) + \tau_1\tau_2(1 - \tau_1)^2(1 - \tau_2)^2 + \ldots$$

$$= \tau_1\tau_2 + \tau_1\tau_2\sum_{i=1}^{\infty}(1 - \tau_1)^i(1 - \tau_2)^i. \tag{3.57}$$

Applying the rule for geometric series to the summation [69], equation (3.57) evaluates to

$$\tau = \frac{\tau_1\tau_2}{1 - (1 - \tau_1)(1 - \tau_2)}. \tag{3.58}$$

If we take the reciprocal on both sides, we obtain

$$\frac{1}{\tau} = \frac{1}{\tau_1} + \frac{1}{\tau_2} - 1. \tag{3.59}$$

You may have seen equation (3.59) in the websites, technical notes, and whitepapers of filter suppliers. In the limit of large transmittances, we let $\tau_1 \rightarrow 1$ and $\tau_2 \rightarrow 1$ in the denominator of equation (3.58), which results in the usual assumption that $\tau = \tau_1\tau_2$. On the other hand, when transmittance is low (such as in the blocking region of bandpass filters), then equations (3.58) and (3.59) should be applied, especially in the case of thin-film filters that block spectra by reflection rather than absorption. The blocking region of filters is often specified in terms of the *optical density* (OD), defined by

$$OD = -\log_{10}\tau = \log_{10}\left(\frac{1}{\tau}\right). \tag{3.60}$$

Therefore,

$$10^{OD} = \frac{1}{\tau}. \tag{3.61}$$

The definition given by equation (3.61) is rather convenient, especially when blocking transmittances are in orders of 0.1, 0.01, ... In such cases, when, for example, $\tau = 0.1$, then it is the same as $\tau = 10^{-1} = 1/10$, so that the logarithm of $1/\tau$ gives the power number, in this case, one. Thus, a filter with transmittance $\tau = 0.001$ has a blocking of $OD = \log_{10}(1/10^3) = 3$. Hence, when we stack two thin-film filters whose blocking levels are due to the rejection of light (which can be multiply reflected between the stacked filters), by way of applying equation (3.59) towards equation (3.61), the OD is

$$OD = \log_{10}\left(\frac{1}{\tau}\right) = \log_{10}\left(\frac{1}{\tau_1} + \frac{1}{\tau_2} - 1\right) = \log_{10}(10^{OD_1} + 10^{OD_2} - 1). \qquad (3.62)$$

3.8.6 Optical fibers versus free-space components for illumination

Optical fibers are quite prevalent in many applications and offer certain benefits when used to guide light for illumination [68]. However, you generally need to *couple* light into a fiber through some focusing means. Also, fibers can obviously only accept rays at angles that meet the total internal reflection (TIR) condition for propagation. For a core index of n_1 and a cladding index of n_2, the acceptance NA is [70]

$$NA = \sqrt{n_1^2 - n_2^2}. \qquad (3.63)$$

If you use fiber bundles for illumination, loss occurs in the gaps between the fibers. You can use a liquid light guide (LLG) instead, whose diameters can be 3 mm or more, but there are other nuances. For instance, all light guides that depend on TIR possess a *minimum bend radius* [68, 71]. Not only does this mean that you cannot bend them too tightly while guiding light, but it also means that such light guides do take up a minimum space given by the bend radius. Another nuance is that unless the light guide is a rectangular or non-cylindrical pipe, the global uniformity of irradiance at the exit is generally unsatisfactory [71]. One therefore has to reshape the output using auxiliary optics, such as any of the methods discussed in chapter 2. If the exit of the light guide is locally uniform, then using the relative illumination equation for a lens system, one can tune the irradiance on the ROI of a screen to be globally uniform.

One common but unobvious light guide nuance that non-optical engineers forget about is that the output rays must possess the same NA as the input rays, by way of étendue conservation. In one experience, a team desired to use a light guide to send light through a small container holding a bio-sample. The same action could have been performed using free-space components, but the team thought that a light guide would *guide* the rays through the sample nicely and would therefore do away with the need for a lens. But at the exit of the light guide, the output rays possess an NA that needs to be collected by a lens for collimation. This collimation would have been performed well by free-space optical components.

Free-space components do not have many of the drawbacks described for light guides, but they are generally not as convenient as light guides when it comes to relaying illumination over long distances. By 'free space,' we mean the use of lenses

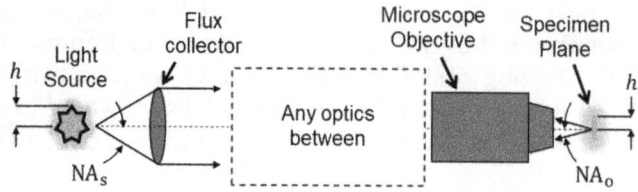

Figure 3.34. The definitions of the quantities in equation (3.64).

and mirrors to channel light from one location to another. The advantage of a light guide is that it can relay light from one end of a room to another without disturbance from ambient light. This is helpful when your source is high-powered, hot, and bulky (you can therefore also mount the source above or below your instrument and use an LLG to channel the light where you want it).

3.8.7 Fundamental limitations to illumination in microscopy

In any microscope system, by virtue of the Lagrange invariant (see section 1.1.3), the following product is constant:

$$h\text{NA}_s = h'\text{NA}_o = \text{constant} \tag{3.64}$$

(the quantities are defined in figure 3.34). **Note that in this formula, the specimen plane need not be conjugate to the source.** To understand this important fact, review the logic in section 1.1.3 that led to equation (1.5). It is the key to understanding the limits to illumination in microscopy, as I will now show.

Since the square of the Lagrange invariant is proportional to the étendue, the square of $h'\text{NA}_o$ sets the upper limit to the fraction of light from the source that is accepted by the objective. Thus, the maximum NA at the source side at which light can be coupled into the objective is given by solving equation (3.64) for NA_s. Taking 100% times the square of NA_s yields the percentage efficiency of the lighting for the microscope:

$$\text{Maximum \% flux efficiency} = 100\% \times \text{NA}_s^2 = 100\% \times \left(\frac{h'\text{NA}_o}{h}\right)^2. \tag{3.65}$$

There is no way around the above. So, as long as there are standardized fields of view (FOVs) and NAs for microscope objectives, then it follows that there are 'standardized % flux efficiencies' for coupling light into microscopes. This means that if you want brighter images at a fixed objective FOV, then you need brighter sources with smaller sizes and a higher NA for the objective. In other words, you need to maximize the numerator and minimize the denominator in equation (3.65).

3.8.8 Drift: an enemy of statistics and the reason for calibration

In real life, all manufactured parts can experience subtle to drastic changes over time, depending on the environmental conditions and external influences on the part in the course of time. A common influence is temperature. Due to these influences, the performance of an instrument—which is an integrated system of individual parts subject to changes over time—can vary over time. This is known as *drift*.

Drift is the reason that is it often necessary to perform routine calibration of an instrument. For instance, a spectrophotometer must often be *baselined*, meaning that a measurement of its transmittance is taken using a covered detector. In this condition, the presence of any mean signal greater than zero should be subtracted so that 'true' zero transmittance takes place at the zero mark in the output. If drift occurs, this zero baseline can suffer drifts and become either elevated or reduced to below the zero-transmittance value.

Another common example of drift is the output fluxes and spectra of 'hot' light sources, such as arc lamps, tungsten halogen bulbs, and even LEDs. Due to temperature variations and the natural life of the source, the emitted flux can drop over time. An LED often possesses a characteristic curve that achieves peak output in roughly the first 15 min after being turned on, followed by a gradual reduction to a *steady-state* output [72, 73]. Beyond this state, the output may still drop over time or change and be elevated, depending on the LED's junction temperature.

It is difficult to control drift. Drift also may change the statistics of parameter distributions, which is troublesome when applying parametric statistical calculations to quantities of interest, because such computations require the distributions of parameters to be stable [16, 37]. Recall from section 3.3.6 that in order to rely on statistical quantities to describe random events, those events must have repeatable histories called probability distributions. Thus, in order to compute probabilities, one needs such distributions to be stable. An earthquake might influence the output of dice throwing, but significant earthquakes (and meteors striking the Earth) are infrequent and completely out of the ordinary. Therefore, under stable conditions, the outcomes of dice throwing are governed by a fixed distribution known as the binomial distribution. But if you lived in a region of the world where slight trembles from the Earth often occur, then your dice throws might be influenced by such movements, resulting in possible drift. Thus, drift that is caused by external influences may need to be controlled through some means of isolation and perhaps calibration. On the other hand, drift that is caused by internal influences (such as temperature) may be controlled by design and provisions for feedback using sensors. For instance, in the design of optical systems influenced by temperature, it is possible to select materials and optimize lens variables such that the optical system become *athermalized* to changes in temperature [74]. In the case of the influence of temperature on the output of a source's flux, a sensor can provide feedback to the drive electronics and electrical power input to change according to the monitored flux.

3.8.9 Should manufacturing processes be easy or hard?

It is reasonable for an organization to want their product to be highly manufacturable. A simple process saves cost and time. But in the highly competitive world of modern product development, an easy manufacturing process would be easy for a competitor to replicate, even if the competitor had no knowledge of the original manufacturer's trade secrets (in cases where the manufacturing process is not published as a patent). It is therefore also not unreasonable to accept a manufacturing process that is hard, especially if the product represents a technological advancement. Besides, a new product that is based on an advanced novel technology can often be difficult to produce at the beginning. If the new product presents unique features to its customers, then perhaps challenges in its production may be worth pursuing. I am of the personal view that a new product with special features should, in fact, be hard to produce—with the provision that the challenge is *masterable* (capable of being mastered).

Anything worth pursuing is often difficult—at first. Playing basketball in the park is easy. Making it to the National Basketball Association (NBA) is hard. Very hard. But it is not impossible, because it is masterable. Over time, you become good at something by doing it over and over, until you master it. However, perhaps the caveat is that a process that is hard is only worth pursuing if it is of significant value. Recall in section 1.2.9 (and appendix A.1) that the MTF of an imaging system need not always be measured, because it is hard to measure it (if one thinks it is easy, then one ought to think hard about what it means to have a MTF computed in an optical design program that is a good match for the MTF determined through actual measurement in a stable lab bench, with all of the necessary filters and calibration steps required to ensure that the MTF being sought through measurement is the same quantity modeled in the program). In this case, if a product's performance can be qualified through a simpler quantity that more accurately reflects how the product is used, then of course, it is sensible to select the simpler quantity. But even this apparently simpler process can require ingenuity that is hard at first. Pursue the challenge of a difficult but worthy process. Master it so that only your organization can achieve the result.

3.8.10 Optical design for robustness

An optical system product can be made robust in different ways. It can be made rugged for reliability, or it can be made less sensitive to production tolerances, or both. Sometimes, even without involving the issue of tolerances, a nominal design may be considered robust if its performance is unaffected by certain influences (such as in the case of an athermalized optical design [74]). In some other cases, the choice of components or optical architecture can make the design robust to certain conditions. For instance, an optical system architecture may be developed in which the filter mount is located in a space that has collimated rays, thereby making the filter's spectral transmittance robust to ray angles. Or perhaps specific aberrations may be imposed on an optical imaging system such that the image focus becomes approximately invariant with respect to the object distance, yielding extended depth of field and/or focus (see section 1.2.6).

In lens design, as discussed in section 1.2.6 (and noting the cited references there), elements may be *desensitized* to production tolerances by way of optimizing surface radii and thicknesses such that aberrations are minimized without high degrees of cancellation among surfaces. This means that the desensitized optical design is achieved through the use of a merit function during design optimization. A rule is to minimize the AOIs of rays at each surface, because the magnitude of aberration contributed by a surface is a function of the AOI. The higher the AOI, the larger the aberration. It is often cumbersome to set merit function operands to minimize these AOIs, so experienced designers often develop special algorithms to perform this task. OS provides two merit function operands as choices for use in desensitizing an optical imaging system. An example is discussed in appendix C.1.

3.8.11 Innovation tip—ask the question: 'How bad is it?'

Many organizations either develop unique methodologies to stimulate innovation or adopt specific techniques that they feel have proven effective for innovation, such as the method known as TRIZ (a Russian acronym for the *Theory of Inventive Problem Solving*) [33, 75]. Everyone therefore has their favorite innovation process, and it is no different for this author. What I often find useful in problem solving and developing new products is to ask how bad it is going to be if I tweak a certain design variable or if I add a component that is usually not meant for a specific condition. One example of this can be seen in the use of a single illumination subsystem at an oblique angle [76]. Other examples included the solar filter I attempted in high school (see figure 3.25) and the LED beam-shaping example in section 2.18.2.

The idea of asking 'How bad is it?' can be justifiably traced to moments in history when certain technologies were invented. Conceivably, when airplanes were first invented, someone surely asked how bad it would be to lift heavy metal using speed and the consequential differential in air pressure generated between two surfaces. The results were surely bad at first, with many failures. But eventually, the technique was mastered. Someone might have also asked how bad it would be to fit a phone into a pocket. And perhaps someone might have asked 'how bad' it would be to send people to the Moon—and it worked. By logical induction, I do believe that asking 'How bad is it?' is an effective way to innovate. If a crazy idea is not too bad on a first trial, then maybe it can be mastered and improved until it actually works.

3.9 Simple conceptual case studies

3.9.1 A compact optical system for virus detection

Let us take some of the principles and concepts discussed in this book and put them into the context of optical system product development. Suppose we are part of a core team in a company whose direction is to develop a compact instrument for virus detection, which is to be used as a *point-of-care* (POC) device [77]. This means that it should be a small portable instrument that medical scientists can use outside the laboratory to test for the presence or absence of specific viruses in patient samples. No one knows how compact the instrument can be, but since the team is in the early stage of product ideation and concept development (see figure 1.1), it is 'ok' to brainstorm some crazy ideas.

Evidence of the presence of a virus in a human body may be detected by several means. One may sense the presence of antibodies, antigens, or specific nucleic acid

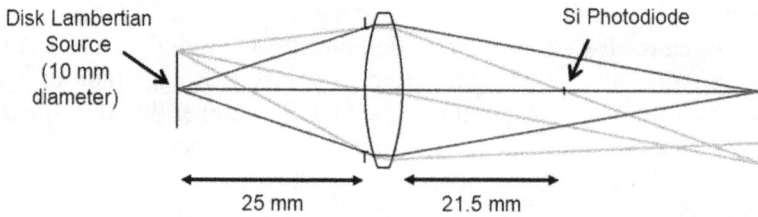

Figure 3.35. A plausible single-well virus detection system optical layout.

sequences in fragments of the virus's RNA (using the PCR method) or even determine the entire sequence (gene sequencing) of the virus's genetic signature [77]. Out of these, gene sequencing costs the most, but it is also the most accurate method. During the COVID-19 pandemic that began in 2020, many antigen tests were developed, which have been quite effective, and also not very costly. Since the PCR method is less costly than sequencing and is also more accurate than antigen tests, let us suppose that we wish to develop a compact virus detection system based on the method of *real-time* qPCR. The method was discussed in some detail in section 3.2, and we first encountered the optical layout of an analytic real-time qPCR instrument in figure 1.2, which is generally not a compact instrument. What can be done to make it smaller?

Applying the innovation tip from section 3.8.11, perhaps we may ask ourselves how bad it would be to convert a virus detection instrument of the form illustrated in figure 1.2 into a much smaller version of it by way of eliminating certain parts of the instrument. For instance, how bad would it be to have just one well rather than a plate of wells? The consequence would be that we would no longer have replicated samples leading to a statistically accurate result. But we also note that rapid antigen tests also do not have replicated tests—unless you simply take another test. Therefore, perhaps we could apply the same process to qPCR testing. That is, if we want to perform another PCR test, then just do another PCR test. One way to obtain a compact optical system with just a single well is to recall from section 2.11 that there is a position of peak irradiance (POP) behind a lens (see figures 2.50 and 2.51). As a starting design concept, we can propose the layout shown in figure 3.35, whose prescription is provided in figure 3.36. The singlet in this figure is a COTS component from Edmund Optics, (part # 63-614), and the Silicon (Si) photodiode is a small sensor with a size of 1.1 × 1.1 mm, manufactured by Hamamatsu (part # S1226-18BK). The ray trace is monochromatic, at 550 nm.

The rationale for the layout and prescription above is as follows: recall from figure 1.2 and section 3.2 that a *well* is a reaction vessel (very much like a test tube in chemistry). Fluids and other reagents are added to the well in order to test for the presence or absence of viral DNA in samples of blood, saliva, etc. taken from a medical patient. Let us suppose that viral DNA is placed into a single well homogeneously filled with PCR reagents, so that fluorescent light may be emitted from the reagent's probes (dye molecules) located everywhere in the fluid during the PCR process. In this case, the surface of the well may appear as if it is a Lambertian

	Surface Type	Comment	Radius	Thickness	Material	Coating	Clear Semi-Dia	Chip Zone
0	OBJECT Standard ▾		Infinity	25.000			5.000	0.000
1	STOP Standard ▾		Infinity	0.000			8.000 U	0.000
2	(aper) Standard ▾		63-614 30.250	5.000	N-SF11		10.000 U	0.000
3	(aper) Standard ▾		-30.250	21.500			10.000 U	0.000
4	Standard ▾	POP	Infinity	26.500 P			0.550 U	0.000
5	IMAGE Standard ▾		Infinity	-			9.409	0.000

Figure 3.36. The prescription for the layout shown in figure 3.35.

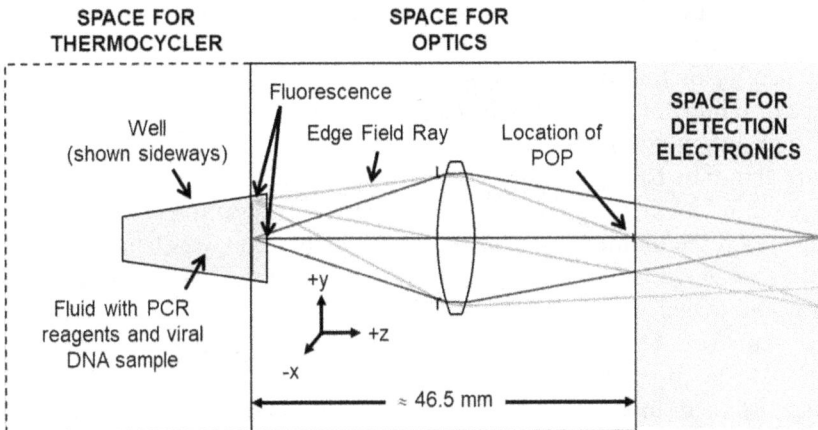

Figure 3.37. A conceptual layout with wells and spaces for other components.

disk source, which is what the layout in figure 3.35 is assuming. Pictorially, this condition is illustrated in figure 3.37, where the well is shown sideways (note that in reality, $+z$ points vertically). The manager of our core team asks, 'Why is the photodiode in figure 3.36 located between the lens and the image, and why is it so small? Shouldn't the sensor be made larger so as to collect all the flux from the emitted fluorescent light?' The answer is that, based on our knowledge of section 2.11 in this book, we know that a POP exists between the lens and the image, which is roughly determined by the intersection of the edge field ray with the lens's axis. Therefore, we can concentrate flux at the POP to (hopefully) obtain a high signal and release space behind the POP to mount electronics for a small sensor.

The size of the well is assumed to be that of a typical so-called *48-well microtiter plate*. The wells in such plates are known to typically possess the shape of a cone that is roughly 10 mm in diameter at the top, 9 mm (or more) at the bottom, and roughly

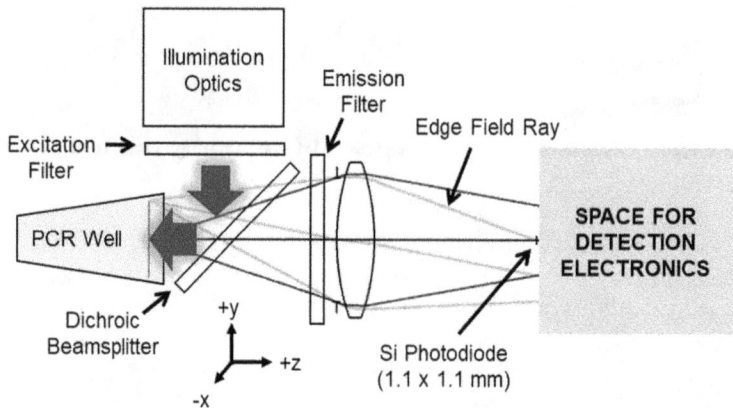

Figure 3.38. A conceptual layout for fluorescence excitation.

18–20 mm tall. For this reason, the well in figure 3.37 is depicted as a cone lying sideways. The next thing to consider is how to illuminate the reagents so as to excite fluorescence. To do this, one could apply a conceptual layout as shown in figure 3.38. Any type of luminaire design may be considered for the illumination optics, as long as the beam output floods the entire volume of the well. The *excitation* and *emission* filters are thin-film bandpass filters, and the dichroic beam splitter should have a multilayer thin-film coating with long-pass spectral transmittance in the green and beyond, so that the illumination can be provided at a blue wavelength, such as that produced by a blue LED. Many dyes can be made to absorb in the blue and emit in the green and longer wavelengths.

Given that the filters in the layout of figure 3.38 are of the multilayer thin-film type, all of the nuances described in section 3.8.4 should be considered in implementing them in the compact virus detection system. However, consider that since the small Si photodiode only receives a fraction of the rays from the well, it may not necessarily be the case that the total spectral transmittance of cone beams is of great importance. For example, out of all the rays arriving from the edge of the well, only the edge field ray strikes the sensor. This ray does not have a high AOI with respect to the emission filter, neither does it depart very far from the nominal 45-degree angle that is required to be incident onto the dichroic beam splitter. At field heights lower than the point of origin of the edge field ray, the AOIs of rays that pass through the beam splitter (which eventually strike the sensor) are even closer to their required nominal 45 degrees.

A similar argument can be made for rays that pass through the emission filter and reach the sensor—their AOIs at the emission filter are less than the AOI of the edge field ray. To see this, we have already noted that the only ray striking the Si photodiode from the full field point is the edge field ray, while we also note that the only ray striking this sensor at the axis of the well is the axial ray. Therefore, in between these two field points, rays striking the Si photodiode must have AOIs at the emission filter that are less than the AOI subtended at that filter by the edge field ray.

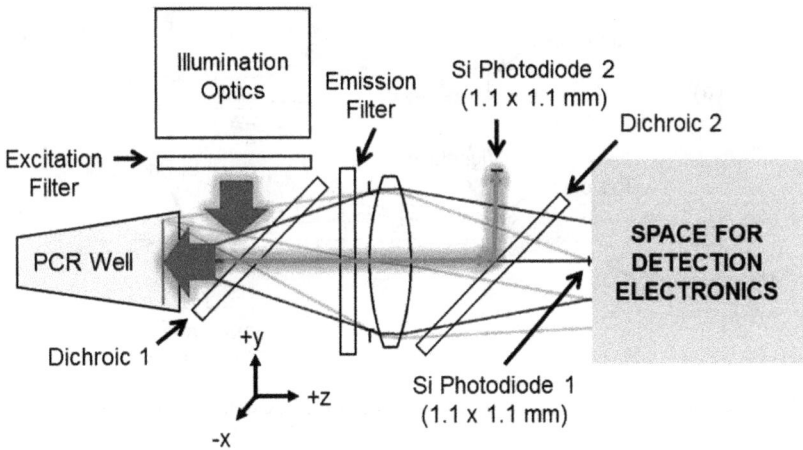

Figure 3.39. A plausible concept for a second fluorescence detection channel.

Accordingly, these rays also have AOIs closer to 45° at the dichroic beam splitter. It may also be possible to include an additional beam splitter and photodiode, as indicated by *Dichroic 2* in figure 3.39. In the FRET process [7], fluorescent probes may be designed to be excited by a blue wavelength and emit at longer wavelengths than usual. Therefore, two probes can be used in the same PCR well to detect different viruses, one emitting in the green, the other emitting in the red, the latter being detected by the sensor labeled as Si Photodiode 2 in figure 3.39. However, the risk here is in considering the amount of stray light from the red leaking through the second dichroic onto the first photodiode in the straight path. This is known as *spectral crosstalk*. Clearly, this nuance must be studied during the actual design, development, and prototyping phases of the project.

Things are not always so simple. The SC ray trace in figure 3.35 assumes that the top surface of the fluid in the well is the only emitter of light. If the well were a cylindrical volume, then the fluorescent fluid in the well would indeed likely appear as if only the surface emitted light. But the well is more like a cone, with a smaller diameter bottom. Hence, from the point of view of the lens, it can see the bottom of the well. We recall from section 1.1.1 that optical modeling begins with the simplest plausible layout and proceeds by way of including further aspects of reality such that there is progressive improvement in the model. A progressive improvement in the case of the compact virus detection system is to maintain ray tracing in the SC mode but now include a multi-configuration setup comprising two configurations: one for rays originating from the surface of the well and the other for rays originating from the bottom, as shown in the multi-configuration layout in figure 3.40. As indicated in figure 3.40(b), rays from the edge of the bottom of the well are vignetted at the edge of the top surface of the well, thereby causing the POP location for the edge field ray for the bottom portion of the well to be shifted slightly to the left of the sensor location indicated in figure 3.35. In this case, we might expect to have a small range of POP locations along the axis, as indicated in figure 3.40(b), but we could not

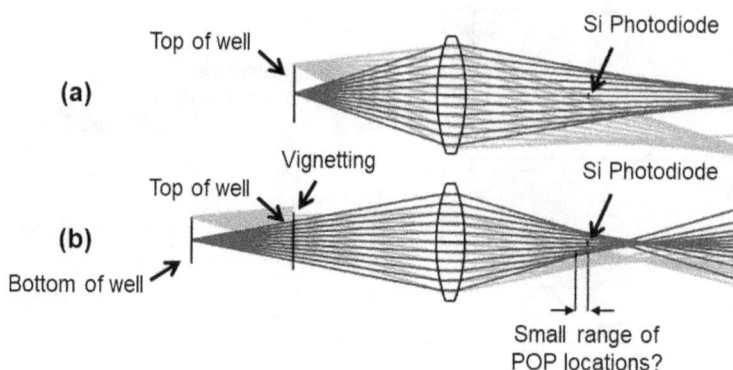

Figure 3.40. A multi-configuration layout. (a) Rays from the top of the well. (b) Rays from the bottom of the well.

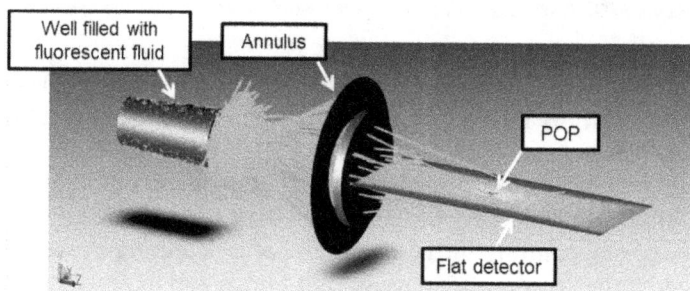

Figure 3.41. The NSC layout for the compact virus detection system.

know this for sure unless were to do an actual test on a lab bench or perform an NSC ray trace.

While the SC ray trace layout with added sketches in the figures above appears quite plausible, an NSC model may reveal further details and nuances. For instance, it would be of interest to check the location of the POP behind the bi-convex lens for the case of rays originating just from the surface of the well and for the case of rays originating from a full volume of fluorescent fluid. A simple NSC model that can test these cases is shown in figure 3.41, where a flat detector reveals the POP for a source modeled as a volume that emits rays in a well. For comparison, figure 3.42 shows the irradiance on the flat detector for the middle column of pixels for the case of a disk Lambertian source (figure 3.42(a)) and for the volume emitter (figure 3.42(b)). Note that the plots are rather similar and contain no clear indication of the range of POP locations that we had thought could exist, as was implied in figure 3.40(b). The reason is because there is an averaging effect for the rays emitted by the entire volume of fluid. This volume emitter is modeled in OS by using a combination of a cylinder *parent object* and a *source object* in the NSC editor, as indicated in rows 2 and 3 in figure 3.43. The parent object models the shape of the volume emitter (the object on row 3), while the source object (i.e. the volume emitter) takes on the shape of the parent object.

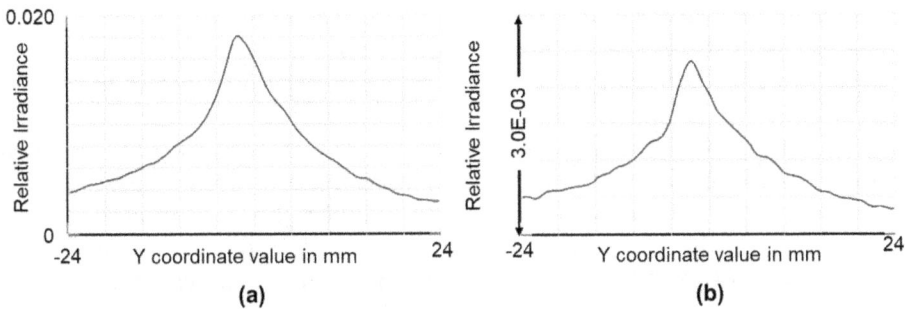

Figure 3.42. The irradiance profile for the middle column of pixels on the flat detector. (a) The rays originating from just the top of the well (modeled as a disk Lambertian source). (b) The rays emitted by a full volume of fluid in the well.

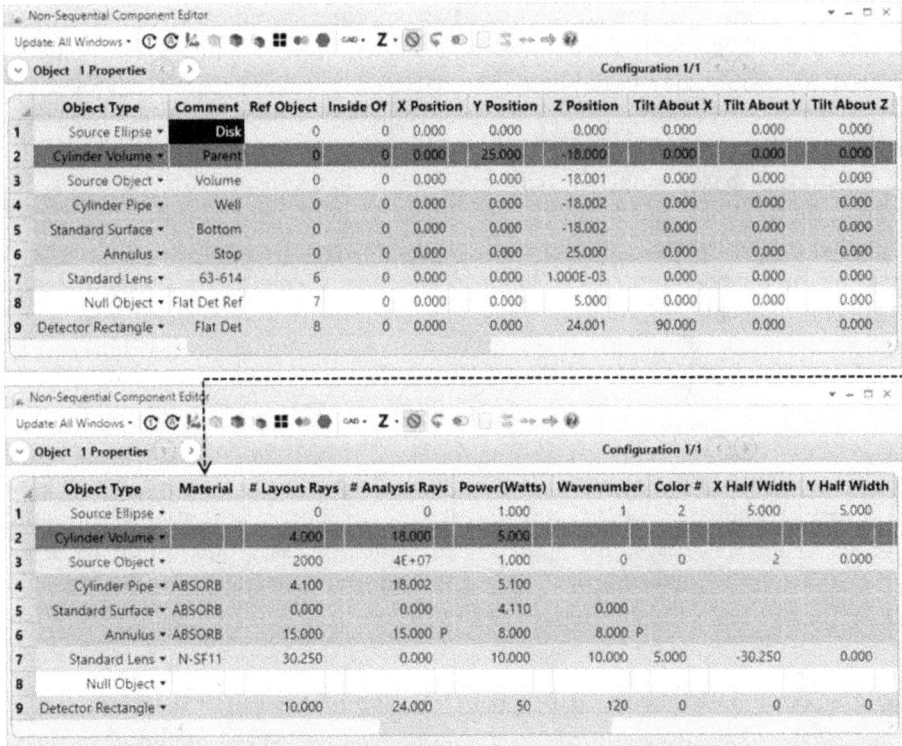

Figure 3.43. The prescription in pure NSC mode for the model in figure 3.41.

In figure 3.43, the row for the parent object is darkened because it is set to be *ignored* by rays, and it is also not drawn in the NSC layout of figure 3.41. As usual, note that the column labels are triggered to have different names when the cursor is placed in the relevant cell of the NSC Editor. If you create this model using the prescription given, you will note that if your cursor is in the cell of row 3 and the column labeled 'X Half Width,' the column's title is changed to 'Parent Object #.' I have inserted '2' in that cell because the parent object for that source object is on row 2, which is the cylinder volume object. The volume emitter (object on row 3) is then enclosed by a cylinder pipe (row 4) with a bottom cover (row 5). Care has been taken to ensure that the cylinder enclosing the volume emitter is slightly larger than the volume emitter. For instance, note that the cylinder pipe's Z Position starts at 18.002 mm from the top of the volume emitter, while the volume emitter's Z Position begins at −18.001 mm. Also, the volume emitter's surface semi-diameter is 5 mm (row 1 in the bottom figure of figure 3.43), while the well containing it has a surface semi-diameter of 5.1 mm.

Everything that has been discussed above has concerned mainly the optical system portion of the detection system (and for this, we would also need to study the achievable SNR and LOD of the system (see, for example, sections 3.4 and 3.5)). There remains the thermocycler development, the electronics, the signals processing unit, and perhaps also the development of specific ingredients required by the biochemistry of the test assays, which would involve biochemists, biomedical scientists, and molecular biologists. Thus, many other team members would have to be involved. An important point for discussion with a broader team concerns our crazy idea of doing a PCR test with only a single well. We note from section 3.2 that in a laboratory-based analytic real-time qPCR instrument, using a plate with multiple wells enables not only the replication of PCR tests but also the use of a population of control wells and reference wells for accurate quantification. We need to ask our biomedical and system engineering colleagues what this would mean to the greater medical community. Thus, the complexity of developing a medical product that involves optical detection becomes quite apparent when you consider all of these other components for product development. And we are not even close to the finish line. At this point, if we were to apply the PCP chart shown in figure 1.1, we would still only be at the ideation and concept development stage. A lab bench breadboard would need to be built to test the feasibility of this concept, which would lead to the feasibility phase. Of course, we shall stop here. Our goal in this simple conceptual case study is to highlight the thought processes that would usually be applied to developing an optical system into a product starting from the beginning, with nothing but an idea and its possibilities.

3.9.2 The Texas Instruments DLP® chip (DMD) projection optical system

Modern projection systems use electronic/digital spatial light modulators (SLMs) in place of transparencies [78]. Examples of such SLMs include digital micromirror devices (DMDs) from Texas Instruments [79], liquid crystal displays (LCDs) [80], the grating light valve (GLV) from Silicon Light Machines [4, 81], and liquid crystal on silicon (LcoS) displays [82]. All of these SLMs have found applications ranging from cinema and lithography [79] to even biology, in which light patterns are projected onto specific areas of cells for in-depth study [83, 84]. Therefore, optical

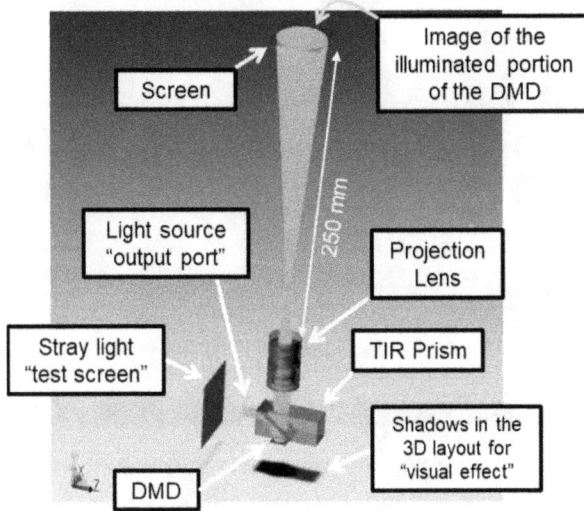

Figure 3.44. A 3D shaded model of a DMD projection optical system.

systems that utilize digital SLMs are often quite complex, involving many elements and parts from different fields.

To get a feel for the complexity of modern optical projection, let us consider the layout of a conceptual optical system involving the use of a DMD. Consider the pure NSC 3D shaded model shown in figure 3.44, whose prescription is provided in figures 3.45–3.47. While complexities exist for a real system, the initial conceptual layout is intended to be kept simple in order to represent a generic DMD projection optical system so that specific nuances of the complexities may be examined one step at a time. For instance, there is a *stray light test screen* indicated at the bottom left, whose purpose shall be explained shortly. We will also explain how the projection lens was conceived and integrated into this layout, which involves a certain nuance related to the discussion about imaging an intermediate transmissive or reflective surface in section 3.8.1.

Many of the NSC objects in the prescription shall be understood in due course. The wavelength settings are 400, 550, and 700 nm, each of which has a weight of one, and the primary is set to 550 nm. The first step in laying out a DMD-based projection optical system is to consider how to set up the path of rays towards the DMD and through a so-called *total internal reflection* (TIR) prism. To do this, we refer to figure 3.48 and note that a ray from the illumination must be reflected by DMD micromirrors (which are the pixels) at a specific angle.

In figure 3.48, the DMD (object 12 in figure 3.45) is assumed to be a Texas Instruments (TI) DLP® chip whose micromirrors tilt by ±12° to create on/off states. Not all of TI's DMDs have micromirror tilts at these angles, but we shall assume 12° for our case study. An example of an existing DMD of the size modeled here is the DLP4500 *chipset* (this is what TI calls their DMD products), whose micromirror pixels do tilt at the ±12° angle. While the DLP54500 DMD has 912 × 1140

Non-Sequential Component Editor
Update: All Windows
Object 3 Properties
Configuration 1/1

	Object Type	Comment	Ref Object	Inside Of	X Position	Y Position	Z Position	Tilt About X	Tilt About Y	Tilt About Z
1	Source Ellipse	Source (B)	0	0	0.000	2.500	0.000	0.000	0.000	0.000
2	Source Ellipse	Source (G)	0	0	0.000	2.500	0.000	0.000	0.000	0.000
3	Source Ellipse	Source (R)	0	0	0.000	2.500	0.000	0.000	0.000	0.000
4	Source Point	Point Source	1	0	0.000	0.000	1.000E-03	0.000	0.000	0.000
5	Source Ray	Ray (400 nm)	0	0	0.000	2.500	0.000	0.000	0.000	0.000
6	Source Ray	Ray (550 nm)	0	0	0.000	2.500	0.000	0.000 P	0.000	0.000
7	Source Ray	Ray (700 mn)	0	0	0.000	2.500	0.000	0.000 P	0.000	0.000
8	Rectangular Volume	Back Check	1	0	0.000	0.000	-25.000	0.000	0.000	0.000
9	Rectangular Volume	TIR Prism1	0	0	0.000	0.000	12.500	0.000	0.000	0.000
10	Rectangular Volume	TIR Prism2	0	0	0.000	0.394	27.000	0.000	0.000	0.000
11	Rectangular Volume	DMD Ref	0	0	0.000	-15.000	30.000	90.000	0.000	0.000
12	MEMS	DMD	11	0	0.000	0.000	-0.250	0.000	0.000	0.000
13	Sphere	Air Gap Check	9	0	0.000	9.984	1.871	0.000	0.000	0.000
14	Null Object	Lens Ref	0	0	0.000	31.000	30.000	-90.000	0.000	0.000
15	Standard Lens	6	14	0	0.000	0.000	0.000	0.000	0.000	0.000
16	Standard Lens	5	15	0	0.000	0.000	2.961	0.000	0.000	0.000
17	Standard Lens	4	16	0	0.000	0.000	4.412	0.000	0.000	0.000
18	Annulus	Aperture	17	0	0.000	0.000	9.796	0.000	0.000	0.000
19	Standard Lens	3	17	0	0.000	0.000	15.217	0.000	0.000	0.000
20	Standard Lens	2	19	0	0.000	0.000	2.051	0.000	0.000	0.000
21	Standard Lens	1	20	0	0.000	0.000	7.154	0.000	0.000	0.000
22	Cylinder Pipe	"Housing"	15	0	0.000	0.000	-1.000	0.000	0.000	0.000
23	Annulus	"Housing Back"	22	0	0.000	0.000	0.000	0.000	0.000	0.000
24	Null Object	Screen Ref	21	0	0.000	-0.800	4.360	0.000	0.000	0.000
25	Detector Rectangle	Screen	24	0	0.000	0.000	250.000	0.000	0.000	0.000

Figure 3.45. The prescription in OS for the pure NSC layout shown in figure 3.44 (the columns to the right of 'Tilt About *Z*' are continued in figures 3.46 and 3.47).

micromirrors, our model has just 25 × 25 such mirrors. For our purpose, we only need to visualize the on/off effect of the micromirror tilts. Figure 3.48 shows the micromirrors in their ON state, where each has been tilted by −12° (this is the required direction for the tilt in our model for the ON state). In order for a ray from the illumination to reflect orthogonally with respect to the plane of the DMD, it must therefore be at an AOI of 24° relative to the normal of that plane. This ray must therefore reflect downwards from the TIR prism in such a way that its ray exiting the prism is at an AOI of 24° to the normal of the DMD plane.

A TIR prism comprises a pair of prisms separated by a thin air gap (roughly 10 μm thick, in order to minimize astigmatism in the imaging path [85]) so that TIR occurs at the air gap for the illumination ray. In our model, object 9 (TIR Prism 1) models the lower prism and object 10 (TIR Prism 2) models the upper prism. The air gap is determined by simple trial and error, guided by a sphere 10 μm in diameter (figure 3.49). This sphere is object 13 in the prescription. Guided by this sphere, the upper prism's location was adjusted until it barely touched the sphere, thereby creating an air gap nearly 10 μm thick.

The two prism pairs are made of N-BK7 material, which has an index of roughly 1.519 at 550 nm. According to Snell's law, if n is the index of the prism, then for TIR

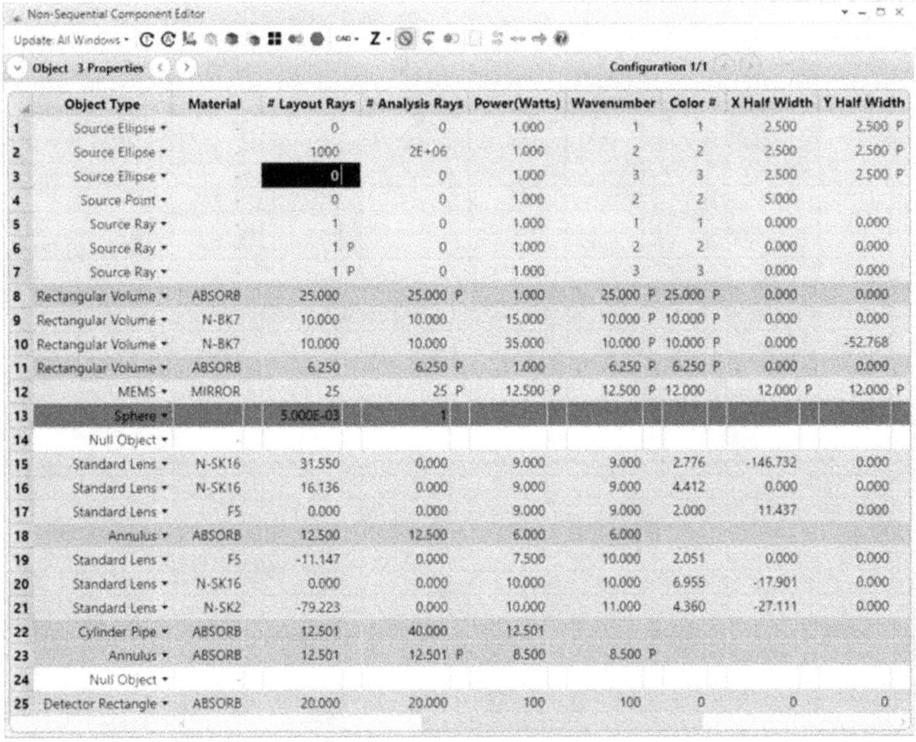

	Object Type	Material	# Layout Rays	# Analysis Rays	Power(Watts)	Wavenumber	Color #	X Half Width	Y Half Width
1	Source Ellipse		0	0	1.000	1	1	2.500	2.500 P
2	Source Ellipse		1000	2E+06	1.000	2	2	2.500	2.500 P
3	Source Ellipse		0	0	1.000	3	3	2.500	2.500 P
4	Source Point		0	0	1.000	2	2	5.000	
5	Source Ray		1	0	1.000	1	1	0.000	0.000
6	Source Ray		1 P	0	1.000	2	2	0.000	0.000
7	Source Ray		1 P	0	1.000	3	3	0.000	0.000
8	Rectangular Volume	ABSORB	25.000	25.000 P	1.000	25.000 P	25.000 P	0.000	0.000
9	Rectangular Volume	N-BK7	10.000	10.000	15.000	10.000 P	10.000 P	0.000	0.000
10	Rectangular Volume	N-BK7	10.000	10.000	35.000	10.000 P	10.000 P	0.000	-52.768
11	Rectangular Volume	ABSORB	6.250	6.250 P	1.000	6.250 P	6.250 P	0.000	0.000
12	MEMS	MIRROR	25	25 P	12.500 P	12.500 P	12.000	12.000 P	12.000 P
13	Sphere		5.000E-03	1					
14	Null Object								
15	Standard Lens	N-SK16	31.550	0.000	9.000	9.000	2.776	-146.732	0.000
16	Standard Lens	N-SK16	16.136	0.000	9.000	9.000	4.412	0.000	0.000
17	Standard Lens	F5	0.000	0.000	9.000	9.000	2.000	11.437	0.000
18	Annulus	ABSORB	12.500	12.500	6.000	6.000			
19	Standard Lens	F5	-11.147	0.000	7.500	10.000	2.051	0.000	0.000
20	Standard Lens	N-SK16	0.000	0.000	10.000	10.000	6.955	-17.901	0.000
21	Standard Lens	N-SK2	-79.223	0.000	10.000	11.000	4.360	-27.111	0.000
22	Cylinder Pipe	ABSORB	12.501	40.000	12.501				
23	Annulus	ABSORB	12.501	12.501 P	8.500	8.500 P			
24	Null Object								
25	Detector Rectangle	ABSORB	20.000	20.000	100	100	0	0	0

Figure 3.46. Prescription continued from figure 3.45.

to occur at a glass–air interface, we must have $n \sin \theta_c = 1$, where θ_c is the critical angle for TIR. For an index of 1.519, this critical angle is roughly 41.188°, so that all rays from the illumination must have AOIs *greater* than this angle. It so happens that, by trigonometry, it was determined that the AOI required to reflect a horizontal ray from the illumination onto the DMD with an AOI of 24° is 52.768°. This is also the tilt angle for the air gap (see figure 3.48). Therefore, the horizontal ray entering the lower prism from the left side has an AOI at the air gap that satisfies the critical angle. But illumination rays often have a divergence. We must then ask what half-cone angle for the illumination satisfies the TIR condition at the air gap.

To address the above concern about the TIR condition for a divergent (or convergent) set of illumination rays, we utilize objects 5, 6, and 7 in the prescription, which are individual rays, set at wavelengths of 400, 550, and 700 nm, respectively. Note in figure 3.48 that these three rays were used to visualize their reflections from the ON state of the micromirrors. In that figure, the rays enter the lower prism horizontally and suffer no dispersion until they emerge and become incident on the micromirror, upon which their dispersion becomes apparent as they continue their propagation vertically through the prism pair. We now wish to use these rays again for the purpose of determining whether a cone of rays at these three wavelengths

Figure 3.47. Prescription continued from figure 3.46.

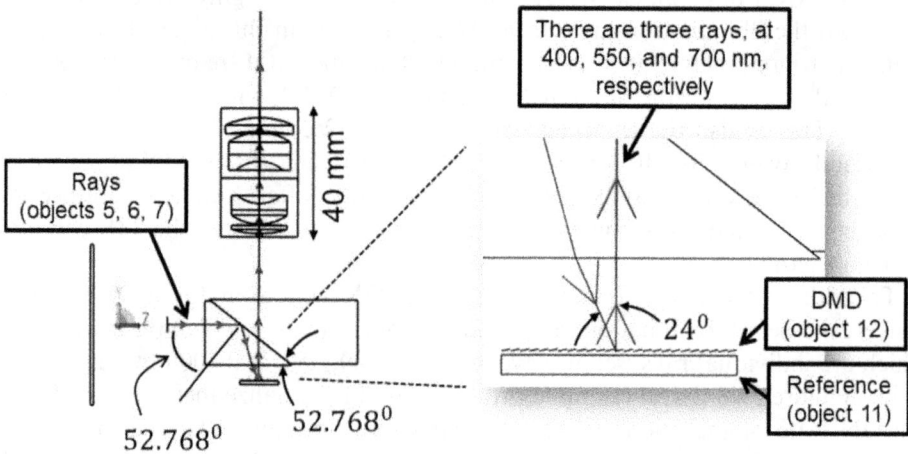

	Object Type	Source Distance	Cosine Exponent	Gauss Gx	Gauss Gy	Source X	Source Y	Minimum X Half Width	Minimum Y
1	Source Ellipse ▾	57.200	0.000	0.000	0.000	0.000	0.000	0.000	
2	Source Ellipse ▾	57.200	0.000	0.000	0.000	0.000	0.000	0.000	
3	Source Ellipse ▾	57.200	0.000	0.000	0.000	0.000	0.000	0.000	
4	Source Point ▾								
5	Source Ray ▾	0.000	0						
6	Source Ray ▾	0.000	0						
7	Source Ray ▾	0.000	0						
8	Rectangular Volume ▾	0.000	0.000						
9	Rectangular Volume ▾	0.000	-52.768						
10	Rectangular Volume ▾	0.000	0.000						
11	Rectangular Volume ▾	0.000	0.000						
12	MEMS ▾	0.000	0	0	0				
13	Sphere ▾								
14	Null Object ▾								
15	Standard Lens ▾	9.000	9.000						
16	Standard Lens ▾	9.000	9.000						
17	Standard Lens ▾	7.000	9.000						
18	Annulus ▾								
19	Standard Lens ▾	10.000	10.000						
20	Standard Lens ▾	10.000	10.000						
21	Standard Lens ▾	11.000	11.000						
22	Cylinder Pipe ▾								
23	Annulus ▾								
24	Null Object ▾								
25	Detector Rectangle ▾	0	0.000	0	0	-90.000	90.000	-90.000	

Figure 3.48. The ray path for a DMD and TIR prism setup.

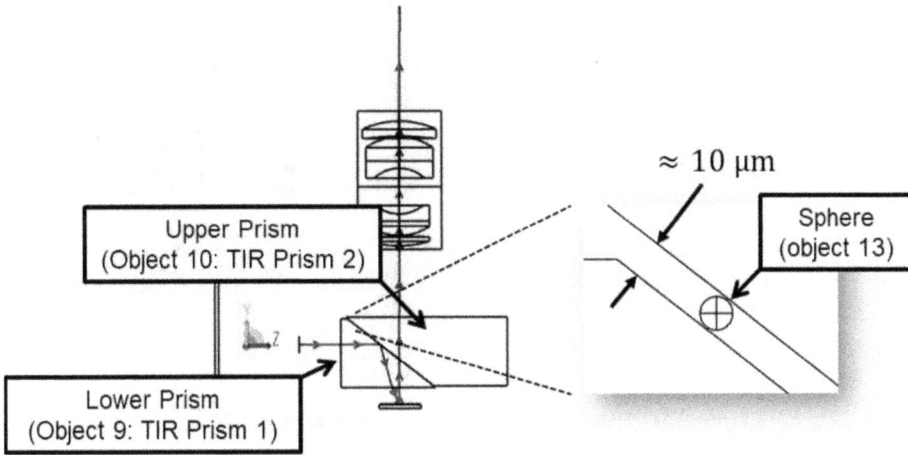

Figure 3.49. The ray path for a DMD and TIR prism setup.

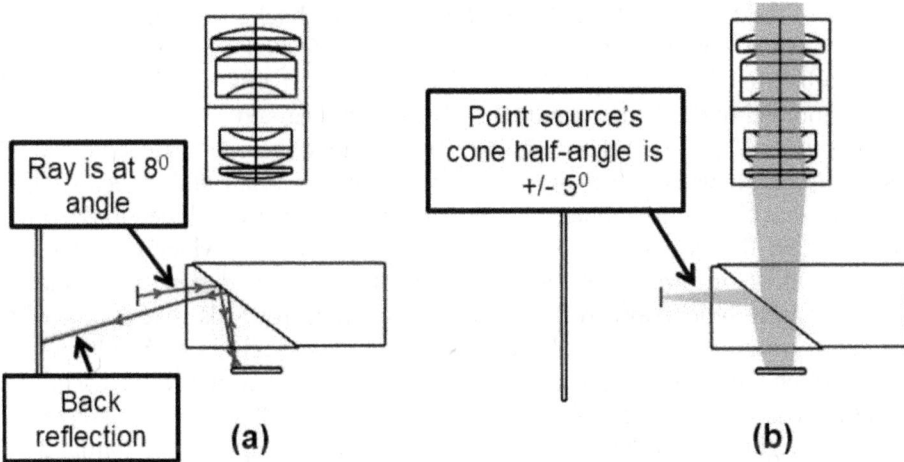

Figure 3.50. TIR condition tests. (a) Using rays. (b) Using a point source.

satisfies the TIR condition at the air gap. To do this, we impose some trial angles and find that when they tilt upwards by 8°, none of them satisfy the TIR condition, resulting in back reflections onto a reference surface (labeled 'stray light test screen' in figure 3.44) behind the rays, as shown in figure 3.50(a). This was the reason for creating the 'stray light test screen.' Through trial and error, it was determined that no back reflections occur for rays ⩽5°. As a further test for this, a point source (object 4) was used, which emitted rays up to a half-cone angle of 5° (figure 3.50(b)). When this point source's wavelength was varied between 400 and 700 nm, the TIR condition held. Thus, this process determined the useable illumination divergence.

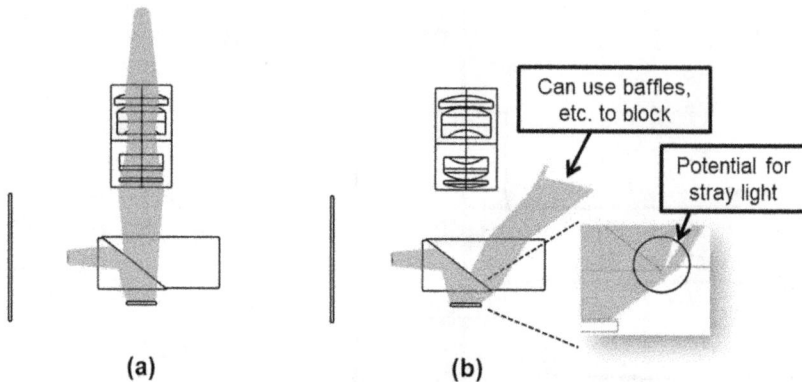

Figure 3.51. The light switching test. (a) ON state. (b) OFF state.

The next step is to let an extended source illuminate the DMD within the determined *full* cone angle of 10° in the ON and OFF states of micromirror tilt, as shown in figure 3.51. Assuming ideal AR coatings on all prism surfaces (with the exception of the air gap surface of the lower prism), no back reflections are observed in either the ON or OFF states, but some of the reflected rays in the OFF state appear to be travelling through a sharp edge of the lower prism, as indicated by the pop-out zoomed-in diagram in figure 3.51(b). This appears to have created split rays at that corner. While the rather simplistic optical model here indicates that the split rays do not go anywhere other than away from the prism pair and the lens, there is no telling what else could happen if some further realism were included in the optical model. Hence, this is a potential source of a stray light issue, which must either be dealt with now or through some benchtop lab tests. By design, we could, for instance, simply block rays from the DMD travelling towards that corner of the prism pair. Alternatively, we could reduce the half-cone angle from the illumination to 2.5°, which seems fruitful, as suggested by the ray trace in figure 3.52. This does not mean it is the best solution, because the rays are still somewhat dangerously close to the right edge of the lower prism. But it is one example of a potential design solution to avoid stray light.

We have not yet discussed how the TIR prisms were modeled in the current prescription. The approach was by way of a quick trick in using the Rectangular Volume NSC object in OS, which is intended only as an initial model rather than a full detailed design (for a complete prism design, one can use the Polygon Object in OS to develop intricate details for prism angles and shapes). The idea here was to simply tilt the right surface of the rectangular object for the lower prism and to do the same for the left surface of the upper prism. Such tricks are useful during the optical architecture phase of the optical system design, which is the assumed phase in our current concept case study.

Now, having reduced the half-cone angle for the illumination beam to 2.5°, we return to the ON state of the ray trace and view the layout horizontally, as shown in figure 3.53. Under this condition, the layout provides visual information about how

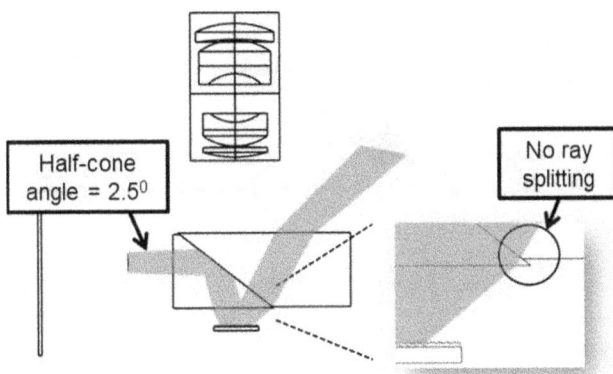

Figure 3.52. A reduced illumination cone angle may help to prevent stray light.

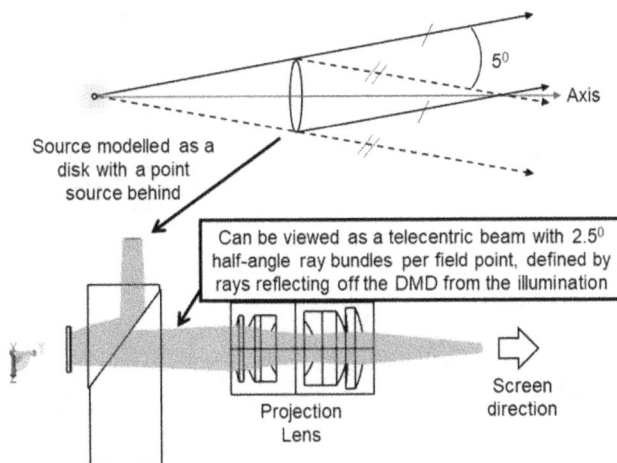

Figure 3.53. The horizontal layout for the projection system.

to design or integrate a projection lens with the DMD-TIR-prism assembly. In particular, we note that if the illumination system is involved, then the plane of the DMD should be regarded as an *intermediate plane* in the path of the rays, because the DMD is essentially a mirror (more on this later, as the situation is made complicated by involving diffraction effects), so that the 2.5° half-angle divergence of rays from the illumination is retained upon reflection from the DMD towards the projection lens. We recall from section 3.8.1 that in this case, should the projection lens be modeled in a separate optical layout, while the object plane would be the plane of the DMD, the rays from this object would not be Lambertian, so that the projection lens's aperture stop would not define the ray path through the lens system. This is illustrated in the HNSC (hybrid NSC) ray trace layout shown in figure 3.54, whose SC and NSC prescriptions are provided in figures 3.55 and 3.56, respectively.

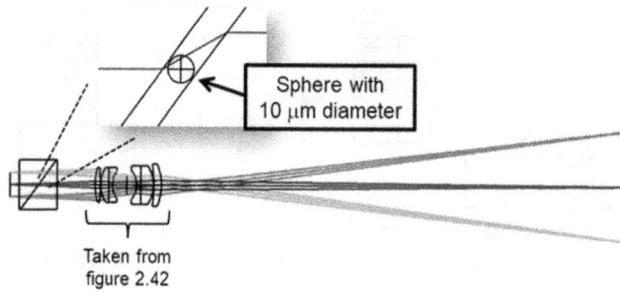

Figure 3.54. The HNSC layout for the projection lens.

	Surface Type		Radius	Thickness	Material	Semi-Diameter	Exit Loc X	Exit Loc Y	Exit Loc Z
0	OBJECT	Standard ▾	Infinity	5.000		6.250			
1		Non-Sequential Component ▾	Infinity	·		15.000 U	0.000	0.000	20.500
2		Standard ▾	Infinity	20.667 V		15.000 U			
3	(aper)	Standard ▾	31.550	2.776	N-SK16	9.000 U			
4	(aper)	Standard ▾	-146.732	0.184		9.000 U			
5	(aper)	Standard ▾	16.136	4.412	N-SK16	9.000 U			
6	(aper)	Standard ▾	Infinity	2.000	F5	9.000 U			
7	(aper)	Standard ▾	11.437	7.796		7.000 U			
8	STOP	Standard ▾	Infinity	5.421		4.925			
9	(aper)	Standard ▾	-11.147	2.051	F5	7.500 U			
10	(aper)	Standard ▾	Infinity	6.955	N-SK16	10.000 U			
11	(aper)	Standard ▾	-17.901	0.199		10.000 U			
12	(aper)	Standard ▾	-79.223	4.360	N-SK2	10.000 U			
13	(aper)	Standard ▾	-27.111	250.000		11.000 U			
14	IMAGE	Standard ▾	Infinity	·		28.942			

Figure 3.55. The prescription (SC portion) for the layout shown in figure 3.54.

Under the current conditions, should the projection lens be designed specifically for the DMD, it must be optimized for rays that are telecentric in object space and have a half-angle of 2.5°, as illustrated in figure 3.53. If the illumination is not telecentric, then the lens must be optimized according to an appropriate system aperture setting that defines the characteristic geometry of the illumination rays.

The lens system in figure 3.55 was taken from the double-Gauss lens that was used in figure 2.42 (see section 2.10) without performing any further optimization. The fields at the object plane are set at 0, 6.25 mm, and −6.25 mm, and the wavelength is monochromatic at 550 nm. Notice that while the aperture stop is at surface 8, its diameter is not set using 'Float by Stop Size.' For this reason, you do not see the *User Defined* symbol 'U' for that surface under the Semi-Diameter column of the prescription. All other surfaces have this 'U' symbol, because they are user-defined apertures. The stop is left with a free semi-diameter, because the beam size is, in

Non-Sequential Component Editor: Component Group on Surface 1

Update: All Windows ▾

Object 1 Properties

Configuration 1/1

	Object Type	Comment	Ref Object	Inside Of	X Position	Y Position	Z Position	Tilt About X	Tilt About Y	Tilt About Z
1	Rectangular Volume ▾	TIR Prism1	0	0	0.000	0.000	0.000	0.000	0.000	0.000
2	Rectangular Volume ▾	TIR Prism2	0	0	0.000	0.000	10.214	0.000	0.000	0.000
3	Sphere ▾	Air Gap Check	0	0	0.000	0.000	10.207	0.000	0.000	0.000

Non-Sequential Component Editor: Component Group on Surface 1

Update: All Windows ▾

Object 1 Properties

Configuration 1/1

	Object Type	Material	X1 Half Width	Y1 Half Width	Z Length	X2 Half Width	Y2 Half Width	Front X Angle
1	Rectangular Volume ▾	N-BK7	10.000	13.000	10.200	10.000	13.000	0.000
2	Rectangular Volume ▾	N-BK7	10.000	13.000	10.200	10.000	13.000	0.000
3	Sphere ▾		5.000E-03	1				

Non-Sequential Component Editor: Component Group on Surface 1

Update: All Windows ▾

Object 1 Properties

Configuration 1/1

	Object Type	Z Length	X2 Half Width	Y2 Half Width	Front X Angle	Front Y Angle	Rear X Angle	Rear Y Angle
1	Rectangular Volume ▾	10.200	10.000	13.000	0.000	0.000	0.000	37.000
2	Rectangular Volume ▾	10.200	10.000	13.000	0.000	37.000	0.000	0.000
3	Sphere ▾							

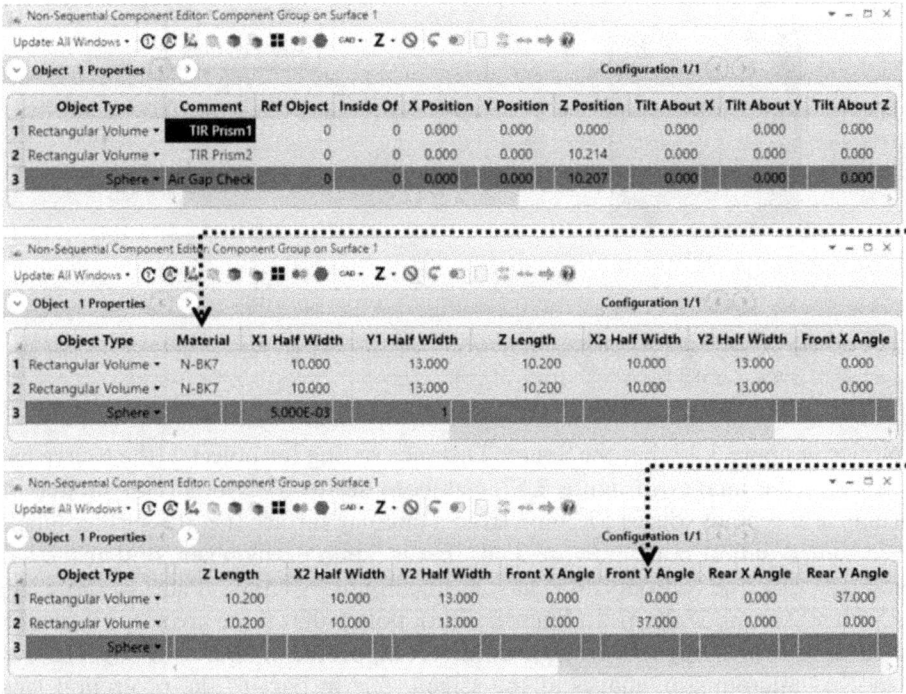

Figure 3.56. The prescription (NSC portion) for the layout shown in figure 3.54.

reality, defined by the illumination incident on the object, which is the DMD surface. In the SC layout, therefore, the system aperture is set to be telecentric in object space, with a half-cone angle of 2.5°, as determined in the analysis steps leading to figure 3.53. Notice that the thickness of surface 2 in figure 3.55 has a variable, which is has been optimized so that the image on the screen (surface 14) is focused. Figure 3.56 shows the NSC Editor settings used to model the TIR prism. No attempt has been made to ensure exact correspondence between this model and that shown in the pure NSC layout. At this stage, the idea is to conceptualize the system as a whole rather than perform detailed design. Note that a sphere object (object 3, with no defined material so that rays ignore it) ten microns in diameter was included in the NSC prescription, which was used to guide the location of the upper prism (TIR Prism 2, object 2) in order to create an approximately ten micron air gap in the HNSC model.

As soon as the lens in figure 3.54 was focused, it was created manually as objects 15–21 in the pure NSC layout (figure 3.45). Then, a housing and cover (objects 22, 23) were added to the lens. The image of the illuminated area of the DMD shown at the top of figure 3.44 was the result of performing an NSC ray trace with 2E+06 rays for just the 550 nm wavelength (object 2). The false color plot and profile for the middle column for this image are shown in figures 3.57(a) and (b), respectively.

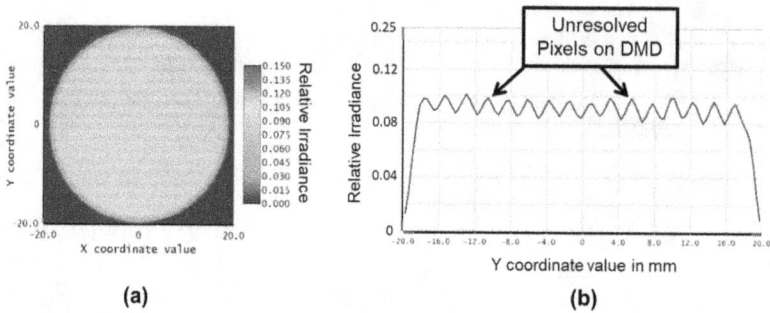

Figure 3.57. The image projected on the screen at the top of figure 3.44. (a) False color. (b) Irradiance profile for the middle column of the image.

Notice in figure 3.47 that the Source Distance setting for object 2 (the source used to generate the images in figure 3.57) has been set to 57.2 mm. This models the placement of a point source 57.2 mm from a circular surface (i.e. the surface labeled 'Light source output port' in figure 3.44) that is 5 mm in diameter, so that the half-cone-angle for the output rays has a maximum of 2.5°. That is, note that $\tan(2.5^0) \approx 2.5mm/57.2mm$. In figure 3.57(a), notice that there are some horizontal line patterns, which appear as a jagged 'sawtooth' profile in figure 3.57(b). These are unresolved micromirror images at the screen, due in part to the projection lens's quality and in part to the aliasing effect of the detector pixels. All micromirrors are in their ON state. In the optical model, the ON and OFF states of the DMD model are controlled by the column labeled 'Color #' for object 12 in figure 3.46. In reality, this column label appears as 'Angle 0' when the computer cursor is placed in that cell, as indicated in figure 3.58.

There are further nuances to consider with regards to the operation and properties of the DMD. In reality, most of TI's DLP® chips (DMDs) have micromirrors that produce ON/OFF states by **tilting about the diagonal axis of the micromirror**, as illustrated in figure 3.59). But the micromirrors in the DMD model for the layout in figure 3.44 do not pivot about the diagonal. Rather, they pivot about their widths. Hence, one would need to account for this in an actual design. Alternatively, in the layout, you can hide the DMD model from view and simply show a rectangular object rotated about the optic axis (provided that the projection optics also rotates about the axis such that the screen is square to the DMD).

Another nuance to consider is that the DMD exhibits diffractive effects, as it is considered to be a reflective blazed diffraction grating in the ON state of the pixels [86–89]. This is quite evident, considering that there are gaps between the micromirrors and that each micromirror tilts—like a reflective blazed grating—to deflect the illumination into the projection lens. But we recall from basic diffraction physics that blazing of the incident beam towards a preferred order of diffraction only occurs for a single wavelength [70]. Therefore, we do not necessarily expect to achieve high flux for illumination rays that are diverted towards the projection lens by the DMD. The analysis is made more complicated by the fact that the usual

Non-Sequential Component Editor

Update: All Windows ▾   CAD ▾ Z ·

Object 12 Properties   Configuration 1/1

	Object Type	Material	# X Pixels	# Y Pixels	X-Width	Y-Width	Angle 0	Angle 1	Angl
1	Source Ellipse ▾		0	0	1.000	1	1	2.500	2.
2	Source Ellipse ▾		1000	2E+06	1.000	2	2	2.500	2.
3	Source Ellipse ▾		0	0	1.000	3	3	2.500	2.
4	Source Point ▾		0	0	1.000	2	2	5.000	
5	Source Ray ▾		0	0	1.000	1	1	0.000	0.
6	Source Ray ▾		0 P	0	1.000	2	2	0.000	0.
7	Source Ray ▾		0 P	0	1.000	3	3	0.000	0.
8	Rectangular Volume ▾	ABSORB	25.000	25.000 P	1.000	25.000 P	25.000 P	0.000	0.
9	Rectangular Volume ▾	N-BK7	10.000	10.000	15.000	10.000 P	10.000 P	0.000	0.
10	Rectangular Volume ▾	N-BK7	10.000	10.000	35.000	10.000 P	10.000 P	0.000	-52.
11	Rectangular Volume ▾	ABSORB	6.250	6.250 P	1.000	6.250 P	6.250 P	0.000	0.
12	MEMS ▾	MIRROR	25	25 P	12.500 P	12.500 P	12.000	12.000 P	12.
13	Sphere ▾		5.000E-03	1					
14	Null Object ▾								
15	Standard Lens ▾	N-SK16	31.550	0.000	9.000	9.000	2.776	-146.732	0.
16	Standard Lens ▾	N-SK16	16.136	0.000	9.000	9.000	4.412	0.000	0.

Figure 3.58. The blackened cell under the 'Angle 0' column controls the ON/OFF states of the micromirrors.

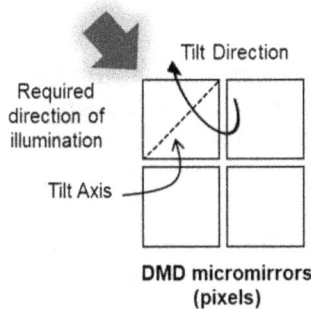

Figure 3.59. The proper direction for illumination incident on a DMD.

far-field Fraunhofer diffraction integral is a paraxial approximation, so that blazing only enters into the resulting formula for the far-field diffraction irradiance distribution at normal incidence for the illumination. At any other AOI for the illumination, the far-field diffraction distribution does not change in its diffraction efficiency [90]. But the illumination incident on the DMD is required to have a non-normal AOI, so that non-paraxial diffraction analysis must be applied [91]. Evidently, DMD diffraction behavior is rather complicated. From experience, it is best to perform experimental tests for a DMD-based optical system on a lab bench; the results of such tests provide accurate feedback for the optical design. Furthermore, scattering effects and other anomalies at the DMD can contribute to reduced image contrast [92, 93], and these effects are best investigated using a real system.

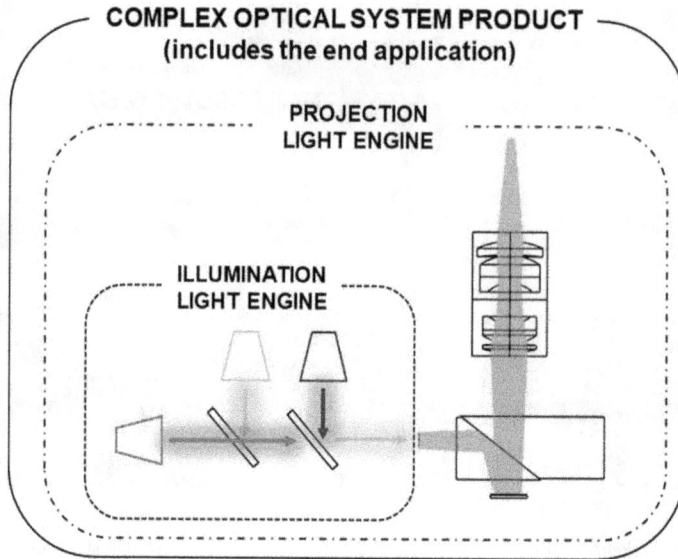

Figure 3.60. Levels of complexity for an optical projection system as a product.

In the context of optical system product development, a DMD-based projection system may turn into a complex instrument, depending on the required end application, as depicted by the hierarchical diagram in figure 3.60. This diagram resembles the levels of complexity for instrument development first introduced in figure 1.2. In the current example, electronics and thermal management are not only relevant to the sources used in the *illumination light engine* but also to the DMD, because of the potential heat load imposed on the DMD by the illumination. Of course, there is also the circuitry, firmware, and software used to control the DMD.

The end application also defines the overall architecture for a *projection light engine*. For instance, the use of a TIR prism is only required for the purpose of enlarging the projected image of the DMD, because in this case, the projection lens must be at the *short conjugate* for imaging (i.e. it must be closer to the object plane, which is the plane of the DMD). Under this condition, since the DMD requires incident oblique illumination, there is often insufficient space to shine rays onto the DMD at a glancing angle of incidence without the use of a TIR prism. But in other applications, such as in microscopy, the projection lens is combined with a microscope objective to project a demagnified image of the DMD onto a biological sample, such as a cell [83, 84]. Under this condition, when the microscope objective is an infinity-corrected objective, the projection lens is a tube lens [83], so that its back focal length can be rather long (roughly the focal length of the tube lens, which often ranges from 160 mm to 200 mm). This eliminates the need for a TIR prism, thereby possibly reducing some stray light nuances (however, diffracted orders from the DMD may not be captured by the tube lens if the tube lens is too far from the DMD).

Certain conditions in the end application may also define the type of illumination required, such as the flux of the sources and the level of achievable uniformity. Accordingly, many of the techniques discussed in chapter 2 of this book may then be applied. Others have also considered various methods, depending on the type of source used, such as arc sources [94], laser diodes [95], and LEDs [96, 97]. For any method of illumination applied, the illumination designer and projection lens designer must communicate. The projection lens designer must know the characteristics of the beam that is incident on the DMD so that the system aperture for the lens design can be made to account for these characteristics, as highlighted above. In particular, suppose that the optical system architect has defined the overall projection light engine to be of the form given in figure 3.44, so that there is an 'output port' for the illumination. If the assumption is that rays from this port behave as if they come from a point source somewhere behind the port (as depicted in figure 3.53), then the illumination designer must develop a luminaire subsystem that provides this characteristic. How may this be done? One method could be to just place a small LED sufficiently far behind the porthole, as depicted in figure 2.1(b). This is not a bad approach in terms of being able to produce good local and global uniformity (see section 2.4 for a review of these terms). The reason for this is that the small LED is not conjugate to the DMD's surface, so there is no visible image of the source at the screen. This is implied in figure 3.44, where it can be seen that there is a focal point for rays between the projection lens and the screen. That focal point is the image of the point source used in that ray trace (i.e. object 2 in the prescription, which has a defined 'Source Distance' 57.2 mm behind the 'output port').

Alternatively, if we wished to increase the amount of flux through the output port, then we would first have to apply the Lagrange invariant (see sections 1.1.3 and 3.8.7) to the rays exiting the output port. In this case, given that the port's semi-diameter is 2.5 mm (see the X and Y Half Width for objects 1–3 in figure 3.46) and the half-cone angle is 2.5° (0.0436 rad), the Lagrange invariant can be roughly calculated by multiplying 2.5 mm by the angle converted to radians. Doing the conversion, the invariant quantity is roughly 0.11 mm rad. Applying this to the portion of a luminaire where flux is being collected by a lens from the source and assuming that the source is an LED whose semi-diameter is 0.5 mm, then the largest possible half-angle subtending a lens that collects this flux is 0.22 rad, which is roughly 12.6°. According to equation (2.6) and assuming a Lambertian LED emitter, the fraction of flux collected at an angle of 12.6° is $\sin^2(12.6°) \approx 0.048$, which is 4.8%. This is not a lot, but it is the limit for the size of LED considered and if the LED emits in a Lambertian manner. If this is an issue, then the optical system architect may need to change the architectural layout for the projection system. On the other hand, if this level of flux efficiency is acceptable, then the illumination design can be a simple aplanatic lens of the form shown in figure 2.81. Alternatively, you may also consider the so-called Villuminator™ [98]. Both luminaire types are good choices, because their output rays are very nearly collimated, resulting in a focal plane for the source that is located between the projection lens and screen, as depicted in figure 3.44 (in other words, they do not provide critical illumination at the DMD). However, as highlighted in section 2.10, an aplanatic luminaire requires that the projection lens must also be aplanatic, which may not always be the case. In this case, the Villuminator™ could be a good alternative, because it does not contribute to local nonuniformities and it can *tune* the global uniformity of illumination if the projection lens is not aplanatic [98].

3.10 Wrap up, README, and I wish you all the best!

So, this is all that can be mustered for this book, given the space, time, and energy available for writing. The basic structure of the book is intended to take you from the basics of imaging (chapter 1) and illumination (chapter 2) towards their application in complex optical systems (chapter 3). Interestingly, in my career, I started in almost the reverse of that path. As I shared in the beginning of the current chapter, I was thrown into the deep end of the pool as soon as I got out of university, having to deal with a large complex instrument involving high-powered near-infrared lasers, imaging relay optics, a new type of spatial light modulator, and a huge rotating drum with infrared sensitive print media. I did not know how to design optical systems using a program, which was frustrating. So, in writing this book, I hope that it will help make things easier for you in transitioning from the academic world into the practical.

We have, of course, left out many other points and topics. In terms of lens design, we did not touch on gradient index optics, diffractive optics, zoom lenses, and others. In terms of illumination, we left out nonimaging concentrators used in reverse and tailored freeform optics. In terms of complex optical systems and instruments, there are plenty of products in the market worth analyzing and modeling (including systems for optical coherence tomography, laser projectors, LCoS spatial light modulators, and many others). We have also left out a discussion of reliability theory and its practical applications, which is an important subject for product development. Last, but certainly not least, we have not covered optical systems in the ultraviolet and infrared. All of these topics continue to be entries on my list of things I still need to understand (or rather, things I need to spend more time on). When I have done enough work on these, perhaps I may write about them.

In a book that hopes to share certain fundamentals, techniques, tips, and tricks, there is a tendency to skip certain details, which may result in something I call a *principle of knowledge dilution*, which is that *every step in simplifying the explanation of a highly technical concept is likely to result in a proportional reduction in accuracy*. For example, in providing a condensed overview of lens design principles in section 1.4, I showed you just enough material to get you started on designing lenses. But there is a little more to be said about the act of designing lenses for specific applications and conditions. For instance, certain applications require special consideration of the use of weights in the merit function editor in tracing rays to minimize spot size and wavefront error. In some cases, a judicious selection of weights that progressively decrease from the center of the pupil outward can be desirable; this is intended to ensure better focus (and therefore a higher concentration of flux) for rays within the central zone [99, 100]. This does not mean that you should weight the rays in this manner all the time. Rather, it is meant to highlight that the ray weights are considered specific to different applications and conditions. This specific example of a subtle detail in lens design also highlights the need for you to broaden your search for answers by not limiting yourself to this book. In practice, one considers as many sources of information as possible before arriving at a conclusion on anything. On this note, please do not forget about the appendices in this book, where I share further notes on specific topics of each chapter, including some brief notes on some advanced topics in appendix D.

To conclude, what further advice may I share here? Well, there is one item. MCOSD is a resource book for the aspiring optical system designer who has been *thrown* into today's modern and capitalistic world of fast-paced product development. Due to this pace, in the course of my career, I have seen some early-career engineers and designers struggle to get things done. Now, I root for the underdog, because I once struggled myself. What I have learned is that in order to make progress in such a rapid work environment in the real world, one has to take leaps and accept that there are risks involved, provided that we consider those risks. For example, if you are doing an exercise in designing a lens, even if you have not reviewed basic aberration theory, maybe you could just begin by placing variables on surface radii and then optimizing? The risk is that you could end up with a terrible design. But it would only take seconds or minutes for you to confirm this. Modern optical design programs are so powerful that you can try anything with them and they will spit out the result before your coffee gets cold. After that, you can return to the books (such as this one). Product development progresses by taking leaps, accepting risks, and creating a fallback plan when issues arise. If the instrument your team is developing comprises an imaging lens, an illumination unit, and a detector, then rather than spending months modeling each subassembly in detail, why not make a best guess at some preliminary choices of COTS components and just string them together in the lab? Then, take a measurement. Get an initial result and therefore an initial feeling for the system. This measurement is good feedback for your models. You just have to tell the managers that the lab jig is not yet the product (make sure you do this, because managers can get optimistic too early in the process). Or maybe do not tell anyone. Just let your best friends at work know about your secret experiment. When you are ready with a fairly reliable result, then it is time for show and tell. What could be the worst-case scenario? Messing up and getting blamed, I guess. But making mistakes earlier is better than later. Also, as Hugh MacLeod put it, 'If you can accept the pain, it cannot hurt you' [101]. I wish you all the very best!

References and further reading

[1] Bunch R M 2021 *Optical Systems Design Detection Essentials* (IOP Series in Emerging Technologies in Optics and Photonics, ed R Barry Johnson) (Bristol: IOP Publishing) ch 1

[2] Kasunic K J 2011 *Optical Systems Engineering* (New York: McGraw-Hill) pp 1–21

[3] Hu C Y 2023 Private communication regarding his work-in-progress book entitled Robotics and Autonomous Systems Computing Architecture: A Practical Approach for Ground, Air, and Space Autonomous Systems

[4] Siew R H 2002 Partial coherent imaging using the grating light valve *Proc. SPIE* **4773** 92–101

[5] Edwards K, Logan J and Saunders N (ed) 2004 *Real-Time PCR: An Essential Guide* (Norfolk: Horizon Bioscience)

[6] Gevertz J L, Dunn S M and Roth C M 2005 Mathematical model of real-time PCR kinetics *Biotechnol. Bioeng.* **92** 346–55

[7] Lakowicz J R 2006 *Principles of Fluorescence Spectroscopy* (New York: Springer) pp 58–9 331–48

[8] Huggett J F, French D, O'Sullivan D M, Moran-Gilad J and Zumla A 2022 Monkeypox: another test for PCR *Euro Surveill.* **27** pii=2200497

[9] World Health Organization *Laboratory Testing for the Monkeypox Virus: Interim Guidance* (https://www.who.int/publications/i/item/WHO-MPX-laboratory-2022.1) (Accessed 22 September 2023)

[10] Livak K J and Schmittgen T D 2001 Analysis of relative gene expression data using real-time quantitative PCR and the 2^(-$\Delta\Delta$Ct) method *Methods* **25** 402–8

[11] Pfaffl M W 2001 A new mathematical model for relative quantification in real-time RT-PCR *Nucleic Acids Res.* **29** e45

[12] Rogers-Broadway K-R and Karteris E 2015 Amplification efficiency and thermal stability of qPCR instrumentation: current landscape and future perspectives *Exp. Ther. Med.* **10** 1261–4

[13] Kim Y H, Yang I, Bae Y-S and Park S-R 2008 Performance evaluation of thermal cyclers for PCR in a rapid cycling condition *Biotechniques* **44** 495–505

[14] Ross S M 2009 *Introduction to Probability and Statistics for Engineers and Scientists* 4th edn (Burlington, MA: Elsevier) pp 107–9, 206–7, 215–6, 274, 517, 598

[15] Wasserstein R L and Lazar N A 2016 The ASA statement on *p*-values: context, process, and purpose *Am. Stat.* **70** 129–33

[16] Squires G L 2001 *Practical Physics* 4th edn (Cambridge: Cambridge University Press) pp 14–5 106–9, 138–42

[17] Taylor J R 1997 *An Introduction to Error Analysis: The Study of Uncertainties in Physical Measurements* 2nd edn (Sausalito, CA: University Science Books) pp 3–14, 25–7, 97–104, 211–2, 227–36, 294–8

[18] Stewart J 2003 *Multivariable Calculus* 5th edn (Belmont, CA: Thomson Learning) p 1000

[19] Rimmer M P 1970 Analysis of perturbed lens systems *App. Opt.* **9** 533–7

[20] Koch D G 1978 A statistical approach to lens tolerancing *Proc. SPIE* **0147** 71–82

[21] Adams G 1988 Selection of tolerances *Proc. SPIE* **0892** 173–85

[22] Youngworth R N 2006 Twenty-first century optical tolerancing: a look at the past and improvements for the future *Proc. SPIE* **6342** 634203

[23] Youngworth R N 2011 Statistical truths of tolerance assignment in optical design *Proc. SPIE* **8131** 81310E

[24] Siew R 2019 *Monte Carlo Simulation and Analysis in Modern Optical Tolerancing* (Bellingham, WA: SPIE) pp 9–18, 20–5

[25] Andrews L C and Phillips R L 2012 *Field Guide to Probability, Random Processes, and Random Data Analysis* (Bellingham, WA: SPIE)

[26] Walpole R E and Myers R H 1985 *Probability and Statistics for Engineers and Scientists* 3rd edn (New York: MacMillan) pp 163–222

[27] Boyd R W 1983 *Radiometry and the Detection of Optical Radiation* (New York: John Wiley and Sons) pp 122, 128–32, 150–6

[28] Sahai H and Ageel M I 2000 *The Analysis of Variance: Fixed, Random, and Mixed Models* (Boston, MA: Birkhäuser)

[30] Pishro-Nik H 2014 *Introduction to Probability, Statistics, and Random Processes* (Amherst, MA: Kappa Research) pp 170–5

[29] Ross S M 2005 *Introductory Statistics* 2nd edn (San Diego, CA: Elsevier) chs 7, 8, and 14

[31] Bevington P R and Robinson D K 2003 *Data Reduction and Error Analysis for the Physical Sciences* 3rd edn (New York: McGraw-Hill) pp 2, 267–8

[32] Baird D C 1995 *Experimentation: An Introduction to Measurement Theory and Experiment Design* 3rd edn (Englewood Cliffs, NJ: Prentice-Hall) pp 62–4

[33] Creveling C M, Slutsky J L and Antis D 2003 *Design for Six Sigma in Technology and Product Development* (Upper Saddle River, NJ: Pearson Education) pp 137, 495–504, 601–14

[34] Sykes C 1994 *No Ordinary Genius: The Illustrated Richard Feynman* (New York: W. W. Norton) ch 3

[35] Montgomery D C and Runger G C 2007 *Applied Statistics and Probability for Engineers* (Hoboken, NJ: Wiley) pp 245–6

[36] Bajorski P 2012 *Statistics for Imaging, Optics, and Photonics* (Hoboken, NJ: Wiley) p 55

[37] Montgomery D C 2009 *Introduction to Statistical Quality Control* 6th edn (Hoboken, NJ: Wiley) pp 28–9

[38] O'Connor P D T and Kleyner A 2012 *Practical Reliability Engineering* 5th edn (West Sussex: Wiley)

[39] Liu D 2016 *Systems Engineering: Design Principles and Models* (Boca Raton, FL: CRC) pp 159–231

[40] Goodman J W 1985 *Statistical Optics* (New York: Wiley) pp 64–8

[41] Castleman K R 1996 *Digital Image Processing* (Upper Saddle River, NJ: Prentice Hall) pp 217–9

[42] Miyaguchi K, Suzuki H, Dezaki J and Yamamoto K 1999 CCD developed for scientific application by Hamamatsu *Nucl. Instrum. Methods Phys. Res., Sect. A* **436** 24–31

[43] Schneider C A, Rasband W S and Eliceiri K W 2012 NIH Image to ImageJ: 25 years of image analysis *Nat. Methods* **9** 671–5

[44] Bohndiek S E, Blue A, Clark A T, Prydderch M L, Turchetta R, Royle G J and Speller R D 2008 Comparison of methods for estimating the conversion gain of CMOS active pixel sensors *IEEE Sens. J.* **8** 1734–44

[45] Pain B and Hancock B 2003 Accurate estimation of conversion gain and quantum efficiency in CMOS sensors *Proc. SPIE* **5017** 94–103

[46] Nakamoto K and Hotaka H 2022 Efficient and accurate conversion-gain estimation of a photon-counting image sensor based on maximum likelihood estimation *Opt. Express* **30** 37493–506

[47] Klymus K E *et al* 2019 Reporting the limits of detection and quantification for environmental DNA assays *Environ. DNA* **2** 271–82

[48] Burns M and Valdivia H 2008 Modelling the limit of detection in real-time quantitative PCR *Eur. Food Res. Technol.* **226** 1513–24

[49] Forootan A, Sjöback R, Björkman J, Sjörgreen B, Linz L and Kubista M 2017 Methods to determine limit of detection and limit of quantification in quantitative real-time PCR (qPCR) *Biomol. Detect. Quantif.* **12** 1–6

[50] Ahmed W, Bivins A, Metcalfe S, Smith W J M, Verbyla M E, Symonds E M and Simpson S L 2022 Evaluation of process limit of detection and quantification variation of SARS-CoV-2 RT-qPCR and RT-dPCR assays for wastewater surveillance *Water Res.* **213** 118132

[51] Shonkwiler R W and Mendivil F 2009 *Explorations in Monte Carlo Methods* (New York: Springer)

[52] Andersen T B and Schweiger P F 2008 Assessment of NIRCam alignment tolerances by Monte Carlo simulations *Proc. SPIE* **7068** 70680H

[53] Schwertz K and Burge J H 2012 *Field Guide to Optomechanical Design and Analysis* (Bellingham, WA: SPIE)

[54] Benton D M 2021 *Alignment of Optical Systems Using Lasers* (Bellingham, WA: SPIE)

[55] Hobbs P C D 2000 *Building Electro-Optical Systems: Making It All Work* 1st edn (Danvers, MA: Wiley) ch 12

[56] Bayar M 1981 Lens barrel optomechanical design principles *Opt. Eng.* **20** 181–6

[57] Lamontagne F and Desnoyers N 2019 New solutions in precision mounting *The 11th Int. Conf. on Optics-Photonics Design and Fabrication (ODF'18) (Hiroshima, Japan); Opt. Rev.* **26** 396–405

[58] Siew R 2017 *Perspectives on Modern Optics and Imaging: With Practical Examples using Zemax® OpticStudio®* (Independently Published) pp 150–5

[59] Grant G B 2011 *Field Guide to Radiometry* (Bellingham, WA: SPIE) pp 28–9

[60] 2014 Theoretical Concepts in Spectrophotometric Measurements *Spectrophotometry: Accurate Measurement of Optical Properties of Materials* (Experimental methods in the physical sciences) ed T A Germer, J C Zwinkels and B K Tsai (Amsterdam: Elsevier) 46 2

[61] Renhorn I G E, Hallberg T and Boreman G D 2023 *Accurate Physics-Based Polarimetric BRDF* (Bellingham, WA: SPIE)

[62] Williams T L 2009 *Thermal Imaging Cameras: Characteristics and Performance* (Boca Raton, FL: CRC Press) pp 28–31, 59–60, 83–4, 119–20, 157

[63] Lau A S 1977 The Narcissus effect in infrared optical scanning systems *Proc. SPIE* **0107** 57–62

[64] ISO 9358 1994 *Optics and Optical Instruments —Veiling Glare of Image Forming System— Definitions and Methods of Measurement* (Geneva: International Organization for Standardization)

[65] Macleod H A 2018 *Thin-Film Optical Filters* (Boca Raton, FL: CRC Press) pp 276–85

[66] Wiley R R 2006 *Field Guide to Optical Thin Films* (Bellingham, WA: SPIE) p 38

[67] Siew R 2023 *Multiple-Field Multispectral Imaging Using Wide-Angle Lenses* (Bellingham, WA: SPIE)

[68] Liang R 2010 *Optical Design for Biomedical Imaging* (Bellingham, WA: SPIE) pp 67–100 184–219

[69] Gradshteyn I S and Ryzhik 2000 *Table of Integrals, Series, and Products* 6th edn ed A Jeffrey and D Zwillinger (San Diego, CA: Academic) p 8

[70] Pedrotti F L S J and Pedrotti L S 1993 *Introduction to Optics* 2nd edn (Englewood Cliffs, NJ: Prentice-Hall) pp 356–9, 505–6

[71] Koshel R J 2013 *Illumination Engineering* (Hoboken, NJ: Wiley) pp 214–6, 220–6

[72] Sakar S, Rönnberg S and Bollen M H J 2018 Light intensity behavior of LED lamps within the thermal stabilization period *IEEE Xplore* pp 1–6

[73] Chung T-Y 2017 *Thermal Management: Component to Systems Level in Handbook of Advanced Lighting Technology* ed R Karlicek, C-C Sun, G Zissis and R Ma (Cham: Springer) pp 239–67

[74] Riedl M J 2001 *Optical Design Fundamentals for Infrared Systems* 2nd edn (Bellingham, WA: SPIE) ch 7

[75] Altshuller G 1999 *The Innovation Algorithm: TRIZ, Systematic Innovation and Technical Creativity* (Worcester, MA: Technical Innovation Center)

[76] Siew R, Wei S Y and Hung J-S 2021 Optical system, and method of illuminating a sample plane *US Patent* 11567007

[77] Martín-Palma R J 2021 *Field Guide to Optical Biosensing* (Bellingham, WA: SPIE)

[78] Rosales-Guzmán C and Forbes 2017 *A How to Shape Light with Spatial Light Modulators* (Bellingham, WA: SPIE)

[79] Gmuender T 2016 *DLP Using Digital Micromirror Devices: A Primer* (Bellingham, WA: SPIE)

[80] Restaino S R and Teare S 2015 *Introduction to Liquid Crystals for Optical Design and Engineering* (Bellingham, WA: SPIE)

[81] Payne A, DeGroot W, Monteverde R and Amm D 2004 Enabling high-data-rate imaging applications with Grating Light Valve technology *Proc. SPIE* **5348** 76–88

[82] Lyu B-H, Wang C and Tsai C-W 2017 Integration of LCoS-SLM and LabVIEW based software to simulate fundamental optics, wave optics, and Fourier optics *Proc. SPIE* **10452** 104521P

[83] Dan D *et al* 2013 DMD-based LED-illumination super-resolution and optical sectioning microscopy *Sci. Rep.* **3** 1116

[84] Allen J 2017 Application of patterned illumination using a DMD for optogenetic control of signaling *Nat. Methods* **14** 1114

[85] Pan J-W and Wang H-H 2013 High contrast ratio prism design in a mini projector *Appl. Opt.* **52** 8347–54

[86] Rice J P, Neira J E, Kehoe M and Swanson R 2009 DMD diffraction measurements to support design of projectors for test and evaluation of multispectral and hyperspectral imaging sensors *Proc. SPIE* (Bellingham, WA: SPIE) 7210 72100D

[87] Deng M-J, Zhao Y-Y, Liang Z-X, Chen J-T, Zhang Y and Duan X-M 2022 Maximizing energy utilization in DMD-based projection lithography *Opt. Express* **30** 4692–705

[88] Liang J, Wu S-Y, Kohn R N, Becker M F and Heinzen D J 2012 Grayscale laser image formation using a programmable binary mask *Opt. Eng.* **51** 108201

[89] Xiong Z, Liu H, Tan X, Lu Z, Li C, Song L and Wang Z 2014 Diffraction analysis of digital micromirror device in maskless photolithography system *J. Micro/Nanolithogr. MEMS MOEMS* **13** 043016

[90] Harvey J E and Pfisterer R N 2019 Understanding diffraction grating behavior: including conical diffraction and Rayleigh anomalies from transmission gratings *Opt. Eng.* **58** 087105

[91] Dong X, Shi Y, Xiao X, Zhang Q, Chen F, Sun X, Yuan W and Yu Y 2022 Non-paraxial diffraction analysis for developing DMD-based optical systems *Opt. Lett.* **47** 4758–61

[92] Dewald D S, Segler D J and Penn S M 2003 Advances in contrast enhancement for DLP projection displays *J. Soc. Inf. Disp.* **11** 177–81

[93] Vorobiev D, Travinsky A, Quijada M A, Ninkov Z, Raisanen A D, Robberto M and Heap S 2016 Measurements of the reflectance, contrast ratio, and scattering properties of digital micromirror devices (DMDs) *Proc. SPIE* **9912** 99125U

[94] Chang C-M and Shieh H-P D 2000 Design of illumination and projection optics for projectors with single digital micromirror devices *Appl. Opt.* **39** 3202–8

[95] Xiong Z, Liu H, Chen R, Xu J, Li Q, Li J and Zhang W 2018 Illumination uniformity improvement in digital micromirror device based scanning photolithography system *Opt. Express* **26** 18597–607

[96] Bai X, Jing X and Liao N 2021 Design method for the high optical efficiency and uniformity illumination system of the projector *Opt. Express* **29** 12502–15

[97] Sun W-S, Chiang Y-C and Tsuei C-H 2011 Optical design for the DLP pocket projector using LED light source *Phys. Procedia* **19** 301–7

[98] Siew R and Tan L 2023 Top hat illumination provides even light distribution across samples *Biophotonics* **30** 32–7

[99] Kingslake R and Johnson R B 2010 *Lens Design Fundamentals* 2nd edn (Burlington, MA: Elsevier) pp 194–7 522–3

[100] Smith G H 1998 *Practical Computer-Aided Lens Design* (Richmond, VA: William-Bell) pp 174–6

[101] MacLeod H 2009 *Ignore Everybody: and 39 Other Keys to Creativity* (New York: Penguin) pp 52–3

[102] Siew R 2021 Basics of optical systems in real-time PCR instruments for virus detection *XLIV OSI Symposium on Frontiers in Optics and Photonics 2021*

Modern Classical Optical System Design
Fundamentals, techniques, tips, and tricks
Ronian Siew

Appendix A

Further notes on imaging

A.1 The relationship between optical crosstalk and MTF

Here, by *optical crosstalk* (or simply crosstalk) we mean the spread of optical flux from an image to an adjacent image on the same image plane, as depicted in figure A.1.

In some cases of practical interest, we may not be able to overcome the challenge of producing perfectly sharp images of objects at some spatial frequency, but we also cannot allow excessive crosstalk between those images. So, the lens that is responsible for doing the imaging may be required to possess an MTF at some spatial frequency ν such that the level of crosstalk is tolerable. We can gain intuition about the relationship between the MTF and crosstalk as follows. For simplicity (and for a reason that will become clear), instead of expressing the two images in figure A.1(a) as rectangular functions with sharp boundaries, let us represent them as *ideal* images given by Gaussian functions with a very narrow width w, expressed as

$$G_1(x) = \frac{\phi_1}{w\sqrt{\pi}} \exp\left\{-[(x - x_1)/w]^2\right\}, \qquad (A.1)$$

and

$$G_2(x) = \frac{\phi_2}{w\sqrt{\pi}} \exp\left\{-[(x - x_2)/w]^2\right\}, \qquad (A.2)$$

where x_1 and x_2 are their positions along the positive x-axis in figure A.1 and ϕ_1 and ϕ_2 are their fluxes, respectively. These images (and a blurred version of $G_1(x)$) are depicted in figure A.2.

Suppose the lens that produces the two narrow-width Gaussian images has a space-invariant point spread function (PSF). Then, the blurred version of $G_1(x)$ is the result of its convolution with that PSF. Suppose that this PSF is also Gaussian but its width is w_p, expressed as

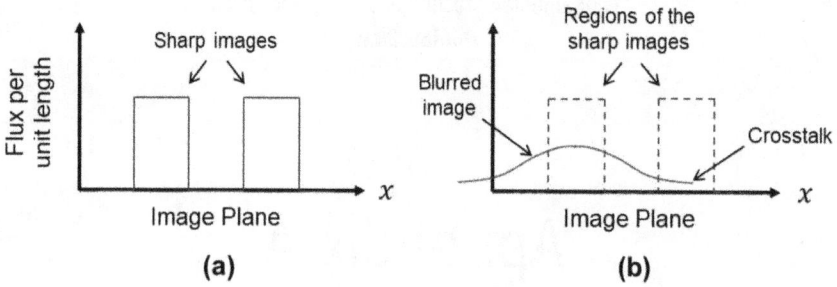

Figure A.1. The definition of crosstalk. (a) Two sharply focused images. (b) The left-hand image has been aberrated or defocused, resulting in the spread of flux into the right-hand image.

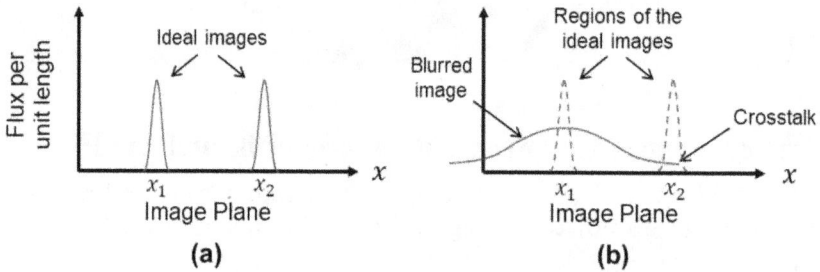

Figure A.2. Crosstalk from a blurred Gaussian image. (a) Two ideal narrow Gaussian images described by equations (A.1) and (A.2). (b) The left-hand image has been aberrated or defocused, resulting in the spread of flux into the right-hand image.

$$P(x) = \exp\left[-\left(x/w_p\right)^2\right]. \tag{A.3}$$

The image distribution $I_1(x)$ (in units of flux per unit length) for the blurred version of $G_1(x)$ is then given by the convolution

$$I_1(x) = \frac{\phi_1}{\pi w w_p} \int_{-\infty}^{\infty} P(x - x')G_1(x')dx', \tag{A.4}$$

where ϕ_1 is the flux in the image and $1/(\pi w w_p)$ is a normalization factor that ensures that the integral of $I_1(x)$ over all linear space is equal to the flux ϕ_1 (see section 2.2.6). Substituting equations (A.1) and (A.3) into (A.4) and regrouping terms, we obtain

$$I_1(x) = \frac{\phi_1}{\pi w w_p} \int_{-\infty}^{+\infty} \exp\left\{-\left[(x - x')/w_p\right]^2 - \left[(x' - x_1)/w\right]^2\right\}dx'. \tag{A.5}$$

To perform the integration shown in equation (A.5), rearrange terms in the exponent into the form $a(x')^2 + bx' + c$, then apply the rule $\int_{-\infty}^{+\infty} \exp\{-[a(x')^2 + bx' + c]\}dx' = \sqrt{\pi/a} \exp[(b^2 - 4ac)/4a]$ (see, for example, Spiegel M R 1994 *Mathematical Handbook of Formulas and Tables* (New York: McGraw-Hill) page 98). Doing this, equation (A.5) may be expressed as

$$I_1(x) = \frac{\phi_1}{\sqrt{\pi\left(w^2 + w_p^2\right)}} \exp\left[-\frac{(x - x_1)^2}{w^2 + w_p^2}\right]. \tag{A.6}$$

In equation (A.6), note that in the limit $w_p \to 0$, $I_1(x)$ takes the form of equation (A.1), which we would expect (in this limit, it is the same as regarding the PSF in equation (A.4) as a Dirac delta function impulse response). The amount of crosstalk at position x_2 from $I_1(x)$ may be regarded as a small amount of flux $d\phi_1 \approx I_1(x_2)dx$. Since dx is just an elemental length, we may disregard it for analytical purposes and simply consider what happens to $I_1(x_2)$ as the PSF's half-width w_p increases. Therefore, the crosstalk into x_2 may be expressed as

$$I_1(x_2) = \frac{\phi_1}{\sqrt{\pi\left(w^2 + w_p^2\right)}} \exp\left[-\frac{(x_2 - x_1)^2}{w^2 + w_p^2}\right]. \tag{A.7}$$

Equation (A.7) makes sense, as the crosstalk depends not only on the size of the PSF's half-width but also how close $G_1(x)$ and $G_2(x)$ are to each other, as given by $(x_2 - x_1)$. Further, a careful examination of equation (A.7) reveals that the amount of crosstalk $I_1(x_2)$ cannot increase indefinitely, because in the limit $w_p \gg w$, the exponent approaches one, while the factor outside the exponent tends to zero. This is sensible, because as the PSF's half-width increases, the flux in the image $I_1(x)$ spreads over an infinite space, thereby reducing the flux per unit length everywhere. Indeed, a plot of equation (A.7) using some selected values for the variables reveals this trend, as shown in figure A.3.

In the current example, the relationship between the crosstalk and the MTF is in the width of the PSF, w_p, which is a function of the width of the MTF curve. Since the Fourier transform of the PSF is proportional to the optical transfer function (OTF), we may take the Fourier transform of equation (A.3) and obtain

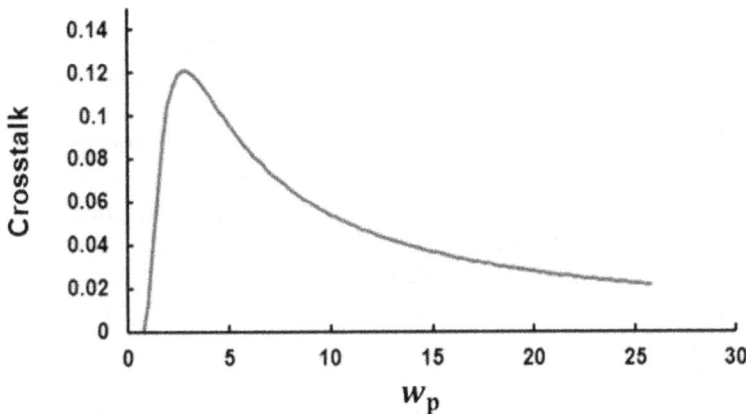

Figure A.3. Crosstalk $I_1(x_2)$ vs the PSF's half-width w_p, obtained by applying equation (A.7) with $w = 0.25$ mm, $x_1 = 5$ mm, and $x_2 = 7$ mm. The unit is mm in the horizontal axis.

$$\text{OTF} \propto \int_{-\infty}^{+\infty} \exp\left[-(x/w_{\text{p}})^2\right] \exp\left(-i2\pi\nu_x x\right) \mathrm{d}x$$

$$\propto w_{\text{p}}\sqrt{\pi} \, \exp\left(-\pi^2 w_{\text{p}}^2\nu_x^2\right), \tag{A.8}$$

where $i = \sqrt{-1}$ and ν_x is spatial frequency in cycles/mm. Note that the integration in equation (A.8) has been performed, once again, by applying the rule for rearranging terms in the exponent that I mentioned above. The MTF is just the normalized modulus (magnitude) of the OTF, so that taking only the exponent in equation (A.8) gives

$$\text{MTF} = \exp\left(-\pi^2 w_{\text{p}}^2\nu_x^2\right). \tag{A.9}$$

This is just another Gaussian function (you can now see why I used Gaussian functions in this example, as the Fourier transform of a Gaussian is a Gaussian, and the convolution of two Gaussians is a Gaussian). More interestingly, if we let w_{MTF} denote the half-width of the MTF curve, then we note from equation (A.9) that this width is given by

$$w_{\text{MTF}}^2 = 1/\left(\pi^2 w_{\text{p}}^2\right). \tag{A.10}$$

So, applying equation (A.10) to equation (A.7) gives

$$I_1(x_2) = \frac{\phi_1}{\sqrt{\pi[w^2 + 1/(\pi^2 w_{\text{MTF}}^2)]}} \exp\left[-\frac{(x_2 - x_1)^2}{w^2 + 1/(\pi^2 w_{\text{MTF}}^2)}\right]. \tag{A.11}$$

Equation (A.11) is about as explicit as one can get in showing the relationship between crosstalk and MTF. However, we can go even further by taking the natural logarithm of equation (A.9), which gives

$$-\ln(\text{MTF}) = \pi^2 w_{\text{p}}^2\nu_x^2. \tag{A.13}$$

The width w_{p} in equation (A.12) can be substituted back into equation (A.7), as was done for equations (A.10) and (A.7), thereby providing a direct link between crosstalk and MTF, but we will save some paper in this book and leave it at that. Of course, the current example uses Gaussian functions to represent images and PSFs, but there is no reason to believe that the general characteristics implied by the present analysis would be any different if rectangular functions were used for the images or if PSFs with lens aberrations were used. In fact, Gaussian PSFs are possible if the lens is apodized by a Gaussian transmittance function (see, for example, pp 66–74 and pp 151–4 in references [8] and [34] of chapter 1, respectively).

The point of this exercise is to address the idea that if a lens's MTF can be related to a specific performance metric (e.g. crosstalk in this case), then it may not be necessary to qualify the lens by its MTF when the lens is received from a supplier and integrated into a product, which can be time-consuming. The current exercise is an analytical approach to show the relationship between crosstalk and MTF, but it can also easily be simulated in OS using the image simulation and GIA features discussed in section 1.2.10. So, correlating measured MTF curves from the lens supplier with a more easily measurable quantity (such as crosstalk) is sensible if they are relatable by some formula in closed form. It is not like correlating two completely separate things (such as the occurrence of rainfall and winning the lottery). A practical approach is to spend some time thinking about the physics that relates the quantities of interest, do some analysis and modeling, then perform some measurements to obtain the correlation.

A.2 A Bravais optical system design example

Based on the equations presented in section 1.3.11, we form the first-order layout and prescription shown in figure A.4, whose starting merit function is provided in figure A.5. As it is a paraxial system model, the wavelength settings are inconsequential, but for consistency throughout this exercise, we will maintain a single wavelength at 550 nm. In the prescription (figure A.4), notice that there is a 200 mm EFL tube lens. The idea of the current example is to design a suitable Bravais lens for the microscope system shown in figure 1.131. So, in the current first-order setup, a simple paraxial thin lens is made to represent that tube lens. Suppose that we have no prior knowledge of the prescription for the microscope system in figure 1.131, other than the field angle (roughly 3.6 degrees) from the pupil, the tube lens's EFL of 200 mm, and that tube lens's BFL, which is 174.25 mm. We then decide to allow a clearance of 4.25 mm behind the tube lens, leaving 170 mm for the distance L, which represents the dimension indicated for L in figure 1.139b. Thus, notice in figure A.4 that a reference surface labeled *Bravais Ref* has been included. The idea is to let that surface be the reference at 4.25 mm from the back of the actual tube lens, so that in the current Bravais design, the total distance $L = 170$ mm is maintained. In the

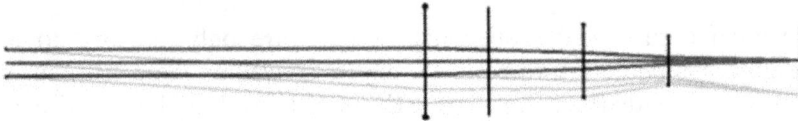

	Surface Type	Comment	Radius	Thickness	Semi-Diameter	Focal Length	OPD Mode
0	OBJECT Standard ▼		Infinity	Infinity	Infinity		
1	STOP Standard ▼		Infinity	200.00000	6.00000 U		
2	Paraxial ▼	"Tube"		30.00000	25.00000 U	200.00000	1
3	Standard ▼	Bravais Ref	Infinity	45.00000 V	25.00000 U		
4	Paraxial ▼	Lens 1		40.00000 V	16.33293	150.00000 V	1
5	Paraxial ▼	Lens 2		64.58333 M	10.77748	-50.00000 V	1
6	IMAGE Standard ▼		Infinity	-	15.72867		

Figure A.4. The Bravais starting first-order layout and its prescription.

	Type	Surf1	Surf2		Target	Weight	Value	% Contrib
1	BLNK ▾	f1						
2	EFLY ▾	4	4		0.00000	0.00000	150.00000	0.00000
3	BLNK ▾	f2						
4	EFLY ▾	5	5		0.00000	0.00000	-50.00000	0.00000
5	BLNK ▾	|f2|						
6	ABSO ▾	4			0.00000	0.00000	50.00000	0.00000
7	BLNK ▾	x						
8	TTHI ▾	5	5		0.00000	0.00000	64.58333	0.00000
9	BLNK ▾	d						
10	TTHI ▾	4	4		0.00000	0.00000	40.00000	0.00000
11	BLNK ▾	y						
12	REAY ▾	4	1 0.00000 0.00000 0.00000 1.00000		0.00000	0.00000	3.75000	0.00000
13	BLNK ▾	y'						
14	REAY ▾	5	1 0.00000 0.00000 0.00000 1.00000		0.00000	0.00000	1.55000	0.00000
15	BLNK ▾	m, by equation (1.29)						
16	DIVI ▾	12	14		0.00000	0.00000	2.41935	0.00000
17	SUMM ▾	8	10		0.00000	0.00000	104.58333	0.00000
18	DIVI ▾	8	17		0.00000	0.00000	0.61753	0.00000
19	PROD ▾	16	18		1.25000	1.00000	1.49402	0.01428
20	BLNK ▾	m, by equation (1.30)						
21	DIVI ▾	2	6		0.00000	0.00000	3.00000	0.00000
22	PROD ▾	18	18		0.00000	0.00000	0.38134	0.00000
23	PROD ▾	21	22		1.25000	1.00000	1.14403	2.69356E-03
24	BLNK ▾	L (constant for a Bravais)						
25	TTHI ▾	3	5		170.00000	1.00000	149.58333	99.98302

Merit Function Editor — Wizards and Operands — Merit Function: 11.7885686889046

Figure A.5. The starting merit function for the layout shown in figure A.4.

merit function (figure A.5), notice that weights are only assigned to the two magnification computations on rows 19 and 23 as well as the constraint on total Bravais system length (row 25). The EFLs for Lens 1 and Lens 2 have been arbitrarily chosen, as per the suggestion made in section 1.3.11 to make initial guesses and then let them be optimized. After a single local damped least-squares optimization run, the result is the layout and prescription shown in figure A.6, whose merit function is provided in figure A.7.

According to the optimized first-order merit function in figure A.7, the EFLs for Lens 1 and Lens 2 are roughly 170 mm and −81 mm, respectively. To turn these paraxial thin lens models into real lenses, we can begin by selecting COTS elements from an online catalog and re-optimizing them. Reasonable starting selections are part numbers 47–740 and 63–764 from Edmund Optics. We then replace the thin lens models in figure A.6 by these COTS elements to obtain the layout and prescription shown in figure A.8 and set up a new merit function for them (figure A.9). Notice the change in construction of the operands in the new merit function, which is meant for real elements. There are no longer distances given by x, y, y', and d, which only apply to thin lenses. Moreover, the magnification is now

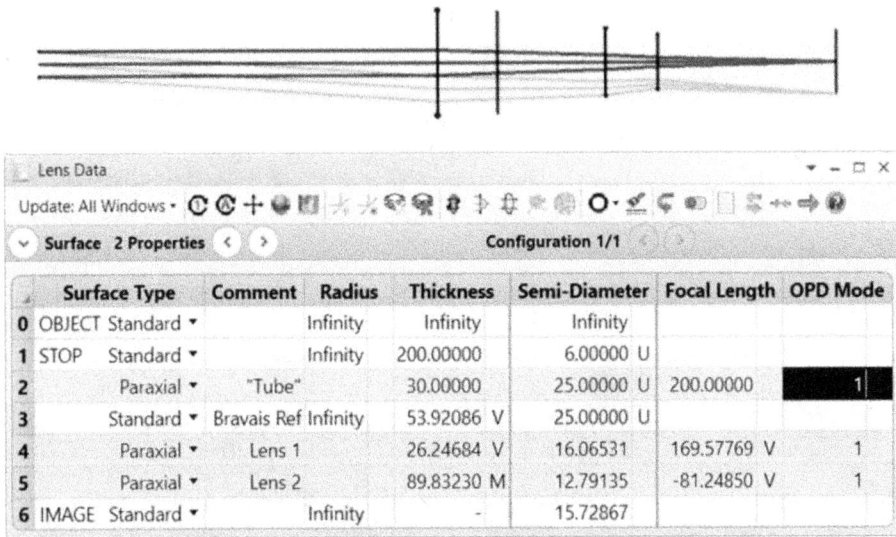

Figure A.6. The optimized first-order layout and prescription.

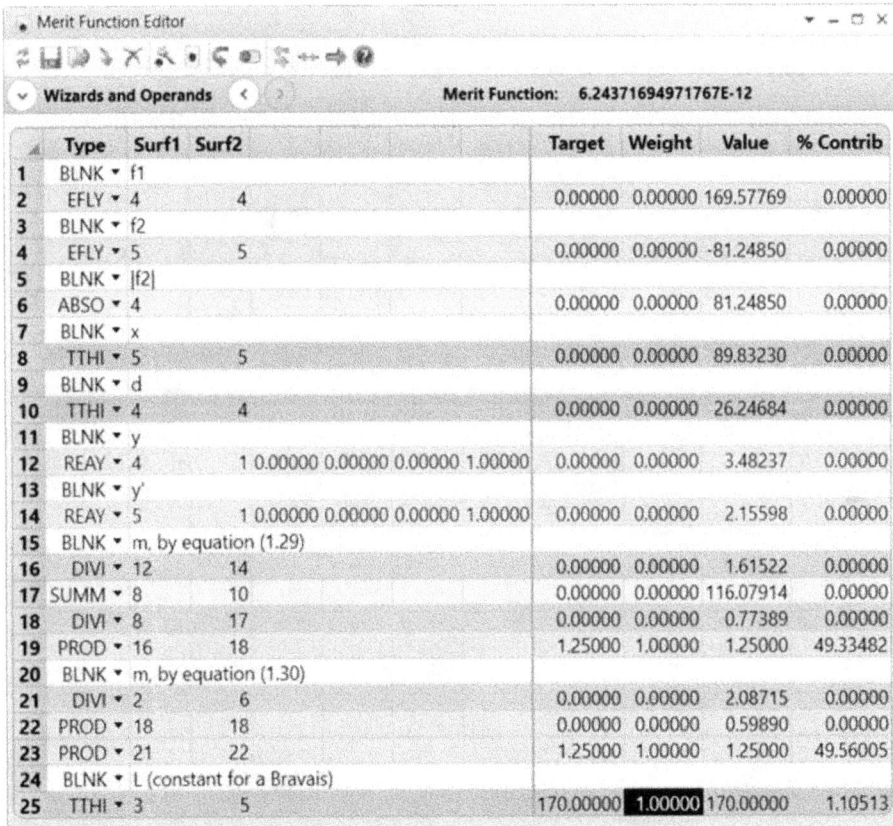

	Surface Type	Comment	Radius	Thickness	Semi-Diameter	Focal Length	OPD Mode
0	OBJECT Standard ▾		Infinity	Infinity	Infinity		
1	STOP Standard ▾		Infinity	200.00000	6.00000 U		
2	Paraxial ▾	"Tube"		30.00000	25.00000 U	200.00000	1
3	Standard ▾	Bravais Ref	Infinity	53.92086 V	25.00000 U		
4	Paraxial ▾	Lens 1		26.24684 V	16.06531	169.57769 V	1
5	Paraxial ▾	Lens 2		89.83230 M	12.79135	-81.24850 V	1
6	IMAGE Standard ▾		Infinity	-	15.72867		

Figure A.6. The optimized first-order layout and prescription.

Merit Function Editor

Wizards and Operands   Merit Function: 6.24371694971767E-12

	Type	Surf1	Surf2		Target	Weight	Value	% Contrib
1	BLNK ▾	f1						
2	EFLY ▾	4	4		0.00000	0.00000	169.57769	0.00000
3	BLNK ▾	f2						
4	EFLY ▾	5	5		0.00000	0.00000	-81.24850	0.00000
5	BLNK ▾	|f2|						
6	ABSO ▾	4			0.00000	0.00000	81.24850	0.00000
7	BLNK ▾	x						
8	TTHI ▾	5	5		0.00000	0.00000	89.83230	0.00000
9	BLNK ▾	d						
10	TTHI ▾	4	4		0.00000	0.00000	26.24684	0.00000
11	BLNK ▾	y						
12	REAY ▾	4	1 0.00000 0.00000 0.00000 1.00000		0.00000	0.00000	3.48237	0.00000
13	BLNK ▾	y'						
14	REAY ▾	5	1 0.00000 0.00000 0.00000 1.00000		0.00000	0.00000	2.15598	0.00000
15	BLNK ▾	m, by equation (1.29)						
16	DIVI ▾	12	14		0.00000	0.00000	1.61522	0.00000
17	SUMM ▾	8	10		0.00000	0.00000	116.07914	0.00000
18	DIVI ▾	8	17		0.00000	0.00000	0.77389	0.00000
19	PROD ▾	16	18		1.25000	1.00000	1.25000	49.33482
20	BLNK ▾	m, by equation (1.30)						
21	DIVI ▾	2	6		0.00000	0.00000	2.08715	0.00000
22	PROD ▾	18	18		0.00000	0.00000	0.59890	0.00000
23	PROD ▾	21	22		1.25000	1.00000	1.25000	49.56005
24	BLNK ▾	L (constant for a Bravais)						
25	TTHI ▾	3	5		170.00000	1.00000	170.00000	1.10513

Figure A.7. The optimized merit function for figure A.6.

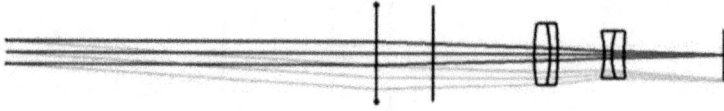

Figure A.8. The starting layout and prescription for a pair of real Bravais elements.

Figure A.9. The starting merit function for the prescription shown in figure A.8.

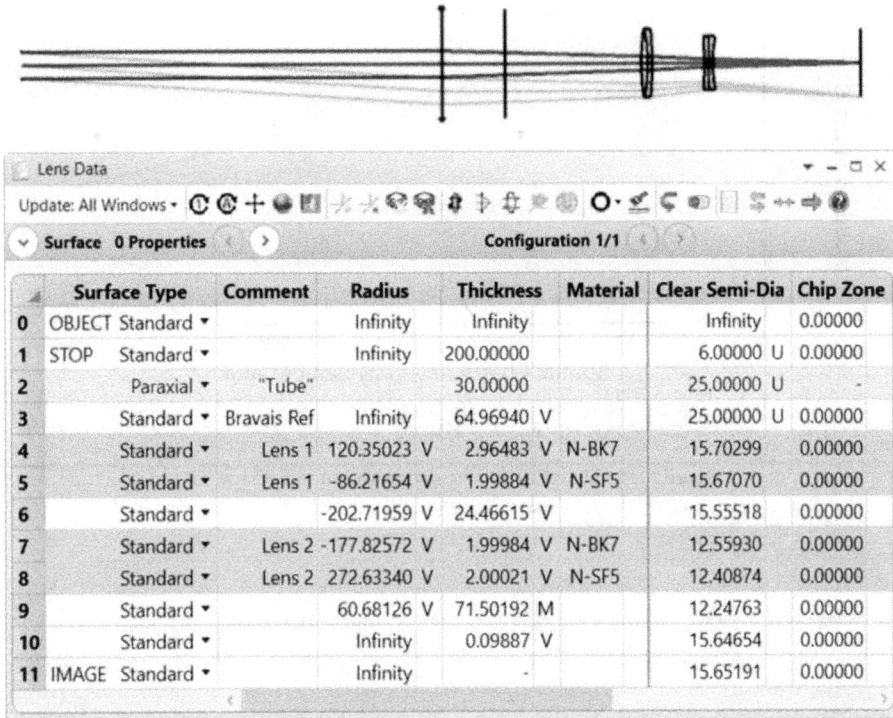

	Surface Type		Comment	Radius		Thickness		Material	Clear Semi-Dia		Chip Zone
0	OBJECT	Standard ▾		Infinity		Infinity			Infinity		0.00000
1	STOP	Standard ▾		Infinity		200.00000			6.00000	U	0.00000
2		Paraxial ▾	"Tube"			30.00000			25.00000	U	-
3		Standard ▾	Bravais Ref	Infinity		64.96940	V		25.00000	U	0.00000
4		Standard ▾	Lens 1	120.35023	V	2.96483	V	N-BK7	15.70299		0.00000
5		Standard ▾	Lens 1	-86.21654	V	1.99884	V	N-SF5	15.67070		0.00000
6		Standard ▾		-202.71959	V	24.46615	V		15.55518		0.00000
7		Standard ▾	Lens 2	-177.82572	V	1.99984	V	N-BK7	12.55930		0.00000
8		Standard ▾	Lens 2	272.63340	V	2.00021	V	N-SF5	12.40874		0.00000
9		Standard ▾		60.68126	V	71.50192	M		12.24763		0.00000
10		Standard ▾		Infinity		0.09887	V		15.64654		0.00000
11	IMAGE	Standard ▾		Infinity		-			15.65191		0.00000

Figure A.10. The optimized Bravais layout and prescription, which started in figure A.8.

computed by way of dividing the real ray height of the current principal ray (row 22) by the real ray height expected of the tube lens (row 21). On row 6, the constraint for L remains, because it must be a Bravais system. On rows 13 to 16, there are now constraints on glass thickness and an air space. RMS spot size operands on rows 8 and 9 now constrain the image quality.

The optimization is first run using constraints on the RMS spot sizes, followed by switching to RMS wavefront errors (rows 10 and 11 in the merit function). The resulting layout/prescription, MTF, and merit function are shown in figures A.10–A.12, respectively. The Bravais lens pair is then integrated with the microscope system in figure 1.131; the resulting system MTF is shown in figure A.13. The integrated layout and prescription are provided in figure A.14. For the monochromatic case, their MTF is comparable to the original microscope system's MTF (figure 1.131). Re-optimization of the Bravais lens pair for a broader wavelength band is possible, as the pair consists of achromats.

In section 1.3.11, I highlighted that if an optical system is diffraction limited, then a Bravais lens system increases magnification at a loss of resolution, due to the natural reduction in the NA of the image. In the case of the current design exercise, there was no loss because the original microscope system (figure 1.131) was not diffraction limited. Since the Bravais lens pair was optimized as much as possible on

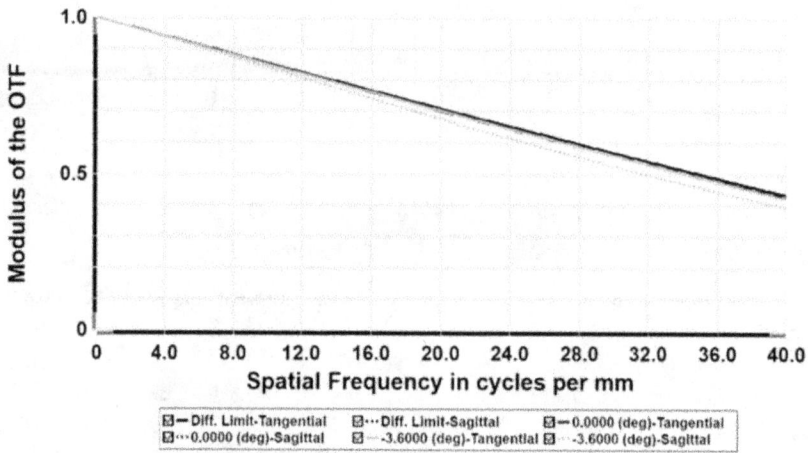

Figure A.11. The MTF for the system shown in figure A.10.

Merit Function Editor

Wizards and Operands — Merit Function: 0.0178335195077503

	Type	Op#1	Op#2					Target	Weight	Value	% Contrib
1	BLNK	f1									
2	EFLY	4	6					0.00000	0.00000	172.81238	0.00000
3	BLNK	f2									
4	EFLY	7	9					0.00000	0.00000	-73.72244	0.00000
5	BLNK	L (constant for a Bravais)									
6	TTHI	3	10					170.00000	1.00000	170.00007	3.17847E-04
7	BLNK	Focus									
8	RSCE	3	1	0.00000	0.00000			0.00000	0.00000	2.32408E-03	0.00000
9	RSCE	3	1	0.00000	1.00000			0.00000	0.00000	2.99272E-03	0.00000
10	RWCE	3	1	0.00000	0.00000			0.00000	1.00000	0.01949	24.87852
11	RWCE	3	1	0.00000	1.00000			0.00000	1.00000	0.03332	72.73850
12	BLNK	Opto-mech									
13	MNCG	4	9					2.00000	0.20000	1.99869	0.02251
14	MXCG	4	9					15.00000	0.20000	15.00000	0.00000
15	MNEG	4	9	1.00000	1			0.50000	0.20000	0.49988	1.84132E-04
16	MNCA	5	7					0.20000	0.20000	0.20000	0.00000
17	BLNK	Magnification									
18	RAID	1	1	0.00000	1.00000	0.00000	0.00000	0.00000	0.00000	3.60000	0.00000
19	TANG	18	1					0.00000	0.00000	0.06291	0.00000
20	EFLY	2	2					0.00000	0.00000	200.00000	0.00000
21	PROD	20	19					0.00000	0.00000	12.58293	0.00000
22	REAY	11	1	0.00000	1.00000	0.00000	0.00000	0.00000	0.00000	15.65314	0.00000
23	DIVI	22	21					1.25000	1.00000	1.24400	2.35997

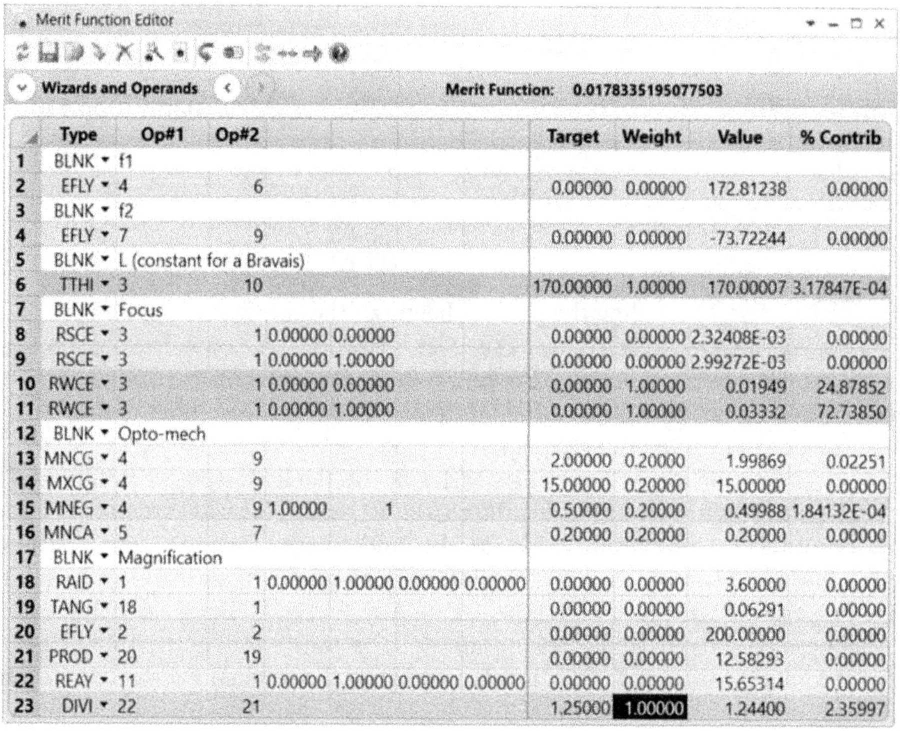

Figure A.12. The optimized merit function for the system shown in figure A.10.

its own, there was no significant change in the integrated system's monochromatic MTF (figure A.14). For a polychromatic condition, the Bravais system would just need to be optimized over a broader band. It should also be noted that the final image quality is significantly improved if all elements (i.e. the objective, tube lens,

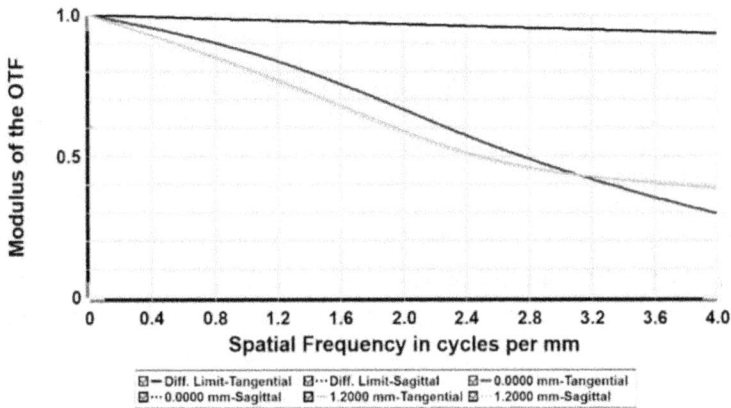

Figure A.13. The MTF (550 nm) for the system shown in figure A.13.

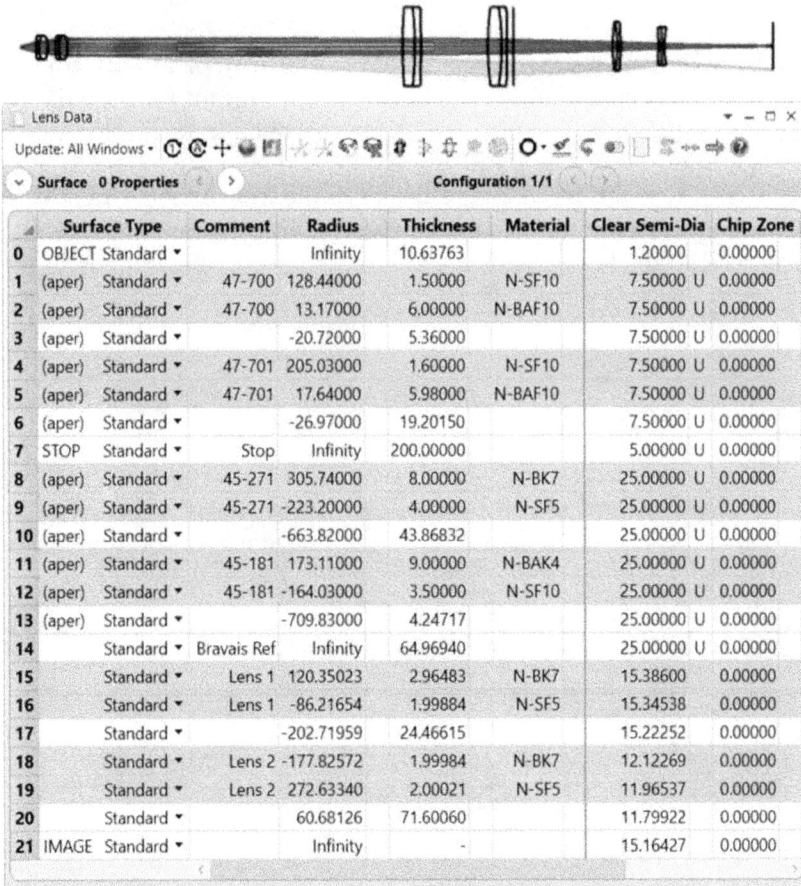

Figure A.14. The microscope system from figure 1.131 integrated with the optimized Bravais lens pair from figure A.10. System wavelength = 550 nm.

and Bravais) in the system are optimized together. However, in most situations, a Bravais system is optimized on its own because it is intended to be inserted into an existing optical system with no change in the back focal distance (or, at most, a slight re-focusing of the image plane if it is accessible). Hence, the current design exercise is intended to mimic that condition.

A.3 Designing an aplanatic singlet by optimizing optical path lengths

In section 1.4.6, the aplanatic aspheric singlet in figure 1.225 was optimized using the RSCE (RMS wavefront error) operand to achieve its MTF. I then highlighted that, alternatively, since the optical path length (OPL) of a ray can be controlled by way of using the OPTH operand in the merit function of OS, it follows that one can achieve similar results for the same lens by constraining ray OPLs to be equal (because after all, this is what lens design is all about, by virtue of Fermat's principle). However, because the oblique rays arrive at the stop at an angle (2°), the stop must be tilted at that angle for those rays, which is the only way to equalize the OPLs for the oblique rays passing through various pupil zones. To see how this is done, figure A.15 displays the optimized layout by way of a multi-configuration approach, in which configuration 1 (figure A.15(a)) is meant for the axial rays and configuration 2 (figure A.15(b)) is meant for the oblique rays. The prescription, multi-configuration editor, MTF (at 587.56 nm), and merit function are provided in figures A.16–A.19, respectively. Notice the tilt of the stop in figure A.15(b), which is at the 2° setting given by the second configuration in figure A.17. Note also the manner in which OPTH operands are set for the two configurations in the merit function (figure A.19).

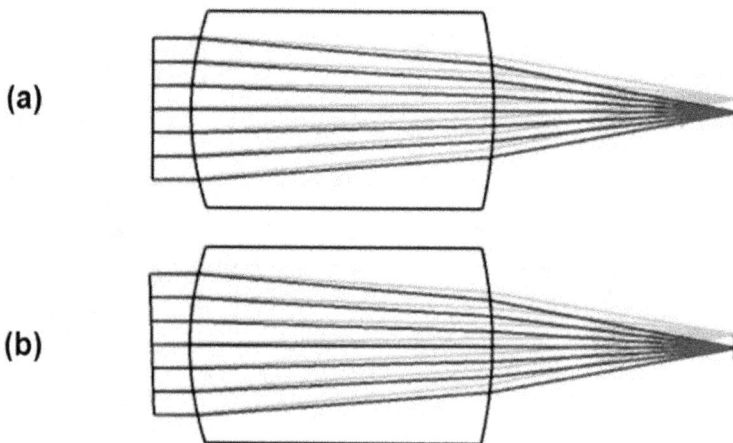

(a)

(b)

Figure A.15. An optimized layout for an aspheric aplanatic singlet obtained using a multi-configuration approach. (a) Axial configuration. (b) Oblique configuration.

Figure A.16. The prescription for the system shown in figure A.15.

Figure A.17. The multi-configuration editor setup for the system shown in figure A.15.

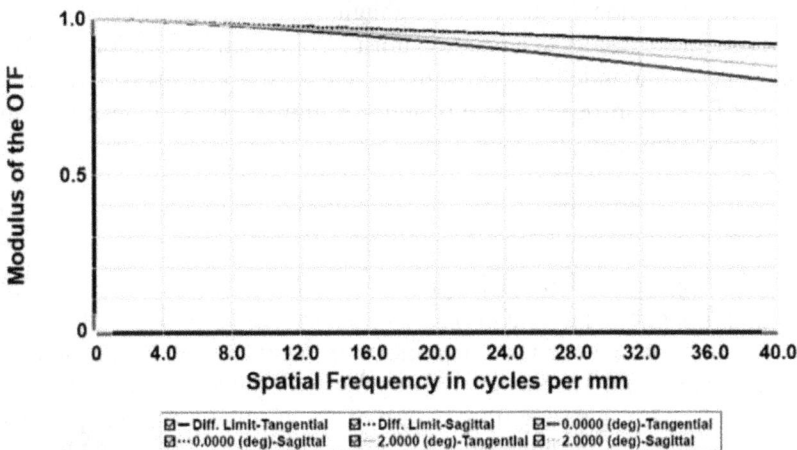

Figure A.18. The MTF (at 587.56 nm) for the system shown in figure A.15.

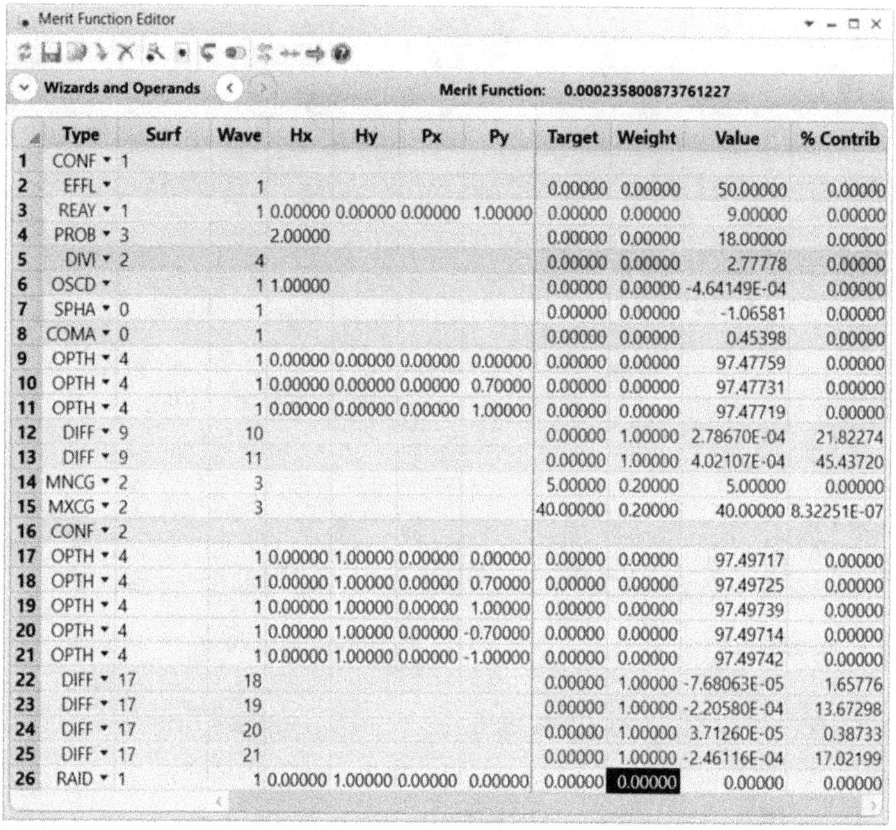

	Type	Surf	Wave	Hx	Hy	Px	Py	Target	Weight	Value	% Contrib
1	CONF	1									
2	EFFL		1					0.00000	0.00000	50.00000	0.00000
3	REAY	1	1	0.00000	0.00000	0.00000	1.00000	0.00000	0.00000	9.00000	0.00000
4	PROB	3		2.00000				0.00000	0.00000	18.00000	0.00000
5	DIVI	2	4					0.00000	0.00000	2.77778	0.00000
6	OSCD		1	1.00000				0.00000	0.00000	-4.64149E-04	0.00000
7	SPHA	0	1					0.00000	0.00000	-1.06581	0.00000
8	COMA	0	1					0.00000	0.00000	0.45398	0.00000
9	OPTH	4	1	0.00000	0.00000	0.00000	0.00000	0.00000	0.00000	97.47759	0.00000
10	OPTH	4	1	0.00000	0.00000	0.00000	0.70000	0.00000	0.00000	97.47731	0.00000
11	OPTH	4	1	0.00000	0.00000	0.00000	1.00000	0.00000	0.00000	97.47719	0.00000
12	DIFF	9	10					0.00000	1.00000	2.78670E-04	21.82274
13	DIFF	9	11					0.00000	1.00000	4.02107E-04	45.43720
14	MNCG	2	3					5.00000	0.20000	5.00000	0.00000
15	MXCG	2	3					40.00000	0.20000	40.00000	8.32251E-07
16	CONF	2									
17	OPTH	4	1	0.00000	1.00000	0.00000	0.00000	0.00000	0.00000	97.49717	0.00000
18	OPTH	4	1	0.00000	1.00000	0.00000	0.70000	0.00000	0.00000	97.49725	0.00000
19	OPTH	4	1	0.00000	1.00000	0.00000	1.00000	0.00000	0.00000	97.49739	0.00000
20	OPTH	4	1	0.00000	1.00000	0.00000	-0.70000	0.00000	0.00000	97.49714	0.00000
21	OPTH	4	1	0.00000	1.00000	0.00000	-1.00000	0.00000	0.00000	97.49742	0.00000
22	DIFF	17	18					0.00000	1.00000	-7.68063E-05	1.65776
23	DIFF	17	19					0.00000	1.00000	-2.20580E-04	13.67298
24	DIFF	17	20					0.00000	1.00000	3.71260E-05	0.38733
25	DIFF	17	21					0.00000	1.00000	-2.46116E-04	17.02199
26	RAID	1	1	0.00000	1.00000	0.00000	0.00000	0.00000	0.00000	0.00000	0.00000

Figure A.19. The merit function (optimized state) for the system shown in figure A.15.

A.4 The effective focal length of a lens system in a medium

Near the end of section 1.4.6, it was highlighted that in OS, the EFL for a lens system whose image resides in a medium must be obtained by multiplying the EFFL operand value by the index of the medium. The reason for this is that the relation between the EFL f in object space of index n and the EFL f' in image space of index n' is given by

$$\frac{f}{f'} = \frac{n}{n'}. \tag{A.12}$$

Following the logic provided by Kingslake and Johnson (see, for example, pp 68–9 in reference [5] of chapter 1), equation (A.12) is derived by noting that the principal planes of a lens system have unit magnification, regardless of the medium that the object and image are in. Therefore, applying the construction in figure A.20, we note that the pair of rays originating from the tip of an object at h emerges parallel in the manner shown on the right-hand side. Note that $h/f = \theta$ and $h/f' = \theta'$. From Snell's law (to the first order), $n\theta = n'\theta'$. Combining these relations, we obtain equation

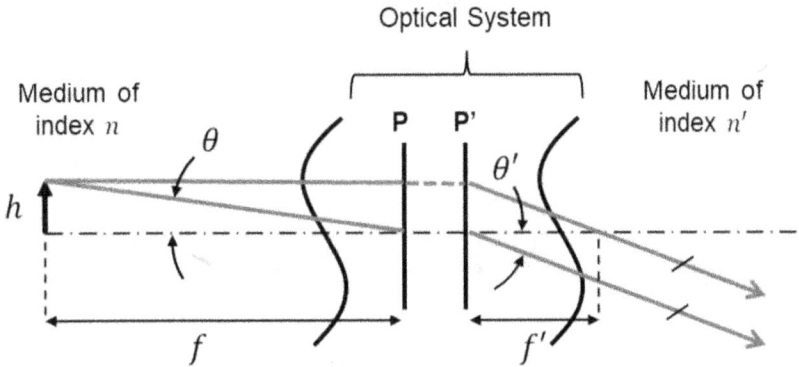

Figure A.20. The construction used to derive equation (A.12).

(A.12). Notice also that if Snell's law is multiplied by h on both sides, then the equality $hn\theta = hn'\theta'$ is the Lagrange invariant (see section 1.1.3), which can be said to contribute to the derivation of equation (A.12).

Based on equation (A.12), it becomes clear that if the EFFL operand in OS gives f for the EFL in object space, then the EFL in image space is $f' = n'f/n$, so that if $n = 1$ in air, then we simply have $f' = n'f$, which is the result mentioned near the end of section 1.4.6. Equation (A.12) is quite useful when considering a lens system in a medium, such as when using an underwater camera. In this case, it is the object side that must be given the index. In the opposite case, the human eye has an image in a medium of index greater than unity. This means that **for an object at infinity** subtending the field angle θ, the image height h' formed in the eye is simply given by $f \tan \theta$, because this product is equal to $f' \tan \theta'$, where θ' is **the angle for the oblique ray emerging from the center of the second principal plane, as in figure A.20**. That is, do not confuse it with the principal ray, which emerges from the center of the exit pupil. Additionally, since the image in the eye is in a medium, the eye's nodal points are not coincident with the location of the principal planes. This introduces other nuances (see, for example, Simpson M J 2022 Nodal points and the eye *Appl. Opt.* **61** 2797–804 and Simpson M J 2023 Focal length, EFL, and the eye *Appl. Opt.* **62** 1853–7).

IOP Publishing

Modern Classical Optical System Design
Fundamentals, techniques, tips, and tricks
Ronian Siew

Appendix B

Further notes on illumination

B.1 The derivation of image irradiance for paraxial thin lenses

In section 2.6.2, it was stated that the axial image irradiance for a **paraxial thin lens model** is given by equation (2.23). There are two ways to get there. The short way is to note that, for a small Lambertian source of elemental area dA, the total flux entering the thin lens is $\pi L dA \sin^2 U$, where U is the object space marginal ray at the full zone of the thin lens and L is the radiance at the source. The image irradiance on the axis is this flux divided by the image area dA'. If the ratio given by dA'/dA is the square of the image magnification given by a paraxial thin lens, then the axial irradiance is

$$E_p = \frac{\pi L dA \sin^2 U}{dA'} = \frac{\pi L \sin^2 U}{m_p^2}, \tag{B.1}$$

which is precisely equation (2.23). Accordingly, inserting equation (2.24) into equation (B.1) yields equation (2.25).

Due to the presence of the paraxial magnification m_p in equation (B.1) rather than the magnification given by the Abbe sine condition (equation (1.52)), there is a hint that something has happened to the radiance. To see this explicitly, we first return to figure 2.21 and now include the elemental areas dA, dA', and da as well as dimensions r, r', s, and s', which are shown in figure B.1:

According to this construction, the element of flux from dA that is contained in a ray pencil subtended by da is given by

$$d^2\phi = \frac{L dA da \cos^2 u}{r^2}. \tag{B.2}$$

On the image side, we allow the possibility that the radiance is different (if it is the same as on the object side, the math will tell us). Denoting the image-side radiance by L', and, by symmetry and flux conservation, letting the elemental flux on the left of da equal the elemental flux on the right of da, we obtain

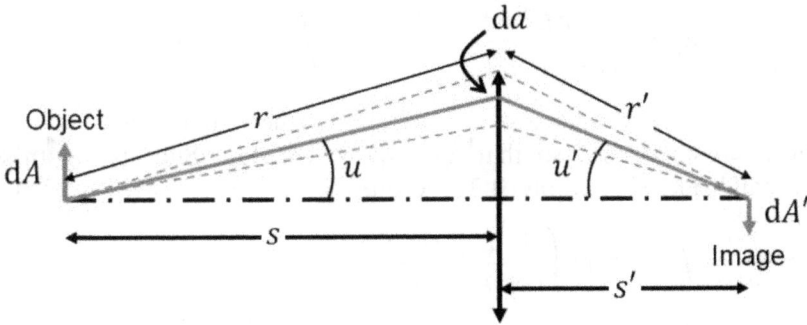

Figure B.1. The construction used to derive radiance and image irradiance for a paraxial thin lens.

$$\frac{L\,dA\,da\,\cos^2 u}{r^2} = \frac{L'\,dA'\,da\,\cos^2 u'}{r'^2}. \tag{B.3}$$

Next, noting that $\cos u = s/r$ and $\cos u' = s'/r'$, inserting these into equation (B.3) and simplifying gives

$$\frac{L\,dA\,\cos^4 u}{s^2} = \frac{L'\,dA'\,\cos^4 u'}{s'^2}. \tag{B.4}$$

But for a paraxial thin lens, $dA/s^2 = dA'/s'^2$; therefore,

$$L\,\cos^4 u = L'\,\cos^4 u'. \tag{B.5}$$

Equation (B.5) is precisely the result stated in section 2.6.2 for the radiance through a paraxial thin lens. We now apply this to the derivation of equation (B.1). To do so, we note that an element of the axial image irradiance contributed by da is given by dividing the right-hand side of equation (B.3) by dA'. Doing this and noting again that $\cos u' = s'/r'$ yields

$$dE_{\mathrm{p}} = \frac{L'\,da\,\cos^4 u'}{s'^2}. \tag{B.6}$$

Substituting L' from equation (B.5) into (B.6), we get

$$dE_{\mathrm{p}} = \frac{L\,\cos^4 u\,da}{s'^2}. \tag{B.7}$$

For a paraxial thin lens, $s'/s = m_{\mathrm{p}}$. So, by substituting this into equation (B.7), we get

$$dE_{\mathrm{p}} = \frac{1}{m_{\mathrm{p}}^2}\frac{L\,\cos^4 u\,da}{s^2}. \tag{B.8}$$

So, interestingly, the irradiance at the image depends on the quantities on the object side. This should remind you of the problem discussed in section 2.6.3. Integrating over the surface of the thin lens from the object side gives

B-2

$$E_p = \frac{1}{m_p^2} \int \frac{L \cos^4 u}{s^2} da = \frac{2\pi L}{m_p^2 s^2} \int_0^\rho \cos^4 u \rho d\rho, \tag{B.9}$$

where ρ is the radial coordinate on the thin lens. Note that $\rho = s \tan u = s \sin u / \cos u$, so that $d\rho = s(1 + \tan^4 u) du$. Substituting these into the right-hand side of equation (B.9), we get

$$
\begin{aligned}
E_p &= \frac{2\pi L}{m_p^2 s^2} \int_0^U \cos^4 u s^2 \left(\frac{\sin u}{\cos u} + \frac{\sin^3 u}{\cos^3 u} \right) du \\
&= \frac{2\pi L}{m_p^2} \int_0^U \cos^4 u \left(\frac{\sin u \cos^2 u}{\cos^3 u} + \frac{\sin^3 u}{\cos^3 u} \right) du \\
&= \frac{2\pi L}{m_p^2} \int_0^U \cos u (\sin u \cos^2 u + \sin^3 u) du \\
&= \frac{2\pi L}{m_p^2} \int_0^U \sin u \cos u (\cos^2 u + \sin^2 u) du \\
&= \frac{2\pi L}{m_p^2} \int_0^U \sin u \cos u du = \frac{2\pi L}{m_p^2} \int_0^{\sin U} \sin u d(\sin u) = \frac{\pi L \sin^2 U}{m_p^2}.
\end{aligned}
\tag{B.10}
$$

This is precisely the result in equation (B.1). Therefore, once again, inserting equation (2.24) into equation (B.10) or (B.1) yields equation (2.25), which is what we intended to prove.

B.2 The derivation of irradiance along the axis behind a lens

Near the end of section 2.11, we discussed the fact that a position of peak irradiance (POP) is located at a point between a lens and the image. Here, we derive two expressions for the axial irradiance behind a lens, which indicate the location of the POP for a **thin real lens** (i.e. not a paraxial thin lens model). To do this, we first apply the construction shown in figure B.2:

The object is assumed to be a flat Lambertian disk source symmetric about the lens's axis, and the lens is in air. Based on the ray geometry in figure B.2, an element

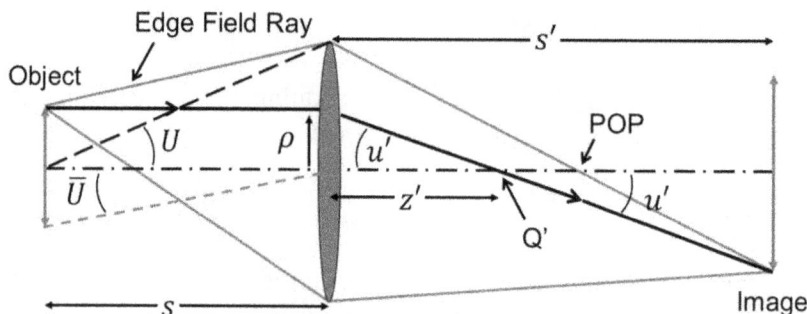

Figure B.2. The construction used to derive axial irradiance for $0 \leqslant z' \leqslant$ POP.

of flux at the point Q' is given by a ray pencil from the tip of the object passing through the lens at radial height ρ. Because this ray ends at the tip of the image, it is the final ray from the lens area making flux contributions at Q'. Other rays contribute at Q' from locations at the axis of the lens up to ρ. Therefore, the total irradiance at Q' must be given by setting up an integral similar to equation (B.9) on the right-hand side of the lens. But since we have a real lens (i.e. it is not a paraxial thin lens), the radiance is conserved on both sides. So, the integral may then be expressed as

$$E = \frac{2\pi L}{z'^2} \int_0^\rho \cos^4 u' \rho d\rho. \tag{B.11}$$

By following the process of integration leading up to equations (B.10), equation (B.11) evaluates to

$$E = \pi L \sin^2 u'. \tag{B.12}$$

Note in equation (B.12) that u' is subtended by the image's full height as long as $0 \leqslant z' \leqslant$ POP, as has been pointed out by Stewart and Johnson (see, for example, pp 149–151 in reference 43 of chapter 2). On the other hand, on the right-hand side of the POP (figure B.3), the position at Q' receives flux from rays over the full radial height of the lens. Therefore, the same integral in equation (B.11) may be applied to compute the irradiance at Q', for which u' is subtended by the full radial height of the lens. Note that this angle is always subtended by the full lens height for positions between the POP and the image. Since u' decreases when Q' is between the POP and the image, and since u' increases when Q' is between the lens and the POP, it follows that the POP is indeed a position of the peak irradiance behind the lens. Notice also that when the object is at the front focal position of the lens (the classical searchlight condition), u' is subtended by the object's height from the vertex of the lens, and it remains constant for a distance behind the lens up to the intersection of the edge field ray with the axis. In this case, there is no POP, just a constant axial irradiance from the lens to that intersection. Beyond that intersection, u' is subtended by the lens, and the axial irradiance begins to drop until it reaches zero at an infinite distance from the lens.

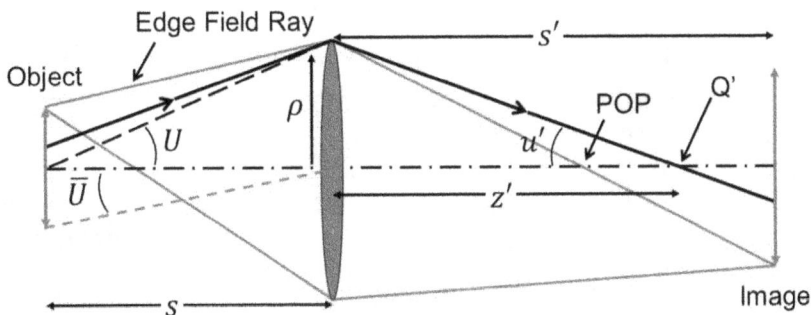

Figure B.3. The construction used to derive the axial irradiance for POP $\leqslant z' \leqslant$ Image.

As for a **paraxial thin lens**, it has been shown (see, for example, references [29] and [44] in chapter 2) that the first-order equations are

$$E_p(z') = \frac{\pi L s'^2}{[m_p z' - (s' - z')]^2} \left[1 - \frac{1}{1 + \tan^2 \overline{U}\left(\frac{m_p z'}{s' - z'} - 1\right)^2} \right], \tag{B.13}$$

for $0 \leqslant z' \leqslant \text{POP}$, and

$$E_p(z') = \frac{\pi L s'^2}{[m_p z' - (s' - z')]^2} \left[1 - \frac{1}{1 + \tan^2 U\left(\frac{s' - z'}{m_p z'} - 1\right)^2} \right], \tag{B.14}$$

for $\text{POP} \leqslant z' \leqslant s'$, where $m_p = s'/s$, which is the paraxial image magnification. In section 2.13, we calculated the axial irradiance on a screen for the critical and Köhler layouts. Suppose we also wanted to know the irradiance for the classical searchlight layout. In this case, the object would be at the front focal position, so that $s' \to \infty$ and also $m_p \to \infty$. In this case, it is more convenient to replace m_p by s'/s, and re-express equation (B.14) as

$$E_p(z') = \frac{\pi L}{[1 - (z'/s)]^2} \left[1 - \frac{1}{1 + \tan^2 U\left(1 - \frac{s}{z'}\right)^2} \right]. \tag{B.15}$$

Now, note that $z'/s = \tan U / \tan U'$. Substituting this into equation (B.15), we get

$$E_p(U, U') = \frac{\pi L}{\left(1 - \frac{\tan U}{\tan U'}\right)^2} \left[1 - \frac{1}{1 + \tan^2 U\left(1 - \frac{\tan U'}{\tan U}\right)^2} \right]. \tag{B.16}$$

In section 2.13, for the conditions in figure 2.53(a), $U' = 2.8624^0$ and $U = 14.0362^0$. If we apply these values to equation (B.16), we obtain $E_p \approx \pi L 0.0024$ watts/area, which is close to the peak in the plot given by OS in figure 2.54(a).

B.3 The derivation of irradiance at the screen in Köhler illumination

Applying the construction shown in figure B.4, for a system of **thin real lenses**, the axial position Q' receives elemental flux from source rays ranging from all annuli centered about the axial point of the source. At the projector's surface, these rays span heights from zero to the full height. This means that the integral in equation (B.11) is applicable to those rays, resulting in equation (B.12) for the axial irradiance at the screen (just replace u' by $\bar{u}'$). At all off-axis points on the screen, if the projector lens has no image distortion for the full field, then the irradiance profile on the screen is given by the irradiance profile on the condenser caused by illumination from the

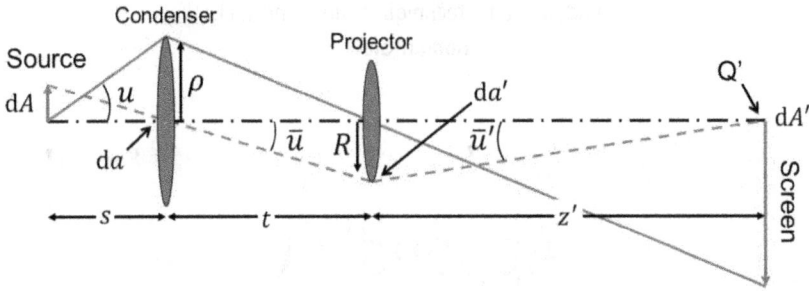

Figure B.4. The construction used to derive the axial irradiance at the screen in Köhler illumination.

source. Therefore, if the source is a small Lambertian emitter, then the irradiance profile on the screen should, according to equation (2.17), be proportional to $\cos^4 u$.

If we have **paraxial thin lens models** for the condenser and projector, applying equation (B.6), a ray pencil with a central ray defined by the dashed line from the tip of the source and ending at Q' contributes an element of irradiance given by

$$dE_\mathrm{p} = \frac{L' da' \cos^4 \bar{u}'}{z'^2}. \tag{B.17}$$

Applying equation (B.5) for the radiance through paraxial thin lenses, equation (B.17) becomes

$$dE_\mathrm{p} = \frac{L da' \cos^4 \bar{u}}{z'^2}. \tag{B.18}$$

But note that $(z'/t)^2 = dA'/da$ and $(t/s)^2 = da'/dA$. Substituting these into equation (B.18), we obtain

$$dE_\mathrm{p} = \frac{L da' da \cos^4 \bar{u}}{t^2 dA'} = \frac{da}{dA'} \frac{L dA \cos^4 \bar{u}}{s^2}. \tag{B.19}$$

The quantity given by $L dA \cos^4 \bar{u}/s^2$ in equation (B.19) is of a form similar to that of equation (B.8). Therefore, integrating it over the source's area yields

$$E_\mathrm{p} = \frac{da}{dA'} \pi L \sin^2 \bar{u}. \tag{B.20}$$

Note that $da/dA' = \tan^2 \bar{u}'/\tan^2 \bar{u}$. Substituting this into equation (B.20), we get

$$E_\mathrm{p} = \frac{\pi L \sin^2 \bar{u} \tan^2 \bar{u}'}{\tan^2 \bar{u}} = \pi L \cos^2 \bar{u} \tan^2 \bar{u}'. \tag{B.21}$$

Equation (B.21) was used to compute the axial irradiance for the layout in figure 2.53(c) for the case of paraxial thin lenses for the condenser and projector. Since, for a Köhler system comprised of paraxial thin lens models for the condenser and projector, there are no aberrations, the irradiance profile on the screen takes on whatever irradiance profile is present on the condenser, provided by the source.

Appendix C

Further notes on optical system product development

C.1 Desensitizing optical systems by design optimization

In section 1.2.6, I discussed some points regarding how some optical imaging systems may be designed to be less sensitive to tolerances. In section 1.2.13, I mentioned that OS provides two merit function operands for achieving this. In particular, they are the TOLR and HYLD operands. In section 3.8.7, I again highlighted that robust optical systems may be either defined by how they are produced for ruggedness and reliability, or that their design possesses an intrinsic property of being *desensitized* to some production tolerances for element variables, such as radii, thickness, tilts, and decenters. Here, I provide a simple example of the latter, using the HYLD operand, which was originally developed by Moore (Moore K E 2019 Optimization for as-built performance *Proc. SPIE* **10925**) and implemented in OS (see also Joonha J and Alisafee 2020 Study of tolerancing optimization approaches in Zemax for as-built performance *Proc. SPIE* **11488**, and Croce A 2020 Next generation optical design methodology *Proc. SPIE* **11287**).

Suppose we return to the toleranced 50 mm EFL double-Gauss lens that was discussed in section 1.2.13, and we want to determine the influence of the tolerances shown in figures 1.79 and 1.80 on the two RMS spot size operands in figure 1.78 (rows 10 and 12). To do this, we remove the weight that is assigned to the EFFL operand and instead assign a weight of one to the RSCE operands on rows 10 and 12 (so, the computed RMS spot size is the RMS average of the two). After performing a Monte Carlo run with some number of trials, we get a certain histogram with a spread of average RMS spot sizes. We then want to know whether we can reduce the spread but use the same tolerances. How can this be achieved? The solution is to desensitize the design through re-optimization, using either the TOLR operand or the HYLD operand. Of these two choices, the HYLD operand is the quicker and more efficient approach. The TOLR operand causes the program to perform tolerancing *live* as the optimization proceeds, while the HYLD operand applies

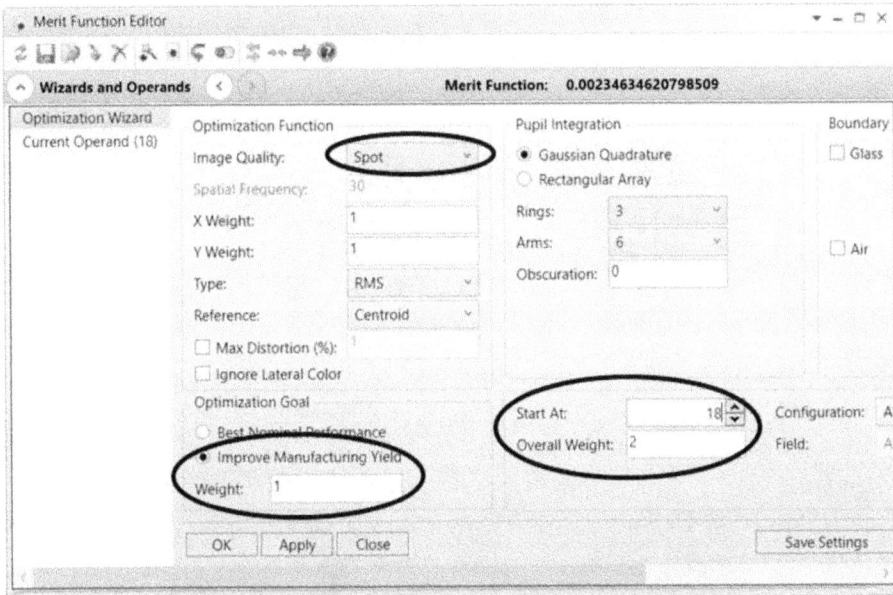

Figure C.1. The settings needed to use the HYLD operand to re-optimize the lens shown in figure 1.70.

an algorithm that minimizes ray angles of incidence (AOIs) on surfaces. Here, we shall apply the HYLD operand to the double-Gauss lens, as follows.

We return to the optimized lens (figure 1.70) and, in the merit function (figure 1.71), remove the weights from the RSCE operands (rows 2 and 4). Then, add two blank rows below the last, so that the last blank row is row #18. Go to the top left of the Merit Function Editor and click on the *optimization wizard* icon to open up its dialog box and enter the settings shown in figure C.1. Notice, in particular, the settings circled in black. The idea is to let the optimization wizard generate RMS spot operands but to involve HYLD functionality (click on the Improve Manufacturing Yield button under Optimization Goal). The Overall Weight setting is meant for the RMS spot size operands, while the weight of one under the Improve Manufacturing Yield setting applies to the HYLD operands. Click Apply, then OK. Then run a local damped least-squares optimization and wait until it is done. It should only take a minute or so.

The resulting layout I obtained is shown in figure C.2, whose prescription and MTF are shown in figures C.3 and C.4, respectively. Notice that the nominal performance is quite good. Next, we apply the same merit function as before (figure 1.78) and assign weights of one for the two RSCE operands only. We then apply the tolerances from figures 1.79 and 1.80 to this lens and perform a Monte Carlo simulation of 500 trials. We do the same thing to the original lens from figure 1.70. The results are shown in figures C.5(a) and (b) for the original and desensitized lens, respectively.

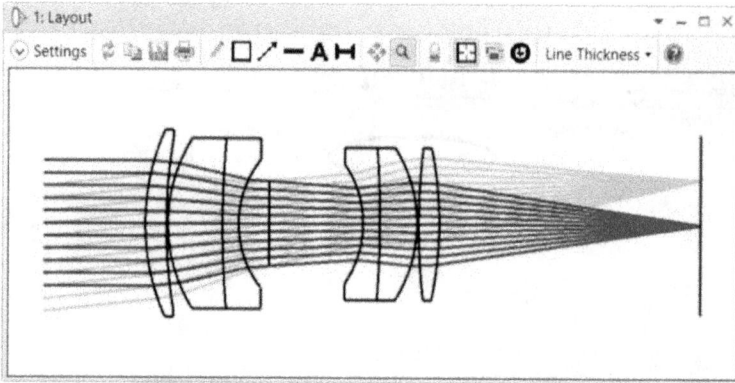

Figure C.2. The re-optimized layout for a desensitized version of the lens shown in figure 1.70.

	Surface Type		Comment	Radius	Thickness	Material	Coating	Clear Semi-Dia
0	OBJECT	Standard ▾		Infinity	Infinity			Infinity
1		Standard ▾	Dummy 1	Infinity	12.50000			15.00000 U
2	(aper)	Standard ▾	1	26.03379 V	2.58480 V	N-SK2		11.00000 U
3	(aper)	Standard ▾		59.02502 V	0.19999 V			10.00000 U
4	(aper)	Standard ▾	2	18.54406 V	6.64665 V	N-SK16		10.00000 U
5	(aper)	Standard ▾	3	109.30179 V	2.00000 V	F5		10.00000 U
6	(aper)	Standard ▾		12.01171 V	3.62921 V			7.50000 U
7	STOP (aper)	Standard ▾	STOP	Infinity	11.45414 V			5.00000 U
8	(aper)	Standard ▾	4	-12.14874 V	2.03436 V	F5		7.00000 U
9	(aper)	Standard ▾	5	-111.06875 V	4.47699 V	N-SK16		9.00000 U
10	(aper)	Standard ▾		-16.56307 V	0.20001 V			9.00000 U
11	(aper)	Standard ▾	6	75.08871 V	2.52147 V	N-SK16		9.00000 U
12	(aper)	Standard ▾		-41.81255 V	31.66909 V			9.00000 U
13	IMAGE (aper)	Standard ▾	Img	Infinity	-			10.62600 U

Figure C.3. The prescription for the lens shown in figure C.2.

Notice the slight compression in the spread for the histogram of the desensitized lens. The HYLD operand really did its job and reduced the sensitivities. As highlighted in section 1.2.6, the desensitization of optical systems has been a subject of great interest for some time, and various methods are discussed in the cited references in that section. The general idea is that sensitivities to perturbations are functions of the magnitude of aberrations contributed by surfaces, and these aberrations are functions of the ray AOIs on those surfaces. Commonly, even if aberrations from surfaces are canceled by those from other surfaces, if the cancellations are numerous, then the system tends to become quite sensitive to

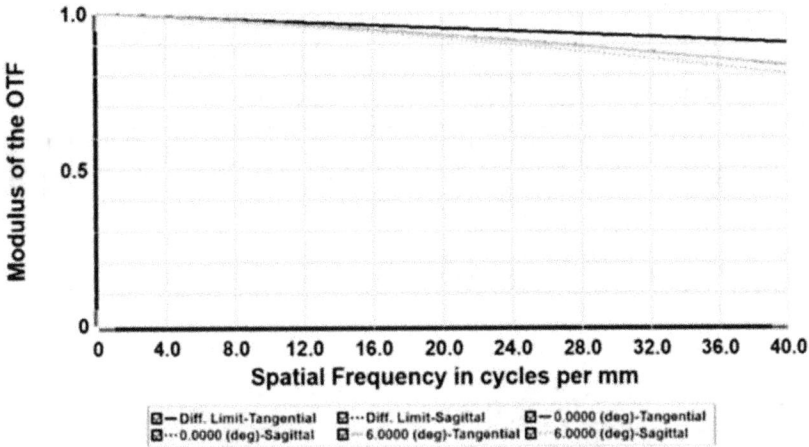

Figure C.4. The MTF for the lens shown in figure C.2.

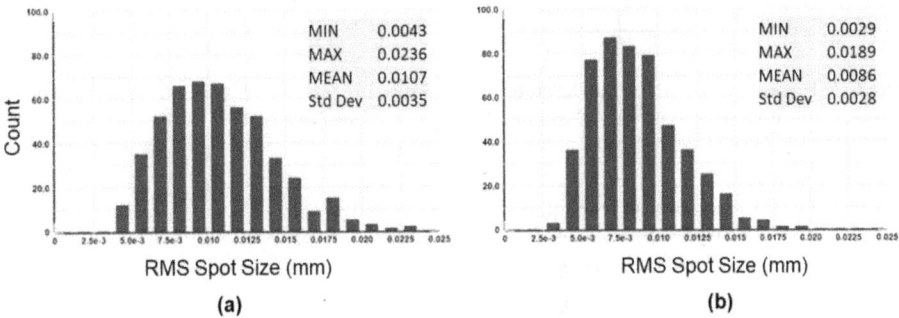

Figure C.5. Histograms of the RMS spot size for the original lens (a) and the desensitized lens (b).

changes in their nominal state. The HYLD operand targets these AOIs and attempts to make them small enough that refraction still occurs (otherwise, no image would be formed) but the aberrations are kept minimal.

C.2 Beam deviation and image quality

Following on from the example provided in appendix C.1, let us examine the angular deviation of the axial ray leaving the lens relative to the axis of the image plane, which is parallel to the object plane. If the object is at infinity, the image plane can be made parallel to a reference *front datum plane* (such as the dummy surface on surface 1 in the prescription given in figure C.3). The angular deviation may be considered a *beam deviation* θ of the axial ray from the object, and it is precisely equal to the AOI at the image plane, as illustrated in figure C.6 for a lens with exaggerated tilts and decenters in its elements. We want to know whether there is any correlation between the beam deviation and the RMS spot size (averaged over the two fields described in the example in appendix C.1). To find out, we plot the

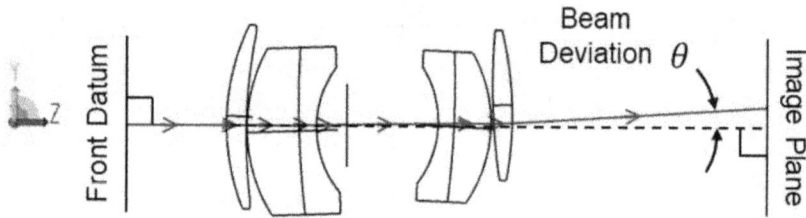

Figure C.6. The definition of beam deviation.

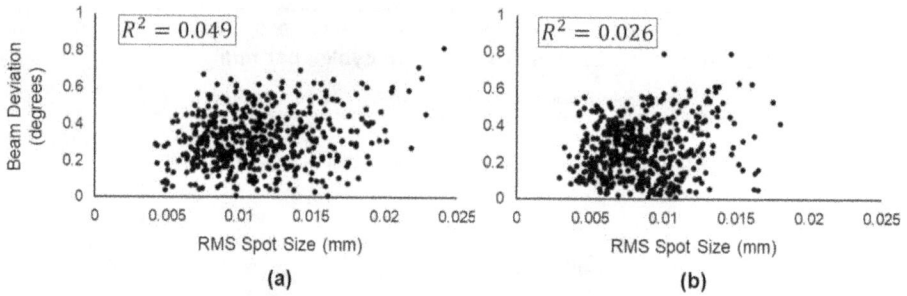

Figure C.7. A scatter plot of the beam deviation vs RMS spot size (averaged across two fields) in 500 Monte Carlo trials. (a) For the original lens of figure C.2. (b) For the lens in figure C.2.

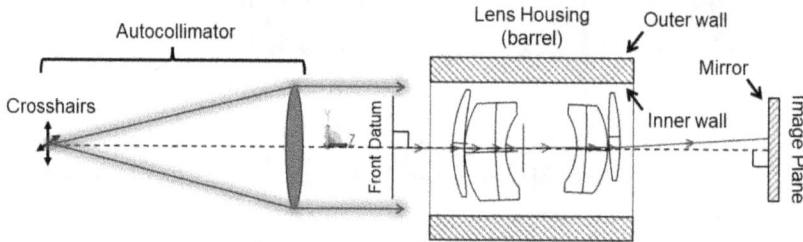

Figure C.8. Lens assembly with an autocollimator.

beam deviation and RMS spot size on a scatter plot. Figure C.7 compares the scatter plots for beam deviation vs the averaged RMS spot size between the desensitized lens of appendix C.1 and the original lens (i.e. the lens shown in figure 1.70).

In figure C.7, it is evident that the spread in RMS spot size for the desensitized lens (figure C.7(b)) is less than for the original lens (figure C.7(a)), which is expected, as this is consistent with the histograms for their RMS spot sizes, as shown in figure C.5. Also, for both lenses, there is no obvious correlation between the beam deviation and RMS spot size across all 500 trial lenses in the Monte Carlo simulation. What this means is that once the elements of these lenses have been mounted (by the 'drop-in' method) into their housings (e.g. a cylindrical barrel), there is no way to simultaneously achieve good image quality and a state of *axial autocollimation* for these assemblies if attempts are made to align them to the axis of an autocollimator, as depicted in figure C.8.

By axial autocollimation, it is meant that when the image of the crosshairs from the autocollimator is projected out and through the lens assembly, it returns to its original axial position in the autocollimator upon being reflected from a mirror that is mounted at the lens's image plane. The situation could be that an image sensor may ultimately be mounted at the mirror plane, so that an engineer may attempt to align the lens assembly relative to that plane by first aligning an autocollimator to a flat mirror located at the image sensor plane. The engineer would ensure that the autocollimator's axis is orthogonal to the mirror, which would be the case if the projected image of the crosshairs from the autocollimator returned to their original axial position upon being reflected by the mirror. Upon achieving this result, the lens assembly would be mounted in some way between the autocollimator and mirror, and the engineer might attempt to decenter and tilt the lens assembly until the return image of the crosshairs returned to its axial location. This could only be achieved if the lens assembly were tilted in such a way that the exiting axial beam was not deviated relative to the image plane. In other words, this alignment process requires zero axial beam deviation. But we know from the scatter plots in figure C.7 that beam deviation does not correlate with good image quality. Thus, the alignment philosophy must change.

This situation is not so simple. Note in figure C.7 that the data clusters are within a beam deviation of about 1°. This means that when the lens housing (which we currently assume is *ideal* in that its outer and inner wall cylinders are coaxial) is tilted by some amount, the beam deviation could be larger than 1°, and the resulting image quality is unlikely to fall within the range shown in the horizontal axes of the scatter plots in figure C.7. This means that for certain degrees of the tilt applied to the whole lens assembly, the image quality could correlate well with the beam deviation, so that the autocollimator could be used in such cases of *coarse tilt adjustment*. In other words, the alignment process for manufactured samples of the current lens should be such that the beam deviation does not exceed about 1° (which we assume is measurable by the autocollimator). Once the returned crosshair image in the autocollimator is brought within 1° of beam deviation, no further tilts to the lens housing are beneficial, and one is then relying on the statistical tolerance stack and interaction effects of the elements in the housing to produce the distribution of image qualities (defined by the averaged RMS spot sizes) in the histograms of figure C.5. The only variable left is focus, which is achieved through autocollimation and examination of the quality of the crosshair image (or any suitable target).

Appendix D

Notes on some advanced topics

D.1 Hamilton's formulation and the optical cosine rule for lens design optimization

In lens design, we constantly try different approaches to optimize the image quality. This involves trying different merit function operands, such as the RMS spot size (e.g. the RSCE operand in OS), RMS wavefront error (e.g. the RWCE operand in OS), the offense against the sine condition (e.g. the OSCD operand in OS), equalizing optical path lengths (e.g. the OPTH and PLEN operands in OS), and sometimes directly optimizing the MTF. Even though all of these operands can be traced to the same fundamental problem of satisfying Fermat's principle, optimizing using different operands can often result in different design solutions. For instance, I often begin optimization with RMS spot size operands, but I eventually switch to RMS wavefront error operands near the end. The latter is more closely related to achieving high MTF (see, for example, Smith G H 1998 *Practical Computer-Aided Lens Design* (Richmond, VA: William-Bell) pp 174–6).

It is quite evident that merit function operands originate from fundamental quantities in optics, and since each operand can lead to different results, it seems worthwhile to explore the other fundamental quantities that may be used for lens design optimization. Consider, for example, a *differential* of Hamilton's point characteristic (see, for example, Walther A 1995 *The Ray and Wave Theory of Lenses* (New York: Cambridge University Press) pp 20–2, 51–2), given by

$$dV = \overline{QQ'} - \overline{PP'} = n'dr' \cos \theta' - ndr \cos \theta, \tag{D.1}$$

where $\overline{QQ'}$ and $\overline{PP'}$ are ray optical path lengths (OPLs) between the points depicted in figure D.1, and the quantity V (known as Hamilton's point characteristic) is an OPL that is a function of the space coordinates x, y, z, x', y', and z', where superscript primes denote the image space (see, for example, Conway A W and Synge J L (ed.) 1931 *The Mathematical Papers of Sir William Rowan Hamilton* vol. 1 (Cambridge: Cambridge University Press)).

doi:10.1088/978-0-7503-6059-3ch7

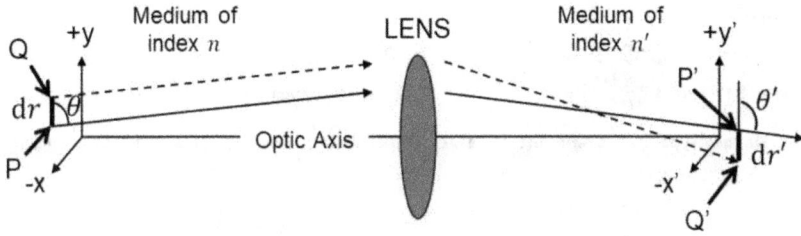

Figure D.1. The construction used for equation (D.1).

Figure D.2. The double-Gauss lens used for the optical cosine rule example.

According to the so-called *optical cosine law* (Smith T 1922 The optical cosine law *Trans. Opt. Soc.* **24** 31–40), which is today actually considered an optical cosine *rule* in that it is not really a very general law, perfect imaging of the differential segment dr to dr' occurs if equation (D.1) is constant for all rays traced from P to P′ and from Q to Q′. Following Walther's notation, we may divide both sides of equation (D.1) by dr to obtain

$$n'm' \cos \theta' - n \cos \theta = \text{constant}, \tag{D.2}$$

where $m' = dr'/dr$, which is the transverse magnification of the image at the specified field point. Equation (D.2) implies that if we can trace rays from an oblique field point that are made to satisfy the constancy of the optical cosine rule, then we should be able to design a lens to achieve good image quality for that field point. As a trivial example, consider the double-Gauss lens shown in figure D.2, whose prescription and merit function are provided in figures D.3 and D.4, respectively. The lens has been taken from figure C.2 and scaled down to have an EFL of 1 mm, so that near-perfect imaging occurs due to the scaled-down aberrations across all fields. The image is purposely defocused by 0.5 mm, and the goal is to let the merit function construction for the optical cosine rule constraint determine the best focus for the oblique field.

The ray traces are performed at 550 nm and the oblique field is at a height of 0.2 mm. The merit function and prescription shown are for the lens's starting state, i.e. prior to optimization. Row 9 in the merit function provides the estimated oblique magnification at the field point of 0.2 mm. The operand REAB computes the direction cosine of the ray under inspection. Thus, rows 11 and 13, for example, compute the respective quantities $\cos \theta'$ and $\cos \theta$ in equation (D.2) for the principal ray. Rows 14, 19, and 24 compute the constant on the right-hand side of equation (D.2) at the same field point, but using different tangential rays. Due to the defocus, each computation has yielded a different 'constant.' As a measure of this variation in

Lens Data

Update: All Windows

Surface 0 Properties — Configuration 1/1

	Surface Type		Comment	Radius	Thickness	Material	Clear Semi-Dia	Conic
0	OBJECT	Standard ▾		Infinity	3.000		0.200	0.000
1		Standard ▾	Dummy 1	Infinity	0.250		0.300 U	0.000
2	(aper)	Standard ▾	1	0.521	0.052	N-SK2	0.220 U	0.000
3	(aper)	Standard ▾		1.181	4.000E-03		0.200 U	0.000
4	(aper)	Standard ▾	2	0.371	0.133	N-SK16	0.200 U	0.000
5	(aper)	Standard ▾	3	2.186	0.040	F5	0.200 U	0.000
6	(aper)	Standard ▾		0.240	0.073		0.150 U	0.000
7	STOP (aper)	Standard ▾	STOP	Infinity	0.229		0.100 U	0.000
8	(aper)	Standard ▾	4	-0.243	0.041	F5	0.140 U	0.000
9	(aper)	Standard ▾	5	-2.221	0.090	N-SK16	0.180 U	0.000
10	(aper)	Standard ▾		-0.331	4.000E-03		0.180 U	0.000
11	(aper)	Standard ▾	6	1.502	0.050	N-SK16	0.180 U	0.000
12	(aper)	Standard ▾		-0.836	0.978		0.180 U	0.000
13		Standard ▾	Defocus	Infinity	0.500 V		0.070	0.000
14	IMAGE	Standard ▾	Img	Infinity	-		0.149	0.000

Figure D.3. The prescription for the lens shown in figure D.2.

the constant, rows 26–28 compute the ratios between each constant and the other. Assigning a weight to each and targeting the weights at unity provides the means to satisfy the optical cosine rule. The thickness of surface 12 in the prescription provides the ideal focal plane location of the image, so that the 0.5 mm thickness of surface 13 represents a defocus from that plane. Notice that a variable has been assigned only for the thickness of surface 13. After performing a single local damped least-squares optimization, that thickness becomes 2.97E−03, which is nearly zero. The merit function values on rows 26 and 27 equal 1.001, while on row 28, the merit function value equals 1.000.

It is conceivable that a more elaborate merit function may be constructed around this cosine rule for appropriately weighted rays in various zones and a variety of fields. In particular, if a standard deviation (std dev) is also computed for the systematic variation in all constants in the cosine rule for a number of rays at specific fields, then for each field, a std dev of constants for that field may be considered a measure of the *offense against the cosine rule*, which could serve a purpose similar to that of the OSC derived by Conrady (see section 1.4.4, and also Conrady A E 1960 *Applied Optics and Optical Design, Part II* (New York: Dover) pp 367–401). That is, values for which the std dev is greater than zero would be the offense, so that optimization serves to target the std dev to zero. Incidentally, when the optical cosine rule is applied to the axial field, then the Abbe sine condition is obtained. Thus, an offence against the cosine rule or some form of optimization involving the cosine rule is not an unreasonable concept for lens design optimization. In fact, when

	Type	Surf	Wave	Hx	Hy	Px	Py	Target	Weight	Value	% Contrib
1	BLNK ▾	Obtain oblique magnification...									
2	REAY ▾ 0		1	0.000	1.000	0.000	0.000	0.000	0.000	0.200	0.000
3	REAY ▾ 0		1	0.000	0.999	0.000	0.000	0.000	0.000	0.200	0.000
4	REAY ▾ 14		1	0.000	1.000	0.000	0.000	0.000	0.000	-0.088	0.000
5	REAY ▾ 14		1	0.000	0.999	0.000	0.000	0.000	0.000	-0.088	0.000
6	DIFF ▾ 2	3						0.000	0.000	2.000E-04	0.000
7	DIFF ▾ 4	5						0.000	0.000	-8.768E-05	0.000
8	BLNK ▾	Oblique magnification									
9	DIVI ▾ 7	6						0.000	0.000	-0.438	0.000
10	BLNK ▾	Optical cosine rule for principal ray...									
11	REAB ▾ 14		1	0.000	1.000	0.000	0.000	0.000	0.000	-0.038	0.000
12	PROD ▾ 9	11						0.000	0.000	0.017	0.000
13	REAB ▾ 0		1	0.000	1.000	0.000	0.000	0.000	0.000	-0.056	0.000
14	DIFF ▾ 12	13						0.000	0.000	0.072	0.000
15	BLNK ▾	Optical cosine rule for upper oblique ray...									
16	REAB ▾ 14		1	0.000	1.000	0.000	1.000	0.000	0.000	-0.157	0.000
17	PROD ▾ 9	16						0.000	0.000	0.069	0.000
18	REAB ▾ 0		1	0.000	1.000	0.000	1.000	0.000	0.000	-0.015	0.000
19	DIFF ▾ 17	18						0.000	0.000	0.084	0.000
20	BLNK ▾	Optical cosine rule for lower oblique ray...									
21	REAB ▾ 14		1	0.000	1.000	0.000	-1.000	0.000	0.000	0.082	0.000
22	PROD ▾ 9	21						0.000	0.000	-0.036	0.000
23	REAB ▾ 0		1	0.000	1.000	0.000	-1.000	0.000	0.000	-0.097	0.000
24	DIFF ▾ 22	23						0.000	0.000	0.061	0.000
25	BLNK ▾	Constrain the rule that n'm'cos(theta') - nmcos(theta) = constant (cosine rule)...									
26	DIVI ▾ 14	19						1.000	1.000	0.865	9.258
27	DIVI ▾ 14	24						1.000	1.000	1.191	18.565
28	DIVI ▾ 19	24						1.000	1.000	1.376	72.177

Figure D.4. The merit function for the lens shown in figure D.2.

Modern Classical Optical System Design was just submitted for copyediting and production in December 2023, a new publication was found on this topic (see, e.g., Wang M and Zhu J 2023 Multi-field cosine condition in the design of wide-field freeform microscope objectives *Opt. Express* **31** 43362–71).

D.2 Skew invariance

In the design of symmetrical optical systems (i.e. systems in which there is rotational symmetry about the optic axis, such as the majority of systems discussed in this book), we apply several invariant (conservative) quantities—such as Snell's law (sometimes called the *refraction invariant*), the Lagrange invariant, and the étendue —to ray tracing and analysis. These quantities are, of course, ultimately traced back to flux conservation in the transport of optical radiation. Snell's law gives the invariance $n \sin \theta = n' \sin \theta'$ for a ray in a plane of incidence and refraction at angles of θ and θ', respectively, but for a skew ray outside this plane, there is also an invariant quantity called the *skewness* of the ray, and the quantity related to a *symmetry of translation* of skewness is called the *skew invariance*. By this, it is meant that the operation of taking a skew ray from a position in the object space and propagating it towards a position in the image space does not result in changes in the skewness of that ray. So, the skewness of the skew ray is invariant.

Skew invariance has an interesting practical application in imaging, which is that the skewness is violated if more than a single conjugate for perfect imaging is attempted. This implies that it is impossible to obtain a perfect imaging condition over a wide object range. A rather pedagogically appealing example of this has been provided by Goodman (e.g. Goodman D S 2001 Imaging limitations related to the skew invariant *Proc. SPIE* **4442** 67–77). However, this constraint is not necessarily a strict one in that many camera lenses are optimized (but perhaps not necessarily at equal performance) over a range of object distances. Further, very near-diffraction-limited performance is maintained even for violations in the invariant when a lens possesses a large depth of field, such as in miniature lenses. We recall from section 1.2.12 that aberrations (with the exception of % image distortion) scale with size of the lens system. Moreover, for a finite conjugate system at a fixed object size, the magnification is reduced through scaling, which results in a significant increase in the depth of field for the miniature lens. To see why, we apply equation (1.20) and solve for the object distance, yielding $s = (f^{-1} - s'^{-1})^{-1}$. Differentiating with respect to the image distance gives $ds = -[(f^{-1} - s'^{-1})^{-2}]s'^{-2}ds'$. But the quantity $f^{-1} - s'^{-1}$ is the reciprocal of the object distance, so that the formula becomes $ds = -(s'/s)^{-2}ds'$. Ignoring the minus sign (which is a consequence of a sign convention for the thin lens formula), the result may also be stated as $ds = m^{-2}ds'$, where m is the image magnification. If we regard ds' as the depth of focus of the lens, then ds is the depth of field in object space, which increases significantly for very small m, due to the negative second power in the magnification.

It is a useful exercise to compute the skew invariance for a miniature lens and see how far the skewness is violated at two different conjugates. To begin, let us derive this invariant quantity. Many authors have provided derivations of the skew invariant, but a particularly simple construction for its derivation has been provided by Welford (see, for example, Welford W T 1986 *Aberrations of Optical Systems* (Bristol: IOP Publishing) pp 84–6, and also Welford W T 1968 A note on the skew invariant of optical systems *Opt. Acta* **15** 621–623), which we shall base our derivation on. To that end, consider the construction shown in figure D.5. Let R, P, and Q be points at the locations shown. Points P and Q lie on a disk at the object plane. Let there be corresponding points P′ and Q′ at the image plane (not shown). Let ε be a small angle that results from rotating ray PP′ along the disk of the object so that the point P reaches Q along the circumference of the disk. The ray PP′ is a skew ray, due to the angle θ. It is then important to note that under perfect imaging conditions, the OPL of the ray segment PP′ is equal to the OPL of the ray segment QQ′, by virtue of Fermat's principle. In addition, due to the line segment connecting P to R at the object plane, there is also a ray RR′, where R′ is the corresponding

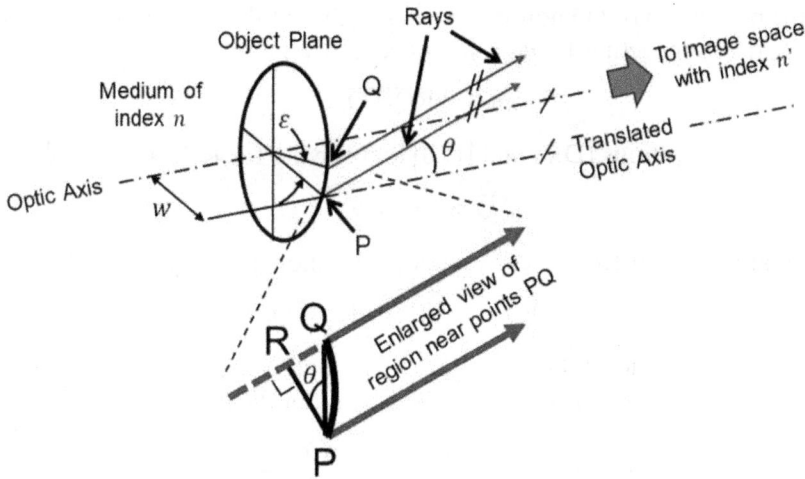

Figure D.5. The construction used to derive the skew invariant.

point for R near the image plane (note that R is not at the object plane). Since the segment RP is perpendicular to the rays PP′ and QQ′, it follows that the OPL of the ray segment RR′ is equal to both the OPLs given by segments PP′ and QQ′. These facts shall be applied to derive the skew invariant in the discussion that follows.

Now, suppose that there is a thin lens between the object and image planes, so that this thin lens is also the pupil of the system. Let A be a point at the pupil joining P such that $n\overline{PA}$ is the OPL of the ray from P to A. Let B be a point at the pupil that joins Q such that $n\overline{QB}$ is the OPL of the ray from Q to B. Then, in the image space, there would be continuing OPLs $n'\overline{AP'}$ and $n'\overline{BQ'}$. There would also be OPLs given by $n\overline{RB}$ and $n'\overline{BR'}$ at the object and image spaces, respectively. Thus, we have two relations given by

$$n\overline{PA} + n'\overline{AP'} = n\overline{RB} + n'\overline{BR'} = n\overline{RQ} + n\overline{QB} + n'\overline{BR'}, \qquad (D.3)$$

$$n\overline{PA} + n'\overline{AP'} = n\overline{QB} + n'\overline{BR'} + n'\overline{R'Q'}. \qquad (D.4)$$

If we take only the left and right sides of equation (D.3) and subtract equation (D.4), we obtain

$$0 = n\overline{RQ} - n'\overline{R'Q'}. \qquad (D.5)$$

Therefore, we have the invariance given by

$$n\overline{RQ} = n'\overline{R'Q'}. \qquad (D.6)$$

It will now be revealed that the quantities on the left and right sides of equation (D.6) are the skew invariant. To see this, we note from figure D.5 that for small ε, we have

$$\varepsilon \approx \overline{PQ}/w, \qquad (D.7)$$

where the bar on top of PQ now denotes a length (not the OPL and it is also not the arc at the disk connecting P and Q). Notice that we also have

$$\sin \theta \approx \overline{RQ}/\overline{PQ}. \tag{D.8}$$

By combining equations (D.7) and (D.8) and solving for $\overline{RQ}$, we obtain

$$\overline{RQ} \approx \varepsilon w \sin \theta. \tag{D.9}$$

At the image plane, there is a corresponding relation given by

$$\overline{R'Q'} \approx \varepsilon w' \sin \theta'. \tag{D.10}$$

Thus, applying equation (D.6) (by multiplying both sides of equation (D.10) by n and both sides of equation (D.11) by n'), we obtain the skew invariant, S:

$$S \equiv nw \sin \theta \approx n'w' \sin \theta'. \tag{D.11}$$

In deriving equation (D.12), Welford replaced the approximation between the left and right sides with an equal sign, but it is unclear why that can be done under the current conditions for derivation. However, in published texts, the skew invariant has been shown to be generalizable, so that we may as well equate the two sides. Let us then check this invariance for two conjugate object-image locations for the diffraction-limited infinite conjugate miniature lens of figure D.2, but used at two

	Surface Type		Comment	Radius	Thickness	Material	Clear Semi-Dia
0	OBJECT	Standard ▾		Infinity	3.00000		0.07338
1		Standard ▾	Dummy 1	Infinity	0.25000		0.13519
2	(aper)	Standard ▾	1	0.52068	0.05170	N-SK2	0.22000 U
3	(aper)	Standard ▾		1.18050	3.99987E-03		0.20000 U
4	(aper)	Standard ▾	2	0.37088	0.13293	N-SK16	0.20000 U
5	(aper)	Standard ▾	3	2.18604	0.04000	F5	0.20000 U
6	(aper)	Standard ▾		0.24023	0.07258		0.15000 U
7	STOP (aper)	Standard ▾	STOP	Infinity	0.22908		0.10000 U
8	(aper)	Standard ▾	4	-0.24297	0.04069	F5	0.14000 U
9	(aper)	Standard ▾	5	-2.22137	0.08954	N-SK16	0.18000 U
10	(aper)	Standard ▾		-0.33126	4.00027E-03		0.18000 U
11	(aper)	Standard ▾	6	1.50177	0.05043	N-SK16	0.18000 U
12	(aper)	Standard ▾		-0.83625	0.10000		0.18000 U
13		Standard ▾	Dummy 2	Infinity	0.87495 M		0.11827
14	IMAGE	Standard ▾	Img	Infinity	-		0.02537

Figure D.6. The prescription for the lens shown in figure D.2, but with object distance controlled by the multi-configuration settings shown in figure D.7.

Active : 1/2		Config 1*	Config 2
1 MOFF	-	Good conjugate	"Bad" conjugate
2 THIC	0	3.000	0.800

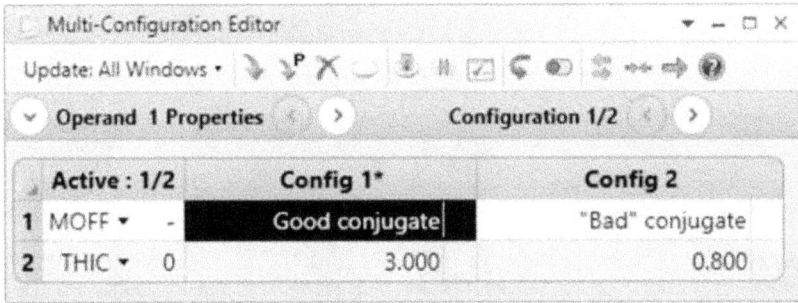

Figure D.7. The multi-configuration settings for the object distance (the thickness of surface 0 in the prescription shown in figure D.6).

	Type	Surf	Wave	Hx	Hy	Px	Py	Target	Weight	Value
1	CONF	1								
2	BLNK	First. find an unvignetted skew ray...								
3	REAY	0	1	0.00000	1.00000	0.00000	0.00000	0.00000	0.00000	0.07338
4	RAID	1	1	0.00000	1.00000	0.00000	0.50000	0.00000	0.00000	8.23787E-06
5	REAB	1	1	0.00000	1.00000	0.00000	0.50000	0.00000	0.00000	-1.43778E-07
6	REAA	1	1	0.00000	1.00000	0.00000	0.50000	0.00000	0.00000	0.00000
7	BLNK	So, a ray at Hy = 1, Px = 0, Py = 0.5 is ray of closest approach at object.								
8	BLNK	This means that giving this ray a dir. cosine in x is a skew ray.								
9	REAB	1	1	0.00000	1.00000	1.00000	0.50000	0.00000	0.00000	1.15739E-04
10	REAA	1	1	0.00000	1.00000	1.00000	0.50000	0.00000	0.00000	0.04119
11	REAR	1	1	0.00000	1.00000	1.00000	0.50000	0.00000	0.00000	0.14398
12	BLNK	So, a ray at Hy = 1, Px = 1, Py = 0.5 is a skew ray.								
13	CONF	1								
14	BLNK	Compute skew at object...								
15	REAY	0	1	0.00000	1.00000	0.00000	0.00000	0.00000	0.00000	0.07338
16	RAID	1	1	0.00000	1.00000	1.00000	0.50000	0.00000	0.00000	2.36056
17	SINE	16	1					0.00000	0.00000	0.04119
18	PROD	15	17					0.00000	0.00000	3.02236E-03
19	BLNK	Compute skew at image...								
20	REAY	14	1	0.00000	1.00000	0.00000	0.00000	0.00000	0.00000	-0.02525
21	RAID	14	1	0.00000	1.00000	1.00000	0.50000	0.00000	0.00000	8.07067
22	SINE	21	1					0.00000	0.00000	0.14039
23	PROD	20	22					0.00000	0.00000	-3.54431E-03
24	ABSO	23						0.00000	0.00000	3.54431E-03
25	BLNK	Config 1: Ratio of object to image skew (good if = 1)								
26	DIVI	18	24					0.00000	0.00000	0.85274

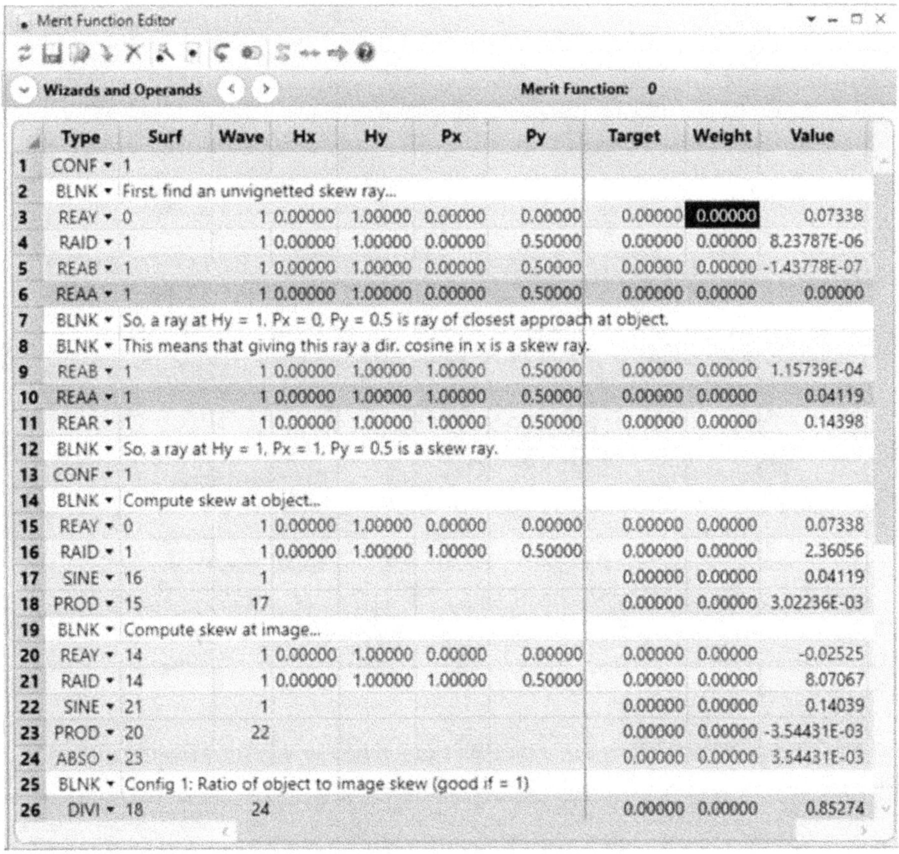

Figure D.8. The merit function for the system shown in figure D.6 (continued in figure D.9).

finite conjugate imaging conditions without re-optimization. The prescription, multi-configuration settings (used to set the two conjugate positions), and merit function are shown in figures D.6–D.9, respectively. The resulting multi-configuration layout is shown in figure D.10. The multi-configuration setting controls the

Figure D.9. The merit function for the system shown in figure D.6 (continued from figure D.8).

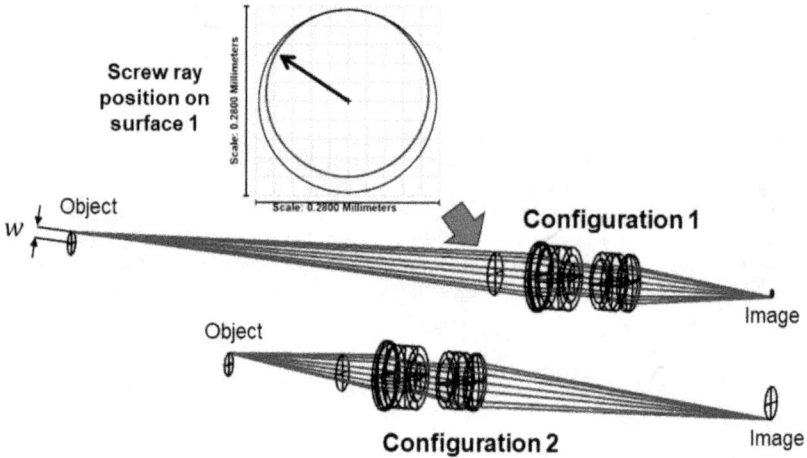

Figure D.10. The multi-configuration 3D layout with a ring of rays for the system shown in figure D.6. The top figure indicates the position of the skew ray being analyzed.

thickness of surface 0, so that this thickness is 3 mm and 0.8 mm for the first and second configurations, respectively. The prescription is shown for the lens in the first configuration, resulting in a marginal ray solve thickness of 0.874 95 mm on surface 13.

The marginal ray solve on surface 13 is set to be for the focal distance at the full zone (i.e. Height = 0, Zone = 1), which is as good a focus as possible for the current lens. When configuration 2 is activated, this solve thickness becomes 1.948 47 mm.

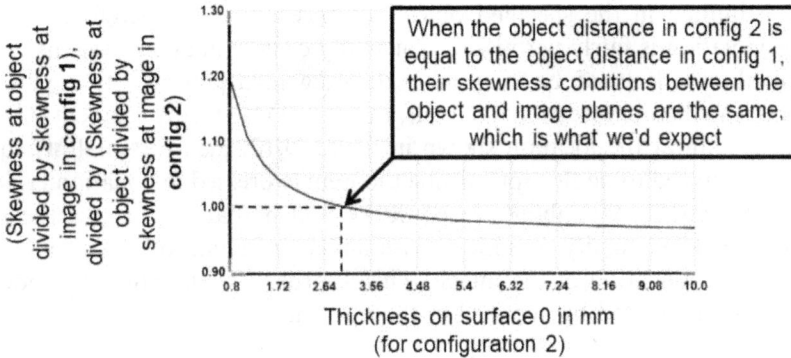

Figure D.11. The skewness ratio of object to image in config 1 divided by the skewness ratio of object to image in config 2, versus the object distance in configuration 2.

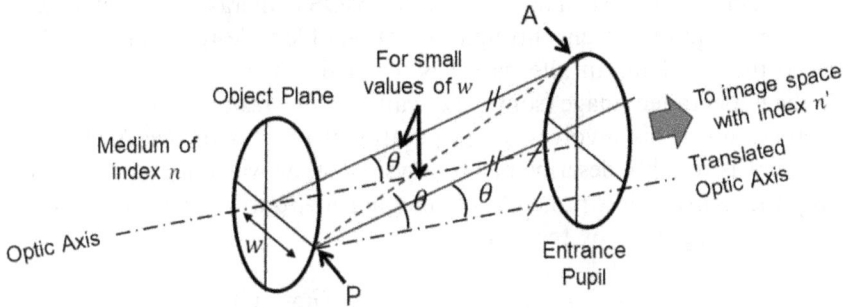

Figure D.12. A construction relating the skew invariant to the optical sine theorem.

In the merit function, at the first configuration, the skewnesses for the skew ray (indicated in figure D.10) at the object and image planes are computed in rows 18 and 24, respectively. In the second configuration, they are computed on rows 32 and 38. Notice that on row 26, the merit function indicates a value of 0.852 74 for the ratio of the skewness at the object to the image. Ideally, it should be near unity under perfect imaging conditions. In the second configuration, the object is closer to the lens and therefore farther from its optimized state (which is for an object at infinity), yielding skewness of 0.715 74, which is further from unity.

At row 42 in the merit function, the ratio of the skewnesses of the first and second configurations is computed. This compares the violation in skewness between the state of the lens in the first configuration and that in the second configuration. Ideally, this ratio should be unity. The violation in skewness condition is made more apparent if we plot this ratio (i.e. the quantity in row 42 of the merit function) as a function of the object distance (the thickness of surface 0) **in configuration 2 only** (this means we leave the thickness of surface 0 in configuration 1 unchanged at 3 mm). Upon doing this, we obtain the plot shown in figure D.11. As indicated, the skewness in configuration 2 departs from that in configuration 1, even though the lens is near diffraction limited in the specified object distance range. It therefore seems that skewness violation is fairly permissible for many optical imaging systems.

As a final remark, it can be shown that the skew invariant may be used to derive the optical sine theorem (section 1.4.5). If, in figure D.5, the ray segment PA is such that A is a point at the location shown in figure D.12, then for small w, equation (D.12) is recognized to be the optical sine theorem expressed in equation (1.58) (just rotate the system by 90° about the optic axis such that θ becomes the sagittal marginal ray). The equality is made precise under the condition that $n = n'$ so that w'/w is the magnification of the image. This approach to deriving the optical sine theorem was highlighted by Welford in the references cited above.

D.3 Deconvolution and noise in imaging

Sometimes, we may wish to take a different approach to the problem of achieving high image quality in the development of optical imaging systems involving electronic image sensors (such as CCDs and CMOS cameras). For instance, instead of providing an optimized lens through the action of lens design, which is a *hardware solution* to the problem, an alternative is to have a *software solution*, which is to *retrieve* the undegraded image (sometimes called *image restoration*) by way of digital image processing. This involves a computational process of *deconvolution* whose basic concept is roughly described as follows. First, as was done in equation (A.4), we apply the convolution principle of imaging in Fourier optics and express the image as being degraded by the PSF:

$$I(x, y) = \int_{-\infty}^{\infty} \int_{-\infty}^{\infty} P(x - x', y - y')G(x', y')\mathrm{d}x'\mathrm{d}y' \tag{D.12}$$

$$\equiv P \otimes G, \tag{D.13}$$

where we regard $I(x, y)$ as an *intensity distribution* of the image (with arbitrary units), $P(x, y)$ is the point spread function (PSF), $G(x', y')$ is the ideal geometric image defining the bounds for integration, and the symbol $\otimes$ is the operator for the convolution. If P is known through some means of calibration, then G may be retrieved through first taking the Fourier transform

$$\mathcal{F}\{I\} = \mathcal{F}\{P \otimes G\} = \mathcal{F}\{P\}\mathcal{F}\{G\}. \tag{D.14}$$

Then, dividing through by $\mathcal{F}\{P\}$ and solving for G we have

$$G = \mathcal{F}^{-1}\left\{\frac{\mathcal{F}\{I\}}{\mathcal{F}\{P\}}\right\}, \tag{D.15}$$

where $\mathcal{F}^{-1}$ denotes taking the inverse Fourier transform. Thus, regardless of the quality of P (thereby relieving the lens designer of the design optimization task), G may always be obtained by the computational steps above.

However, the problem in the above is that random noise (such as photon shot noise and sensor dark current shot noise) in the captured digital image is always present, so that the deconvolved image ends up having even more noise (by virtue of error propagation, each computational step involving quantities with noise would only increase the noise contribution in the quantities). Moreover, from the fact that $\mathcal{F}\{P\}$ is the optical transfer function (OTF) (whose magnitude is often low at high spatial frequencies and can have zero crossings), having $\mathcal{F}\{P\}$ in the denominator would amplify the noise at those spatial frequencies, which is a mess. To see how these issues arise, we must include a noise term into the right side of equation (D.13). Suppose that the random noise intensities fluctuate by amount $\pm\Delta$, then the detected image intensity may be expressed as

$$I = P \otimes G + \Delta. \tag{D.16}$$

Applying the Fourier transform on both sides of equation (D.16), we obtain

$$\mathcal{F}\{I\} = \mathcal{F}\{P \otimes G + \Delta\} = \mathcal{F}\{P\}\mathcal{F}\{G\} + \mathcal{F}\{\Delta\}. \tag{D.17}$$

Solving for G, we obtain

$$G = \mathcal{F}^{-1}\left\{\frac{\mathcal{F}\{I\}}{\mathcal{F}\{P\}} - \frac{\mathcal{F}\{\Delta\}}{\mathcal{F}\{P\}}\right\}. \tag{D.18}$$

In equation (D.18), note that the ratio $\mathcal{F}\{\Delta\}/\mathcal{F}\{P\}$ introduces unwanted noise, which is amplified by low values and zero crossings of $\mathcal{F}\{P\}$ at high spatial frequencies.

Luckily, experts in digital image processing know how to deal with the noise issue (see, e.g., Castleman K R 1996 *Digital Image Processing* (Upper Saddle River, NJ: Prentice Hall) pp 216–29, and Gonzalez R C and Woods R E 2002 *Digital Image Processing* 2nd edn (Upper Saddle River, NJ: Prentice Hall) ch 5; see also, e.g., Khare K, Butola M and Rajora S 2023 *Fourier Optics and Computational Imaging* 2nd edn (Switzerland: Springer) pp 59–70). It is worthwhile to get a basic understanding of the deconvolution process when noise is involved. Also, as convolution is the fundamental process of image formation in optical systems, its inverse (deconvolution) is a problem worth gaining some familiarity with.

To begin, we must first require that $\mathcal{F}\{P\}$ has positive and nonzero values. Then, let us re-express equation (D.17) as

$$\mathcal{F}\{I\} = \mathcal{F}\{P\}\mathcal{F}\{G\}\left(1 + \frac{\mathcal{F}\{\Delta\}}{\mathcal{F}\{P\}\mathcal{F}\{G\}}\right). \tag{D.19}$$

Now, it seems reasonable that if the noise term $\mathcal{F}\{\Delta\}/(\mathcal{F}\{P\}\mathcal{F}\{G\})$ can be somehow eliminated (cancelled) in some way, then what is left is $\mathcal{F}\{P\}\mathcal{F}\{G\}$, which would enable the retrieval of G through the step given by equation (D.15). Based on the digital processing methods discussed in the references cited above, it turns out that the noise term can be effectively *filtered out* if we multiply the right side of equation (D.19) by a *Wiener filter*, $\mathcal{W}$, which is a function in Fourier space expressed as

$$\mathcal{W} = \frac{\mathcal{F}\{P\}*}{|\mathcal{F}\{P\}|^2 + \frac{\langle|\mathcal{F}\{\Delta\}|^2\rangle}{\langle|\mathcal{F}\{G\}|^2\rangle}}, \tag{D.20}$$

where $\mathcal{F}\{P\}*$ is the complex conjugate of $\mathcal{F}\{P\}$, and the angled brackets enclosing $|\mathcal{F}\{\Delta\}|^2$ and $|\mathcal{F}\{G\}|^2$ denote taking an *ensemble average* of the respective squared Fourier transforms. We will return to the concept of the ensemble average shortly. Meanwhile, to see the cancellation effect of the noise, the Wiener filter $\mathcal{W}$ may be algebraically re-arranged by multiplying the top and bottom by $\mathcal{F}\{P\}$, yielding

$$\mathcal{W} = \frac{1}{\mathcal{F}\{P\}} \frac{\mathcal{F}\{P\}\mathcal{F}\{P\}*}{|\mathcal{F}\{P\}|^2 + \frac{\langle|\mathcal{F}\{\Delta\}|^2\rangle}{\langle|\mathcal{F}\{G\}|^2\rangle}} = \frac{1}{\mathcal{F}\{P\}}\left[\frac{1}{1 + \frac{\langle|\mathcal{F}\{\Delta\}|^2\rangle}{|\mathcal{F}\{P\}|^2\langle|\mathcal{F}\{G\}|^2\rangle}}\right]. \qquad (D.21)$$

If we then multiply the right side of equation (D.19) by $\mathcal{W}$, we obtain the filtered transform of the image:

$$\mathcal{F}\{I\} = \mathcal{F}\{G\}\mathcal{F}\{P\}\left(1 + \frac{\mathcal{F}\{\Delta\}}{\mathcal{F}\{P\}\mathcal{F}\{G\}}\right)\frac{1}{\mathcal{F}\{P\}}\left[\frac{1}{1 + \frac{\langle|\mathcal{F}\{\Delta\}|^2\rangle}{|\mathcal{F}\{P\}|^2\langle|\mathcal{F}\{G\}|^2\rangle}}\right] \qquad (D.22)$$

$$= \mathcal{F}\{G\}\left(\frac{1 + \frac{\mathcal{F}\{\Delta\}}{\mathcal{F}\{P\}\mathcal{F}\{G\}}}{1 + \frac{\langle|\mathcal{F}\{\Delta\}|^2\rangle}{|\mathcal{F}\{P\}|^2\langle|\mathcal{F}\{G\}|^2\rangle}}\right), \qquad (D.23)$$

where the superscript prime in $\mathcal{F}\{I\}'$ denotes that $\mathcal{F}\{I\}'$ is not $\mathcal{F}\{I\}$. In equation (D.23), if, for some reason, the quantity $\mathcal{F}\{\Delta\}/(\mathcal{F}\{P\}\mathcal{F}\{G\})$ resembles $\langle|\mathcal{F}\{\Delta\}|^2\rangle/(|\mathcal{F}\{P\}|^2\langle|\mathcal{F}\{G\}|^2\rangle)$, then indeed, some degree of cancellation can occur, enabling the retrieval of G through the inverse Fourier transform operation.

You would have every right to wonder how random noise is cancellable. Error analysis tells us that random noise propagates through each step in an arithmetic computation, such as the division of two noisy quantities. Despite this, there is some theory behind how random noise looks in Fourier space, especially if the noise is shot noise from photon detection and in the pixels (such as from dark current). The basic point in the theory is as follows. In sections 3.4 and 3.5, we talked about the degradation of signals by random noise, and that averaging the degraded signals would minimize the noise contribution. Therefore, if we average infinite ensembles of the acquired noisy signal, then we should be able to obtain an accurate estimate of the undegraded original signal. But in reality, we would never afford the time to perform such averaging. Instead, when capturing digital images in electronic image sensors, a single image capture is usually taken over a specified camera integration period, T. Therefore, some residual noise often exists in any signal captured through the use of image detection systems. Now, suppose that we regard the step in integrating photons over a period T as being equivalent to truncating a noisy signal by that period, then provided that the signal is **ergodic**, the truncated signal has a Fourier transform whose ensemble average of the squared absolute value is known as the noise's **spectral density**, which is also called its **power spectrum** (see the above references on digital image processing, as well as reference [27] in chapter 3, pp 119–41, and also Goodman J W 1985 *Statistical Optics* (New York: Wiley) pp 60–73). It turns out that $\langle |\mathcal{F}\{\Delta\}|^2 \rangle$ is the power spectrum of most random noise sources (including shot noise) and that its ensemble average is a constant of some magnitude (think of a stream of detected photons as a sum of delta function *pulses*, then take the Fourier transform of that sum). In statistical theory, a measurement (such as the captured image signal at integration time T) is an estimate of a quantity, and the average of the estimate (if it is an unbiased estimate) tends towards the *true* value of the quantity. By analogy, think of the constancy in $\langle |\mathcal{F}\{\Delta\}|^2 \rangle$ as the *true* function for the estimate of the Fourier spectrum in the noise. In digital image processing, the Wiener filtering step assumes that $\mathcal{F}\{\Delta\}$ is fairly constant with spatial frequency so that its form is approximately equal to $\langle |\mathcal{F}\{\Delta\}|^2 \rangle$. As for $\mathcal{F}\{G\}$, the expected value (mean) given by the ensemble average $\langle |\mathcal{F}\{G\}|^2 \rangle$ is done for a *class* of images of the object of interest. In this way, we hope that $\langle |\mathcal{F}\{G\}|^2 \rangle$ resembles $\mathcal{F}\{G\}$. We also hope that $\mathcal{F}\{P\}$ resembles $|\mathcal{F}\{P\}|^2$, so that the quantities $1 + [\mathcal{F}\{\Delta\}/(\mathcal{F}\{P\}\mathcal{F}\{G\})]$ and $1 + [\langle |\mathcal{F}\{\Delta\}|^2 \rangle/(|\mathcal{F}\{P\}|^2 \langle |\mathcal{F}\{G\}|^2 \rangle)]$ in equation (D.23) may cancel, which would enable the retrieval of G (to be precise, an *estimate* of G in the statistical sense). Furthermore, note that the ratios given by $\langle |\mathcal{F}\{\Delta\}|^2 \rangle/\langle |\mathcal{F}\{G\}|^2 \rangle$ and $\mathcal{F}\{\Delta\}/\mathcal{F}\{G\}$ are the reciprocals of the signal-to-noise ratio (SNR). Hence, when the SNR is high, equation (D.23) reduces to (D.14) but with the quantity $\mathcal{F}\{P\}$ eliminated, thereby enabling the restored quantity, G.

Incidentally, Gonzalez and Woods (see the reference cited above) state that if, in Wiener deconvolution, $\langle |\mathcal{F}\{G\}|^2 \rangle$ cannot be obtained, then the ratio $\langle |\mathcal{F}\{\Delta\}|^2 \rangle/(|\mathcal{F}\{P\}|^2 \langle |\mathcal{F}\{G\}|^2 \rangle)$ may often be replaced by some constant that is iteratively guessed through trial and error, yielding sufficiently deconvolved and restored images. Indeed, restored images have been reported in much of the literature on deconvolution. In applications involving the problem of extended depth of field, the problem involves restoring a degraded image that is invariant to changes in the object distance for a fixed plane of the camera in the image. This is made possible through the use of phase masks that create a defocus invariant PSF (see section 1.2.6 as well as the references cited there). Here and in many other applications involving deconvolution, the Wiener filter approach is not the only method of image restoration, and digital image processing is a rich subject with plenty of possibilities and applications for modern imaging. The reader is encouraged to consult the references cited above for further study.